MODERN MICROBIAL GENETICS

Second Edition

MODERN MICROBIAL GENETICS

Second Edition

E D I T E D   B Y

Uldis N. Streips

Department of Microbiology and Immunology
School of Medicine
University of Louisville
Louisville, Kentucky

Ronald E. Yasbin

Program in Molecular Biology
University of Texas at Dallas
Richardson, Texas

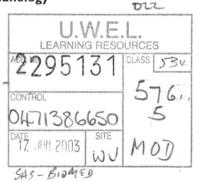

WILEY-LISS

A JOHN WILEY & SONS, INC., PUBLICATION

Library of Congress Cataloging-in-Publication Data

Modern microbial genetics / edited by Uldis N. Streips, Ronald E. Yasbin.—2nd ed.
　　p. ; cm.
　Includes bibliographical references and index.
　ISBN 0–471–38665–0 (cloth : alk. paper)
　　　1. Microbial genetics. I. Streips, Uldis N., 1942–II. Yasbin, Ronald E.
　[DBLM: 1. Genetics, Microbial. QW 51 M689 2002]

　QH434 .M634 2002
　579′.135—dc21

2001045534

Contents

Preface

The impetus for this updated edition of *Modern Microbial Genetics* came from many discussions among the authors and editors with the leadership and participants at the lovely Wind River Conference on Prokaryotic Biology held in Estes Park, Colorado every June. The first edition, though comprehensive, had become outdated and the need for an up-to-date, advanced textbook for microbial genetics was palpable. With the able encouragement and cooperation of our editor Luna Han, at John Wiley & Sons, Inc., the agreement was reached to publish this text. So, we welcome you to *Modern Microbial Genetics II*.

We have maintained the same model for chapter authorship. Even though in some ways it would be optimal to have a single author for the entire textbook, we felt that this in-depth material could be handled far better by enlisting experts in their fields to put together chapters of their own respective insights. Moreover, we chose authors who are also excellent teachers so that the textbook could be easily adapted to classrooms in advanced undergraduate and graduate courses.

A quick comparison of the two editions should point out a universal truth about scientific publications: namely, a published book may advance information a step, or at most a few steps, ahead of other existing books, but the moment it is published, the book is miles behind where the information will ultimately lead. Because of this, in *Modern Microbial Genetics II* the chapters are extensively revised and updated, some are removed, and others added. This happens to be the most complete and relevant information at this point in time from our perspective. Publication on the Web will further allow for more facile updating and diminish the inevitable dissipation of current information.

As we stated in the first edition, this book presents a vibrant field of knowledge with many areas anxiously awaiting new investigators. After going through this text, one or another of the chapters may beguile you, the reader, enough to willingly immerse yourself in the wonderful discipline of microbial genetics. Again we say—Welcome!

We wish you success in adding the extensive knowledge presented in this textbook to your previous experience in microbial genetics and applying it to your own future goals and objectives. We look forward to many of you joining us in generating information, and perhaps even chapters, for future editions and updates to this textbook.

Uldis N. Streips
Ronald E. Yasbin

Preface to the First Edition

The information presented in this book represents the best efforts by a select group of authors, who are not only productive in research but who are also excellent teachers, to delineate the limits of knowledge in the various areas of microbial genetics. We feel the use of multiple authors provides not only for depth of material, but also enriches the perspectives of this textbook. The limits of knowledge need to be stretched continuously for science to remain exciting and meaningful. It should be obvious that this then leaves a vast field for future work, where some of you readers will find a lifetime of productive research. Moreover, it should also be obvious that many of the areas discussed in this book still contain pathways and byways which sometimes have never been explored, and sometimes have side roads waiting for eager minds to map and meld within the pool of knowledge which we call modern microbial genetics. We expect that you will have had some previous exposure to microbial genetics and will use this text to build on that experience. As you probe in depth the thought processes and experiments which were used to formulate the fundamental concepts in modern microbial genetics, one or another of the included chapters may spark the interest in your mind to become a traveler within this vast and exciting discipline. If that is the case—Welcome!

We wish you success in adding the knowledge presented in this textbook to your previous experience in microbial genetics and applying it to your future goals and objectives. We thank the many reviewers who helped to enhance the accuracy and presentation of this material. In this regard, Marti Kimmey was most helpful in correlating the various chapters.

Uldis N. Streips
Ronald E. Yasbin

Introduction

ULDIS N. STREIPS AND RONALD E. YASBIN

The initial studies, which presaged the emergence of the capabilities for the complete sequencing of genomes and the study of whole organism proteomics in addition to various aspects of molecular biology, are now almost 90 years old. The early reports on bacteriophage by Twort (1915), d'Herelle (1917), and Ellis and Delbruck (1939) and the initial description of the pneumococcal type "transformation" by Griffith (1928) preshadowed this explosion of information by laying down a solid foundation on which to build layer upon layer of new ideas and facts. Even though these early workers had no basis for concluding more than their time in the flow of events allowed them to conjecture, we can envision that an unbreakable thread was formulated by their work. The scientists in the many subsequent decades have woven this initially thin thread into an extensive and mutlicolored tapestry in which are embedded the stories of the research that is described in *Modern Microbial Genetics II*. It is fascinating that for the first years the major debate was on the existence and function of DNA. Entering the New Millenium, not only can we reproducibly obtain DNA, deliver it to any cell we choose, but we can also unlock every secret in that molecule.

In the 1940s and 1950s two major research thrusts permanently changed the perspectives on microbial genetics and provided the basis for the explosion of information in the field of molecular biology. These were, first, the documentation of DNA as the carrier of an organism's genetic information by Avery and coworkers (1944) and the subsequent deciphering of the chemical structure of this molecule by Watson and Crick (1953), and second, the discovery of mobile genetic elements by McClintock (1956).

The seminal work on proving that DNA is the stuff of heredity, can be manipulated, and indeed is self-manipulating, rapidly led to the description in 1950s and 1960s of genetic exchange in bacteria and in subsequent years to modern microbial genetics. In this textbook there are detailed descriptions of three major areas. The first is DNA Metabolism: how DNA replicates (Firshein), how DNA is repaired (Yasbin), how DNA is transcribed and the transcription regulated (Helmann) and how DNA recombines (Levene and Huffman). This section also includes the genetics of bacteriophage including the T-even phages (Guttman and Kutter), the lambdoid phages (Hendrix), the phages with nucleic acids other than double stranded DNA (Leclerc), and how restriction and modification directs microbial existence (Blumenthal and Cheng). A chapter (Geoghegan) on DNA manipulation techniques and application to molecular biology completes the DNA Metabolism section.

The second section is on Genetic Response and includes several chapters on how microorganisms interact with the environment. The role and mechanism of bacteria in establishing disease states is discussed by Hassett and coauthors. How cells react to environmental stress is shown in the chapters by Moran on sporulation and Streips on stress shock. Two environmental organisms that depend on genetic versatility are discussed in the chapters on *Myxococcus* by Hartzell and *Agrobacterium* by Ream. The ability of microorganisms to constantly sense their environment is revealed in chapters on two-component sensing by Bayles and Fujimoto and quorum sensing by Parsek and Fuqua.

The last section on Genetic Exchange includes the latest information on the classic exchange mechanisms (see the Chapters by Streips on transformation, Porter on conjugation, and Weinstock on transduction). Perlin discusses the genetics of plasmids that do not belong to the F family. In addition this section also includes recent information about transposons and their ability to move from cell to cell (Whittle and Salyers). Finally, the molecular study of bacteria which have no standard genetic systems is described by Haller and DiChristina and concludes this book.

The elucidation of global regulatory systems, which control everything from DNA uptake to emergency responses and overall microbial development, are widely discussed in various chapters in this book and they help to bring the study of molecular biology full circle. As described by Helmann, Streips, and Moran, there are genes and operons in bacteria which are coordinately regulated and defined as regulons. So, from the initial consideration about the existence and nature of DNA, now assumptions are made about how genes network and cooperate in multigene regulons to suit the needs of the bacterial cell.

McClintock's early work showed that DNA was not merely a static chemical molecule, but rather a dynamic structure which can be amplified to a myriad of genetic possibilities. So it is once the fundamental aspects of bacterial genes and their exchange were elucidated, it became apparent that bacteria, bacteriophage, and also eukaryotes, through mutation, evolution, and genetic exchange have arranged and rearranged their genetic material to take an optimal advantage of their niche in the environment. This theme is the constant thread that connects the various sections and subject areas of *Modern Microbial Genetics II*.

This textbook is our approach to link the pioneering work of the past to the modern technology available today and to start answering some of the major questions about the molecular mechanisms operating in microbial cells.

REFERENCES

Avery OT, MacLeod CM, McCarty M (1944): Studies on the chemical nature of the substance inducing transformation of pneumococcal types. Induction of transformation by a desoxyribonucleic acid fraction isolated from pneumococcus type III. J Exp Med 79:137–158.

D'Herelle F (1917): Sur un microbe invisible antagoniste des bacilles dysenteriques. CR Acad Sci 165:373.

Griffith F (1928): The significance of pneumococcal types. J Hyg 27:113–159.

McClintock B (1956): Controlling elements in the gene. Cold Spring Harbor Symp Quant Biol 21:197–216.

Twort FW (1915): An investigation on the nature of the ultramicroscopic viruses. Lancet 11:1241.

Watson JD, Crick FHC (1953): Molecular structure of nucleic acids. Nature 171:737–738.

Contributors

Kenneth W. Bayles, Department of Microbiology, Molecular, and Biochemistry, The College of Agriculture, University of Idaho, Moscow, ID 83844–3052

Robert M. Blumenthal, Department of Microbiology and Immunology, Medical College of Ohio, Toledo, OH 43614–5806

Xiaodong Cheng, Biochemistry Department, Emory University, Atlanta, GA 30322–4218

Thomas J. DiChristina, School of Biology, Georgia Institute of Technology, Atlanta, GA 30332

William Firshein, Department of Molecular Biology and Biochemistry, Wesleyan University, Middletown, CT 06459

David F. Fujimoto, Biology Department LS–416, San Diego State University, San Diego, CA 92182

Clay Fuqua, Department of Biology, Indiana University, Bloomington, IN 47405

Thomas Geoghegan, Department of Biochemistry and Molecular Biology, University of Louisville School of Medicine, Louisville, KY 40292

Burton S. Guttman, The Evergreen State College, Olympia, WA 98505

Carolyn A. Haller, School of Biology, Georgia Institute of Technology, Atlanta, GA 30332

Patricia L. Hartzell, Department of Microbiology, Molecular Biology, and Biochemistry, University of Idaho, Moscow, ID 83844–3052

Daniel J. Hassett, Department of Molecular Genetics, Biochemistry, and Microbiology, University of Cincinnati, College of Medicine, Cincinnati, OH 45267–0524

John D. Helmann, Department of Microbiology, Cornell University, Ithaca, New York 14853–8101

Roger W. Hendrix, Pittsburgh Bacteriophage Institute, Department of Biological Sciences, University of Pittsburgh, Pittsburgh, PA 15260

Kenneth E. Huffman, Department of Molecular and Cell Biology, University of Texas at Dallas, Richardson, TX 75083–0688

Barbara H. Iglewski, Department of Microbiology and Immunology, University of Rochester School of Medicine, Rochester, NY 14642

Teresa de Kievit, Department of Microbiology and Immunology, University of Rochester School of Medicine, Rochester, NY 14642

Elizabeth M. Kutter, The Evergreen State College, Olympia, WA 98505

J. Eugene LeClerc, Molecular Biology Division, Center for Food Safety and Applied Nutrition, US Food and Drug Administration, Washington, DC 20204

Stephen D. Levene, Department of Molecular and Cell Biology, University of Texas at Dallas, Richardson, TX 75083–0688

Thomas S. Livinghouse, Department of Chemistry and Biochemistry, and Department of Land Resources and Environmental Sciences, Montana State University, Bozeman, MT 59717

Timothy R. McDermott, Department of Land Resources and Environmental Sciences, Montana State University, Bozeman, MT 59717

Charles P. Moran Jr., Department of Microbiology and Immunology, Emory University School of Medicine, Atlanta, GA 30322

Urs A. Ochsner, Department of Microbiology, University of Colorado Health Sciences Center, Denver, CO 80262

Matthew R. Parsek, Department of Civil Engineering, Northwestern University, Evanston, IL 60208

Luciano Passador, Department of Microbiology and Immunology, University of Rochester, School of Medicine, Rochester, NY 14642

Michael H. Perlin, Department of Biology, University of Louisville, Louisville, KY 40292

Ronald D. Porter, Department of Biochemistry and Molecular Biology, The Pennsylvania State University, University Park, PA 16802

Walt Ream, Department of Microbiology, Oregon State University, Corvallis, OR 97331

John J. Rowe, Department of Biology, University of Dayton, Dayton, OH 45469

Abigail A. Salyers, Department of Microbiology, University of Illinois, Urbana, IL 61801

Uldis N. Streips, Department of Microbiology and Immunology, School of Medicine, University of Louisville, Louisville, KY 40292

George M. Weinstock, Department of Biochemistry and Molecular Biology, University of Texas Medical School, Houston, TX 77225

Jeffrey A. Whitsett, Division of Pulmonary Biology, Children's Hospital Medical Center, Cincinatti, OH 45229–3039

Gabrielle Whittle, Department of Microbiology, University of Illinois, Urbana, IL 61801

Ronald E. Yasbin, Program in Molecular Biology, University of Texas at Dallas, Richardson, TX 75083

Section 1: DNA METABOLISM

I

Prokaryotic DNA Replication

Department of Molecular Biology and Biochemistry, Wesleyan University, Middletown,
Connecticut 06459

I. Introduction 3
II. General Concepts of DNA Replication 4
   A. Semiconservative Synthesis.................... 4
   B. The Replicon Model 5
III. Replication Operations 6
   A. Initiation............................... 6
   B. Elongation 7
      1. Fine Details of Elongation 8
   C. Termination 12
   D. Precursors in DNA Replication............... 16
      1. Introduction......................... 16
      2. Types of Metabolic Pathways.............. 16
      3. Multienzyme Complexes 18
IV. The Replicon Membrane Interaction 18
   A. Introduction............................. 18
   B. Specific Organisms........................ 19
      1. *E. coli* 19
      2. *B. subtilis* 20
      3. Plasmid RK2.......................... 21
V. General Conclusions 22

I. INTRODUCTION

Ultimately DNA structure must be understood in terms of its function just as function requires knowledge of structure. Each function must be resolved and reconstituted in complete detail in order to connect it to a structure in the cell. In the case of DNA, three hierarchical functions—storage of genetic information, replication of this information from generation to generation, and ultimate control of the functions of cellular activities—have been elucidated in exquisite detail, although our understanding of those details is far from complete. Much of the success was made possible after Watson and Crick (1953) proposed that the structure of DNA existed as a double helix of sugar-phosphates held together by two purine and pyrimidine base pairs, adenine-thymine and guanine-cytosine, respectively. It was the sequence of these base pairs that determined the exact composition of the DNA molecule and the molecular structure of the gene (storage of genetic information).

Modern Microbial Genetics, Second Edition. Edited by
Uldis N. Streips and Ronald E. Yasbin. ISBN 0–471–38665–0
Copyright © 2002 Wiley-Liss, Inc.

Replication of the double helix was proposed by Watson and Crick to be based upon the separation of two helices which acted as templates for the precise copying of complementary strands to form two progeny double helices according to the sequence of the base pairs (termed *semiconservative replication*). However, in attempting to identify the components (enzymes, control factors) responsible for this precise duplication, it became obvious that the process was interdependent with other related phenomena such as repair and recombination of DNA. Some of the enzymes could be used for all of the processes. In fact there is a growing body of knowledge that not only are the pathways intimately related, but many of the proteins may be part of a "superfamily" in which all of them share a highly conserved DNA-binding motif as determined by X-ray crystallography or electron microscopy (Engelman, 2000).

The difficulty (and complexity) of elucidating these interactions is further underscored by two additional characteristics of the replicative process. First, unlike RNA and protein synthesis, DNA replication occurs at discrete times during the cell cycle. The many components involved must be assembled and disassembled after each round of replication. Second, unlike the organelle involved in protein synthesis (the ribosome) which is held together with strong forces, those that maintain the DNA replisome (the components involved in DNA replication) involve weak electrostatic forces which can be dissociated under mild salt conditions. Thus in vitro studies that have formed the bases for understanding many of the intricacies of replication are subject to artifacts because extraction of the replisome from cells may be disruptive and not represent the in vivo condition as fully as possible.

Nevertheless, much has been revealed by classic in vitro studies of prokaryotes using single stranded DNA viruses that infect *Escherichia coli* and sequester many of the host's components (Kornberg and Baker, 1992) and recombinant plasmids containing

the beginning (origin) of replication (*oriC*) for this and other organisms such as *Bacillus subtilis* (Kornberg and Baker, 1992; Moriya et al., 1994).

II. GENERAL CONCEPTS OF DNA REPLICATION

A. Semiconservative Synthesis

How could Watson and Crick's model be proven that replication occurred in a semiconservative manner? In fact two additional possibilities existed besides such a mechanism. These included conservative (both strands replicated simultaneously) or dispersive (each strand was fragmented, copied and joined to form a completely new parental and progeny strand).

The most important and definitive experiments that proved that DNA was replicated semiconservatively were carried out by Meselson and Stahl (1958). They adapted *E. coli* to a growth medium containing $N^{15}H_4Cl$ ensuring that every molecule in the cell containing nitrogen (including DNA) would have the N^{15} heavy density label. When these cells were shifted to a medium containing the normal light density $N^{14}H_4Cl$, the resulting progeny double helices after one generation consisted of a hybrid density DNA species containing presumably one strand of N^{15}-DNA and one strand of N^{14}-DNA. After a second generation in light density medium, the double helices consisted equally of both the hybrid density species and a complete light density species. This is seen in Figure 1 where the various DNA species are separated by centrifugation in a neutral cesium chloride equilibrium density gradient.

The other hypotheses could not be supported by these results. Further proof of the mechanism was obtained by separating the hybrid density species in an alkaline cesium chloride density gradient which denatured the DNA into two single stranded forms on the gradient, one consisting of N^{15}-DNA, the other of N^{14}-DNA (Meselson and Stahl, 1958).

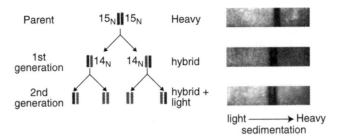

Fig. 1.   Semiconservative replication of *E. coli* DNA ^{15}N (heavy) parental, ^{14}N (light) progeny and hybrid first-generation DNAs are separated by sedimentation in a cesium chloride equilibrium density gradient. (Reproduced from Kornberg and Baker, 1992, with permission of the publisher.)

B. The Replicon Model

DNA replication is divided into three parts or phases. These include initiation (or beginning of replication), elongation, and termination. Early studies had demonstrated that initiation required the synthesis of new proteins (Maaloe and Hanawalt, 1961) while elongation did not. However, it was the autoradiographic studies of Cairns (1963) with *E. coli* and the genetic studies of Yoshikawa and Sueoka (1963) with *B. subtilis* that demonstrated that bacterial chromosomes had a fixed origin of replication. Cairns results further confirmed earlier genetic studies (Hayes, 1962) that the *E. coli* chromosome consisted of one double helical species of DNA without a free end, namely a circular molecule. During replication this molecule was split into two "replicating" forks that traveled along the template in opposite directions migrating away from the site where elongation began, namely the origin, until they met, approximately 180° opposite the origin where elongation terminated.

Based on these studies, Jacob et al. (1963) proposed a model that envisioned one circular chromosome as a genetic unit of replication (the replicon) in which replication began at a defined small region on the chromosome, the origin, and proceeded, usually bidirectionally to the terminus via two replication forks seen in the theta circles of Cairns. This model is depicted in Figure 2.

A number of control features were proposed based on a positive activation

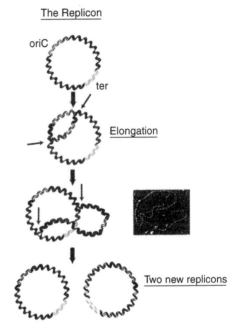

Fig. 2.   Bidirectional replication of the *E. coli* chromosome. The thin black arrows identify the advancing replication forks. The micrograph is of a bacterial chromosome in the process of replication, comparable to the figure next to it. (Modified from Klug and Cummings, 2000, with permission of the publisher.)

mechanism in which a series of genes near the origin region controlled the synthesis of initiation proteins (the initiator) that activated the "replicator" (the origin region), which then replicated (elongated) any DNA that was part of the replicon. The replicon was thought to be anchored to a specific site in the cells (probably the membrane) where the various enzymes and initiator proteins could be sequestered and where segregation of the newly synthesized chromosome could occur.

Even after 35 years, the replicon model still represents a good conceptual framework (with modifications) to explain these important events.

III. REPLICATION OPERATIONS

A. Initiation

Many complex problems must be resolved to replicate a complete chromosome. Among the first of these is the localized unwinding of the duplex at a specific site (the origin) in order that each stand can begin its function as a template. In addition this open configuration must be stabilized so that synthesis can actually occur. The process is called initiation and is at the heart of the entire replicative process.

Although a number of origin regions have been studied, the most complete analysis to date has been that of the *E. coli* origin (or *oriC*). It was detected and localized by several different strategies, among them its inability to be deleted from the chromosome without destroying the cell, gene dosage experiments in which genes near the origin were replicated first, and the construction of an *oriC* plasmid which required the same functions as replication of the entire chromosome including bidirectional replication (Bird et al., 1972; Von Meyenburg et al., 1979; Meijer and Messer, 1980).

The minimum *oriC* region (absolutely required for in vitro synthesis) (Oka et al., 1980) consists of 245 base pairs in a negative supercoil state that can be subdivided on the basis of function into two parts. The first part contains four repeating sequences of 9 base pairs (9 mers), while the second part contains three repeating sequences of 13 base pairs (13 mers). The latter are highly AT rich, while the former contains sequences (consensus 5$^-$-TTAT C/A CAC/AA-3′) that

recognize an important 52 kDa initiation protein (DnaA) (Fuller and Kornberg, 1983). This protein is not only a central player in initiation of *oriC* in *E. coli* but in the initiation of many other bacteria (Zyskind et al., 1983; Zyskind and Smith, 1986; Bramhill and Kornberg, 1988) and even of some plasmids (Masai et al., 1987; Konieczny and Helsinski, 1997), suggesting its ubiquity and fulfilling one of the tenets of the replicon model (Jacob et al., 1963).

The remarkable structure of the origin predicted how the localized denaturation was effected, and a variety of experimental approaches (genetic, biochemical, and ultrastructural) confirmed the predictions (Funnell et al., 1987; Bramhill and Kornberg, 1988; Echols, 1990). The process was initiated in *E. coli* by the specific binding of the DnaA protein (20–40 monomers) to the 9 base pair DnaA boxes.

As a result of this binding and in the presence of ATP, a basic instability of the AT-rich region is heightened and approximately 45 base pairs are denatured to mark it for recruitment of other essential proteins into the bubble (e.g., DnaB and DnaC) that further open and destabilize the complex (see next section). Additional accessory proteins also are involved in the process. These include the HU protein, a small double stranded DNA-binding protein (Rouviere-Yaniv and Gros, 1975) that may be involved in bending the DNA, and the single-stranded DNA-binding protein (SSB) that may stabilize the single-stranded regions when they are present (Meyer and Laine, 1990).

Figure 3a, b depicts the initiation process as currently envisaged with electron micrographs that visualize the complex.

There are additional important aspects concerning the regulation and activity of the DnaA initiation protein in the initiation process that point to its probable location (and that of the replicon) in the cell, namely the cell membrane. DnaA functions primarily in a membrane environment (Yung and Kornberg, 1988), is activated by anionic phospholipids (Sekimizu and Kornberg,

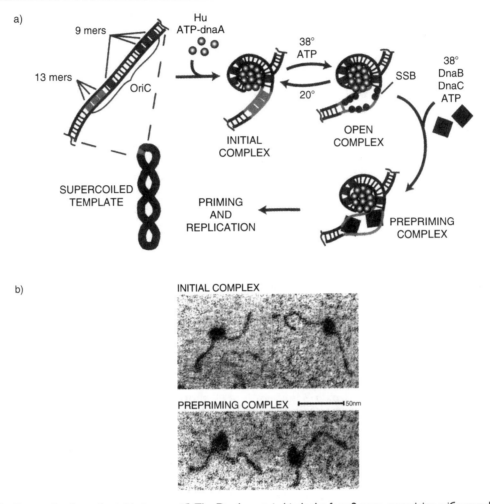

Fig. 3.   a: A scheme for initiation at *oriC*. The DnaA protein binds the four 9 mers, organizing *oriC* around a protein core to form the initial complex. The three 13 mers are then melted serially by DnaA protein to create the open complex. The DnaB–DnaC complex can now be directed to the 13-mer region to extend the duplex opening and generate a prepriming complex, which unwinds the template for priming and replication. (Modified from Kornberg and Baker, 1992, with permission of the publisher.) **b:** Electron micrographs of protein complexes at *oriC*. Initial complexes (above) were formed on a supercoiled *oricC* plasmid with DnaA protein only. Prepriming complexes (below) were formed with DnaA, DnaB, DnaC, and HU proteins. Complexes were cross-linked and the DNA was cut with a restriction endonuclease. Protein complexes are seen at the *oriC* site asymmetrically situated on the DNA fragments. (Taken from Kornberg and Baker, 1992, with permission of the publisher.)

1988), and has been found in living cells to be located at the cell membrane (Newman and Crooke, 2000). This will be discussed further in a separate section (Section IV).

B. Elongation

The elongation of DNA from the initiation site bubble is truly a remarkable and efficient process. Enzymes (see below) interact with each template strand polymerizing new DNA at approximately 6×10^4 base pairs/ min, completing a typical prokaryotic chromosome of approximately 4.8 megabases bidirectionally in 40 minutes (Helmstetter and Leonard, 1987). The amazing aspect of this feat is that there are many other essential proteins (more than 40 at the latest count) acting in coordination and that each strand is elongated in opposite directions, although it appears that they are being synthesized simultaneously. At the heart of the elongation process are the activities of two out of the three known DNA polymerases (I and III) that are absolutely required for elongation. The latter polymerase is the principal replicative enzyme in most prokaryotes, consisting of at least 10 distinct subunits organized as two complete units, one for each strand (see below).

Three inherent problems embody the complexity of the process. The first is that the two DNA strands of the helix are mirror images of each other (or antiparallel) and that one strand runs in the 5' to 3' direction, while the other strand runs in the opposite direction 3' to 5'. These notations refer to the chemical structure of each DNA strand as shown in Figure 4a.

The second is that all DNA polymerases, as far as is known, extend a growing DNA chain by the addition of a deoxyribonucleoside triphosphate precursor (dNTP) to an open 3' OH group in a 5' to 3' direction as shown in Figure 4b. Thus only one strand (termed the *leading strand*) can be extended continuously in the same direction as the replication fork. The other strand (termed the *lagging strand*) can not be extended in this way because there is no open 3' OH group at the same end as its complementary strand (see Figs. 4a, b above). Therefore it is necessary to elongate DNA, literally, in the direction opposite from that of the leading strand and replication fork, at least for a short distance, in order for it to appear as if elongation of both strands is occurring simultaneously in the same direction.

The third is that no DNA polymerase is capable of starting a DNA chain de novo. It can only extend a chain already initiated. Therefore another mechanism is required to provide a "primer" with an open 3' OH group for extension on both the leading and lagging strands).

After considerable genetic and biochemical analysis of these three problems (summarized in Kornberg and Baker, 1992; Marians, 1992, 1996; and Ogawa and Okazaki, 1980) the concept of continuous and discontinuous synthesis was proposed in which one strand (3'–5') serves as a template for continuous DNA synthesis (leading strand) while the other strand (5'–3') serves as a template for discontinuous DNA synthesis (the lagging strand). In the former only one point of initiation is required whereas in the latter many separate initiation points are necessary. Involvement of RNA as the primer for DNA chain extension was inferred initially from the sensitivity of such replication to rifampicin, an inhibitor of RNA polymerase activity (Brutlag et al., 1971). However, it appears that this particular requirement is related to the phenomenon of transcriptional activation of regions upstream from the initiation site which aid the DnaA protein to open the DNA duplex (Baker and Kornberg, 1988). Instead, the primer that does consist of a short 10 to 15 bp segment of RNA is synthesized by another type of polymerizing enzyme, the primase, which actually can polymerize both DNA and RNA precursors (see next section). The existence of an RNA-DNA single-stranded molecule during elongation was demonstrated by a number of techniques, among them the detection of a covalent phosphodiester bond between a deoxynucleotide (DNA precursor) and a ribonucleotide (RNA precursor) (described in Kornberg and Baker, 1992). The entire process is illustrated in Figures 5a and 5b.

I. Fine details of elongation

Knowledge of the details of elongation is still emerging, and although there is general agreement concerning the basic mechanisms, much still remains to be settled. Genetic analysis has played a leading role in explaining most of what is known. One of the classic mutations (polA1) revealed that the first and most abundant DNA polymerase discovered (DNA pol I; Lehman et al., 1958) was not

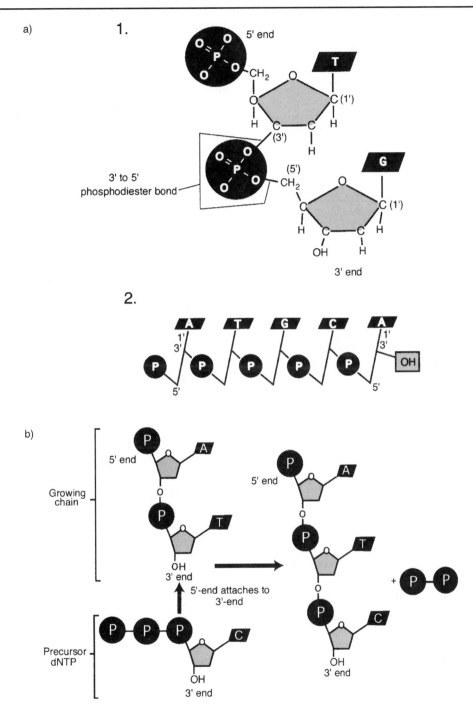

Fig. 4. a: (1) The linkage of two nucleotides by the formation of a C-3′-5′ (3′–5′) phosphodiester bond, producing a dinucleotide. **(2)** A shorthand notation for a polynucleotide chain. (Reproduced from Klug and Cummings, 2000, with permission of the publisher.) **b:** Demonstration of 5′ to 3′ synthesis of DNA.

a)

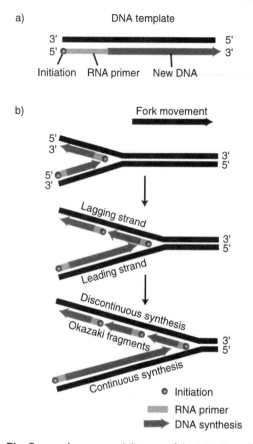

Fig. 5. a: A conceptual diagram of the initiation of DNA synthesis. A complementary RNA primer is first synthesized to which DNA is added. All synthesis is in the 5′–3′ direction. (Reproduced from Klug and Cummings, 2000, with permission of the publisher.) b: Illustration of the opposite polarity of DNA synthesis along the two strands, necessary because the two strands of DNA run antiparallel to one another and DNA polymerase III synthesis occurs only in one direction (5′–3′). On the lagging strand, synthesis must be discontinuous. On the leading strand, synthesis is continuous. RNA primers are used to initiate synthesis on both strands. (Reproduced from Klug and Cummings, 2000, with permission of the publisher.)

the only polymerase present in bacteria but also was not the true replicative polymerase since polA1 mutants could still replicate DNA (Delucia and Cairns, 1969). Nevertheless, an essential function of DNA pol I was revealed because of defects in the mutant's ability to repair DNA. Since then, a variety

of other mutants, in particular, conditional mutants (those whose products are inhibited under restrictive, but not permissive, conditions, e.g., high [42 °C] and low [30 °C] temperatures respectively) provided significant insights into whether a particular component (enzyme or control protein) was present and active in a particular complex. Table 1 depicts a number of genes that were discovered in this and other ways involved in many aspects of DNA replication or repair in *E. coli*. Although such genes are presumably present in other organisms, only *B. subtilis* has been investigated to any great extent and there are some significant differences between these two organisms as well (Yoshikawa and Wake, 1993; Imai et al., 2000).

The steps in elongation of *E. coli* can be described best as a series of points (summarized in Kornberg and Baker, 1992; Marians, 1992, 1996, 2000; Kelman and O'Donnell, 1995).

1. After the melting of the AT rich regions by the DnaA-DNA complex in the origin to provide a 45 base pair bubble during initiation, the separated single-stranded DNAs are coated with SSB (Meyer and Laine, 1990) to (a) prevent their degradation, (b) keep each strand rigid, and (c) possibly direct priming of DNA synthesis to specific unbound regions in the open bubble.

2. The DnaA protein directs the DnaB–DnaC protein complex into the open region at *oriC* to extend the duplex opening and generate a prepriming complex by a protein-protein interaction with DnaB (Marszalek and Kaguni, 1994). DnaB performs two functions. It acts as a helicase to unwind the duplex in front of the replication fork (in the presence of ATP) (see step 4 below) and as a marker or activator for the primase (DnaG) that begins the synthesis of the RNA primers both on the leading and lagging strands (Rowen and Kornberg, 1978). Because replication is bidirectional, two DnaB–DnaC complexes are positioned at the beginning of each replication fork. The DnaC protein also has two functions (Kobori and Kornberg, 1982; Allen and

TABLE 1. A Partial List of Genes Involved in DNA Replication and Repair of *E. coli*

Gene	Protein
polA	DNA polymerase I (repair and replication)
polB	DNA polymerase II (repair of UV damage)

dnaE, N, Q, X, holA, holB,
    holC, holD, holE
	DNA polymerase III subunits (main replicative
	    enzyme)

dnaG	Primase (initiates RNA primers)
priA, priB, priC, dnaT	subunits of the primosome (with DnaG, DnaB and DnaC)
dnaB, C	Helicase and helicase binder, respectively
dnaA	Initiation
gyrA, B	Gyrase subunits (relaxes supercoils)
lig	DNA ligase (joining enzyme)
ssb	Single-stranded binding proteins
rnha	Ribonuclease H (degrades single-stranded RNA in RNA-DNA hybrid molecule)

Sources: Compiled from Kornberg and Baker, 1992; Marians, 1992, 1996, 2000.

Kornberg, 1991). It keeps the DnaB protein in an inactive state until the latter is positioned via a cryptic DnaC–DNA binding site onto the SSB-free denatured DNA. Once binding occurs, DnaC is dissociated from *oriC* to allow for DnaB function as a helicase.

3. At this instant, two different but interacting complexes form at each replication fork. One contains the DNA polymerase III ten subunit replicase (called the *holoenzyme*) that synthesizes both nascent leading and lagging strands in a coordinated manner (one holoenzyme for each strand), and the other contains the primosome. The latter consists of a seven subunit multienzyme complex that is positioned along the lagging strand template and unwinds both the parental template and synthesizes the RNA primers for multiple initiations of the small 2 kb DNA fragments (termed *Okazaki fragments*, after the scientist who discovered them) that characterize discontinuous DNA synthesis (Ogawa and Okazaki, 1980; Marians, 1992). The leading strand presumably only requires one priming event with the DnaG protein, and so synthesis is continuous.

4. As unwinding mediated by the DnaB helicase proceeds, supercoiling intensifies ahead of the replication fork; that is, the double helix becomes twisted more tightly. Such supercoiling must be relieved and that is accomplished by the action of special enzymes called *topoisomerases* (in bacteria, *DNA gyrase*) (Wang, 1987). This bacterial enzyme acts by nicking the supercoil on both strands to "relax" the supercoil and then acts to reseal the nick or nicks ahead of the DnaB helicase (all of this occurring in the presence of ATP).

5. DNA pol III is the principal replication enzyme required for elongation of a duplex template in *E. coli* and probably most other prokaryotes although different subunits might replace some of the *E. coli* components in these other organisms. Present only in very small amounts (10–20 molecules/cell with a molecular weight of 900 kDa), it is a remarkably efficient "replicating machine" (Kelman and O'Donnell, 1995). How the holoenzyme is actually recruited to the initial replication forks is unknown. It may not be due to any specific protein to protein interaction but rather to an extremely efficient recognition of a primer terminus (O'Donnell, personal communication), which then acts to bind the holoenzyme as follows: One of the subunits is a complex of

six monomers (β-subunit) that encircles the DNA as a clamp, while another of the subunits acts to load the clamp on to the DNA (clamp loader). The clamploader is quite complex and consists of five different proteins (γ-complex). A two-stage process is envisaged in which the γ-complex first recognizes a primed template (on the leading or lagging strand) and in the presence of ATP assembles the clamp on to the template. In a second step the catalytic core of the polymerase (consisting of three subunits including a 3′–5′ endonuclease [and its stimulator] to excise the occasional mismatched nucleotide in base pairing [proofreading] and the polymerizing [catalytic] enzyme that recognizes the correct deoxyribonucleoside triphosphate precursor as well as the correct template base to which it will be paired) is assembled behind the clamp. Although it was originally assumed that the clamp slid along the DNA template in a processive manner (i.e., the maintenance of enzyme activity over a relatively long sequence of template for both the leading and lagging strands), it is most probable, instead, that the template migrates through a fixed holoenzyme site (factory model; Lemon and Grossman, 1998) and that this site is probably the cell membrane (Firshein, 1989, Firshein and Kim, 1997) (see Section IV). Other important features of this remarkable process include the dimerization of the catalytic core by the *dnaX* gene. Most of these points are depicted in Table 2 and Figures 6 to 10.

C. Termination

The most significant event to occur after elongation is initiated in a bidirectional manner from a fixed origin on a circular chromosome, is termination approximately 180° from the initiation site. The movement of the replication forks, like two express trains, must be inhibited from "crashing" into each other, and at the same time some mechanism must exist to separate the two chromosomes after they have been completed. Only two bacterial species, *E. coli* and *B. subtilis*, have been examined in detail in this respect, and although the overall features are similar (inhibition of replication forks, and separation of completed chromosomes), many differences exists in specific details (for reviews, see Yoshikawa and Wake, 1993; and Baker, 1995). In both organisms the main target for inhibition may be the DNA helicases (DnaB in *E. coli*; DnaC in *B. subtilis*) (Lee et al., 1989; Khatri et al., 1989; Imai et al., 2000).

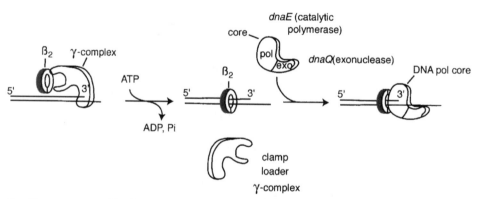

Fig. 6. Two-stage assembly of a processive polymerase. The γ-complex recognizes a primed template and couples hydrolysis of ATP to assemble β on DNA. The γ-complex easily dissociates from DNA and can resume its action in loading β clamps on other DNA templates. In a second step, core assembly with the β clamp to form a processive polymerase. (Taken from Kelman and O'Donnell, 1995, with permission of the publisher.)

TABLE 2. DNA Polymerase III Holoenzyme Subunits and Subassemblies

Subunit	Gene	Mass (kDa)	Function	Subassembly
α	dnaE	129.9	DNA polymerase	core / pol III′ / pol III*
ε	dnaQ, mutD	27.5	Proofreading 3′–5′ exonuclease	core / pol III′ / pol III*
φ	holE	8.6	Stimulates ε exonuclease	core / pol III′ / pol III*
τ	dnaX†	71.1	Dimerizes core. DNA-dependent ATPase	pol III′ / pol III*
γ	dnaX†	47.5	Binds ATP	γ-complex / pol III*
δ	holA	38.7	Binds to β	γ-complex / pol III*
δ′	holB	36.9	Cofactor for γ ATPase and stimulates clamp loading	γ-complex / pol III*
χ	holC	16.6	Binds SSB	γ-complex / pol III*
ψ	holD	15.2	Bridge between χ and γ	γ-complex / pol III*
β	dnaN	40.6	Clamp on DNA	

Note: † This gene contains two coding sequences due to a frameshift.
Source: (modified from Kelman and O'Donnell, 1995, with permission of the publisher).

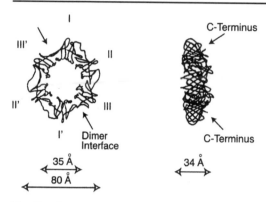

Fig. 7. The molecular structure of the β clamp. A "doughnut" structure consisting of a head to tail dimer containing six domains with a central opening large enough to accommodate duplex DNA. (Taken from Kelman and O'Donnell, 1995, with permission of the publisher.)

The terminus regions (Ter) of both organisms contain multiple DNA replication terminators consisting of short DNA sequences that bind specific terminator proteins (replication terminator protein—RTP, in *B. subtilis* and terminus utilization substance; Tus, in *E. coli*). Thus far, 9 such terminators have been detected in *B. subtilis* and 10 in *E. coli* (Coskun-Ari and Hill, 1997; Griffiths et al., 1998). There is no relationship between the proteins involved or the sequences in the terminator regions of both organisms (Baker, 1995). Thus, in *E. coli*, Ter sites are 22 bp in length, while the *B. subtilis* Ter sites consist of 30 bp imperfect inverted repeats. The *E. coli* regions are recognized by a monomer of the Tus proteins (MW 36,000), while the *B. subtilis* terminators are recognized by two dimers of the RTP (MW 14,500). The Ter sites in *E. coli* are spread over a long distance on the chromosome (approximately 50 kb), while the Ter sites in *B. subtilis* encompass only 59 bp.

These multiple terminator sites have been thought to act as a series of trip wires to slow down the replication forks, with the outer regions acting as backups to those more centrally located within the terminus region (Griffith and Wake, 2000). However, it is important to point out that some of the Ter sites are oriented to stop the clockwise replication fork, while others are oriented to inhibit the anticlockwise replication fork in both organisms. For example, in *E. coli*, the clockwise fork passes through three Ter sites that are in an inactive orientation until it contacts one oriented in the right direction (Baker, 1995).

A model of helicase inactivation by the Ter complexes presupposes that there is an

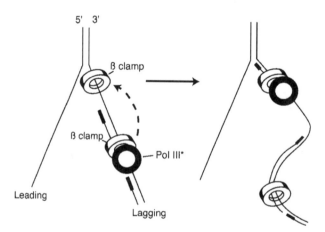

Fig. 8. Scheme of polymerase cycling on the lagging strand. Pol III holoenzyme is held to DNA by the β clamp for continuous (processive) extension of Okazaki fragment (left). At the end of polymerization, pol III (every subunit except the clamp) is dissociated from the fragment and reattaches to a new RNA primer (right) with the original clamp remaining on the finished Okazaki fragment. (Taken from Stukenberg et al., 1994, with permission of publisher.)

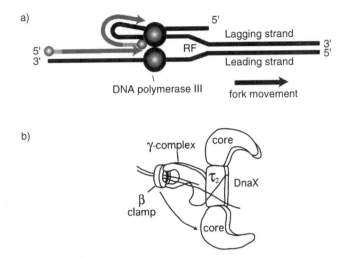

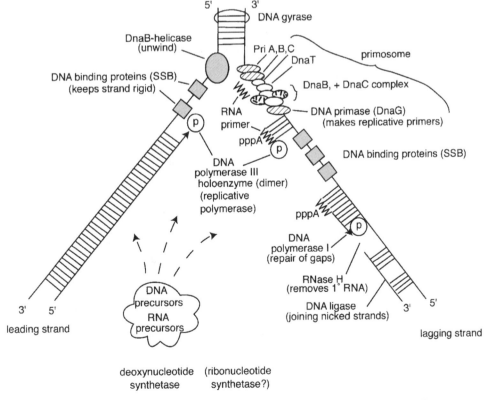

Fig. 9. Illustration of how concurrent DNA synthesis may be achieved on both the leading and lagging strands at a single replication fork. The lagging template strand is "looped" in order to invert the physical direction of synthesis, but not the biochemical direction. The enzyme functions as a dimer with each core enzyme achieving synthesis on one or the other strands. **a:** Conceptual diagram (taken from Klug and Cummings, 2000, with permission of the publisher). **b:** Two pol III cores interacting with *dnaX* which also interacts with the γ-complex clamp loader (taken from Kelman and O'Donnell, 1995, with permission of the publisher).

Fig. 10. Conceptual model of DNA replication fork without looping of the tagging strand.

inhibitory surface of the protein-DNA complex that can be oriented away or toward the helicase. In the nonpermissive orientation (toward), helicase translocation or activity is blocked, while in the permissive (away) orientation, the helicase displaces the protein and continues to unwind the DNA. Whether a protein-protein interaction between other replication fork proteins and the Ter complexes occurs is not known (Baker, 1995). One interesting consideration in this respect, however, is the lack of any role for DNA gyrase, the enzyme that precedes even the helicase (see Section IIIC) in replication arrest. It is curious that this enzyme would not interact with the terminator complexes in some manner (Yoshikawa and Wake, 1993).

Figure 11 depicts the organization of the replication and termination sites on the chromosomes of *E. coli* and *B. subtilis*.

It has been known for many years that the terminator regions of both *E. coli* and *B. subtilis* can be deleted genetically without affecting viability, or in the case of *B. subtilis* sporulation (Yoshikawa and Wake, 1993; Baker, 1995). Does this indicate that such regions are superfluous? There is in fact evidence that termination via the normal mechanism is advantageous for the organisms. Such a system may prevent "overreplication" of DNA that could generate multimeric forms of the double helix by the continuing activity of the DnaB (or DnaC) helicase after a round of replication (Hiasa and Marians, 1994). Preventing the formation of multimeric forms Is important because it could interfere with normal chromosome segregation and cell division. Indeed, a specific recombination site (dif) exists in *E. coli* (in addition to the Ter sites) precisely to prevent the formation of such structures (Baker, 1995).

D. Precursors in DNA Replication

I. Introduction

DNA replication, of course, cannot occur without the presence (really the sequestration) of the immediate DNA precursors (deoxyribonucleoside triphosphates) at the replication fork. In addition a similar sequestration must occur for the ribonucleoside triphosphates (the immediate RNA precursors) in order for synthesis of the RNA primers to occur. Among the many factors involved in this sequestration, two stand out as highly important. First, the precursors are not simply floating around in the cytoplasm, they must be brought to the replication fork simultaneously and in a balanced concentrated form. Second, since replication is such a rapid process, simple diffusion can not explain how the precursors are concentrated; rather, it is probable that some type of multienzyme complex must be kinetically coupled to replication in order for them to be available.

2. Types of metabolic pathways

Basically two types of metabolic pathways exist for the synthesis of DNA precursors in

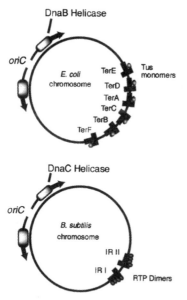

Fig. I I. Organization of the replication initiation and termination sites on the *E. coli* and *B. subtilis* chromosomes. The origins of bidirectional replication are labeled *oriC*, and the two replication forks are represented by the DnaB and DnaC helicases. The termination sites are labeled Ter for *E. coli* and IR for *B. subtilis*. The T shape denotes the polarity of the site; replication forks meeting the flat side (the top of T) are arrested. (Modified from Baker, 1995, with permission of the publisher.)

all cells, salvage and de novo (for review, see Kornberg and Baker, 1992). They are, however, interconnected with each other and with the metabolic pathways for RNA precursor synthesis and the synthesis of coenzymes as shown in Figure 12.

Salvage pathways use degradation products of DNA (and RNA) derived either from the extracellular environment or internally (purine and pyrimidine bases, deoxy(ribo)nucleosides, and deoxy(ribo)nucleotides) to recycle them back to the immediate deoxyribonucleoside triphosphate precursors via the mediation (primarily) of a group of enzymes, the deoxy(ribo)nucleoside and -tide kinases. They use ATP as a phosphoryl donor to add successive phosphate groups to the deoxy(ribo)nucleoside or deoxy(ribo)-nucleotide. (The bases must first be condensed with deoxyribose or ribose before the kinases can act.) Many interconversions of the purine and pyrimidine bases can occur when they are by themselves or part of the nucleoside-nucleotide structures to satisfy the needs for DNA synthesis.

De novo synthetic pathways supply the immediate DNA precursors by an extensive series of enzymatic reactions beginning with the formation of the purine and pyrimidine skeletons from several different amino acids (glutamic acid, glycine, and aspartic acid), formate, CO_2, and NH_3. The next step, interestingly enough, is the formation *not* of deoxyribonucleosides, but of ribonucleotides (ribonucleoside monophosphates), using phosphoribosyl pyrophosphate (PRPP) a ribose derivative that is formed by the condensation of ribose with ATP under the control of the enzyme PRPP synthase. All four ribonucleotides are produced after a complex series of additional enzymatic reactions, which are then phosphorylated by different kinases in the presence of ATP to form the ribonucleoside diphosphate derivatives.

It is at this level that perhaps the most important step occurs, the reduction of the ribo derivative to the deoxy-derivative simply by reducing ribose to deoxyribose (i.e., removing the hydroxyl group on the second carbon of ribose and replacing it with a hydrogen atom). The enzyme controlling this reaction (ribonucleotide reductase, or more properly, the ribonucleoside diphosphate reductase) consists of two subunits (R1 and R2) encoded by two genes (*nrdA* and *nrdB*) within the same operon (Jordan and Reichard, 1998). They associate in the presence of Mg^{++}. Each subunit contains two identical polypeptides (85 kDa for R1 and 43 kDa for R2). The heavier polypeptide is involved primarily in binding the substrates and in regulating their activities by binding allosteric effectors notably ATP (which activates the enzyme) and dATP (which inhibits it). Other specific deoxynucleoside triphosphate effectors are also in-

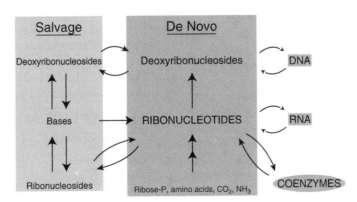

Fig. 12. Salvage and de novo pathways of nucleotide biosynthesis. (Taken from Kornberg and Baker, 1992, with permission of the publisher.)

volved in regulation of specific substrate activations, resulting in a strikingly fine adjustment to the needs for DNA synthesis (Fig. 13). The lighter polypeptide is involved in electron transfer, which leads to the actual catalytic reduction of the ribo- to deoxyribo-derivative.

The thymine moiety of DNA is not made directly from the reductase pathway (see Fig. 13). Instead, the product of uridine diphosphate (UDP) reduction, dUDP, is phosphorylated to dUTP by a diphosphate kinase (see below) and then degraded by a strong phosphatase (dUTPase) (Shlomai and Kornberg, 1978; Hoffman et al., 1987) to dUMP. Mutants deficient in this enzyme (dut) (Hajj et al., 1988) permit incorporation of dUTP into DNA resulting in the activation of repair systems to excise the uracil moiety. It is at the level of dUMP that dTMP (thymidine monophosphate) is formed by the addition of a methyl group. This latter step is mediated by another enzyme subject to extensive regulation, thymidylate synthetase (Belfort et al., 1983; Climie and Santi, 1990).

One more important enzyme, nucleoside diphosphate kinase, phosphorylates all eight of the deoxy(ribo)nucleoside diphosphates to the triphosphate immediate DNA and RNA precursor level. This enzyme is a powerful nonspecific kinase (Roisin and Kepes, 1978; Munoz-Dorado et al., 1990) and is used in both the salvage and de novo pathways.

3. Multienzyme complexes

As discussed in the Introduction to this section, the precursors are probably synthesized by a multienzyme complex, somehow kinetically coupled to the polymerizing activities of the DNA pol III holoenzyme. Evidence for such multienzyme complexes in prokaryotes have been presented by a number of investigators (Firshein, 1974; Lunn and Piaget, 1979; Chiu et al., 1982; Mathews et al., 1989; Laffan et al., 1990). As many as 10 or more enzymes have been found to be associated with a deoxynucleotide synthetase complex, including deoxynucleoside and -tide kinases, ribonucleoside diphosphate reductase, thymidylate synthetase, and the nucleoside diphosphokinase. Evidence for the existence of this complex includes the following: coelution after fractionation of whole cell extracts by affinity chromatography or comigration after gel electrophoresis, the detection of mutants defective in complex formation, and catalytic facilitation in which earlier precursors are more efficiently processed than later ones as a substrate for a particular end product. However, in only two studies have these precursor activities been coupled to DNA polymerase and DNA ligase activity, namely by a membrane associated DNA complex extracted from Pneumococci (Firshein, 1974) and *B. subtilis* (Laffan et al., 1990).

A model for kinetic coupling and catalytic facilitation is depicted in Figure 14.

IV. THE REPLICON MEMBRANE INTERACTION

A. Introduction

The replicon model of Jacob et al. (1963) (Section IIB) proposed that the membrane was the site of DNA replication in the prokaryotic cell, as well as the site through

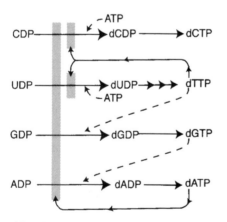

Fig. 13. Allosteric effects of ATP and specific deoxyribonucleoside-triphosphates on ribonucleoside diphosphate reductase activity. Solid bars indicate inhibitions; dashed lines indicate activations. (Taken from Kornberg and Baker, 1992, with permission of the publisher.)

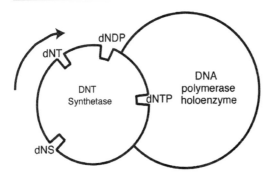

Fig. 14. Model for kinetic coupling and catalytic facilitation. Adding dNS (deoxynucleosides), dNT (deoxynucleotides), or dNDP (deoxynucleoside diphosphates) enables the precursors to be channeled better than dNTP (deoxynucleoside triphosphates, the immediate DNA precursors) because they can enter the complex more easily.

which the newly synthesized chromosome was segregated into a daughter cell. It made evolutionary sense that because of the lack of a nucleus in prokaryotes, it was necessary to sequester the many components involved in these events so that they could interact, even though most of them are not, by themselves, membrane proteins. Nevertheless, since the model was proposed, many proteins involved in DNA replication of prokaryotes have been found to require a membrane environment to function, to be activated by components of the membrane, or to be membrane associated (for reviews, see Firshein, 1989; Firshein and Kim, 1997; Sueoka, 1998). Of particular importance in this respect are a significant number of initiation proteins that function in a wide variety of bacteria, plasmids, and bacterial viruses. They include the all important DnaA protein of *E. coli* (Sekimizu and Kornberg, 1987; Yung and Kornberg, 1988), the DnaB protein of *B. subtilis* (not to be confused with the DnaB helicase of *E. coli*) (Hoshino et al., 1987), the TrfA initiation proteins of the broad host range plasmid RK2 (Kim et al., 2000), and the gene 69 product of bacteriophage T4 (Mosig and MacDonald, 1986). The requirements for these types of proteins to be concentrated at the origin of replica-

tion, coupled with relatively low rates of synthesis in the cell (e.g., Durland and Helinski, 1990) probably accounts for their membrane affinity, and in fact two such proteins contain domains that enable them to interact with the membrane (Kim et al., 2000; Newman and Crooke, 2000).

Proof for the membrane as the site for replication has been subject to many difficulties such as confirming that macromolecules observed in a cell lysate represent a natural complex before lysis. In addition the existence of weak DNA-membrane interactions and the probable temporary nature of many such interactions, as well as a lack of knowledge of membrane receptors for DNA, has caused conflicting interpretations of specific results. Nevertheless, the total weight of support for the replicon-membrane interaction is compelling, consisting of genetic, molecular, ultrastructural, and biochemical analysis. However, as will be discussed below, both positive and negative models for control of DNA replication by the membrane have been proposed. Therefore it is not surprising that different organisms display differences in their DNA-membrane interactions. Indeed, it appears that basic differences exist between gram-negative and gram-positive bacteria in this respect, and that other modifications may be present in plasmids and bacteriophages.

B. Specific Organisms

1. E. coli

Early attempts to elucidate how *E. coli* DNA bound to the membrane involved the identification of membrane proteins that could affect DNA replication (e.g., Gudas et al., 1976; Heidrich and Olsen, 1975). However, it was not until the isolation or detection of the origin of replication that further progress was possible. A number of groups reported that origin region DNA was not only enriched in membrane fractions extracted from *E. coli*, but that it was the outer membrane fraction that was involved (Wolf-Watz and Masters, 1979; Jacq et al., 1983;

Hendrickson et al., 1982). An important new detail was added in that origin DNA bound to the outer membrane only when it was in a hemimethylated state (methylation of only one strand of double-stranded DNA), whereas fully methylated or unmethylated DNA did not (Ogden et al., 1988). This region contains many GATC sites recognized by adenine methylase (dam sites) that are apparently important in mediating the origin-membrane interaction.

When additional studies by Landoulsi et al. (1990) demonstrated an inhibition of DNA replication in vitro in a crude *oriC* plasmid system by outer cell membrane preparations only when the *oriC* plasmid was in a hemimethylated state (but not unmethylated, or fully methylated), the possibility of a negative control mechanism for the membrane (instead of positive!) was raised. Such a mechanism in *E. coli* has in fact proved to possess some validity (Marians, 1992), although many questions still remain. To be sure, it is a membrane event in which newly replicated origins are hemimethylated and transiently sequestered by the outer membrane perhaps with the aid of a 20.5 kDa protein (SeqA) (Lu et al., 1994) to prevent premature reinitiation. Initiation would occur again by release of the sequestered hemimethylated origin from the outer membrane. The *dnaA* gene is also involved in this model in that in order for it to be transcribed to produce the DnaA protein, the promoter must be fully methylated (Braun and Wright, 1986) which it is at the time of initiation (Ogden et al., 1988). However in a hemimethylated form, the *dnaA* promoter would be sequestered by the outer membrane and unable to transcribe the gene.

Recent studies, however, have suggested that the SeqA protein may have other inhibitory functions on both methylated and hemimethylated origins. These include (a) the interference with open complex formation during initiation (Torheim and Skarstad, 1999) and (b) the displacement of the DnaA protein from its origin binding (Skarstad et al., 2000). Clearly, the relationship between outer membrane binding of *oriC*, the state of methylation of *oriC*, and the role of the SeqA protein remains to be elucidated.

Perhaps the most important question in this respect is where initiation is occurring when hemimethylated *oriC* DNA is remethylated and released from the outer membrane or when the SeqA protein is rendered inactive? In fact, numerous studies in vivo and in vitro have suggested strongly that the DnaA initiation protein functions as a membrane associated protein in a membrane environment (Yung and Kornberg, 1988) and that anionic phospholipids are vital in maintaining its activity (Sekimizu and Kornberg, 1987; Newman and Crooke, 2000). Could it be that replication still occurs in a membrane environment, but in association with the inner membrane and that transfer of *oriC* between these two domains would control initiation, as suggested by Landoulsi et al. (1990)? Some support for this possibility comes from Chakraborty et al (1992) who observed *oriC* binding activity in a small inner membrane subfraction, although the DNA was still in a hemimethylated form.

However, until an in vitro membrane associated replicating system is developed that can replicate *E. coli* DNA (or an *oriC* plasmid) endogenously (i.e., without the addition of exogenous template or enzymes) as in other organisms (Laffan and Firshein, 1987; Kim and Firshein, 2000), it will be difficult to elucidate the exact mechanisms involved in control of initiation.

2. *B. subtilis*

The situation in *B. subtilis* with respect to the role of the membrane in DNA replication is significantly different than *E. coli*. First, *B. subtilis* has one membrane as compared to the two membranes found in *E. coli*. Second, there are no methylation sites in *oriC* (Yoshikawa and Wake, 1993) as there are in *E. coli*. Third, *B. subtilis* contains a gene, *dnaB*, whose counterpart does not exist in *E. coli* and whose product may be involved in the attachment of *oriC* to the membrane (Winston and Sueoka, 1980).

Early genetic studies by Sueoka and colleagues suggested that both the origin and terminus regions were membrane associated (Yoshikawa and Sueoka, 1963). The unique initiation *dnaB* gene was first detected as a temperature sensitive mutation (Karamata and Gross, 1970; White and Sueoka, 1973) and was subsequently shown to be involved in binding of the origin region to the membrane, but not the terminus (Winston and Sueoka, 1980). The *dnaB* gene was cloned (Hoshino et al., 1987; Ogasawara et al., 1986) and found to be the first gene of an operon containing three or four other genes including one that is very hydrophobic (ORFZ), and the *dnaI* gene which is part of the *B. subtilis* primosome and resembles the *dnaC* gene of *E. coli* (helicase loader). The DnaB protein itself (55 kDa) has a hydrophobic region, a possible ATP-binding site, and binds tightly but nonspecifically to single-stranded DNA (Sueoka, 1998).

Despite the extensive characterization of the DnaB protein and its possible role in mediating a membrane-*oriC* interaction, there is no definitive mechanism to explain its function. It may in fact be that the presence of the *dnaB* gene in the same operon as the *dnaI* gene (the probable functional equivalent of the helicase loader gene in *E. coli*) could point to another role, namely as an aid to the *dnaI* gene product in loading the *B. subtilis* helicase (Imai et al., 2000).

In another approach to characterize the membrane DNA-interaction, a membrane associated DNA replication complex was developed in which a membrane fraction was used as the sole source of template and enzymes (termed *endogenous replication*) to synthesize DNA. The principle was that if the site of the replicon was indeed in the membrane, it should be possible to use an extracted membrane fraction to detect initiation and elongation, simply by adding soluble precursors, cofactors, cations, and an energy-generating system. Much success has in fact been achieved with this system *in B. subtilis* (for review, see Firshein, 1989), although it is possible that the initiation was due to activation or continuation of preexisting initiating complexes that were formed prior to the extraction of the membrane fraction. Nevertheless, one interesting new observation was the detection of a membrane associated protein that bound double-stranded DNA near the origin region and acted as an inhibitor of initiation (Laffan and Firshein, 1988). The identification of this protein as a subunit of the pyruvate dehydrogenase complex (dihydrolipoamide acetyltransferase) was remarkable in that it constituted a new and direct link between the metabolic state of the cell and gene expression, in this case DNA replication (Stein and Firshein, 2000).

3. Plasmid RK2

Of all the studies concerning the replicon membrane interaction, most of the positive results have been obtained with this broad host range plasmid (for review, see Firshein and Kim, 1997). Not only has a unique membrane subfraction been detected that binds the plasmid origin (oriV), but the binding is apparently mediated by initiation proteins encoded by the plasmid, and the subfraction synthesizes plasmid DNA uniquely.

Early studies with miniplasmid deivatives of the 60 kb plasmid (cultured in its *E. coli* host) demonstrated that a mini plasmid DNA/membrane complex could be extracted from *E. coli* that synthesizes the entire supercoil DNA product in a semiconservative manner (Firshein et al., 1982; Kornacki and Firshein, 1986; Michaels et al., 1994). Of interest was the observation that the plasmid initiation proteins, TrfA-33_kDa and 43 kDa (encoded by overlapping genes), were strongly associated with both the inner and outer membrane fractions, although they are not integral membrane proteins and have only one short hydrophobic amino acid domain which is too small to traverse the membrane (Kostyal et al., 1989).

Further results demonstrated that despite the binding of the initiation proteins to both membrane fractions, it was only the inner membrane that synthesized plasmid specific

DNA that was inhibited by anti-TrfA antibody (Michaels et al., 1994). The outer membrane was mostly inactive in this respect as was the soluble (cytoplasmic) fraction. Similar results were obtained with four other gram negative species that harbored the plasmid (Banack et al., 2000). Fractionation of the inner and outer membrane fractions into a series of six subfractions by flotation sucrose gradient centrifugation (Ishidate et al., 1986) revealed that the initiation proteins are bound in vivo and in vitro to two specific subfractions, one derived from the inner membrane fraction (representing only 10% of the entire membrane) and one derived from the outer membrane. However, the specific activity of binding was much greater in the former than the latter. In addition the same inner membrane subfraction was found to bind *oriV* much more strongly than the outer membrane fraction as judged by nitrocellulose filter binding assays (Mei et al., 1995). Binding of *oriV* to the membrane was mediated solely by the TrfA initiation proteins, since it occurred only when the subfractions were extracted from plasmid containing cells. If the membrane subfractions were derived from plasmid-free cells, *oriV* binding was nonspecific and erratic. To further characterize the subfractions, each was assayed for their ability to synthesize plasmid DNA. Of great significance was the finding that only the subraction derived from the inner membrane that contained the TrfA proteins and bound *oriV* synthesized such DNA (Kim and Firshein, 2000). Thus these results bring together for the first time synthetic capability, presence of the TrfA initiation proteins, and *oriV* binding into one relatively small membrane domain, suggesting that it is the site of the plasmid replicon in the bacterial cell.

V. GENERAL CONCLUSIONS

The last decade has witnessed enormous progress in understanding the details of DNA replication in prokaryotes and the structures of many of the important enzymes, in particular, the DNA polymerase III holoenzyme, the true replicative polymerase of most prokaryotes. In addition the almost universal mechanism for the initiation of DNA replication involving the denaturation of AT-rich regions at a unique site on the chromosome induced by a number of initiation proteins acting alone or in tandem has been reinforced and expanded. Most important, the wisdom of the replicon model enunciated more than 38 years ago by Jacob et al., (1963) has, with modifications, proved to be of utmost value in conceptualizing the circular chromosome as a genetic unit of replication. A particularly unifying theme that has received significant new support has been the probable identification of the site of DNA replication (and the replicon) as the cell membrane, although there may be differences in control features between the two large groups of prokaryotes (gram-negative and gram-positive bacteria). Nevertheless, this confirmation has induced a more significant analysis of the coordination of cell division with chromosome replication, since the cell membrane is also the site for many partitioning proteins.

ACKNOWLEDGMENTS

Support from the Army Research office is greatly appreciated.

REFERENCES

Allen GC, Kornberg A (1991): Fine balance in the regulation of DnaB helicase by DnaC protein in replication in *Escherichia coli*. J Biol Chem 266:22096–22101.

Baker TA (1995): Replication arrest. Cell 80:521–524.

Baker TA, Kornberg A (1988): Transcriptional activation of initiation of replication from the *E. coli* chromosomal origin: an RNA-DNA hybrid near *oriC*. Cell 55:113–123.

Banack T, Kim, PD, Firshein, W. (2000): TrfA-dependent inner membrane-associated plasmid RK2 DNA synthesis and association of TrfA with membranes of different gram-negative hosts. J Bacteriol 182:4380–4383.

Belfort M, Maley GF, Maley, F (1983): Characterization of the *Escherichia coli* ThyA gene and its amplified thymidylate synthetase product. Proc Natl Acad Sci USA 80:1858–1862.

Bird RE, Louarn JM, Martuscelli J, Caro L (1972): Origin and sequence of chromosome replication in *Escherichia coli* J Mol Biol 70:549–566.

Bramhill D, Kornberg A (1988): A model for initiation at origins of DNA replication. Cell 54:915–918.

Braun RE, Wright A (1986): DNA methylation differentially enhances the expression of one of the two *E. coli dnaA*, promoters in vivo and in vitro. Mol Gen Genet 202:246–250.

Brutlag D, Schekman R, Kornberg A (1971): A possible role for RNA polymerase in the initiation of M13 DNA synthesis. Proc Natl Acad Sci USA 68:2826–2830.

Cairns J (1963): The bacterial chromosome and its manner of replication as seen by autoradiography. J Mol Biol 6:208–231.

Chakraborti A, Gunji S, Shakibai N, Cubedda J, Rothfield L (1992): Characterization of the *Escherichia coli* membrane domain responsible for binding *oriC* DNA. J Bacteriol 174:7202–7206.

Chiu CS, Cook KS, Greenberg GR (1982): Characterization of bacteriophage T4-induced complex synthesizing deoxyribonucleotides. J Biol Chem 257: 15087–15097.

Climie S, Santi DV (1990): Chemical synthesis of the thymidylate synthase gene. Proc Natl Acad Sci USA 87:633–637.

Coskun-Ari, FF, Hill, TM (1997): Sequence-specific interactions in the Tus-Ter complex and the effect of base pair substitutions on arrest of DNA replication in *Escherichia coli*. J Biol Chem 272:26448–26456.

Delucia P, Cairns J (1969): Isolation of an *E. coli* strain with a mutation affecting DNA polymerase. Nature 224:1164–1166.

Durland RH, Helinski DR (1990): Replication of the broad host-range plasmid RK2: direct measurement of intracellular concentrations of essential TrfA proteins and their effect on plasmid copy number. J Bacteriol 172:3849–3858.

Echols H (1990): Nucleoprotein structures initiating DNA replication, transcription and site specific recombination. J Biol Chem 265:14697–14700.

Engleman E (2000): A common structural core in proteins active in DNA recombination and replication. Trends in Biochem Sci 25:180–182.

Firshein W (1974): In situ activity of enzymes on polyacrylamide gels of a DNA-membrane fraction extracted from pneumococci. J Bacteriol 126:777–784.

Firshein W, Strumph P, Benjamin P, Burnstein K, Kornacki J (1982): Replication of a low-copy-number plasmid by a plasmid DNA-membrane complex extracted from minicells of *Escherichia coli*. J Bacteriol 150:1234–1243.

Firshein W (1989): Role of the DNA/membrane complex in prokaryotic DNA replication. Annu Rev Microbiol 43:89–120.

Firshein W, Kim PD (1997): Plasmid replication and partition in *Escherichia coli*. Is the cell membrane the key? Mol Microbiol 23:1–10.

Fuller RS, Kornberg A (1983): Purified dnaA protein in initiation of replication at the *Escherichia coli* chromosomal origin of replication. Proc Natl Acad Sci USA 80:5817–5821.

Funnell BE, Baker TA, Kornberg A (1987): In vitro assembly of a prepriming complex at the origin of the *Escherichia* chromosome. J Biol Chem 262: 10327–10334.

Griffiths AA, Andersen, PA, Wake RG (1998): Replication terminator protein-based replication fork–arrest systems in various *Bacillus* species. J Bacteriol 180: 3360–3367.

Griffiths AA, Wake RG (2000): Utilization of subsidiary chromosomal replication terminators in *Bacillus subtilis*. J Bacteriol 182:1448–1451.

Gudas LJ, James R, Pardee AB (1976): Evidence for the involvement of an outer membrane protein in DNA initiation. J Biol Chem 251:3740–3479.

Hajj HH, Zhang H, Weiss B (1988): Lethality of a *dut* (deoxyuridine triphosphatase) mutation in *Escherichia coli*. J Bacteriol 170:1069–1075.

Hayes W (1962): Conjugation in *Escherichia coli*. British Med Bull 18:36–40.

Heidrich HG, Olson WL (1975): Deoxyribonucleic acid-envelope complexes from *Escherichia coli*: A complex-specific protein and its possible function for the stability of the complex. J Cell Biol 67:444–460.

Helmstetter CE, Leonard AC (1987): Mechanisms for chromosome and minichromosome segregation in *Escherichia coli*. J Mol Biol 197:195–204.

Hendrickson WE, Kusano T, Yamaki H, Balakrishnan R, King M, Benson A, Schaechter M (1982): Binding of replication of *Escherichia coli* to the outer membrane. Cell 30:915–923.

Hiasa H, Marians KJ (1994): Tus prevents over replication of *oriC* plasmid DNA. J Biol Chem 269: 26959–26968.

Hoffman I, Widstrom J, Zeppezouer M, Nyman PO (1987): Overproduction and large-scale preparation of deoxyuridine triphosphate nucleotidohydrolase from *Escherichia coli*. Eur J Bioch 164:45–51.

Hoshino T, McKenzie T, Schmidt S, Tanaka T, Sueoka N (1987): Nucleotide sequence of *Bacillus subtilis dnaB*; an essential gene for DNA replication initiation and membrane attachment. Proc Natl Acad Sci USA 84:653–657.

Imai Y, Ogasawara N, Ishigo-oka D, Kadoya R, Daito D, Moriya S (2000): Subcellular localization of Dna-initiation proteins of *Bacillus subtilis*: Evidence that chromosome replication begins at either edge of nucleoids. Mol Microbiol 36:1037–1048.

Ishidate E, Creegar ES, Zrike J, Deb S, Glauner B, MacAlister TJ, Rothfield LI (1986): Isolation of differentiated membrane domains from *Escherichia coli* and *Salmonella typhimurium*, including a fraction containing attachment sites between the inner and outer membranes and the murein skeleton of the cell envelope. J Biol Chem 261:428–443.

Jacq A, Kohiyama M, Lother H, Messer W (1983): Recognition sites for a membrane-derived DNA binding protein preparation in the *E. coli* replication origin. Mol Gen Genet 191:460–465.

Jacob F, Brenner S, Cuzin F (1963): On the regulation of DNA replication in bacteria. Cold Spring Harbor Symp Quant Biol 28:289–348.

Jordan A, Reichard P (1998): Ribonucleotide reductases. Annu Rev Biochem 67:71–98.

Karamata D, Gross J (1970): Isolation and genetic analysis of temperature sensitive mutants of *B. subtilis* defective in DNA synthesis. Mol Gen Genet 108:277–287.

Kelman Z, O'Donnell M (1995): DNA polymerase III holoenzyme: structure and function of a chromosomal replicating machine. Annu Rev Biochem 64:171–200.

Khatri GS, MacAllister T, Sista PR, Bastia D (1989): The replication terminator protein of *E. coli* is a DNA sequence-specific contra-helicase. Cell 59:667–674.

Kim PD, Rosche TM, Firshein W (2000): Identification of a potential membrane targeting region of the replication initiation protein (TrfA) of broad host range plasmid RK2. Plasmid 43:214–222.

Kim PD, Firshein W (2000): Isolation of an inner membrane derived subfraction that supports in vitro replication of a mini RK2 plasmid in *Escherichia coli*. J Bacteriol 182:1757–1760.

Klug WS, Cummings MR (2000): "Concepts of Genetics," 6th ed. UpperSaddle River: Prentice Hall.

Kobori JA, Kornberg A (1982): The *Escherichia coli* dnaC gene product. II. Purification, physical properties and role in replication. J Biol Chem 257:13763–13769.

Konieczny I, Helinski DR (1997): Helicase delivery and activation by DnaA and TrfA proteins during the initiation of replication of the broad host range plasmid RK2. J Biol Chem 272:33312–33318.

Kornacki JA, Firshein W (1986): Replication of Plasmid RK2 in vitro by a DNA/membrane complex: Evidence for initiation and its coupling to transcription and translation. J Bacteriol 167:319–336.

Kornberg A, Baker TA (1992): "DNA Replication," 2nd ed. San Francisco: WH Freeman.

Kostyal DA, Farrell M, McCabe A, Firshein W (1989): Replication of an RK2 miniplasmid derivative in vitro by a DNA membrane complex extracted from *Escherichia coli*: Involvement of the dnaA but not dnaK host protein and association of these and plasmid-encoded proteins with the inner membrane. Plasmid 21:226–237.

Laffan J and Firshein W (1987): DNA replication by a DNA/membrane complex extracted from *Bacillus subtilis*. Site of initiation *in vitro* and analysis of initiation potential of subcomplexes. J Bacteriol 169:2819–2827.

Laffan J, Firshein W (1988): Origin specific DNA binding membrane associated protein may be involved in repression of initiation in *Bacillus subtilis*. Proc Natl Acad Sci USA 85:7452–7456.

Laffan JJ, Skolnik IL, Hadley DA, Bouyea M, Firshein W (1990): Characterization of a multienzyme complex derived from a *Bacillus subtilis* DNA-membrane extract that synthesizes RNA and DNA precursors. J Bacteriol 172:5724–5731.

Landoulsi A, Malki M, Kern R, Kohiyama M, Hughes P (1990): The *E. coli* cell surface specifically prevents the initiation of DNA replication at *oriC* on hemimethylated DNA templates. Cell 63:1053–1060.

Lee EH, Kornberg A (1992): Features of replication fork blockage by the *Escherichia coli* terminus-binding proteins. J Biol Chem 267:8778–8784.

Lehman IR (1974): DNA ligase: Structure, mechanisms and function. Science 186:790–797.

Lehman IR, Bessman MJ, Simms ES, Kornberg A (1958): Enzymatic synthesis of deoxyribonucleic acid I. Preparation of substrates and partial purification of an enzyme from *Escherichia coli*. J Biol Chem 233:163–170.

Lemon KD, Grossman AD (1998): Localization of bacterial DNA polymerase: Evidence for a factory model of replication. Science 282:1516–1519.

Lu M, Campbell JL, Boye E, Kleckner N (1994): SeqA: a negative modulator of replication initiation in *E. coli*. Cell 77:413–426.

Lunn CA, Pigiet V (1979): Characterization of a high activity form of ribonucleoside diphosphate reductase from *E. coli*. J Biol Chem 254:5008–5014.

Maaloe O, Hanawalt PC (1961): Thymine deficiency and the normal DNA replication cycle. J Mol Biol 3:144–155.

Marians KJ (1992): Prokaryotic DNA replication. Annu Rev Biochem 61:673–719.

Marians KJ (1996): Replication fork propagation. In Neidhardt FC, Curtiss III R, Ingraham JL, Lin ECC, Low KB, Magasanik B, Reznikolf WS, Riley M, Schaechter M, Umbarger, HE (eds): "*Escherichia coli* and *Salmonella*: Cellular and Molecular Biology," 2nd ed. Washington, DC: ASM Press, pp 749–763.

Marians KJ (2000): Pri A-directed replication fork restart in *Escherichia coli*. Trends in Biochem Sci 25:185–188.

Marszalek J, Kaguni JM (1994): DnaA protein directs the binding of DnaB protein in initiation of DNA replication in *Escherichia coli*. J Biol Chem 269:4883–4890.

Masai H, Arai K (1987): RepA and DnaA proteins are required for initiation of R1 plasmid replication in vitro and interact with the *oriR* sequence. Proc Natl Acad Sci USA 84:4781–4785.

Mathews CK, Thylen C, Wang Y, Ji J, Howell ML, Slabaugh MB, Mun B (1989): Intercellular organization of enzymes of DNA precursor biosynthesis. In Srere PA, Jones ME, Mathews CK (eds): "Structural and Organizational Aspects of Metabolic Regulation." UCLA Symp on Molecular and Cell Biology, Vol 133. New York: Wiley, pp 139–152.

Mei J, Benashki S, Firshein W (1995): Interactions of the origin of replication (oriV) and initiation proteins (TrfA) of plasmid RK2 with submembrane domains of Escherichia coli. J Bacteriol 177:6766–6772.

Meijer M, Messer W (1980): Functional analysis of minichromosome replication: Bidirectional and unidirectional replication from the Escherichia coli replication origin, oriC. J Bacteriol 143:1049–1053.

Meselson M, Stahl FW (1958): The replication of DNA in Escherichia coli. Proc Natl Acad Sci USA 44:671–682.

Meyer RR, Laine PS (1990): The single-stranded DNA binding protein of Escherichia coli. Microbiol Rev 54:342–380.

Michaels K, Mei J, Firshein W (1994): TrfA-dependent inner-membrane associated plasmid RK2 DNA synthesis in Escherichia coli maxicells. Plasmid 32:19–31.

Moriya S, Firshein W, Yoshikawa H, Ogasawara N (1994): Replication of a Bacillus subtilis oriC plasmid in vitro. Mol Microbiol 12:469–478.

Mosig G, MacDonald P (1986): A new membrane-associated DNA replication protein, the gene 69 product of bacteriophage T4 shares a patch of homology with the Escherichia coli dnaA protein. J Mol Biol 189:243–248.

Munoz-Dorado J, Inouye S, Inouye M (1990): Nucleoside diphosphate kinase from Myxococcus xanthus. J Biol Chem 265:2707–2712.

Newman G, Crooke E (2000): DnaA, the initiator of Escherichia coli chromosomal replication, is located at the cell membrane. J Bacteriol 182:2604–2610.

Ng JY, Marians KJ (1996): The ordered assembly of the ϕX174–type primosome I. Isolation and identification of intermediate protein-DNA complexes. J Biol Chem 271:15642–15648.

Ogasawara J, Moriya S, Mazza PG, Yoshikawa H (1986): Nucleotide sequence and organization of dnaB gene and neighboring genes on the Bacillus subtilis chromosome. Nucl Acids Res 14:9989–9999.

Ogawa T, Okazaki T (1980): Discontinuous DNA synthesis. Annu Rev Biochem 49:421–457.

Ogden GB, Pratt MJ, Schaechter M (1988): The replicative origin of the Escherichia coli chromosome binds to cell membranes only when hemimethylated. Cell 54:127–135.

Oka A, Sugimoto K, Takanami M, Hirota Y (1980): Replication origin of the Escherichia coli K-12 chromosome: The size and structure of the minimum DNA segment carrying the information for autonomous replication. Mol Gen Genet 178:9–20.

Roisin MB, Kepes A (1978): Nucleoside diphosphate kinase of Escherichia coli, a periplasmic enzyme. Biochim Biophys Acta 526:418–428.

Rouviere-Yaniv J, Gros F (1975): Characterization of a novel low molecular-weight DNA-binding protein. Proc Natl Acad Sci USA 72:3428:3432.

Rowen L, Kornberg A (1978): Primase, the dnaG protein of Escherichia coli. J Biol Chem 253:758–764.

Shlomai J, Kornberg A (1978): Deoxyuridine triphosphatase of Escherichia coli. J Biol Chem 253: 3305–3312.

Sekimizu K, Kornberg A (1988): Cardiolipin activation of DnaA protein, the initiation protein of replication in Escherichia coli. J Biol Chem 263:7131–7135.

Skarstad K, Lueder G, Lurz R, Speck C, Messer W (2000): The Escherichia coli SeqA protein binds specifically and cooperatively to two sites in hemimethylated and fully methylated oriC. Mol Microbiol 36:1319–1326.

Stein A, Firshein W (2000): The probable identification of a membrane associated repressor of Bacillus subtilis DNA replication as the E2 subunit of the pyruvate dehydrogenase complex. J Bacteriol 182:2119–2124.

Stukenberg TP, Turner J, O'Donnell M (1994): An explanation for lagging strand replication polymerase hopping among DNA sliding clamps. Cell 78: 877–887.

Sueoka N (1998): Cell membrane and chromosome replication in Bacillus subtilis. Prog Nucl Acid Res Mol Biol 59:35–53.

Torheim NK, Skarstad, K (1999): Escherichia coli SeqA protein affects DNA topology and inhibits open complex formation at oriC. EMBO J 180:4882–4888.

Von Meyenburg K, Hassen FG, Riise E, Bergmans HE, Meijer M, Messer W (1979): Origin of replication, oriC of the Escherichia coli K-12 chromosome: genetic mapping and minichromosome replication replication. Cold Spring Harbor Symp Quant Biol 43: 121–128.

Wang JC (1987): Recent studies of DNA topoisomerases. Biochim Biophys Acta 909:1–9.

Watson JD, Crick FC (1953): Molecular structure of nucleic acids: A structure for deoxyribose nucleic acids. Nature 171:737–738.

White K, Sueoka N (1973): Temperature-sensitive DNA synthesis mutants of Bacillus subtilis—Appendix: Theory of density transfer for symmetric chromosome replication. Genetics 73:185–214.

Winston S, Sueoka J (1980): DNA membrane association is necessary for initiation of chromosomal and plasmid replication in Bacillus subtilis. Proc Natl Acad Sci USA 77:2834–2838.

Wolf-Watz H, Masters M (1979): DNA and outer membrane strains diploid for the oriC region show elevated levels of a DNA binding protein and evidence for specific binding of the oriC region to the outer membrane. J Bacteriol 140:50–58.

Yoshikawa H, Sueoka N (1963): Sequential replication of *Bacillus subtilis* chromosome I. Comparison of marker sequences in experimental and stationary growth phases. Proc Natl Acad Sci USA 49:559–566.

Yoshikawa H, Wake RG (1993): Initiation and termination of chromosome replication. In Sonenshein AL, Hoch JA, Losick R (eds): "*Bacillus subtilis* and Other Gram Positive Bacteria." Washington, DC: ASM Press, pp 507–528.

Yung BY, Kornberg A (1988): Membrane attachment activates DnaA protein, the initiation protein of chromosome replication in *Escherichia coli*. Proc Natl Acad Sci USA 85:7202–7205.

Zyskind JHW, Cleary JM, Brusilow WS, Harding NE, Smith DW (1983): Chromosome replication origin from the marine bacterium *Vibrio harvey* functions in *Escherichia coli oriC* consensus sequence. Proc Natl Acad Sci USA 80:1164–1168.

Zyskind JHW, Smith DW (1986): The bacterial origin, *oriC*. Cell 46:489–490.

2

DNA Repair Mechanisms and Mutagenesis

RONALD E. YASBIN

Program in Molecular Biology, University of Texas at Dallas, Richardson, Texas 75083

I. INTRODUCTION

All living cells are constantly exposed to chemical and physical agents that have the ability to alter the primary structure of DNA. Such alterations, if not corrected, would result in mutations. While many of these mutations would be neutral (i.e., no changes in the amino acid sequences of peptides) or would be insignificant (no involvement of regulatory regions for the DNA and RNA or in the case of proteins no alteration of active sites), the accumulation of significant mutations has the potential to increase the genetic diversity of a species. Such genetic diversity is an essential component of evolution and the ability of species to survive in changing environments. However, there does come a critical point in the accumulation of mutations (genetic load) at which time the species can no longer exist (Dobshansky, 1950). Thus it would seem obvious that living systems must maintain mechanisms for the repairing of DNA damage. It would also seem obvious that

these same systems must balance the removal of DNA damage with the accumulation of a finite number of mutations. In this chapter we discuss the diversity, as well as the exciting intricacy, of the DNA repair systems found in the paradigm *Escherichia coli*. In addition we consider the dramatic changes that have occurred within the last 10 years to our understanding of the processes of DNA repair and mutagenesis. Many of these changes have been brought about by the information gained through the various genome projects.

Beginning to understand the processes associated with DNA repair and mutagenesis requires visiting the debate over whether mutations arise spontaneously or are directed by environmental conditions—Darwin versus Lamarck. In 1942 Luria and Delbrück seemed to answer this question following the publication of their fluctuation tests (Luria and Delbrück, 1943). By combining statistics with

Modern Microbial Genetics, Second Edition. Edited by Uldis N. Streips and Ronald E. Yasbin. ISBN 0-471-38665-0

an elegant investigation of mutation numbers, these pioneers demonstrated that under their laboratory conditions bacterial mutations arose spontaneously during growth. While these and related results (Lederberg and Lederberg, 1952; Newcomb, 1949) clearly supported the view that mutations are non-directed and arise spontaneously, the debate has never really ended. The last 10 years has seen a dramatic increase in the interest shown in "directed" and stress-induced mutagenesis (Wright, 2000). While this is not a new concept, the very mention of naturally occurring "directed" mutagenesis invokes the passions associated with Lamarck's views on the inheritance of acquired characteristics. In all fairness, Lamarck should also be remembered for having articulated the need for a gradual evolution from the simplest species to the most complex. Evolutionists (Dobshansky, 1950) and mathematicians have consistently questioned the probability that evolution could have proceeded as rapidly as demonstrated had true random mutagenesis been the only factor in providing genetic diversity (Wright, 2000). The validity of these questions is attested to, since today we know of the impact that transposons and transpositions can have on diversity and on the evolutionary process (Labrador and Corces, 1997; see Whittle and Salyers ch. 17). Furthermore there are data that strongly support the existence of stress-related "directed" mutagenesis mechanisms (Wright, 2000). However, it is important to note that in all of these cases there is no evidence found to support the Lamarckian concept of the inheritance of acquired characteristics.

Consistently spontaneous mutations were thought to arise almost exclusively as a consequence of growth (either errors in replication, unrepaired DNA damage or as a result of errors during the process of repairing damaged DNA). As described in the chapter by Frishein, this volume, prokaryotic DNA replication is the primary responsibility of the replicating complex. This complex of DNA polymerases and accessory proteins perform the normal semiconservative replication with

a great deal of accuracy (Friedberg et al., 2000; Friedberg et al., 1995; Ohashi et al., 2000). Without the involvement of any factors contributed by the bacteria, the potential error frequency associated with the pairing of bases would be between 1 to 10% per nucleotide. However, the actual mutation frequency for newly replicated *E. coli* DNA is six to nine orders of magnitude less frequent than the prediction based solely on energetics. At least three to six orders of magnitude of this enhanced fidelity is due to inherent properties associated with the replication machinery including the $3'$ to $5'$ exonuclease function that has editing or proofreading activity. Further reduction in the replication errors occur as a result of the functioning a protein systems involved in mismatch correction (described below).

Recently a family of error-prone polymerases that lack the $3'$ to $5'$ exonuclease editing function have been identified in eubacteria, archaea, and eukaryotes (Friedberg et al., 2000; Gerlach et al., 1999). In *E. coli* these designated DNA polymerases IV (DinB) and V (UmuD'C) have been associated with translesion processing of DNA (replication past a noninstructional lesion) and consequently with the potential generation of mutations. In eukaryotes, homologs of these polymerases have been associated with human diseases including cancer and potentially with the functioning of the diversity associated with the immune system. The existence of these polymerases and their stress-related regulation has spawned an intensive re-investigation into the nature of the mutagenesis process(es).

In 1988 John Cairns and his collaborators published a controversial and exciting article that forced rethinking about how spontaneous mutations might arise when cells are under a stress-induced selection (Cairns et al., 1988). Although there were some problems with this first report (Prival and Cebula 1996), Cairns and Foster (1991) confirmed that mutations arise in nondividing or stationary phase bacteria when the cells are subjected to nonlethal selective pressure such as nutrient-limited environments. The authors termed the accumulation of these types of mutants as adaptive mutagenesis. Since this report, additional data have accumulated supporting the existence of mutations generated while cells are in stationary phases (Bridges, 1998; Cairns and Foster, 1991; Hall, 1997) (Foster, 1998) (Rosenberg et al., 1995). While most of the research has involved the *E. coli* model system, similar observations have been made for other prokaryotes (Kasak et al., 1997) as well as for eukaryotic organisms (Steele and Jinks-Robertson, 1992). Regardless of the organism utilized, these types of mutations (called either *adaptive-* or *stationary-phase induced*) and the processes that generate them are of real interest because of their implications to evolution and the generation of diversity across the domains of life.

In the frameshift-reversion assay system that has been studied in *E. coli*, stationary-phase or adaptive mutations can be distinguished from normal growth-dependent spontaneous mutations. Specifically, the mutations generated in stationary-phase cells require a functional homologous recombination system (see chapter by Levene and Huffman, this volume), F' transfer functions (see chapter by Porter, this volume), and a component(s) of the SOS system (see below). Genetic evidence suggests that DNA polymerase III and DNA polymerase IV are responsible for the synthesis errors that lead to these mutations (Foster 1999; McKenzie et al 2000). The mechanism(s) responsible for this stationary-phase mutagenesis have not yet been delineated. However, studies have suggested that in a starving or "stressed" culture a small subpopulation of the cells seem to have an overall increased mutation frequency (Hall, 1990; Bridges, 1997; Foster, 1998; Torkelson et al., 1997; Lombardo et al., 1999). Theoretically bacteria may differentiate a hypermutable subpopulation when cells are under stressed conditions, and these hypermutable cells generate mutations randomly. The existence of a hypermutable subpopulation(s) responsible for generating genetic diversity in a stressed population raises fascinating questions concerning the nature of the molecular mechanisms that control this process as well as the potential involvement of quorum sensing systems (see chapter by Fuqua and Parsek) and prokaryotic differentiation and development regulons (see chapters by Moran, Streips, Hartzell and Ream, in this volume). Significantly in the *Bacillus subtilis* model prokaryotic system a subpopulation has already been characterized that enhances diversity through differentiation and the development of natural competence (see chapter by Streips-Transformation, in this volume) and the induction of DNA repair systems (Bol and Yasbin, 1991; Cheo et al., 1993; Yasbin et al., 1992).

II. DNA DAMAGES

Each time DNA is synthesized (see chapter by Firshein, this volume), either following semiconservative chromosome replication or following repair-replication, there is the possibility that mispairing of bases will occur. The rate of mispairing can be significantly affected by cellular metabolism, chemical alterations of the bases, and by the presence of base analogues (Friedberg et al., 1995). A transient rearrangement of bonding among the bases—this process is called a *tautomeric shift*—can occur during normal cellular metabolism. Such a rearrangement results in the production of a structural isomer of a base. These tautomers will enhance mispairing. For instance, guanine and thymine can shift from their normal keto form to an enol form. When either is in its enol form, these bases will now be able to bond to each other rather than their normal bonding partners, cytosine and adenine, respectively. Similarly, when either cytosine or adenine shift from their amino form to the imino tautomer they can now bind with each other. Following the next round of replication, the improper bonding caused by the tautomeric shifts will result in the fixation of mutations in at least one of the newly replicated DNA strands.

The exocyclic amino groups that can be found on some of the bases in DNA can be lost spontaneously in reactions that are dependent on temperature and pH. This deamination process results in cytosine, adenine, and guanine being converted to uracil, hypoxanthine, and xanthine, respectively. The products of some of these deaminations can give rise to mutations due to incorrect pairing following DNA replication (Friedberg et al., 1995). The significance of this problem is demonstrated by the existence of repair systems specific for the removal of these deamination products from DNA (described below). One deamination product that potentially represents a very serious problem is the conversion of 5-methylcytosine (a common modification) to thymine. While the other deamination products mentioned are not normally found in DNA, thymine is a natural component and its recognition as being in an incorrect location is certainly not straightforward.

In addition to deaminations, environmental and metabolic factors result in the loss of purines and pyrimidines from the DNA (apurinic and apyrimidinic sites) as well as nonenzymatic methylations of bases within the DNA. Again, these chemical changes can result in mispairing during DNA replication.

Because of its charged nature, components of the DNA are subjected to attack by reactive oxygen species. Such interactions represent a major source of spontaneous damage to DNA. Normal by-products of oxidative metabolism as well environmental factors such as ionizing radiation, near-UV light (UVA), and heat. (Friedberg et al., 1995) generate a variety of these reactive oxygen species. These species include peroxides, as well as superoxide and hydroxyl radicals that all can react with DNA directly or indirectly (Balasubramanian et al., 1998; Imlay and Linn, 1988) to produce strand breaks as well as altered bases such as 8-Oxo-7,8-dihydrodeoxyguanine (8-oxoG). This particular altered base often pairs with adenine instead of cytosine resulting in a GC to TA transversion following replication and will be dis-

cussed in a subsequent section (Friedberg et al., 1995).

Substantial evidence has been presented that hydrogen peroxide and the superoxide radicals do not react directly with DNA. Instead, the primary source of DNA damage caused by the presence of these reactive species seems to be the result of the generation of hydroxyl radicals (*OH) through Fenton-like reactions (Imlay and Linn, 1988). With respect to strand breaks and base loss, it is known that the hydroxyl radicals can abstract protons from the deoxyribose of the DNA (Balasubramanian et al., 1998).

Classically the study of DNA damage and repair systems has primarily involved the effects of ultraviolet (UV) radiation (Friedberg et al., 1995). From a research view, UV can be easily administered to cells under defined conditions. From an evolutionary view, living systems have been continually exposed to UV from the very beginning of life on the planet. From a public health view, UV and its effects are important with respect to human disease (especially cancer; Setlow 1978) as well as the maintenance of our ecosystem (Pienitz and Vincent, 2000).

The wavelengths that comprise the ultraviolet spectrum have been divided into three bands: UVA (400–320 nm), UVB (320–290 nm) and UVC (290–100 nm). While most of the early laboratory work involved UVC, it is actually UVA and UVB that constitute the majority of solar radiation that reaches the surface of the planet since wavelengths below 320 nm do not penetrate well the atmospheric ozone layer (Friedberg et al., 1995). Nevertheless, all of the research performed using UV radiation has been instrumental in our understanding of a myriad of processes including DNA repair, replication, recombination, mutagenesis, cancer biology, and cell cycling.

Following exposure to UV, the bases in the DNA strongly absorb photons that energize and lead to rearrangements of the chemical bonds. The first type of UV damage that was extensively studied was the pyrimidine dimer. In this damage product, the rings of two adjacent pyrimidines fuse. A cyclobutane ring is formed when the 5-carbon atoms and the 6-carbon atoms of adjacent pyrimidines join. Another type of dimer results when the 6-carbon of one pyrimidine is joined to the 4-carbon of an adjacent pyrimidine. This photoproduct is referred to as a 6–4 lesion or the pyrimidine-pyrimidone (6–4) photoproduct. Additional photoproducts are found less frequently in DNA following UV irradiation or only under special conditions. These products include 5, 6-dihydroxydihydrothymine (thymine glycol; Demple and Linn, 1982); the spore photoproduct (5′-thyminyl-5,6-dihydrothymine) (Varghese, 1970) and pyrimidine hydrates (Fisher and Johns, 1976).

While only a sampling of the types of DNA damage and base changes have been presented, the diversity of the sample highlights the importance and need of living cells to protect and repair their genetic material. While this point should have been obvious, interest in this important area of research did not really assume high priority until almost the start of the 1950s. One of the milestones was a report by Dr. Evelyn Witkin in 1947. Essentially Dr. Witkin (1947) observed that a mutant of *E. coli* could be shown to have decreased resistance to DNA-damaging agents. The conclusion that could be drawn from this result was that the bacterium had genetic information that determined how sensitive it was to the killing effects of DNA-damaging agents. Thus there must be a DNA repair mechanism(s). For the past 50 years there has been an extensive delineation of the mechanisms responsible for maintaining the integrity of DNA.

III. PHOTOREACTIVATION

The first DNA repair mechanism discovered was photoreactivation (Dulbecco, 1949; Kelner, 1949). Photoreactivation reduces the deleterious effects of UV irradiation (200–300 nm) by means of a light-dependent process in which the cis-syn cyclobutyl pyrimidine dimers are enzymatically monomerized (Setlow et al., 1965; Wulff and Rupert,

1962). This process (Fig. 1) involves a single enzyme, a DNA photolyase. In *E. coli*, this photolyase is a flavoprotein that functions by a two-step mechanism (Sancar et al., 1985). Initially DNA photolyase binds to pyrimidine dimers in a light-independent reaction. Upon subsequent exposure to light of wavelengths greater than 300 nm, the enzyme cleaves the dimer and dissociates from the substrate, leaving the original primary structure of the DNA.

DNA photolyase activity has been detected in a wide variety of microorganisms, plants, and animals (Rupert, 1975). The exceptions to this near-universal distribution are in naturally competent eubacteria (Campbell and Yasbin, 1979), and in animals higher on the evolutionary tree than marsupials (Friedberg et al., 1995). Since many or the organisms harboring a DNA photolyase never, or very rarely, come in contact with the necessary wavelengths of light, it remains a question as to why this genetic information would have been maintained through evolution of these organisms. It has been suggested that these enzymes might have additional functions. For instance, it has been demonstrated that the presence of a functional photoreactivation gene in strains of *E. coli* that are deficient in recombination (*recA*) decreases the sensitivity of these strains to UV irradiation even without any exposure to photoreactivating light (Yamamoto et al., 1984). Furthermore the *E. coli* photolyase, under nonphotoreactivating conditions, stimulates in vitro both the rate and cutting by the excision nuclease (see below) of UV-irradiated DNA (Sancar et al., 1984).

IV. NUCLEOTIDE EXCISION REPAIR

Bulky, noncoding lesions that produce a block to DNA replication can be removed from damaged DNA through the action of a nucleotide excision repair (NER) system. A noncoding lesion constitutes some alteration of a nucleotide(s) contained within the DNA such that the replication machinery of the

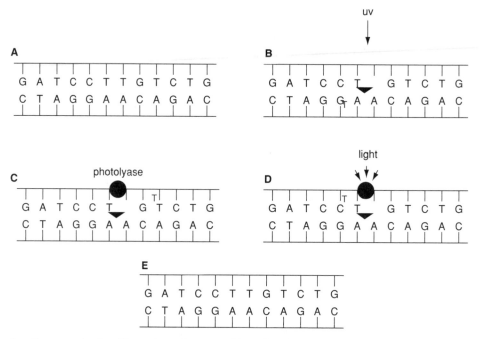

Fig. 1. Photoreactivation. Shown is a schematic of how the photolyase enzyme with the help of species-specific wavelengths of light enzymatically cleaves pyrimidine dimers and thus restores the integrity of the DNA. (**A**) A DNA sequence (**B**) That sequence following exposure to UV (approximately 254 nm). The triangle represents the pyrimidine dimer that was formed. (**C**) A molecule of photolyase recognizes the dimer, binds to it, and sits there until it is activated by specific wavelengths of light (**D**). Once activated, the dimer is cleaved and the DNA sequence is restored (**E**).

cell can no longer use this nucelotide as a template for normal base-pairing purposes. The general properties of NER include a five-step mechanism (Friedberg et al., 1995):

1. Recognition of the bulky lesion in the DNA.
2. Hydrolyzing a phosphodiester bond in the deoxyribose backbone on the 5' side of the lesion.
3. Excising the lesion (along with a limited number of nucleotides on its 3' side).
4. Filling in the resultant gap using the information from the complementary strand.
5. Closing the nicked DNA to generate intact strand (Schendel 1981).

The best characterized model system for NER involves the removal of the pyrimidine dimer by the *E. coli* UvrA, B, C exonuclease (Sancar and Rupp, 1983). The first three steps of the NER in *E. coli* are accomplished through the cooperative functioning of the UvrA, UvrB, and UvrC proteins (Fig. 2). These three proteins comprise the UvrABC endonuclease (Sancar and Rupp, 1983). Two copies of the UvrA protein, and one copy of the UvrB protein form a complex that binds to DNA even in the absence of damage. The complex moves along the DNA, apparently with the DNA wrapped around the A2-B complex (Verhoeven et al., 2001) until a helix distortion (bulky lesion) is identified. The complex will stop at the damage (in this case the pyrimidine dimer), the UvrA protein will exit and be replaced by the UvrC protein. The binding of the UvrC protein to the UvrB causes the UvrB to make a cut in the DNA usually 4 nucleotides 3' of the damage. Then the UvrC protein cuts the DNA 7 nucleotides 5′ of the damaged base. Following the cuts in the DNA, the UvrD protein (a DNA helicase) removes the oligonucleotides that contain the damage, while DNA polymerase I resynthesizes the removed strand using the opposite strand as the template. Finally, the ligase reseals the newly synthesized strand.

It has been determined that low levels of the UvrA, B, C,and D proteins are found in normal cells. However, the levels of the UvrA, B, and D proteins are significantly enhanced following the introduction of certain types of DNA damage. The genes encoding these proteins have been shown to be part of the SOS regulon (see below).

Research into the NER systems in both prokaryotes and eukaryotes have shown that this type of repair process can be directed to specific regions of the chromosome(s). In particular, NER systems have evolved to treat specific regions of the chromosome(s) differently with respect to what genes should be repaired first or even repaired at all. The most dramatic example of this directed repair can be seen in higher eukaryotes where there is a repair bias for expressed genes as compared to the nonexpressed genes in each cell type (Friedberg et al., 1995). In prokaryotes there is also a mechanism that directs the NER system to preferentially function on transcribed regions (Selby and Sancar, 1994). A protein, the product of the *mfd* gene, called the *transcription coupling repair factor* (TRCF) causes the RNA polymerase to be

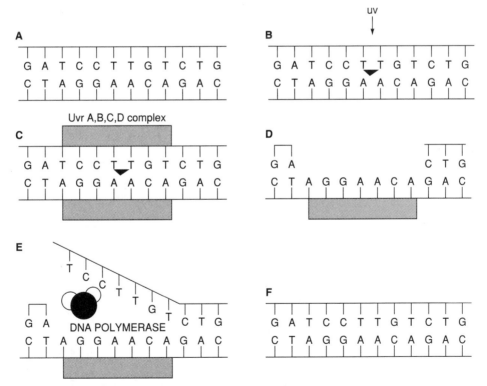

Fig. 2. Nucleotide excision repair. The DNA (**A**) has been exposed to the far UV (**B**) and the pyrimidine dimer has been formed. In (**C**) the UvrA,B,C complex has formed around the damaged DNA and with the help of UvrD this complex cuts the DNA and opens up the DNA (**D**) in order for the DNA polymerase (most often DNA polymerase I) to begin to resynthesize the damaged strand using the opposite strand as a template (**E** and **F**).

displaced when the transcription complex stalls at the site of DNA damage. Because of the damage the RNA polymerase cannot proceed. TRCF also binds to the UvrA protein. Following the TRCF mediated displacement of the RNA polymerase, the UvrA2B complex then binds to the DNA that contains the damage. This interaction accelerates the repair of actively transcribed regions of the genome. A phenomenon related to the functioning of the TRCF is the process called *mutation frequency decline*. This process is the rapid and irreversible decline in suppressor mutation frequency that occurs when the cells are kept in nongrowth media immediately following mutagenic treatment and requires the functioning of the TRCF or the *mfd* gene product.

V. BASE EXCISION REPAIR

Base excision repair (BER) is a second method by which bulky, noncoding lesions can be removed from DNA. In addition BER represents an efficient mechanism for the removal of many base alterations that are the result of metabolic factors (deaminations, alkylations, oxygen radicals, etc.). BER differs from NER in that damaged or incorrect bases are excised as free bases (Fig. 3) rather than nucleotides or oligonucleotides (Friedberg et al., 1995). Total removal of the DNA lesion requires a two-step process. First, BER involves the hydrolysis of the N-glycosylic bond that links the base to the deoxyribose-phosphate backbone of the DNA. This is performed by the action of a class of DNA repair enzymes called

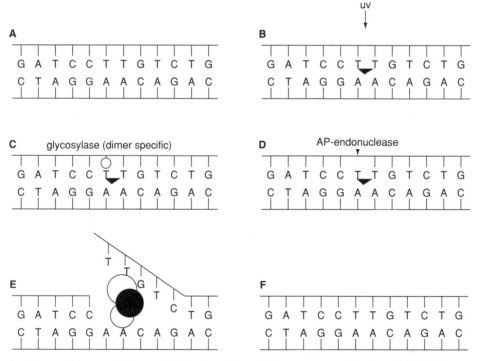

Fig. 3. Base excision repair. The pyrimidine dimer formed after exposure of the bacteria to UV is removed by the activities of a damage-specific glycosylase and AP-endonuclease(s). The DNA sequence (**A**) is exposed to UV and the dimer is formed (**B**). The damage-specific glycosylase recognizes the pyrimidine dimer, attaches to it, and cleaves one of the N-glycosylic bonds (**C**) and (**D**) In D, the apyrimidinic site is recognized by an AP-endonuclease, a cut is made in the DNA backbone, some bases are excised, and DNA polymerase resynthesizes the strand using the opposite strand as a template (**E** and **F**).

glycosylases. Once a damaged base, an incorrect base pair or an inappropriate base is recognized by a specific glycosylase, the N-glycosylic bond is cut, leaving an apurinic or apyrimidinic (AP) site in the DNA. The second step in BER in the removal of this AP site via the action of one or more nucleases. Sites of base loss in DNA are specifically recognized by enzymes known as AP endonucleases (Lindahl, 1979). Repair synthesis and ligation in BER proceed as discussed for NER.

Bacteria posses many different DNA glycosylases, and as stated above, each is specific fo a particular type of lesion in the DNA. Included in this group are glycosylases that recognize uracil, hydroxymethyl uracil, 5-methylcytosine, hypoxanthine, 3-methyladenine, 7-methylguanine, 3-methylguanine, and 8-hydroxyguanine (Friedberg et al., 1995). In addition glycosylases have been identified that recognize DNA containing 5,6-hydrated thymine moieties and DNA containing pyrimidine dimers.

As can be seen from this partial list, BER can function on altered bases as well as on bases that are not normally present in DNA (uracil, hypoxanthine, etc.). Another aspect of BER is the removal of bases involved in mispairing. Essentially there are glycosylases that recognize very specific types of mispairing events. Some interesting examples are the enzymes involved in handling the potential problems caused by the generation of 8-oxoG (described above). A DNA glycosylase has been identified that recognizes 8-hydroxyguanine residues in DNA as well as some imidiazole ring-opened forms. Subsequent evaluations determined that this glycosylase was product of the *mutM* gene. This loss of this gene had been shown to cause an increase in the GC to TA transversion rate and a decreased ability to handle the mutagenic effect of 8-oxoG. The biochemical analysis demonstrated that the MutM glycosylase also recognizes 8-oxoG residues in DNA. However, if all of the 8-oxoG residues are not removed by this mechanism, the glycosylase specified by the *mutY* gene functions to

help reduce the potential problems caused by the presence of this mutagenic lesion in the DNA. Essentially the MutY glycosylase recognizes 8-oxoG-adenine mispairs in the DNA and removes the adenine. BER will now function to restore an 8-oxoG-cytosine pairing. This pairing could then be the substrate for the MutM glycosylase (to remove the 8-oxoG) (Friedberg et al., 1995).

With respect to the 8-oxoG lesion, there is one additional type of repair that operates but is not part of the BER. The product of the *mutT* gene is a phosphatase that specifically degrades 8-oxodGTP to 8-oxodGMP. This action prevents the DNA polymerases from incorporating 8-oxoG into the DNA. Collectively the *mutT*, *mutY*, and *mutM* gene products function to reduce the mutagenic impact of 8-oxoG. (Michaels et al., 1992).

As mentioned earlier, BER functions via the combined mechanisms of damage-specific DNA glycosylases and AP endonucleases. AP endonucleases produce incisions in the duplex DNA by hydrolysis generally of the phosphodiester bond that is 5' to the AP site. The result of this incision is the generation of a 5' terminal deoxyribose-phosphate residue. These residues can be removed by the action of either exonucleases or DNA-deoxyribophosphodiesterases. In the former case there is a removal of tracts of nucleotides followed by DNA polymerase and ligase activity while in the later case, a single nucleotide gap is generated and that gap is replaced by DNA polymerase and ligase activities (Friedberg et al., 1995). In addition to the separate AP endonucleases that have been characterized, several of the glycosylases have associated AP-lyase activity that may or may not play important roles in the actual removal of DNA damage (Friedberg et al., 1995; Vasquez et al., 2000).

VI. MISMATCH EXCISION REPAIR

Classically *E. coli* mismatch excision repair (mismatch repair) was defined as a methyl-directed postreplication repair system which eliminated replicative errors within newly synthesized DNA (Harfe and Jinks-Robertson, 2000; Modrich and Lahue, 1996). These replicative errors or mismatches are distinguished from the preexisting correct base in the parental strand due to the undermethylated state of the newly synthesized daughter strand. The repair of these mismatches involves localized excision and resynthesis of nucleotides at the site of the mismatch.

The methyl-directed mismatch repair system has been extensively characterized both genetically and biochemically. Mutations in the *dam, mutH, mutL, mutS,* and *uvrD* genes lead to increases in the spontaneous mutation frequencies between 10-and 1000-fold. This increase in the spontaneous mutation frequency is due to a deficiency in the mismatch repair system (Friedberg et al., 1995) (Modrich and Lahue, 1996). The product of the *dam+* gene is a DNA adenosine methylase that methylates the adenine in the site GATC (Herman and Modrich, 1982). The presence of hemimethylated DNA seems to trigger the enzymes involved in this repair system to search for mismatches. The hemimethylated state would tend to be more prevalent in the newly replicated DNA adjacent to the replication fork. This mechanism would imply that the mismatch would be corrected in favor the parental strand. This does in fact occur. As discussed later, this particular methyl-directed mismatch repair system is not as common as nonmethyl-directed mismatch repair systems.

As one would expect, strains carrying a mutant allele for the *dam* gene perform undirected mismatch excision repair (loss of preferential repair of newly replicated DNA strand). This undirected mismatch repair is the reason behind these strains having an increased mutation frequency or a so-called mutator phenotype. Also, as expected, *dam* mutants are more sensitive to agents that cause strand breaks either directly or as a result of attempted repair of the base damage inflicted by the agent. These mutants exhibit increased recombination frequency and an increased induction of prophage. In addition these mutations are in viable when strains also carry mutations in *recA, recB, recC, recJ, lexA,* or *polA* (Bale et al., 1979; Friedberg et al., 1995). These phenotypes can all be explained by the model for the methyl-directed mismatch repair system that has been advanced. Specifically, strains carrying a mutant *dam* gene would excise relatively long patches due to the undirected nature of the repair. This accumulation of long patches of single-stranded DNA would lead to an increase in the generation of double-strand breaks. These double-strand breaks would enhance an promote recombination, prophage induction and lethality (Section VII). As expected, suppressors of the *dam rec* double-mutation combinations have been found to be mutant alleles of *mutH, mutL,* and *mutS* (mutations that prevent the excision patches).

The *E. coli* mismatch repair system does not identify and correct all potential mismatches with equal efficiency (Radman and Wagner, 1986). In general, the extent of the repair depends on the type of mismatch as well as the neighboring nucleotide sequences. Specifically, transition mismatches (G-C to G-T and A-T to A-C) appear to be repaired more readily than are transversion mismatches. G-G and A-A mismatches seem to be repaired efficiently, while T-T, G-A, and C-T are repaired less efficiently. There seems to little repair of the C-C mismatch. Furthermore, increasing the G-C content in the neighboring nucleotide sequences enhances the probability that a given mismatch will be repaired.

Besides the mismatches listed above, the *E. coli* system can recognize and repair frameshift heteroduplexes. These types of heteroduplexes (the result of either additions or deletions to one strand of the heteroduplex) do not technically contain a mismatch. Rather, there is an extra, and therefore unpaired, base in one of the strands. Mismatch works equally well on both strands when the DNA is nonmethylated. In the presence of methylated DNA, the heteroduplex repair is directed and would therefore seem to function in the region of the replication fork.

The mismatch repair system described above is the classic example of a type of repair process that has been designated long patch mismatch repair (LPMR). While the LPMR system in *E. coli* requires or is dependent on the activity of the dam^+ gene, in *Streptococcus pneumoniae* an LPMR directed by the hex^+ genes is independent of DNA methylation (Radman, 1988). This apparent paradox seems to be resolved by the observations that in *E. coli* the product of the $mutH^+$ gene nicks the nonmethylated GATC sequence and persistent nicks in heteroduplex DNA can effectively substitute for the functions of both the MutH protein and the nonmethylated GATC sequence (Radman, 1988). Thus LPMR is not necessarily dependent on a DNA methylation system (as is the case in *S. pneumoniae* and most other organisms) but instead could be directed against strands that have single-strand ends.

In the non-methyl-directed LPMR systems, homologues of MutS and MutL have been identified (Harfe and Jinks-Robertson, 2000). However, MutH homologues are not found in these systems. This again indicates that the MutH protein is specifically involved in the interaction with the methylated sequence. For the *E. coli* system, a complex of MutH, MutL, and MutS bind to the mispaired region, and then the DNA apparently forms an alpha loop (which requires ATPase activity). The excision of the mispaired base or region occurs once the unmethylated GATC site (or single-strand break) is reached. DNA helicase II (the product of the $uvrD$ or $mutU$ gene), DNA polymerase III, and ligase are required to complete the repair. It is interesting to note that this is one of the few cases in which DNA pol III is the preferred repair polymerase.

In addition to LPMR, another type of mismatch correction system has been characterized in prokaryotes and eukaryotes by short spans of DNA being repair replicated (Coic et al., 2000; Lieb and Bhagwat, 1996; Lieb and Rehmat, 1995; Turner and Con-

nolly, 2000). An example of such a short-patch repair system would by the one controlled by the $mutY$ gene (Section IV). Another well-studied short-patch repair system is the one termed VSPMR (very short-patch mismatch repair; Radman, 1988). This system repair those G-T mismatches that apparently originate by the deamination of 5-methylcytosine to thymine in the sequence 5′-CC (A or T) GG-3′. The enzyme encoded by the dcm gene methylates the second C in the sequence. The repair of these G-T mismatches to the correct G-C pairing by VSPMR significantly reduces the mutation "hot spots" generated by the presence of 5-methylcytosine. In this *E. coli* system the products of the dcm^+ gene (cytosine methyl transferase) and the $mutS^+, mutL^+$, and $polA^+$ are essential. However, the products of the $mutH^+$ or $mutU^+$ are not required.

In the case of the *S. pneumoniae*, the VSPMR system acts on the sequence 5′-ATTAAT-3′, and the repair pattern involves the correction of G-A to G-C (Sicard et al., 1985, 2000) and seems to be involved in the efficiency of some markers during the transformation process (see chapter by Streips-Transformation, this volume).

VII. POSTREPLICATION REPAIR OR DAMAGE BYPASS

In *E. coli* treated with UV or other agents that cause the production of bulky, noncoding lesions in the DNA, damages that are not removed before being encountered by the replication machinery constitute a block to further DNA synthesis (Setlow et al., 1963). DNA replication can be resumed if the DNA polymerase dissociates from the DNA when it encounters a noncoding lesion and then initiates replication on the other side of the lesion (see chapter by Firshein, this volume). Such a mechanism was first proposed in 1968 (Howard-Flanders et al., 1968). This type of mechanism would result in gaps in the newly synthesized daughter strand, which subsequently become filled-in by some process. In support of this model it was observed that the daughter strands are much smaller than the parental template following UV irradiation. This size of this newly synthesized DNA approximates the average interdimer distance in the template (Sedgwick, 1975). Upon continued incubation, daughter strands become longer until they eventually reach the same size as the parental strands.

The daughter-strand gaps are filled in by a recombination event (Ganesan, 1974). Hence this process has been called postreplication repair, daughter strand gap repair and recombination repair (Fig. 4). Regardless of the name that is applied, this type of mechanism exemplifies tolerance of DNA damage rather than a true repair process since the actual damage is night physically removed from the DNA. Rather, the damage is bypassed by this process.

The evidence for the involvement of recombinational events in this process is as follows: In UV-irradiated *E. coli*, newly synthesized DNA was found in both the daughter and parental strands, physically demonstrating that strand exchange had occurred. From these data it was estimated that one genetic exchange occurred per pyrimidine dimer. The reciprocal of these data was also observed in that dimers were found to be equally distributed be parental and progeny strands (Ganesan, 1974). In addition UV-irradiated DNA that has replicated is highly recombinogenic

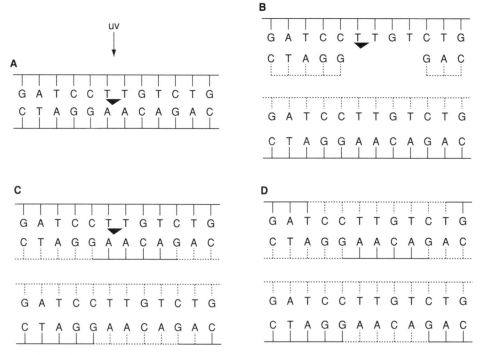

Fig. 4. Recombination or translesion bypass. This type of repair is actually a tolerance mechanism. Following the introduction of damage into the DNA. **A:** the DNA will now have problems being replicated. When the DNA is replicated, the two daughter strands cannot be completely finished. Opposite a site of damage in a parental strand there will be a gap in the daughter strand. **B:** As long as the gaps are not overlapping in the two daughter strands, recombination can be utilized remove the single-strand gaps. **C:** DNA from the other parental strand can be recombined into one of the daughter strands. This will now result in parental DNA being found in a daughter strand and newly synthesized DNA being found in the parental strand that was a donor of DNA to the gapped daughter strand. **D:** The damaged DNA (in this case a pyrimidine dimer) can be repaired and actually removed via any of the mechanisms previously discussed (photoreactivation, nucleotide excision repair, base excision repair).

(Howard-Flanders et al., 1968) and the presence of the photoproducts in the replicating DNA is responsible for the increase in recombination (Lin and Howard-Flanders, 1976). Finally, strains of *E. coli* carrying the *recA1* allele do not convert short, newly synthesized, DNA strands into high molecular weight DNA (Smith and Meun, 1970).

As mentioned above, the *recA*[+] gene product is required for postreplication repair. It is not too surprising that other genes known to be involved with general recombination events would also influence postreplication repair (see chapter by Levene and Huffman, this volume). In general, there is a strong correlation between the levels of recombination proficiency and UV resistance in various mutant strains. Specifically, mutations in *recA, recB,C,D, recE, recF, recJ, recN, recQ, ruvB,C,* and *ssb* reduce UV resistance in genetic backgrounds where they reduce recombination proficiency (Mahajan, 1988). Postreplication repair proceeds by two major *recA*[+] dependant processes. One pathway repairs most of the DNA daughter-strand gaps via the *recF*[+]-mediated process, while the other repairs double-strand breaks produced by the cleavage of unrepaired gaps. This second pathway is dependent on the functions of the *recBCD* gene products (see chapter by Levene and Huffman, this volume). Finally, the repair of cross-links in the DNA presents a situation where both recombination (postreplication) repair and excision repair must function together. In this type of damage, adducts are covalently attached to both strands of the DNA. The UvrABC complex removes the adduct from one strand, producing a gap, while the opposite strand still contains the adduct. Recombinational repair would allow for the filling-in of the gap, permitting the adduct to now be removed from the opposite strand (Cole, 1973).

VIII. TRANSLESION DNA SYNTHESIS

Postreplication repair illustrates DNA damage tolerance via a discontinuous mode of DNA synthesis. However, DNA damage tolerance could occur via a continuation of DNA synthesis, opposite a noncoding lesion, without gap formation. This is termed *translesion DNA synthesis* (Friedberg et al., 1995). While the former type of postreplication repair should be relatively error free (not produce mutations), translesion DNA synthesis should result in the production of errors or mutations.

Translesion DNA synthesis is one of a myriad of coordinately induced cellular responses observed in *E. coli* and collectively known as the SOS system or regulon (Little and Mount, 1982; Radman, 1974; Walker, 1987; Witkin, 1976). As part of this SOS system, translesion DNA synthesis usually has been called *error-prone* repair or inducible DNA repair. This nomenclature arose because an increased mutation frequencies was observed in *E. coli* populations induced for SOS functions following some type of DNA damage. However, as with postreplication repair, this "error-prone" repair was postulated to result in a dilution out of DNA lesions rather than a true repair of the DNA. Therefore it was suggested (Miller, 1982) that mutations should be thought of as occurring by replication across from altered bases rather than as a result of a true repair process. Hence the term *translesion DNA synthesis* arose.

A partial list of *E. coli* SOS responses is given in Table 1. These phenomena are coordinately induced in *E. coli* cells that have been exposed to UV radiation, chemicals that produce bulky lesions, or agents that arrest DNA synthesis (Walker, 1984; Witkin, 1976). Radman (1974) first formalized the SOS hypothesis by suggesting that DNA damage or the consequence of this damage initiates some sort of regulatory signal that simultaneously causes the derepression of a number of genes. He further speculated that this "danger" signal might be a temporary block in DNA replication. Expression of these SOS phenomena have traditionally been described as depending upon the functioning of the RecA and LexA proteins. However, some recent results in *E. coli* as

TABLE 1. Phenomena That Are Components of the SOS Regulon

Phenomenon	Description
Prophage induction	Resident prophage are induced to enter lytic cycle (i.e., λ)
W reactivation	Enhanced survival or irradiated phage
W mutagenesis	Enhanced mutation rate of W-reactivated phage
UV mutagenesis	Ability of UV to cause mutations
Filamentation	Bacteria grow as long filaments
Induction of Din genes	Genes that are DNA damage inducible such as *recA*, *lexA*, *himA*, *uvrA, B, dinA, B, D, F*
Cessation of respiration	Loss of active aerobic growth
Alleviation of restriction	Decrease in the effect of restriction enzymes
Stable DNA replication	New rounds of DNA replication begin

well as studies in other organisms indicate that the SOS system might involve more than one type of gene regulation (Humayun, 1998; Cheo et al., 1993; Yasbin et al., 1992). Despite this potential diversity in regulation, the general working model for the control of the primary SOS regulon is as shown in Figure 5 (Little and Mount, 1982; Walker, 1984; Witkin, 1976). Essentially, in the undamaged wild-type cell, the SOS regulon genes are repressed by the LexA protein. The products of these genes (including LexA and RecA) are synthesized at low constitutive levels (or not produced at all). An SOS-inducing signal is generated by DNA damage. There is convincing evidence that the major signal for this induction are the regions of single-stranded DNA that are generated when the molecular machinery attempts to replicate a damaged DNA template or when the normal process of DNA replication is blocked (Friedberg et al., 1995). RecA binds to these single-stranded regions, in the presence of nucleoside triphosphates, and allosterically converts (reversibly) to a form that has been called RecA*. LexA protein comes in contact with the RecA* nucleoprotein complex, resulting in the autoproteolysis of LexA at a specific

Ala-Gly bond. In this sequence of events the RecA* functions as a coprotease and the proteolytic activity actually resides within the LexA protein itself. Following cleavage, the LexA protein can no longer function as a cellular repressor. In addition to LexA, the RecA* can cause similar activation of proteolytic activity in certain prophage repressors (i.e., λ) and the UmuD protein (discussed below). There have been recent reports that the activation of this proteolytic activity might involve interactions with polyamines (Kim and Oh, 2000). However, the complete nature of the activation process requires additional studies.

LexA has been shown to be the repressor of over 20 genes, including *recA*, *lexA*, *uvrB*, and *umuD,C*. It is possible that other cellular repressors in *E. coli* may exist that are sensitive to auto-proteolytic cleavage following activation RecA*. As mentioned above, RecA* also leads to the auto-proteolytic digestion of λ repressor and the UmuD protein. In any event the pools of LexA protein decrease very rapidly after inducing treatment (activation of RecA to RecA*), and the end result is the derepression of the SOS regulon and the expression of the SOS phenomena. This expression will continue as

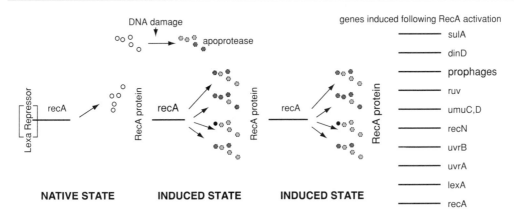

Fig. 5. Induction of the SOS system. The LexA protein is a repressor of at least 20 different genes on the *Escherichia coli* chromosome. This replicon can be induced following the introduction of certain types of damage into the bacterial DNA. Once the DNA has been damaged, a signal is produced that activates the RecA protein into becoming a apoprotease. This activated form of RecA causes LexA to cleave itself, thus inducing all of the genes under its control. Interestingly, *lexA* and *recA* are two of the genes under the control of the LexA repressor. Thus, when the system is induced, large quantities of RecA and LexA proteins are produced. The LexA protein is inactivated as long as there is an activated form of the RecA protein present. Once the damage has been removed, the signal no longer exists and then the RecA protein is no longer activated. When this occurs there is sufficient LexA present to shut down the SOS regulon. In addition to causing the LexA protein to cleave itself, the activated form of the RecA protein also causes the UvrD protein and many different types of prophage repressors to cleave themselves.

long as sufficient inducing signal persists. As the level of this signal subsides, less RecA* is available and the cellular concentration of the LexA protein will rise. Eventually the entire SOS regulon will again be repressed, and the cell will return to its normal, uninduced state. This return to steady-state level of LexA repressor following the removal of the inducing signal occurs rapidly.

This mechanism for regulation of the SOS response offers many opportunities for fine-tuning of the system. First, RecA* exhibits varied efficiencies in causing the auto-proteolysis of different proteins. For instance, the LexA protein is activated to auto-cleave itself more readily than is the λ repressor (Little and Mount, 1982). Therefore one would expect that the SOS regulon genes that are repressed by the LexA protein would be preferentially induced when compared to the induction of prophage λ. Second, the LexA protein has different binding efficiencies for the various operator regions of the SOS regulon genes. The *recA* operator binds LexA more strongly than

do the operators of the *uvrB* or *lexA* genes (Brent and Ptashne, 1981). The binding strength of LexA protein is greatest for *dinD*, somewhat weaker for *umuD, C*, and weaker yet for *uvrA*, *dinA (polB)*, and *dinB (polIV)* (Kreuger et al., 1983). This indicates that the potential exists for intermediate induction of the SOS system and for the production of mutations (Walker 1984).

The operator regions of a number of SOS genes have been sequenced and protein protection experiments have resulted in the identification of DNA sequences that have been called SOS boxes or Lex boxes (Friedberg et al., 1995). The operator regions have similar base sequences, about 20 bp long, that are binding sites for the LexA protein. All the binding sites include inverted repeat sequences that contain as a minimum 5'CTG~10N-CAG3'. The *lexA* gene has two nearly identical SOS boxes in its operator region, again adding another dimension of control.

Finally, the fact that the repression of the *lexA* gene is autoregulated (Friedberg et al.,

1995) markedly influences SOS induction. This autoregulation allows for the expression of only a subset of the SOS responses, depending on the strength of the inducing signal. It also guards against full induction of the system in response to a mildly damaging situation, since the LexA protein has a greater affinity for the *recA* operator than its own operator. In addition autoregulation of the production of the LexA protein allows for a speedy return to the repressed state, which is observed when the inducing signal subsides.

It has long been established that mutagenesis of the *E. coli* chromosome by UV, as well as certain chemicals such as methylmethanesulfonate (MMS) and 4-nitroquinoline-1-oxide (4NQO) is dependent on the *recA* and *lexA* gene products (Witkin, 1976). Therefore induced mutagenesis is one of the SOS responses. The mechanism for this mutagenesis has only recently begun to be elucidated. In *E. coli* three DNA polymerases have been shown to be under SOS regulation; Pol II, Pol IV, and Pol V (Wagner et al., 1999; O'Grady et al., 2000; Sutton et al., 2000). Pols IV (product of the *dinB* gene) and V (product of the *umuD, C* genes) belong to a superfamily of DNA polymerases that have been found in eubacteria, archaea, and eukaryotes (Friedberg et al., 2000; Gerlach et al., 1999). Pols IV and V are nonprocessive polymerases that can perform translesion bypass. In addition to their roles in translesion bypass, the UmuDC proteins have also been associated with prokaryotic cell cycle control (Sutton et al., 2001a, 2001b; Sutton and Walker, 2001), which represents another important survival aspect of the SOS system.

As mentioned above, variations on the SOS regulon have begun to be identified. In addition to regulation by a Lex-like cellular repressors, the SOS genes have been shown to be under the control of prokaryotic development and differentiation factors (Cheo et al., 1992, 1993; Lovett et al., 1989; McVeigh and Yasbin, 1996; Yasbin and Miehl-Lester, 1990). In addition the binding sites for the cellular repressors have shown divergence among the gram-negative bacteria and between the gram-positive and gram-negative kingdoms (Winterling et al., 1997). There is also a tremendous diversity of the types of genes that are grouped into SOS regulons in different organisms. These genes range from ones whose products are involved in DNA repair to genes that play essential roles in virulence, metabolism, growth, and development (Friedberg et al., 1995). Thus the SOS regulon developed early in evolution and has been conserved as well as modified to play important roles in the survival of species.

IX. ADAPTIVE RESPONSE

E. coli posses an inducible repair system that protects against the lethal and mutagenic effects of alkylation damage (Jeggo et al., 1977; Landini and Volkert, 2000; Samson and Cairns, 1977). This repair system has been termed the adaptive response, due to its particular mode of functioning. Specifically, *E. coli* cultures exposed to low levels of an alkylating agent such as N-methyl-N'-nitro-N-nitosoguanidine (MNNG) and then subsequently challenged by a much higher dose of this agent are able to withstand both the cytotoxic and mutagenic effects of such an exposure. Hence these cultures have "adapted" to the deleterious effects of MNNG.

The adaptive response is regulated by the product of the *ada* gene and also during stationary phase by rpoS-dependent gene expression (see Moran; Streips-Stress Shock, Landini and Volkert, 2000). This interesting protein has a molecular weight of 37,000 daltons and has at least three known functions. First, this protein is a positive regulatory element that is involved in the increased transcription of at least four genes (*ada*, *alkA*, *alkB*, and *aidB*). The *ada* and *alkB* genes are in an operon, while the other two known genes of this regulon are dispersed on the chromosome. The enzymatic function of alkB has yet to be clearly defined. However, it is known that bacteria deficient in this product are more sensitive to some alkylating agents and that this protein is needed to remove certain damages (Landini and Volkert, 2000). The alkA gene encodes a glycosylase that repairs several alklyation caused lesions including N7-methylguanine, N3-methyl purines, and O2-methyl pyrimidines. The aidB gene is homologous to the mammalian isovaleryl coenzyme A dehydrogenase (IVD), and appears to have IVD activity to function and inactivate nitrosoguanidines or their reactive intermediates. However, its exact enzymatic activity has not been completely established.

In addition to its activity as a regulatory element, the Ada protein is a methyltransferase. Ada has two active methyl acceptor cysteine residues, Cys-69 and Cys-321, that are required for demethylation of damaged DNA (Friedberg et al., 1995). Both sites can be methylated but are utilized to repair different types of damages. The Cys-321 is the methyl acceptor site required for the removal of two very mutagenic lesions: methyl groups from either O6-methylguanine or O4-methylthymine. The Cys-69 is involved in the removal of methyl groups from the phosphomethyltriesters in the sugar-phosphate backbone. The Ada protein is not turned over following its acceptance of alkyl groups, and thus it can be classified as a "suicide" protein. Furthermore the transfer of a methyl group from the triester, rather than from the guanine or thymine, is responsible for causing the Ada protein to become a positive effector molecule for transcription. Importantly, the Ada protein can function as both a positive and negative effector of transcription (Saget and Walker, 1994) (Landini and Volkert, 2000).

X. UNIVERSALITY OF DNA REPAIR MECHANISMS

While E. coli has functioned as the principal model for investigations into DNA repair mechanisms, by no means is it a unique organism. The DNA repair systems identified in this paradigm have been discovered in most other organisms studied. While not all organisms may have all of the same systems identified in E. coli, it is clear that DNA

repair systems are an important evolutionary advantage and as such they have been conserved in both prokaryotic and eukaryotic systems. This fact has been made even more evident by the results of the genome-sequencing efforts (Wood et al., 2001); Ronen and Glickman, 2001). Not only have the genes and the proteins discussed above been shown to be involved in survival and mutagenesis, but homologues of these proteins play essential roles in disease prevention, cell-cycle regulation as well as normal development and differentiation (Aquilina and Bignami, 2001; Khanna and Jackson, 2001; Modrich and Lahue, 1996; Sutton et al., 2001b). Clearly, the pioneering investigations into DNA repair mechanisms in E. coli and other prokaryotes have greatly enhanced our understanding of the ability of life systems to survive, adapt, and evolve.

REFERENCES

Aquilina G, Bignami M (2001): Mismatch repair in correction of replication errors and processing of DNA damage. J Cell Physiol 187:145–154.

Balasubramanian B, Pogozelski WK, Tullius TD (1998): DNA strand breaking by the hydroxyl radical is governed by the accessible surface areas of the hydrogen atoms of the DNA backbone. Proc Natl Acad Sci USA 95:9738–9743.

Bale A, d'Alarcao M, Marinus MG (1979): Characterization of DNA adenine methylation mutants of Escherchia coli K-12. Mutat Res 59:157–165.

Bol DK, Yasbin RE (1991): The isolation, cloning and identification of a vegetative catalase gene from Bacillus subtilis. Gene 109:31–37.

Brent R, Ptashne M (1981): Mechanism of action of the lexA gene product. Proc Natl Acad Sci USA 78:4204–4208.

Bridges BA (1997): Hypermutation under stress. Nature 387:557–558.

Bridges BA (1998): The role of DNA damage in stationary phase ("adaptive") mutation. Mutat Res 408:1–9.

Cairns J, Foster PL (1991): Adaptive reversion of a frameshift mutation in Escherichia coli. Genetics 128:695–701.

Cairns J, Overbaugh J, Miller S (1988): The origin of mutants. Nature 335:142–145.

Campbell LA, Yasbin RE (1979): DNA repair capacities of Neisseria gonorrhoeae: Absence of photoreactivation. J Bacteriol 140:1109–1111.

Cheo DL, Bayles KW, Yasbin RE (1992): Molecular characterization of regulatory elements controlling expression of the *Bacillus subtilis recA* gene. Biochimie 74:755–762.

Cheo DL, Bayles KW, Yasbin RE (1993): Elucidation of regulatory elements that control damage induction and competence induction of the *Bacillus subtilis* SOS system. J Bacteriol 175:5907–5915.

Coic E, Gluck L, Fabre F (2000): Evidence for short-patch mismatch repair in *Saccharomyces cerevisiae*. EMBO J 19:3408–3417.

Cole RS (1973): Repair of DNA containing interstand cross-links in *Escherichia coli*: Sequential excision and recombination. Proc Natl Acad Sci USA 70: 1064–1068.

Demple B, Linn S (1982): 5,6 Saturated thymine lesions in DNA: production by ultraviolet light or hydrogen peroxide. Nucleic Acids Res 10:3781–3789.

Dobshansky T (1950): The genetic basis of evolution. Sci Am 182:32–41.

Dulbecco R (1949): Reactivation of ultraviolet inactivated bacteriophage by visible light. Nature 163: 949–950.

Fisher GJ, Johns HE (1976): Pyrimidine hydrates. In Wang SY (ed): "Photochemistry and Photobiology of Nucleic Acids", Vol. 1. New York: Academic Press, pp 169–294.

Foster PL (1998): Adaptive mutation: Has the unicorn landed? Genetics 148:1453–1459.

Foster PL (1999): Mechanisms of stationary phase mutation: A decade of adaptive mutation. Annu Rev Genet 33:57–88.

Friedberg EC, Feaver WJ, Gerlach VL (2000): The many faces of DNA polymerases: Strategies for mutagenesis and for mutational avoidance. Proc Natl Acad Sci USA 97:5681–5683.

Friedberg EC, Walker GC, Siede W (1995): "DNA Repair and Mutagenesis." Washington, DC: ASM Press.

Ganesan AK (1974): Persistence of pyrimidine dimers during post-replication repair in ultraviolet light-irradiated *Escherichia coli* K-12. J Mol Biol 87:102–119.

Gerlach VL, Aravind L, Gotway G, Schultz RA, Koonin EV, Friedberg EC (1999): Human and mouse homologs of *Escherichia coli dinB* (DNA polymerase IV), members of the UmuC/DinB superfamily. Proc Natl Acad Sci USA 96:11922–11927.

Hall BG (1990): Spontaneous point mutations that occur more often when advantageous than when neutral. Genetics 126:5–16.

Hall BG (1997): On the specificity of adaptive mutations. Genetics 145:39–44.

Harfe BD, Jinks-Robertson S (2000): DNA mismatch repair and genetic instability. Annu Rev Genet 34:359–399.

Herman GE, Modrich P (1982): *Escherichia coli dam* methylase: Physical and catalytic properties of the homogenous enzyme. J Biol Chem 257:2605–2612.

Howard-Flanders P, Rupp WD, Wilkins BM, Cole RS (1968): DNA replication and recombination after UV irradiation. Cold Spring Harb Symp Quant Biol 33:195–207.

Humayun MZ (1998): SOS and mayday: Multiple inducible mutagenic pathways in *Escherichia coli*. Mol Microbiol 30:905–910.

Imlay JA, Linn S (1988): DNA damage and oxygen radical toxicity. Science 240:1302–1309.

Jeggo P, Defais M, Samson L, Schendel P (1977): An adaptive response of *E. coli* to low levels of alkylating agent: Comparison with previously characterized DNA repair pathways. Mo Gen Genet 157:1–9.

Kasak L, Horak R, Kivisaar M (1997): Promoter-creating mutations in *Pseudomonas putida*: A model system for the study of mutation in starving bacteria. Proc Natl Acad Sci USA 94:3134–3139.

Kelner A (1949): Effect of visible light on the recovery of *Streptomyces griseus* conidia from ultraviolet light irradiation injury. Proc Natl Acad Sci USA 35:73–79.

Khanna KK, Jackson SP (2001): DNA double-strand breaks: Signaling, repair and the cancer connection. Nat Genet 27:247–254.

Kim IG, Oh TJ (2000): SOS induction of the *recA* gene by UV-, gamma-irradiation and mitomycin C is mediated by polyamines in *Escherichia coli* K-12. Toxicol Lett 116:143–149.

Kreuger JH, Elledge SJ, Walker GC (1983): Isolation and characterization of Tn5 mutations in the lexA gene of *Escherichia coli*. J Bacteriol 153:1368–1378.

Labrador M, Corces VG (1997): Transposable element-host interactions: Regulation of insertion and excision. Annu Rev Genet 31:381–404.

Landini P, Volkert MR (2000): Regulatory responses of the adaptive response to alkylation damage: A simple regulon with complex regulatory features. J Bacteriol 182:6543–6549.

Lederberg J, Lederberg EM (1952): Replica plating and indrect selection of bacterial mutants. J Bacteriol 63:399–406.

Lieb M, Bhagwat AS (1996): Very short patch repair: Reducing the cost of cytosine methylation. Mol Microbiol 20:467–473.

Lieb M, Rehmat S (1995): Very short patch repair of T:G mismatches in vivo: Importance of context and accessory proteins. J Bacteriol 177:660–666.

Lin P-F, Howard-Flanders P (1976): Genetic exchanges caused by ultraviolet photoproducts in phage lambda DNA molecules: The role of DNA replication. Mol Gen Genet 146:107–115.

Lindahl T (1979): DNA glycosylases, endonucleases for apurinic/apyrimidinic sites and base excision-repair. Prog Nucleic Acid Research Mol Biol 22:135–192.

Little JM, Mount D (1982): The SOS regulatory system of *Escherichia coli*. Cell 29:11–22.

Lombardo MJ, Torkelson J, Bull HJ, McKenzie GJ, Rosenberg SM (1999): Mechanisms of genome-wide hypermutation in stationary phase. Annu NY Acad Sci 870:275–289.

Lovett CM, Jr, Love PE, Yasbin RE (1989): Competence-specific induction of the *Bacillus subtilis* RecA protein analog: Evidence for dual regulation of a recombination protein. J Bacteriol 171:2318–2322.

Luria SE, Delbrück M (1943): Mutations of bacteria from virus sensitive to virus resistance. Genetics 28:491–511.

Mahajan SK (1988): Pathways of homologous recombination in *Escherchia coli*. In Kucherlapatic R, Smith GR (eds): "Genetic Recombination", Washington, DC: ASM Press, pp 87–140.

McKenzie GJ, Harris RS, Lee PL, Rosenberg SM (2000): The SOS response regulates adaptive mutation. Proc Natl Acad Sci USA 97:6646–6651.

McVeigh R, Yasbin RE (1996): The smart phages of *B. subtilis*: Type 4 SOS Response. J Bacteriol 178:3399–3401.

Michaels ML, Cruz C, Grollman AP, Miller JH (1992): Evidence that MutY and MutM combine to prevent mutations by an oxidative damaged form of guanine in DNA. Proc Natl Acad Sci USA 89:7022–7025.

Miller JH (1982): Carcinogens induce targeted mutations in *Escherichia coli*. Cell 31:5–7.

Modrich P, Lahue R (1996): Mismatch repair in replication fidelity, genetic recombination and cancer. Annu Rev Biochem 65:101–133.

Newcomb HB (1949): Origin of bacterial variants. Nature 164:150.

O'Grady PI, Borden A, Vandewiele D, Ozgenc A, Woodgate R, Lawrence CW (2000): Intrinsic polymerase activities of UmuD′(2)C and MucA′(2)B are responsible for their different mutagenic properties during bypass of a T-T *cis*-syn cyclobutane dimer. J Bacteriol 182:2285–2291.

Ohashi E, Bebenek K, Matsuda T, Feaver WJ, Gerlach VL, et al. (2000): Fidelity and processivity of DNA synthesis by DNA polymerase kappa, the product of the human DINB1 gene. J Biol Chem 275:39678–39684.

Pienitz R, Vincent WF (2000): Effect of climate change relative to ozone depletion on UV exposure in subarctic lakes. Nature 404:484–487.

Prival M, Cebula T (1996): Adaptive mutation and slow-growing revertants of an *Escherichia coli lacZ* amber mutant. Genetics 144:1337–1341.

Radman M (1974): Phenomenology of an inducible mutagenic DNA repair pathway in *Escherichia coli*: SOS repair hypothesis. In Prakash L, Sherman F, Miller M, Lawrence C, Tabor HW (eds): "Molecular and Environmental Aspects of Mutagenesis", Springfield, IL: Charles C. Thomas, pp 128–142.

Radman M (1988): Mismatch repair and genetic recombination. In Kucherlapatic R, Smith GR (eds): "Genetic Recombination", Washington, DC: ASM Press, pp 169–192.

Radman M, Wagner R (1986): Mismatch repair in *Escherichia coli*. Annu Rev Genet 20:523–538.

Ronen A, Glickman BW (2001): Human DNA repair genes. Environ Mol Mutagen 37:241–283.

Rosenberg SM, Harris RS, Torkelson J (1995): Molecular handles on adaptive mutation. Mol Microbiol 18:185–189.

Rupert LS, (ed) (1975): "Enzymatic Photoreactivation: An Overview". New York: Plenum Press, pp 73–87.

Saget B, Walker GC (1994): The Ada protein acts as both a positive and negative modulator of *Escherichia coli's* response to methylating agents. Proc Natl Acad Sci USA 91:9730–9734.

Samson L, Cairns J (1977): A new pathway for DNA repair in *Escherichia coli*. Nature 267:281–283.

Sancar A, Franklin KA, Sancar GB (1984): *Escherichia coli* photolyase stimulates UvrABC excision nuclease *in vitro*. Proc Natl Acad Sci USA 81:7397–7401.

Sancar A, Rupp WD (1983): A novel repair enzyme: UVRABC excision nuclease of *Escherchia coli* cuts a DNA strand on both sides of the damaged region. Cell 33:249–260.

Sancar GB, Smith FW, Sancar A (1985): Binding of *Escherichia coli* DNA photolyase to UV-irradiated DNA. Biochem 24:1849–1855.

Schendel PF (1981): Inducible repair systems and their implications for toxicology. CRC Crit Rev 8:311–362.

Sedgwick SG (1975): Genetic and kinetic evidence for different types of post-replication repair in *Escherichia coli* B. J Bacteriol 123:154–161.

Selby CP, Sancar A (1994): Mechanisms of transcription-repair coupling and mutation frequency decline. Microbiol Rev 58:317–329.

Setlow JK, Boling ME, Bollum FJ (1965): The chemical nature of photoreactivable lesions in DNA. Proc Natl Acad Sci USA 53:1430–1436.

Setlow RB (1978): Repair deficient human disorders and cancer. Nature (London) 271:713–717.

Setlow RB, Swenson PA, Carrier WL (1963): Thymine dimers and inhibition of DNA synthesis by ultraviolet irradiation of cells. Science 142:1464–1466.

Sicard M, Gasc AM, Giammarinaro P, Lefrancois J, Pasta F, Samrakandi M (2000): Molecular biology of *Streptococcus pneumoniae*: An everlasting challenge. Res Microbiol 151:407–411.

Sicard M, Lefevre JC, Mostachfi P, Gasc AM, Mejean V, Claverys JP (1985): Long- and short-patch gene conversions in *Streptococcus pneumoniae* transformation. Biochimie 67:377–384.

Smith KC, Meun DHC (1970): Repair of radiation-induced damage in *Escherichia coli*. Effect of *rec* mutations on postreplication repair of damage due to ultraviolet radiation. J Mol Biol 51:459–477.

Steele DF, Jinks-Robertson S (1992): An examination of adaptive reversion in *Saccharomyces cerevisiae*. Genetics 132:9–21.

Sutton MD, Kim M, Walker GC (2001a): Genetic and biochemical characterization of a novel *umuD* mutation: Insights into a mechanism for UmuD self-cleavage. J Bacteriol 183:347–370.

Sutton MD, Murli S, Opperman T, Klein C, Walker GC (2001b): *umuDC-dnaQ* interaction and its implications for cell cycle regulation and SOS mutagenesis in *Escherichia coli*. J Bacteriol 183:1085–1089.

Sutton MD, Smith BT, Godoy VG, Walker GC (2000): The SOS response: Recent insights into *umuDC*-dependent mutagenesis and DNA damage tolerance. Annu Rev Genet 34:479–499.

Sutton MD, Walker GC (2001): *umuDC*-Mediated cold sensitivity is a manifestation of functions of the UmuD2C complex involved in a DNA damage checkpoint control. J Bacteriol 183:1215–1123.

Torkelson J, Harris RS, Lombardo MJ, Nagendran J, Thulin C, Rosenberg SM (1997): Genome-wide hypermutation in a subpopulation of stationary-phase cells underlies recombination-dependent adaptive mutation. EMBO J 16:3303–3311.

Turner DP, Connolly BA (2000): Interaction of the *E. coli* DNA G:T-mismatch endonuclease (Vsr protein) with oligonucleotides containing its target sequence. J Mol Biol 304:765–778.

Varghese AJ (1970): 5-Thyminyl-5,6-dihydrothymine from DNA irradiated with ultraviolet light. Biochem Biophys Res Commun 38:484–490.

Vasquez DA, Nyaga SG, Lloyd RS (2000): Purification and characterization of a novel UV lesion-specific DNA glycosylase/AP lyase from *Bacillus sphaericus*. Mutat Res 459:307–316.

Verhoeven EE, Wyman C, Moolenaar GF, Hoeijmakers JH, Goosen N (2001): Architecture of nucleotide excision repair complexes: DNA is wrapped by UvrB before and after damage recognition. EMBO J 20:601–611.

Wagner J, Gruz P, Kim SR, Yamada M, Matsui K, et al. (1999): The *dinB* gene encodes a novel *E. coli* DNA polymerase, DNA pol IV, involved in mutagenesis. Mol Cell 4:281–286.

Walker GC (1984): Mutagenesis and inducible responses to deoxyribonucleic acid damage in *Escherichia coli*. Microbiol Rev 48:60–93.

Walker GC (1987): The SOS response of *E. coli*. In Neidhardt FC (ed): "*Escherichia coli* and *Salmonella typhimurium*." Washington, DC: ASM Press, pp 1346–1357.

Winterling KW, Levine AS, Yasbin RE, Woodgate R (1997): Characterization of DinR, the *Bacillus subtilis* SOS repressor. J Bacteriol 179:1698–1703.

Witkin E (1947): Genetics of resistance to radiation in *Escherichia coli*. Genetics 32:221–.

Witkin EM (1976): Ultraviolet mutagenesis and inducible DNA repair in *Escherichia coli*. Bacteriol Rev 40:869–907.

Wood RD, Mitchell M, Sgouros J, Lindahl T (2001): Human DNA Repair Genes. Science 291:1284–1289.

Wright BE (2000): A biochemical mechanism for non-random mutations and evolution. J Bacteriol 182:2993–3001.

Wulff DL, Rupert CS (1962): Disappearance of thymine photodimer in ultraviolet irradiated DNA upon treatment with a photoreactivating enzyme from baker's yeast. Biochem Biophys Res Commun 7:237–240.

Yamamoto K, Satake M, Shinagawa H (1984): A multicopy *phr*-plasmid increased the ultraviolet resistance of a *recA* strain of *Escherichia coli*. Mutat Res 131:11–18.

Yasbin RE, Cheo DL, Bayles KW (1992): Inducible DNA repair and differentiation in *Bacillus subtilis*: Interactions between global regulons. Mol Microbiol 6:1263–1270.

Yasbin RE, Miehl-Lester R (1990): DNA repair and mutagenesis. In Streips UN, Yasbin RE (eds): "Modern Microbial Genetics." New York: Alan R. Liss, pp 77–90.

3

Gene Expression and Its Regulation

JOHN D. HELMANN

Department of Microbiology, Cornell University, Ithaca, New York 14853-8101

I. INTRODUCTION

Bacteria have a remarkable ability to adapt to a rapidly changing environment. In most cases adaptation requires that new proteins be synthesized to adjust the metabolic capacity of the organism to the available nutrients or to defend against chemical or physical toxins. In this chapter we will survey the diverse ways that bacteria have evolved to coordinate gene expression with environmental signals.

Gene expression begins with the copying of discrete segments of the DNA into RNA, a process known as *transcription*. The products of transcription include four classes of RNA molecules: messenger RNA (mRNA),

Modern Microbial Genetics, Second Edition. Edited by Uldis N. Streips and Ronald E. Yasbin. ISBN 0–471–38665–0 Copyright © 2002 Wiley-Liss, Inc.

ribosomal RNA (rRNA), transfer RNA (tRNA), and regulatory RNA. For protein-coding genes, the corresponding mRNA molecule binds to ribosomes and directs the synthesis of one or more specific proteins in a process called *translation*. These processes are so central to all living things that the informational transfer of "DNA makes RNA makes protein" has been referred to as the "central dogma" of molecular biology.

In bacteria, gene expression is most frequently regulated at the level of transcription. That is, bacteria only transcribe the subset of their genes that are necessary for growth and survival under the existing environmental conditions. The remaining regions of the genome are silent. In some cases, however, genes are transcribed into mRNA even when their protein products may not be needed. In these cases the process of translation is likely to be the focus of regulation. To appreciate the diverse mechanisms that allow bacteria to regulate transcription and translation, we will first review these processes and the enzymes that catalyze them.

II. RNA POLYMERASE AND THE PROCESS OF TRANSCRIPTION

A. Structure of RNAP

The first step in gene expression is the transcription of an RNA molecule complementary to the DNA template catalyzed by DNA-dependent RNA polymerase (RNAP). As befits its central role in the cell, RNAP is highly conserved and very complex (McClure, 1985; Young, 1991). All cells contain a multisubunit RNAP with two large subunits and a variable number of smaller subunits. In bacteria, a catalytically active RNAP core enzyme has minimal subunit composition $\beta\beta'\alpha_2$ (indicated as "E") (Burgess et al., 1969). Additional small proteins, including the omega (ω) polypeptide (Gentry and Burgess, 1993; Mukherjee and Chatterji, 1997) and, in some, gram-positive bacteria, the delta (δ) subunit (Juang and Helmann, 1994; Lopez de Saro et al., 1995; Lopez de

Saro et al., 1999), are also often present. The core enzyme can faithfully copy DNA into RNA over many thousands of base pairs, but by itself is incapable of recognizing promoter elements.

Promoter recognition requires a separate, dissociable specificity protein known as σ (Gross et al., 1992; Helmann, 1994). The complex formed by binding of σ to the core enzyme is called *holoenzyme* (Fig. 1) and is often identified by the associated σ factor. In *Escherichia coli*, for example, the primary σ factor is 70 kDa in size and is referred to as σ^{70}. The complex formed by the binding of σ^{70} to the core RNAP is the σ^{70} holoenzyme ($\beta\beta'\alpha_2\sigma^{70}$ or $E\sigma^{70}$). As we will see, substitution of one σ factor by an alternative specificity subunit is a powerful mechanism for activating the transcription of new sets of genes (Table 1).

Eukaryotes have three nuclear RNAP forms that have between 10 and 12 protein subunits each, including two similar in sequence to the large β and β' subunits that make up the bulk of the bacterial core enzyme. Recently the three-dimensional structures of the RNAP from the thermophilic bacterium *Thermus aquaticus* (Zhang et al., 1999) and from the yeast *Saccharomyces cerevisiae* (Cramer et al., 2000; Fu et al., 1999) have been determined at atomic resolution. The resulting structures reveal that those regions that are highly similar in sequence between the bacterial and eukaryotic RNAP subunits are closely clustered around the active site for RNA synthesis.

B. The Bacterial Transcription Cycle—Overview

In general, processes of macromolecular synthesis can be divided into three major phases: initiation, elongation, and termination. In the case of RNA synthesis (Fig. 2A), the initiation phase involves the interaction of RNAP with specific *promoter* sites that identify the start point of an RNA molecule (deHaseth et al., 1998). Once bound to the promoter, initially as a *closed complex*, RNAP locally separates the two DNA

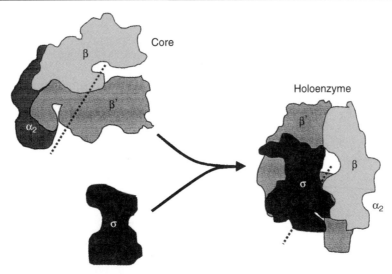

Fig. 1. Structure of RNA polymerase core and holoenzyme. In bacteria, RNA synthesis is carried out by the core enzyme containing the β, β', and 2α subunits (abbreviated E). Promoter recognition requires that the core enzyme by associated with a σ factor to form the holoenzyme (Eσ). Recent X-ray crystallography studies have allowed the overall molecular dimensions of both core enzyme (left) and holoenzyme (right) to be visualized (Finn et al., 2000; Zhang et al., 1999). Binding of the σ subunit to the RNAP core enzyme leads to a significant conformational change in both core subunits and σ. As a result of these changes, the "clawlike" features of the core enzyme close to form a channel, thought to bind DNA. Note that the holoenzyme in this example is viewed from a different angle than the core, and the two α subunits are largely hidden behind the much larger β and β' subunits. The dashed line indicates a possible DNA trajectory, but this is not yet clear (for a discussion see Naryshkin et al., 2000; Nudler, 1999).

strands over a span of about 12 bp to generate the *open complex* (deHaseth and Helmann, 1995). This process is catalyzed, in part, by interactions between the σ subunit of RNAP and the nontemplate strand of the −10 element (Helmann and deHaseth, 1999).

RNA synthesis commences when the open complex binds to the first two nucleoside triphosphate (NTP) substrates, as specified by the template DNA. The template-directed polymerization of NTPs into an RNA product then proceeds. During the early stages of elongation, short RNA products frequently dissociate from the complex and are released. These *abortive products* are typically only a few nucleotides in length and can be produced in large amounts from some promoter sites (Hsu, 1996; Hsu et al., 1995). Another phenomenon that sometimes accompanies the initiation phase is *slippage synthesis*. In this case the synthesis of short repeated sequences can lead to a misalign-

ment of the RNA product on the DNA template. For example, transcription initiation in the (nontemplate) sequence *A*TTTTTTG may lead to RNA molecules starting with A followed by 15 or more U residues rather than a complete RNA transcript (Qi and Turnbough, 1995). Although initially viewed as a curiosity of relevance only to afficionados of RNAP enzymology, it is now clear that both abortive initiation and slippage synthesis can be used by the cell to regulate transcription at selected promoter sites (Uptain et al., 1997).

Once the RNA product passes approximately 10 nt in length, the σ subunit is usually released and the RNAP-DNA complex undergoes a substantial structural rearrangement to generate a highly stable *elongation complex* (Nudler, 1999). This complex, sometimes called a *ternary complex* to denote the presence of DNA-RNAP-RNA, can synthesize RNA of many thousands of nucleotides

TABLE 1.   Sigma Factors of *E. coli* and *B. subtilis*

Organism	σ	Gene	Function (s)
E. coli	σ^{70}	*rpoD*	Housekeeping genes
	$\sigma^{32}(\sigma^{H})$	*rpoH (htpR)*	Heat shock
	$\sigma^{24}(\sigma^{E})$	*rpoE*	Extreme heat shock, periplasmic stress
	$\sigma^{28}(\sigma^{F})$	*fliA*	Flagellar-based motility
	$\sigma^{38}(\sigma^{S})$	*rpoS*	Stationary phase adaptive response
	σ^{54}	*rpoN*	Nitrogen-regulated genes
	σ^{fecI}	*fecI*	Iron-citrate transport
B. subtilis	σ^{A}	*sigA*	Primary σ
	σ^{B}	*sigB*	General stress response
	σ^{D}	*sigD*	Flagella, chemotaxis, autolysins
	σ^{E}	*sigE*	Sporulation—early mother cell
	σ^{F}	*sigF*	Sporulation—late mother cell
	σ^{G}	*sigG*	Sporulation—late forespore
	σ^{H}	*sigH*	Sporulation, competence
	σ^{K}	*sigK*	Sporulation, late forespore
	σ^{L}	*sigL*	Levanase, amino acid catabolism
	σ^{ykoZ}	*ykoZ*	Unknown
	ECF σs	*sigM, sigV, sigW, sigX,*	Extracytoplasmic functions (details
		sigY, sigZ, ylaC	largely unknown)

in length without dissociation—that is, it is highly processive (Nudler et al., 1996).

At specific sequences, or in response to specific protein factors, the processivity of RNA synthesis can be interrupted and the completed RNA chain released in an event called *termination*. The two major classes of termination events in bacteria are those mediated by Rho-independent terminator sites and those catalyzed by the Rho termination factor (Landick, 1997; Mooney et al., 1998; Uptain et al., 1997). Rho-independent terminators (also called *factor-independent sites*) are sequences that encode GC-rich RNA stem-loop (hairpin) structures often followed immediately by a U-rich sequence (Fig. 2B). These RNA structures interact with sites on RNAP to trigger a conformational change leading to release of the transcript from the ternary complex and dissociation of RNAP from the template. Rho protein acts to terminate the synthesis of transcripts for protein coding genes that are not being translated. In the absence of translation, Rho is able to bind to unstructured regions of RNA, particularly in regions en-

riched in cytosine, and then translocate along the RNA to interact with RNAP and trigger dissociation of the ternary complex (Platt, 1994). As a result of Rho action, an inability to *translate* one gene in an operon (e.g., due to a nonsense mutation) will often lead to a failure to even *transcribe* the downstream genes of the same operon: a phenomenon known as *polarity* (Peters and Benson, 1995; Stanssens et al., 1986).

To complete the transcription cycle, the core enzyme must be released from the template DNA and then rebind to a σ subunit to reform holoenzyme before it can again begin the cycle of promoter recognition, RNA chain initiation, elongation, and termination (Fig. 2A). As we will see, each of these various steps can be regulated.

C. Promoter Structure

In bacteria, the promoter is often identified with two conserved sequence elements located approximately 35 and 10 bp upstream of the transcription start point (TSP). These consensus sequences, known as the −35 and −10 elements, are major determinants of

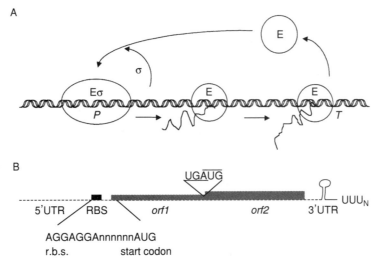

Fig. 2. The bacterial transcription cycle and structure of a generic operon. **A:** Transcription initiation begins when the holoenzyme (Eσ) binds to the promoter site, establishes the strand-separated open complex, and begins the tempate-directed polymerization of NTPs. The RNA chain is elongated and σ is released. Elongation continues until RNAP encounters a Rho-independent terminator (stem-loop) structure. Dissociation of the core enzyme, followed by rebinding of σ factor completes the cycle. **B:** The RNA product illustrated encodes two protein products. An initial 5′ UTR (untranslated region) may contain regulatory signals (for translational control or attenuation mechanisms). The RBS and start codon define the beginning of the first gene. In this case, ribosomes completing synthesis of gene product 1 will immediately begin translation of gene product 2 by the mechanism of *translational coupling*. The 3′ UTR contains the structure corresponding to the Rho-independent transcription terminator.

promoter strength. In nearly all bacteria, most promoters recognized by the predominant form of RNAP (equivalent to the *E. coli* σ⁷⁰ holoenzyme) have similarity to the classic consensus elements: TTGACA N(16–18) TATAAT (Helmann, 1995; Lisser and Margalit, 1993). These two hexamer sequences, separated by about 17 bp (the spacer region) contact the σ subunit during the process of promoter recognition. It is important to appreciate that, on average, promoters may match these sequences at only 7 or 8 of the 12 conserved positions. In general, those with a closer match to consensus tend to be stronger promoters (they initiate transcription rapidly), while those with fewer matches to consensus are often weaker. However, many promoter sites have been identified that do not closely match these consensus sequences. Often these are sites that require an activator protein or a different σ factor for recognition (Gross et al., 1992; Helmann,

1994). In addition, a subset of σ⁷⁰ promoters lack a −35 region altogether, having instead an extended −10 element with consensus sequence TGnTATAAT. These are designated "extended −10" promoters (reviewed in Bown et al., 1997).

While the −35 and −10 consensus elements are certainly defining features of bacterial promoters, it has become increasingly clear that promoter strength depends on many factors in addition to the −35 and −10 elements (Fig. 3). To appreciate the complexity of the initiation process, consider that RNAP is a very large protein (Mr ~450 kDa) and interacts with as much as nearly 110 bp of DNA when bound at a promoter site. Interactions throughout this region (extending from nearly −90 to +20 relative to the TSP) can and do affect promoter strength. The major functional regions of a bacterial promoter can be arbitrarily divided into the upstream promoter region (−90 to −40), the consensus

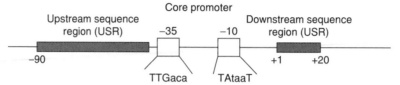

Fig. 3. Bacterial promoter structure. The structure of a bacterial promoter is defined relative to the transcription start point (TSP), designated +1. The two critical conserved regions for σ factor recognition are located near −35 and −10. The upstream promoter region (−40 to −90) may contain an UP element (for interaction with the α-CTD and/or activator binding sites. The initial transcribed region (+1 to +20) may also contain regulatory elements, and it can affect the efficiency of RNAP clearance of the promoter region by affecting the processes of abortive initiation, transcriptional slippage, and pausing. While activator proteins bind most frequently to the upstream promoter region, repressor proteins bind most frequently to sites overlapping the consensus elements, the TSP, or in the initial transcribed region. The consensus sequences are shown, based on the *E. coli* model, with highly conserved positions in upper case and more weakly conserved bases in lower case.

element/spacer region (−40 to −1), and the TSP and downstream sequence region (DSR) (+1 to +20). Note that by convention the TSP is designated +1 and the immediately preceding base is designated −1 (there is no 0 near the TSP!).

The role of the upstream promoter region in modulating transcription initiation is complex. First, this is often a site where activator proteins can bind to DNA and stimulate transcription initiation by RNAP bound to DNA at adjacent −35 and −10 elements (Rhodius and Busby, 1998). Second, sequence elements in this region can stimulate transcription independent of any bound transcription factors (Gourse et al., 2000). Often these stimulatory sequences include short runs of repeated adenine or thymine residues, such as AAAA or TTTT, that are known to lead to intrinsic DNA bends. It was therefore postulated that such intrinsically bent DNA could facilitate the wrapping of the promoter DNA around the large RNAP molecule during promoter recognition and subsequent initiation (Perez-Martin and de Lorenzo, 1997). Third, this region can provide an additional sequence-specific interaction site for RNAP (in addition to the −35 and −10 elements). Specific AT-rich sequences, called UP elements, are often present between −40 and −60 and interact favorably with the carboxyl-terminal domains of the two α subunits of RNAP (α-CTD) (Gourse et al., 2000).

Interactions of α with the UP element can greatly increase promoter strength (as much as 200-fold) by increasing the rate and affinity of the interaction of RNAP with the promoter region. The best characterized UP element is found just upstream of a promoter for the *E. coli* ribosomal RNA operon B (*rrnB* P2), but related sequences are widespread in many bacterial promoters including those requiring an alternative σ subunit for activation (Fredrick et al., 1995; Ross et al., 1998). The interaction of the α-CTD regions with the UP element DNA likely occurs in the minor groove and is favored by AT-rich sequences. Indeed, it has been shown that the ability of short A- and T-rich sequences to stimulate transcription also depends on the presence of α-CTD (Aiyar et al., 1998). Thus at least two classes of promoter region sequences, oligo-A directed DNA bends and UP elements (that may or may not be bent) probably stimulate transcription by a similar mechanism. In addition to its important role in contacting promoter DNA, the α-CTD is also a frequent target for protein-protein interactions with upstream activator proteins (see below). As these results make clear, the process of transcription initiation can be very complex and involves many different proteins interacting with different parts of the promoter region.

Transcription in the Archaea and in eukaryotes is catalyzed by a multisubunit enzyme considerably more complex than its bacterial homologue (Young, 1991; Bell and Jackson, 1998; Soppa, 1999). Archaea have a single RNAP, while eukaryotes typically have three discrete species known as RNAP I, II, and III. Transcription of mRNA is mediated by RNAP II. In these organisms, promoter recognition follows a decidedly different pathway from the bacteria that does not involve a σ factor and does not require conserved −35 and −10 recognition elements. Instead, a multisubunit transcription factor complex recognizes a conserved TATA element upstream of the TSP. A key player in this process is the TATA-binding protein (TBP) which, at least in eukaryotes, is in an assembly with other transcription factors (together known as TFIID) (Green, 2000a). After binding of TBP (and associated proteins) to the TATA box element, another transcription factor, TFIIB (TFB in Archaea), binds to the adjacent DNA. RNAP (which in its simplest form contains 10–12 subunits) then binds to the DNA-protein complex that has defined the initiation region (Buratowski, 2000). The bound RNAP is then capable of initiating transcription, often in response to bound activator proteins.

Current models for Archaeal promoters include three key sequence elements: an initiator (INR) region near the TSP, a TATA box near −26, and upstream pair of adenine residues (−34, −33) that define a TFB recognition element (Soppa, 1999). Clearly, this structure is very different from the classic −35, −10 element architecture associated with Bacterial promoters. Thus, studies of gene expression and its regulation in the Archaea must often take cues from eukaryotic, rather than classic prokaryotic (bacterial) paradigms.

III. RIBOSOMES AND THE PROCESS OF TRANSLATION

A. Structure of Ribosomes

While RNA synthesis is a relatively straightforward, template-directed copying of one type of nucleic acid (DNA) into another (RNA), the process of translating the resulting mRNA sequences into protein is considerably more complex. The process of translation occurs on the ribosomes, which provide a scaffold for the alignment of specific adaptor molecules that translate the nucleic acid sequence of the mRNA into the amino acid sequence of the protein product (Fig. 4A). These adaptors, of course, are the charged aminoacyl tRNA (aa-tRNA) molecules which each recognize one or more triplet codons in the mRNA and carry, covalently attached at their 3′ end, the corresponding amino acid. During the process of translation, the ribosome binds two different tRNA molecules at a time and catalyzes the bond-forming reaction (peptidyltransferase reaction) between the amino acids.

The ribosome itself is an enormously complex macromolecular assembly consisting of dozens of proteins bound to highly structured rRNA molecules (Green and Noller, 1997). By weight, the ribosome is about half RNA and half protein. Altogether, the bacterial ribosome has a sedimentation coefficent of 70 S (S = Svedberg units) in a ultracentrifugation experiment. Indeed, most components of the ribosome are defined, for largely historical reasons, by their sedimentation values. The ribosome can be reversibly dissociated into two functional subassemblies, the small "subunit" (30 S) and the large "subunit" (50 S). Each subunit still contains at least one large rRNA molecule and numerous proteins. Recently high-resolution structural techniques have allowed the bacterial ribosome to be visualized at near atomic resolution (Fig. 4A; Ban et al., 2000; Carter et al., 2000; Cech, 2000; Puglisi et al., 2000; Wimberly et al., 2000).

The ribosome can be most simply viewed as a two part molecular machine (Fig. 4B; see Frank, 1998, for review). The small 30 S subunit plays a primary role in binding to the mRNA and is the site of decoding. It is on this subunit that the anticodon loops of the tRNA molecules will base-pair with the codons on the mRNA (Carter et al., 2000). The large subunit is the site of peptide bond

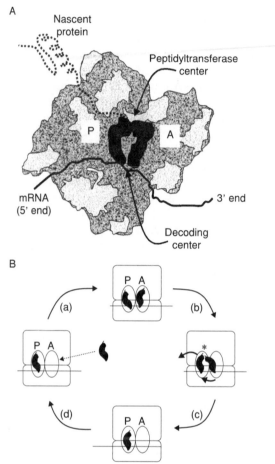

Fig. 4. The translation cycle. **A:** The structure of a ribosome elongation complex is illustrated as modeled from X-ray crystallography (Nissen et al., 2000a). The lighter colored regions correspond to ribosomal proteins. The bulk of the ribosome (darker stippling) is made up of RNA. The large subunit corresponds to the upper half of the ribosome, while the small subunit is the lower half. Note that the two tRNAs (black) are aligned along the mRNA in the decoding center on the small ribosomal subunit. Note that the nascent polypeptide chain, attached to the tRNA in the P site, is already starting to fold into its three-dimensional structure. **B:** The elongation cycle of translation. In step (a), the incoming aminoacyl-tRNA is delivered to the empty A site by elongation factor Tu. If the tRNA contains the correct anticodon to pair with the codon in the mRNA, EF-Tu hydrolyzes GTP and releases the tRNA into the A site. In step (b), the amino acid attached to the A site tRNA moves into the P-site and forms a peptide bond with the nascent polypeptide chain. This reaction, peptidyltransferase (*), is catalyzed by the large subunit 23 S rRNA. Note that the anticodon loop remains bound to the A site on the small subunit) to generate a "hybrid state" tRNA. In step (c) translocation occurs when EF-G repositions the two tRNAs and the bound mRNA in the small subunit by three nucleotides. Note that after step (c), the tRNA has been completely moved into the P site, leaving an empty A site, and the mRNA has translocated within the decoding center. The ribosome is now ready to accept the next tRNA corresponding to the next codon (d).

formation (peptidyltransferase), a reaction actually catalyzed by the 23S rRNA (Green and Noller, 1997). The tRNA molecules bind to two distinct sites on the surface of the ribosome (the A and P sites), each site bridging from the decoding site (on the small subunit) to the peptidyltransferase center (on the large subunit). The part of the tRNA that binds to the decoding site is the anticodon loop, while the part that carries the amino acid, and binds to the large subunit peptidyltransferase center, is the acceptor end. As the mRNA is channeled through the ribosome, three nucleotides at a time, the corresponding aa-tRNA molecules are loaded into the aminoacyl (A) site where their bound amino acid becomes linked to the growing polypeptide chain. Concomitant with peptide bond formation, the acceptor end of the tRNA moves into the peptidyl (P) site, displacing the now empty (uncharged) tRNA from that site.

B. The Bacterial Translation Cycle

In bacteria, a typical mRNA molecule contains an initial 5'-untranslated region (5'UTR), one or more coding segments, and a final 3'-untranslated region (3'UTR) (Fig. 2B). The 5'UTR contains signals that define the start site of the first gene (coding sequence), while the 3'UTR will often has a stem loop structure that functions both as a transcription terminator and to stabilize the mRNA against exonucleolytic degradation (Grunberg-Manago, 1999). Since transcription and translation are closely coupled in prokaryotes, ribosomes may bind to the nascent mRNA molecules as soon as the 5'UTR is extruded from the transcribing RNAP.

The initial interaction between mRNA and the ribosome involves an RNA-RNA annealing reaction. The small, 30 S ribosomal subunit contains a large RNA molecule (16 S). The 3'-end of the 16 S rRNA can anneal to specific, complementary sequences located just upstream of a protein-coding region. These ribosome-binding sites (RBS or Shine-Dalgarno region) have a consensus sequence, in E. coli, of AGGAGGA. Recognition of the RBS by the 30 S subunit, facilitated by protein initiation factors, allows the binding of a specific initiator tRNA that is complementary to the initiation codon for protein synthesis (Schmitt et al., 1996; Brock et al. 1998). The initiation codon is most commonly AUG, but GUG is also sometimes used. The initiator tRNA carries formyl-Methionine (fMET) at its 3' end and binds to the peptidyl (P) site on the ribosome.

Once the 30 S initiation complex has assembled, the large 50 S ribosomal subunit binds to form the 70 S initiation complex. For translation to begin, another tRNA molecule is loaded onto the A site of the ribosome (by the action of the elongation factor, EF-Tu) as specified by the identity of the second codon triplet in the mRNA. Note that the two tRNA binding sites on the ribosome, the A and the P sites, are biochemically distinct. The A site is specific for tRNA molecules carrying (unmodified) amino acids, while the P site carries the growing polypeptide chain (or the initiating amino acid, fMET).

Once the A and P site tRNAs are bound, the free amino group of the A site aa-tRNA becomes linked, in an RNA-catalyzed reaction, to the fMET on the P site initiator tRNA (Cech, 2000; Nissen et al., 2000a). This reaction is quite complicated as it involves the movement of the acceptor end of the tRNA into the P site on the large subunit, while the anticodon end of the tRNA, bound to the small subunit, remains bound to the A site (Fig. 4B). This leads to a hybrid state that involves a motion of the large subunit of the ribosome relative to the smaller subunit (Green and Noller, 1997). In this hybrid state, the initiator tRNA is still retained, by virtue of its interaction with the decoding center in the P site—and a transient interaction with an exit (E) site on the large subunit—but it no longer carries an amino acid.

Subsequent to peptidyltransfer, the ribosome translocates along the mRNA by three nucleotides so that the next codon is brought into the decoding center on the small subunit and the next tRNA can then be bound. The process of ribosome translocation is catalyzed by a protein, elongation factor G (EF-G) (Green, 2000b). In a remarkable example of molecular mimicry, EF-G has a three-dimensional shape that closely resembles the shape of EF-Tu when bound to an aa-tRNA (Nissen et al., 2000b). Thus it is proposed that EF-G binds to the same surface of the ribosome used by EF-Tu when delivering a tRNA into the decoding pocket (A site) on the small subunit of the ribosome. However, instead of delivering a tRNA, EF-G inserts a protein domain into this site and thereby moves the anticodon loop of the peptide carrying tRNA (together with the annealed mRNA) from the small subunit A site into the small subunit P site (the acceptor end already moved into the large subunit P site during the peptidyltransferase reaction). This motion leads to the ejection of the initiator tRNA (retained by its small subunit P site interaction) and regenerates a vacant A site for the next tRNA. The process of polypeptide elongation involves a repeating cycle of EF-Tu catalyzed aa-tRNA binding (to the A site), peptidyltransferase to form a peptide bond and generate tRNAs bound in hybrid states, binding of EF-G and translocation of the ribosome relative to the mRNA with regeneration of an empty A site.

Polypeptide elongation continues until the ribosome encounters one of the three termination (stop) codons. These sites are not normally recognized by a cognate aa-tRNA so the ribosome stalls. The stop codon is then recognized by specific protein release factors that trigger the release of the completed polypeptide from the tRNA and the ribosome (Wilson et al., 2000).

The overall process of protein synthesis is a dominant one within the cell (Neidhardt et al., 1990). The ribosomes are a major fraction of the cell mass, accounting for nearly 50% of cell dry weight in rapidly growing cells. In addition translation requires a lot of energy. ATP hydrolysis drives the coupling of amino acids to tRNA and GTP hydrolysis is coupled both to the loading of the correct aa-tRNA onto the ribosome (by EF-Tu) and to translocation (EF-G). During rapid growth the process of translation consumes approximately half of all energy generated by metabolism. If ATP generation is blocked, by chemical poisons, the ongoing translation can completely deplete the cell of ATP (and GTP) energy reserves in a matter of seconds.

Because translation is such a central feature of metabolism, the cell tightly regulates its total capacity for translation by controlling the number of ribosomes per cell in response to growth rate. Thus rapidly growing cells need proportionally more ribosomes per cell than slowly growing cells. This phenomenon, growth rate control, results in rates of transcription for rRNA, ribosomal protein, and tRNA genes that vary as the square of the growth rate (Gourse et al., 1996). The mechanisms that contribute to growth rate control eluded investigators for many years, but recent insights suggest a surprisingly simple mechanism that couples rates of transcription to cellular energy charge (Gaal et al., 1997). This transcriptional control mechanism will be described in more detail below.

IV. TRANSCRIPTIONAL REGULATION—REPRESSORS AND ACTIVATORS

In bacteria, groups of functionally related genes are often clustered on the chromosome into *operons* that can be cotranscribed into a single mRNA molecule carrying the information for multiple proteins (Fig. 2B). The operon arrangement allows for the coordinate regulation of related functions (Salgado et al., 2000). Control of operon expression is often mediated by regulatory proteins acting at defined binding sites in or near the promoter region (Collado-Vides et al., 1991; Struhl, 1999). Negative acting *repressor* proteins bind to sites called *operators* (Rojo, 1999), while positive acting *activators* bind to *activator binding sites* (Rhodius and Busby, 1998). In this and subsequent sections examples will be largely drawn (except where noted) from the extensive literature on *E. coli*.

Many regulatory proteins regulate more than one operon. The collection of operons that respond to a common regulator define a *regulon*. A final level of organization is represented by the *stimulon*: all those genes that respond to a particular stimulus or signal (Neidhardt et al., 1990). In bacteria, for example, there is a complex heat shock stimulon representing a large set of genes that are strongly activated in response to high-temperature stress (see Streips, this volume). This stimulon involves the coordinate transcriptional induction of numerous regulons, controlled by several different transcription factors (Narberhaus, 1999; Yura and Nakahigashi, 1999).

The hierarchical organization of genes into regulatory units is made more complex by the fact that many genes and operons belong to multiple regulons (and hence to multiple stimulons). Some genes may be induced by both heat and oxidative stress, and others by heat and osmotic stress, but not by oxidative stress. At the level of regulation, these complexities are often reflected in the presence of promoters that can respond to multiple regulators, multiple promoter elements preceding an operon, promoter elements within an operon, or other regulatory inputs that affect gene expression (Neidhardt and Savageau, 1996).

A. Regulation by Repressors

Conceptually the simplest mechanism of transcriptional control is repression by a repressor protein binding to an operator site (Rojo, 1999). This is also the first mechanism

discovered, having been identified as the basis for the induction of the lactose operon in response to lactose (and related gratuitous inducers). Indeed, the repressor model of gene regulation was so dominant in the early days of molecular biology that the first documented example of positive control, induction of the arabinose operon by arabinose, met with fierce resistance (see Beckwith, 1996, for an interesting account).

In the simplest cases repression can be explained by steric occlusion. That is, the binding of a repressor protein to its DNA target site occludes, or blocks, the interaction of RNAP with the promoter. Analysis of large numbers of operator sites suggests that this is a very common mechanism of action for repressor proteins: operator sites frequently overlap the promoter region (Fig. 5; see Collado-Vides et al., 1991). The activity of repressors is itself regulated in response to specific chemical or physical signals. This regulation can be at the level of repressor synthesis or activity. For example, many repressors are regulated by the reversible binding of small molecules. In some cases, these signal molecules may act as *co-repressors* to alter the conformation of the repressor to favor DNA binding (Fig. 6).

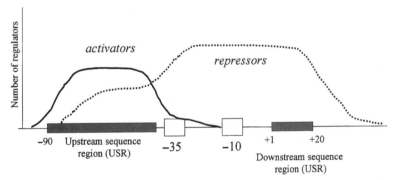

Fig. 5. Distribution of repressor and activator binding sites on *E. coli* promoters. A compilation of known repressor and activator binding sites on *E. coli* promoters reveals a distinct spatial distribution (Gralla and Collado-Vides, 1996). Activator sites typically occur in the upstream promoter region while operator sites are found most frequently overlapping the −35 and −10 consensus elements, the TSP, and in the early transcribed region. In addition to the sites summarized here, some activator binding sites (particularly those for the σ^{54} holoenzyme) are located much farther upstream or even downstream of the promoter; some auxilliary repressor binding sites (e.g., those involved in looping) are also located at a greater distance from the promoter.

A

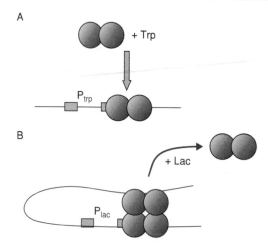

B

Fig. 6. Bacterial repressors: TrpR and LacI. **A:** Repression by the *E. coli* TrpR protein occurs when the TrpR protein binds tryptophan as a corepressor. The resulting protein–amino acid complex represses synthesis of tryptophan biosynthesis genes by binding to a site overlapping the promoter for the tryptophan operon. The −35 and −10 elements are represented by the boxes. **B:** The LacI repressor is active in the absence of its signal molecule, lactose, and is thought to form a repression loop involving interaction between proteins bound just downstream of the promoter (+11) and a site at −82. Upon binding lactose, the LacI-lactose complex loses the ability to bind tightly to its cognate operator sites, the repression loop opens up, and the genes for lactose utilization are induced.

One example is the repressor of the tryptophan biosynthesis operon (TrpR). In the presence of tryptophan, TrpR binds to operator sites to prevent further transcription of the tryptophan biosynthesis genes (Somerville, 1992). In other instances, the signal may serve as an *inducer* and inhibit repressor binding to DNA. A well-characterized example of this mechanism is the ability of lactose to prevent repression of the *lac* operon by the LacI repressor (Matthews and Nichols, 1998). Finally, many regulatory proteins can be controlled by covalent modification. A particularly widespread example of this mechanism is the large family of two-component regulatory proteins in which a sensor kinase phosphorylates a response regulator that can then serve as either a repressor or an activator of transcription. Note that consistent with accepted conventions, gene names are written in italics with an initial lowercase letter (e.g., *trpR*), while the corresponding protein product is not italicized and has an initial capital letter (e.g., TrpR).

In general, repression can occur by any mechanism that reduces the frequency with which a promoter initiates transcription (Choy and Adhya, 1996; Rojo, 1999). As reviewed above, transcription initiation begins with promoter binding to form the closed complex, isomerization of the closed complex to the strand-separated open complex, and then initiation of the RNA chain. Since early stages of transcription can sometimes be limiting, for example, due to abortive initiation, this phase is often referred to as *promoter clearance*. While repressors frequently block the binding of RNAP to the promoter, repression can also occur when a protein impedes later steps, including open complex formation or promoter clearance.

At the *gal* operon, for example, repression can involve either of two distinct types of mechanisms (Choy and Adhya, 1996). The binding of two repressor molecules (GalR) to sites both upstream and downstream of the promoter, together with a DNA-bending protein (HU), forms a DNA loop that prevents RNAP binding to the intervening promoter sequence. Under conditions when GalR cannot form a repression loop (e.g., if HU is absent or if the downstream binding site is mutated), GalR can still bind to the upstream site and repress initiation from the *gal* P1 promoter (there are actually two closely spaced promoters for the *gal* operon). In this case GalR does not prevent RNAP binding but inhibits the progression of the RNAP from a closed to an open complex.

Finally, some repressors may allow RNAP to initiate short, abortive products but inhibit the promoter clearance process. An example of this mechanism is the *Bacillus subtilis* bacteriophage φ29 p4 protein which represses an early viral promoter (A2c) (Monsalve et al., 1996; Rojo, 1999). A related phenomenon occurs when promoters are engineered to have "optimal" consensus elements both in the upstream region (e.g., A tract containing sequences) as well as in the −35 and −10 elements. In this case RNAP binds so tightly to the promoter that it has trouble escaping into a productive elongation mode (it actively synthesizes abortive products, however). As a result the "optimized" promoter actually becomes weaker rather than stronger (Ellinger et al., 1994a; Ellinger et al., 1994b). This serves to illustrate that promoter strength is a complex phenomenon that requires that all steps in initiation be optimized to facilitate both rapid binding *and* rapid clearance.

B. Regulation by Activators

Transcription activator proteins accelerate the rate of RNA synthesis from promoter sites. For most promoters, there is a low (basal) level of transcription in the absence of activation, and the activator protein serves to greatly increase promoter efficiency. Unlike repression, which can occur by blocking any of the many steps of transcription initiation, activators have to accelerate the slowest (rate-limiting) step in order to stimulate transcription (Roy et al., 1998). Many activators stimulate the rate of RNAP binding to the promoter, while others act by accelerating the rate of open complex formation or promoter clearance.

Since activator proteins typically bind to promoter DNA at the same time as RNAP, activator-binding sites are most frequently found just upstream of the −35 consensus element (Fig. 3). An activator bound at this position can make favorable protein-protein interactions with bound RNAP. These interactions most commonly include contacts to the α-CTD, as mentioned above, but can also involve contacts to the N-terminal domain of α, or to σ, β, or β′ (Geiduschek, 1997; Lonetto et al., 1998; Miller et al., 1997; Rhodius and Busby, 1998).

One exceptionally well-studied activator protein is the cyclic AMP receptor protein

(CRP), also known as the catabolite activator protein (CAP), which binds as a dimer to sites centered either 41 or 61 bp upstream of the transcription initiation site (Fig. 7). The detailed interactions between CAP and RNAP that serve to enhance transcription initiation have been very well defined at these two classes of binding sites. When bound at the upstream position (centered near −61; class I sites), CAP interacts with a specific cluster of amino acids on the surface of the α-CTD region of one of the α subunits of RNAP (Busby and Ebright, 1994). These interactions, in turn, bring the α-CTD into

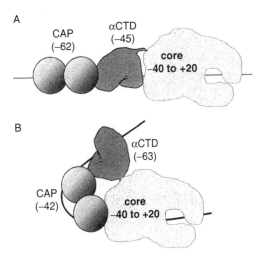

Fig. 7. Models for activation complexes formed by CAP with RNAP at class I and class II promoter sites. Activation of transcription by the catabolite activator protein (CAP; also known are the cyclic-AMP receptor protein, CRP) require the dimeric CAP protein bound to the co-activator, cyclic-AMP. Operons regulated by CAP fall into two main groups. **A:** At those operons with a CAP-binding site centered upstream of the RNAP binding site (class I) activation occurs when CAP contacts the C-terminal domains of the two a subunits, which themselves interact with the intervening DNA. **B:** At class II sites, the CAP binding site site is centered near −41.5 and CAP can interact directly with either the N-terminal domain of α or the portion of the sigma subunit that contacts the −35 element (region 4 of σ). At these promoters the αCTD can still interact with upstream DNA regions (near −63 in this example), particularly if these regions are AT rich (Lloyd et al., 1998).

contact with DNA and increase the affinity of RNAP for the promoter. When bound at class II sites (centered near −42), CAP directly contacts the amino-terminal domain of the alpha subunit of RNAP to stimulate initiation. The displaced α-CTD bridges over the bound CAP and interacts with DNA upstream of the bound CAP. Such straddling is facilitated by the flexible linker that connects the α-CTD to the remainder of the α polypeptide and by the ability of the DNA to bend around the RNAP molecule. For detailed discussions of CAP-mediated activation at class I and class II promoters, the reader is referred to Busby and Ebright (1997, 1999) and Rhodius and Busby (1998).

In some cases activator proteins function from other positions. For example, some activator proteins bind far upstream, or less commonly downstream, of RNAP bound at the promoter. Another unusual class of activators stimulates RNAP from a binding site located within the spacer region of the promoter. The best characterized example of this mechanism is the Tn501 MerR protein which activates transcription of a mercury inducible promoter with an abnormally long (19 bp) spacer region (Fig. 8). MerR binds to the spacer region of a mercury resistance operon together with RNAP, which is bound primarily to the opposed face of the DNA. In the absence of the inducer, mercuric ion, MerR functions to prevent initiation. It is a repressor that acts by blocking the closed to open complex transition. However, upon binding to mercuric ion, MerR undergoes a conformational change that results in a distortion of the spacer DNA. This in turn serves to realign the −35 and −10 promoter elements to facilitate a productive interaction with RNAP. In essence the DNA distortion imposed by the MerR-Hg(II) complex compensates for the abnormally long spacer region (Summers, 1992). Other regulators in the MerR family, which likely use a similar DNA distortion mechanism, include the superoxide responsive SoxR regulator (Hidalgo et al., 1998), the TipA regulator of antibiotic synthesis in *Streptomyces coelicolor*

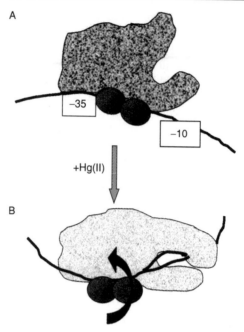

Fig. 8. Mechanism of action of MerR in transcriptional repression and activation. **A:** In the absence of MerR, the *mer* operon is weakly active due to an overly long spacer region (19 or 20 bp, depending on the operon). In the presence of MerR, but the absence of mercuric ion, RNAP binds to the promoter but is unable to establish an open complex. As a result transcription is repressed. **B:** In the presence of both MerR and mercuric ion, the spacer region DNA is distorted by DNA twisting to facilitate productive initiation by RNAP (Ansari et al., 1992, 1995).

(Chiu et al., 1999), and the Mta regulator of multidrug resistance in *B. subtilis* (Baranova et al., 1999).

As noted for repressor proteins, the activity of activators is tightly regulated. Some activators are regulated at the level of synthesis, but in many cases the activator is regulated by either the reversible binding of a small molecule (e.g., binding of cAMP to CRP or Hg(II) to MerR) or by covalent modification (e.g., two component response regulator proteins). In the case of regulation of the *E. coli* arabinose operon, activation is accomplished by conformational changes in the AraC regulator that affect the nature of the DNA complex, rather than binding itself. In the absence of arabinose, AraC binds to two distant half-sites and a loop is formed.

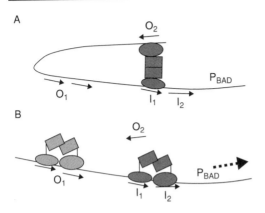

Fig. 9. Model for activation by the AraC arabinose regulatory protein. **A:** In the absence of arabinose, the AraC regulatory protein binds to the I_1 and O_2 half-sites leading to formation of a repression loop. Note that AraC has both a DNA-binding domain (oval) and a dimerization domain (rectangle). **B:** When arabinose binds to the AraC protein (in the dimerization domain) a conformational change takes place, leading to the preferential binding of the protein subunits to two adjacent half-sites (I_1 and I_2) rather than to distant half-sites. In addition another protein dimer binds to the O1 operator. Occupancy of the I_2 site allows productive interaction between AraC and RNAP bound at the *araBAD* promoter, leading to transcription activation. For more details on this "light switch" mechanism, see (Harmer et al. 2001).

Binding of arabinose leads to a conformational change such that AraC now binds preferentially to adjacent half-sites. Only in this conformation is the I2 half-site occupied, which allows productive contact with RNAP (Fig. 9).

V. TRANSCRIPTIONAL REGULATION—OTHER MECHANISMS

A. Alternative Sigma Factors

In 1969, when the role of σ factor in allowing promoter recognition was first reported (Burgess et al., 1969), it was suggested that substitution of one σ factor by an alternative σ could be a mechanism of gene regulation. Indeed, regulation by alternative σ factors has proven to be widespread. Most bacterial genomes encode multiple σ factors (Table 2).

These typically include one essential (primary) σ factor (σ^{70} equivalent) that controls transcription of the vast majority of genes during logarithmic growth and as many as several dozen alternative σ factors that control sets of genes activated in response to particular stress conditions. In a formal sense, an alternative σ factor is an activator protein for RNAP. However, instead of stimulating transcription by holoenzyme, the alternative σ factor binds to core to generate a new holoenzyme with a distinct promoter selectivity.

Alternative σ factors are commonly designated by a superscript reflecting their molecular weight (in kDa) or by a letter or gene name. The unfortunate nonuniformity in σ nomenclature reflects the historical processes of discovery. For example, in *E. coli* alternative σ factors include σ^{32}, σ^{54} (σ^{N}), σ^{E}, σ^{F}, σ^{S}, and σ^{fecl}. The σ^{32} heat shock σ factor is the product of the *htpR* (high-temperature protein regulator) gene, which is now called *rpoH*. Similarly σ^{E} is the product of the *rpoE* gene, while σ^{fecl} is the product of the *fecI* (ferric citrate transport) gene. In this system the primary σ factor (σ^{70}) is the product of the *rpoD* gene.

Because many alternative σ factors have similar molecular weights, and this system becomes cumbersome for organisms with a great many σ factors, the use of letters as superscripts is preferred. In this system, originally introduced for *B. subtilis* (Losick et al., 1986), the corresponding gene name is also implicit. Thus in *B. subtilis* (which has 17 σ factors), the primary σ factor is σ^{A} (encoded by the *sigA* gene), and alternative σ factors include σ^{B}, σ^{D}, σ^{E}, σ^{F}, and so forth. The use of *sig* (rather than *rpo*) is preferred as the genetic prefix for σ factors because it avoids confusion with the other RNAP subunits encoded by the *rpoA* (α), *rpoB* (β), *rpoC* (β'), and, in *B. subtilis*, the *rpoE* (δ) genes. Unfortunately, there is little to no correspondence between gene names and function for σ factors found in different organisms. For example, *E. coli* σ^{F} (the product of the *fliA* gene) is functionally

TABLE 2. Examples of NTP-Mediated Regulation in *E. coli*

Operon	Physiological Function	Initial Transcribed Region	NTPs	Mechanism
pyrBI	Pyrimidine biosynthesis	TATAATGCCGGACAATTGCCG	UTP	High UTP leads to nonproductive slippage synthesis (AAUUUUU$_N$).
pyrC	Pyrimidine biosynthesis	TATCCTTTGTGTCCGGCAAAAA	CTP	High CTP allows initiation with CTP; the resulting longer transcript has stem-loop structure that blocks r.b.s. (translational control). In low [CTP], transcription initiates with GTP two bases downstream and translation is efficient.
carAB	Pyrimidine biosynthesis	CAGAATGCCGCCGTTTGCCAGA	UTP	High UTP promotes slippage synthesis.
codBA	Cytosine uptake and utilization	TAGAATGCGGGCGGATTTTTGG	UTP	Low UTP, transcription initiates with GA and escapes slippage synthesis. High UTP, initiation occurs with AU and RNAP enters nonproductive slippage synthesis (AUUUUUU$_N$).
upp	Pyrimidine salvage	TATAATCCGTCGATTTTTTTG	UTP	Same as for *codBA*.

Note: The mechanisms referred to are described in Cheng et al., 2001; Han and Turnbough, 1998; Liu et al., 1994; Qi and Turnbough, 1995; Wilson et al., 1992.

In the initial transcribed region the −10 element is underlined and the transcriptional start site(s) are in bold type.

equivalent to *B. subtilis* σ^D (the product of the *sigD* gene) as both control transcription of flagellar biosynthesis and chemotaxis genes, and they have closely related promoter selectivity (the *B. subtilis* σ^F participates in sporulation control).

As a group the alternative σ factors can be divided into two evolutionarily distinct groups. Most σ factors have amino acid sequences related to *E. coli* σ^{70} and *B. subtilis* σ^A, the primary σ factors of these two model organisms (Lonetto et al., 1992). These proteins define the σ^{70} superfamily and include subfamilies of factors involved in regulating heat shock, flagellar motility, sporulation (in *B. subtilis*), and extracytoplasmic (ECF) functions (Lonetto et al., 1994). In contrast, many bacteria have one (rarely two) member of a distinct class of σ factor related to *E. coli* $\sigma^{54}(\sigma^N)$ (Studholme and Buck, 2000a,b). These regulators recognize promoters with conserved sequence elements at -24 and -12, rather than the typical -35 and -10 position characteristic for σ factors related to σ^{70}. The σ^{54} family of proteins are also distinct in that they form holoenzymes with an obligate requirement for a positive activator protein, and these activators can function from atypically large distances from the promoter region.

Production of an alternative σ factor is a powerful mechanism for redirecting the transcriptional program of the cell. In some cases, transcription by alternative holoenzymes can dominate RNA synthesis in the cell leading to a large-scale switch in protein production. This occurs during conditions of extreme heat shock and during *B. subtilis* sporulation. In other cases, alternative σ factors may be active at a low level to redirect a minor sub-population of RNAP to new promoter sites (Ishihama, 2000).

A remarkable diversity of mechanisms have been described that act to control σ factor activity. As for any other type of gene regulation, chances are that if a mechanism can be envisioned, it has developed in some organism during evolution. Thus alternative σ factors are known to be regulated at the level of transcription, translation, stability, and by post-translational events such as protein processing and interaction with specific inhibitor proteins (anti-σ factors). Indeed, all of these mechanisms are operative on the σ factors controlling sporulation in *B. subtilis*, which provides an outstanding example of the complexities of alternative σ factors and their regulation (Haldenwang, 1995; Kroos et al., 1999) (see Moran, this volume).

One particularly interesting group of alternative σ factors is the extracytoplasmic function (ECF) subfamily of the σ^{70} family. This family is represented by seven σ factors *in B. subtilis* (Table 1), 17 in *Pseudomonas aeruginosa* (Stover et al., 2000), and 10 in *Mycobacterium tuberculosis* (Cole, 1998). In general, these σ factors control responses having to do with the cell surface, such as secretion, synthesis of extracellular factors, uptake of nutrients, and transport (Missiakas and Raina, 1998). In most cases these σ factors are held inactive by binding to a membrane-bound anti-σ. The gene for the anti-σ is often encoded in the same operon as the σ factor. Thus, the synthesis of a σ-anti-σ pair can be thought of as forming a signaling complex localized to the membrane and poised to receive extracellular signals. This system in analogous to the more familiar two-component regulatory systems, which are also well positioned to activate gene expression in response to signals external to the cell (Fig. 10) (see Bayles and Fujimoto, this volume).

B. Direct Regulation of RNAP Activity by NTPs

The processes connecting a chemical or physical signal to changes in gene expression can be enormously complex, involving cascades of many regulators. At the other extreme, some regulatory processes involve a single protein that acts both to sense a signal and to regulate transcription. In some cases it is actually RNAP itself that functions as the sensor: there is no classic activator or repressor component. Specifically, RNAP is able to sense fluctuations in NTP levels within the cell and convert this information directly

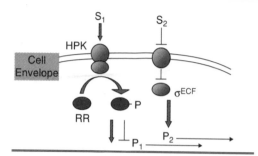

Fig. 10. Two mechanisms of transcriptional control by extracellular signals. Two component regulatory systems typically contain a membrane-localized histidine protein kinase (HPK) that binds to small molecules (signal I; SI) present outside the cell (in the periplasm in gram-negative bacteria). This binding regulates the activity of the HPK which phosphorylates (and often also dephosphorylates) a specific response regulator (RR). The phosphorylated RR often functions as a transcription factor, either as an activator (arrow) or a repressor (bar) of transcription. A second mechanism of sensing external signals involves a membrane-localized anti-σ factor that binds, and thereby inactivates, a σ factor (of the extracytoplasmic function, or ECF, subfamily). In the presence of an inducing signal (S2), the σ factor is released and can bind to core RNAP to direct transcription of specific target operons (e.g., promoter P2).

into a regulatory response. In this section we will consider several examples of how NTP levels can regulate gene expression.

One particularly dramatic example of a regulatory mechanism that takes advantage of the NTP-sensing ability of RNAP is the growth rate-dependent control of ribosomal RNA transcription (Roberts, 1997). As noted above, rRNA synthesis varies as the square of the growth rate and considerable effort, over more than two decades, sought to define the molecular basis of this control mechanism. One especially well-studied example of growth rate control is the *rrnB* P1 promoter. As noted previously, this very strong promoter contains a stimulatory UP element (Roberts, 1997), and it also binds three copies of the FIS activator protein to sites upstream of the promoter (Roberts, 1997). However, deleting both the FIS binding sites and the UP element leads to a much weakened pro-

moter that is nevertheless still subject to growth rate regulation. Further dissection of this system revealed that the only sequences obviously required for growth rate regulation were the core promoter elements (-35 and -10 consensus sites), and no evidence could be obtained for a binding site for a regulatory protein (Bartlett and Gourse, 1994; Gourse et al., 1996).

Studies of the *rrnB* P1 promoter in vitro revealed that this site requires an unusually high concentration of the initiating NTP to begin transcription (Fig. 11). In general, transcription initiation is strongly affected by the concentration of the first two NTPs (the K_m for the initiating NTPs is higher than that for elongation). Transcription preferentially initiates with purines (A or G), and as a result initiation rates are most sensitive to ATG and GTP levels. Since growth rate regulated promoters, such as *rrnB* P1, appear to have unusually low affinities for the initiating NTP, it is suggested

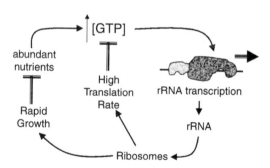

Fig. 11. Proposed mechanism for growth rate regulation *in E. coli*. Initiation of transcription of the *rrnB* operon is highly sensitive to the intracellular level of the first (initiating) NTP in the transcript (in this case, GTP). If nutrients are abundant, GTP levels will be high (GTP is in equilibrium with ATP), and initiation of ribosomal RNA operons is stimulated. This increased rRNA synthesis supports the increased synthesis of ribosomes. Since higher levels of ribosomes allow a more rapid growth rate, nutrients will be more rapidly consumed. In addition translation is the major consumer of GTP in the cell. Increased rates of translation will act to lower intracellular GTP levels. These mechanisms may account for the observation that rapidly growing cells contain far more ribosomes per cell than slowly growing cells (Gaal et al., 1997; Roberts, 1997).

that they will be particularly sensitive to changes that affect the cellular energy charge (Gaal et al., 1997). As noted previously, one of the largest single drains on cellular energy reserves is the process of translation itself. Thus, when the translation capacity of the cell is insufficient relative to its metabolic capacity, a high-energy charge will result, leading to increased expression of growth rate regulated promoters. These promoters in turn directly increase the cellular capacity for translation by exressing the RNA and proteins needed for ribosome assembly and function. Conversely, when a cell becomes nutrient and energy limited, the first promoters affected by the declining levels of ATP and GTP will be those that control ribosome production. While it is not yet clear whether or not this is the major mechanism for growth rate control in bacteria, it nevertheless serves to illustrate how changing NTP levels can selectively regulated a subset of promoters.

Several other examples of NTP-mediated regulation have come to light from analysis of purine and pyrimidine biosynthetic operons in *E. coli* (Table 2). Here I discuss two cases that illustrate how the cell has evolved regulatory mechanisms that take advantage of certain, intrinsic properties of RNAP. The first concerns the regulation of start site selection during initiation of transcription at the *pyrC* gene encoding an enzyme in the pathway for pyrimidine biosynthesis (Wilson et al., 1992). In general, RNAP prefers to initiate transcription with a purine (ATP or GTP) rather than a pyrimidine and the preferred distance from the −10 element to the start site is about 7 bp. In the initiation region for the *pyrC* promoter, these two preferences are in competition with one another. RNA synthesis can either start at the preferred distance, but must use CTP, or at a slightly longer distance (9 nt), with the more preferred initiating substrate, GTP. The regulatory signal that affects start site choice, then, is the relative level in the cell of the two possible initiating NTPs, GTP and CTP. If the ratio of CTP to GTP is high (i.e.,

pyrimidines are abundant), initiation occurs with CTP, whereas if the ratio is low (pyrimidines are scarce), initiation occurs with GTP. How then is transcriptional start site selection coupled to gene regulation? Inspection of the initial transcribed region reveals that initiation with CTP leads to a stable stem-loop structure in the mRNA which sequesters the RBS, thereby blocking efficient translation of the mRNA. Conversely, initiation at the more downstream position yields a more unstructured mRNA that is efficiently translated. Thus a subtle switch in start site selection, mediated by the competing preferences of RNAP for NTP substrates, contributes directly to a translational control mechanism for expression of the genes for pyrimidine biosynthesis (Wilson et al., 1992).

The mechanism of NTP-mediated regulation of the *codBA* operon is conceptually related to that just described (Qi and Turnbough, 1995). In this case the *codBA* gene products are involved in the uptake of cytosine and are more highly expressed under conditions of pyrimidine limitation. Initiation, in this system, initiates within the sequence GATTTTT with either GTP or ATP. In this case selection of the transcription start site is sensitive to the level of UTP, since RNAP must bind the first two NTPs in order to initiate transcription. Under pyrimidine limiting conditions, initiation occurs with GTP, since then only GTP and ATP need to be bound to RNAP. In contrast, high levels of pyrimidines favor initiation with ATP, for which both ATP and UTP must be bound to RNAP. If pyrimidines are low, and initation occurs with G, the RNA transcript begins with the sequence GAUUUUUGG... and is efficiently elongated into full-length mRNA. In contrast, when pyrimidine levels are high, the initial transcript begins with AUUUU... and enters into slippage synthesis, producing long transcripts of repeating U residues. For reasons that are not entirely clear, a switch in transcription start site of a single base has a large influence on the ability of RNAP to transcribe through the run of six U residues without slipping.

This example illustrates how a subtle change in start site selection, mediated by relative NTP levels, can control the efficiency with which RNAP escapes from a promoter site into a productive elongation mode. Similar mechanisms control other pyrimidine-sensitive operons in *E. coli*, including the *carAB* (Han and Turnbough, 1998) and *upp* operons (Cheng et al., 2001) (Table 2). A final example of regulation by NTP levels is attenuation control of the *pyrBI* operon, which will be considered in more detail below.

C. RNAP Substitution or Modification during Phage Infection

During the lytic growth of bacteriophage, the transcriptional capacity of the cell is often completely redirected toward the production of phage mRNA. Bacteriophage have evolved several remarkable mechanisms to ensure this transition. In the case of coliphage T7 (and related phages), early after infection the host RNAP transcribes a gene for a new RNAP, the phage T7 RNAP (Chamberlin et al., 1970). In addition other phage-encoded genes inhibit the activity of the host RNAP (Nechaev and Severinov, 1999). Since the phage T7 RNAP has a rapid elongation rate and only recognizes phage promoters (and ignores the host chromosome), the bulk of transcription is redirected towards the phage DNA.

The strategy adopted by phage T4 (and other T even phages) is somewhat different. Rather than producing a new RNAP, the phage modifies the host RNAP to redirect the enzyme to transcribe phage DNA. This modification includes a covalent ADP-ribosylation of the α-CTD of the core enzyme. ADP-ribosylation of Arg 265 residue prevents the interaction of the α-CTD with DNA, particularly important for those promoters that are largely dependent on UP elements for their high transcriptional activity (Gourse et al., 2000). In this way the phage effectively shuts down transcription of the strongly transcribed *rrn* promoters of the host. Subsequently middle gene transcrip-

tion is activated by a transcription factor, MotA (Hinton et al., 1996), working in concert with AsiA (Adelman et al., 1997; Colland et al., 1998), which modifies the interactions of σ^{70} with DNA. Finally, late transcription is activated when phage T4 specifies a new σ factor (σ^{gp55}) that redirects the host enzyme to phage late promoters (Brody et al., 1995; Tinker-Kulberg et al., 1996) in a process activated by concurrent DNA replication (see also Guttman and Kutter, this volume).

In addition to reprogramming RNAP, many phage have evolved DNA genomes that are modified. Typical modifications include the use of uracil instead of thymine, the methylation of bases (5-Me-C), or glucosylation. By altering RNAP so that modified DNA is preferentially recognized or transcribed after infection, the phage efficiently redirects RNAP away from the host promoters. One remarkable example is the phage T4 Alc protein, which efficiently terminates transcription on the unmodified host DNA, but not on modified phage DNA (Kashlev et al., 1993). These are just a few of the many remarkable examples of control mechanisms evolved by bacteriophage to subvert the host transcriptional program for efficient production of phage.

D. Regulation of Transcription Termination

1. Overview

While significantly less common than regulation at the level of initiation, the processes of transcription elongation and termination are also regulated for many operons. There are two general classes of mechanisms that serve to regulate termination by elongating RNAP. The first class, referred to as *antitermination*, involves modification of RNAP into a termination resistance state, usually by association with one or more protein factors. The second class, *attenuation*, refers to the regulated formation of a Rho-independent terminator structure (or an alternative structure, called an *antiterminator*) in the leader region of an operon, usually by the regulated binding of molecules to the nascent RNA chain during the process of transcription.

One of the best studies examples of antitermination control occurs in late gene expression during phage lambda infection (Roberts, 1988). Transcription from the lambda P_L and P_R promoters initially produce relatively short transcripts due to efficient termination at nearby terminator sites. The lambda N protein functions by interaction with RNAP to modify terminator recognition and allow readthrough into the downstream transcription units (Greenblatt et al., 1998). One of the gene products of the extended P_R transcript is the lambda Q protein which also functions as an elongation factor. RNAP that initiates transcription from the $P_{R'}$ promoter enters into the elongation phase but pauses (stalls) for a minute or more near position +16 (Roberts et al., 1998). This paused RNAP complex (which still retains the σ^{70} subunit; Ring et al., 1996) interacts with Q protein which binds to DNA in the promoter region (Yarnell and Roberts, 1992). The interaction of Q with RNAP leads to the formation of a termination-resistance form of elongating RNAP that essentially ignores downstream terminator sites. This allows efficient readthrough transcription into the lambda late genes (Roberts et al., 1998).

Another form of antitermination control is important for transcription of ribosomal RNA operons (Condon et al., 1995; Squires et al., 1993). As noted earlier, transcription of large regions of RNA that are not being translated, particularly unstructured regions rich in C residues, leads to efficient Rho-dependent termination (Stanssens et al., 1986). Since *rrn* operons encode structural RNA that is not translated, it is proposed that RNAP is modified during initiation from these sites into a termination-resistant form. The modified RNAP elongates rRNA at about twice the rate of elongation typical for mRNA genes (Vogel and Jensen, 1997). This *rrn* antitermination system involves a conserved DNA site (box A) and several

elongation factors including NusA, NusB, and NusG (see Hendrix, this volume).

2. Attenuation

Attenuation refers to a large class of regulatory mechanisms in which a regulatory protein (or process) affects the formation of a transcription terminator (attenuator) located prior to a coding region (Table 3). Under conditions that lead to operon expression, an alternative RNA structure is favored that prevents formation of the terminator and thereby allows "read-through" of RNAP into the operon (Landick et al., 1996; Yanofsky, 2000). In the classic example of attenuation control in the *trp* operon, whether the nascent RNA folds into a terminator structure or the alternative antiterminator structure depends on the translation of a short leader peptide that is rich in Trp (Fig. 12A). Ribosomes bind to the nascent RNA and initiate translation of the leader peptide while RNAP is still elongating through the leader region. If the ribosomes stall over the Trp codons, because of an insufficient supply of charged Trp-tRNA, the formation of the antiterminator structure is favored, the terminator cannot form, and RNAP proceeds to transcribe the Trp biosynthetic genes. Conversely, when Trp is abundant, the ribosomes do not stall, and the terminator structure can form and transcription terminate prior to the structural genes. Attenuation is an effective mechanism for gene control because transcription and translation are closely coupled in prokaryotes. In the case of the Trp operon, the RNAP senses the relative rate of movement of RNAP down the template DNA and the ribosomes down the mRNA. Since the motion of the ribosomes is sensitive to the availability of charged tRNA, this system (and similar systems for other amino acid biosynthetic operons) is sensitive to amino acid levels (Landick et al., 1996).

An alternative type of attenuation mechanism governs the transcription of the *pyrBI* operon. As for the *trp* operon, the leader region can form two alternative secondary

TABLE 3. Selected Examples of Attenuation Mechanisms

Organism	Operon	Signal	Mechanism	Reference
E. coli	trp	uncharged trp-tRNA^{Trp}	Pausing of ribosome translating a leader peptide affects RNA secondary structure	Landick et al., 1996; Yanofsky, 2000
E. coli	*pyrBI*	UTP, CTP	Pyrimidine levels affect elongation rate of RNAP; slow elongation allows close coupling with ribosome and prevents formation of the terminator structure	Landick et al., 1996
B. subtilis	*tyrS*	uncharged tRNA^{Tyr}	Uncharged tRNA binds to the mRNA leader region stabilizing the formation of an anti-terminator structure	Grundy and Henkin, 1993; Henkin, 1994
B. subtilis	trp	Tryptophan	The TRAP protein-tryptophan complex binds the leader region and prevents formation of the anti-terminator, leading to formation of the transcription terminator	Babitzke, 1997
B. subtilis	*pyr*	UMP	PyrR binds RNA in response to UMP and prevents formation of an antiterminator; leading to formation of the transcription terminator	Switzer et al., 1999

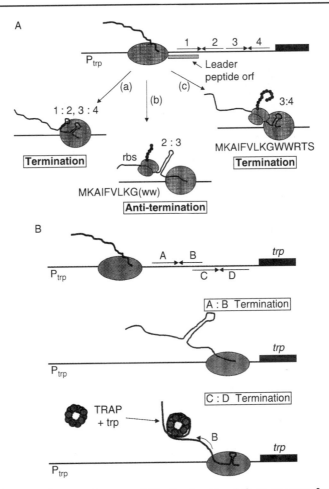

Fig. 12. Two mechanisms of attenuation control: The *E. coli* and *B. subtilis trp* operons. **A:** Attenuation control of the *trp* operon in *E. coli* involved translation of a leader peptide encoded in the mRNA upstream of the *trp* operon structural genes (*trp*) (Landick et al., 1996). In the absence of translation (**a**), the leader region forms a secondary structure containing two stem-loop features: designated 1 : 2 and 3 : 4. The 3 : 4 stem-loop functions as a rho-independent transcription terminator. This greatly reduces the amount of transcription that continues into the trp operon. Usually ribosomes bind to the leader region and initiate translation of the leader peptide. If Trp is scarce (**b**), then translation of the leader peptide will tend to pause when the ribosome encounters the two tandem Trp codons (ww) (if Trp is scarce, then there will also be a shortage of charged trp-tRNA). The paused ribosome covers segment 1 of the leader region, allowing segment 2 to pair with 3. The resulting 2 : 3 stem-loop structure is called the antiterminator. If the 2 : 3 stem loop is formed, then transcription through region 4 does not lead to formation of the 3 : 4 terminator structure, and transcription continues into the *trp* structural genes. If Trp is abundant (**c**) translation will proceed to the end of the leader peptide and the ribosome will be released. A ribosome paused at the end of the leader peptide open reading frame will sequester the portion of the mRNA corresponding to both regions 1 and 2, and as a result the 3 : 4 terminator will be able to form. **B:** In *B. subtilis* the trp operon is also regulated by transcription attenuation. However, the leader region does not encode a leader peptide. Instead of sensing tryptophan by the ability of the ribosome to translate Trp codons, in this organism there is a regulatory protein: tryptophan attenuation protein (TRAP). In the absence of TRAP, or under low Trp conditions, the leader region forms a very stable A : B stem loop that prevents formation of the C : D terminator. However, when Trp is abundant, the TRAP protein binds to Trp and binds to the trp leader region RNA at a specific sequence. This binding covers region A, thereby preventing formation of the A : B antiterminator. Instead, the C : D terminator hairpin can form. Note that TRAP is an unusual protein of 11 identical subunits that forms a donutlike structure. It wraps the target RNA around its surface. For details of the TRAP structure and its interaction with RNA, see Antson et al., (1995, 1999).

structures. However, in this case it is the motion of RNAP that is regulated rather than the ribosome (Turnbough et al., 1983). Under conditions of pyrimidine starvation, RNAP will transcribe more slowly, and the closely following ribosome will prevent formation of the terminator structure, allowing expression of the pyrimidine biosynthetic genes. If pyrimidines (CTP, UTP) are abundant, close coupling is not favored, and a terminator structure can form.

While in both of the preceding examples, attenuation is associated with translation of a leader peptide, this is not always the case. Many examples are now known of attenuation mechanisms in which the formation of the terminator structure is prevented by binding of a regulatory protein (an RNA-binding protein) to the nascent mRNA. In one well-studied system (Fig. 12B), the *B. subtilis* TRAP protein binds to Trp and then binds to the 5′UTR of the *trp* operon to favor formation of a terminator structure (Antson et al., 1999; Babitzke, 1997).

In gram-positive bacteria, including *B. subtilis*, there is yet another class of attenuation mechanism. In these systems, which include many genes for amino acid synthesis, attenuation control is regulated by the tRNA species corresponding to the given amino acid (Grundy and Henkin, 1993; Henkin, 1994). For example, the operon that encodes tryosyl-tRNA aminoacyl synthetase (*tyrS*) is preceded by a 274 nt long 5′UTR that contains a Rho-independent terminator. When uncharged tyrosyl-tRNA accumulates, this tRNA interacts specifically by RNA-RNA annealing reactions with the 5′UTR of the *tyrS* operon. This interaction involves base-pairing between one part of the leader region and the anticodon loop (thereby ensuring that the correct tRNA is bound), and between another part of the leader region and the 3′end (thereby ensuring that only uncharged tRNA is bound). The binding of tRNA alters the RNA structure to prevent formation of the terminator and thereby facilitates efficient read-through transcription into the downstream *tyrS*

gene. Thus, under conditions leading to inefficient charging of tyrosyl-tRNA, the enzyme responsible for this step is up-regulated. Inspection of leader region sequences for many different tRNA synthetase and amino acid biosynthetic operons reveals conserved sequences indicating that tRNA-mediated attenuation is a widespread control mechanism in gram-positive bacteria (Grundy and Henkin, 1994).

VI. TRANSLATIONAL REGULATION

As noted for the process of transcription, translation can be conveniently divided into the initiation, elongation, and termination phases. While most translational control mechanisms act on the initiation step, in some systems the elongation and termination phases are also subject to regulatory control.

A. Regulation of Translation Initiation

Translation initiation begins with the recognition of the RBS by the 30 S ribosomal subunit. One widespread, and conceptually simple, class of regulatory mechanisms involves factors that influence whether or not a RBS is accessible to ribosomes. As we have seen in the case of the *pyrC* operon, accessibility of the RBS can be influenced by RNA secondary structures. Accessibility can also be influenced by RNA-binding proteins that function as translational repressors. To ensure specificity, these regulators must recognize some unique sequence or structural feature of their target mRNA.

One classic example of a translational repressor protein is the phage T4 encoded single-stranded DNA binding protein (gp32). This protein plays an accessory role during phage replication by binding tightly and cooperatively to ssDNA. Once the available ssDNA sites in the cell are saturated, it then binds selectively to its own mRNA to repress translation. Recognition of the gp32 mRNA is thought to require a unique structure (a pseudoknot) located near the 5′end (Shamoo et al., 1993).

A second example of a translational repressor is the ribosomal protein S4. This small RNA-binding protein interacts specifically with a site on the 16S rRNA during ribosome assembly. Once the available rRNA sites are saturated, the protein then binds to a similar structure in the leader region of the operon encoding the S4 protein. The mRNA binding site is likely to fold into a structure that closely mimics the binding site recognized by S4 during ribosome assembly (Tang and Draper, 1989).

A third example of translational repression is the action of antisense RNA. RNA transcribed from the antisense strand of a gene (or an RNA with similar sequence encoded elsewhere in genome) can anneal to an mRNA at sites adjacent to or overlapping the RBS and thereby block translation. A number of regulatory RNAs that function in this manner have been described (Wassarman et al., 1999). For example, the *micF* RNA anneals to the divergently transcribed *ompF* mRNA to block translation (Coleman et al., 1984). The *oxyS* RNA binds as an antisense message to the mRNA for a transcription activator (FhlA) and also binds to a translational activator protein (Hfq) needed for efficient translation of the σ^S message (Altuvia et al., 1997; Altuvia et al., 1998; Zhang et al., 1998). Thus OxyS acts as a pleiotropic regulator ultimately affecting the expression of at least 40 proteins.

Other regulatory factors act to determine whether or not a RBS is accessible by influencing mRNA structure. This likely contributes to the widespread phenomenon of *translational coupling* (Fig. 2B). In bacterial operons it is common for the termination codon of one gene to overlap the initiation codon of another gene (in another reading frame). For example, the bases TGATG encode both a termination codon (UGA) and a start codon (AUG). At such sites translation of the downstream gene may be coupled to translation of the upstream gene. Perhaps the same ribosome that terminates translation of one protein can immediately reassemble an initiation complex for the next protein (Oppenheim and Yanofsky, 1980). In some cases translation may involve a RBS for the downstream protein that is not normally accessible in the mRNA unless translation first serves to melt away secondary structures or displace translational repressors (Chiaruttini et al., 1997). In any event the result of translational coupling is that adjacent, and frequently overlapping, genes can be translated at equal rates. This is particularly advantageous for proteins that are needed in stoichiometric amounts for assembly of a multiprotein complex.

Another example of a signal that can affect RNA structure is temperature. Translation of the gene for σ^{32}, an alternative σ factor controlling heat shock genes, is enhanced at elevated temperatures. In this case it is the mRNA itself that serves as a heat sensor (Morita et al., 2000; Yura et al., 1993). When the temperature is elevated, an mRNA structure that sequesters the RBS becomes unstable and translation can respond directly to the temperature change. Another process that can affect mRNA structure is nucleolytic processing (Petersen, 1992). Although many genes are cotranscribed into long mRNA molecules, subsequent processing of these polycistronic mRNAs into smaller transcripts can either activate or inactivate translation of subsets of proteins (Mattheakis et al., 1989).

One final example of regulation at the level of translation initiation is illustrated by the autoregulation of the initiation factor 3 gene (*infC*). The product of this gene, IF3, acts during the process of translation initiation to help dock the initiator fMET-tRNA with the AUG or GUG initiation codons. In the presence of IF3, initiation at other codons is efficiently prevented. However, the *infC* gene itself begins with an AUU codon (Fig. 13). Efficient translation of this gene responds directly to the absence of IF3 (Butler et al., 1987). As we will see, an analogous mechanism regulates the translation of release factor 2.

A

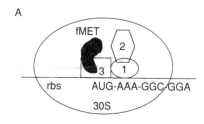

B

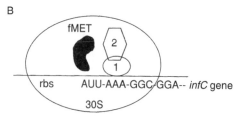

Fig. 13. Autoregulation of translation of translation initiation factor 3 (IF3). **A:** The process of translation begins when the small subunit of the ribosome (30 S) binds to the ribosome-binding site (rbs) near the 5′ end of the mRNA. Assembly of a functional initiation complex requires a specific initiator tRNA (fMET-tRNA) and three additional initiation factors, IF1, IF2, and IF3. IF3 is thought to interact with the initiator tRNA to help ensure accurate recognition of the initiation codon (usually AUG or occassionally GUG or UUG). IF2 and IF3 are proposed to bind to the A site of ribosome. Upon association with the 50 S subunit to form the intact 70 S ribosome, the three initiation factors are released. **B:** The gene encoding initiation factor 3 (infC) is unique in *E. coli* in that it begins with an AUU codon, instead of AUG. Normally messages starting with AUU are not translated. However, if IF3 levels are limiting, then translation initiation from an AUU codon is permitted. Thus IF3 regulates its own translation by inhibiting translation initiation at the unusual AUU codon (Brock et al., 1998; Butler et al., 1986).

B. Regulation of Translation Elongation and Termination

Most of the time translation of mRNA occurs in an orderly manner with each nucleotide triplet specifying a unique amino acid. At first glance this process would not appear a suitable target for regulation. In recent years, however, we have seen numerous examples wherein the process of translation occurs in unexpected ways and, at least in some cases these unusual processes could be subject to regulatory control (reviewed in

Baranov et al., 2001). Examples include programmed co-translational insertion of selenocysteine (Bock et al., 1991), translational frameshifting (Engelberg-Kulka and Schoulaker-Schwarz, 1994; Farabaugh, 1996), translational bypass (Herr et al., 2000), and regulated tagging of proteins by tmRNA (Karzai et al., 2000; Keiler et al., 1996; Muto et al., 1998).

A handful of enzymes, both in prokaryotic and eukaryotic systems make use of the unusual amino acid selenocysteine. The insertion of selenocysteine occurs at specific UGA codons, which normally function as a stop codon (Bock et al., 1991). The insertion of selenocysteine at UGA codons requires a downstream mRNA stem-loop structure which may act to pause the elongating ribosome(Huttenhofer et al., 1996). Charged selenocysteinyl-tRNA is then delivered to the ribosome by a specialized elongation factor, SelB (Commans and Bock, 1999). The mechanisms that allow the stop codon, UGA, to be recognized in an alternative manner, and only in certain mRNA contexts, are not yet clear.

Programmed translational frameshifting occurs when ribosomes shift the reading frame during translation (Table 4). Typically, the reading frame is maintained with high fidelity during translation. However, some gene products are actually encoded by two, overlapping reading frames and their expression requires a precise frameshift (usually +1 or −1 relative to the initial reading frame). These frameshifting events can occur with reasonably high efficiency in response to specific nucleotide sequences or structures in the mRNA (Farabaugh, 1996). Translational frameshifting occurs in several *E. coli* genes including *prfB, dnaX,* and *trpR* and several other examples have been described in a variety of other organisms and viruses. Some of the signals that can contribute to translational frameshifting include the presence of a stop codon or codon for a rare tRNA (which may lead to ribosomal pausing), a downstream secondary structure (hairpin or pseudoknot),

TABLE 4. Examples of Programmed Translational Frameshifting and Hopping

Organism	Gene	Type of Event	Site	Regulation
E. coli	*prfB*	+1 frameshift[a]	CUU UGA C	Low levels of protein release factor 2 stimulate
E. coli	*dnax*	−1 frameshift	A AAA AAG	Ribosome-binding site located 10 bp upstream; 3′ stem loop
B. subtilis	*cdd*	−1 frameshift	A CGA AAG	Ribosome-binding site located 14 bp upstream
phage T4	60	Bypass	GGA (47) GGA	Unusual translational bypass stimulated by RNA secondary structure

Source: Baranov et al., 2001.
[a] This mechanism is found in most bacteria.

and RBS like sequences positioned to facilitate ribosome realignment (Larsen et al., 1995).

One of the best characterized examples of a programmed translational frameshift, and one with clear regulatory significance, occurs in the gene for protein release factor 2, RF2 (*prfB*). This protein is encoded by two overlapping partial reading frames (Fig. 14). The first reading frame terminates with a UGA codon which is normally recognized by RF2. When RF2 is limiting, the ribosome fails to efficiently terminate at this site, and the paused ribosome efficiently shifts into the +1 reading frame allowing expression of full length RF2 (Kawakami and Nakamura, 1990). This +1 frameshift can occur with efficiencies approaching 50%. A key feature allowing the +1 frameshift to occur is the pausing of the ribosome as judged by experiments in which the UGA codon is replaced by various sense codons. If UGA is replaced by the Trp codon, UGG, frameshifting is still quite frequent (11%). This is probably due to the low abundance of the cognate tryptophanyl-tRNA since overexpression of this tRNA from a plasmid could reduced the frequency of translational frameshifting (Sipley and Goldman, 1993). Translational frameshifting is also thought to be used as a regulatory mechanism controlling expression of the mammalian ornithine decarboxylase antizyme gene (Ivanov

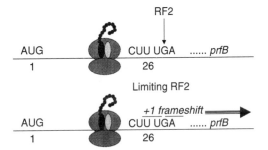

Fig. 14. Autoregulation of translation of release factor 2 (RF2). The release factor 2 gene, *prfB*, contains an in-frame stop codon (UGA) at codon 26. The UGA codon is specifically recognized by RF2. When RF2 levels are sufficient, translation is terminated at the stop codon and a 25 amino acid, presumably nonfunctional peptide is produced (top). When RF2 levels are limiting in the cell, the ribosome pauses at the stop codon, which allows for a +1 frameshift to occur. In this instance, the P-site bound Leucyl-tRNA (bound to the CUU codon) slips into the +1 reading frame (UUU). Translation of the remaining portion of the RF2 protein then continues in the +1 reading frame (Craigen and Caskey, 1986). Genome sequencing reveals that an internal stop codon, and therefore a frameshifting event, is important in autoregulation of translation of RF2 in many bacteria.

et al., 2000). Antizyme production, which is induced by polyamines, inhibits ornithine decarboxylase, the rate-limiting enzyme for polyamine biosynthesis. Production of antizyme in response to polyamines appears to involve a polyamine-mediated stimulation of a +1 frameshift at efficiencies approaching 20%.

Even more dramatic than translational frameshifting is the phenomenon of translational hopping (Herr et al., 2000). In hopping a segment of an mRNA is simply skipped over during translation. Note that this is quite distinct from RNA splicing, as commonly occurs in eukaryotic cells, because the intervening RNA segment is not removed, it is merely ignored. Translational hopping was first discovered in a bacteriophage T4 gene for a topoisomerase II subunit. In this example a 47 nt stretch of RNA is bypassed. Analysis of predicted RNA secondary structures suggests that the RNA efficiently folds into a structure that juxtaposes the 46th and 47th codons despite the presence of the intervening segment. A similar translational hop contributes to expression of a TrpR-LacZ +1 frameshifted translational fusion protein (Benhar and Engelberg-Kulka, 1993; Benhar et al., 1992).

As we have seen, translation normally terminates efficiently when the elongating ribosomes encounter a stop codon and the resulting complex is recognized by a release factor (Buckingham et al., 1997). The release factor then triggers hydrolysis of the tRNA-peptide bond leading to release of the completed protein. However, the cell has evolved a separate mechanism to process translation complexes that cannot be released by this pathway due to the lack of a termination codon (Karzai et al., 2000; Muto et al., 1998). These can arise if mRNA molecules are released prematurely from RNAP or cleaved inappropriately by enzymes or chemical reactions in the cell. While translation can initiate normally on such partial messages, the ribosome will stall upon reaching the 3′end of the mRNA and, in the absence of a release factor, will be unable to release the (incomplete) polypeptide. Such stalled ribosomes are recognized by a specific RNA molecules called *tmRNA*, to indicate that it has properties of both tRNA and mRNA (Fig. 15). The tmRNA interacts at the vacant A site of ribosomes that have

come to the 3′end of a damaged mRNA. The tmRNA is folded into a structure closely resembling alanyl-tRNA and is efficiently charged by the alanyl-tRNA synthetase. Upon interaction with the ribosome, alanine is added to the growing polypeptide chain by the "tRNA-like" function of the tmRNA. At the same time a large loop region inserted into the "tRNA-like" structure binds as a surrogate mRNA to the ribosome, and the ribosome hops off the end of the broken mRNA onto the "mRNA-like" portion of the tmRNA. Then, in a conventional translation process, a 11 amino acid polypeptide segment is then added to the partial polypeptide and a stop codon is encountered, allowing efficient peptide release. By this mechanism the partial polypeptide encoded by the broken mRNA is not only efficiently released from the ribosome, it is tagged at its carboxyl-terminus with an 11 amino acid segment. This hydrophobic peptide segment is a recognition signal for proteolysis so that the partial, and presumably nonfunctional, proteins can be efficiently degraded. A pathway for the regulated release and degradation of proteins encoded by broken mRNA molecules is highly conserved as judged by the presence of tmRNA genes in most sequenced bacterial genomes (Williams, 1999; Williams and Bartel, 1996).

VII. REGULATION BY DNA MODIFICATIONS

Regulation of gene expression necessarily involves changes in transcription or translation. However, in some cases the regulatory mechanism actually operates by affecting the covalent structure of the DNA. In this section we consider three classes of DNA structural change: changes in the linear arrangement of the genetic information (by segment inversion or excision), local mutational changes that affect reading frame choice, and modification of nucleotides. In some cases these changes are readily reversible, in others they are not. In addition some types of changes can be inherited (they are mutations), whereas the modification of nucleotides cannot.

Salmonella enterica serovar Typhimurium is a motile bacterium that expresses a proteinaceous flagella. These flagella are antigenic in infected hosts and the resulting immune response helps to clear the infection. However, Salmonella can actually express two antigenically distinct flagella, depending on which of two flagellin structural genes is transcribed (Silverman and Simon, 1980; Zieg et al., 1978). Regulation of gene expression is controlled by an invertible DNA segment carrying the promoter for expression of

the H2 flagellin (Fig. 16). In the ON orientation, this DNA segment drives transcription of the H2 flagellin gene and a cotranscribed repressor that blocks transcription of the H1 flagellin gene. In the OFF orientation, the H2 gene is silent, while the h1 gene becomes active. The frequency of segment inversion is fairly low, so that at any given time most of the cells of the population will express the same class of filament. A strong immune response against this class of cell will allow the small subpopulation that has experienced (stochastic) segment inversion to multiply. This general class of mechanism, in which a

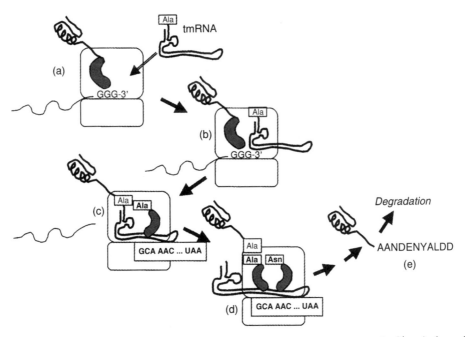

Fig. 15.   Mechanism of action of tmRNA. (**a**) If a translating ribosome encounters the 3′ end of an mRNA prior to successful termination of translation at a stop codon, the ribosome is stalled and cannot proceed farther. This might happen if an mRNA is cleaved by a nuclease or if the ribosome reads through a stop codon (nonsense suppression). (**b**) The charged tmRNA^{Ala} molecule can then bind to the empty A site. (**c**) The tmRNA^{Ala} serves as a substrate for the peptidyltransferase action of the ribosome. As a result the nascent polypeptide is transfered onto the Ala residue at the 3′ end of the tmRNA, and the tmRNA is translocated into the P site (just like any other tRNA). Concurrent with this translocation, the large, looplike region of tmRNA enters into the mRNA binding site to serve as a surrogate mRNA. This portion of the tmRNA functions like an mRNA and encodes 10 amino acids followed by a stop codon. In panel (**c**) the next codon (GCA) is recognized by the anticodon loop of the charged Ala-tRNA that has bound in the A site. Ala is then added to the growing polypeptide chain and the tRNA-like portion of the tmRNA is released from the ribosome P site. The remaining steps of translation occur normally (**d**) using the tmRNA as message. In the end the protein is "tagged" by a unique 11 amino acid sequence terminating in two aspartate residues. This sequence targets the protein for rapid degradation by proteases within the cell (Keiler et al., 1996).

A

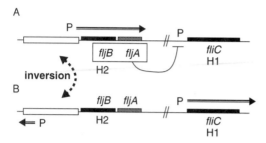

B

Fig. 16. Regulation of phase variation in *Salmonella enterica* serovar Typhimurium. In the H2 "on" orientation the promoter contained within the invertible DNA segment drives expression of the H2 flagellin gene (*fljB*) and a linked gene (*fljA*) encoding a repressor of the H1 flagellin gene (*fliC*). When the DNA segment inverts (catalyzed by the H-invertase, or Hin protein), repression of the H1 flagellin is relieved, and H2 is no longer expressed (Henderson et al., 1999b).

cell periodically varies its surface antigens to foil a host immune response, is called *phase variation* (Table 5). In Salmonella, phase variation is catalyzed by the H-antigen invertase (Hin) protein which binds to short, inverted repeat sequences flanking the invertible DNA segment (Merickel et al., 1998).

A related mechanism contributes to the expression of the gene for the late, mother cell-specific σ^K factor in *B. subtilis*. This protein is encoded by two partial genes separated by an intervening 48 kb segment of DNA (Kunkel et al., 1990). Late during the process of spore formation, this large intervening DNA segment is excised by recombination between two directly repeated segments. This assembles the *sigK* gene allowing expression of an mRNA that encodes σ^K. The intervening DNA segment is lost (see Moran, this volume). Since this event occurs late during spore formation, and the mother cell lyses shortly thereafter, the loss of genetic information contained in this segment of DNA is not a problem. A similar phenomenon occurs in the cyanobacterium Anabaena sp. strain PCC 7120 during heterocyst differentiation (Lammers et al., 1986; Ramaswamy et al., 1997).

Another common mechanism of phase variation involves changes in the structure of gene segments that affect reading frame. For example, in *Neisseria gonorrhoeae* cell surface antigens (opacity proteins) are encoded by gene with a repeating 5 nt motif (CTCTT) in a signal peptide region (Table 5). The number of repeats in the gene can vary, presumably due to unequal crossing-over between different gene copies. Repeat expansion and contraction, in steps of 5 nt, alters the reading frame of the *opa* genes (Stern and Meyer, 1987; van Belkum et al., 1999). By this mechanism the expression of this gene can be turned on and off in a stochastic fashion, governed primarily by the frequency of recombination. A related mechanisms of phase variation involves the expansion and contraction of dinucleotide repeats with the spacer region of the promoter for fimbriae in *Haemophilus influenzae* (which thereby affects promoter strength). Thus expansion and contraction in simple repeat sequences can either affect translation (by altering reading frame) or affect promoter strength (Henderson et al., 1999).

Not all regulatory mechanisms that act on DNA actually change the sequence of the DNA. In some cases, the DNA is modified by methylation, or other changes to the nucleotide bases. Since these changes are not heritable, they are not mutations, and the related regulatory phenomena are consider *epigenetic* changes. One example is the modification N6-methylation of adenine residues within the sequence GATC as catalyzed by the product of the Dam (DNA-adenine methylase) (Palmer and Marinus, 1994). Sites of Dam methylation serve to distinguish the newly synthesized daughter strand from the parent strand immediately after replication: information that is important for methyl-directed mismatch repair pathways (see Yasbin, this volume). However, some sites of Dam methylation overlap promoter regions and can affect promoter activity. In one such example, expression of a surface pilus is activated by transcription factors that bind to DNA regions containing recognition sequences

TABLE 5. Selected Mechanisms of Bacterial Phase Variation

Organism (Product)	Operon	Repeat	Number	Result	Mechanism
Neiserria (opacity)	opa	CTCTT	9, 12, 15, etc.	On	Variable number of 5 nt repeats in coding region controls active reading frame for cell surface opacity proteins
			7, 8, 10, 11, etc.	Off	
Haemophilus (fimbriae)	hifA	TA	10	On	Variable number of TA repeats in the spacer region of the hifA promoter. Ten repeats gives an optimal spacer length of 16 bp, and the promoter is on. Other values lead to less or no activity.
			8, 9	Off	
Bordetella (fimbriae)	fim	C	14	On	Promoter activity is regulated by the length of the poly(C) tract.
			11, 12, 13,	Off	

Source: Saunders, 1999.

for Dam gene (van der Woude et al., 1992). If these sites are methylated, transcription is prevented. In *Salmonella*, mutants deficient in Dam methylation were found to be severely impaired in the expression of virulence genes and it has been proposed that these attenuated strains may be good vaccine candidates (Garcia-Del Portillo et al., 1999). Thus, in this, and possibly other pathogens, Dam methylation may serve a global regulatory function.

Regulation of gene expression by DNA modification is also a widely employed strategy among bacterophage. As alluded to previously, many phage have DNA genomes that are chemically distinct from the host genome. Common examples include the incorporation of uracil in place of thymine (which is normally removed from DNA by uracil N-glycosase) or the use of methylated or glucosylated bases. Subsequent to phage infection, RNAP is modified to selectively transcribe DNA with modified nucleosides, while ignoring the host genome (which in the case of lytic phages is often targeted for degradation).

VIII. CONCLUSIONS

In this chapter we have reviewed the key steps in gene expression, transcription and translation, and I have introduced some of the more baroque elaborations of these processes that have developed during evolution: phenomena like abortive initiation, pausing, transcriptional slippage, and translational frameshifting and hopping. We have surveyed, in a cursory manner, a wide range of regulatory mechanisms including those both familiar (repressors and activators) and those that may be less well known (Baumberg, 1999, provides further reading on many of these mechanisms). If there is one obvious lesson that emerges from this broad viewpoint, it is that evolution has taken advantage of the regulatory potential at each and every step in the process of gene expression. Often regulatory mechanisms seem to surprise investigators with their simplicity and logic: the direct regulation of

growth rate controlled promoters by ATP and GTP, the autoregulation of IF3 and RF2, and the feedback control of ribosomal protein synthesis. In other cases the regulatory mechanisms are much less direct, and yet retain an undeniable logic. No doubt the future holds new surprises and yet undiscovered mechanisms that have yet to be imagined.

REFERENCES

Adelman K, Orsini G, Kolb A, Graziani L, Brody EN (1997): The interaction between the AsiA protein of bacteriophage T4 and the sigma70 subunit of *Escherichia coli* RNA polymerase. J Biol Chem 272: 27435–27443.

Aiyar SE, Gourse RL, Ross W (1998): Upstream A-tracts increase bacterial promoter activity through interactions with the RNA polymerase alpha subunit. Proc Natl Acad Sci USA 95:14652–14657.

Altuvia S, Weinstein-Fischer D, Zhang A, Postow L, Storz G (1997): A small, stable RNA induced by oxidative stress: Role as a pleiotropic regulator and antimutator. Cell 90:43–53.

Altuvia S, Zhang A, Argaman L, Tiwari A, Storz G (1998): The *Escherichia coli* OxyS regulatory RNA represses fhlA translation by blocking ribosome binding. EMBO J 17:6069–6075.

Ansari AZ, Bradner JE, O'Halloran TV (1995): DNA-bend modulation in a repressor-to-activator switching mechanism. Nature 374:371–375.

Ansari AZ, Chael ML, O'Halloran TV (1992): Allosteric underwinding of DNA is a critical step in positive control of transcription by Hg-MerR. Nature 355:87–89.

Antson AA, Dodson EJ, Dodson G, Greaves RB, Chen X, Gollnick P (1999): Structure of the trp RNA-binding attenuation protein, TRAP, bound to RNA. Nature 401:235–242.

Antson AA, Otridge J, Brzozowski AM, Dodson EJ, Dodson GG, Wilson KS, Smith TM, Yang M, Kurecki T, Gollnick P (1995): The structure of trp RNA-binding attenuation protein. Nature 374: 693–700.

Babitzke P (1997): Regulation of tryptophan biosynthesis: Trp-ing the TRAP or how *Bacillus subtilis* reinvented the wheel. Mol Microbiol 26:1–9.

Ban N, Nissen P, Hansen J, Moore PB, Steitz TA (2000): The complete atomic structure of the large ribosomal subunit at 2.4 A resolution. Science 289:905–920.

Baranov PV, Gurvich OL, Fayet O, Prere MF, Miller WA, Gesteland RF, Atkins JF, Giddings MC (2001): RECODE: A database of frameshifting, bypassing and codon redefinition utilized for gene expression. Nucleic Acids Res 29:264–267.

Baranova NN, Danchin A, Neyfakh AA (1999): Mta, a global MerR-type regulator of the *Bacillus subtilis* multidrug-efflux transporters. Mol Microbiol 31:1549–1559.

Bartlett MS, Gourse RL (1994): Growth rate-dependent control of the rrnB P1 core promoter in Escherichia coli. J Bacteriol 176:5560–5564.

Baumberg S (ed) (1999): "Prokaryotic Gene Expression." New York: Oxford University Press.

Beckwith J (1996): The Operon: An Historical Account. In Neidhardt FC (ed): "*Escherichia coli* and *Salmonella:* Cellular and Molecular Biology", Vol 1. Washington, DC: ASM Press, pp 1227–1231.

Bell SD, Jackson SP (1998): Transcription and translation in Archaea: A mosaic of eukaryal and bacterial features. Trends Microbiol 6:222–228.

Benhar I, Engelberg-Kulka H (1993): Frameshifting in the expression of the *E. coli trpR* gene occurs by the bypassing of a segment of its coding sequence. Cell 72:121–130.

Benhar I, Miller C, Engelberg-Kulka H (1992): Frameshifting in the expression of the *Escherichia coli trpR* gene. Mol Microbiol 6:2777–2784.

Bock A, Forchhammer K, Heider J, Leinfelder W, Sawers G, Veprek B, Zinoni F (1991): Selenocysteine: The 21st amino acid. Mol Microbiol 5:515–520.

Bown JA, Barne KA, Minchin SD, Busby SJW (1997): Extended -10 Promoters. In Eckstein F, Lilley DMJ (eds): "Nucleic Acids and Molecular Biology," Vol 11. Berlin: Springer, pp 41–52.

Brock S, Szkaradkiewicz K, Sprinzl M (1998): Initiation factors of protein biosynthesis in bacteria and their structural relationship to elongation and termination factors. Mol Microbiol 29:409–417.

Brody EN, Kassavetis GA, Ouhammouch M, Sanders GM, Tinker RL, Geiduschek EP (1995): Old phage, new insights: two recently recognized mechanisms of transcriptional regulation in bacteriophage T4 development. FEMS Microbiol Lett 128:1–8.

Buckingham RH, Grentzmann G, Kisselev L (1997): Polypeptide chain release factors. Mol Microbiol 24:449–456.

Buratowski S (2000): Snapshots of RNA polymerase II transcription initiation. Curr Opin Cell Biol 12:320–325.

Burgess RR, Travers AA, Dunn JJ, Bautz EK (1969): Factor stimulating transcription by RNA polymerase. Nature 221:43–46.

Busby S, Ebright RH (1994): Promoter structure, promoter recognition, and transcription activation in prokaryotes. Cell 79:743–746.

Busby S, Ebright RH (1997): Transcription activation at class II CAP-dependent promoters. Mol Microbiol 23:853–859.

Busby S, Ebright RH (1999): Transcription activation by catabolite activator protein (CAP). J Mol Biol 293:199–213.

Butler JS, Springer M, Dondon J, Graffe M, Grunberg-Manago M (1986): *Escherichia coli* protein synthesis initiation factor IF3 controls its own gene expression at the translational level in vivo. J Mol Biol 192:767–780.

Butler JS, Springer M, Grunberg-Manago M (1987): AUU-to-AUG mutation in the initiator codon of the translation initiation factor IF3 abolishes translational autocontrol of its own gene (infC) in vivo. Proc Natl Acad Sci USA 84:4022–4025.

Carter AP, Clemons WM, Brodersen DE, Morgan-Warren RJ, Wimberly BT, Ramakrishnan V (2000): Functional insights from the structure of the 30S ribosomal subunit and its interactions with antibiotics. Nature 407:340–348.

Cech TR (2000): Structural biology: The ribosome is a ribozyme. Science 289:878–879.

Chamberlin M, McGrath J, Waskell L (1970): New RNA polymerase from *Escherichia coli* infected with bacteriophage T7. Nature 228:227–231.

Cheng Y, Dylla SM, Turnbough CL, Jr. (2001): A long T: A tract in the upp initially transcribed region is required for regulation of upp expression by UTP-dependent reiterative transcription in *Escherichia coli.* J Bacteriol 183:221–228.

Chiaruttini C, Milet M, Springer M (1997): Translational coupling by modulation of feedback repression in the IF3 operon of *Escherichia coli.* Proc Natl Acad Sci USA 94:9208–9213.

Chiu ML, Folcher M, Katoh T, Puglia AM, Vohradsky J, Yun BS, Seto H, Thompson CJ (1999): Broad spectrum thiopeptide recognition specificity of the *Streptomyces lividans* TipAL protein and its role in regulating gene expression. J Biol Chem, 274:20578–20586.

Choy H, Adhya S (1996): Negative Control. In Neidhardt FC (ed): "*Escherichia coli* and *Salmonella:* Cellular and Molecular Biology," Vol. 1. Washington, DC: ASM Press, pp 1287–1299.

Cole ST (1998): Deciphering the biology of *Mycobacterium tuberculosis* form the complete genome sequence. Nature 393:537–544.

Coleman J, Green PJ, Inouye M (1984): The use of RNAs complementary to specific mRNAs to regulate the expression of individual bacterial genes. Cell 37:429–436.

Collado-Vides J, Magasanik B, Gralla JD (1991): Control site location and transcriptional regulation in *Escherichia coli.* Microbiol Rev 55:371–394.

Colland F, Orsini G, Brody EN, Buc H, Kolb A (1998): The bacteriophage T4 AsiA protein: A molecular switch for sigma 70–dependent promoters. Mol Microbiol 27:819–829.

Commans S, Bock A (1999): Selenocysteine inserting tRNAs: An overview. FEMS Microbiol Rev 23:335–351.

Condon C, Squires C, and Squires CL (1995): Control of rRNA transcription in *Escherichia coli*. Microbiol Rev 59:623–645.

Craigen WJ, Caskey CT (1986): Expression of peptide chain release factor 2 requires high-efficiency frameshift. Nature 322:273–275.

Cramer P, Bushnell DA, Fu J, Gnatt AL, Maier-Davis B, Thompson NE, Burgess RR, Edwards AM, David PR, Kornberg RD (2000): Architecture of RNA polymerase II and implications for the transcription mechanism. Science 288:640–649.

deHaseth PL, Helmann JD (1995): Open complex formation by *Escherichia coli* RNA polymerase: The mechanism of polymerase-induced strand separation of double helical DNA. Mol Microbiol 16:817–824.

deHaseth PL, Zupancic M, Record MT, Jr. (1998): RNA polymerase-promoter interaction: The comings and goings of RNA polymerase. J Bact 180:3019–3025.

Ellinger T, Behnke D, Bujard H, Gralla JD (1994a): Stalling of *Escherichia coli* RNA polymerase in the +6 to +12 region in vivo is associated with tight binding to consensus promoter elements. J Mol Biol 239:455–465.

Ellinger T, Behnke D, Knaus R, Bujard H, Gralla JD (1994b): Context-dependent effects of upstream A-tracts. Stimulation or inhibition of *Escherichia coli* promoter function. J Mol Biol 239:466–475.

Engelberg-Kulka H, Schoulaker-Schwarz R (1994): Regulatory implications of translational frameshifting in cellular gene expression. Mol Microbiol 11:3–8.

Farabaugh PJ (1996): Programmed translational frameshifting. Microbiol Rev 60:103–134.

Finn RD, Orlova EV, Gowen B, Buck M, van Heel M (2000): *Escherichia coli* RNA polymerase core and holoenzyme structures. EMBO J 19:6833–6844.

Frank J (1998): How the ribosome works. Am Scientist 86:428–439.

Fredrick K, Caramori T, Chen YF, Galizzi A, Helmann JD (1995): Promoter architecture in the flagellar regulon of *Bacillus subtilis*: High-level expression of flagellin by the sigma D RNA polymerase requires an upstream promoter element. Proc Natl Acad Sci USA 92:2582–2586.

Fu J, Gnatt AL, Bushnell DA, Jensen GJ, Thompson NE, Burgess RR, David PR, Kornberg RD (1999): Yeast RNA polymerase II at 5 A resolution. Cell 98:799–810.

Gaal T, Bartlett MS, Ross W, Turnbough CL, Jr, Gourse RL (1997): Transcription regulation by initiating NTP concentration: rRNA synthesis in bacteria. Science 278:2092–2097.

Garcia-Del Portillo F, Pucciarelli MG, Casadesus J (1999): DNA adenine methylase mutants of *Salmonella typhimurium* show defects in protein secretion, cell invasion, and M cell cytotoxicity. Proc Natl Acad Sci USA 96:11578–11583.

Geiduschek EP (1997): Paths to activation of transcription. Science 275:1614–1616.

Gentry DR, Burgess RR (1993): Cross-linking of *Escherichia coli* RNA polymerase subunits: Identification of beta' as the binding site of omega. Biochem 32:11224–11227.

Gourse RL, Gaal T, Bartlett MS, Appleman JA, Ross W (1996): rRNA transcription and growth rate-dependent regulation of ribosome synthesis in *Escherichia coli*. Annu Rev Microbiol 50:645–677.

Gourse RL, Ross W, Gaal T (2000): UPs and downs in bacterial transcription initiation: The role of the alpha subunit of RNA polymerase in promoter recognition. Mol Microbiol 37:687–695.

Gralla JD, Collado-Vides J (1996): Organization and function of transcription regulatory elements. In Neidhardt FC (ed): "*Escherichia coli* and *Salmonella*: Cellular and Molecular Biology," Vol. 1. Washington, DC: ASM Press, pp 1232–1245.

Green MR (2000a): TBP-associated factors (TAFIIs): Multiple, selective transcriptional mediators in common complexes. Trends Biochem Sci 25:59–63.

Green R (2000b): Ribosomal translocation: EF-G turns the crank. Curr Biol 10:R369–373.

Green R, Noller HF (1997): Ribosomes and translation. Annu Rev Biochem 66:679–716.

Greenblatt J, Mah TF, Legault P, Mogridge J, Li J, Kay LE (1998): Structure and mechanism in transcriptional antitermination by the bacteriophage lambda N protein. Cold Spring Harb Symp Quant Biol 63:327–336.

Gross CA, Lonetto M, Losick R (1992): Bacterial sigma factors. In McKnight SL, Yamamoto KR (eds): "Transcriptional Regulation," Vol. 1. Cold Spring Harbor, NY: Cold Spring Harbor Press, pp 129–176.

Grunberg-Manago M (1999): Messenger RNA stability and its role in control of gene expression in bacteria and phages. Annu Rev Genet 33:193–227.

Grundy FJ, Henkin TM (1993): tRNA as a positive regulator of transcription antitermination in B. subtilis. Cell 74:475–482.

Grundy FJ, Henkin TM (1994): Conservation of a transcription antitermination mechanism in aminoacyl-tRNA synthetase and amino acid biosynthesis genes in gram-positive bacteria. J Mol Biol 235:798–804.

Haldenwang WG (1995): The sigma factors of *Bacillus subtilis*. Microbiol Rev 59:1–30.

Han X, Turnbough CL, Jr (1998): Regulation of carAB expression in *Escherichia coli* occurs in part through UTP-sensitive reiterative transcription. J Bacteriol 180:705–713.

Harmer T, Wu M, Schleif R (2001): The role of rigidity in DNA looping-unlooping by AraC. Proc Natl Acad Sci USA 98:427–431.

Helmann JD (1994): Bacterial sigma factors. In Conaway RC, Conaway J (eds): "Transcription: Mechanisms and Regulation", Vol 3. New York: Raven Press, pp 1–17.

Helmann JD (1995): Compilation and analysis of *Bacillus subtilis* sigma A-dependent promoter sequences: Evidence for extended contact between RNA polymerase and upstream promoter DNA. Nucleic Acids Res 23:2351–2360.

Helmann JD, deHaseth PL (1999): Protein-nucleic acid interactions during open complex formation investigated by systematic alteration of the protein and DNA binding partners. Biochem 38:5959–5967.

Henderson IR, Owen P, Nataro JP (1999): Molecular switches—The on and off of bacterial phase variation. Mol Microbiol 33:919–932.

Henkin TM (1994): tRNA-directed transcription antitermination. Mol Microbiol 13:381–387.

Herr AJ, Atkins JF, Gesteland RF (2000): Coupling of open reading frames by translational bypassing. Annu Rev Biochem 69:343–372.

Hidalgo E, Leautaud V, Demple B (1998): The redox-regulated SoxR protein acts from a single DNA site as a repressor and an allosteric activator. EMBO J 17:2629–2636.

Hinton DM, March-Amegadzie R, Gerber JS, Sharma M (1996): Characterization of pre-transcription complexes made at a bacteriophage T4 middle promoter: Involvement of the T4 MotA activator and the T4 AsiA protein, a sigma 70 binding protein, in the formation of the open complex. J Mol Biol 256:235–248.

Hsu LM (1996): Quantitative parameters for promoter clearance. Methods Enzymol 273:59–71.

Hsu LM, Vo NV, Chamberlin MJ (1995): *Escherichia coli* transcript cleavage factors GreA and GreB stimulate promoter escape and gene expression in vivo and in vitro. Proc Natl Acad Sci USA 92:11588–11592.

Huttenhofer A, Heider J, Bock A (1996): Interaction of the *Escherichia coli* fdhF mRNA hairpin promoting selenocysteine incorporation with the ribosome. Nucleic Acids Res 24:3903–3910.

Ishihama A (2000): Functional modulation of *Escherichia coli* RNA polymerase. Annu Rev Microbiol 54:499–518.

Ivanov IP, Gesteland RF, Atkins JF (2000): Survey and summary. Antizyme expression: A subversion of triplet decoding, which is remarkably conserved by evolution, is a sensor for an autoregulatory circuit. Nucleic Acids Res 28:3185–3196.

Juang YL, Helmann JD (1994): The delta subunit of *Bacillus subtilis* RNA polymerase: An allosteric effector of the initiation and core-recycling phases of transcription. J Mol Biol 239:1–14.

Karzai AW, Roche ED, Sauer RT (2000): The SsrA-SmpB system for protein tagging, directed degradation and ribosome rescue. Nat Struct Biol 7:449–455.

Kashlev M, Nudler E, Goldfarb A, White T, Kutter E (1993): Bacteriophage T4 Alc protein: A transcription termination factor sensing local modification of DNA. Cell 75:147–154.

Kawakami K, Nakamura Y (1990): Autogenous suppression of an opal mutation in the gene encoding peptide chain release factor 2. Proc Natl Acad Sci USA 87:8432–8436.

Keiler KC, Waller PR, Sauer RT (1996): Role of a peptide tagging system in degradation of proteins synthesized from damaged messenger RNA. Science 271:990–993.

Kroos L, Zhang B, Ichikawa H, Yu YT (1999): Control of sigma factor activity during *Bacillus subtilis* sporulation. Mol Microbiol 31:1285–1294.

Kunkel B, Losick R, Stragier P (1990): The *Bacillus subtilis* gene for the development transcription factor sigma K is generated by excision of a dispensable DNA element containing a sporulation recombinase gene. Genes Dev 4:525–535.

Lammers PJ, Golden JW, Haselkorn R (1986): Identification and sequence of a gene required for a developmentally regulated DNA excision in Anabaena. Cell 44:905–911.

Landick R (1997): RNA polymerase slides home: Pause and termination site recognition. Cell 88:741–744.

Landick R, Turnbough CL, Jr, Yanofsky C (1996): Transcription attenuation. In Neidhardt FC (ed): "*Escherichia coli* and *Salmonella:* Cellular and Molecular Biology," Vol 1. Washington, DC: ASM Press, pp 1263–1286.

Larsen B, Peden J, Matsufuji S, Matsufuji T, Brady K, Maldonado R, Wills NM, Fayet O, Atkins JF, Gesteland RF (1995): Upstream stimulators for recoding. Biochem Cell Biol 73:1123–1129.

Lisser S, Margalit H (1993): Compilation of E. coli mRNA promoter sequences. Nucleic Acids Res 21:1507–1516.

Liu C, Heath LS, Turnbough CL, Jr (1994): Regulation of pyrBI operon expression in *Escherichia coli* by UTP-sensitive reiterative RNA synthesis during transcriptional initiation. Genes Dev 8:2904–2912.

Lloyd GS, Busby SJ, Savery NJ (1998): Spacing requirements for interactions between the C-terminal domain of the alpha subunit of *Escherichia coli* RNA polymerase and the cAMP receptor protein. Biochem J 330:413–420.

Lonetto M, Gribskov M, Gross CA (1992): The sigma 70 family: Sequence conservation and evolutionary relationships. J Bacteriol 174:3843–3849.

Lonetto MA, Brown KL, Rudd KE, Buttner MJ (1994): Analysis of the *Streptomyces coelicolor* sigE gene reveals the existence of a subfamily of eubacterial RNA polymerase sigma factors involved in the regulation of extracytoplasmic functions. Proc Natl Acad Sci USA 91:7573–7577.

Lonetto MA, Rhodius V, Lamberg K, Kiley P, Busby S, Gross C (1998): Identification of a contact site for different transcription activators in region 4 of the

Escherichia coli RNA polymerase sigma70 subunit. J Mol Biol 284:1353–1365.

Lopez de Saro FJ, Woody AY, Helmann JD (1995): Structural analysis of the *Bacillus subtilis* delta factor: A protein polyanion which displaces RNA from RNA polymerase. J Mol Biol 252:189–202.

Lopez de Saro FJ, Yoshikawa N, Helmann JD (1999): Expression, abundance, and RNA polymerase binding properties of the delta factor of *Bacillus subtilis*. J Biol Chem 274:15953–15958.

Losick R, Youngman P, Piggot PJ (1986): Genetics of endospore formation in *Bacillus subtilis*. Annu Rev Genet 20:625–669.

Mattheakis L, Vu L, Sor F, Nomura M (1989): Retro-regulation of the synthesis of ribosomal proteins L14 and L24 by feedback repressor S8 in *Escherichia coli*. Proc Natl Acad Sci USA 86:448–452.

Matthews KS, Nichols JC (1998): Lactose repressor protein: Functional properties and structure. Prog Nucleic Acid Res Mol Biol 58:127–164.

McClure WR (1985): Mechanism and control of transcription initiation in prokaryotes. Annu Rev Biochem 54:171–204.

Merickel SK, Haykinson MJ, Johnson RC (1998): Communication between Hin recombinase and Fis regulatory subunits during coordinate activation of Hin-catalyzed site-specific DNA inversion. Genes Dev 12:2803–2816.

Miller A, Wood D, Ebright RH, Rothman-Denes LB (1997): RNA polymerase beta' subunit: A target of DNA binding-independent activation. Science 275:1655–1657.

Missiakas D, Raina S (1998): The extracytoplasmic function sigma factors: Role and regulation. Mol Microbiol 28:1059–1066.

Monsalve M, Mencia M, Salas M, Rojo F (1996): Protein p4 represses phage phi 29 A2c promoter by interacting with the alpha subunit of *Bacillus subtilis* RNA polymerase. Proc Natl Acad Sci USA 93:8913–8918.

Mooney RA, Artsimovitch I, Landick R (1998): Information processing by RNA polymerase: Recognition of regulatory signals during RNA chain elongation. J Bacteriol 180:3265–3275.

Morita MT, Kanemori M, Yanagi H, Yura T (2000): Dynamic interplay between antagonistic pathways controlling the sigma 32 level in *Escherichia coli*. Proc Natl Acad Sci USA 97:5860–5865.

Mukherjee K, Chatterji D (1997): Studies on the omega subunit of *Escherichia coli* RNA polymerase—Its role in the recovery of denatured enzyme activity. Eur J Biochem 247:884–889.

Muto A, Ushida C, Himeno H (1998): A bacterial RNA that functions as both a tRNA and an mRNA. Trends Biochem Sci 23:25–29.

Narberhaus F (1999): Negative regulation of bacterial heat shock genes. Mol Microbiol 31:1–8.

Naryshkin N, Revyakin A, Kim Y, Mekler V, Ebright RH (2000): Structural organization of the RNA polymerase-promoter open complex. Cell 101:601–611.

Nechaev S, Severinov K (1999): Inhibition of *Escherichia coli* RNA polymerase by bacteriophage T7 gene 2 protein. J Mol Biol 289:815–826.

Neidhardt FC, Ingraham JL, Schaechter M (1990): "Physiology of the Bacterial Cell: A Molecular Approach." Sunderland, MA: Sinauer Assoc.

Neidhardt FC, Savageau MA (1996): Regulation beyond the operon. In Neidhardt FC (ed): "*Escherichia coli* and *Salmonella:* Cellular and Molecular Biology," Vol 1. Washington, DC: ASM Press, pp 1310–1324.

Nissen P, Hansen J, Ban N, Moore PB, Steitz TA (2000a): The structural basis of ribosome activity in peptide bond synthesis. Science 289:920–930.

Nissen P, Kjeldgaard M, Nyborg J (2000b): Macromolecular mimicry. EMBO J 19:489–495.

Nudler E (1999): Transcription elongation: structural basis and mechanisms. J Mol Biol 288:1–12.

Nudler E, Avetissova E, Markovtsov V, Goldfarb A (1996): Transcription processivity: Protein-DNA interactions holding together the elongation complex. Science 273:211–217.

Oppenheim DS, Yanofsky C (1980): Translational coupling during expression of the tryptophan operon of *Escherichia coli*. Genetics 95:785–795.

Palmer BR, Marinus MG (1994): The dam and dcm strains of *Escherichia coli*—A review. Gene 143:1–12.

Perez-Martin J, de Lorenzo V (1997): Clues and consequences of DNA bending in transcription. Annu Rev Microbiol 51:593–628.

Peters JE, Benson SA (1995): Characterization of a new rho mutation that relieves polarity of Mu insertions. Mol Microbiol 17:231–240.

Petersen C (1992): Control of functional mRNA stability in bacteria: Multiple mechanisms of nucleolytic and non-nucleolytic inactivation. Mol Microbiol 6:277–282.

Platt T (1994): Rho and RNA: Models for recognition and response. Mol Microbiol 11:983–990.

Puglisi JD, Blanchard SC, Green R (2000): Approaching translation at atomic resolution. Nat Struct Biol 7:855–861.

Qi F, Turnbough CL, Jr (1995): Regulation of codBA operon expression in *Escherichia coli* by UTP-dependent reiterative transcription and UTP-sensitive transcriptional start site switching. J Mol Biol 254:552–565.

Ramaswamy KS, Carrasco CD, Fatma T, Golden JW (1997): Cell-type specificity of the Anabaena fdxN-element rearrangement requires xisH and xisI. Mol Microbiol 23:1241–1249.

Rhodius VA, Busby SJ (1998): Positive activation of gene expression. Curr Opin Microbiol 1:152–159.

Ring BZ, Yarnell WS, Roberts JW (1996): Function of *E. coli* RNA polymerase sigma factor sigma 70 in promoter-proximal pausing. Cell 86:485–493.

Roberts J (1997): Control of the supply line. Science 278:2073–2074.

Roberts JW (1988): Phage lambda and the regulation of transcription termination. Cell 52:5–6.

Roberts JW, Yarnell W, Bartlett E, Guo J, Marr M, Ko DC, Sun H, Roberts CW (1998): Antitermination by bacteriophage lambda Q protein. Cold Spring Harb Symp Quant Biol 63:319–325.

Rojo F (1999): Repression of transcription initiation in bacteria. J Bacteriol 181:2987–2991.

Ross W, Aiyar SE, Salomon J, Gourse RL (1998): *Escherichia coli* promoters with UP elements of different strengths: Modular structure of bacterial promoters. J Bacteriol 180:5375–5383.

Roy S, Garges S, Adhya S (1998): Activation and repression of transcription by differential contact: Two sides of a coin. J Biol Chem 273:14059–14062.

Salgado H, Moreno-Hagelsieb G, Smith TF, Collado-Vides J (2000): Operons in *Escherichia coli*: Genomic analyses and predictions. Proc Natl Acad Sci USA 97:6652–6657.

Saunders JR (1999): Switch systems. In Baumberg S (ed): "Prokaryotic Gene Expression." New York: Oxford University Press, pp 229–252.

Schmitt E, Guillon JM, Meinnel T, Mechulam Y, Dardel F, Blanquet S (1996): Molecular recognition governing the initiation of translation in *Escherichia coli*: A review. Biochimie 78:543–554.

Shamoo Y, Tam A, Konigsberg WH, Williams KR (1993): Translational repression by the bacteriophage T4 gene 32 protein involves specific recognition of an RNA pseudoknot structure. J Mol Biol 232:89–104.

Silverman M, Simon M (1980): Phase variation: Genetic analysis of switching mutants. Cell 19:845–854.

Sipley J, Goldman E (1993): Increased ribosomal accuracy increases a programmed translational frameshift in *Escherichia coli*. Proc Natl Acad Sci USA 90:2315–2319.

Somerville R (1992): The Trp repressor, a ligand-activated regulatory protein. Prog Nucleic Acid Res Mol Biol 42:1–38.

Soppa J (1999): Transcription initiation in Archaea: Facts, factors and future aspects. Mol Microbiol 31:1295–1305.

Squires CL, Greenblatt J, Li J, Condon C (1993): Ribosomal RNA antitermination in vitro: Requirement for Nus factors and one or more unidentified cellular components. Proc Natl Acad Sci USA 90:970–974.

Stanssens P, Remaut E, Fiers W (1986): Inefficient translation initiation causes premature transcription termination in the *lacZ* gene. Cell 44:711–718.

Stern A, Meyer TF (1987): Common mechanism controlling phase and antigenic variation in pathogenic neisseriae. Mol Microbiol 1:5–12.

Stover CK, Pham XQ, Erwin AL, Mizoguchi SD, Warrener P, Hickey MJ, Brinkman FS, Hufnagle WO, Kowalik DJ, Lagrou M, Garber RL, Goltry L, Tolentino E, Westbrock-Wadman S, Yuan Y, Brody LL, Coulter SN, Folger KR, Kas A, Larbig K, Lim R, Smith K, Spencer D, Wong GK, Wu Z, Paulsen IT (2000): Complete genome sequence of *Pseudomonas aeruginosa* PA01, an opportunistic pathogen. Nature 406:959–964.

Struhl K (1999): Fundamentally different logic of gene regulation in eukaryotes and prokaryotes. Cell 98:1–4.

Studholme DJ, Buck M (2000a): The biology of enhancer-dependent transcriptional regulation in bacteria: Insights from genome sequences. FEMS Microbiol Lett 186:1–9.

Studholme DJ, Buck M (2000b): Novel roles of sigmaN in small genomes. Microbiol 146:4–5.

Summers AO (1992): Untwist and shout: A heavy metal-responsive transcriptional regulator. J Bacteriol 174:3097–3101.

Switzer RL, Turner RJ, Lu Y (1999): Regulation of the *Bacillus subtilis* pyrimidine biosynthetic operon by transcriptional attenuation: control of gene expression by an mRNA-binding protein. Prog Nucleic Acid Res Mol Biol 62:329–367.

Tang CK, Draper DE (1989): Unusual mRNA pseudoknot structure is recognized by a protein translational repressor. Cell 57:531–536.

Tinker-Kulberg RL, Fu TJ, Geiduschek EP, Kassavetis GA (1996): A direct interaction between a DNA-tracking protein and a promoter recognition protein: Implications for searching DNA sequence. EMBO J 15:5032–5039.

Turnbough CL, Jr, Hicks KL, Donahue JP (1983): Attenuation control of pyrBI operon expression in *Escherichia coli* K-12. Proc Natl Acad Sci USA 80:368–372.

Uptain SM, Kane CM, Chamberlin MJ (1997): Basic mechanisms of transcript elongation and its regulation. Annu Rev Biochem 66:117–172.

van Belkum A, van Leeuwen W, Scherer S, Verbrugh H (1999): Occurrence and structure-function relationship of pentameric short sequence repeats in microbial genomes. Res Microbiol 150:617–626.

van der Woude MW, Braaten BA, Low DA (1992): Evidence for global regulatory control of pilus expression in *Escherichia coli* by Lrp and DNA methylation: Model building based on analysis of pap. Mol Microbiol 6:2429–2435.

Vogel U, Jensen KF (1997): NusA is required for ribosomal antitermination and for modulation of the transcription elongation rate of both antiterminated RNA and mRNA. J Biol Chem 272:12265–12271.

Wassarman KM, Zhang A, Storz G (1999): Small RNAs in *Escherichia coli*. Trends Microbiol 7:37–45.

Williams KP (1999): The tmRNA website. Nucleic Acids Res 27:165–166.

Williams KP, Bartel DP (1996): Phylogenetic analysis of tmRNA secondary structure. Rna 2:1306–1310.

Wilson HR, Archer CD, Liu JK, Turnbough CL, Jr (1992): Translational control of pyrC expression mediated by nucleotide-sensitive selection of transcriptional start sites in *Escherichia coli*. J Bacteriol 174:514–524.

Wilson KS, Ito K, Noller HF, Nakamura Y (2000): Functional sites of interaction between release factor RF1 and the ribosome. Nat Struct Biol 7:866–870.

Wimberly BT, Brodersen DE, Clemons WM, Jr, Morgan-Warren RJ, Carter AP, Vonrhein C, Hartsch T, Ramakrishnan V (2000): Structure of the 30S ribosomal subunit. Nature 407:327–339.

Yanofsky C (2000): Transcription attenuation: Once viewed as a novel regulatory strategy. J Bacteriol 182:1–8.

Yarnell WS, Roberts JW (1992): The phage lambda gene Q transcription antiterminator binds DNA in the late gene promoter as it modifies RNA polymerase. Cell 69:1181–1189.

Young RA (1991): RNA polymerase II. Annu Rev Biochem 60:689–715.

Yura T, Nagai H, Mori H (1993): Regulation of the heat-shock response in bacteria. Annu Rev Microbiol 47:321–350.

Yura T, Nakahigashi K (1999): Regulation of the heat-shock response. Curr Opin Microbiol 2:153–158.

Zhang A, Altuvia S, Tiwari A, Argaman L, Hengge-Aronis R, Storz G (1998): The OxyS regulatory RNA represses rpoS translation and binds the Hfq (HF-I) protein. EMBO J 17:6061–6068.

Zhang G, Campbell EA, Minakhin L, Richter C, Severinov K, Darst SA (1999): Crystal structure of *Thermus aquaticus* core RNA polymerase at 3.3 A resolution. Cell 98:811–824.

Zieg J, Hilmen M, Simon M (1978): Regulation of gene expression by site-specific inversion. Cell 15:237–244.

4

Bacteriophage Genetics

BURTON S. GUTTMAN AND ELIZABETH M. KUTTER

The Evergreen State College, Olympia, Washington 98505

I. INTRODUCTION

Bacteriophages, or bacterial viruses, have been major research tools for molecular biology, and the history of research with them is virtually a history of molecular biology itself

(see Cairns et al., 1966). In this chapter we focus primarily on the large, virulent T-even

Modern Microbial Genetics, Second Edition. Edited by
Uldis N. Streips and Ronald E. Yasbin. ISBN 0–471–38665–0
Copyright © 2002 Wiley-Liss, Inc.

coliphages (viruses of *Escherichia coli*) because of the central role they have played in the development of our understanding of many fundamental processes and control mechanisms. For example, they were first used to demonstrate that viruses can direct the synthesis of enzymes that the host was not previously capable of making, and thus they carry their own genetic information. Other important advances included demonstrations of DNA as the genetic material; the colinearity of gene and protein; the nonoverlapping triplet nature of the genetic code, with specific triplets used to signal the end of the protein; the existence and properties of messenger RNA; the processes leading to the assembly of complex functional structures; the mechanism of DNA replication; and the occurrence of DNA restriction and modification. Karam et al. (1994) give a thorough review of the work with these phages up to that date. We will therefore focus particularly on work since then, as well as on some of the most classical experiments. We will also look briefly at some *Bacillus subtilis* phages as parallel examples in gram-positive bacteria. In addition, temperate phages are considered by Hendrix (this volume) and small single-stranded phages by LeClerc (this volume).

II. THE CONCEPT OF A VIRUS

Viruses are too often discussed as if they are merely small, simple organisms. They are not. The distinction between the concepts of "organism" and "virus" was drawn clearly, and with great good humor, by Lwoff (1953) and Lwoff and Tournier (1966). Only confusion results from any attempt to meld them into a single category.

1. An organism is always a cell or collection of cells (or, sometimes, a multinucleated cytoplasm that is simply not divided by cell membranes). No virus has such a structure. A virus is a particle, called a *virion*, which consists of a nucleic acid genome enclosed in a protein covering, or *capsid*, of distinctive geometry (Fig. 1). The nucleic acid and

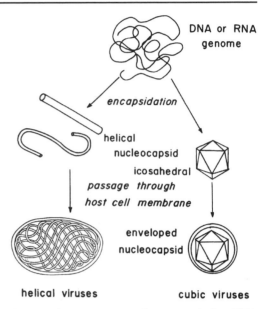

Fig. 1. A virion consists of a genome (either DNA or RNA) enclosed in a protein capsid, which has either a helical or icosahedral form. Some viruses mature by passage through a host-cell membrane, thus acquiring an outer envelope that encloses the nucleocapsid.

capsid together are known as the *nucleocapsid*. The nucleocapsid of some viruses is surrounded by an enclosing membrane, derived from the membranes of the host cell in which it was formed, but this in no way gives such a virion the properties of a cell.

a. The virion contains only one kind of nucleic acid—either DNA or RNA—whereas every cell needs both kinds to function. Viruses reproduce solely using the information from this one nucleic acid (using additional machinery from the host), whereas organisms, including infectious organisms, reproduce by means of an integrated action of their nucleic acid constituents.

b. Cells grow by enlargement and binary fission. No virus grows in this way. The virion is merely a vehicle for transporting the nucleic acid genome to another host cell. The genome enters the host cell and begins an infection, which results in production of a large number of new virions; the capsid is not reused.

2. Viral genomes do not contain the information for any kind of apparatus to generate high potential energy—what Lwoff called a Lipmann system. The virus is thus totally dependent on its host cell for a chemiosmotic potential, for ATP, and for any other source of energy.

3. A virus makes use of its host's protein-synthesizing apparatus: its ribosomes, transfer RNAs, and other factors. Some viral genomes encode special tRNAs, but no virus supplies the entire protein-synthetic system. Again, it is absolutely dependent on its host.

Some bacteria (rickettsias, chlamydias, *Bdellovibrio*) are obligate intracellular parasites, some even within other bacteria, and some of them have degenerated to the point of needing energy supplied by their host cells. But none of these agents have the properties of viruses. There are no organisms usefully seen as "quasi-viruses" or transition states between viruses and organisms.

III. HISTORICAL BACKGROUND: BASIC METHODOLOGY

Bacteriophages were first identified by Twort (1915) and d'Herelle (1917) as agents that caused clearing in cultures of bacteria. A number of microbiologists pursued this phenomenon during the following years, but were unable to obtain clear-cut results and to understand the nature of phage. This was largely due to the phenomenon of lysogeny (see Hendrix, this volume), in which the genetic material of the phage takes up residence within host bacteria and the bacteria then produce new phage irregularly. Sorting out these phenomena required better techniques, first developed with phage that have simple growth cycles not involving lysogeny. The required methods were developed by Ellis and Delbrück (1939), who performed experiments that are now the basis of all phage work.

The study of phage begins with plaque formation (Fig. 2). A sample of liquid that

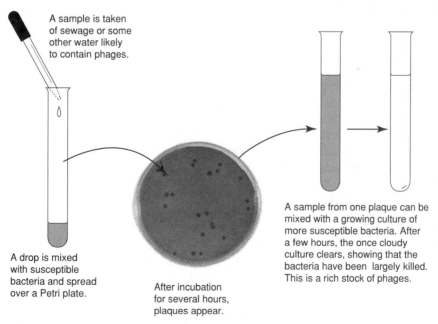

A sample is taken of sewage or some other water likely to contain phages.

A drop is mixed with susceptible bacteria and spread over a Petri plate.

After incubation for several hours, plaques appear.

A sample from one plaque can be mixed with a growing culture of more susceptible bacteria. After a few hours, the once cloudy culture clears, showing that the bacteria have been largely killed. This is a rich stock of phages.

Fig. 2. A method of isolating bacteriophage from a new source. The photograph shows a plate with a lawn of bacteria interrupted by clear holes, or plaques. Each plaque is a focus of infection where bacteria have been killed. The material from a plaque, which is rich in phages, may be used to start a new infection in a fresh cell culture to make a concentrated phage stock.

is likely to contain phage, such as sewage or a specimen of human or animal stools, is filtered, centrifuged, or treated otherwise to remove all bacteria and other organisms. Then dilutions of the filtrate are mixed with a few drops of a culture of some susceptible bacterium, and the mixture is spread over the surface of a petri plate containing nutrient agar. After incubation for several hours, when appropriate dilutions are made, the added "plating" bacteria form a continuous layer, or *lawn*, over the plate, but the lawn is interrupted by clear, round areas of various sizes. Each of these is a *plaque*, which represents an area where the bacteria have been infected by phage and killed. (Plaques for most phage do not grow in size indefinitely because the phage grow well only in bacteria in exponential phase. As bacteria enter the stationary phase, further infection is limited for most phage, and the size of the plaque is therefore defined.) Plaques made by different phages differ in size, in degree of clearing, and in characteristic circular zones of clarity or turbidity.

In general, each plaque is initiated by a single phage. It was the phenomenon of plaque formation that first indicated that phage should be thought of as particulate entities, rather than some kind of "poison," and Ellis and Delbrück (1939) demonstrated a linear relationship between dilution of a phage stock and number of plaques obtained under optimum conditions. (Phage were not actually observed as particles until several years later, with the development of electron microscopy.) Thus the titer of a phage stock (number of phage particles per milliliter) may be obtained by plating appropriate dilutions to obtain plaques, counting the number of plaques, and multiplying by the dilution factor, just as bacteria are enumerated by counting colonies. Similarly a single strain of phage may be purified out of a mixture by carefully removing a sample from one plaque (with a sterilized bacteriological needle, capillary tube, or toothpick) and regrowing in a fresh bacterial culture.

Using the plaque enumeration method, Ellis and Delbrück performed the experiment shown in Figure 3. At time zero, phage are mixed with appropriate host bacteria. After a few minutes, the mixture is diluted, and samples are removed at various times and plated. The result is that the number of plaques remains constant for about 25 minutes; it then rises sharply and levels off at about 100 times its initial value. (d'Herelle demonstrated this result in 1929; see Summers, 1999, p. 87. However, his work did not become widely known.) Ellis and Delbrück interpreted this result as showing that after bacteria are infected by phage, they remain intact for about 25 minutes; then each cell suddenly bursts, or *lyses*, liberating about 100 new phage on average. The ratio between number of plaques obtained after and before lysis is called the *burst size*. This, again, tends to be characteristic of each phage strain.

Advances in the biology of phage were made over the next few years by Delbrück and his colleagues, particularly S.E. Luria and A.D. Hershey, and their students. Until 1944 various laboratories had used different kinds of phage, making it impossible to

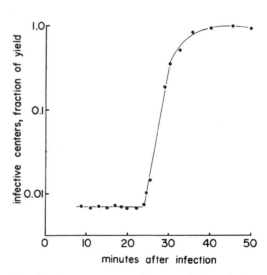

Fig. 3. A one-step growth curve of phage T4 on rapidly dividing bacteria at 37 °C. The difference between the initial and final numbers show a burst size of close to 200 phages per cell.

compare their results. Delbrück therefore arranged the "phage truce," whereby the community of investigators agreed to use only a set of seven phages selected by Demerec and Fano (1945) and numbered T1 through T7 (T for "type"), growing on *E. coli* strain B in nutrient broth at 37 °C. These are all "well-behaved" phages, in that they give easily countable plaques and show no confusing phenomena such as lysogeny. T2, T4, and T6 (the "T-even" phages) happen to be very closely related; their large icosahedral heads and contractile tails now assign them to the family Myoviridae (Murphy et al., 1995). (An icosahedron is a solid with 20 equilateral triangular faces; as discussed in Section V, it is the typical form of many virus particles.) Much of the early work focused on them and then became even more focused on T4, our main example.

For comparison, we should note that T5 belongs to the Siphoviridae, the phages with a long, flexible, noncontractile tail and an icosahedral head (90 nm in diameter); its genome is about two-thirds the size of the T4 genome. T3 and T7 are Podoviridae, with short stubby tails. They are about a quarter the size of the T-even phages and are distinguished particularly by producing their own phage-directed RNA polymerase to transcribe their late genes; this polymerase and its associated distinct promoters have been useful for cloning work, particularly when potentially very toxic gene products are involved. The last of the group, T1, is also a member of the Siphoviridae; it has a 60 nm icosahedral head and a genome size of about 48.5 kbp, and looks much like the temperate bacteriophage lambda. T1 has been studied much less, mainly because it is so difficult to contain in the lab; unlike the other T phages, it survives drying and thus often turns up in unexpected and undesired places.

T2, T4, and T6 are so closely related that they can recombine with one another in a mixed infection, producing mixed particles with the capsid of T4 and much of the genome of T2 (Novick and Szilard, 1951).

The T-even phages show dominance over other T-series phages and virtually all other phages in mixed infections, inhibiting their synthesis just as they do that of the host even when the other phages are already well through their infectious cycle. T4 further distinguishes itself by using an odd base in its DNA: 5-hydroxymethylcytosine rather than cytosine. This substitution is instrumental in many aspects of the mechanisms the phage uses to dominate the host.

Early electron microscopy (Anderson et al., 1945) showed that the various T phages are minute tadpole-shaped particles with large, rounded heads and thin tails. Anderson (1952) later demonstrated that these particles attach to the host-cell surface by their tails. Upon mixing phage with susceptible bacteria, the process of attachment, or *adsorption*, occurs rapidly and with second-order kinetics; that is, the rate of adsorption is proportional to the concentrations of both phage and cells, indicating that the process is nothing more than a specific molecular interaction between structures on the phage and others on the cell surface. Adsorption requires only a collision between the two in the right orientation.

A variation on the one-step growth experiment was performed by Doermann (1952). It was already known that infection of bacteria with a high ratio of phage to cells (a high *multiplicity of infection*, abbreviated MOI) would result in almost immediate lysis (*lysis from without*), as if the cell walls were suddenly weakened by so many infecting particles. Doermann infected cells with T4 and used a different phage, such as T6, plus cyanide, to lyse the infected cells at various times. He thus discovered that new T4 phage cannot be detected intracellularly until about 11 to 12 minutes after infection; the period before this time, known as the *eclipse period*, thus became a mystery, for it could not be explained why cells that would shortly contain hundreds of phage contained none at all for a time. To understand eclipse, the nature of the phage particle had to be determined.

The typical phage particle is made of about equal amounts of protein and DNA. Before the genetic role of DNA had been firmly demonstrated, Hershey and Chase (1952) separated the roles of the protein and the DNA in phage by a classic series of experiments. They grew one stock of phage T2 in medium containing ^{32}P, to label its DNA, and another stock in medium containing ^{35}S, to label its protein. They then followed the fates of the labeled components. Anderson's observation of phage attached to the bacterial cell surface after infection suggested that this component might be stripped off by violent agitation; Hershey and Chase therefore looked for the release of radioactivity from infected cells vortexed in a blender for various times. They showed that DNA (^{32}P label) remained almost entirely in the infected cells, which can be collected by centrifugation, but that protein, labeled with ^{35}S, was easily released into the supernatant by blending. Thus they concluded that only the DNA of the phage is actually injected into the cell, the protein remaining outside. When phage were mixed with bacterial cell wall fragments, they could be made to adsorb to these fragments and release their DNA into the medium. Furthermore the labeling pattern of newly made phage showed that large amounts of labeled DNA are passed on to the next generation, while little or no parental protein is contained in the new phage.

The Hershey-Chase experiment was the classical demonstration that DNA is the stuff of heredity, so for this reason it is important to all of biology. But it also clearly established the general pattern of phage growth, and it explained the eclipse period. The first event following adsorption of the phage particle must be injection of its DNA. The DNA takes over the cellular apparatus and initiates the synthesis of new phage proteins, but the first whole phage particles are not made until about 11 to 12 minutes, and then their numbers increase rapidly.

IV. OVERVIEW OF THE BACTERIOPHAGE T4 INFECTIOUS CYCLE

T4 is a large, complex bacteriophage that infects *E. coli*. Its spaceshiplike capsid (Fig. 4) carries about 169,000 bp worth of genetic information, coding for about 300 genes, 130 of which have been mapped and characterized in some detail (Fig. 5). As indicated in Figure 5, genes of related function are largely clustered. About 40% of the genome codes for the phage's complex structural assembly: 24 genes for head morphogenesis, 10 of them encoding structural components, and 26 proteins in the tail and fibers, with 5 additional ones needed for assembly. Thus it is not surprising that T4 serves extensively as a model system for studying self-assembly and mediated-assembly processes. The DNA is a linear molecule, but the genomes packaged into various phage particles are circularly permuted, ending at many different sites in the genome, and each genome has a terminal redundancy of about 6%; as a result the genetic map is circular. The physical basis for this phenomenon is discussed in section V.D below.

T4 rapidly directs the bacterial cell to stop making all of its own macromolecules—that is, DNA, RNA, and protein—and turns it into a factory for making more T4. This transition involves a carefully orchestrated series of developmental steps, and it has been used as a model system for better understanding the changes in gene expression that occur during embryological development in complex organisms. The steps of phage development are as follows:

1. As soon as the T4 DNA is injected, the host RNA polymerase binds to several strong promoter regions on the T4 DNA, leading to transcription of a group of so-called *immediate-early genes* (Fig. 6a). The products of these genes are mainly small proteins, primarily involved with shut-off of host functions or initiation of phage

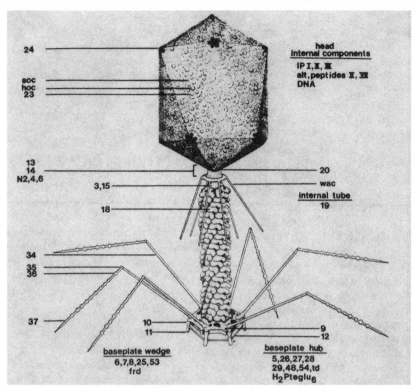

Fig. 4. Structure of the phage T4 virion, based on electron microscopic analysis. The locations of major proteins, named primarily after their genes, are shown. The baseplate is made of a central plug and six wedges; proteins whose locations are not yet certain are listed. The internal tail tube, which is not visible, is made of gp19. The collar and whiskers are apparently made of one protein species, gpwac. (Reproduced from Eiserling, 1983, with permission of the publisher.)

infection. They are only made for the first 3 to 5 minutes after infection at 37°C. (These proteins are generally quite stable and remain throughout infection, but no more of these proteins are produced after 5 minutes, no matter what else is going on in the infection process.)

2. Synthesis of a second group of early proteins starts about 3 minutes after infection (see Fig. 6b). Some of these *delayed-early proteins* form the complexes of enzymes that replicate T4 DNA and provide the precursors for DNA replication. Others are nucleases that degrade the host DNA, and some are proteins that further modify the host RNA polymerase to allow recognition of the genes producing the proteins for new phage capsids.

3. Phage DNA synthesis starts about 5 minutes after infection, mediated by a *replisome*, a complex of eight proteins that polymerizes nucleotides. Nucleotides are efficiently fed into the replisome by a complex of nucleotide-synthesizing enzymes (Mathews and Allen, 1983). The daughter DNA molecules recombine extensively, in a process apparently mediated by yet another complex multiprotein "machine," producing a complicated, multibranched ball of replicating DNA.

4. Synthesis of late phage proteins, mostly those that form the phage capsid, starts about 7 minutes after infection (Fig. 6c). Meanwhile synthesis of the second group of T4 early enzymes gradually stops. If anything blocks phage DNA synthesis, such as

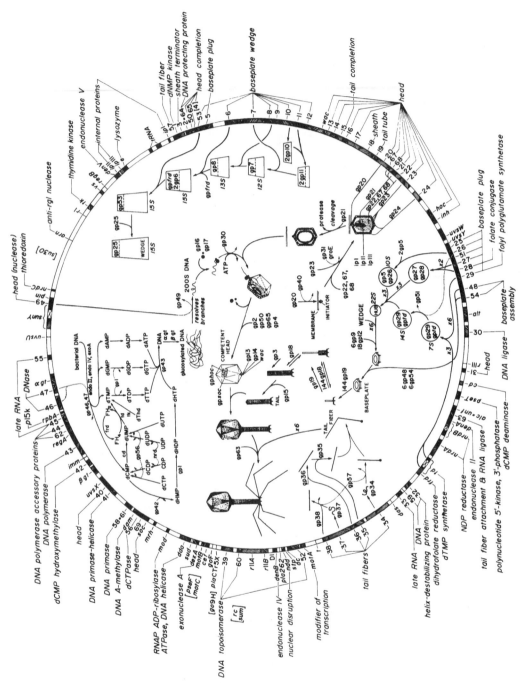

Fig. 5. The genomic map of phage T4. Numbered genes are defined on the basis of amber and temperature-sensitive mutations, as explained in the text; thus their absence is lethal under normal conditions for infection. Other genes are designated by mnemonics that reflect their function. Genes that are not securely mapped are located approximately in brackets. Genes are highly organized by function, especially those that encode portions of the capsid. The functions of other gene products, primarily enzymes, are specified where these are known. The major pathways of nucleotide metabolism and the capsid assembly pathways are summarized inside the map.

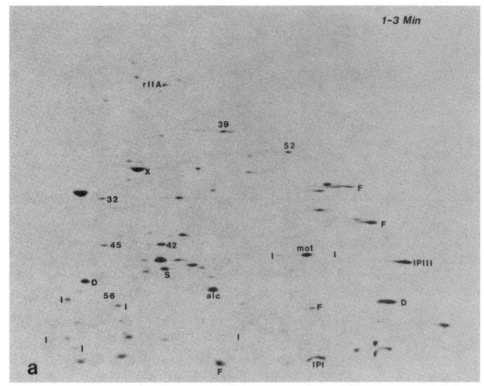

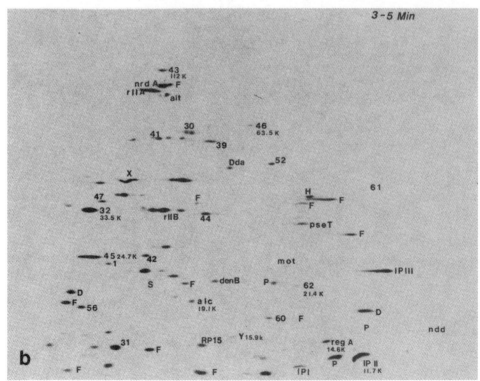

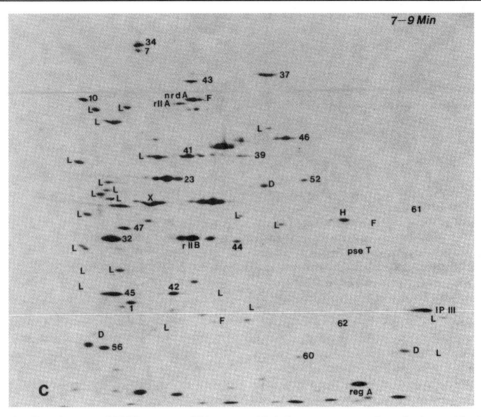

Fig. 6. Two-dimensional PAGE patterns of T4 proteins, labeled at various times after infection. **a:** 13 minutes. **b:** 3 to 5 minutes. **c:** 7 to 9 minutes. Many proteins have been identified by comparisons of mutant and nonmutant phage. Others, not fully identified yet, are partly correlated with the genetic map because they are missing in certain long deletions (F, I, P, D).

an antibiotic or a mutation in a gene essential for phage DNA replication, synthesis of many T4 early enzymes continues, as if trying to get around the block by sheer numbers. Furthermore no structural proteins are made, reflecting the direct link between replication and late transcription; that is, a regulatory mechanism blocks capsid synthesis until phage DNA molecules are available to be packaged in them.

5. Phage heads, tails, and tail fibers assemble via independent pathways. Heads assemble while bound to the cell membrane. Then each of them is filled with a headful of DNA from the replicating complex, while any single-stranded breaks are repaired and any branches are resolved in the process. Tails and tail fibers are then added.

6. Cell lysis occurs normally about 25 to 30 minutes after infection at 37°C. Oxidative metabolism suddenly stops, and lysis is mediated by the combined action of T4 lysozyme, which seems to act like known eukaryotic lysozymes, working in conjunction with at least one other, less well understood protein, encoded by the *t* gene. The released phage, about 100 to 200 per cell, are then ready to start another cycle of infection. T4 virions, like those of many other viruses, can remain viable for many years, waiting for another susceptible *E. coli* to show up—unless they dry out, their DNA is damaged by radiation, their tail fibers get knocked off, or their DNA is released by osmotic shock.

During its intracellular phase, T4 can switch into another growth strategy, termed *lysis inhibition*, in response to a signal that there is a bacterial shortage at the moment. If another T4 tries to *superinfect* a cell—that is, to get into a cell already infected by T4—this event is taken to indicate an overabundance of phage relative to cells, so the best strategy for reproduction is to delay lysis. Instead of lysing the cell after only half an hour, the virus—through some unknown mechanism—maintains the cell intact for at least 4 to 6 hours, squeezing out every last phage particle it can make, sometimes over 400 phage per cell. Expansion of the phage population is clearly slower under lysis-inhibition conditions than when a new round of infection is initiated every half hour, but this is a more effective strategy when the bacterial population is limited, and it gives any remaining bacteria more opportunity to reproduce while the phage are developing.

One T4 particle is enough to cause a normal infection. When several T4 infect *E. coli* at the same time, they peacefully coexist, mutually complement any genetic defects they may have, recombine avidly with each other, and produce progeny with all possible combinations of the available genetic information. However, if more than 25 to 30 phage try to infect the same cell simultaneously, they may seriously damage the bacterial membrane so much that the cell just disintegrates, because of *lysis from without*.

V. FOUNDATIONS OF PHAGE GENETICS

A. Mutant Phages

Modern molecular biology has grown rapidly through the development of a number of important techniques. No technical innovation, however, has been more productive than the development of genetic analysis, which depends on finding mutants and using them to elucidate the normal structure and operation of specific systems.

Phage genetics began with the recognition of phage mutants. In any stock one always finds a few phage that make plaques with unusual morphologies. Turbid (*tu*) mutants make somewhat cloudy plaques minute (*mi*) mutants make small plaques and rapid lysis (*r*) mutants make somewhat larger-than-normal plaques with sharp edges (Fig. 7a). Luria (1945) also recognized host-range (*h*) mutants. Wild-type T4 cannot grow on a phage-resistant strain of *E. coli* B (B/4); *h* mutants have altered adsorption properties, so they can grow on B/4. *h* mutants are distinguished from h^+ (wild-type) phage because they form clear plaques on mixed indicator bacteria (B and B/4), a condition in which h^+ make turbid plaques since they do not infect the B/4 bacteria. The ability to find such phage and bacterial mutants shows how specific the attachment of the phage to the cell surface is. It can be shown that the resistant bacteria do not adsorb the phage in question because the requisite surface structures have been altered and that the phage mutants likewise have altered adsorption structures suited to the new bacterial morphology.

Hershey and Rotman (1948) used mutants of all these types to demonstrate recombination in phage T4 and to develop the first genetic map of the virus. In these experiments bacteria are infected at relatively high MOI with a mixture of two phages, say, an *r* mutant and a *mi* mutant. A large fraction of the progeny phage will naturally be either *r* or *mi*; these are the *parental types*. In addition two *recombinant types* of phage also appear. One is wild-type in all its characteristics; the other makes plaques with a combination of *r* and *mi* features and can be shown to carry both *r* and *mi* mutations (Fig. 7a). These recombinant types must have been formed through complex interactions among the replicating phage DNA molecules inside a cell so when a cell is mixedly infected with phage carrying two different mutations, some genomes are created that carry both, or neither (Fig. 7b).

The general principles of genetic mapping of phage, derived from classical Mendelian genetics, are fairly simple. Every mutation

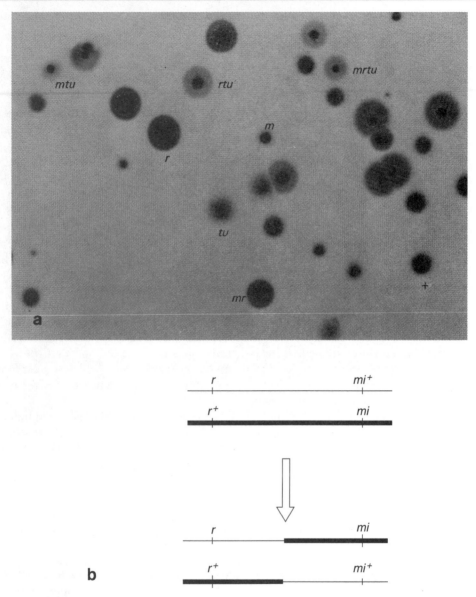

Fig. 7. a: A plate containing plaques made by several mutants with distinctive plaque morphologies and various recombinants between them. An investigator can learn to distinguish the various kinds of plaques; their genotypes are confirmed by picking phage from a plaque and performing additional genetic tests. **b:** Recombination occurs when genomes carrying two different markers interact and engage in exchange (crossover), so new genomes are generated with different combinations of the markers. The complex processes occurring at the point of crossover are not specified, and are not yet fully understood; they are discussed in more detail by Fishel (this volume). (Photograph courtesy of Dr. J. Foss.)

may be used as a *genetic marker*, a tag that marks a certain point on the phage genome. Every point on the genome is considered to have at least two alternative states: crudely, either normal (wild-type) or mutant. The objective of mapping experiments is to establish the physical relationships among all these points and, eventually, to establish the boundaries of all genes or other functional units.

Suppose that in an infection with equal multiplicities of two parental phages, a total of n progeny phage are produced and that m of these are recombinant types. Then $R = m/n$ is the *frequency of recombination* between the markers involved. R may then be taken as a measure of the distance between the markers, making the assumption of classical genetics that the probability of a crossover between the markers should increase with increasing distance. The maximum value of R should theoretically be 0.5 (50%) for the case in which the markers in question are completely unlinked to each other—that is, on different "chromosomes" or pieces of DNA, or at least far enough apart on one piece of DNA to recombine at random. In this event the probability of incorporating either allele of one marker (e.g., a or a^+) is 0.5, and there is the same probability of incorporating either allele of the other marker (b or b^+). Thus the four possible genetic combinations will be formed with equal probabilities of $0.5 \times 0.5 = 0.25$, and since two of these combinations are recombinant and two are parental, $R = 0.5$.

Using these principles, Hershey and Rotman established a simple genetic map with three linkage groups. The frequency of recombination between any two markers on different linkage groups is 0.5, indicating that these groups are not (detectably) linked to one another. This map was based on relatively few mutations, but it did include several independent r mutations in the second linkage group (rII mutations) that mapped in a cluster and might represent a single gene; this idea is explored below.

Early studies on the DNA of T4 were consistent with this map, since they indicated that the phage might contain several independent DNA molecules. However, Streisinger and Bruce (1960) performed more sensitive mapping experiments using additional genetic markers and demonstrated that an extensive series of genetic markers are in fact linked in a single group.

Since plaque-morphology and host-range mutants represent a small number of possible genes, they have limited genetic uses. R.S. Edgar and R.H. Epstein recognized that T4 could only be properly explored genetically if one could collect a large number of mutants and if the mutations might, in principle, affect any gene. They therefore searched for, and found, *temperature-sensitive* (*ts*) mutants, defined as mutants able to grow at 30°C but not at 42°C. Since any protein of the phage might become inactivated at high temperature through a change in one of its amino acids, *ts* mutations might be obtained in any gene; however, they can only be observed if the absence of that gene product is lethal, or at least very deleterious, to the phage under the growth conditions being used.

At about the same time, a second general type of mutant was discovered, a *host-dependent* type that is able to grow in certain bacterial hosts but not others. Benzer (1955) had already shown that the *rII* mutants are host dependent. They grow readily in *E. coli* strain B but not in strain K, although it soon became apparent that the critical factor is really that strain K is usually lysogenic and carries a totally unrelated phage, lambda (λ). Epstein and C. Steinberg decided to search for "anti-*rII*" mutants that could grow in strain K but not in B, and they found a number of them. But in contrast to *rII* mutants, these mutants mapped at many places, not just in one gene. It thus became evident that these mutations were of a general type that might occur in any gene, and they were searched for actively, along with *ts* mutants. (They were subsequently named *amber* [*am*] in honor of Harris Bernstein, the graduate student who helped to isolate them; the German word *Bernstein* means "amber.") *Amber* mutants have been identified in many organisms and viruses; they involve mutation to a "stop" codon within gene and thus to premature termination.

We can now understand the results of genetic studies with this variety of phage mutants.

B. Topology and Topography of the Phage Genome

Benzer carried out a now-classical series of experiments with the *rII* mutants that revealed basic facts about genetic structure. As stated above, these experiments depended on the fact that *rII* mutants will grow in B but not in K strains. When any two *rII* mutants are crossed, any wild-type recombinants they might produce are easily detectable because they alone will plate on K. Since distances between genetic markers are measured by *R* values, the fine structure of a gene can only be investigated if one can find very small numbers of recombinants among large numbers of progeny.

Most of the *rII* mutants carry point mutations, which appear to be changes at one point on the DNA and which revert (back-mutate) to wild-type at measurable rates. However, Benzer found some *rII* mutants

that appear to be deletions: they do not revert to wildtype, and when mapped against other mutations they fail to recombine with two or more point mutants whose defects map at distinct sites, indicating that the deletion covers a short stretch of the genome and is unable to interact with mutations at any site within that stretch. If any two deletions overlap—that is, delete some common stretch of the genome—they also will be unable to recombine and produce wild-type phage. If all these deletions affect a simple linear structure—presumably a DNA molecule—it should then be possible to arrange them in a linear sequence on the basis of their pattern of overlaps. Figure 8 shows Benzer's arrangement of one set of deletions on this basis. Eventually he was able to arrange 145 deletions in an unambiguous sequence, with no need to postulate anything more complicated than the simple linear structure.

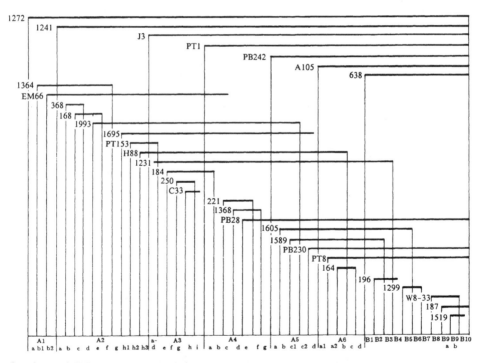

Fig. 8. A set of *rII* deletions placed in the proper relationship to one another on the basis of their patterns of recombination. Two mutations that do not recombine to produce wild-type recombinants must overlap, so the approximate relationships of a set of deletions can be determined uniquely. Lengths of deletions and the exact positions of end points can only be located with more refined mapping.

His experiments thus showed that the phage genome is topologically linear, as it should be if its information is simply encoded in a DNA molecule.

Given the deletion map, it is then easy to map any point mutation by crossing it against the deletions. It is first localized roughly by crossing against a set of long deletions (Fig. 9); then its location is narrowed down by crossing against shorter deletions. Finally, its position relative to other nearby point mutations is determined by standard crosses. Using this procedure, Benzer determined that the *rII* point mutations map at a large number of sites, some so close to each other that they must be changes at neighboring nucleotide pairs in the DNA. Thus these

experiments showed that the phage genome can be understood as a simple DNA molecule, with mutations being changes in its nucleotides.

C. Complementation and the Operational Definition of a Gene

Mapping a series of mutations, even those very close to one another, still leaves open the question of where boundaries between genes occur. Theoretically, a gene is best understood as the region that encodes a single polypeptide; operationally, there must be some way to delimit such a region.

Following Benzer, imagine that the *rII* mutations actually fall into two neighboring genes (Fig. 10). Suppose both gene products

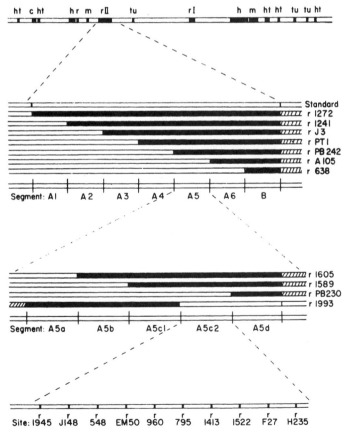

Fig. 9. A set of deletions is used to map other mutations. The unknown mutation is confined to shorter and shorter segments by crossing it with selected deletions; when it has been located in the shortest segment defined by the deletions, it is located relative to other mutations by standard crosses.

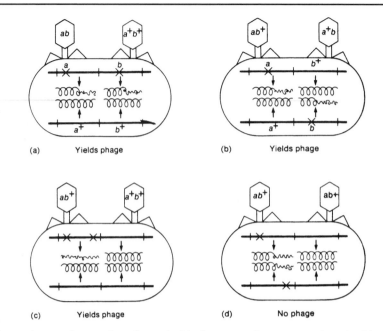

Fig. 10. A complementation test is performed with phages carrying two mutations, in either a *cis* arrangement (both in the same genome) or a *trans* arrangement (in different genomes). Protein products are represented by helices, and a defective protein by an interrupted helix. If the mutations are in the same gene, there will be no functional proteins (and therefore no phage) in the *trans* position; if they are in different genes, the mutants will complement each other and produce phages.

are required for growth in strain K. Any two mutants could represent changes in the same gene or in different genes. Now infect strain K cells simultaneously with any two mutants. If their defects are in different genes, then each one can supply a function that the other lacks; they are able to *complement* each other and produce an infection yielding viable phage. If, on the other hand, both mutations fall in the same gene, the two phages together are no better off than each one by itself, and they cannot grow. The complementation test is thus an operational definition of a gene. When applied to the *rII* mutants, it showed that there are in fact two neighboring genes. The same test can be used with *amber* or *ts* mutants, and it is in this way that the known genes of T4 have been defined and mapped.

It is important, incidentally, to see that complementation is quite different from recombination. In recombination tests, the question is whether, and with what fre-

quency, two genomes can recombine their information to produce a new genome; one must wait until the next generation to determine the answer. In complementation tests, the question is whether two genomes, each missing some functional unit, can mutually supply gene products (generally proteins) to produce a normal function and thus viable phage under otherwise nonpermissive conditions.

Occasionally allele-specific apparent anomalies have been observed that reflect the unusual ability of two mutant proteins to interact and form a functional dimer. For example, two temperature-sensitive mutants of the enzyme hydroxymethylase, LB-1 and LB-3, can complement to make a functional protein. There is little general difficulty in distinguishing this allele-specific *intragenic* complementation from the *intergenic* complementation discussed above, but it is important to be aware of the possibility.

D. General Genomic Structure: Circularity and Gene Arrangement

Streisinger and Bruce's (1960) mapping data had given some slight indication of circularity: that is, the end markers of the single linkage group appeared to be linked to each other, making the map a circle. Given additional markers, in the form of *am* mutations, Streisinger and coworkers (1964) were able to demonstrate that in fact a series of markers spread over the entire genome are all linked circularly.

Genetic circularity could result from several quite different phenomena—most obviously that the phage chromosome is simply a physical circle of DNA. However, Streisinger postulated a more complex, and more interesting, basis for the circularity of phage T4. This explanation was precipitated by the observation that in a cross of an *rII* mutant by *r⁺*, a small fraction of progeny produce mottled plaques, with interspersed sectors of the two phenotypes; they appear to have come from a "het" (for heterozygote) that carried both markers within its genome. A het could have either of two structures (Fig. 11): a *heteroduplex*, in which the two DNA strands at some point are mismatched and are carrying the two markers, or a duplication of two double-stranded DNA molecules limited to the stretch in question. Doermann and Boehner (1963) adduced evidence for the latter model; they created phage that were simultaneously

heterozygous for six closely linked markers and showed that these phage then segregated their markers in a gradient, with the largest numbers of mutation 1 and the smallest of mutation 6. This result suggested that the marked region was being affected by a nearby physical end. However, at about the same time Berns and Thomas (1961) presented evidence that the T4 genome is a single unbroken structure.

In a brilliant theoretical gambit, Streisinger sought to resolve this apparent contradiction with the model shown in Figure 12. The genome of each phage is considered to have a small terminal redundancy, with a few percent of its length duplicated. (This terminal redundancy is one solution to the het problem, since a phage can be a het by carrying different markers in the duplicated genes at its ends.) After an infecting genome has replicated even a few times, there will be a pool of molecules with identical duplicated ends that can "mate" with each other, and recombination within the hybridized regions could then produce a long *concatemer*, containing the equivalent of several genomes. When new phage particles are made, each of them will contain a "headful" of DNA cut from such a concatemer. However, each phage will contain a different terminal redundancy, and the phages will be related to one another by circular permutations of the gene sequence. Genetic crosses with any population of phage will then yield a circular

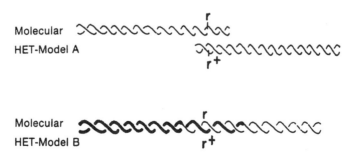

Fig. 11. A heteroduplex might be a phage carrying two double-stranded DNA molecules with partially duplicated information (model A) or a phage with a heteroduplex structure (model B) in which each strand of the DNA carries different information.

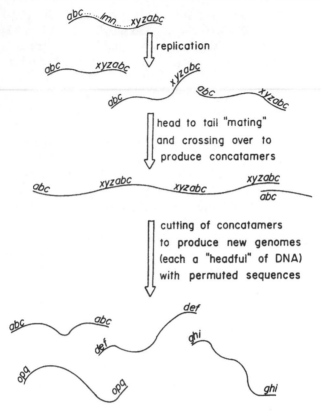

Fig. 12. Streisinger's model of T4 structure and replication. Each phage genome is terminally redundant. After replication begins (or if more than one phage initiates the infection), the pool of DNA molecules can "mate" with another and, through crossing-over, produce long concatamers. Eventually new phage genomes are cut from these large molecules by removing a "headful" of DNA, but since these genomes are cut with terminal redundancies, the whole population bears circularly permuted sequences.

map, even though the genome of each phage is linear. This model was well confirmed through a series of critical experiments (Streisinger et al., 1967), and molecules of concatemer length have been identified physically.

By using a large collection of *am* and *ts* mutants, Epstein et al. (1963) identified many T4 genes, designated simply by numbers, and outlined the general structure of the T4 genome (Fig. 5). The genome extends over 169,000 nucleotide pairs and is standardly drawn with its arbitrary zero point at the junction between the *rIIA* and *rIIB* genes, at 9 o'clock. The genome is highly organized by function, with the late or capsid-related genes falling predominantly into a large block between about 73 and 120 kb and another be-

tween about 150 and 160 kb. The genes for a few late proteins are interspersed in early regions; transcription and translation of such genes is subjected to unusual transcriptional and translational controls (discussed in Section VIIIG). The other regions contain primarily early genes, many of which are not essential under ordinary laboratory conditions and thus cannot be defined by *am* or *ts* mutations in the usual way. Many such "nonessential" genes have been identified by a variety of other mutations, obtained by special methods. Their names of one to four letters are mnemonics for their function (e.g., *e* for lysozyme ["endolysin"], *denA* and *denB* for DNA endonucleases, or *rpbA* and *rpbB* for RNA polymerase binding proteins).

The gene products (in general, proteins) of T4 genes are all designated by "gp" plus the name or number of the gene. This is especially useful for genes whose products are known only as capsid components or for those whose specific function has not yet been determined. Note that "gp" is used differently here than it is for eukaryotic systems, where "gp" means "glycoprotein"; none of the T4 proteins are known to be glycosylated. The still-unidentified open reading frames (ORFs) are named with regard to the preceding characterized gene in the clockwise direction—for example, ORF 60.3 or nrdC.8—until a specific function or property is assigned.

The general function of each late gene was originally outlined by electron microscopy of lysates, which contain large amounts of incomplete capsids, even if they are missing some critical protein essential to formation of mature phage. More specific functions have been revealed primarily by in vitro studies of capsid assembly (see Section VI), and the functions of many early genes have been determined in studies of host shutoff, nucleotide metabolism, and DNA replication (see Section VIIIE) or of gene regulation (see Section VIIIQ).

VI. STRUCTURE OF THE PHAGE PARTICLE

The major components of the capsid were identified by Brenner et al. (1959) with electron microscopy. Eiserling and Black (1994) and Mosig and Eiserling (1988) present excellent reviews. The particle consists of a large head and a tail with six tail fibers attached to its end. The tail consists of several components. At its distal end is a hexagonal baseplate, bearing the tail fibers. A thin tail tube is built up on the baseplate, and this is surrounded by a sheath, which contracts upon infection and becomes demonstrably shorter and thicker. The baseplate also undergoes a conformational change at the same time.

The details of capsid structure have been elucidated through electron microscopic studies, along with studies of capsid assembly. Edgar and Wood (1966) demonstrated that phage will assemble themselves in vitro if the lysates of a "headless" and a "tailless" mutant are mixed. This observation opened up several series of very fruitful investigations in which the steps in self-assembly were determined, and these have naturally led to detailed information about the structures being assembled.

In considering the structures of small viruses, Crick and Watson (1957) noted that the genome of such a virus is too small to encode all the protein in its capsid. They suggested therefore that the genome encodes only a subunit and that the whole capsid is made by assembly of many subunits into a regular structure. Caspar and Klug (1962) showed that only a few symmetrical structures are possible and that the so-called spherical viruses should actually be icosahedrons. They also proposed that protein structures should be governed by the general principle of self-assembly: that once the polypeptide components of the structure are formed, they should automatically assemble themselves into a stable, least-energy form which will be the proper functional structure. The principle applies generally to virus assembly, and virtually all of the T4 capsid structures assemble themselves when their components are mixed. The general experimental method is to begin with lysates from two different mutant infections, each one missing a different protein. The lysates complement in vitro and produce functional structures. However, in a few cases assembly requires some enzymatic modification by a phage encoded protein; this is a slight variation on self-assembly.

Wood and his associates (Bishop et al., 1974) determined the comparatively simple pathway of tail fiber assembly. The fiber is made primarily of two very long proteins, gp34 and gp37, with three much smaller proteins, gp35, gp36, and gp38. The whole pathway is shown in Figure 13a. Notice that at two points gp57 catalytically modifies the structural proteins. Gene 57 is the only gene remote from the others. The five structural proteins are encoded by a linked block of genes, illustrating the general principle that genes for components that must be connected should be tightly linked. A somewhat more complex pathway mediates tail assembly (Fig. 13b). Six wedges are assembled and joined to a preassembled core; this then is used as a platform on which to build both the tail tube and the sheath. A specific baseplate protein, gp29, acts as a sort of "ruler" to determine tail length.

In accordance with the Caspar-Klug principles, the head of T4 (and of all the known phages that have a head and a tail) is an icosahedron, but it is somewhat elongated rather than isometric. Its principal subunit is gp23, with gp24 at the vertices and several other proteins involved in its assembly (see drawing in Fig. 5).

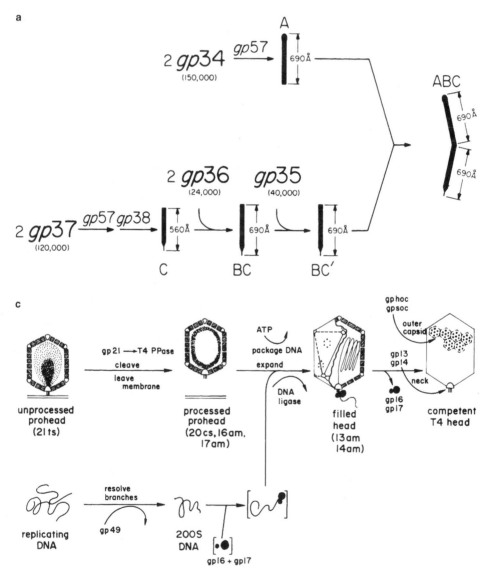

Fig. 13. The pathways of T4 capsid assembly. **a:** The tail fiber assembly pathway. **b:** The tail assembly pathway. **c:** The head assembly pathway.

13b

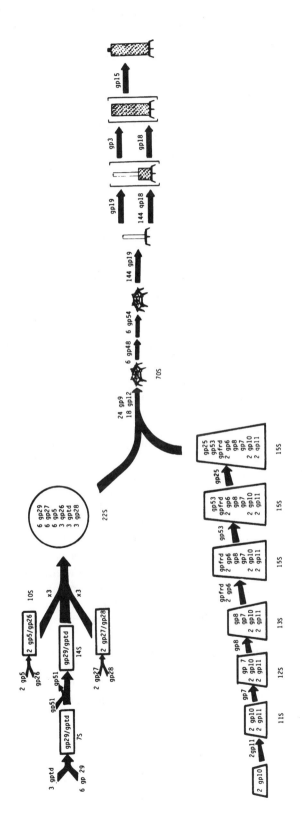

The bacterial cell membrane forms a foundation on which the T4 head (as well as those of several other phages) is assembled (Fig. 13c). At least one bacterial protein, groEL, is known to be essential. An initiator is first formed on the membrane by gp20 and gp40, and a large prohead, which is seen by electron microscopy as a rather spherical complex with two or three layers of density, is assembled on this foundation by gp21, gp22, gp67, gp68, and the internal head proteins. gp23 forms a shell around the internal core, and gp24 is added to the vertices; at this point the GroEL protein and gp31 are required. Both gp23 and gp24 are initially added in the form in which they were synthesized, but in a subsequent step the head matures into its final form by cleavage of these two proteins to somewhat smaller molecules (gp23* and gp24*) by an internal protease, gp21. In this maturation step, the head is released from the cell membrane.

The completed head attaches to the end of one of the many branches of the massive replicating DNA complex and spools in a headful by means of gp16 and gp17, with endonuclease VII (gp49) resolving branches and the DNA ligase repairing nicks in the process. Some additional proteins, some of them nonessential, are added to the outside of the head and to the neck. Finally the head and tail self-assemble, and tail fibers are then added with the assistance of gp63. (The entire process, indicating associated genes, is summarized in Fig. 5.)

VII. SPECIAL PROPERTIES OF T4

Some of the most interesting properties of phage T4 stem from the fact that it has an unusual base, 5-hydroxymethylcytosine (hmdC), in its DNA in place of the normal cytosine. In the DNA helix, the hydroxymethyl groups, like the methyl groups of thymine, are located in the major groove, where they do not affect base pairing but can be used as recognition signals. The 5-methylcytosine formed at specific sites after DNA synthesis acts as a control signal in both prokaryotic and eukaryotic systems.

The use of this new base facilitates the viral domination of the host in several ways:

1. T4 is immune to most bacterial restriction systems. Bacteria protect their own DNA from restriction endonucleases by marking it with methyl groups at the cleavage site, and these enzymes are also blocked by the hydroxymethyl groups and do not attack T4 DNA. E. coli does have a nuclease that specifically recognizes 5-hydroxymethylcytosine as foreign, but T4 blocks this nuclease by glucosylating the hmdC residues in its DNA. No E. coli enzyme has yet evolved that can attack the sugar-coated DNA,

2. T4 makes cytosine-specific nucleases that degrade the host DNA but do not attack their own DNA.

3. T4 inhibits transcription of bacterial DNA by producing a small protein, gpalc, which interacts with both the RNA polymerase and (cytosine-containing) DNA to block transcription of cytosine containing DNA.

4. T4 has no mechanism to ensure that it only encapsidates hmdC DNA; it can therefore package host DNA and carry it to a new cell, acting as a transducing phage, as long as the degradation of host DNA is blocked by eliminating the genes that encode the specific T4 nucleases (Wilson et al., 1979). Reasonably efficient transduction only occurs with phage mutants that use cytosine rather than hmdC in their DNA and do not block host transcription, since alc is defective.

T4 also makes several enzymes that are particularly useful in genetic engineering work, even though their value to T4 itself is still unclear:

1. A DNA ligase that can join two blunt-ended pieces of DNA. The other known DNA ligases will only join DNA pieces that have complementary single-stranded ends and thus can be held in register.

2. An RNA ligase that seems to be involved in splicing a tRNA that is cleaved by an unusual sort of restriction system encoded by a cryptic element, *prr*, in certain *E. coli* strains. Strangely it also aids the joining of the tail fibers to the tail, a process that involves only proteins, not RNA. Though at least three T4 genes contain introns (the first to be demonstrated in eubacterial systems), the RNA ligase apparently plays no role in their splicing, which occurs autocatalytically. (This is discussed in Section VIIIF below).

3. A 3'-phosphatase, 5'-kinase that acts on DNA, RNA, a number of vitamins and cofactors, and a variety of other molecules. Mutations in this gene have no observable deleterious effects on the phage.

VIII. SOME DETAILS OF THE PROCESS OF INFECTION

A. Adsorption and Injection

T4 phage particles initially adsorb to the surfaces of sensitive cells through specific contact between the distal ends of the tail fibers and specific outer-membrane receptors. The potential versatility of this interaction is emphasized by the fact that these are diglucosyl residues of the lipopolysaccharides of the *E. coli* B cell surface (Simon and Anderson, 1967) or OmpC on the surface of K strains. This binding leads to an allosteric hexagon to star transition in the arrangement of tail baseplate proteins, quickly followed by irreversible attachment by means of the now-exposed gp12, on the tip of the baseplate, and thence to rearrangements in the tail sheath. The sheath contracts while the baseplate stays bound to the cell surface, forcing the central, noncontracting core of the tail fiber through the membrane of the cell.

A few internal proteins are injected into the bacterial cell along with the DNA; one of these, gp2, protects the ends of the DNA from exonucleolytic degradation, while another, gp*alt*, ADP-ribosylates Arg 285 of one of the two α subunits of each host RNA polymerase.

B. Shutoff of Host Functions

T4 efficiently shuts off host transcription, translation and replication and substantially alters a number of other host pathways, as reviewed by Kutter, White, et al. (1994).

The process of adsorption itself appears to trigger cellular metabolic changes that would be irreversible if DNA injection and subsequent phage-induced changes did not occur. This has been shown by studying the effects of phage *ghosts*, the empty capsids made by osmotically shocking phage particles so that they release their DNA. Ghosts adsorb to cells just as whole phages do, and in so doing, they kill. The action is similar to the killing activity of certain colicins (see Perlin, this volume), which appear to produce a general inactivation of the membrane-bound metabolic apparatus (cytochromes, etc.), but the mechanism of ghost-mediated killing is poorly understood; leakage of ions and other small molecules may be involved.

It has been clear for a long time that T4 infection quickly stops all synthesis of host proteins. Monod and Wollman (1947) showed that the enzyme β-galactosidase cannot be induced in infected cells, and Levinthal et al. (1967) showed that no synthesis of host proteins can be detected after about 2 to 3 minutes. By hybridizing labeled RNA to specific DNAs, Nomura et al. (1960) and Hall and Spiegelman (1961) demonstrated that within a few minutes after infection essentially all of the newly synthesized RNA is transcribed from T4 DNA. A further level of complexity was introduced, however, by the report by Nomura et al. (1966) that there are actually at least two modes of inhibition. One of them is multiplicity dependent and insensitive to chloramphenicol (i.e., not dependent on protein synthesis after infection), while the other requires protein synthesis and is independent of multiplicity.

The rate of host DNA replication also decreases sharply over the first 5 minutes. At the same time the host nucleoid, which is normally a compact structure in the center of the cell attached to the membrane at only a few points, is disrupted; electron microscopy shows that it becomes strongly associated with the cell membrane at many points (Fig. 14). This process is the result of the gene *ndd* (nuclear disruption defective) that maps near the *rII* region (see Kutter, White, et al., 1994); *ndd* mutants shut off host transcription at the normal rate but are somewhat defective in shutting off host replication.

In a completely independent process, the host DNA is attacked and degraded by several T4-encoded enzymes. Because, as discussed above, T4 DNA contains hydroxymethylcytosine and is also coated with glucose residues, enzymes can distinguish between host and phage DNA. Host DNA is attacked by at least two endonucleases: endonuclease II, the product of the *denA* gene, makes single-strand nicks in cytosine-

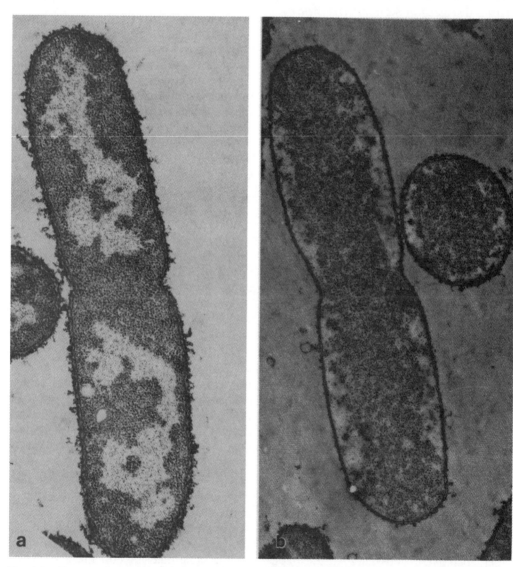

Fig. 14. Electron micrographs showing the effect of nuclear disruption on the arrangement of the *E. Coli* nucleoid after T4 infection. **a:** An uninfected cell in which the nucleoid (white material) is located centrally. **b:** A cell infected with wild-type T4 in which the nuclear material is marginalized. (Courtesy of Dr. D.P. Snustad.)

containing DNA, and endonuclease IV, encoded by the *denB* gene, attacks single-stranded cytosine-containing regions, including those opposite these nicks. (Endo IV is not required, however, for host DNA degradation, implying the potential involvement of some other still-unidentified host or phage endonuclease as well.) The products of genes 46 and 47 form an exonuclease that then degrades the fragmented DNA to mononucleotides, which are efficiently used for T4 DNA synthesis. However, this whole degradative process is relatively slow. By 8 minutes after infection, fragments with a molecular weight of 5×10^7 daltons predominate, in addition to some mononucleotides; the 20% still in acid-insoluble form has fallen to an average of 2×10^6 daltons by 25 minutes. But transcription of the genes in this DNA was already terminated quite early, due to the action of other T4 genes. The degradation of host DNA is not crucial to T4 infection; mutants lacking endo II have no phenotype unless the production of new nucleotides by ribonucleotide reductase is blocked, as by the addition of hydroxyurea. Mutations in genes 46 and 47 are very deleterious, but only because the gp46-47 exonuclease also participates in T4 recombination, which is essential to late initiation of DNA replication (see Section VIIIE).

The major regulatory process affecting transcription of host DNA involves a gene called *alc* or *unf*. Its two names reflect the somewhat circuitous way it was discovered as well as its dual functions. The substitution of hmdC for C in T4 DNA depends on one enzyme, dCTPase, which converts dCTP to dCMP, and a second enzyme, dCMP hydroxymethylase, which adds the hydroxymethyl group to dCMP. Kutter et al. (1975) showed that T4 DNA containing at least 95% cytosine (dC-DNA) is made by mutants that lack endonucleases II and IV and the dCTPase. However, such mutants make no late proteins and, hence, no phage. (The mutants are propagated by using *amber* dCTPase mutants, which grow perfectly well in Su+ hosts.) Phage are easily selected that have one additional mutation that bypasses the transcriptional

block of the T4 dC-DNA, so they make late proteins and therefore grow on the otherwise nonpermissive (Su−) host. All mutations thus selected map in a single gene named *alc*, for allows lates on C-DNA (Snyder et al., 1976)—or, better: attenuates elongation on dC-DNA. Alc is an 18 kb neutral protein.

Kutter, White, et al. (1994) followed the turnoff of specific host transcription by hybridizing RNAs labeled after infection to specific cloned DNAs, comparing *alc* and *alc+* phage. The *alc* mutants show a significant delay in shutoff of both host mRNA and rRNA synthesis (Fig. 15).

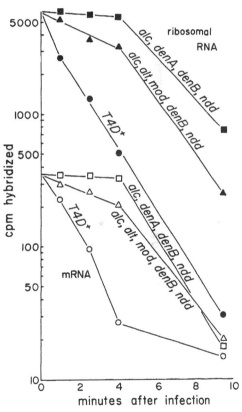

Fig. 15. Shutoff of host transcription in vivo by phage T4. RNA was labeled in 1 minute pulses at various times after infection with wild-type T4 or various mutants. The relative amounts of specific transcripts were measured by hybridization to nitrocellulose filters bearing certain DNAs (in this case, for ribosomal RNA and for one of the ribosome proteins). There is a clear delay in shutoff of host transcription with *alc* mutants.

A protein that inhibits transcription of dC-DNA might conceivably act on the DNA directly, on the RNA polymerase (RNAP), or on a complex of the two at cytosine-rich sequences. Snustad et al. (1986) and McKinney and Kutter (unpublished) have shown that gpalc can indeed bind to DNA, albeit weakly. To test the possibility of direct polymerase effects, Drivdahl and Kutter (1990) investigated the activity of RNAP from lysates infected with T4 or with various mutants. They found that transcription (of a T7 DNA template) is reduced in two stages (Fig. 16). An early sharp decrease appears to result from the action of the *alt* protein, a noncapsid protein found in the phage particle and injected with the DNA, which is responsible for *alteration* of the RNAP: the addition of an ADP-ribosyl unit to one of its β subunits. The slower second-stage decrease in RNAP activity is due to gpalc. The fact that partially purified RNAP from *alc*+ cells shows this change and that the difference disappears when more highly purified RNAP is used suggests that gpalc binds, although rather weakly, to the

RNAP. The functional nature of this interaction is supported by the observation of the 18-kDa gpalc on gels of partially purified polymerase from wild-type T4 but not from an *alc* missense mutant making a protein that is only slightly more basic than the wild-type gpalc. Drivdahl and Kutter (1990) have shown that the effect is at the level of *elongation* of the transcript, rather than at the level of promoter recognition or initiation of transcription. Gpalc has no detectable effect until the polymerase has lost the sigma factor, left the initiation site, and gone into "elongation" mode. Snyder and Jorissen (1988) provided genetic evidence that gpalc interacts with the polymerase: they isolated *E. coli* mutants reducing gpalc/*unf* action, which map in the gene for the β subunit of the RNA polymerase. The effect of *gpalc* is also reversible. Late-protein synthesis is detectable by 1 to 2 minutes after inactivation of a temperature-sensitive gpalc by shifting up to 41°C from 27°C, at which temperature no late-protein synthesis occurs with this mutant when the progeny DNA contains cytosine rather than hmdC.

The *alc* gene was also identified as being responsible for the reported *unfolding* of the host nucleoid. However, that unfolding appears to be an artifact of the high-salt isolation procedure standardly used, which dissociates many bound proteins from the DNA; stabilization of the nucleoid structure during such isolation seems to depend on entanglement in nascent RNA strands. The folded nucleoid actually involves about 50 separate supercoiled domains; a single nick in the DNA releases the supercoiling of a single domain and rifampicin, which blocks transcription, makes little difference in this domain structure (Sinden and Pettijohn, 1981). Using the low-salt method of isolating nucleoids developed by Kornberg and coworkers (1974), followed by sucrose density gradient centrifugation, we have evidence that the nucleoid is actually still largely in a folded state in vivo up to at least 7 minutes after infection with wild-type T4, by which time it is unfolding due to nuclease attack.

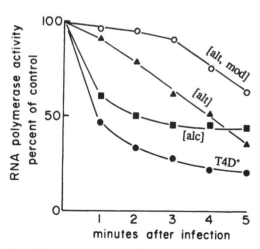

Fig. 16. Patterns of in vitro shutoff of host transcription. RNA polymerase was extracted from cells at various times after infection with wild-type T4 or various mutants. Polymerase activity was measured by following transcription of a standard T7 DNA template. The results indicate that the early sharp decline in activity is due to the Alt protein and the second slower decline is due to the Alc protein.

It appears that not only the ionic strength but also the specific nature of the ions involved may be very important experimental parameters. As reported by Leirmo et al. (1987), *E. coli* normally has almost no intracellular chloride ion. Substituting glutamate for chloride in vitro has a substantial effect on such properties as transcription. Leirmo's evidence indicates, however, that the primary effect is on *initiation* of transcription and may be related to the much higher degree of bound water around the glutamate. She found little difference in elongation rates under otherwise standard conditions, but there may well be effects on at least some DNA-binding proteins and factor-specific termination events.

C. Regulation of T4 Gene Expression: Transcriptional Controls

One of the most important uses for T4 has been in elucidating mechanisms of gene regulation. The phage expresses several identifiable classes of genes at different times, and since the overall patterns seem to be common to all viruses, one might expect studies of this relatively simple, easily controlled system to provide insights into mechanisms of viral gene regulation that might then be extended to complex developmental systems. The observed general pattern is presented above (Figs. 6a-c) and the timing details are summarized in Fig. 17.

The mechanisms that produce this pattern are complex. Some of them are transcriptional and some translational; we deal with the translational mechanisms in the next section. There seem to be two major types of transcriptional controls: changes in the RNA polymerase which direct it to different classes of promoters and changes in termination that extend the lengths of transcription units (see Brody et al., 1983; Rabussay, 1983). (A transcription unit is the space between a promoter and a termination site; it may contain one gene or several.)

One set of terms is used to describe the classes of genes and another to describe the classes of promoters; the two terminologies must be kept separate, since many genes are transcribed from more than one promoter. Immediately upon infection, a set of *immediate-early* (IE) genes is turned on. They are transcribed by unmodified host RNAP, and there are no controls that can stop their expression. IE transcripts are synthesized even in the presence of chloramphenicol, which inhibits protein synthesis.

A second set of genes called *delayed-early* (DE) is turned on approximately 2 minutes

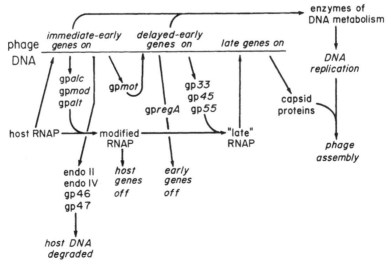

Fig. 17. General mechanisms that regulate transcription of phage T4.

postinfection; they are also transcribed by unmodified host RNAP, but their transcription is inhibited by chloramphenicol, indicating that some protein product of the IE genes is necessary for DE expression. DE expression can be initiated at two different types of promoters (Fig. 18): at new *middle promoters* or at *early promoters*, the same ones used for IE transcription. The latter form of DE expression entails the elimination of termination points, so that the DE genes are transcribed as the more promoter-distal portions of early transcripts. Interestingly both methods of accessing middle-mode transcription are affected by the DNA-binding protein gp*motA*, which appears to play an enhancing role in antitermination as well as strongly stimulating recognition of middle-mode promoters (see Brody et al., 1983; Guild et al., 1988).

Finally, there are the late genes, whose transcription is always initiated at *late promoters* (see Geiduschek et al., 1983; Christensen and Young, 1983). This phase of transcription requires new RNAP-binding proteins.

IE transcription thus begins at early promoters, but is terminated at rho-dependent sites downstream. The initial transcripts are relatively short; this is shown by separation of labeled messengers by molecular weight on gels and identification of their informational contents by either hybridizing them to cloned T4 genes or by in vitro translation and identification of the protein products by gel electrophoresis (Christensen and Young, 1983). Transcripts for several different genes occur in multiple forms. For instance, gene 39 is initially transcribed onto a short messenger that apparently contains no other gene copies, but later in infection gene 39 also appears on a much longer transcript containing gene 60 and the *rIIA* and *rIIB* genes. The *rII* genes are also transcribed by themselves on a DE messenger. This pattern of transcription indicates that the gene 39 message, which terminates just downstream of that gene in IE transcription, is later extended into the *rII* genes and that there is also a middle promoter upstream of the *rII* genes from which a separate message is transcribed (Fig. 18).

Transcription of T4 late genes requires a new sigma factor, gp55 (Malik and Goldfarb, 1984), and the products of at least two other genes, 33 and 45. This sigma factor

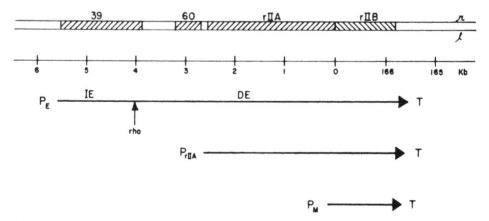

Fig. 18. Patterns of early transcription illustrated with the gene 39-*rII* region. Transcription begins immediately from a PE promoter upstream from gene 39; these transcripts are terminated at a rho-dependent site at the end of this gene. A second promoter lies at the beginning of the *rIIA* gene. Delayed-early transcription then occurs in two ways. First, the rho site is bypassed so that transcription beginning at PE continues downstream through *rIIB*. Second, a new middle promoter (PM) becomes available, under the influence of gp*mot*, so that a transcript for *rIIB* alone is made. There is apparently a single termination site downstream of *rIIB*. (Reproduced from Brody et al., 1983, with permission of the publisher.)

uses a different promoter, with the consensus sequence TATAAATACTATT spanning the position of the *E. coli* −10 consensus sequence but showing no consensus sequence in the −35 region (see Christensen and Young, 1983). This fits with the observation that the polymerase coupled to sigma-T4 (gp55) only spans nucleotides −30 to +20, not the −50 to +20 span observed with the bacterial sigma-70 (Malik and Goldfarb, 1984). Late transcription normally also requires DNA replication, with the transcription apparatus functioning as a "moving enhancer"; however, this requirement can be bypassed if the phages are mutated in DNA ligase, DNA polymerase, and the gp46 exonuclease (see Williams et al., 1994).

D. Translational Controls and Autoregulation

At least three T4 proteins act as specific repressors of translation; gp32 and gp43 control translation of their own mRNAs, while gp*regA* exerts translational control over the expression of 10 to 15 separate mRNAs, including its own (see Wiberg and Karam, 1983). The three proteins clearly recognize different classes of targets, but all see determinants in the ribosome-initiation domains of their substrates. For example, Andrake et al. (1988) have described footprinting of the DNA polymerase and polymerase transcript, showing that the enzyme specifically binds to, and can protect, a sequence of about 35 nucleotides, including the ribosome-binding site and a stem-loop structure that may be involved in recognition; the protected sequence ends just before the initiating AUG.

The mRNA for the single-strand binding protein gp32 has a 40-bp stretch just before the initiating AUG, which forms no secondary structure but clearly contributes to the ability of the protein to inhibit its own translation. However, the major responsible element is a stretch some 40 bases away that can form a "pseudoknot." This involves a stem-loop with seven perfectly matching base pairs and a sequence of four bases in the loop that are complementary to four bases

shortly before the stem, perfectly spaced to pair and form the pseudoknot. The gp32 protein nucleates on the pseudoknot and thence cooperatively binds across the structureless region to block gene-32 translation. Putting this structure in front of other genes similarly allows gp32 to block their translation.

The mechanism of *regA* repression is much less well understood. Mutations blocking the responsiveness of *rIIB* to *regA* have been studied extensively (see Karam et al., 1981) and all map in the same general vicinity, on both sides of the AUG. Certain sequence elements seem to be prevalent (with varying spacings) in *regA*-responsive genes, but there is still no clear picture of the precise mechanism of action of this small 12-kDa protein in inhibiting transcription of so many early genes late in infection.

E. DNA Replication and the Nucleotide Precursor Complex

Since T4 uses 5-hydroxymethylcytosine rather than cytosine in its DNA, the substrate for its DNA polymerase, in addition to dATP, dTTP, and dGTP, is normally hmdCTP: deoxy-5-hydroxymethylcytosine triphosphate. The normal dCTP has to be destroyed to keep it out of the way, since the phage DNA polymerase uses dCTP and hmdCTP indiscriminately. The whole pathway of aerobic nucleotide biosynthesis is as follows:

$$\text{ADP} \xrightarrow{1} \text{dADP} \xrightarrow{6} \text{dATP}$$

$$\text{CDP} \xrightarrow{1} \text{dCDP} \xrightarrow{2} \text{dCMP} \xrightarrow{3} \text{hmdCMP} \xrightarrow{5} \text{hmdCDP} \xrightarrow{6} \text{hmdCTP}$$

$$\text{GDP} \xrightarrow{1} \text{dGDP} \xrightarrow{6} \text{dGTP}$$

$$\text{UDP} \xrightarrow{1} \text{dUDP} \xrightarrow{2} \text{dUMP} \xrightarrow{4} \text{dTMP} \xrightarrow{3} \text{dTDP} \xrightarrow{6} \text{dTTP}$$

T4 makes a ribonucleotide reductase (1) that parallels the function of the *E. coli* enzyme. It also encodes a new enzyme, a dCTPase-dCDPase (2), which removes dCTP as a substrate for DNA synthesis and at the same time provides dCMP as the substrate for another new phage-directed enzyme, dCMP hydroxymethylase (HMase) (3); this produces deoxy-5-hydroxymethylcytosine monophosphate, in parallel with the way (4) that dTMP is made from dUMP. (It was the identification of this unprecedented enzyme that established the fact that viruses do encode at least some of their own proteins.) A new T4 dNMP kinase (5) then phosphorylates hmdCMP to hmdCDP (along with dTMP and, incidentally, dGMP), while an abundant and active host kinase (6) can phosphorylate all of these diphosphates (as well as ribose diphosphates) to the triphosphate level, taking the phosphate from ATP. Thus the pathway for making the odd base, 5-hydroxymethylcytosine, parallels the pathway (4) for making thymidine, which is 5-methyluracil.

Genetic and biochemical evidence indicates that these enzymes are all organized into a nucleotide precursor complex (Fig. 19), which is normally coupled to the multiprotein DNA polymerase complex, so one

funnels nucleotides into the other as they are needed (Mathews and Allen, 1983; Greenberg et al., 1994). This is the only system to date where such a complex has been shown to operate. It is also the system where DNA replication in general is best understood, due to the availability of mutants and the ability to assemble the system in a test tube. As discussed by Nossal (1994) and by Selick et al. (1987), the marriage of genetics and biochemistry has been a particularly fruitful one in studying T4 DNA replication.

DNA replication itself entails at least eight proteins. The current model for the replisome, or replicating "machine," they form is shown in Figure 20. Its main features are the following:

1. Two DNA polymerase molecules work simultaneously, one synthesizing the leading strand and one the lagging strand, clamped to the DNA by a complex of gp44/62 and gp45 to give a highly processive enzyme (i.e., one that repeats its action while remaining attached to its substrate).

2. The DNA helix is rapidly unwound in front of the polymerase making the leading strand by the combined actions of the helix-destabilizing protein gp32, the polymerase itself, and gp41, a DNA helicase that uses GTP hydrolysis to force open the template

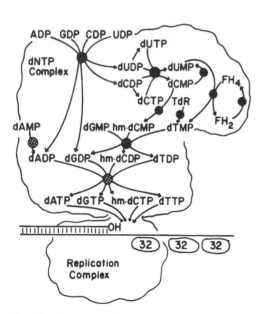

Fig. 19. Structure of the nucleotide precursor complex formed by T4-encoded enzymes. It is shown connected to the T4 replication complex, so that one feeds nucleotides directly to the other. (Reproduced from Matthews and Allen, 1983, with permission of the publisher.)

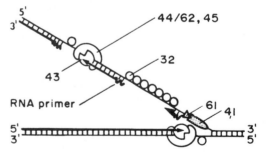

Fig. 20. Structure of the DNA replication complex of phage T4. The polymerase itself is gp43; gp44, 62, and 45 are accessory proteins. Proteins gp41 and 61 form a primase complex that unwinds the helix, and gp32 is a helix-stabilizing protein that binds to single-stranded DNA. (Reproduced from Nossal and Alberts, 1983, with permission of the publisher.)

helix and remains tightly bound to the template for the lagging strand.

3. New RNA-primed Okazaki fragments are generated about every 4 seconds (1500 bp) on the gp32-coated template of the lagging strand; each fragment starts with a pentanucleotide primer (pppApCpNpNpN) synthesized by a primase-helicase composed of gp41 and gp61 (giving a second key role for gp41, which may be involved in keeping leading-and lagging-strand synthesis synchronized).

4. The DNA polymerase synthesizing the lagging strand, like that synthesizing the leading strand, remains with its replication fork for a prolonged time, so the lagging-strand template must be folded to bring the 5'-hydroxyl end of a completed Okazaki fragment adjacent to the start site for the next Okazaki fragment, as in Figure 20. This allows the same polymerase to move processively to the next Okazaki fragment. (This coupling generates few Okazaki fragments shorter than 500 nucleotides, even though potential primer start sites occur quite frequently.)

5. The T4 replication apparatus lays down a series of Okazaki fragments still containing their intact RNA primers on the 5' end. In vitro, these primers are removed by a T4-encoded RNase H and replaced with DNA, and the fragments are joined by DNA ligase. It is not clear whether RNase H has the same role in vivo.

6. One additional protein, the product of gene dda (DNA-dependent ATPase), has been identified as a DNA helicase that facilitates movement of the replicating complex past DNA-bound RNA polymerase or regulatory proteins (Bedinger et al., 1983). The loss of this protein is not lethal, so there is presumably a second phage or host protein that can substitute for gpdda.

Surprisingly, the T4 replication complex can easily pass an RNA-polymerase complex transcribing in the same or opposite direction without disrupting either process, as long as the replication complex contains gp41 (Liu and Alberts 1995).

The normal initiation of replication has not yet been reconstructed in vitro. The host RNA polymerase and several topoisomerase components seem to be involved in the usual initial process, while later initiation seems to happen mainly at recombination sites (once the T4 transcriptional program has altered the RNA polymerase) and thus requires the gp46/47 exonuclease (see Mosig, 1987; Mosig and Eiserling, 1988). An additional mode of initiation, independent of gp46 and gp47 but also rifampicin resistant, has also been reported by Kreuzer and Alberts (1985).

F. Introns in T4 Genes and Novel Homing Endonucleases

A large fraction of eucaryotic genes are now known to be fragmented by the insertion of one or more nontranslated *intervening sequences*, or *introns*, within their coding sequences, which must then be excised from the primary transcripts. However, such complexities were considered a purely eukaryotic phenomenon until the report by Chu et al. (1984) of a 1-kb intron within the thymidylate synthase (*td*) gene of bacteriophage T4. Two additional T4 genes have since been shown to also contain introns: the nucleotide reductase gene *nrdB* and a gene initially termed *sunY*. *SunY* is now known to encode an *anaerobic* ribonucleotide reductase that functions at the nucleotide triphosphate level and is the T4 gene most closely related to a gene of its host.

Additional research in the laboratories of Belfort, Shub, and Chu has shown that the T4 introns are self-splicing and can assume a secondary structure virtually identical to that of the eukaryotic type-I self-splicing introns (Fig. 21) (Shub et al., 1994). The splicing, in all cases, occurs via the same mechanism, involving a series of transesterifications, or phosphodiester bond transfers, with the RNA functioning as an "enzyme" (see Cech, 1986):

1. Nucleophilic attack by a sugar hydroxyl of a guanosine cofactor at the 5' splice

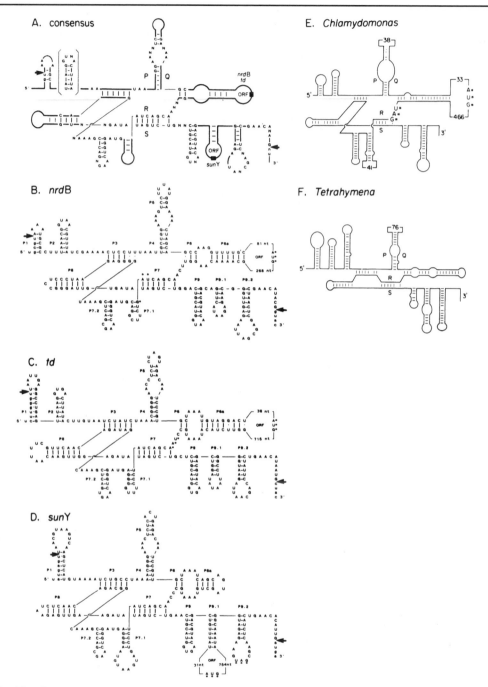

Fig. 21. Structures of the T4 introns. A consensus sequence of these introns is shown in A, along with the predicted structures of the three introns by themselves. The structures of *Chlamydomonas* and *Tetrahymena* class I self-splicing introns are shown for comparison. ORFs are the open reading frames included in the T4 introns. (Courtesy of Dr. D. Shub.)

site cleaves the chain just 3' of a particular uracil residue and adds the G to the new free 5' end of the chain. This ability to label the intron RNA specifically, if transiently, with free labeled GTP, in the absence of protein, was used in combination with Southern blotting to locate new T4 introns after the initial one had been observed and to find introns in a number of other phages of both gram-negative and gram-positive bacteria.

2. The new free hydroxyl end of the upstream exon attacks a bond at the 3' splice site, forming the mature transcript and releasing the intron in linear form.

3. The intron cyclizes via intramolecular nucleophilic attack.

4. The intron is slightly shortened through a cycle of repeated hydrolysis and cyclization.

It is not yet clear whether the structural homology between the T4 and eukaryotic introns reflects an ancient evolutionary origin of such splicing or a later transfer of introns from eukaryotes to T4; in either case, many interesting questions are opened up.

Phage T4 is a system in which the role of various intron and exon sequences in the self-splicing reaction can be studied easily because it is so easy to do with this phage, especially since there are already many useful mutants and because additional mutations, in both directions, can be selected in the *td* gene. Mutants inactivating the gene cloned on a plasmid can be selected in an *E. coli thyA* host by looking for the development of resistance to trimethoprim, which is converted to the active form by thymidylate synthase; thymine must be supplied and ampicillin is used to maintain the plasmid. In the other direction, mutations to an active gene can be recognized by growth of the *thyA* host strain in the absence of thymine. Studies of these mutants have shown the following:

1. Fewer than 166 nucleotides at the 5' end and 226 nucleotides at the 3' end of the intron constitute the catalytic core of the intron.

2. Some of the observed mutations are in phylogenetically conserved elements already shown to be functionally crucial in *Tetrahymena* and fungal mitochondria (P1, P4, and P7), while others imply important roles for structural elements P5, P6, P7.2, P8, P9, and P9.2 (see Fig. 22).

3. No mutations affecting splicing were ever found in other T4 or host genes or in the parts of the introns that protrude from the core, and pseudoreversions of mutations always involved compensatory changes that restored base pairing.

Two other observations raise further interesting questions about the evolutionary origins of the T4 introns and their meaning. First, the various T-even phages have different intron patterns, with the two parts of the gene being contiguous in some of them, and the two major T2 strains even show differences. Second, the *B. subtills* phage SPO1 also contains an intron, in its DNA polymerase gene (Shub et al., 1988a); this intron has many homologies with the others but looks more like chloroplast introns in the positioning of two of its loops. Two of three closely related phages, originally isolated from different parts of the world, have similar introns (Shub, personal communication). These observations encourage the speculation that there may be some relationship between introns and transposable elements, even though no introns have been detected in either of the hosts or their other phages, using the same techniques of in vitro guanosine labeling followed by blotting.

The three known T4 introns are also unusual in that each one contains an open reading frame—that is, a gene that is apparently informationally and functionally unrelated to the gene formed from the two exons spanning the intron. These genes-within-introns are translated only late in infection, under late regulation, even though the intron-containing genes themselves are controlled like typical early genes. Two of the three encode proteins that are quite large (>30 kDa), very basic, and are related to the intron endonucleases in the mitochondria of filamentous fungi. For both *td* and

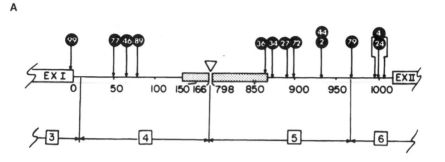

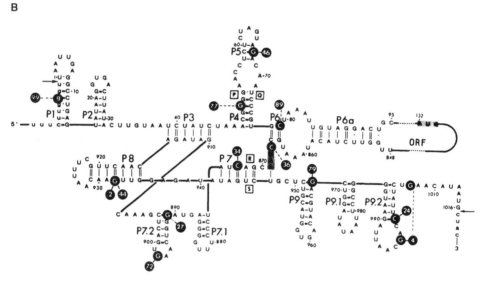

Fig. 22. Structure of one T4 intron, showing the positions of mutations affecting the self-splicing. (Courtesy of Dr. M. Belfort.)

nrdD (but not *nrdB*) these intron-encoded proteins are specifically responsible for intron insertion at their respective cognate sites in phage genomes that do not already contain introns (Quirk et al., 1989), further emphasizing a similarity with transposable elements. All together, the T4 genome contains 13 members of this class of genes, representing all 3 of the families seen in filamentous fungi, all of them located between genes or in introns; a number of them have been shown to indeed be homing endonucleases.

The pattern of late translation of genes transcribed in the early direction is not unique to these homing endonucleases, but has also been observed for at least three other late genes located in early regions. These include gene *e* (lysozyme) (McPheeters et al., 1986); the gene for a small outer-capsid protein, *soc* (Macdonald et al., 1984); and gene 49, which encodes the DNA packaging nuclease (Barth et al., 1988). It appears that, in the early transcript of the latter, the ribosome-binding site is sequestered in a stem-loop structure that blocks translation. A late promoter, however, is located within the stem; in transcripts initiated from this promoter, the ribosome-binding site is accessible and the proteins can be made under all of the usual late-protein controls. This is an interesting, novel form of gene regulation that seems to get around potential problems of proteins being made prematurely from read-through of transcripts from neighboring genes. This model is discussed in some detail by McPheeters et al. (1986).

G. A Novel Form of Gene Splicing

Phage T4 seems to be a never-ending source of novelties for molecular biologists. One of the most interesting of them was described by Huang et al. (1988), who found a mechanism in which a segment of the information in one gene is skipped over during translation. Gene 60 encodes an 18-kDa subunit of the DNA topoisomerase, which is involved in phage DNA replication. While cloning the genes for topoisomerase proteins, Huang and her colleagues discovered that the gene is split, with a sequence of 50 untranslated nucleotides in the middle. However, there is strong evidence that this sequence is not removed as an intron; when the mRNA for the gene is used as a template for reverse transcriptase, the resulting DNA is identical in sequence with the original gene, showing that the untranslated sequence is still in the messenger.

A reasonable model can be drawn in which the interruption, which is bracketed by a direct repeat of five nucleotide pairs, is pushed out in a kind of hairpin loop so that the codons on either side of it are brought together. Though the structure at this point is unusual, a ribosome can presumably move right through it, translating the messenger properly while ignoring all the nucleotides in the loop.

These investigators transferred segments of gene 60, with and without the interruption, into the amino-terminal coding sequence of the β-galactosidase gene of *E. coli*, where it also is bypassed without being either excised or translated. These fused genes show comparable levels of enzyme activity, indicating that the looped-out sequence has little effect on translation of the messenger. This interruption has been found

in the T4 gene 60 and that of six other T-even phages examined by Repoila et al. (1994). There is nothing like it in the comparable gene of phages T2, T6, and most other T-even phages, where it is actually fused with the gene for another topoisomerase subunit, coded in T4 by gp39.

IX. THE T-ODD COLIPHAGES

A. Bacteriophages T7 and T3

Bacteriophage T7 has also played important roles in the development of molecular biology. It was the first of the larger phages to be fully sequenced—by I. J. Dunn and F. W. Studier in 1983—and the functions of most of its genes were soon identified. It has a stubby, external tail (about 10×20 nm) and an intraviral portion that expands after attachment to the target cell, forming a complex organ for DNA transfer into the cell. DNA transfer always begins from the left end on the standard genomic map, aided by host-polymerase transcription of the first 19% of the genome from three strong promoters located within the first 750 bp. The genes in this "first-step-transfer" portion encode several key proteins: (1) inactivators of the host restriction enzyme and of its dGTP triphosphohydrolase, a protein kinase that also shuts off host transcription by a phosphorylation-independent mechanism; (2) a DNA ligase; and (3) a new single-subunit RNA polymerase. This RNA polymerase then transcribes—and thus helps to draw in—the remainder of the genome; it first transcribes a cluster of genes involved primarily in DNA metabolism and then, from stronger promoters, the genes responsible for the phage capsid. T7 has ten promoters for the middle genes, five for the late genes, and one to initiate replication.

The phage-encoded RNA polymerase is particularly interesting. It has significant homology with the *Saccharomyces cerevisiae* mitochondrial RNA polymerase and transcribes 10 times as fast as the host polymerase. The T7 polymerase recognizes promoters with a highly conserved sequence between bases -17 and $+6$ relative to the transcription start site. There is little recognition for noncognate promoters between the polymerases from T7, T3, and related phage of other bacteria, but changes in a single amino acid can interconvert the T3 and T7 specificities. Taking advantage of its speed and specificity, this enzyme has been used to create tightly controlled, high-level expression vectors that so overproduce an encoded protein that it forms as much as half of the total cell protein, a tremendous bonus for gene engineers. The T7 promoter sequences are rare enough that these vectors can be engineered to work even in eucaryotic cells. The crystal structure of the T7 polymerase has recently been determined to high resolution and is providing much insight into the general mechanisms of transcription (Cheetham and Steitz, 2000).

T7 DNA replicates as a linear molecule and then forms concatamers using unreplicated terminal repeats of 160 bp, which are later duplicated during the packaging process. Growth of T7 and many of its relatives (but not T3) is inhibited by F plasmids. This inhibition involves specific interactions with the F-factor *pif* gene and causes inhibition of membrane functions and of all macromolecular synthesis. Several other prophages and resident plasmids can inhibit infection by T7.

B. Bacteriophage T5

The DNA of T5 is about 121 kbp long with 10-kbp terminal repeats, unique ends, and four nicks in one strand at specific sites. The DNA enters the cell in a two-step process. The left terminal repeat enters first, and the "pre-early proteins" that it encodes completely to shut off host replication, transcription, and translation, block host restriction systems, and degrade the host DNA to free bases and deoxyribonucleosides that are ejected from the cell. This first-step-transfer DNA segment also contains genes needed for the rapid entry of the rest of the genome once the initial takeover of the cell is complete; some of these genes shut off the pre-

early genes and then program orderly expression of early and then late genes from the rest of the genome. T5 encodes various enzymes of nucleotide metabolism and DNA synthesis, and modifiers of RNA polymerase.

The genome appears to replicate in a rolling-circle mode; circular DNA molecules with a single copy of the terminal repeat are found inside the cell. Precut genomes containing both terminal repeats are inserted into the preformed heads. About 25 kbp of the genome, in three large blocks, is in principle deletable; these regions includes genes for tRNAs for all 20 amino acids as well as a number of ORFs. However, if more than 13.3 kb is deleted, the DNA cannot be packaged without a compensating insertion.

C. Bacteriophage T1

Phage T1 multiplies rapidly, with a latent period of only about 13 minutes and a burst size of about 100. T1 uses proteins of the iron-transport pathway to enter the cell, first binding reversibly to membrane protein TonA, then irreversibly to TonB. It is the only one of the T-phages requiring an energized cell membrane for irreversible binding, and its DNA entry is effected by a proton symport system involving TonB. The infecting phage creates a transient fall in protonmotive force, ATP, and GTP; this inhibits the initiation of translation of host proteins while allowing phage transcription and translation to proceed. However, host *transcription* continues until the host DNA is degraded, a process that is tightly coupled to phage DNA synthesis (though not required to produce phage DNA). Phage DNA synthesis requires phage proteins, but elongation is carried out by the host Pol III α subunit, a mode of replication that is common for temperate but not for lytic phages. The DNA is packaged from concatamers by a mechanism that measures out headfuls. Early in the cycle (or in the absence of host DNA degradation), it can package host DNA to produce generalized transducing phages. However, transduction can be observed only with double *amber*-mutant strains plated on nonpermissive recipients, since otherwise the transductants are all killed by the large excess of viable virulent phage.

X. *BACILLUS SUBTILIS* PHAGES

Every type of bacterium probably lives amid many types of phage that could infect it. A survey of various phages (some discussed in other chapters) is as fascinating as a trip to the zoo. Among the relatively large virulent phages of bacteria other than *E. coli*, those that infect *B. subtilis* are the best known, and some of them have interesting and unusual properties.

Some *B. subtilis* phages are known to have unusual bases in their DNA, comparable to the T-even situation. For instance, the phages SPO1, SP82G, and SP8 all have genomes comparable in size to those of the T-even coliphages but containing hydroxymethyluracil (hmU). These phages encode enzymes that set up metabolic pathways like those of T4 (Fig. 23): a dTTPase-dUTPase that converts dTTP to dTMP and also hydrolyzes dUTP back to dUMP, and a dUMP hydroxymethylase that adds the hydroxymethyl group to form hmdUMP. Also the normal conversion of dUMP to dTMP, by thymidylate synthase, is blocked. However, the dTTPase-dUTPase is not essential, as it is in T4, and in the absence of this enzyme viable phages are made that contain up to 20% thymine instead of hmU (Marcus and Newton, 1971).

As mentioned above, SPO1 has recently been found to have another feature in common with T4 in possessing at least one intron, located in its DNA polymerase gene.

There are also *B. subtilis* phages, such as PBS1, whose DNA contains uracil instead of thymine. (PBS1 is a generalized transducing phage that can take large chunks of DNA and is excellent for mapping purposes.) The new metabolic pathways set up during infections by these phage are summarized in Fig. 23. One enzyme converts dUMP to dUDP, and another removes dTMP by converting it to thymidine. It is also necessary for the phage to make inhibitors of enzyme systems that usually prevent the synthesis of uracil-containing DNA, including a nuclease that releases deoxyuridine from DNA and an *N*-glycosidase that normally acts as a DNA repair enzyme scouting for cytosine deamination and releases free uracil. These phages are particularly interesting because there is a nearly universal synthesis, in all organisms and almost all other viruses, of DNA containing thymine and of RNA containing uracil. This use of thymine in DNA seems

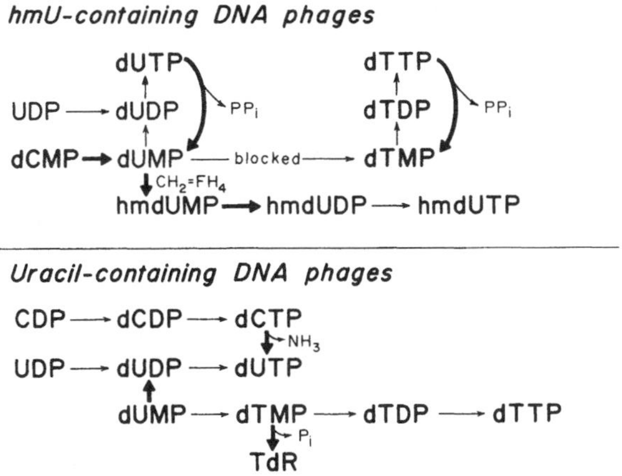

hmU-containing DNA phages

Uracil-containing DNA phages

Fig. 23. Pathways of nucleoticle metabolism in *Bacillus subtilis* infected with phages containing either hmU or U. Heavy arrows show new phage-encoded enzymes.

important to avoid the accumulation of mutations due to the rather high spontaneous level of deamination of cytosine to uracil, and so it will be interesting to see what special problems face a phage that violates the rule.

Phage φ29 is another virus with an icosahedral head and a tail, like the coliphages discussed above, but its tail is comparatively short relative to the head and a complex structure of spikes is associated with the collar region where the head and tail join (Anderson et al., 1966). It is much smaller than T4, having a double-stranded DNA genome of only 19,285 bases; the sequence is known completely (Yoshikawa and Ito, 1982; Garvey et al., 1985). The sequence is identical in all phage particles, with no circular permutation. Twenty-three distinct proteins encoded by the phage have been identified.

φ29 is also one of the viruses that has little effect on the normal synthesis of host macromolecules; it lacks any of the T4-like mechanisms for shutting off its host's activities and simply begins to synthesize its own molecules in competition with them. The early genes of φ29 fall into two clusters, located at the ends of the genome, with the late genes in the middle. There are eight promoters for early genes and only one for the late genes. As infection is initiated, the early genes are transcribed by unmodified RNA polymerase, from the light strand of DNA. The product of one of the early genes, gene 4, then promotes transcription of the late genes from the heavy strand, although it is not yet known whether this protein acts as a sigma-like factor or some other kind of activator (Salas, 1988). However, there is apparently no shutoff of early transcription once late transcription has begun.

The most unusual feature of φ29, and of several related phages, is a protein (the product of its gene 3) that is covalently linked to the 5' ends of its DNA. This is a phosphodiester linkage between the hydroxyl group of one serine residue and the 5'-dAMP found on both ends of the DNA (Hermoso and Salas, 1980). This protein is required for DNA replication and enters into a novel mechanism for initiation of synthesis, in which the primer is a hydroxyl group of one of the residues in the protein, rather than the 3'-OH group of a nucleotide. The products of four phage genes are actually required for replication: two of them (2 and 3) for initiation and two others (5 and 6) for elongation, and the product of a fifth gene (17) is involved but not required. All this suggests a multiprotein replication "machine" comparable to the one described above for T4. The bacterial DNA polymerases I and III are not involved. DeVega et al. (2000) provide an excellent discussion of this unique mode of replication.

XI. CONCLUSIONS AND FUTURE DIRECTIONS

Bacteriophages, especially the large, complex phages discussed here, have long been a major focus of molecular biology. An amazing amount of basic biological information has been uncovered with the T-even coliphages alone. Although much of the excitement of molecular biology has now shifted to eukaryotic systems, many investigators continue to work with phage and continue to astonish their colleagues with discoveries of previously undreamt-of mechanisms and processes. Furthermore phage systems, which are generally easy to handle and control, involving inexpensive materials and short time scales, remain excellent material for working out the details of many kinds of complex mechanisms. The same ease of control makes them excellent training grounds for young investigators, who can make good progress without having to fight the degree of technical problems associated with most eukaryotic systems. One major advantage is the degree of genetic understanding of the phage-host system and the ability to combine genetic, physical, and biochemical tools in attacking a problem. The discussions of each of the different phages in the excellent reference work Encyclopedia of

Virology, edited by Webster and Grannoff (1994), provides a good starting point for going more deeply into any of them.

The T4 genome is now sequenced, but we still do not know the functions of almost half of the nearly 300 genes that has revealed. Very few T4 genes other than those involved in nucleotide metabolism show any significant homologies to genes from any of the sequenced organisms. The similarities that are there emphasize the ancient nature of the phages and the virtual absence of exchange of genetic information between these large, lytic phages and their hosts. The evidence is strong that the T4 thymidylate synthase diverged before the separation of the bacterial and eukaryotic thymidylate synthases; here and in several other enzymes, T4 has many residues unique to eukaryotes but well conserved there interspersed between residues that are unique and conserved for bacteria. The enzyme most similar to that of *E. coli* is the anaerobic ribonucleotide triphosphate reductase, but even that clearly diverged well before *E. coli* and *Haemophilis influenzae*. (Both the genome analysis and the discussion of evolutionary relationships are still being written up.)

This line of work reemphasizes an important point about biological research: that simply knowing the structure of a DNA molecule is not enough, because the sequence of nucleotides tells little about the function of that sequence, even though it may yield important clues. Biology is something more than chemistry. A gene is not merely a segment of a DNA molecule; it is a meaningful segment, which must be expressed and regulated, often through complex mechanisms, and there is no way to know those mechanisms a priori just by doing chemical experiments. This has again been reemphasized with the discovery of the folded-out intron in gene 60. Molecular biology has been fruitful primarily because it combines chemical work with biological—especially genetic—studies. And much of its fascination lies in its promise of another surprise after every experiment. The 1994 ASM book, *The Molecular Biology of Bacteriophage T4*, suggests many directions in which further research is warranted, and also has a number of chapters detailing various experimental techniques for working with phages. Amazingly, for example, almost no work has been carried out exploring coliphage infection in conditions that reasonably approximate those in the real world, such as when the bacteria are in stationary phase or growing very slowly or during anaerobic growth (as would be seen in the mammalian colon). What little has been published in that field is discussed by Kutter, Kellenberger, et al (1994). It seems likely that a number of the still-uncharacterized T4 genes will play significant roles here, with many others being involved in redundant ways in the shift from host to phage metabolism.

Another area that is attracting a great deal of interest recently is the use of phages as antibiotics, to deal with the growing problem of bacteria that are resistant to all available antibiotics. While phages have been little used in the West since the advent of sulfa drugs and penicillin, they were very widely used in the Soviet Union, with the Bacteriophage Institute in Tbilisi, Georgia, leading the research and implementation. Interestingly, virtually all of the therapeutic cocktails used against gram-negative bacteria contain T-even phages, and many in the collection of over 100 were isolated in therapeutic contexts, including T2 and, possibly, T4. In contrast, it is important to avoid temperate phages like lambda for therapeutic purposes. This is true both because lambdoid lysogens become resistant to infection by all related phages and because of the possibility of recombining with resident prophages and carrying around information such as that in pathogenicity islands, which are related to lambdoid prophages in several different pathogenic bacteria. Sulakvelidze et al. (2001) have written an excellent review of phage therapy, and a good deal of historical and current information on the topic is available on our Web site: *www.evergreen.edu/bacteriophage*.

REFERENCES

Alberts BM (1984): The DNA enzymology of protein machines. Cold Spring Harbor Symp Quant Biol 49:1–12.

Anderson DL, Hickman DD, Reilly BE (1966): Structure of *Bacillus subtilis* bacteriophage φ29 and the length of φ29 deoxyribonucleic acid. J Bacteriol 91:2081–2089.

Anderson TF (1952): Stereoscopic studies of cells and viruses in the electron microscope. Am Nat 86:91–100,

Anderson TF, Delbrück M, Demerec M (1945): Types of morphology found in bacterial viruses. J Appl Physiol 16:264.

Andrake M, Guild N, Hsu T, Gold L, Tuerk C, Karam J (1988): DNA polymerase of bacteriophage T4 is an autogenous translational repressor. Proc Natl Acad Sci USA 85:7942–7946.

Barth KA, Powell D, Trupin M, Mosig G (1988): Regulation of two nested proteins from gene 49 (recombination endonuclease VII) and of a λ RexA-like protein of bacteriophage T4. Genetics 120:329–343.

Bedinger P, Hochstrasser M, Jongeneel CY, Alberts BM (1983): Properties of the T4 bacteriophage DNA replication apparatus: The T4 *dds* DNA helicase is required to pass a bound RNA polymerase molecule. Cell 34:115.

Benzer S (1955): Fine structure of a genetic region in bacteriophage. Proc Natl Acad Sci USA 41:344–354.

Berns KI, Thomas CA Jr (1961): A study of single polynucleotide chains derived from T2 and T4 bacteriophage. J Mol Biol 3:289–300.

Bishop RJ, Conley MF, Wood WB (1974): Assembly and attachment of bacteriophage T4 tail fibers. J Supramol Struct 2:196–201.

Brenner S, Streisinger G, Horne RW, Champe SP, Barnett L, Benzer S, Rees MW (1959): Structural components of bacteriophage. J Mol Biol 1:281–292.

Brody E, Rabussay D, Hall DH (1983): Regulation of transcription of prereplicative genes. In Mathews CK, Kutter E, Mosig G, Berget PB (eds): "Bacteriophage T4." Washington, DC: ASM Press, pp 174–183.

Cairns J, Stent GS, Watson JD (eds) (1966): "Phage and the Origins of Molecular Biology." Cold Spring Harbor, NY: Cold Spring Harbor Laboratory.

Caspar DLD, Klug A (1962): Physical principles in the construction of regular viruses. Cold Spring Harbor Symp Quant Biol 27:1–24.

Cech T (1986): The generality of self-splicing RNA: Relationship to nuclear messenger RNA splicing. Cell 44:207–210.

Cheetham GM, Steitz TA (2000): Insights into transcription structure and function of single subunit DNA-dependent RNA polymerases. Curr Opin Struct Biol 10:117–123.

Christensen AC, Young ET (1983): Characterization of T4 transcripts. In Mathews CK, Kutter E, Mosig G, Berget FB (eds): "Bacteriophage T4." Washington, DC: ASM Press, pp 184–188.

Chu FK, Maley GF, Maley F, Belfort M (1984): Intervening sequence in the thymidylate synthase gene of bacteriophage T4. Proc Natl Acad Sci USA Biol P1(10):3049–3053.

Crick FHC, Watson JD (1957): The structure of small viruses. Nature 177:473–475.

Demerec M, Fano U (1945): Bacteriophage-resistant mutants in *Escherichia coli*. Genetics 30:119–136.

deVega M., Lazaro J-M, Salas M (2000) Phage φ29 DNA polymerase residues involved in the proper stabilisation of the primer-terminus at the 3′5′ exonuclease active site. J Mol Biol 304:1–9.

Doermann AH (1952): The intracellular growth of bacteriophages. 1. Liberation of intracellular bacteriophage T4 by premature lysis with another phage or with cyanide. J Gen Physiol 35:645–656.

Doermann AH, Boehner L (1963): An experimental analysis of bacteriophage T4 heterozygotes. 1. Mottled plaques from crosses involving six *rII* loci. Virology 21:551–567.

Drivdahl RH, Kutter EM (1990): Inhibition of transcription of cytosine-containing DNA in vitro by the *alc* gene product of bacteriophage T4. J Bacteriol 172:2716–2727.

Edgar KS, Wood WB (1966): Morphogenesis of bacteriophage T4 in extracts of mutant-infected cells. Proc Natl Acad Sci USA 55:498–505.

Eiserling FA (1983): Structure of the T4 virion. In Mathews CK, Kutter E, Mosig G, Berget FB (eds): "Bacteriophage T4." Washington, DC: ASM Press, pp 11–24.

Ellis EL, Delbrück M (1939): The growth of bacteriophage. J Gen Physiol 22:365–384.

Epstein RH, Bolle A, Steinberg C, Kellenberger E, Boy de la Tour E, Chevalley R, Edgar R, Susman M, Denhardt C, Lielausis I (1963): Physiological studies of conditional lethal mutants of bacteriophage T4D. Cold Spring Harbor Symp Quant Biol 28:375–392.

Garvey KJ, Yoshikawa H, Ito J (1985): The complete sequence of the *Bacillus* phage φ29 right early region. Gene 40:301–309.

Geiduschek EP, Elliott T, Kassavetis GA (1983): Regulation of late gene expression. In Mathews CK, Kutter E, Mosig G, Berget FB (eds): Bacteriophage T4. Washington, DC: ASM Press, pp 189–192.

Greenberg GR, He P, Hilfinger J, Tseng M-J (1994): Deoxyribonucleoside triphosphate synthesis and T4 DNA replication. In Karam JD, Drake JW, Kreuzer KN, Mosig G, Hall DH, Eiserling FA, Black LW, Spicer EK, Kutter E, Carlson K, Miller ES (eds): "Molecular Biology of Bacteriophage T4." Washington, DC: ASM Press, pp 14–27.

Guild N, Gayle M, Sweeney R, Hollingsworth T, Modeer T, Gold L (1988): Transcriptional activation of bacteriophage T4 middle promoters by the *motA* protein. J Mol Biol 199:241–258.

Hall BD, Spiegelman S (1961): Sequence complementarity of T2 DNA and T2–specific RNA. Proc Natl Acad Sci USA 47:137–146.

d'Herelle F (1917): Sur un microbe invisible antagoniste des bacilles dysenteriques. CR Acad Sci 165:373.

Herman RE, Haas N, Snustad DP (1984): Identification of the bacteriophage T4 *unf* (= *alc*) gene product, a protein involved in the shutoff of host transcription. Genetics 108:305–317.

Hermoso JM, Salas M (1980): Protein p3 is linked to the DNA of phage φ29 through a phosphoester bond between serine and 5′-dAMP. Proc Natl Acad Sci USA 77:6425–6428.

Hershey AD, Chase M (1952): Independent functions of viral protein and nucleic acid in growth of bacteriophage. J Gen Physiol 36:39–56.

Hershey AD, Rotman R (1948): Linkage among genes controlling inhibition of lysis in a bacterial virus. Proc Natl Acad Sci USA 34:89–96.

Karam, J. ed (1994). "The Molecular Biology of Bacteriophage T4." Washington, DC: ASM Press.

Huang WM, Ao S-Z, Casjens S, Orlandi R, Zeikus R, Weiss R, Winge D, Fang M (1988): A persistent untranslated sequence within T4 DNA topoisomerase gene 60. Science 239:1005–1012.

Karam JD, Gold L, Singer BS, Dawson B (1981): Translational regulation: Identification of the site on bacteriophage T4 *rIIB* mRNA recognized by the *regA* gene function. Proc Natl Acad Sci USA 78:4669–4673.

Komberg T, Lockwood A, Worcel A (1974): Replication of the *Escherichia coli* chromosome with a soluble enzyme system. Proc Natl Acad Sci USA 71:3189–3193.

Kreuzer KN, Alberts BM (1985): A defective phage system reveals bacteriophage T4 replication origins that coincide with recombination hot spots. Proc Natl Acad Sci USA P2(10):3345–3349.

Kutter E, Beug A, Sluss R, Jensen L, Bradley D (1975): The production of undegraded cytosine-containing DNA by bacteriophage T4 in the absence of dCTPase and endonucleases II and IV, and its effects on T4–directed protein synthesis. J Mol Biol 99:591–607.

Kutter E, Kellenberger E, Carlson K, Eddy S, Neitzel J, Messinger L, North J and Guttman B (1994). Effects of Bacterial Growth Conditions on T4 Infection. In Karam J (ed): "Bacteriophage T4." Washington, DC: ASM Press, pp 406–420.

Kutter E, White T, Kashlev M, Uzan M, McKinney J and Guttman B. Effects on host genome structure and expression. In Karam J (ed): "Bacteriophage T4." Washington, DC: ASM Press, pp 357–368.

Leirmo S, Harrison C, Gayley DS, Burgess RR, Record MT (1987): Replacement of potassium chloride by potassium glutamate dramatically enhances protein-DNA interactions in vitro. Biochem 26:2095–2101.

Levinthal C, Hosoda J, Shub D (1967): The control of protein synthesis after phage infection. In Colter JS, Paranchych W (eds): "The Molecular Biology of Viruses." New York: Academic Press, pp 71–87.

Liu B and Alberts BM (1995): Head-on collision between a DNA replication apparatus and RNA polymerase transcription complex. Science 267:1131–1137.

Luria SE (1945): Mutation of bacterial viruses affecting their host range. Genetics 30:84–99.

Lwoff A (1953): Lysogeny. Bacteriol Rev 17:269–337.

Lwoff A, Tournier P (1966): The classification of viruses. Annu Rev Microbiol 20:45–74.

Macdonald PM, Kutter E, Mosig G (1984): Regulation of a bacteriophage T4 late gene, *soc*, which maps in an early region. Genetics 106:17–27.

Malik S, Goldfarb A (1984): The effect of a bacteriophage T4-induced polypeptide on host RNA polymerase interaction with promoters. J Biol Chem 259:13292–13297.

Marcus M, Newton MC (1971): Control of DNA synthesis in *Bacillus subtilis* phage J. Virology 44:83.

Mathews CK, Allen JR (1983): DNA precursor biosynthesis. in Mathews CK, Kutter E, Mosig G, Berget FB (eds): "Bacteriophage T4." Washington, DC: ASM Press, pp 59–70.

Mathews CK, Kutter E, Mosig G, Berget PB (eds) (1983): "Bacteriophage T4". Washington, DC: ASM Press.

McPheeters DS, Christiansen A, Young EA, Stormo G, Gold L (1986): Translational regulation of expression of bacteriophage T4 lysozyme gene. Nucleic Acids Res 14:5813–5826.

Monod J, Wollman EL (1947): L'inhibition de la croissance et de l'adaption enzymatique chez les bacteries infectees par le bacteriophage. Ann Inst Pasteur 73:937–956.

Mosig G (1987): The essential role of recombination in phage T4 growth. Annu Rev Genet 21:347–371.

Mosig G, Eiserling F (1988): Phage T4: Structure and metabolism. In Calendar R (ed): "The Bacteriophages II." New York: Plenum Press, pp 521–606.

Murphy FA, Fauquet CM, Bishop DHL, Ghabrial SA, Jarvis AW, Martelli GP (1995): "The Classification and Nomenclature of Viruses." New York: Springer.

Nomura M, Witten C, Mantel N, Echols H (1966): Inhibition of host nucleic acid synthesis by bacteriophage T4: Effect of chloramphenicol at various multiplicities of infection. J Mol Biol 17:273–278.

Nomura M, Hall BD, Spiegelman S (1960): Characterization of RNA synthesized in *Escherichia coli* after bacteriophage T2 infection. J Mol Biol 2:306–326.

Nossal NG (1994): The bacteriophage T4 DNA replication fork. In Karam JD, Drake JW, Kreuzer KN, Mosig G, Hall DH, Eiserling FA, Black LW, Spicer EK, Kutter E, Carlson K, Miller ES (eds): "Molecular Biology of Bacteriophage T4." Washington, DC: ASM Press, pp 43–53.

Novick A, Szilard L (1951): Virus strains of identical phenotype but different genotype. Science 113:34–35.

Prehm P, Jann B, Jann K, Schmidt G, Stirm S (1975): On a bacteriophage T3 and T4 receptor region within the cell wall lipopolysaccharide of *Escherichia coli* B. J Mol Biol 101:277–281.

Quirk SM, Bell-Pedersen D, Belfort M (1989): Intron mobility in the T-even phages: High frequency inheritance of group I introns promoted by intron open reading frames. Cell 56:455–465.

Rabussay D (1983): Phage-evoked changes in RNA polymerase. In Mathews CK, Kutter C, Mosig G, Berget FB (eds): "Bacteriophage T4." Washington, DC: ASM Press, pp 167–173.

Repoila, F., Tetart F, Bouet JY, Krisch HM (1994): Genomic polymorphism in the T-even bacteriophages. EMBO J 13:4181–4192.

Salas M (1988): Phages with protein attached to the DNA ends. In Calendar R (ed): "The Bacteriophages I." New York: Plenum Press, pp 169–192.

Selick HE, Barry J, Cha T-A, Munn M, Nakanishi M, Wong ML, Alberts BE (1987): Studies on the T4 bacteriophage DNA replication system. In McMacken R, Kelley TJ (eds): "DNA Replication and Recombination." New York: Alan R. Liss, pp 183–214.

Shub D, Coetzee T, Hall D, Belfort M (1994) The Self-splicing introns of bacteriophage T4. In Karam J (ed): "Bacteriophage T4." Washington, DC: ASM Press, pp 186–192.

Shub DA, Goodrich H, Gott J, Xu M-Q, Scarlato V (1988a): A self-splicing intron in the DNA polymerase gene of the Bacillus subtilis bacteriophage SPO1. J Cell Biochem (Suppl) 12D:30.

Shub D, Gott J, Xu M-Q, Lang BF, Michel F, Tomaschewski J, Pedersen-Lane J, Belfort M (1988b): Structural conservation among three homologous introns of bacteriophage T4 and the group I introns of eukaryotes. Proc Natl Acad Sci USA 85:1151–1155.

Simon LD, Anderson TF (1967): The infection of *Escherichia coli* by T2 and T4 bacteriophage as seen in the electron microscope. 1. Attachment and penetration. Virol 32:279–297.

Sinden R, Pettijohn D (1981): Chromosomes in living *Escherichia coli* cells are segregated into domains of supercoiling. Proc Natl Acad Sci USA 78:224–228.

Snyder L, Jorissen L (1988): *Escherichia coli* mutations that prevent the action of the T4 *unf/alc* protein map in an RNA polymerase gene. Genetics 118:173–180.

Streisinger G, Bruce V (1960): Linkage of genetic markers in phages T2 and T4. Genetics 45:1289–1296.

Streisinger G, Edgar RS, Denhardt GH (1964): Chromosome structure in phage T4. 1. Circularity of the linkage map. Proc Natl Acad Sci USA 51:775–779.

Streisinger G, Emrich J, Stahl MM (1967): Chromosome structure in phage T4. III. Terminal redundancy and length determination. Proc Natl Acad Sci USA 57:292–295.

Summers, WC (1999): "Felix d'Herelle and the Origins of Molecular Biology." New Haven: Yale University Press.

Sulakvelidze A, Alavidze Z, Morris JG, Jr (2001): Bacteriophage Therapy. Antimicrob Agents Chemother 45:649–659.

Tétart F, Desplats C, Kntateladze M, Monod C, Ac-Kermunn H, Krisch HM (2001): Phytogeny of the Major Head and Tail genes of the wide-ranging T4 type phages. J Bact 183:358–366.

Twort FW (1915): An investigation on the nature of the ultramicroscopic viruses. Lancet 11:1241.

Webster RG, Grannoff A (1994): "Encyclopedia of Virology." London: Academic Press.

Wiberg JS, Karam JD (1983): Translational regulation in T4 phage development. In Mathews CK, Kutter E, Mosig G, Berget FB (eds): "Bacteriophage T4." Washington, DC: ASM Press, pp 193–201.

Williams KP, Kassavetis GA, Geiduschek EP (1987): Interactions of the bacteriophage T4 gene 55 product with *Escherichia coli* RNA polymerase: Competition with *E. coli* sigma-70 and release from late T4 transcription complexes following initiation. J Biol Chem 262:2365–2371.

Williams KP, Kassevetis GA, Herendeen DR, Geiduschek EP (1994): Regulation of late-gene expression. In Karam JD, Drake JW, Kreuzer KN, Mosig G, Hall DH, Eiserling FA, Black LW, Spicer EK, Kutter E, Carlson K, Miller ES (eds): "Molecular Biology of Bacteriophage T4." Washington, DC: ASM Press, pp 161–175.

Wilson GG, Young KKY, Edline GJ, Konigsberg W (1979): High frequency generalized transduction by bacteriophage T4. Nature 280:80–82.

Yoshikawa H, Ito J (1982): Nucleotide sequence of the major early region of bacteriophage φ29. Gene 17:323–335.

5

Bacteriophage λ and Its Relatives

ROGER W. HENDRIX

Pittsburgh Bacteriophage Institute, Department of Biological Sciences, University of Pittsburgh,
Pittsburgh, Pennsylvania 15260

I. INTRODUCTION

Bacteriophages, the viruses that infect bacteria, are almost incomprehensibly abundant in the environment. There are, for example, about 10 million bacteriophage particles in a typical milliliter of coastal seawater. Numbers like this lead to the estimate that the global population of phages is somewhere in excess of 10^{30} individuals. And since the number of phage particles in environmental samples is typically 10-fold higher than the number of bacterial cells, it has been suggested that bacteriophages are the most abundant—in fact constitute the majority of—organisms on the planet. Whether or not this is literally true, there is no doubt that phages play a major role in the ecology, genetics, and evolution of their bacterial hosts, as described below and elsewhere in this book. The discussion in this chapter applies to the dsDNA-containing, tailed phages—and particularly the member of that group called *phage λ*.

There is a different aspect of bacteriophage biology, relating to the history of the discipline of molecular biology, that explains why two individual phages (out of the population of $> 10^{30}$) have entire chapters devoted to them in a book such as this one. That is, these two phages of *Escherichia coli* (λ and T4) plus a handful of others, were chosen, rather arbitarily, in the early years of what came to be known as molecular biology, as *model experimental systems* for understanding the molecular basis of life processes, and in particular, the molecular nature of genes and how they work. The early molecular biologists chose phages to work with largely because phages were experimentally tractable, but also because they believed that the basic life processes they could learn about from phages were the same as the basic life processes of cellular organisms such as *E. coli*, humans, sea urchins, mushrooms, and redwood trees. The remarkable degree to which this belief turned out to be correct has meant that

Modern Microbial Genetics, Second Edition. Edited by
Uldis N. Streips and Ronald E. Yasbin. ISBN 0–471–38665–0
Copyright © 2002 Wiley-Liss, Inc.

many of the most fundamental things we know about the molecular basis of life were learned first in studies of phages— prominently λ and T4. It has been said (and it may be true) that per base pair of genome, more scientist-years have been expended studying, and more is known about, bacteriophage λ than about any other organism on Earth. This is a sobering thought, given that much still remains to be learned about phage λ, and contemporary studies on λ regularly yield up new understanding of its life style and how it interacts with its bacterial host. Figure 1 gives picture of λ.

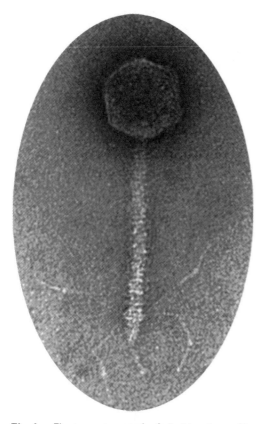

Fig. 1. Electron micrograph of a λ virion, imaged by the negative stain technique. The head, the tail, and the tail fibers are visible. The DNA, which is packed tightly in the head, exits through the tail during infection and into the cytoplasm of the cell. The length of the virion, from tail tip to top of head, is about 200 nm.

II. DISCOVERY OF λ

Bacteriophage λ was discovered in 1951, more or less by accident, by Esther and Joshua Lederberg during their pioneering studies of *E. coli* conjugation. It turns out that the K12 strain of *E. coli*, which the Lederbergs were using in their experiments, carries a quiescent copy of the λ chromosome, known as a prophage (see below), associated with the bacterial chromosome. In the course of mutagenesis to produce nutritional mutants, one of the mutant strains of *E. coli* K12 lost its λ prophage, which made it susceptible to infection and killing by λ phage particles. When the Lederbergs cross-streaked two of their mutant strains to check for nutritional cross-feeding, they got more than they bargained for: the phage particles associated with the strain carrying the prophage infected and lysed the strain that had lost the prophage, revealing the presence of the phage. The world was apparently waiting for a phage that infected *E. coli* and λ's characteristics, because within a few years research on λ was booming and λ had become the exemplar of a *temperate* bacteriophage.

III. THE TEMPERATE PHAGE LIFESTYLE

While phage T4, the *Tyrannosaurus rex* of bacteriophages, is a prototypical example of a *lytic* or *virulent* phage, λ is the prototype of the large group of phages known as *temperate* phages. When a virulent phage infects a cell, the results are always the same: the phage genes co-opt the cellular machinery and turn the cell into a factory for making new phages; phage DNA is replicated, new virions (virus particles) are assembled, and the cell lyses, releasing perhaps 100 to 200 progeny phages into the medium. A temperate phage like λ, on the other hand, every time it infects a cell has a choice between two very different ways of interacting with the cell; these are referred to as the *lytic cycle* and the *lysogenic cycle*. Figure 2 outlines the temperate lifestyle, with its choice between

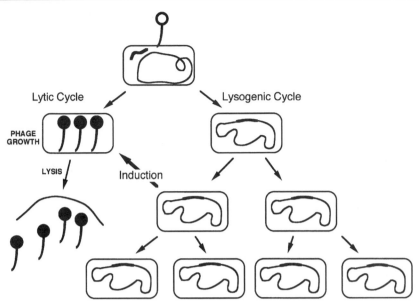

Fig. 2.   Schematic diagram of the λ life cycles, showing the consequences of the lytic/lysogenic decision. Relative sizes are distorted for graphical clarity; thus the length of the bacterial chromosome (thin line) should be about 500 times the length of the cell, the length of the prophage DNA (thick line) should be about 1% the length of the bacterial chromosome, and the length of the virion should be about 5 times less than shown, relative to the cell.

the lytic and lysogenic cycles. At the top of the diagram the phage infects a cell by adsorbing to the cell surface and injecting its DNA into the cytoplasm. Once inside the cell, a subset of the phage genes is expressed, and the decision between the lytic and lysogenic life cycles is made. (The molecular basis of this decision is described below.) If the phage opts for the lytic cycle, it proceeds through an orderly expression of genes, production of progeny virions, and release of the progeny through cell lysis, as described in detail for λ below.

Two major events differentiate the lysogenic cycle from the lytic cycle. First, expression of virtually all the phage genes is shut off in the lysogenic cycle through the action of a repressor protein. To a first approximation, the only gene expressed from the phage under these circumstances is the repressor gene, and the repressor protein binds to two operators that flank the repressor gene and blocks transcription from the two associated promoters (P$_L$ and P$_R$ in λ). Since all the

other genes of the phage depend, either directly or indirectly, on these promoters for their expression, the repressor effectively holds the entire phage (except its own gene) transcriptionally silent. An important consequence is that the genes that would be lethal to the host in a lytic infection are not expressed and the host survives. The second major event of the lysogenic cycle is that the entirety of the phage genome becomes inserted ("integrated") into the continuity of the host genome. The result of integration is that the phage genome becomes part of the bacterial genome; this means that every time the bacterial genome is replicated the phage genome is replicated as part of the bargain, and each daughter cell ends up with its own copy of the phage genome. This situation, with the phage hitchhiking a ride in the bacterium, can persist indefinitely. The phage genome in this state is called a *prophage*. The bacterial cell carrying the prophage is called a *lysogen* (because the low level of phage particles associated with a culture of lysogens can give rise to lysis of other susceptible bacteria, as in the experiment cited above); alternatively, such a cell is said to be *lysogenic for* the phage in question.

A consequence of expression of the repressor by a prophage is that the lysogen carrying the prophage acquires *immunity* to infection by another phage of the same type. Repressor molecules in the cell that are not bound to the operators of the resident prophage are available to bind to the operators of incoming phage DNA, which prevents it from entering either the lytic or lysogenic cycle and effectively aborts the infection.

There is also a way for the prophage to leave the lysogenic cycle and enter lytic growth. This process, called *induction*, is ordinarily a very rate event, with perhaps one lysogenic cell in 10^6 undergoing induction each generation. In the induced lysogen, the prophage becomes detached ("excised") from the bacterial chromosome, goes through the lytic cycle, producing a crop of progeny phages, and lyses the cell to release the progeny into the culture. This low level of "spontaneous induction" accounts for the low level of infectious phage particles in a lysogenic culture. (These phages cannot successfully infect other cells in the culture, because those cells are immune to the infecting phages by virtue of the repressor expressed from their prophages.)

With some phages, including λ, induction can be converted from a rare event into an event that happens in every cell in the culture by giving the cells an appropriate dose of ultraviolet radiation. The UV turns on the cell's SOS response, which activates a number of DNA damage repair mechanisms to counter the effects of the UV on the cell. However, the phage repressor is programmed to respond to the SOS response by inactivating itself (by autoproteolysis), which leads to derepression of the prophage and therefore induction, with consequent phage production and death of the cell. From an evolutionary perspective, this is a sensible thing for the prophage to do. The presence of the SOS response means that the cell has sustained damage and may be in serious trouble; the prophage is following the same logic as the proverbial rats that desert a sinking ship.

IV. LYTIC GROWTH OF PHAGE λ

As with most viruses, expression of λ's genes during lytic growth is organized temporally. For the first 10 minutes following infection or induction, the *early genes* are expressed exclusively. These genes encode the phage proteins responsible for DNA replication, repair, and recombination; the proteins with regulatory roles; and other proteins whose early expression is advantageous to survival of the phage, such as a protein that counteracts the effects of host restriction enzymes. Starting at 10 to 12 minutes after infection, expression of the *late genes* begins and continues at a high level until *cell lysis* at about 50 minutes. The late genes encode the proteins that will make up the structure of the virion—the head, the tail, and the tail fibers, and they also include the genes that cause cell lysis at the end of lytic growth.

The temporal organization of gene expression in the lytic cycle is accomplished by regulation of transcription. Figure 3 shows

Fig. 3. The physical map of the λ genome is shown in the upper part of the figure, divided into halves to fit on the page. The scale bar represents the DNA, and the boxes above it show the positions and sizes of the genes. Shaded boxes represent genes transcribed leftward and open boxes genes transcribed rightward. (The vertical offsets of the boxes are for graphical clarity and have no biological significance.) Arrows below the scale bar show the locations and extent of transcription, with the thin arrow denoting transcription from the repressed prophage, the medium arrows denoting early transcription, and the thick arrows denoting late transcription. Note that in the cell, the two ends of the genome are joined together, and as a result transcription initiating at PR' can continue across the joined ends and into the head and tail genes. The regions around PL and PR are shown in expanded form in the lower part of the figure.

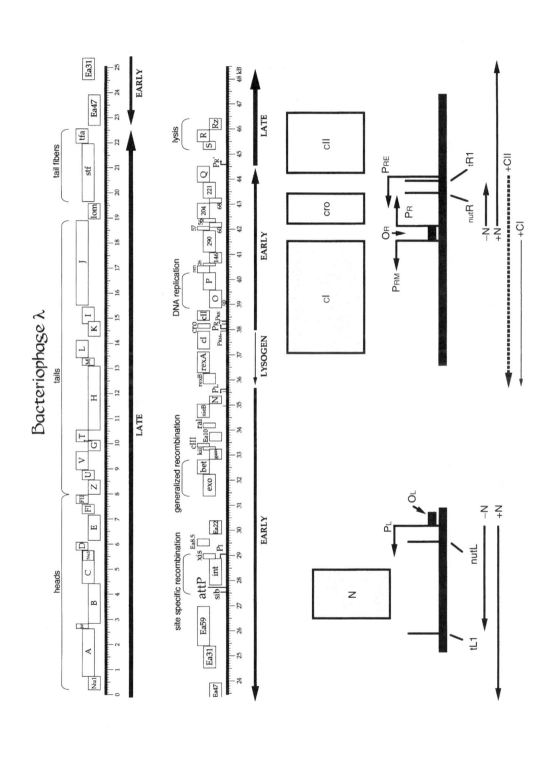

Bacteriophage λ

a map of the λ genome where one basis for this regulation can be seen, namely that the genes are clustered by function and organized into operons. This means that their transcription can be controlled in groups and from a small number of promoters. In λ, all transcription is done by the host (*E. coli*) RNA polymerase, and its orderly progression through the different transcription units is accomplished by a cascade mechanism in which the protein product of a gene in one transcription unit activates the polymerase to read the next. This activation is achieved by a *transcription antitermination* mechanism, as described in detail below.

Let's now follow λ through one cycle of lytic growth, from the initial infection until cell lysis some 50 minutes later.

A λ virion adsorbs to a cell through an interaction between the tail fiber protein at the tip of the tail and an outer membrane protein (LamB) of the host. Successful adsorption triggers injection of the DNA, which passes through the cell envelope into the cytoplasm. Once in the cytoplasm, the first thing that happens to the DNA is that it is converted from a linear double-stranded molecule to a double-stranded circle through annealing of complementary single-stranded 12 base extensions on the two ends, followed by ligation to make a covalently sealed 48,503 bp circle. The second thing to happen to the DNA is that its two early promoters, PL and PR, are recognized by the host RNA polymerase, which initiates transcription. The resulting transcripts are short, since in each case the polymerase encounters a termination signal soon after it has transcribed the first gene. These two genes—*N* transcribed from PL and *cro* from PR—are sometimes called the *immediate early* genes.

Nothing more would happen except for the product of the *N* gene. The action of the N protein modifies the RNA polymerase so that it ignores termination signals (hence "antitermination"). However, the N protein does not modify RNA polymerases indiscriminately; it confines its attention to polymerases that have initiated transcription at one of the early promoters, PL or PR. This is because N protein acts by forming a complex with the RNA polymerase, three host proteins and—crucially—with a special sequence in the mRNA as it is being synthesized by the polymerase. This special sequence, called the "N-utilization" or "*nut*" site, occurs downstream from the PL and PR promoters and nowhere else in the λ genome, which is why only polymerase starting at those promoters can be modified.

Once the N protein is available, then, RNA polymerase reading from the early promoters is not sensitive to termination signals and reads through to the ends of the two early operons. As a result the rest of the early proteins are made (using the host translation apparatus). These include, most important, the O and P proteins, which direct the host DNA replication machinery to replicate the λ DNA. Replication occurs initially by a "theta" mechanism in which one circular molecule is replicated into two circular daughters. At about 12 minutes after infection replication switches to the "rolling circle" mechanism, which produces long head-to-tail linear concatemers of the genome, the appropriate substrate for packaging into the phage head.

Most of the other early genes have either auxiliary roles or roles in the decision between the lytic and lysogenic cycles, which will be discussed below. The one additional early gene with an essential role in lytic growth is the *Q* gene. The Q protein acts to turn on transcription of the late genes in a way that is conceptually very similar to the way the N protein acts—that is, by antitermination of transcription—though the biochemical mechanism is somewhat different. Late transcription starts from the late promoter, PR', located just downstream from the Q gene. PR' is a strong promoter that, like the early promoters, is read by the unmodified host RNA polymerase, but transcription stops soon thereafter at a terminator. This termination is not overcome by the presence of the N protein, since PR' does not have an associated *nut* site. However, it does have sequences that allow Q protein to interact with the RNA polymerase as it is initiating transcription and render it insensitive to termination signals. The polymerase is now able to read through the entire 26 genes of the late operon.

The late proteins include those necessary for assembling the virion—head, tail, and tail fiber proteins—plus the proteins responsible for cell lysis. Once late synthesis starts, heads and tails assemble in separate pathways, the initially empty heads package a genome's worth of DNA by a mechanism that superficially resembles an ill-behaved child eating spaghetti, and tails join to heads to form infectious virions. These accumulate inside the cell, together with the endolysin enzyme (product of gene R), until the holin protein (product of gene S) reaches an appropriate level to form pores in the cytoplasmic membrane, allowing the endolysin to reach and digest its substrate, the cell wall. Deprived of the support of the cell wall, the cell explodes due to the osmotic pressure difference between the cytoplasm and the surrounding medium, allowing the progeny phages to escape to find a fresh cell to infect.

V. THE LYTIC/LYSOGENIC DECISION

In describing the lytic cycle of λ, we gave only brief mention of the early genes that have roles in the decision between lytic and lysogenic growth. Now we will explicitly consider these genes and how they allow the infecting phage to assess the conditions in the cell and to choose the strategy of growth that will maximize its success in propagating its genes.

There are three additional phage genes to think about when we consider the lytic/lysogenic decision; these are cI, cII, and cIII. (The I's in these gene names are Roman numerals, so these genes are pronounced "C-one", "C-two", and "C-three.") The cI gene encodes the repressor protein that we encountered above, which is also frequently called the "CI repressor" or "CI protein." The CI repressor carries out its repressing function in the lysogenic state by binding to two operators, O_L and O_R, which overlap the corresponding P_L and P_R promoters, thereby preventing expression of all the phage genes required for lytic growth of the phage. (Figure 3 shows the locations of O_L and O_R; Fig. 4A shows the detailed organ-

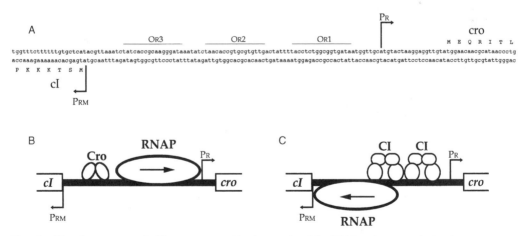

Fig. 4. The O_R operator. **A:** The sequence of both strands of the DNA is shown, with the three operator subsites, O_R1, O_R2, and O_R3, indicated. The bent arrows show the start sites of transcription from P_R and P_RM, and the first several amino acids encoded by the cI and cro genes are shown in the one letter code. **B:** The O_R region early during lytic growth, with a Cro dimer bound to O_R3 and RNA polymerase (RNAP) bound to P_R, ready to transcribe cro and the genes downstream. **C:** The O_R region in the repressed prophage, with CI dimers bound to O_R1 and O_R2, blocking transcription from P_R and activating transcription from P_RM.

ization of O$_R$.) The decision between lytic and lysogenic growth following infection is in essence determined by whether or not enough CI repressor gets made fast enough to clamp down on expression of the genes required for lytic growth before the lytic cycle is irreversibly established. Production of CI repressor is determined in turn by how much CII protein is available, with an auxiliary role played by CIII protein. The CII protein can be regarded as the phage's environmental sensor, inasmuch as the levels of functional CII protein respond to the conditions in the cell, as described below, and thereby transmit information about those conditions to the decision-making process.

The *cII* gene lies just to the right (downstream) of *cro*, and so it begins to be expressed by transcription from P$_R$ as soon as the N protein allows RNA polymerase to read through the terminator between *cro* and *cII* (Fig. 3). CII protein turns out to be a potent transcriptional activator that specifically turns on transcription from a leftward-pointing promoter called P$_{RE}$ ("promoter for repressor establishment"), located right at the beginning of the *cII* gene. Successful high-level transcription from P$_{RE}$, as happens when CII protein levels are high, reads backward through the rightward pointing *cro* gene and then forward through the leftward pointing *cI* gene, resulting in high levels of CI repressor production and the establishment of repression through CI binding at O$_L$ and O$_R$.

The question then becomes, how are the levels of CII protein determined? We know of two important ways that CII levels are influenced by the cellular environment. The first of these derives from the fact that CII is sensitive to degradation by a host protease specified by the *hflA* and *hflB* genes. If the Hfl protease were always fully active, λ would never enter the lysogenic cycle because the CII protein would be degraded as soon as it was synthesized, and CI repressor would never be made. However, the activity of Hfl protease is modulated by the physiological state of the cell in response to levels of the signaling molecule cyclic AMP. Higher levels of cAMP lead to lower Hfl protease activity and therefore slower degradation of CII and higher probability of entering the lysogenic cycle. It is also at the level of Hfl protease activity that the CIII protein acts. CIII inhibits the Hfl activity and therefore pushes the balance of the system in favor of the lysogenic cycle. The other known important influence on the lytic/lysogenic decision is the multiplicity of infection—that is, the number of phages infecting a single cell simultaneously. Higher multiplicity of infection strongly favors the lysogenic cycle, probably because the concentration of CII and CIII proteins in the cell increases as the number of copies of the *cII* and *cIII* genes being expressed in the cell increases, but the Hfl activity remains constant. This way of responding to high multiplicity of infection appears to make biological sense for the phage: if a cell is simultaneously infected by multiple phages, it must mean that the phages outnumber the bacteria in the local environment, and the progeny of any infecting phage that chose the lytic cycle under these conditions would most likely be released into an environment in which there were no host cells left to be infected.

VI. THE SWITCH AT O$_R$

Another important component of the decision between the lytic and lysogenic pathways is a molecular switch centered around the O$_R$ operator region. This region of 101 bp, located between the start points of the diverging *cI* and *cro* genes, contains multiple repressor binding sites plus the promoter P$_R$ (driving rightward transcription of *cro* and *cII*) and a second promoter we have not yet encountered, P$_{RM}$ (driving leftward transcription of *cI*, though under different control from that of P$_{RE}$ transcription discussed above). If we think of the level of CII protein as what determines which way the lytic/lysogenic decision goes, by determining the amount of CI repressor made, then O$_R$ is the place where CI acts to carry out that decision.

The O$_R$ operator contains three "subsites," O$_R$1, O$_R$2, and O$_R$3 (Fig. 4). Each of these subsites is a binding site for CI repressor, and each has twofold symmetry (an inverted repeat) in the sequence, corresponding to the fact that the repressor binds as a twofold symmetric dimer. The Cro protein turns out also to be a repressor, and like the CI repressor, it also binds (as a dimer) to the three subsites of OR. At first sight this seems paradoxical because, as we will see, the effects of CI and Cro are quite different. The answer to this conundrum lies in the fact that while the three subsites of OR are similar in sequence, they are not identical, and the differences in sequence have crucially different effects on how CI and Cro bind to them to carry out their regulatory roles. Thus CI protein binds with highest affinity to OR1 and lowest affinity to OR3, while Cro protein has just the opposite binding appetites.

When CI encounters OR, the first thing it does is to bind to OR1. The presence of a CI dimer at OR1 increases the affinity of the adjacent OR2 site for a second CI dimer, so OR2 rapidly fills up once the first CI dimer has bound to OR1; however, CI does not bind to the low affinity OR3 site until its concentration is considerably higher. The presence of CI dimers bound at OR1 and OR2 has two effects (see Fig. 4C). First, RNA polymerase is denied access to PR, and transcription of *cro* and the genes downstream is blocked. (CI binding at OL has a similar repressing effect on transcription leftward from PL.) Second, CI bound at OR2 acts as a positive transcription factor for transcription from PRM. Similar to what we have seen for CII protein and its role in transcription from PRE, RNA polymerase cannot recognize the PRM promoter unless CI is bound at OR2. A sufficiently high level of CI protein leads to lysogeny, first by shutting down transcription of all the genes involved in lytic growth and second by establishing CI synthesis from PRM.

The synthesis from PRM is the only source of CI repressor once lysogeny is established, since there is no transcription of *cII* from a repressed prophage to allow transcription of *cI* from PRE. (The name of the promoter,

PRM, is an abbreviation for promoter for repressor maintenance.) In a lysogenic cell, the amount of *cI* transcription is regulated in both a positive and a negative sense by the ultimate product of that transcription, CI protein. The positive regulation works as described above, through CI bound to OR2; when CI concentrations begin to get too high, a CI dimer binds to the low affinity OR3 site and blocks further transcription from PRM, with the net effect being rather precise regulation of CI concentration.

As mentioned above, Cro protein's binding preferences for the three OR subsites are opposite to the preferences of CI, so when Cro encounters OR it binds first to OR3 (see Fig. 4B). Under these circumstances Cro may have some effect of pushing the lytic/lysogenic decision in the direction of the lytic cycle—for example, by blocking *cI* transcription from PRM or competing with CI for binding to OR. However, Cro's most important function probably comes after the phage is committed to the lytic cycle. At about 10 minutes after infection in the lytic cycle, Cro concentration builds up to the point that it begins to occupy OR2. The effect is to turn down the rate of transcription from PR and, by the same mechanism operating at OL, to turn down transcription from PL. The apparent logic of this is that by this time in the lytic cycle enough of the early proteins—for example, DNA replication proteins—have been made that a continued high rate of synthesis is not needed, and the phage will do better to devote more of the cellular resources to making the late proteins it needs for constructing virions.

VII. PROPHAGE INTEGRATION AND EXCISION

When lambda enters the lysogenic cycle, the phage DNA must become integrated into the host chromosome to form a prophage. This is accomplished by a *site-specific recombination* event catalyzed by a phage-encoded enzyme, Integrase (or Int), together with a multi-subunit host factor, the integration host factor or

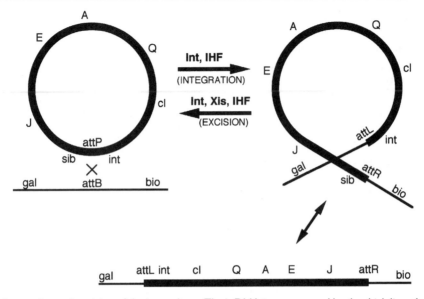

Fig. 5. Integration and excision of the λ prophage. The λ DNA is represented by the thick line; the thin line represents a small fragment of the bacterial chromosome surrounding *attB*. Integration entails a reciprocal, break-and-join recombination between *attP* and *attB* and results in the insertion of the prophage DNA into the continuity of the bacterial chromosome, with a consequent increase in the separation between the flanking bacterial genes.

IHF. The integration reaction is a reciprocal recombination between the "attachment site" on the phage DNA, *attP*, and the corresponding attachment site on the bacterial chromosome, *attB*. Since the phage genome is circular at the time of integration, a reciprocal recombination results in the insertion of a linear version of the phage DNA into the continuity of the bacterial genome (Fig. 5).

The two attachment sites, *attP* and *attB*, share 15 bp of sequence identity called the *core*, and it is here that the recombination takes place. Overlapping the core sequence there are two Integrase binding sites that are bound by a DNA-binding site on the Int protein. That description applies to both attachment sites and in fact completes the description of *attB*. On the other hand, *attP* extends upstream ~150 bp from the core and ~90 bp downstream. This extra sequence includes multiple binding sites recognized by a second DNA binding domain of the Int protein as well as sites for the binding of IHF. During the integration reaction the DNA of both attachment sites is bound and wrapped

up with Integrase and IHF into a compact complex known as the *intasome*. The intasome brings the core sequences of the two attachment sites into juxtaposition with each other and with the catalytic domains of the Integrase enzymes, and it is in this complex that the reaction takes place. The details of the geometrical arrangement of the components of the intasome that make this reaction possible are still being worked out. In contrast, the chemical mechanism of the reaction is rather well understood. Briefly, Integrases cut the "top" strand of each attachment site at a position 4 bp from the start of the 15 bp core, preserving the energy of the phosphodiester bond by transferring the 3'-OH of the cut strand to an ester linkage with a tyrosine in the active site of the enzyme. The enzyme bound strands are swapped between the two attachment sites and rejoined to the free end of the opposite attachment site. These actions from a "Holliday junction" crossover structure, which then moves 7 bp to the right by branch migration. The Holliday junction is finally resolved into the two recombinant products by a repeat of the catalytic action of Integrase, with the cutting of the "bottom" strand of each attachment site, swapping position, and rejoining to the opposite partner strand.

Since the sequences of the *attP* and *attB* sites are different from each other outside the 15 bp core sequences, the recombinant sites that are created at the ends of the prophage are different from either *attP* or *attB*. These are called *attL* ("attachment site on the left") and *attR* ("attachment site on the right"), and these are the substrates for the reciprocal recombination reaction of excision that happens following prophage induction (Fig. 5). In parallel with the fact that the substrates for the integration and excision reactions are different, the co-factor requirements are also different: excision requires not only Integrase and IHF but also Excisionase, a small protein encoded by the phage *xis* gene. Excisionase (also called Xis, pronounced "e<u>xcise</u>") has a binding site in the left arm of *attR*, and its binding to that site, together with Int and IHF binding to their sites, causes formation of a reaction complex in which *attL* and *attR* recombine to release the prophage from the host DNA, recreating the *attP* site in the phage genome and *attB* in the host genome.

VIII. REGULATION OF INTEGRATION AND EXCISION

Since the integration and excision reactions have different requirements for phage-encoded proteins (Int for integration; Int + Xis for excision), the phage could in principle regulate which of the two reactions occurs in any particular situation by differentially regulating the synthesis of Int and Xis. The phage in fact does just that, producing Int exclusively when it needs to integrate and producing both Int and Xis when it needs to excise. The question is then how the phage senses whether its DNA is integrated into the chromosome or not, and having sensed that, how it regulates expression of *int* and *xis* appropriately. To understand that, we need to examine how the *int* and *xis* genes are transcribed.

Figure 6 shows the *int* and *xis* genes, together with the two promoters, PI and PL, that are responsible for their transcription. Note also that the attachment site, *attP* is just downstream from *int*, and that *sib*, which is a regulatory sequence with a central role in regulation of *int* and *xis*, is on the opposite side of *attP* from *int*. The PI promoter is regulated essentially the same way as the PRE promoter that was discussed earlier: it is normally inactive but it is turned on strongly in the presence of the CII protein. Thus, after an infection in which the conditions are favorable for lysogeny, CII is produced at high levels, and transcription at PI is stimulated, ensuring that enough Int protein is made to cause integration of the prophage as it enters the lysogenic state. Since PI is located within the coding sequence of *xis*, a transcript from PI does not encode Xis, which would be deleterious under these circumstances, for it could cause reversal of the prophage's integration into the chromosome. The transcript from PL, on the other hand, encodes both Int and Xis, but this transcript is degraded from its 3′ end soon after it is made, with the result that neither Int nor Xis is made to a significant extent from this transcript. Degradation of the PL transcript is mediated by the *sib* site, which acts as a "poison" signal that marks the PL transcript for destruction as soon as the RNA polymerase has transcribed *sib* into RNA. We discuss below how the *sib* site can differentiate between transcripts that started at PL and ones that started at PI, signaling destruction of the former and not molesting the latter. (Since in the case of *sib*-mediated regulation the expression of the *int* and *xis* genes is controlled by an element downstream from the genes themselves, it is often referred to as *retroregulation*.)

Now consider the case of prophage induction, in which transcription has begun from PL but the prophage DNA is still integrated into the host chromosome. The crucial difference here is that, because of the rearrangement of DNA sequences that took place during the integration reaction, the *sib* site is

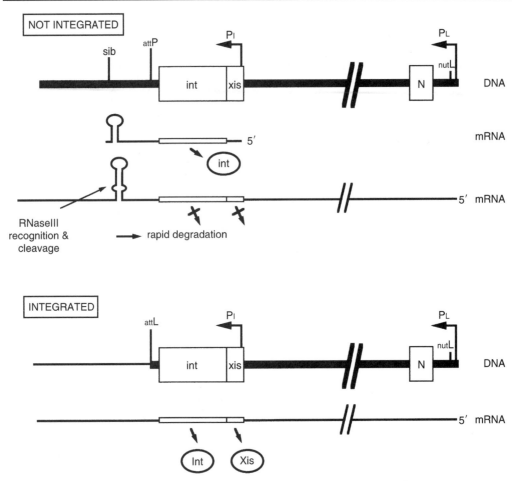

Fig. 6. Regulation of integration and excision. In the nonintegrated prophage DNA (upper diagram), transcription from PI terminates at *sib*, resulting in a mRNA with a stable 3' end that can produce the Int needed for integration. RNA polymerase originating at PL, on the other hand, reads through the terminator, allowing formation of the slightly larger secondary structure that tags the mRNA for destruction. In the integrated prophage (lower diagram) *sib* is no longer downstream from *att*, so the transcript originating at PL does not form the destruction signal, and the mRNA can produce the Int and Xis proteins needed for excision.

no longer downstream from PL, so the PL transcript never encounters *sib* and therefore does not get rapidly degraded, with the result that both Int and Xis proteins are produced and the excision reaction goes ahead. To summarize, the phage uses the *sib* site to sense whether the phage DNA is integrated or not: in the integrated state the *sib* site is removed from the PL transcript, and both Int and Xis synthesis are therefore allowed and excision can occur. In the nonintegrated state the *sib* site is part of the PL transcript and that transcript is destroyed, leaving only

Int produced from the PI transcript to catalyze the integration reaction.

It only remains now to describe how the *sib* site selectively targets transcripts that originated at PL for destruction while leaving transcripts from PI untouched. The reader may have guessed that the crucial difference between these two transcripts is the one already described: an RNA polymerase transcribing from PL has been acted on by the N protein and is consequently insensitive to termination signals, while a polymerase reading from PI is still susceptible to those termination signals. When a transcribing polymerase enters the *sib* region it encounters an inverted repeat in the sequence, which folds into a stem-loop structure in the RNA, followed (in the RNA) by a string of 6 uracils. This is a conventional factor-independent transcription termination signal, and if the polymerase started at PI, then transcription terminates at the end of the run of U's to leave a relatively stable mRNA encoding Int. On the other hand, when a polymerase that initiated at PL encounters the *sib* region, it ignores the termination signal and continues transcribing beyond it. A secondary structure is then able to form in the mRNA which includes the stem-loop of the terminator but also a stem that is made of sequences both upstream and downstream from the terminator. This secondary structure (an interrupted stem-loop) can only form because the polymerase ignored the terminator and made the crucial part of the RNA sequence downstream from the terminator. Once formed, it is specifically recognized by the enzyme RNaseIII as a signal for cleaving the mRNA, and once cleaved by RNaseIII, other nucleases in the cell rapidly degrade the mRNA. Thus the seeds of destruction for the mRNA lie in the property of the RNA polymerase that allows it to read through terminators.

IX. EVOLUTION OF λ AND ITS RELATIVES

Phage λ has been isolated from nature only once, but from the earliest days of research on λ, biologists working on this phage have made use of a group of independently isolated related phages, often called the *lambdoid* phages, to provide comparisons. These phages have the same genome organization as λ—that is, the same kinds of genes in the same order along the genome—and they can recombine with λ to make biologically functional hybrids. It would seem that with enough information about all these phages, for example, the complete DNA sequences of their genomes, it should be possible to deduce phylogenetic relationships among

them and learn something about how they have evolved. This has in fact turned out to be the case, but with an interesting twist. That is, when the sequences of any two of these phages, say λ and the lambdoid phage HK97, are compared, they are clearly seen to be related, but in a much more complex way than might have been imagined; each phage in the lambdoid group can be thought of as a *genetic mosaic* with respect to the rest of the group. Thus in a pairwise comparison of phages, one pair of genes—for example, the *cI* genes—may be very similar in sequence between the two phages, but the adjacent pair of genes may have a very much lower level of similarity. Said another way, if we take the degree of sequence similarity between two phages to be a measure of how long ago they diverged from a common ancestor, we get very different answers when we look at the sequences of different pairs of genes.

The solution to this conundrum is as follows: As in any other evolving population, diversity among the lambdoid phages in the population arises in part as a result of mutational changes in the genome sequences, and further diversity is generated as those mutational differences are reassorted with each other through homologous recombination. (It is this diversity that natural selection acts on.) In phages, however, another very important source of diversity is the process known as *horizontal exchange*. Horizontal exchange refers to the swapping of chunks of DNA sequence between genomes through the process of nonhomologous recombination—that is, recombination between sequences that are different from each other—to create novel juxtapositions of sequence that did not exist in either parent. In the hybrid phage genomes that result, one part of the sequence may have a very different evolutionary history from another.

Nonhomologous recombination is essentially a mistake of the recombination system, since homologous recombination is "supposed" to recombine two identical or nearly identical sequences to produce progeny that are very much like both parents. Nonhomologous recombination occurs quite rarely, but given the enormous numbers of phages in the biosphere and the very long time phages are thought to have been engaging in recombination with each other, it has evidently occurred innumerable times among the lambdoid phages. Note also that because nonhomologous recombination can paste together DNA sequences essentially at random, it is very likely that most of the hybrid phages produced in this way are nonfunctional monsters; the ones that survive natural selection to be examined by us are the rare ones that are as fit or more fit than their parents.

The process of horizontal exchange of genes means that phages can sometimes acquire some rather unexpected, "un-phage-like" genes that can then be carried into a bacterial genome as part of a prophage. When this happens, the novel gene (as is also true for all the other phage genes) becomes part of the bacterial genotype and can therefore potentially affect the bacterial phenotype. In this way, phages, and particularly temperate phages, can have a big impact on the evolution of their hosts. Among the many examples of phage genes that alter the phenotype of their host cells by this mechanism are the genes encoding the toxins of diptheria, botulism, cholera, scarlet fever, the deadly O157:H7 strain of *E. coli*, and ovine footrot.

The lambdoid phages are not the only ones that show abundant horizontal exchange of genes. Examination of the genome sequences of groups of phages different from the lambdoid phages, for example, phages that infect the Mycobacteria, shows that these groups also undergo high levels of horizontal exchange within their own group. More surprisingly, there is evidence for horizontal exchange of sequences at a much reduced frequency even between very different groups of phages, such as the lambdoid phages and the mycobacterial phages. Thus in this sense all of the phages (or at least all of the $> 10^{30}$ dsDNA tailed phages, which is what we are considering here) are part of a single genetic population.

We are just in recent years beginning to get a glimpse of how astoundingly numerous and diverse that population is; suffice it to say that if each of those 10^{30} phages were transformed into a beetle, the surface of the Earth would be covered with a 50,000 km deep layer of beetles. Each of those beetles—I mean phages—is presumably as complex and elegantly regulated as λ, but each one carries out its program of infection and propagation with a different specific combination of genes, gene sequences, and regulatory sequences and with a correspondingly different (sometimes very different!) variation on the themes of lifestyle, genetic regulation and biochemical mechanism that have been investigated so thoroughly in phages like λ and T4. We are coming to realize that the diversity of the global phage population constitutes a rich and largely untapped resource of genes and genetic and biochemical mechanisms, not only for revealing novel mechanisms of biological function but also as a source of raw materials for pharmaceutical and other biotechnological applications. The task ahead for phage biologists is to figure out how to use the extensive knowledge gained over the past 50 years of studying λ and a few other phages to mine the riches of the global phage population as a whole.

SUGGESTED READING

Hershey, AD (ed) (1971): "The Bacteriophage Lambda," Cold Spring Harbor, NY: Cold Spring Harbor Laboratories.

Hendrix RW, Roberts JW, Stahl FW, Weisberg RA (eds) (1983): "Lambda II." Cold Spring Harbor, NY: Cold Spring Harbor Laboratories.

These two books, published just over ten years apart, give comprehensive views of the then-current states of λ biology. In addition to detailed reviews of the topics covered in this chapter (among others), they provide access to the original research papers on these topics.

Ptashne M (1992): "A Genetic Switch: Phage λ and Higher Organisms." Cambridge, MA: Blackwell Scientific.

This book provides a readable summary of work leading to our understanding of how repression and the lytic/lysogenic decision work. There is an emphasis on the logic and progress of the research as well as its results.

Reichardt L (1975): Control of bacteriophage lambda repressor synthesis after phage infection: The role of the *N, cII, cIII* and *cro* products. J Mol Biol 93: 267–288.

This research article gives definitive information about how the lytic/lysogenic decision is made.

Casjens SR, Hendrix RW (1988): Control mechanisms in dsDNA bacteriophage assembly. In Calendar R (ed): "The Bacteriophages." New York: Plenum Press, pp 15–91.

This review covers a topic not discussed in detail in this chapter, namely how virions are assembled from their component macromolecules.

Campbell A (1994): Comparative molecular biology of lambdoid phages, Annu Rev Microbiol 48: 193–222.

Casjens S, Hatfull G, Hendrix R (1992): Evolution of dsDNA tailed–bacteriophage genomes. Seminars in Virology 3: 383–397.

These two reviews summarize much of our current understanding about lambdoid phage evolution and population structure.

Hendrix RW, Smith MCM, Burns RN, Ford ME, Hatfull GF (1999): Evolutionary relationships among diverse bacteriophages and prophages: All the world's phage. Proc Natl Acad Sci USA 96: 2192–2197.

This research article makes use of recently determined phage and prophage genome sequences to derive a broad view of the evolutionary relationships among all tailed phages.

Where do all these claims about how λ works come from? Here's a very abbreviated glimpse at some of the experimental basis for our current understanding of λ repressor and how it is regulated.

The first mutants of λ to be isolated were clear plaque mutants, isolated in the mid-1950s by Dale Kaiser, who was a graduate student at Caltech and then a postdoctoral fellow at the Pasteur Institute. (Plaques are the visible areas of phage growth—and bacterial killing—in a lawn of bacterial growth that are used to assay phages. λ plaques are normally slightly cloudy or "turbid" due to the growth of lysogenic cells that were established during formation of the plaque. Mutants of the phage that are unable to form lysogens make "clear" plaques.) Kaiser carried out genetic complementation experiments to divide his clear plaque mutants into three complementation groups, defining three genes which he named *cI, cII,* and *cIII.* Subsequent genetic experiments showed that the *cI* gene encodes some sort of "repressor substance" responsible both for repression of the prophage and for immunity of the lysogenic cell to superinfection. A decade later, Mark Ptashne, a junior professor at Harvard University, succeeded in isolating the repressor and showing that it was a protein. Biochemical experiments by Ptashne and others worked out the molecular behavior of repressor protein, including how it recognizes and binds specifically to its operator sites. There followed a large number of both genetic and biochemical experiments by many labs around the world directed at the mechanisms of the lytic/lysogenic decision. Particularly notable was work done by Louis Reichardt as part of his Ph.D. dissertation research in the laboratory of Dale Kaiser at Stanford University. Reichardt established the role of CII protein and the Pre promoter in establishing repression.

APPENDIX: SPECIALIZED TRANSDUCTION

In normal prophage excision, the excisive recombination event takes place between the *attL* and *attR* sites at the ends of the prophage, reconstituting the *attP* site and precisely removing the phage DNA from the bacterial chromosome (see Fig. 5). On rare occasions, however, recombination happens by mistake, not at one of the attachment sites but in the adjacent bacterial DNA. As a result the DNA that is excised and packaged into phage particles includes some of the DNA that flanked one end of the prophage. The effects of this process were originally seen when it was noticed that the phages produced by induction of a λ lysogen could transfer genetic information from the genes of the galactose operon of the phages' original host into the genetic makeup of the next host that those phages infected. Such transfer of genetic information is termed *specialized transduction*— "transduction" describes the virus-mediated transfer of genetic information from one cell to another, and "specialized" refers to the fact that λ is only able to transduce genes that lie adjacent to the prophage DNA in the lysogen. (Some phages, in contrast to λ, will transduce any genes of their host. This process, called *generalized transduction*, occurs by a different mechanism from specialized transduction; it is discussed in detail by Weinstick, this volume.) In addition to the *gal* genes, which lie on one side of the prophage, λ is also able to mediate specialized transduction of the genes of the biotin (*bio*) operon from the other side of the prophage.

Specialized transduction results from the rare aberrant excision of the prophage described above and the consequent packaging of some host DNA into the virions. Such virions can transfer their DNA (including the attached host DNA) into a new host by the same efficient DNA injection mechanism that normal virions use to infect a cell. Once in the cell, if the transducing phage enters the lysogenic cycle (and therefore doesn't kill the cell), the genes from the previous host can become integrated into the new host's genome as part of the prophage. If the recipient host was a *gal⁻* mutant, the transduced cell ("transductant") can be selected easily by its ability to grow on galactose-containing medium. If this new lysogen is subsequently induced, all of the virions produced will be transducing virions—that is, they should all carry the host DNA—and the efficiency of transduction with such a preparation is many orders of magnitude higher than with the original one. In reality the

situation is a bit more complicated than described. The fact is that in the original aberrant excision, in order for the host-DNA-containing genome to fit into the phage capsid, the excision must occur in such a way that DNA is lost from the opposite end of the prophage to compensate for the extra host DNA. Thus all transducing virions are missing some phage genes, or said another way, some of their phage genes have been replaced by host genes. Since the phage genes that are missing often include ones that are essential for lytic growth of the phage—tail and sometimes head genes for λ*gal* transducing phages—these phages can only be propagated in the presence of a "helper phage" that can provide the missing functions *in trans*.

Specialized transduction is really a particular example of a more general phenomenon known as *lysogenic conversion*. Lysogenic conversion refers to changes in the phenotype of a cell that result from its acquisition of a prophage. Perhaps the clearest example of lysogenic conversion is that when an *E. coli* cell acquires a λ prophage, it becomes immune to infection by other λ's because of the expression of the CI repressor by the prophage. In addition to the conversion of host phenotype that results when a specialized transducing phage becomes a prophage, other examples of lysogenic conversion include the examples cited above of prophages that carry the toxin and other pathogenicity genes of pathogenic bacteria.

Studies of specialized transduction by λ played a critical role in early λ genetics. Most important, in working out how specialized transduction works, a major contribution was made to deciphering the mechanisms of prophage integration and excision by the wild-type phage. The availability of transducing phages also greatly facilitated studies on the *gal* and *bio* genes of *E. coli*, as well as studies on the relatively few other sets of genes found close to the *attB* sites of different temperate phages. With the advent of recombinant DNA techniques, studies with λ specialized transducing phages were largely eclipsed and are now not common. However, the concepts developed in the early studies of λ specialized transducing phages played an important role in the development of cloning vectors. λ based cloning vectors—which are really just specialized transducing phages that are not restricted to carrying only DNA from near their attachment site, or for that matter to only carrying DNA from a particular organism—were among the very first cloning vectors developed for cloning DNA, and they have remained important to the present time. More generally, the idea of using temperate phages to carry nonphage DNA between cells has been expanded to include other viruses. As an example of recent interest, much of the current work on gene therapy uses viruses in which viral DNA has been replaced with nonviral DNA as vectors to introduce theraputic DNA into cells, an idea with its roots in the early studies of specialized transduction by λ.

6

Single-Stranded DNA Phages

J. EUGENE LECLERC

Molecular Biology Division, Center for Food Safety and Applied Nutrition, US Food and Drug
Administration, Washington, District of Columbia 20204

I. INTRODUCTION

Although the discovery of the smallest bacteriophages can be traced to 1927, interest in them intensified only after R. L. Sinsheimer demonstrated in 1959 that the particles contain a single-stranded DNA genome. The research spurred by these features—the small and single-stranded DNA genome—led over the next 20 years to revelations of fundamen-

Modern Microbial Genetics, Second Edition. Edited by
Uldis N. Streips and Ronald E. Yasbin. ISBN 0–471–38665–0
Copyright © 2002 Wiley-Liss, Inc.

tal information on the frugal uses of nucleotide sequence, of viral and host mechanisms for DNA replication, and of host functions adapted for viral reproduction. The φX174 phage that Sinsheimer studied was an important participant in the first era of molecular biology, largely phage biology, which culminated in the total in vitro synthesis of infectious φX174 DNA. During the second great era, φX174 DNA was the first genome to be entirely sequenced and engineered derivatives of M13 and f1 genomes were the most commonly used vectors for sequencing cloned DNA.

In this chapter we describe the life cycles, genetics, and biochemistry of the two general classes of single stranded DNA phages: the spherical or isometric phages, including φX174 (commonly called φX), S13, and G4, and the rod-shaped or filamentous phages M13, f1, and fd. Two other classes of bacteriophages, the isometric single-stranded and double-stranded RNA phages, will only be briefly discussed. Finally, we review some of the current and novel uses of the single-stranded DNA phages, which continue to have a significant place in genetics and biochemistry research.

II. HISTORY AND CLASSIFICATION

The initial characterization of the smallest phages of enteric bacteria came from measuring the sizes of bacterial viruses by filtration and sedimentation analyses (see review by Hoffman-Berling et al., 1966). S13 was shown to have a particle diameter of 25 mm, less than one-fourth the size of T phages. Subsequent electron micrographs of φX174 showed a similar size; the particles were polyhedral and contained a knob or spike at each of 12 axes of symmetry (Hall et al., 1959). Several φX-like phages have since been identified, differing in serological properties and preferences for hosts among strains of *Escherichia*, *Shigella*, and *Salmonella*. Taking into account their morphology and nucleic acid content, the phages are currently classified together as isometric DNA phages. Members of the isometric group that have been studied to varying degrees include φX174, S13, G4, St-1, φK, and α; they seem to have conserved from a common ancestor a similar morphology, genome organization, and protein functions, while differing significantly in nucleotide sequence. The phages of the isometric group are virulent; that is, they kill and lyse their host bacteria at the end of the phage life cycle.

The initial impetus for studying the small phages was the limited size of their genomes; the potential existed for completely defining the genetic content and organization of homogeneous DNA and for understanding all viral functions upon infection of host cells (see Sinsheimer, 1966, 1991). The realization of these goals started with the early studies of φX174 by Sinsheimer and on S13 by I. and E.S. Tessman. Characterization of the phage DNAs showed unusual features that contrasted sharply with the known properties of double-stranded DNA (Sinsheimer, 1959a,b; Tessman, 1959): φX DNA reacted with formaldehyde and was precipitated with lead ions, indicating that the amino groups of the purine and pyrimidine bases were accessible and not involved in base pairing; ultraviolet absorption of φX DNA was dependent on temperature over a wide range, unlike double-stranded DNA, which shows a sharp transition upon denaturation; and the density of the DNA, its light-scattering properties, and degradation by nuclease were more like denatured than native DNA. An extensive series of radiobiological experiments showed that the inactivation constants of S13 and φX phages for incorporated ^{32}P was near unity, that is, every disintegration breaks a single strand of DNA and causes lethality (Tessman et al., 1957; Tessman, 1959). Finally, determination of the base composition of φX DNA revealed ratios of A : T = 0.75 and G : C = 1.3, unlike A = T and G = C in double-stranded DNA; in retrospect, the genome of the single-stranded DNA phage was the exception that proved Chargaff's rule for complementary base pairing in double-stranded DNA. Since experiments with both intact phage and purified DNA exhibited unusual properties for genomic DNA, it was concluded that each phage particle contains one DNA molecule in single-stranded form.

The description of single-stranded DNA in ϕX and S13 phages led quickly to similar findings for another class of small phages, the filamentous group, specific for *E. coli* that contain the male fertility factor F^+ (see Porter, this volume). One rationale for the search that yielded male-specific phages was that sensitivity to phage infection, among other properties, might distinguish F^- and F^+ or Hfr bacteria (Loeb, 1960). In 1963 the descriptions of three new single-stranded DNA phages were reported: f1 (Zinder et al., 1963), fd (Marvin and Hoffmann-Berling, 1963), and M13 (Hofschneider, 1963), isolated respectively from sewers in New York, Heidelberg, and Munich. It was soon shown that the isolates were closely related and, in fact, they may be considered mutants of the same phage (Salivar et al., 1964). Besides their specificity for male hosts, explained by the requirement for F pili during adsorption, other properties of these phages differ significantly from the isometric group: no cell lysis occurs during their life cycles, and the phage particles are long, thin, and flexible, without heads or spikes. Indeed, it was initially difficult to distinguish the phage particles from the pili of host bacteria. These small phages, measuring about a micron in length, are classified as filamentous DNA phages, or FV, for filamentous viruses; more specifically, M13, f1, and fd are classified as Ff phages, denoting filamentous phages that require the pili encoded by F conjugative plasmids for infection (see review by Marvin and Hohn, 1969). Phages of the filamentous group are ubiquitous and include If and IKe phages, which require, for infection, pili specified by conjugative plasmids of the I and N incompatibility groups, respectively, and phages that infect *Pseudomonas* and *Xanthomonas*.

After infection by Ff phage, progeny phage are extruded into the medium as host cells continue to grow, albeit at a slower rate; phage plaques are visible only because of the slower growth rate of infected cells. Among the samples from which the f1 turbid plaque-formers were discovered (Loeb, 1960), clear plaques were also evident and were separately chosen for study; designated f2, they turned out to be the first RNA phages described (Loeb and Zinder, 1961). Isometric RNA phages, now isolated on every continent, form a group about as diverse as the isometric DNA phages. They are obviously fascinating for study of their mode of replication, particularly exploited in the cases of Qβ and f2. In addition, f2, MS2, R17, and Qβ became the workhorses for studies on the mechanism of translation and the ideal substrates for developing RNA composition and sequence methodology. For comprehensive discussions on all aspects of these phages, the reader is referred to *RNA Phages*, edited by N.D. Zinder (1975), and reviews by Fiers (1979) and Van Duin (1988).

III. THE VIRIONS

The single-stranded DNA of the isometric phages is enclosed in a capsid of *icosahedral* symmetry, its 20 faces requiring 60 identical subunits arranged on a sphere (Fig. 1). It has been noted that such an arrangement is commonly found for virus construction—icosahedral symmetry allows the subunits to enclose a larger volume than other types of symmetry (see review by Denhardt, 1977). Sixty molecules of gene F protein fulfill this role as the major capsid protein. At each of 12 vertices of the icosahedron is a spike or knob that may be considered a short, primitive tail; the five gene G proteins and one gene H protein that compose each spike are likely involved in host recognition and attachment to the cell surface. Within the icosahedral shell of 25 nm (to 36 nm including the spikes), the DNA forms a densely packed core, its phosphate groups neutralized by polyamines and the positive charges of virion proteins. In addition, the core contains a minor virion protein, from gene J, which may be involved in condensation of the viral DNA.

Fig. 1. Schematic representation of the φX174 icosahedron. A spike of gene H and gene G proteins, represented by the filled area, is located at the apex of each fivefold axis of symmetry. It is surrounded by gene F coat proteins, represented by stippled areas. (Reproduced from Hayashi, 1978, with permission of the publisher.)

In sharp contrast to the fixed φX spheres, the size of filamentous phage particles is determined by the length of DNA contained within them, suggesting a fundamentally different mechanism for phage morphogenesis. Approximately 1% of Ff phage preparations is composed of "miniphage" particles (Fig. 2), with 0.2 to 0.5 times the length of normal particles and DNA (Griffith and Kornberg, 1974; Enea and Zinder, 1975; Hewitt, 1975), and "polyphages," which contain multiple genomes in multiple-length particles (Salivar et al., 1967; Scott and Zinder, 1967). A

normal particle of 900 nm length and 6 to 9 nm width contains a circular strand of DNA in a protein coat of five proteins (reviewed by Marvin, 1998). The particle is a protein tube sheathing the DNA, made up of about 2700 molecules of the α-helical gene VIII protein, its subunits overlapping and likened to scales on a fish (Marvin et al., 1974). At one end of the tube are four or five copies each of gene III and gene VI proteins, involved in adsorption of phage at the tip of an F pilus, and at the other end are similar numbers of gene VII and gene IX proteins. The DNA is embedded in the 2 nm core of the particle, its circular single strand lying like a stretched-out loop in the tube. A fascinating result of morphogenesis is that the loop of DNA is always oriented the same way, gene III sequences at the gene III protein-bearing end of the particle and, at the other end, an intergenic region encoding the initiation sites for DNA replication (Webster et al., 1981).

IV. GENETIC ORGANIZATION

The inability to digest ends of φX174 DNA by exonuclease treatment led Sinsheimer to surmise that the DNA is circular. This assumption was confirmed by physical studies (Fiers and Sinsheimer, 1962) and electron microscopy (Freifelder et al., 1964). Genetic

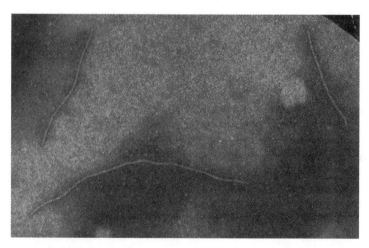

Fig. 2. Electron micrograph of an MI 3 bacteriophage filament and two miniphage particles. (Courtesy of J. Griffith, University of North Carolina.)

recombination tests using phage mutants also established that the genetic map of S13 is circular (Baker and Tessman, 1967). Indeed, determining the number and organization of genes on the circle largely depended on the analysis of phage mutants, which affected host range or plaque morphology or were conditionally lethal. The conditional lethal mutants, either temperature sensitive or suppressible, were the most useful because they identified the gene products essential for phage development; hundreds of mutants for both the isometric and filamentous phages have been collected. In order to enumerate the phage genes, genetic complementation tests were used, wherein mutations are assigned to different genes if two conditional lethal mutants, infecting cells together under nonpermissive conditions (high temperature or in suppressor-free hosts), produced progeny phage. In this way seven genes in both φX174 and S13, and eight genes in the Ff phages, were identified. Although these numbers are now revised with the availability of nucleotide sequence information, it should be noted that the bulk of our knowledge on the viral genes and their protein functions comes from the work on these extensive sets of phage mutants. (See Pratt, 1969, for a review of the early genetic work.)

The new era for analysis of genome organization started with the complete nucleotide sequence determination of φX174 DNA (Sanger et al., 1977). Entire sequences are now known for phages φX174 (revised in Sanger et al., 1978), G4 (Godson et al., 1978) fd (Beck et al., 1978), M13 (van Wezenbeek et al., 1980), fl (Beck and Zink, 1981), Hill and Peterson, 1982), and IKe (Peeters, et al., 1985). Although much remains to be learned about the functions encoded in phage genes, the goal of a complete knowledge of their organization is nearly realized. In addition the sequences provide a rich source of information for evolutionary description. In the isometric group, φX174 with 5386 nucleotides and G4 with 5577 nucleotides show considerable variability; in addition to the different lengths of the

genomes, the coding regions show 33% nucleotide changes. Differences are particularly evident in the region of the genome involved in the regulation of DNA replication; as discussed later, these phages indeed have different mechanisms for initiating replication. There is no significant homology between the phage genomes of the isometric group and the filamentous group. As anticipated, the nucleotide sequences of the Ff phages M13 (6407 nucleotides), fd (6408 nucleotides), and f1 (6407 nucleotides) are nearly identical, showing 97–99% homology, and most of the base substitutions do not result in amino acid changes. Although the gene organization for the Ff phages and the N-specific phage IKe is identical, overall homology is only about 55%. Particular divergence has occurred for the IKe protein required for binding the infecting phage to host pili, providing one molecular basis for the host ranges of F-specific as opposed to. N-specific phages.

The genetic organization for representative phages is depicted in Figures 3 and 4, giving φX174 for the isometric group and M13 for the filamentous group. The maps are drawn to show clockwise transcription of φX DNA and counterclockwise transcription of M13 DNA. The polarities of the mRNAs of both phages in fact correspond to that of the viral DNAs, making the single-stranded DNA phages plus-strand viruses; that is, all transcription for gene expression occurs on the complementary, minus strand of replicated phage DNA. The gene-encoded functions are also summarized in the figures; overall similarities between the phages may be noted, particularly in the clustering the initiation sites for complementary strand synthesis of common functions for DNA replication and phage morphogenesis. For the following discussion, however, it is best to diverge here and treat the groups separately.

A. Isometric Phages

The φX genome contains 11 genes (Fig. 3), roughly grouped corresponding to the functions of phage DNA replication and phage

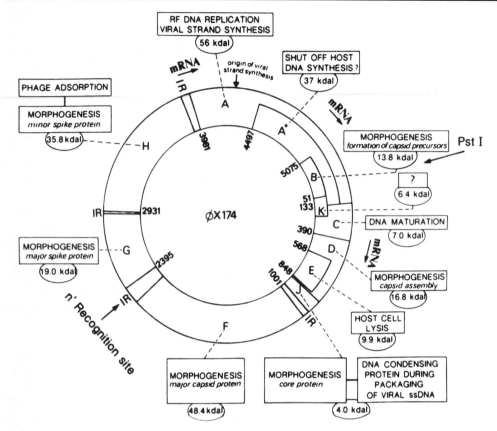

Fig. 3. Genetic organization and gene products of φX174 DNA. The numbers in the inner circle indicate the first nucleotide of the initiation codon for the respective protein, numbered from a unique *Pst* I restriction endonuclease cleavage site (position 5386/1). The protein functions (if known) and their molecular weights (derived from the DNA sequence) are given outside the circle. The direction of transcription for the major transcripts on φX DNA and the approximate positions of the initiation sites for complementary strand synthesis (n′ recognition site) and viral strand synthesis are indicated by arrows. IR, intergenic region. (Reproduced from Baas, 1985, with permission of the publisher.)

morphogenesis. Transcription starts at three promoters preceding clusters of genes whose products are used at different stages, or in different amounts, during the life cycle: genes A and A*, for proteins controlling the early functions of DNA replication and shutting off host DNA synthesis; genes B, C, and K, for proteins involved in early steps of capsid morphogenesis and DNA maturation (gene K function is unknown); and genes D, E, J, F, G, and H, whose products are used in phage morphogenesis and host cell lysis. A main mRNA terminator is located between genes H and A; other termination signals have been mapped, but since readthrough

occurs, the potential exists for more transcription of the morphogenesis genes whose products are needed in greater supply (see Fujimura and Hayashi, 1978). In addition to the coding sequences, untranslated intergenic regions (IR) ranging from 8 to 110 nucleotides are found at the borders of genes J, F, G, H, and A. Although their complete functions are probably not known, they are certainly used efficiently: all contain a ribosome binding site for the proximal gene; the H/A space has the gene A promoter; and the IR between *F* and *G* contains a recognition site for proteins that initiate DNA replication.

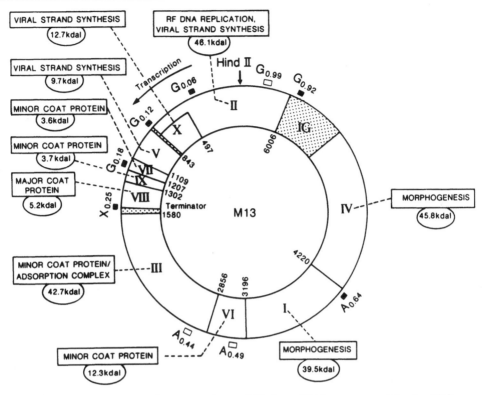

Fig. 4. Genetic organization and gene products of Ff phage DNA, represented by the M13 genome. Proteins and sequence positions are noted as in Figure 3; numbering begins at a unique *Hind*II cleavage site (6407/1). The direction of transcription is indicated by the arrow and the location of promoters is shown by bars outside the circle. (Promoters indicated by open bars are not active in vivo; see text.) IG, the main intergenic region; terminator, the main termination site for transcription. (Reproduced from Baas, 1985, with permission of the publisher.)

The most astonishing result revealed by the φX nucleotide sequence, combined with protein and mutant analyses, was the discovery of overlapping genes, or one stretch of DNA coding for more than one protein (Barrell et al., 1976; Smith et al., 1977; Weisbeek et al., 1977). As shown in Figure 3, gene B lies totally within gene A, and gene E within gene D. Sequence analysis of G4 DNA revealed an eleventh gene, K, that spans both genes A and C; it is also present in φX174 (Shaw et al., 1978). The mRNAs for overlapping genes are translated from ribosome binding sites within the preceding genes and read in different frames. Indeed, for five nucleotides that overlap genes A, C, and K, all three possible reading frames are used! A priori one might expect that the simultaneous use of a coding region in two reading frames would severely restrict the sequence differences between the genes in φX and G4. In that sense the 22–23% nucleotide changes observed for overlapping genes in the two phages are extraordinary; necessarily fewer nucleotide differences are third position or other conservative changes, leading to higher than average amino acid differences (Godson et al., 1978).

Another translational control mechanism serves to expand the use of the φX genome in gene A. The 37 kDa gene A* protein is formed by reinitiation of translation at an AUG codon within gene A mRNA, which encodes the entire 56 kDa protein (Linney and Hayashi, 1974). Proteins of different functions are specified. The same translational phase is used, so the amino acid sequence of the A* protein is identical to the carboxy-terminal half of A protein. Accordingly, nonsense mutations that block synthesis of gene A* protein also terminate gene A protein, but not the overlapping gene B protein read in a different frame (Weisbeek et al., 1977).

Although there can be argument about considering the A and A* sequences as separate genes, it is clear that five proteins (A, A*, B, C, and K) share a DNA sequence through the use of transcriptional and translational control mechanisms. These extensions of the "one gene–one protein" hypothesis display a variety of means for the frugal usage of nucleotide sequence in the small phage genomes for "deriving the most protein from the least DNA" (Pollock et al., 1978). The extra coding capacity is equivalent to nearly 1500 bp, 27% more than if the DNA were used only once (Weisbeek and van Arkel, 1978). Maybe constraints on DNA content imposed by the size of the icosahedral phage particles, or the morphogenesis process, led to the expanded use of available sequence. Mutants of φX174 have been constructed to test the maximum genome size that can be packaged; inserts greater than 3–4% of the genome were highly unstable (Russell and Miller, 1984). It has been noted that in the filamentous phages, with little constraint on genome size, φX-like overlapping genes do not occur and use is made of a prominent intergenic region (Kornberg and Baker, 1992).

Multiple uses for shared DNA sequence also present interesting puzzles for describing the evolution of new protein function. For instance, gene E protein, required for host cell lysis, may have evolved its membrane-related functions by mutagenesis of a preexisting gene D. The third position of a high proportion of gene D codons contains U; the overlapping gene E, its reading frame one nucleotide downstream, then contains the U residues in the second codon position.

Such codons specify amino acids, such as leucine, that are hydrophobic, of the kind found in polypeptides that interact with the cell membrane (Barrell et al., 1976).

B. Filamentous Phages

The ten gene products of the Ff genome can be assigned to three functional groups, their genes arranged counterclockwise on the map of Figure 4: DNA replication proteins, the products of genes II, X, and V; the coat proteins from genes VII, IX, VIII, III, and VI; and morphogenesis proteins from genes I and IV. Promoters have been identified in vitro in front of all genes except VII and IX, although not all of these promoters are used during infection; apparently the in vivo requirements for transcription are more stringent than those in vitro (Smits et al., 1984). Two intergenic regions exist. A small one of 59 nucleotides between genes VIII and III contains the main transcription terminator, so transcription from strong upstream promoters leads to much more expression of gene VIII (major coat protein) than gene III (minor coat protein) from its promoter. The large intergenic region between genes IV and II, referred to as IG, contains multiple regulatory elements for replication, transcription, and phage morphogenesis. Few nucleotides otherwise separate the genes of Ff phages, or overlaps of a few nucleotides occur, so that the 5′ regulatory elements (promoters and ribosome binding sites) for several genes lie within the coding sequences of preceding genes. In phage Ike (55% identity to the Ff genome), the genetic organization is highly similar, yet the controlling elements are quite different (Stump et al., 1997). In Ff phage, for instance, translation of protein VII is coupled to that of protein V, while in Ike it is not coupled. Yet, the same levels of protein are produced (Madison-Antenucci and Steege, 1998). It is suggested that there is a biological or evolutionary significance to maintaining the same basic genetic arrangement in the filamentous bacteriophages, while control mechanisms have evolved.

The sequence of the 508-nucleotide intergenic region, IG, of the Ff phage genomes is given in Figure 5, drawn to depict the extensive secondary structure of the multiregulatory unit. At its ends are the stop site and start site for gene IV and gene II proteins, respectively, and within the region are the origins for both complementary and viral strand replication, a rho-dependent transcription terminator, a sequence required for morphogenesis of the phage, and the gene II promoter. Transcription termination in the presence of the host rho factor occurs in the A hairpin (map position 5565 in Fig. 5) on the 5' side of IG. Hence no mRNA is synthesized in the rest of the intergenic region. That IG is the business end of the genome of phage DNA replication and packaging is evidenced by the propagation for miniphage, containing IG but no intact cistrons; in the presence of wild-type helper phage to provide phage proteins, these sub-particles outgrow the wild-type phage (Griffith and Kornberg, 1974; Enea and Zinder, 1975; Hewitt, 1975). Furthermore insertion of IG into the pBR322 cloning vector yields, again in the presence of helper phage, transducing phage particles that contain single-stranded, chimeric plasmid DNA (Cleary and Ray, 1980; Dotto et al., 1981). Therefore the noncoding region contains the *cis*-acting elements sufficient for controlling the initiation and termination reactions of DNA replication and for initiating DNA packaging (reviewed by Zinder and Horiuchi, 1985).

Although φX-like, out-of-frame overlapping genes are not found in the Ff phages, reinitiation of translation at an internal AUG codon in gene II mRNA does serve to expand the use of the genome in a manner analogous to gene A of φX, and curiously, in a gene encoding a protein of analogous function in DNA replication. The carboxy-terminal 27% of gene II protein is identical to

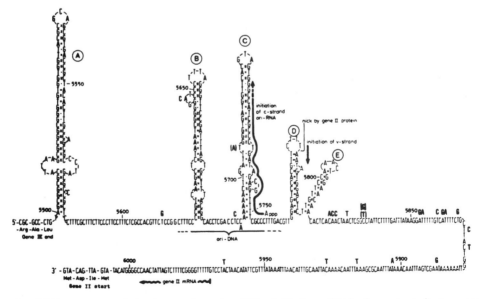

Fig. 5. DNA sequence of the intergenic region (IG) of Ff phage DNA, drawn to indicate potential secondary structures. The fd DNA sequence is given, with numbering from a unique *Hind* II cleavage site (6408/1), and base exchanges in other Ff phages are indicated as follows: parentheses, exchange in fl only; brackets, exchange in M13 only; no parentheses or brackets, exchanges in both fl and M13. See text for discussions of functions encoded in hairpins A–E and initiation sites for complementary strand (c-strand) and viral strand (v-strand) DNA synthesis. (Reproduced from Beck and Zink, 1981, with permission of the publisher.)

gene X protein, so that gene products of 42 kDa and 13 kDa are encoded in the same sequence and reading frame (Yen and Webster, 1981). Studying the in vivo functions of the products from in-frame overlapping genes presents an interesting challenge—in the overlapping region, a mutation in one gene necessarily affects the other gene, unlike out-of-frame overlaps. Fulford and Model (1984) did a clever analysis by site-specifically mutagenizing the gene X initiation codon, AUG, to an amber (termination) UAG codon so that no gene X protein is made. They then propagated the phage in amber-suppressing cells carrying gene X on a plasmid. In the absence of the complementing gene X on the plasmid, the amber mutant could not grow; it produced no single-stranded DNA for progeny virions, indicating a specific requirement for gene X protein function in phage DNA synthesis.

V. PHAGE LIFE CYCLES

The consequences for the host bacteria are drastically different in the cases of infection by isometric phages or filamentous phages. Isometric phages are virulent and lyse the host cell after about 30 minutes of infection, which yields about 200 progeny phage per cell. Filamentous phages grow in their host as parasites, more akin to plasmids, and the host cells continue to grow, divide, and extrude several hundred phage per cell generation. A variety of control mechanisms are used throughout the phage life cycles, either to "manage their own course of development and streamline the exploitation of the hapless host" in the case of isometric phages, or in contrast for the Ff phages, to "stage a rapid coup and then install a stable new regime" (Fulford et al., 1986). Although the phage groups will often be treated together in the following description of life cycles, it will be useful to keep in mind the fundamentally different phage and host relationships that ensure efficient viral reproduction. For purposes of the description, stages of the life cycle will be designated adsorption and penetration, DNA replication, and phage assembly and release, but in reality these stages overlap and contain several steps.

A. Adsorption and Penetration

The extraordinary accomplishment of phages is that one infecting particle so quickly and efficiently gains control of the host synthetic apparatus, in competition with the host genome a thousand times its size. The adsorption and penetration stages at the onset of successful infection unfortunately remain the least well-understood aspects of the infection process for the small phages. A general model was proposed for a "pilot" protein that guides the phage and its DNA through many stages of development, from cell surface interactions through DNA replication (Jazwinski et al., 1975a). Although properties of the proteins encoded by gene H of the isometric phages and gene III of the Ff phages fulfill many of the criteria for a pilot function, evidence does not strongly support such multifunctional roles for one phage protein (see Tessman and Tessman, 1978; Rasched and Oberer, 1986). But it is still useful to consider how the phage genome is delivered to a site available for immediate replication, as a consequence of the phage adhering to the cell surface and penetrating the cell membrane.

I. Isometric phages

For the isometric phages, the 12 spikes of the capsid are the adsorption organelles, each composed of one molecule of H protein surrounded by 5 molecules of G protein. As might be expected, phage mutations that affect the host range and adsorption rates map not only in genes G and H, but also in gene F for the coat protein (see Tessman and Tessman, 1978). The early work that pointed to adsorption occurring at a spike involved some good inference. Hutchison et al. (1967) infected E. coli with both wild-type φX phage and host range mutants (in gene H), yielding progeny phage that contained capsids with varying proportions of wild-type and mutant subunits. Homogeneous fractions of phage from the mixed progeny could be prepared by electrophoresis, so these workers compared the host range phenotype of hybrid capsids with their sub-unit composition, estimated from electro-phoretic mobility. With the assumption that all subunits must be mutant to adsorb to the extended host, the results nicely conformed, on a statistical basis, with 5 subunits at the adsorption site. Since a fivefold axis of symmetry exists at each of the 12 vertices of the φX icosahedral particle, Hutchison et al. (1967) suggested that a spike at the vertex is responsible for adsorption. Their inference was supported by the results of experiments on the adsorption of phages to cell wall fragments and their attachment, viewed by electron microscopy, at the tip of a vertex (Brown et al., 1971). The cellular component for phage interaction is lipopolysaccharide (LPS) in the outer cell membrane (Incardona and Selvidge, 1973; Bruse et al., 1991), and it is the H protein that interacts specifically with LPS (Suzuki et al., 1999). Different bacteriophages have adopted a wide variety of specific cell surface molecules as receptors; even the closely related phages φX174 and S13 differ, in that N-acetylglucosamine on the lipopolysaccharide chain is required for binding φX, but not S13 (Jazwinski et al., 1975b). S13 adsorbs to specific polysaccharides, requiring either hexose with substituted alpha-D-glycopyranosyl groups or heptose with the same groups in a different position (Bruse et al., 1989).

The adsorption stage of infection is reversible; infectious phage particles can be detached from cells, for instance, by EDTA treatment. The next complex steps of "injecting" phage DNA into the cell are poorly understood. For most bacteriophages an eclipse stage has been defined during which infectivity of the phage particles for fresh cells is irreversibly lost, presumably because of conformational changes in the capsid proteins and/or their removal from viral DNA. The removal of coat proteins during eclipse of φX and S13 is coupled to replication of the viral DNA. Curiously, at least one of the gene H proteins remains

attached to the DNA enters the cell and is involved, as assessed by mutant studies, in the replication of phage DNA (Jazwinski et al., 1975a). The possible multifunctional roles of H protein led to the aforementioned proposal for its pilot function, but the function(s) of the wild-type H protein at each step of infection need to be determined in order to adequately evaluate the model. Another intriguing result is that both adsorbed phage and its replicated DNA occur at zones of adhesion between the outer and inner membranes of E. coli (Bayer and Starkey, 1972). Does the cell surface receptor direct infecting DNA to membrane sites with immediate access to the DNA replication machinery? These observations need clarification, not only for understanding phage biology but also because the phage systems provide exceptional models for exploring how viruses so effectively gain control of host metabolism.

2. Filamentous phages

The specificity of filamentous phages for male host bacteria provides the clue that adsorption involves the conjugative pili of transmissible plasmid-bearing strains. Electron micrographs indeed show adherence at the tips of a phage filament and a pilus, while F-specific RNA phages are shown attached to the sides of F pili. It is considered unlikely, however, that the pilus is used to conduct phage DNA in a manner analogous to conjugation (see Marvin and Hohn, 1969). Rather, the whole pilus with attached phage may be retracted into the cell surface, or the phage may be guided down the pilus to a receptor on the cell surface. The ability of cells lacking pili to propagate phage, for instance after transfection using viral DNA, indicates that pili are used solely at the adsorption and/or penetration stages of infection. A cell envelope protein, encoded by the E. coli fii locus, is required for penetration, but not adsorption, by phages f1 and IKe (Sun and Webster, 1986). Since fii mutants are proficient for infection by the isometric DNA phages and RNA phages, as well as for conjugation, analysis of Fii function should aid in understanding the penetration step for the filamentous phages.

Much information is available on the roles of Ff phage proteins during the adsorption and penetration stages; an extensive literature on the subject is thoroughly discussed by Rasched and Oberer (1986). A summary picture that we have, partly based on model studies (Griffith et al., 1981), is as follows. Gene III protein is located at one end of the phage filament, probably in an "adsorption complex" with gene VI protein. During adsorption to the cell, the complex appears with a "knob-on-stem" structure; the knob is the amino terminus of gene III protein attached to the pilus tip and the stem anchoring the phage is the carboxyl terminus attached to the phage coat of gene VIII protein (Gray et al., 1981). During eclipse, a transition occurs in the gene VIII protein (it loses α-helical structure), contracting the phage filament to a spheroid as DNA is ejected, its gene III end first, through a ruptured gene III protein end of the particle. The opposite end of the DNA loop, containing IG, remains associated with the phage coat, possibly attached to the gene VII and/ or gene IX proteins. How DNA release from the phage protein(s) is accomplished is not known. The uncoating of the viral DNA is coupled with its replication to a double-stranded form on the inner cell membrane. Monomers of gene VIII protein are stored in the membrane, so essentially the whole phage is resorbed during infection. Since gene III protein from the phage coat is also found in the inner membrane, associated with the replicative form (RF) of phage DNA, a pilot function for gene III protein has been suggested, analogous to that for φX gene F protein (Jazwinski et al., 1973, 1975a). Some specific objections to the proposed pilot function are that the gene III protein is on the opposite end of the phage from IG, which controls the initiation of DNA replication, and that gene III protein is not needed for phage DNA replication, either in vitro or during transfection with

purified DNA (see Rasched and Oberer, 1986). As with φX, the coordination of events during eclipse and the involvement of the cell membrane need further investigation.

B. DNA Replication

The two special features of the genomes of the single-stranded DNA phages made them fascinating for studying their modes of reproduction. First, that the genomes are small suggests that even when frugally used, they are mostly devoted to genes specifying phage morphogenesis and structure, leaving little room for DNA replication enzymes. Second, their single-stranded nature requires them to be converted to a double-stranded form as a prerequisite for most DNA transactions, in particular, transcription for the expression of any phage-encoded proteins that function in replication. Necessarily the phages then rely on host enzymes for DNA replication. Kornberg and Baker (1992) considered the mechanisms of viral replication as "windows on cellular replication"; indeed, the work carried out using the φX, G4, and M13 (or fd) systems led to the discoveries of most of the proteins used for the replication of E. coli DNA (see Firshein, this volume).

The usefulness of the small, single-stranded, homogeneous, and biologically active phage DNAs as templates for DNA synthesis was demonstrated in a dramatic experiment reported in 1967. The so-called Goulian, Kornberg, and Sinsheimer experiment showed that DNA polymerase I from E. coli, provided primer fragments and deoxyribonucleotide precursors, could totally copy pure φX174 circles. Synthesis was accurate since the products of the in vitro reaction, sealed by polynucleotide ligase, produced progeny phage in transfection assays (Goulian et al., 1967). The important series of enzymological studies that followed were carried out using reconstituted in vitro systems, starting with purified phage DNAs as templates for DNA replication in extracts of uninfected cells (R. B. Wickner et al., 1972; W. Wickner et al., 1973). Cell extracts were then fractionated and the purified components added back together in *reconstitution assays* to achieve replication activity. Alternatively, specific components from extracts of wild-type cells were added to mutant cell extracts, devoid of replication activity, in order to recover activity in *complementation assays*, much like genetic complementation. It was in using these assays to detect activity during the purification and analysis of the DNA replication proteins, and the template DNAs from φX, G4, or M13 phages that have different requirements for the initiation of DNA replication, that the biochemical pathways for the conversion of viral single strands to double-stranded DNA were reconstructed in vitro. For working out the enzymology of later steps in replication, which utilize phage-encoded proteins, similar methods were employed starting with extracts of phage-infected cells.

Preceding the enzymological work, pioneering studies by Sinsheimer and colleagues defined the intermediate structures of replicated φX DNA observed during the course of infection (see review by Sinsheimer, 1968). The studies relied on marking the infecting DNA molecules with density isotopes and radioactivity, and then analyzing the in vivo replication products in cell extracts by centrifugation to equilibrium in CsCl2 gradients, in order to follow the conversion of single-stranded DNA to forms of newly synthesized DNA. Phage DNA was distinguished from the cellular DNA in extracts by transfection of spheroplasts, a method that had recently been worked out for φX DNA (Guthrie and Sinsheimer, 1960). These were the studies that defined the circular, double-stranded "replicative form" (RF) of DNA as the product of replication on an infecting viral single strand (SS). Parental RF, consisting of the viral SS DNA (the plus strand) and a complementary nascent strand (the minus strand), is then replicated in a semiconservative process to produce a pool of progeny RF. No evidence for free progeny SS DNA in the cell was found; its appearance only in phage particles implied

that SS DNA synthesis is coordinated with DNA packaging in the φX system. Ray and coworkers, whose studies defined the replication requirements and intermediates during M13 infection, found a pool of intracellular single strands late in infection, suggesting a different mode of packaging DNA in the Ff phages (Ray and Sheckman, 1969; also see review by Ray, 1977).

From a combination of biophysical, genetic, and enzymological studies, we now recognize three steps in the replication of phage DNA, for both the isometric and filamentous phages: (1) SS→RF, the synthesis of a complementary strand to produce parental RF DNA, is carried out by host enzymes that exist prior to infection; (2) RF→RF, the replication of parental RF to yield progeny RF DNA, requires phage-encoded enzymes in addition to those of the host and produces RF DNA adequate for phage-specific transcription; and (3) RF→SS, the replication of RF DNA to yield the progeny viral strands that are packaged into phage particles. A complicating feature of RF replication is that the process is asymmetric, unlike the concerted leading and lagging strand synthesis of most double-stranded DNAs. That is, replication on an RF molecule uses the complementary strand as template to produce one progeny viral strand, thereby forming another RF molecule; the "old" viral strand is peeled away, to re-enter the RF pool (via the SS→RF mode) or to be sequestered for packaging in phage coats. Which course is followed depends on the stage of infection and is subject to elaborate controls. The following only summarizes the highlights of each step in phage DNA replication, with emphasis on those control mechanisms. All aspects are described beautifully in *DNA Replication* (Kornberg and Baker, 1992) and detailed reviews of the research can be found in Marians (1984) and Baas (1985).

I. SS→RF

The reactions used in common by the single-stranded DNA phages for complementary

158 LECLERC

strand synthesis are chain elongation by DNA polymerase III holoenzyme, gap filling by DNA polymerase I, ligation of the duplex circles by polynucleotide ligase, and formation of the superhelical RFI molecules by DNA gyrase (DNA topoisomerase II). In addition all of the phage DNAs are coated with single-stranded DNA-binding protein (SSB) upon infection; the protein is required for replication, and it may protect the exposed single strands from endonucleolytic degradation. The differences among the phages that made them favorite models for host DNA replication are in the initiation reactions for de novo synthesis on infecting single strands. As depicted in Figure 6, RNA primer formation on the three representative phage DNAs is accomplished by three different enzyme systems. In the case of the filamentous phages, the host RNA polymerase recognizes the duplex regions of hairpins B and C (Fig. 5) and synthesizes approximately 30 nucleotides of RNA primer on hairpin C. The duplex region is the only phage DNA not coated with SSB, but its binding to the DNA released from the hairpin upon primer formation may terminate RNA synthesis. This simple, efficient system makes use of a plentiful enzyme, RNA polymerase holoenzyme, for priming DNA replication, and is

well suited for the parasitic lifestyle of the filamentous phages. In contrast, the virulent isometric phages use enzymes that are in shorter supply. G4 phage (as well as St-1, α3, and φK) use E. coli primase, the DnaG protein, for RNA primer synthesis in a simple reaction like that of RNA polymerase. The hairpin recognized by DnaG primase resides in the IR between genes F and G in G4 DNA, while complementary strand origins of similar sequence are located between genes G and H in the DNAs of St-1, α3, and φK. The most complex reaction is carried out on φX (and S13) DNA (see Marians, 1984). A prepriming complex of host proteins is constructed at the n′ recognition site (now called the primosome assembly site or PAS) in the IR between genes F and G. The prepriming event is necessary for association of the DnaG primase, forming a primosome that functions as a "mobile promoter" for synthesizing RNA chains on φX174 DNA (Arai and Kornberg, 1981). Although the synthesis of one primer on φX DNA probably suffices for the DNA polymerase III-mediated extension to the whole complementary strand, the priming system can generate primers repeatedly during processive movement around the DNA circle. The primosome moves in the template 5′ → 3′ direction, opposite the direction of primer formation and DNA chain growth. Since lagging strand synthesis of chromosomal DNA occurs in discontinuous steps away from the replication fork, the system is an attractive model for the repeated priming needed for initiation of Okazaki fragments (see Firshein, this volume).

The results of in vivo analyses and host mutant studies are in total agreement with the well-defined biochemical pathways for SS→RF conversion. What has not been demonstrated in an in vitro system are functions for the products of Ff gene III and isometric phage gene H, implicated by phage mutant studies to have a role in complementary strand synthesis. Their putative pilot functions could well be dispensable in soluble enzyme systems, or the pilot hypothesis

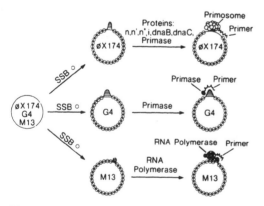

Fig. 6. Pathways for the initiation of complementary strand synthesis by RNA priming on single-stranded phage DNAs. SSB, single-stranded DNA binding protein from E. coli. (Reproduced from Baas, 1985, with permission of the publisher.)

may simply be wrong. It is clear, however, that synthesis of the parental RF occurs at the membrane, coupled to the uncoating of phage particles during penetration. Whether the discovery of membrane-associated functions changes the current picture of the SS-RF replication pathway remains to be seen.

2. RF→RF

Key participants in RF→RF replication are the phage initiator proteins, the products encoded by gene A of the isometric phages and gene II of the filamentous phages. Each of these proteins is a site-specific endonuclease-ligase that acts on the viral strand of RF molecules, first to introduce a nick at the origin of viral strand replication and, after replication of the strand, to religate the "old" viral strand circle. The origin sequence is located in IG of the Ff phage genome and, strikingly, in gene A of the isometric phages (cf. Figs. 3 and 5). The purpose of the nick is to provide a 3'-OH terminus for extension by DNA polymerase III holoenzyme. Other proteins required for RF replication are the single-stranded DNA-binding protein (SSB) and another host protein, the product of the *E. coli rep* gene. As a component of the replication complex, the Rep protein acts as a helicase to unwind the duplex DNA of RF ahead of the replication fork, using its ATPase activity to provide energy for strand separation. SSB then coats the separated strands. Upon completion of one round of replication, the viral strand is acted upon again by the initiator protein, cleaving the regenerated origin sequence to free the viral strand and ligating it to form a viral strand circle. Progeny RF continues asymmetric synthesis in this rolling circle mode (Gilbert and Dressler, 1968), while the displaced single-stranded circle is again incorporated into double-stranded DNA by the SS→RF mode. Hence two mechanisms for initiation are used during RF→RF replication: covalent starts at the cleavage sites on RF viral strands for continuous viral strand synthesis, and one of the processes of RNA priming for complementary strand

synthesis, which is required anew on each viral strand.

Although the φX gene A protein and Ff gene II protein have analogous functions in RF replication, they display several differences in reaction mechanisms. Gene A protein becomes covalently linked to the 5'-phosphate at the origin cleavage site, the energy of the phosphodiester bond being conserved throughout the replication cycle for recircularization of the viral strand. The replicative intermediates are described as looped rolling circles, consistent with the A protein-5' terminus remaining associated with the replication complex at the 3' growing end of the chain. This mechanism allows gene A protein to act processively: the A protein is transferred to progeny RF during the concerted cleavage and ligation reaction, and the replication cycle continues. In contrast, gene II protein action is distributive; the protein is either released after viral strand cleavage or it forms a weak complex with the complementary strand. Hence the subsequent replication step may utilize another gene II protein, and the energy for circularization comes from cleavage of the viral strand after replication. How the 5' end of the linear viral strand becomes associated with its 3' end for closure is not clear.

An event special to the isometric phages is likely a control step late in RF→RF synthesis. About halfway through infection, host DNA synthesis ceases, either because one or more proteins used for replication of both phage and host DNA are in limited supply or because of the specific action of a phage-encoded protein. The best candidate for the latter mechanism would be the gene A* protein, which results from the translational restart within gene A, because some mutants in the distal part of gene A fail to shut off host synthesis (Martin and Godson, 1975). As such, gene A* function may be part of a control system for the switch to RF→SS synthesis; the production of progeny RF from viral strand circles also ceases after a burst of RF synthesis, and then DNA replication is committed to the RF→SS stage for the production of phage. No abrupt cessation of RF→RF synthesis occurs during filamentous phage infection, and RF→RF synthesis occurs simultaneously with RF→SS synthesis, albeit at a reduced rate. Of course, host macromolecular synthesis is continuous for the steady-state production of filamentous phage.

3. RF→SS

Although RF→SS replication uses the enzyme systems for viral strand synthesis already described, phage-encoded proteins have the predominant roles at this stage. For the isometric phages, SS DNA synthesis is coupled to phage maturation, wherein viral strands are immediately encapsidated in a viral protein complex, the prohead, rather than being coated with SSB for subsequent doubling up by complementary strand synthesis. The prohead is comprised of the morphogenesis proteins from genes B, D, F, G, and H, and import of DNA to the prohead utilizes the products of the maturation genes C and J. Since mutant forms of the morphogenesis proteins block SS DNA synthesis, protein-DNA or protein-protein interactions that differ from those in RF→RF synthesis must be operating at the RF→SS stage. In the case of the filamentous phages, SS DNA synthesis and phage morphogenesis are not coupled. Rather than being packed immediately, the nascent SS DNA is coated with a phage-specfic single-stranded DNA-binding protein, the product of gene V. Progeny SS DNA accumulates in the cell in these nucleoprotein complexes, to be transferred to the membrane for DNA packaging.

Since the rolling-circle mode accounts for viral strand synthesis during both the RF→RF and RF→SS stages of DNA replication, what controls the switch that is essential for efficient viral reproduction? The answer probably lies in the absolute and relative amounts of phage-induced proteins in the host cell, which vary during the time course of infection; note, for instance, that for both the isometric and filamentous phages, *E. coli* SSB protein that coats the

nascent SS strands in RF→RF synthesis is replaced with phage-specific proteins during RF→SS synthesis. The most developed model to explain the fate of SS DNA comes from the work of Fulford and Model (1988a, b) on the control of f1 DNA replication (also see review by Fulford et al., 1986). By their model, three multifunctional phage proteins modulate the rates of DNA replication, favoring either RF multiplication or SS DNA destined for phage particles. The gene II protein that initiates all synthesis on RF molecules also enhances the doubling of viral strands, possibly by competing with gene V protein for binding the hairpin origin sites used for SS→RF conversion. Conversely, the cooperative binding of gene V protein to SS DNA inhibits the conversion to RF, perhaps by melting out the hairpin regions, and sequesters viral strands for packaging. In addition unbound gene V protein—at high concentration late during infection—acts as a translational repressor for gene II protein synthesis, further thwarting RF production (Model et al., 1982; Yen and Webster, 1982). Gene V protein therefore appears to gauge the ratio of SS to RF molecules during the course of infection: if low, unbound protein (made from excess RF) inhibits gene II protein synthesis and checks the accumulation of RF; if high, the protein is bound up in nucleoprotein complexes and gene II protein advances SS→RF and RF→RF conversions. Subsequent experiments, however, have shown that protein V control of protein II synthesis is dispensable; it may be ancestral (Zaman et al., 1992). Finally, the gene X protein, its synthesis tied to that of gene II protein and also repressed by gene V protein, appears to antagonize gene II protein actions in RF production. (Identical to the carboxyl third of gene II protein, maybe it binds to the same sequence, but without catalytic activity or the ability to overcome gene V protein.) The interplay of the three phage-encoded proteins involved in DNA replication, from genes II, X, and V, may thus describe the regulatory circuit that keeps phage productionin a long-term steady

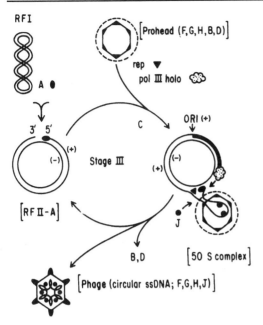

Fig. 7. Model for RF→SS DNA replication and production of φX174 virions. The nicked replicative form of DNA, in a complex with gene A protein (RFII-A), is replicated asymmetrically in the presence of the prohead, DNA polymerase III holoenzyme, and the *E. coli* rep protein. In a concerted reaction, the "old" viral (+) strand is packaged into the phage head to produce the mature φX phage particle. (Reproduced from Aoyama et al., 1983, with permission of the publisher.)

state, implying that Ff phages have evolved means to limit growth appropriate to the capability of their host bacteria. In this regard it may be worthy to recall the surmised function of the φX gene A* protein in shutting off host DNA synthesis, and possibly in the abrupt switch to phage SS DNA synthesis. In genetic organization, gene A* is analogous to the Ff gene X. Since gene X protein function appears to discourage RF production, there may be functional similarities too, except that gene A* protein evolved to fit the virulent life cycle of the isometric phages.

C. Phage Assembly and Release

1. Isometric phages

Figure 7 shows a model for RF→SS replication and packaging reactions on φX174 DNA, based on in vitro studies in which

the process was reconstituted using purified components (Aoyama et al., 1983). The reactions are concerted because the prohead and gene C protein are required for DNA synthesis; therefore the third stage of phage DNA replication is more complex than viral strand synthesis in the second stage. The steps in the process are summarized as follows (Hayashi, 1978; Burch and Fane, 2000):

1. Five gene B (internal scaffolding) proteins bind to the underside of five molecules of gene F capsid protein (9Svedberg) and likely trigger conformational changes in the upper surface of the 9S particle (Dokland et al., 1999). This then favors further interaction with five molecules of gene G spike protein (6Svedberg) into a 12S particle, preventing self-aggregation of the 9S particles.

2. Twelve of the 12S particles, 240 copies of the gene D "external scaffolding" protein, and twelve copies of the gene H protein form the 108S prohead.

3. The displaced viral strand from RF→SS synthesis is associated with 60 copies of the gene J protein and introduced into the prohead to form a compact 50S complex. Gene A and C proteins are required at this step.

4. Gene A protein catalyzes the DNA cleavage and recircularization steps, freeing the viral strand circle from RF DNA.

5. Maturation of the phage particle requires elimination of the internal scaffolding (gene B) protein to form a 132 S complex and subsequent removal of the external scaffolding (gene D) protein, to form the mature phage as a 114S particle.

The mature bacteriophage contains 60 copies each of the F protein (426 amino acids), G protein (175 amino acids), J protein (37 amino acids), and twelve copies of the H protein (328 amino acids). Most of the protein-protein interactions arise from the interfaces of the F and G proteins. Pentameric F and G proteins form the assembly intermediates (9Svedberg and 6Svedberg) and ultimately the pentameric G protein spikes are stabilized, centered at the fivefold vertices of the F protein phage coat (McKenna et al., 1994).

The COOH-terminus of the gene B product (internal scaffolding protein) is necessary for specific coat protein interactions, though the internal scaffolding proteins from the various phages in this group can crossfunction (Burch and Fane, 2000).However, even though the external scaffolding proteins (protein D) have significant homology among the different phages of the group, the stages of morphogenesis outlined above are highly specific for each of the phages and foreign D proteins inhibit morphogenesis (Burch and Fane, 2000). Of two domains in the D protein, one probably blocks formation of the procapsid while the other inhibits DNA packaging. Therefore the individual phages can control self-morphogenesis at this level, as has been noted in other bacteriophages and animal viruses as well (Marvik et al., 1995; Spencer et al., 1998).

The lytic cycle of isometric phage infection ends with the release of about 200 phage particles per cell (reviewed in Young, 1992). Cell lysis specifically requires the gene E product; when cells were infected with a gene E *am* mutant of φX and lysed artificially, the burst size averaged 2000 particles per cell (Hutchison and Sinsheimer, 1966)! The E protein does not function as a lysozyme. The N-terminal region of the protein is very hydrophobic, similar to signal peptides involved in membrane transfer, so E protein interaction with the membrane is likely (Barrell et al., 1976). It is also the N-terminal 35 amino acids that are required for the lysis activity. The lysis reaction has a dependence on the host protein SlyD, a member of the rotamase (*cis-trans*-isomerase) protein family, which may help protein E accumulate in the membrane (Roof et al., 1994). Isolation of phage E mutants that plated on *slyD* hosts demonstrated that SlyD activity is dispensable and led to the identification of the *mraY* gene, encoding translocase I, as the true target of protein E action. Since the translocase forms the first lipid-linked intermediate in cell wall biosynthesis, cell lysis then appears to result from inhibition of cell wall synthesis (Bernhardt et al., 2000). Incidently, it had been noted that the mechanism of protein E lysis is quite similar to that for penicillin (Roof and Young, 1993).

2. Filamentous phages

An overview of the morphogenesis of filamentous phages is that gene V protein, which coats nascent viral DNA, is exchanged for gene VIII protein as the nucleoprotein filament is extruded through the cell membrane (reviewed by Russel, 1991; 1995). Two aspects of this directional membrane process make it a fascinating model for nucleoprotein membrane interactions. The first is the nature of the "morphogenetic signal" encoded in IG of the phage genome: How does a regulatory sequence direct the initiation of the packaging process at the membrane? Second, the major coat protein from gene VIII is found associated only with the membrane of infected bacteria. How the protein is deposited there and then utilized for phage architecture without disrupting the cell are fundamental and yet unanswered questions in membrane biology.

The replacement of gene V protein by capsid proteins during morphogenesis requires a specific nucleotide sequence encompassing much of hairpin A (Fig. 5) (Dotto and Zinder, 1983). The delineation of the morphogenetic signal came from experiments with cells that contain the plasmid vector pBR322 carrying segments of IG (Cleary and Ray, 1980; Dotto et al., 1981). Upon infection with helper phage, progeny particles that transduce ampicillin resistance were assessed as a measure of packaged chimeric DNA. The efficient production of transducing particles required both hairpin A and the complete origins for DNA replication, interpreted to mean that, in the presence of phage proteins, the chimeric plasmid DNA can be replicated in the phage mode as single-stranded DNA and then encapsulated, as is phage DNA. Hairpin A is required for packaging (transducing particles were reduced 100-fold in its absence) and the viral strand sequence (plus strand) must be located in *cis* arrangement to the DNA that is packaged, although it can be separated from the rest of IG by thousands of nucleotides. Finally, the morphogenetic signal is not needed for either viral or complementary strand DNA synthesis (Dotto and Zinder, 1983). How the morphogenetic signal in hairpin A works is not understood. It may be recognized by membrane-bound gene VII and/or gene IX proteins during the initiation of morphogenesis, because (1) the A hairpin is located on the gene VII/IX protein end of the phage filament (Shen et al,, 1979; Webster et al., 1981) and (2) that is the end of the phage extruded first from the host cell (Lopez and Webster, 1983). Three to five copies of the proteins encoded by gene VII and gene IX are present at the leading end of the virus particle (Russel, 1995). There is also evidence that protein VII and protein IX may interact with the packaging signal for bacteriophage DNA and that protein VII interacts with protein VIII, the major coat protein (Endemann and Model, 1995). Indeed, the minor coat proteins seem to define the ends of the phage particle; the end bearing the gene III and VI proteins is these first to penetrate the cell and the last to leave. The functions of the minor coat proteins must be known in order to describe phage assembly and extrusion.

The gene IV protein product is required for phage assembly. It is an integral membrane protein, usually located in the outer membrane of infected cells. It is rich in charged amino acids and has extensive β-sheet structure, similar to many outer membrane proteins. It is produced as a precursor protein, and then cleaved as it is transported to the outer membrane. While membrane integration is independent of phage assembly, this protein may be vital for extruding mature phage (Brisette and Russel, 1990). Not much is known about another phage protein essential for morphogenesis, the product of gene I.

The gene VIII protein has been studied in much more detail than the other coat pro-teins—it is abundant in the infected cell and it has been a popular model for intrinsic membrane proteins (see Wickner, 1983). The protein is synthesized in precursor form with an N-terminal signal peptide of 23 amino acids, which is cleaved after association with the inner membrane by a host peptidase. The mature protein of 50 amino acids is stored as an integral membrane protein when it is not being used as a phage coat. In the membrane the orientation of the gene VIII protein is the same whether it is inserted from the inside (during synthesis and processing) or from the outside (during the penetration step of infection): the basic C-terminus exposed to cytoplasm, the central hydrophobic interior anchoring the protein in the lipid bilayer, and the acidic N-terminus in the periplasm. (The N-termini are also exposed on the phage coat.) The hydrophilic terminus of the protein (about 50% of total protein) is in a-helical conformation when present in the membrane, while almost 100% α-helices exist in the phage coat (Nozaki et al., 1978). The transition may take place during the exchange of gene V protein for gene VIII protein (ca. 1500 molecules for 2800 molecules, respectively). Webster and Cashman (1978) picture the gene VIII molecules stacked in the membrane; a group of the C-terminal chains protruding into the cytoplasm interact with the viral DNA during or after release from gene V protein, taking on total α-helical conformation as they envelope the DNA; another group of C-terminal ends is then available to the next portion of DNA until the particle is extruded from the cell. Such a mechanism would allow the particle size to adjust to different genome lengths, observed in natural infection as the small proportion (1%) of miniphages and polyphages. In this respect it is extraordinary that the normal phages have maintained genome lengths within one nucleotide of each other in M13, fd, and fl.

Another aspect of the assembly of filamentous phages that deserves comment is the curious role of a host factor in the process. An *E.coli* gene called *fipA*, required specifically for the assembly step as assessed by mutant studies, turned out to encode thioredoxin, the cofactor for reduction of ribonucleoside diphosphates to deoxyribonucleotides by ribonucleotide reductase (Russel and Model, 1985; Lim et al., 1985; also see review by Fulford et al., 1986). The host function of thioredoxin in phage assembly apparently does not involve redox reactions, however; it is the reduced form of the protein, or the conformation of the reduced state, that has been adapted for use in the packaging pathway. Some mutants of the phage gene I can grow on *fipA* mutant hosts, taken as evidence for an interaction between thioredoxin and the gene I product, but the latter protein's function is not known either. Since *E. coli* itself grows with mutant thioredoxins, normally an abundant protein, investigations into the role of thioredoxin in phage assembly may reveal an unimagined property of the protein used in host metabolism.

VI. RNA BACTERIOPHAGES

Although this chapter and others in this volume have concentrated on bacteriophages containing DNA genomes, the reader should be aware that other classes of bacteriophages, like mammalian viruses, utilize RNA for their genetic information. In this section we discuss bacteriophages that use RNA genomes in either double-stranded or single-stranded form. The discussion will be brief, but will provide the reader with enough information and references to initiate further study. Comprehensive information can be found in *The Bacteriophages*, Vol. 2, edited by Calendar (1992).

A. Double-Stranded RNA Bacteriophages

Bacteriophage φ6 is a well-characterized RNA bacteriophage virulent for *Pseudomonas* species of bacteria (see de Haas et al., 1999; Mindich, 1999). The phage has a genome consisting of three segments of double-stranded RNA (*s*, small; *m*, medium; and *l*, large) and a lipid envelope surrounding the virus (Li et al., 1993). The virions are spherical, 86 nm in diameter, and have surface projections from the membrane as 8 nm long spikes of protein P3 anchored by protein P6. The nucleocapsids are isometric and 56 nm in diameter. The virus utilizes the spike to attach to male pili of its principal host, *Pseudomonas syringae*. As mentioned above, the strategy for pilus-specific bacteriophages is to depend on the retraction of the pilus to reach the bacterial surface. There the bacterial outer membrane and bacteriophage membranes fuse, releasing the nucleocapsid and protein P5, a lysozymelike protein, into the periplasm of the bacterial cell. Protein P5 then digests peptidoglycan and permits the nucleocapsid to reach the inner membrane (Mindich and Lehman, 1979). The nuclecapsid breaches the inner membrane through a step that requires the energized membrane and protein P8 (Romantschuk et al, 1988; Ojala et al., 1990). P8 releases from the complex and activates a four-protein polymerase complex in the nucleocapsid, which replicates the viral genome. Recently several other dsRNA-containing bacteriophages, differing from φ6, have been isolated (Mindich et al., 1999). Some of these fuse directly with the outer membrane of the susceptible bacteria, bypassing the pilus requirement.

The assembly of the virus is well documented and has been replicated in vitro (Gottlieb et al., 1990). The nucleocapsid enzyme complex shields the replication of the virus from the cytoplasm and utilizes phage products. The framework for the complex is made up of protein P1 (85 kDa) and is the site where RNA binds. The P2 protein (75 kDa) is the RNA-dependent RNA polymerase used for genome duplication. The P4 and P7 proteins are involved in (+) strand packaging and the fidelity of synthesis. The three RNA segments are packaged successively and processively. One model suggests that empty particles can bind only the small segment of RNA; once *s* has been packaged, the sites for *m* and *l* segments appear in succession until all three segments have been packaged. (Onodera et al., 1998). The filled procapsids then are covered by protein P8 to become nucleocapsids and then are enveloped by the lipid membrane (Gottlieb et al., 1990). The morphogenesis of these bacteriophage resembles in many ways that of the mammalian reovirus.

B. Single-Stranded RNA Bacteriophages

This brief review will concentrate on bacteriophage MS2 (see Stockley et al., 1994). MS2 is 275 μm in diameter with no tail and infects F pilus-displaying bacteria. The genome is a (+) sense single strand of RNA containing 3569 nucleotides. This genome encodes four genes coding for the phage capsid protein, the maturation protein (protein A), the replicating RNA-dependent RNA polymerase, and a lysis protein. There are overlapping genes in this phage with the gene for the lysis protein spanning both the capsid protein and replicase-specifying genes (Fiers et al., 1976). Following adsorption to the F pilus, the single A protein in the phage is split proteolytically into fragments of 15 and 24 kDa. The fragments remain attached to the RNA as the pilus is retracted; the complex reaches the bacterial surface and penetrates the membrane. The protein A fragments have no further documented role in phage maturation (Van Duin, 1988). The phage RNA recruits host ribosomes and, by regulating the level of expression by both ribosome binding and translation controls, produces the required level of proteins for phage assembly and the lysis of the infected cell.

The bacteriophage uses interesting genetic means to control the level of expression from the RNA genome during phage maturation. For instance, at the 5′ end of the RNA molecule a stem-loop structure forms, which can interact with a single dimer of the coat protein (Lago et al., 2001). As the level of coat protein rises, this binding is complete and inhibits ribosome binding to sequences in the 3′ end that are needed for the expression of the replicase. Consequently the achievement of the proper level of coat proteins

signals that the replication of RNA molecules can cease and further maturation may ensue. The RNA-coat protein complex then attracts further capsid dimers and a copy of the maturation protein A. This ultimately results in a self-assembled capsid. It is suggested that the free energy change that results from capsid formation may provide the momentum for this assembly.

The replicase for the RNA bacteriophage is inaccurate. This has been studied more extensively in Qβ where the misincorporation rate is between 10^{-3} and 10^{-4} per nucleotide per replication cycle (Van Duin, 1988). This is due to the lack of a 3′ to 5′ proofreading nuclease in the replicase. Frequent deletions are also noted, presumably due to regions of homology or perhaps to hairpin loops in the RNA. At any rate, though there must be selection for a wild-type RNA, the population of Qβ phage contains many variants. The genomes of these RNAs phages exhibit a significant degree of secondary structure, which may prevent annealing between the (+) and (−) strands of RNA.

The genome of MS2 bacteriophage is characterized by an overlapping lysis gene. The protein from this gene is responsible for lysing the host bacterium at the end of the infection. Control of lysis protein expression is coordinated with that of the coat protein. When coat gene translation terminates, translation of the overlapping lysis gene is triggered (Benhardt et al., 2001). A secondary structure in the RNA prevents lysis gene translation in the absence of coat gene translation. It is assumed that the same ribosome switches from coat gene to lysis gene translation. With the lysis of the cell, the cycle for the single-stranded RNA bacteriophage is concluded.

VII. USES IN BIOTECHNOLOGY

A. Cloning and Sequencing Vectors

As far back as the 1978 Cold Spring Harbor monograph *The Single-Stranded DNA Phages* (Denhardt et al., 1978), six reports revealed the development of the M13, f1, and fd genomes as carriers for foreign DNA, that is, cloning vectors. Several features of the life cycle of the filamentous phages commend their use for the cloning and subsequent manipulations of DNA:

1. There are few constraints on the size of insertions, since the Ff phages allow the packaging of larger than unit-length DNA (although inserts < 5 kb are the most stable).
2. The phages do not lyse host bacteria, allowing the expression of cloned genes or selective markers on vector DNA. Hence recombinants can be propagated and handled either as phage or as phage-infected cells.
3. The replication cycle yields RF DNA that is like plasmid DNA for all of its technical uses in genetic engineering, such as in restriction enzyme digestion and ligation.
4. The abundance of viral DNA—up to 200 RF molecules per cell and phage titers > 10^{12} per milliliter—means that adequate DNA for analysis can be obtained from as little as a milliliter of cells.
5. The bonus from the phage systems is that the DNA is packaged as single strands; double-stranded DNA may be cloned into RF in either orientation, yielding the separated strands in unique viral strand progeny. As such, the DNA from phage particles is ideally suited for the chain termination ("dideoxy") method of DNA sequencing (Sanger et al., 1977), so the greatest efforts have gone into constructing phage vectors for rapid and efficient sequencing technology.

Reviews that describe a variety of the early phage vectors and methodologies have been written by Zinder and Boeke (1982), Messing (1983), and Gelder (1986), and useful manuals on cloning and sequence analysis using the phage systems are available from the major biotechnology vendors.

The M13mp (for Max Planck) phages are probably the most popular vectors for sequencing cloned DNA. These are the "blue plaque formers" developed by Messing and

coworkers, in which a color system is used to detect clones, rather than the more usual case of inactivating antibiotic resistance genes (Messing et al., 1977; Messing, 1983). An amusing account of the events leading to the M13mp vectors emphasizes the rich history of phage biology and *lac* genetics that were brought together in their development (Messing, 1988; also see Messing, 1996). The M13mp system is diagrammed in Figure 8. The vector contains an insert of 789 nucleotides from the *E. coli lac* operon: the regulatory region and a *lacZ'* sequence encoding only the amino-terminal end (α-peptide) of β-galactosidase. The *lacZ* gene of the host cell is also defective, containing a deletion of 93 nucleotides (*lacZ*ΔM15), which eliminates amino acids 11 to 41 of β-galactosidase. Functional enzyme is produced upon phage infection (or transfection using phage DNA) by α-complementation, wherein the amino portion of the enzyme, encoded in hybrid phage DNA, complements the *lacZ*M15 protein of the host cell. When the hybrid phage are plated in a lawn of *lacZ*M15 cells in the presence of the histochemical dye 5-bromo-4-chloro-3-indoyl-β-D-galactoside ("Xgal"), α-complemented enzyme hydrolyzes the galactoside, the indoyl moiety causing the plaques of infected cells to turn a deep blue

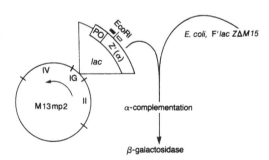

Fig. 8. The M13mp2 vector system for cloning and sequencing DNA. An *EcoR*I restriction enclonuclease site in the *lacZ*(α) gene of hybrid phage DNA provides the cloning site for foreign DNA. The open box represents the universal primer for sequencing cloned DNA; the filled box represents the probe primer for labeling DNA 5′ to the cloning site. The direction of DNA synthesis for primer elongation is indicated by the arrow on phage DNA.

color. The blue phenotype of M13lac-infected cells is ideal for cloning experiments: inactivation of the a-peptide by inserting foreign DNA into the *lacZ'* gene of RF DNA yields colorless plaques of infected cells—a vivid indicator for screening recombinants. The first usable vector, M13mp2, was developed by mutagenizing the fifth codon of the *lacZ'* gene, creating a unique EcoRl site for cloning into hybrid phage DNA. Since the early codons can be interrupted with small in-frame insertions and still encode α-complementing activity (albeit forming lighter blue plaques), multiple restriction endonuclease sites or polylinkers have been engineered at the cloning site in successive generations of M13mp vectors, making them adaptable to virtually any cloning situation. Their use is so routine that authors of published reports have referred to the parent vector as wild-type M13!

The principal use of phage vectors is sequencing recombinant DNA; they have little advantage for the expression of cloned genes. The feature that makes the phage systems advantageous is the provision of single-stranded template DNA for the chain termination method of sequencing. Double-stranded inserts cloned into phage RF DNA are naturally subjected to strand separation by the asymmetric mode of Ff DNA replication; the single-stranded viral DNA carries either strand of insert free of the other

strand, so template DNA can be obtained simply by preparing virions and extracting their DNA. To aid in cloning the desired strand, pairs of M13mp vectors have been designed to contain the same polylinker sequence in opposite orientations (shown in Fig. 9). The vectors can be cut with two different restriction enzymes to produce different ends and then the appropriately cleaved fragments "force-cloned" in either orientation. When two clones contain inserts in opposite orientations, the single-stranded DNAs from lysed virions hybridize in the complementary region; the structure migrates slower than unannealed molecules during gel electrophoresis, providing a simple assay for the orientation of recombinant DNA. Another feature introduced with the use of phage vectors for sequencing is the *universal primer*, an oligonucleotide that anneals to the vector DNA at a site flanking, and 3' to, the cloning site. Since the cloning site is at the same map position in all M13mp vectors, one primer sequence is used for all recombinant clones rather than using many different primers on one recombinant DNA. In order to accommodate the limit of sequencing 500 to 1000 nucleotides from a single primer, numerous methods have been devised for preparing nested deletions in cloned DNA, systematically drawing the distal sequences of long inserts closer to the primer site. Alternatively, small fragments

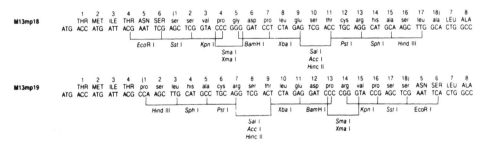

Fig. 9. Multiple cloning sites of the Ff phage cloning vectors M13mp18 and M13mp19. The amino acid sequence, given in uppercase letters, corresponds to the sequence of the amino portion of β-galactosidase encoded in the phage vector. In-frame insertions of restriction endonuclease cleavage sites, and corresponding amino acids given in lowercase letters, are in opposite orientations in the pair of vectors. (Courtesy of Research Products Division, Life Technologies, Inc., Gaithersburg, MD.)

are randomly cloned and sequenced ("shot-gun sequencing"), and the results are sorted out by computer analysis (Messing et al., 1981). Many of these novel methods developed for sequencing with the Ff vectors are now amenable to the plasmid vectors, with the introduction of efficient protocols for dideoxy sequencing of double-stranded DNA (Wallace et al., 1981; Chen and Seeburg, 1985).

The cloning sites for most of the filamentous phage vectors reside in the large intergenic space, IG, one of the few regions of the phage genome that can be interrupted without affecting viral genes. But IG is hardly devoid of function, and an insertion such as the lac insert in M13mp vectors has had interesting consequences for phage viability. The lac DNA in the progenitor of the series was cloned into domain B of the viral strand replication origin (at position 5867 on the sequence of Fig. 5). Disruption of this replication enhancer, which lies downstream from the recognition sequence for gene II protein, normally drops the activity of the viral strand origin to 1% or less. It turns out that the viability of the M13mp hybrid phages depends on a compensatory mutation within gene II; other mutations that overcome domain B defects lie within gene V or in the regulatory region for gene II mRNA, all causing the overexpression of the gene II protein by removing gene V protein–mediated repression of its translation (Dotto and Zinder, 1984a, b). Thus qualitative or quantitative changes in gene II protein make the replication enhancer largely dispensable. Nevertheless, viability is still affected in the case of M13mp vectors, since they produce titers 5- to 10-fold lower than wild-type phages or the f1 cloning vectors that have insertions upstream from the DNA replication origins (Boeke et al., 1979).

B. Hybridization Probes

Nucleic acid hybridization is a sensitive and powerful means to study the structure and function of genes—their numbers and sizes, their transcription, and their relatedness.

The preparation of single-stranded DNA probes for hybridization is another useful application of the natural strand separation afforded by filamentous phage vectors. The recombinant DNA is produced in single-stranded form, available for annealing without denaturation, and more important, it is strand specific. The latter feature is most significant for the analysis of gene transcription, since only the coding strand inserted into the viral DNA hybridizes to mRNA. Labeled probes may be prepared by using DNA polymerase and radioactive deoxyribonucleotides to elongate a probe primer that anneals to vector DNA beyond the 5′ side of the vector cloning site; complementary sequences in phage DNA are synthesized and labeled, leaving the recombinant DNA in single-stranded form. For other applications, strand-specific recombinant DNA (complementary to the insert) can be labeled by extending the universal primer and separating it before use. Specific applications of these methods are given in the review by Gelder (1986).

C. Site-Directed Mutagenesis

Changing the regulatory or coding sequences in DNA at will, once the geneticists' and biochemists' dream, is now made routine by methods of site-directed (or site-specific) mutagenesis. This technology relies on the solid phase-supported synthesis of oligodeoxyribonucleotides, so that nucleotide sequences can be designed to contain the desired changes. Smith and colleagues pioneered the methods for incorporating the changes into a genome; they used φX174 viral DNA as a model system (Hutchison et al., 1978). The oligonucleotide, made complementary to its target sequence except for the mutation site, was used as a primer for in vitro synthesis on the single-stranded DNA so that the mutant sequence could be incorporated into the complementary strand of duplex phage DNA. Upon transfection of E. coli the two strands of the heteroduplex segregate by replication; the two types of progeny phage produced were the wild-type and the desired mutant clones. The technology was quickly adopted for mutating recombinant DNA in the Ff cloning vectors when their use became widespread. In addition methods were developed to improve the efficiency of the process and to detect mutant clones by probing with labeled oligonucleotides (see Zoller and Smith, 1983).

Transfection studies with heteroduplex f1 molecules have shown that the genotypes among the progeny phage strongly favor the complementary (minus) strand of the original RF molecule; the complementary strand acts as an *"master template"* so that its genetic information dominates in the production of progeny molecules (Enea et al., 1975). Although the reasons for this phenomenon are not entirely clear, it implies that the progeny RF molecules from transfecting DNA are not used equally during RF→RF synthesis—replication must involve the asymmetric rolling-circle replication of Ff DNA. In practice, the heteroduplexes derived from site-directed mutagenesis protocols usually yield only 10–20% progeny from the mutant (complementary) strand, owing to the inefficiency of the in vitro reactions, repair or mismatch correction during transfection, or unknown causes. Therefore both biochemical and genetic approaches have been developed to trick host cells into favoring the mutant strand. Kunkel (1985) devised one of the most effective protocols, which yields 60–80% mutant progeny. The selection relies on the host excision repair system for uracil residues in DNA. Uracil in DNA (e.g., from deamination of cytosine) is normally removed by uracil DNA-glycosylase (encoded by the ung^+ gene), which excises the base and leaves an abasic site on deoxyribose as the first step in base excision repair of double-stranded DNA (see Yasbin, this volume). In single-stranded DNA, subsequent phosphodiester bond cleavage at the abasic site destroys the single strand. Uracil-containing phage DNA can be obtained by growing recombinant phage in the presence of uridine in an *E. coli ung dut* mutant strain;

the uracil residues are not excised because of the *ung* mutation and the coding properties of the DNA are unaffected, since uracil substitutes for thymine. (The additional *dut* mutation inactivates dUTPase, which cleaves dUTP, also keeping uracil out of DNA.) Used as template DNA in the site-directed mutagenesis procedure, the uracil-substituted viral strand in the resulting heteroduplex resides opposite a nonsubstituted complementary strand synthesized in vitro. By transfecting an *E. coli ung*$^+$ strain, the host uracil glycosylase provides a strong selection for the desired mutant strand, both by inactivating the viral strand of the heteroduplex and by destroying unextended template DNA. Such a high proportion of transfectants are site-specifically modified that no screening step is required—a few clones are sequenced to verify the mutant change.

D. Plasmids with Phage Origins of Replication

Although the availability of single-stranded DNA from the phage vectors is advantageous for specific applications, the smaller plasmid vectors are favored for several reasons, including the stability of large inserts, means for amplification to obtain large amounts of recombinant DNA, RNA, or protein from cloned genes, and simply because of their familiarity (see Geoghegan, this volume). The advantages of both systems have been achieved by the construction of plasmid vectors that contain the origins of DNA replication and morphogenetic signal (ca. 500 bp) from IG of the Ff phage genome (Dente et al., 1983; Levinson et al., 1984; Zagursky and Berman, 1984; review by Cesareni and Murray, 1987). The IG segment carried on the plasmid is normally silent in the cell; it is double-stranded, so the complementary strand origin is not functional, and the absence of phage-specific proteins precludes RF replication and morphogenesis. Hence the plasmid-IG chimera is treated in all ways like plasmid DNA. When an application benefits from the provision of single-stranded DNA, the cells are infected with helper phage, resulting in the extrusion of particles that contain the single-stranded plasmid DNA. Although helper phage are also produced (exceeding the yield of packaged plasmid when the plasmid insert is large), the phage DNA may not interfere with particular applications or the plasmid single strands may be purified. In addition an M13 derivative has been specially constructed for use as a helper (Vieira and Messing, 1987). Owing to inserts in domain B of the viral strand origin and the altered gene II protein from the M13mp phages, replication from the helper phage origin is less efficient than from the wild-type phage origin on chimeric plasmid DNA, resulting in helper phage titers 10- to 100-fold lower than plasmid particles.

The orientation of the IG segment in plasmid DNA determines the plasmid strand that is packaged, since it is the viral (plus) strand that is replicated during RF→SS synthesis and encodes the packaging signal. Therefore either strand of recombinant DNA can be packaged by cloning inserts in opposite orientations. Use of pairs of plasmid vectors constructed with IG inserted in opposite orientations achieves the same end. A clever biotechnological use has been made of the finding that the gene II proteins from the Ff and IKe phages are origin specific; that is, the Ff gene II protein does not act on the viral strand origin of IKe, and vice versa (Peeters et al., 1986). The plasmid vector pKUN9 contains the M13 replication origins and packaging signal on one strand and those from IKe on the other (i.e., they were cloned in opposite orientations). In cells that bear both F-and N-specific episomes, the pKUN9 strand and the recombinant strand linked to it are packaged according to which helper phage is used for infection—either M13 or IKe.

E. Phage Display Technology Using Filamentous Bacteriophage

M13 bacteriophage have been modified for use in displaying foreign proteins on the bac-

teriophage surface (Smith, 1985; reviewed by Wilson and Finlay, 1998; and see Geoghegan, this volume). Both protein VIII (major coat protein) and protein III (at tip of particle) have been used for this purpose (see Davies et al., 2000). As an example, investigators were able to display antibodies linked to protein VIII and then resultant bacteriophage could be "biopanned" with antigens of interest. Kang and coworkers created a combinatorial library of functional antibodies by combining random (kappa) light chain DNA and Fd heavy chain DNA fused to gene VIII (Kang et al., 1991). They used protein VIII since that would allow display all along the phage particle, rather than just at the tip. The light and heavy chains assemble in the periplasm with the Fd chain anchored in the membrane together with protein VIII. Phage morphogenesis then assembles a coat with the antibody on the outer surface. The phage had 1 to 24 copies of the antibody per bacteriophage. The presence of functional Fab on the surface of the phage could be confirmed by ELISA assay, and the isolated phage could interact with antigens. Waterhouse and coworkers utilized a site-specific recombination system from bacteriophage P1 to fuse heavy and light chain genes on separate replicons prior to packaging (Waterhouse et al., 1993), and then used fd filamentous bacteriophage for the display.

VIII. CONCLUDING REMARKS

The current and novel uses of the filamentous phage genomes for recombinant DNA technology are striking examples of the application of basic research. A thorough knowledge of the life cycles of the phages led to most of the developments; they were devised and constructed with predictable outcomes. In other cases a chance occurrence led to unexpected results, and we learned something new about phage biology. It is the natural progression of science that the basic research on the processes of infection by the single-stranded DNA phages has declined; the research groups actively pursuing fundamental problems have dwindled to perhaps a dozen. The irony is that the number of laboratories using the phages—as tools of research—is many times that in the heyday of phage research. Several questions remain unanswered. The molecular events at the beginning and at the end of the phage life cycles are not understood, particularly where the host cell membranes are involved. The functions of a few phage gene products are a matter of conjecture. Much can be learned about the properties of host proteins by discovering the functions that the phage systems have adopted for their reproduction. The reasons for which Sinsheimer studied the small phages, encompassed in his description of ΦX174 as *multum in parvo* (Sinsheimer, 1966), are still applicable for the fundamental problem of how genes encode information for morphogenesis and architecture. These problems will be solved experimentally; others, like describing the evolution of the phage groups (convergent or divergent?), will continue to be debated. Even if the current great era of molecular genetics, that of the human genome, seems far removed from the small phages, they are involved there too, in new biotechnological uses devised for analysis of human DNA.

REFERENCES

Aoyama A, Hamatake RK, Hayashi M (1983): In vitro synthesis of bacteriophage ΦX174 by purified components. Proc Natl Acad Sci USA 80:4195–4199.

Arai K, Kornberg A (1981): Unique primed start of phage ΦX174 DNA replication and mobility of the primosome in a direction opposite chain synthesis. Proc Natl Acad Sci USA 78:69–73.

Baas PD (1985): DNA replication of single-stranded *Escherichia coli* DNA phages. Biochim Biophys Acta 825:111–139.

Baker R, Tessman I (1967): The circular genetic map of phage S13. Proc Natl Acad Sci USA 58:1438–1445.

Barrell GB, Air GM, Hutchison CA III (1976): Overlapping genes in bacteriophage ΦX174. Nature 264:34–41.

Bayer ME, Starkey TIN (1972): The adsorption of bacteriophage ΦX174 and its interaction with *Escherichia coli*, a kinetic and morphological study. Virol 49:236–256.

Beck E, Sommer R, Auerswald EA, Kurz C, Zink B, Osterburg G, Schaller H, Sugimoto K, Sugisaki H,

Okamoto T (1978): Nucleotide sequence of bacteriophage fd DNA. Nucleic Acids Res 5:4495–4503.

Beck E, Zink B (1981): Nucleotide sequence and genome organization of filamentous bacteriophages f1 and fd. Gene 16:35–58.

Bernhardt TG, Roof WD, Young R (2000): Genetic evidence that the bacteriophage ΦX174 lysis protein inhibits cell wall synthesis. Proc Natl Acad Sci USA 97:4297–4302.

Bernhardt TG, Wang I-N, Stuck DK, Young R (2001): A protein antibiotic in the phage Qβ Virion: Diversity in lysin targets. Science, 292:2326–2329.

Boeke JD, Vovis GF, Zinder ND (1979): Insertion mutant of bacteriophage f1 sensitive to *EcoR*l. Proc Natl Acad Sci USA 76:2699–2701.

Brissette JL, Russel M (1990): Secretion and membrane integration of a filamentous phage-encoded morphogenetic protein. J Mol Biol 211:565–580.

Brown DT, MacKenzie JM, Bayer ME (1971): Mode of host cell penetration by bacteriophage ΦX174. Virol 7:836–846.

Bruse GW, Wollin R, Oscarson S, Jansson PE, Lindberg AA (1991): Studies of the binding activity of phage S13 to synthetic trisaccharides analagous to binding structures in *Salmonella typhimurium* and *Escherichia coli* C core saccharide. Correlation between conformation and binding activity. J Mol Recognit 4:121–128.

Burch AD, Fane BA (2000): Foreign and chimeric external scaffolding proteins as inhibitors of Microviridae morphogenesis. J Virol 74:9347–9352.

Burch AD, Fane BA (2000): Efficient complementation by chimeric Microviridae internal scaffolding proteins is a function of the COOH-terminus of the encoded protein.Virol 270:286–290.

Calendar R (1992): "The Bacteriophages". Vol 1. New York: Plenum Press.

Cesareni G, Murray JAH (1987): Plasmid vectors carrying the replication origin of filamentous single-stranded phages. In Setlow JK (ed): "Genetic Engineering". New York: Plenum Press, pp 135–153.

Chen EY, Seeburg PH (1985): Supercoil sequencing: A fast and simple method for sequencing plasmid DNA. DNA 4:165–170.

Cleary JM, Ray DS (1980): Replication of the plasmid pBR322 under the control of a cloned replication origin from the single-stranded DNA phage M13. Proc Natl Acad Sci USA 77:4638–4642.

Davies JM, O'Hehir RE, Suphioglu C (2000): Use of phage display technology to investigate allergen-antibody interactions. J Allergy Clin Immunol 105:1085–1092.

deHaas F, Paatero AO, Mindich L, Bamford DH, Fuller SD (1999): A symmetry at the site of RNA packaging in the polymerase complex of dsRNA bacteriophage Φ6. J Mol Biol 294:357–372.

Denhardt DT (1975): The single-stranded DNA phages, CRC Crit Rev Microbiol 4:161–223.

Denhardt DT (1977): The isometric single-stranded DNA phages. Compr Virol 7:1–104.

Denhardt DT, Dressler D, Ray DS (1978): "The Single-Stranded DNA Phages". Cold Spring Harbor, NY: Cold Spring Harbor Laboratory.

Dente L, Cesareni G, Cortese R (1983): pEMBL: A new family of single-stranded plasmids. Nucleic Acids Res 11:1645–1655.

Dokland T, Bernal RA, Burch A, Pletnev S, Fane BA, Rossman MG (1999): The role of scaffolding proteins in the assembly of the small, single stranded DNA virus φX174. J Mol Biol 288:595–608.

Dotto GF, Enea V, Zinder ND (1981): Functional analysis of bacteriophage f1 intergenic region. Virol 114:463–473.

Dotto GP, Zinder ND (1983): The morphogenetic signal of bacteriophage f1. Virol 130:252–256.

Dotto GF, Zinder ND (1984a): Increased intracellular concentration of an initiator protein markedly reduces the minimal sequence required for initiation of DNA synthesis. Proc Natl Acad Sci USA 81:1336–1340.

Dotto GP, Zinder ND (1984b): The minimal sequence for initiation of DNA synthesis can be reduced by qualitative or quantitative changes of an initiator protein. Nature 311:279–280.

Endemann H, Model P (1995): Location of the filamentous phage minor coat proteins in phage and in infected cells. J Mol Biol 250:496–506.

Enea V, Vovis GF, Zinder ND (1975): Genetic studies with heteroduplex DNA of bacteriophage f1. Asymmetric segregation, base correction and implications for the mechanism of genetic recombination. J Mol Biol 96:495–509.

Enea V, Zinder N (1975): A deletion mutant of bacteriophage f1 containing no intact cistrons. Virol 68:105–114.

Fiers W (1979): Structure and function of RNA bacteriophages. Compr Virol 13:69–179.

Fiers W, Contreras R, Duerink F, Haegaman G, Iserentant D, Merregaert J, Min Jou W, Molemans F, Raemaekers A, Van den Berghe A, Volckaert G, Ysebaert M (1976): Complete nucleotide sequence of bacteriophage MS2 RNA: Primary and secondary structure of the replicase gene. Nature 260:500–507

Fiers W, Sinsheimer RL (1962): The structure of the DNA of bacteriophage φX174. III. Ultracentrifugal evidence for a ring structure. J Mol Biol 5:424–434.

Freifelder D, Kleinschmidt AK, Sinsheimer RL (1964): Electron microscopy of single-stranded DNA: Circularity of DNA of bacteriophage φX174. Science 146:254–255.

Fujimura FK, Hayashi M (1978): Transcription of isometric single-stranded DNA phage. In Denhardt DT, Dressler D, Ray DS (eds): "The Single-Stranded DNA Phages". Cold Spring Harbor, NY: Cold Spring Harbor Laboratory, pp 485–505.

Fulford W, Model P (1984): Gene X of bacterlophage f1 is required for phage DNA synthesis. Mutagenesis of in-frame overlapping genes. J Mol Biol 178:137–153.

Fulford W, Model P (1988a): Regulation of bacteriophage f1 DNA replication. 1. New functions for genes II and X. J Mol Biol 203:49–62.

Fulford W, Model P (1988b): Bacteriophage f1 DNA replication genes. II. The roles of gene V protein and gene II protein in complementary strand synthesis. J Mol Biol 203:39–48.

Fulford W, Russel M, Model P (1986): Aspects of the growth and regulation of the filamentous phages. Prog Nucleic Acid Res Mol Biol 33:141–168.

Geider K (1986): DNA cloning vectors utilizing replication functions of the filamentous phages of *Escherichia coli*. J Gen Virol 67:2287–2303.

Gilbert W, Dressler D (1968): DNA replication: The rolling circle model. Cold Spring Harbor Symp Quant Biol 33:473–484,

Godson GN, Barrell BG, Staden R, Fiddes JC (1978): Nucleotide sequence of bacteriophage G4 DNA. Nature 276:236–247.

Gottlieb P, Strassman J, Quao X, Frucht A, Mindich L (1990): In vitro replication, packaging, and transcription of the segmented, double-stranded RNA genome of bacteriophage φ6: Studies with procapsids assembled from plasmid-encoded proteins. J Bacteriol 172:5774–5782.

Goulian M, Kornberg A, Sinsheimer RL (1967): Enzymatic synthesis of DNA. XXIV. Synthesis of infectious phage φX174 DNA. Proc Natl Acad Sci USA 58:2321–2325.

Gray C, Brown R, Marvin D (1981): Adsorption complex of filamentous fd virus. J Mol Biol 146:621–627.

Griffith J, Kornberg A (1974): Mini M13 bacteriophage: Circular fragments of M13 DNA are replicated and packaged during normal infections. Virol 59:139–152.

Griffith J, Manning M, Dunn K (1981): Filamentous bacteriophage contract into hollow spherical particles upon exposure to a chloroform-water interface. Cell 23:747–753.

Guthrie GD, Sinsheimer RL (1960): Infection of protoplasts of *Escherichia coli* by subviral particles of bacteriophage φX174. J Mol Biol 2:297–305.

Hall CE, Maclean EC, Tessman I (1959): Structure and dimensions of bacteriophage φX174 from electron microscopy. J Mol Biol 1:192–194.

Hayashi M (1978): Morphogenesis of the isometric phages. In Denhardt DT, Dressler D, Ray DS (eds): *The Single-Stranded DNA Phages*. Cold Spring Harbor, NY: Cold Spring Harbor Laboratory, pp 531–547.

Hewitt JA (1975): Miniphage—A class of satellite phage to M13. J Gen Virol 26:87–94.

Hill D, Peterson G (1982): Nucleotide sequence of bacteriophage f1 DNA. J Virol 44:32–46.

Hoffman-Berling H, Kaerner HC, Knippers R (1966): Small bacteriophages. Adv Virus Res 12:329–370.

Hofschneider PH (1963): Untersuchungen uber "kleine" *E. coli* K12 bacteriophagen. 1 und 2 mitteilung. Z Naturforsch 18b:203–210.

Hutchison CA III, Marshall EH, Sinsheimer RL (1967): The process of infection with bacteriophage φX174. XII. Phenotypic mixing between electrophoretic mutants of φX174. J Mol Biol 23:553–575.

Hutchison CA III, Phillips S, Edgell MH, Gillam S, Jahnke P, Smith M (1978): Mutagenesis at a specific position in a DNA sequence. J Biol Chem 253:6551–6560.

Hutchison CA III, Sinsheimer RL (1966): The process of infection with bacteriophage φX174. X. Mutations in a φX lysis gene. J Mol Biol 18:429–447.

Incardona NL, Selvidge L (1973): Mechanism of adsorption and eclipse of bacteriophage φX174. II Attachment and eclipse with isolated *Escherichia coli* cell wall lipopolysaccharide. J Virol 11:775–782.

Jazwinski M, Marco R, Komberg A (1973): A coat protein of the bacteriophage M13 virion participates in membrane-oriented synthesis of DNA. Proc Natl Acad Sci USA 70:205–209.

Jazwinski MS, Marco R, Kornberg A (1975a): The gene H spike protein of bacteriophage φX174 and S13. 11. Relation to synthesis of parental replicative form. Virol 66:294–305.

Jazwinski SM, Lindberg AA, Kornberg A (1975b): The lipopolysaccharide receptor for bacteriophages φX174 and S13. Virol 66:268–282.

Kang, AS, Barbas, CF, Janda KD, Benkovic SJ, Lerner RA (1991): Linkage of recognition and replication functions by assembling combinatorial antibody Fab libraries along phage surfaces. Proc Natl Acad Sci USA 88:4363–4366.

Kornberg A, Baker TA (1992): *DNA Replication.* New York: Freeman.

Kunkel TA (1985): Rapid and efficient site-specific mutagenesis without phenotypic selection. Proc Natl Acad Sci USA 82:488–492.

Lago H, Parrott, AM, Moss T, Stonehouse NJ, Stockley PG (2001): Probing the kinetics of formation of the bacteriophage MS2 translational operator complex: identification of a protein conformer unable to bind RNA. J Mol Biol 305:1131–144.

Levinson A, Silver D, Seed B (1984): Minimal size plasmids containing an M13 origin for production of single strand transducing particles. J Mol Appl Genet 2:507–517.

Li T, Bamford DH, Bamford JKH, Thomas GJJ(1993): Structural studies of the enveloped dsRNA bacteriophage φ6 of *Pseudomonas syringae* by Raman spectroscopy. I. The virion and its membrane envelope. J Mol Biol. 230:461–472.

Lim C, Haller B, Fuchs J (1985): Thioredoxin is the bacteria protein encoded by *fip* that is required for filamentous bacteriophage f1 assembly. J Bacteriol 161:799–802.

Linney E, Hayashi M (1974): Intragenic regulation of the synthesis of φX174 gene A proteins. Nature 249:345–348.

Loeb T (1960): Isolation of a bacteriophage specific for the F+ and Hfr mating types of *Escherichia coli* K12. Science 131:932–933.

Loeb T, Zinder ND (1961): A bacteriophage containing RNA. Proc Natl Acad Sci USA 47:282–289.

Lopez J, Webster R (1983): Morphogenesis of filamentous bacteriophage f1: orientation of extrusion and production of polyphage. Virol 127:177–193.

Madison-Antenucci S, Steege DA (1998): Translation limits synthesis of an assembly-initiating coat protein of filamentous phage IKe. J Bacteriol 180:464–472.

Marians KJ (1984): Enzymology of DNA in replication in prokaryotes. CRC Crit Rev Biochem 17:153–215.

Martin DO, Godson GN (1975): Identification of a φX174 coded protein involved in the shut off of host DNA replication. Biochem Biophys Res Commun 65:323–330.

Marvik OJ, Dokland T, Nokling RH, Jacobsen E, Larsen T, Lindqvist BJ (1995): The capsid size-determining protein sid forms an external scaffold on phage P4 procapsids. J Mol Biol 251:59–75.

Marvin D (1998): Filamentous phage structure, infection and assembly. Curr Opin Struct Biol 8:150–158.

Marvin D, Pigram W, Wiseman R, Wachtel E, Marvin F (1974): Filamentous bacterial viruses. XII. Molecular architecture of the class I (fd, f1, IKe) virion. J Mol Biol 88:581–598.

Marvin DA, Hoffmann-Berling H (1963): A fibrous DNA phage (fd) and a spherical RNA phage specific for male strains of *E coli.* Z Naturforsch 18b:884–893.

Marvin DA, Hohn B (1969): Filamentous bacterial viruses. Bacteriol Rev 33:172–209.

McKenna R, Ilag LL, Rossman MG (1994): Analysis of the single stranded DNA bacteriophage φX174 at a resolution of 3.0A. J Mol Biol 237:517–543.

Messing J (1983): New M13 vectors for cloning. Methods Enzymol 101:20–79.

Messing J (1988): M13, the universal primer and the polylinker. Focus 10:21–26.

Messing J (1996): Cloning single-stranded DNA. Molecular Biotechnology 5:39–47.

Messing J, Crea R, Seeburg PH (1981): A system for shotgun DNA sequencing. Nucleic Acids Res 9:309–321.

Messing J, Gronenborn B, Müller-Hill B, Hofschneider PH (1977): Filamentous coliphage M13 as a cloning vehicle: Insertion of a *Hind*II fragment of the *lac* regulatory region in M13 replicative form in vitro. Proc Natl Acad Sci USA 74:3642–3646.

Mindich L, Lehma JF (1979): Cell wall lysin as a component of the bacteriophage φ6 virion. J Virol 30:489–496.

Mindich L (1999): Precise packagaing of the three genomic segments of the double-stranded-RNA bacteriophage φ6. Microbiol Mol Rev 63:149–160.

Mindich L, Qiao X, Qiao J, Onodera S, Romantschuk M, Hoogstraten, D (1999): Isolation of additional bacteriophages with genomes of segmented double-stranded RNA. J Bacteriol 181:4505–4508.

Model P, McGill C, Mazur B, Fulford W (1982): The replication of bacteriophage fl: Gene 5 protein regulates the synthesis of gene 2 protein. Cell 29:329–335.

Nozaki Y, Reynolds J, Tanford C (1978): Conformational states of a hydrophobic protein: the coat protein of fd bacteriophage. Biochem 17:1239–1246.

Ojala PM, Romantschuk M, Bamford DH (1990): Purified φ6 nucleocapsids are capable of productive infection of host cells with partially disrupted outer membranes. Virol 178:364–372.

Onodera S, Qiao X, Qiao J, Mindich L (1998): Directed changes in the number of double-stranded RNA segments in bacteriophage φ6. Proc Natl Acad Sci USA 95:3920–3924.

Peeters BPH, Peters RM, Schoenmakers JGG, Konings RNH (1985): Nucleotide sequence and genetic organization of the genome of the N-specific filamentous bacteriophage 1Ke. Comparison with the genome of the Ff-specific filamentous phages M13, fd and fl. J Mol Biol 181:27–39.

Peeters BPH, Schoenmakers JGG, Konings RNH (1986): Plasmid pKUN9, a versatile vector for the selective packaging of both DNA strands into single stranded DNA-containing phage-like particles. Gene 41:39–46.

Pollock TJ, Tessman 1, Tessman ES (1978): Potential for variability through multiple gene products of bacteriophage φX174. Nature 274:34–37.

Pratt D (1969): Genetics of single-stranded DNA bacteriophages. Ann Rev Genet 3:343–361.

Rasched L, Oberer E (1986): Ff coliphages: Structural and functional relationships. Microbiol Rev 50:401–427.

Ray DS (1968): The small DNA-containing bacteriophages. In Fraenkel-Conrat H (ed): "Molecular Basis of Virology". New York: Reinhold, pp 222–254.

Ray DS (1977): Replication of filamentous bacteriophages. Compr Virol 7:105–178.

Ray DS, Sheckman RW (1969): Replication of bacteriophage M13 1. Sedimentation analysis of crude lysates of M13–infected bacteria. Biochim Biophys Acta 179:398–407.

Romantschuk M, Olkkonen VM, Bamford DH (1988): The nucleocapsid of bacteriophage φ6 penetrates the host cytoplasmic membrane. Embo J 7:1821–1829.

Roof WD, Horne SM, Young KD, Young R (1994): slyD, a host gene required for φX174 lysis, is related to the FK506-binding protein family of peptidyl-prolyl cis-trans-isomerases. J Biol Chem 269:2902–2910.

Roof WD, Young R (1993): φX174 E complements lambda S and R dysfunction for host cell lysis. J Bacteriol 175:3909–3912.

Russel M (1991): Filamentous phage assembly. Mol Microbiol 5:1607–1613.

Russel M (1995): Moving through the membrane with filamentous phages. Trends Microbiol 3:223–228.

Russel M, Model F (1985): Thioredoxin is required for filamentous phage assembly. Proc Natl Acad Sci USA 82:29–33.

Russell PW, Muller UR (1984): Construction of bacteriophage φX174 mutants with maximum genome sizes. Virol 52:822–827.

Salivar WO, Henry T, Pratt D (1967): Purification and properties of diploid particles of coliphage M13. Virol 32:41–51.

Salivar WO, Tzagoloff H, Pratt D (1964): Some physical-chemical and biological properties of the rod-shaped coliophage M13. Virol 24:359–371.

Sanger F, Air G, Barrell BG, Brown NL, Coulson AR, Fiddes JC, Hutchison CA, Slocombe PM, Smith M (1977): Nucleotide sequence of bacteriophage φX174 DNA. Nature 265:687–695.

Sanger F, Coulson AR, Friedmann T, Air GM, Barrell BG, Brown NL, Fiddes JC, Hutchison CA III, Slocombe PM, Smith M (1978): The nucleotide sequence of bacteriophage φX174. J Mol Biol 125:225–246.

Sanger F, Nicklen S, Coulson AR (1977): DNA sequencing with chain terminating inhibitors. Proc Natl Acad Sci USA 74:5463–5468.

Scott JR, Zinder ND (1967); Heterozygotes of phage fl. In Colter JS, Faranchych W (eds): The Molecular Biology of Viruses. New York: Academic Press, pp 211–218.

Shaw DC, Walker JE, Northrop FD, Barrell BG, Godson GN, Fiddes JC (1978): Gene K, a new overlapping gene in bacteriophage G4. Nature 272:510–515.

Shen CK, Ikoku A, Hearst JE (1979): A specific DNA orientation in the filamentous bacteriophage fd as probed by psoralen crosslinking and electron microscopy. J Mol Biol 127:163–175.

Sinsheimer RL (1959a): Purification and properties of bacteriophage φX174. J Mol Biol 1:37–42.

Sinsheimer RL (1959b): A single-stranded deoxyribonucleic acid from bacteriophage φX174. J Mol Biol 1:43–53.

Sinsheimer R (1966): φX: Multum in parvo. In Cairns J, Stent GS, Watson JD (eds): Phage and the Origins of Molecular Biology. Cold Spring Harbor, NY: Cold Spring Harbor Laboratory, pp 258–264.

Sinsheimer RL (1968): Bacteriophage φX174 and related viruses. Prog Nucleic Acid Res Mol Biol 8:115–169.

Sinsheimer RL (1991): The discovery of a single-stranded, circular DNA genome. BioEssays 13:89–91.

Smith M, Brown NL, Air GM, Barrel BG, Coulson AR, Hutchison CA, Sanger F (1977): DNA sequence at the C termini of the overlapping genes A and B in bacteriophage φX174. Nature 265:702–705.

Smits M, Jansen J, Konings R, Schoenmakers J (1984): Initiation and termination signals for transcription in bacteriophage M13. Nucleic Acids Res 12:4071–4081.

Spencer JV, Newcomb WW, Thomsen DR, Homa FL, Brown JC (1998): Assembly of herpes simplex virus capsid: preformed triplexes bind to nascent capsid. J Virol 72:3944–3951.

Stockley PG, Stonehouse NJ, Valegard K (1994): The molecular mechanism of RNA phage morphogenesis. Int J Biochem 26:1249–1260.

Stump MS, Madison-Antenucci S, Kokoska RJ, Steege DA (1997): Filamentous phage IKe mRNAs conserve form and function despite divergence in regulatory elements. J Mol Biol 266:51–65.

Suzuki R, Inagaki M, Karita S, Kawaura T, Kato M, Nishikawa S, Kashim N, Morita J (1999): Specific interaction of fused H protein of bacteriophage φX174 with receptor lipopolysaccharides. Virus Res 60:95–99.

Sun T, Webster R (1986): fii, a bacterial locus required for filamentous phage infection and its relation to colicin-tolerant tolA and tolB. J Bacteriol 165:107–115.

Tessman ES, Tessman 1 (1978): The genes of the isometric phages and their functions. In Denhardt DT, Dressler D, Ray DS (eds): "The Single-Stranded DNA Phages". Cold Spring Harbor, NY: Cold Spring Harbor Laboratory, pp 9–29.

Tessman 1 (1959): Some unusual properties of the nucleic acid in bacteriophage S13 and φX174. Virol 7:263–275.

Tessman 1, Tessman ES, Stent GS (1957): The relative radiosensitivity of bacteriophages S13 and T2. Virol 4:209–215.

Van Duin J (1988): Single-stranded RNA bacteriophages. In Calendar R (ed): The Bacteriophages. New York: Plenum Press, Vol. 1, pp 117–167.

van Wezenbeek PMGF, Hulsebos TJM, Schoenmakers JGG (1980): Nucleotide sequence of the filamentous bacteriophage M13 DNA genome: Comparison with phage fd. Gene 11:129–148.

Vieira J, Messing J (1987): Production of single-stranded plasmid DNA. Methods Enzymol 153:3–11.

Wallace RB, Johnson MJ, Suggs SV, Miyoshi K, Bhatt R, Itakura K (1981): A set of synthetic oligodeoxyribonucleotide primers for DNA sequencing in the plasmid vector pBR322. Gene 16:21–26.

Waterhouse P, Griffiths AD, Johnson KS, Winter G (1993): Combinatorial infection and in vivo recombin-

ation: A strategy for making large phage antibody repertoires Nucleic Acids Res 21:2265–2266.

Webster RE, Cashman JS (1978): Morphogenesis of the filamentous single-stranded DNA phages. In Denhardt DT, Dressler D, Ray DS (eds): The Single-Stranded DNA Phages. Cold Spring Harbor, NY: Cold Spring Harbor Laboratory, pp 557–569.

Webster BE, Grant RA, Hamilton LW (1981): Orientation of the DNA in the filamentous bacteriophage fl. J Mol Biol 152:357–374.

Weisbeek PJ, Borrais WE, Langeveld SA, Baas FD, van Arkel GA (1977): Bacteriophage φX174: Gene A overlaps gene B. Proc Natl Acad Sci USA 74:2504–2508.

Weisbeek PJ, van Arkel GA (1978): The isometric phage genome: Physical structure and correlation with the genetic map. In Denhardt DT, Dressler D, Ray DS (eds): "The Single-Stranded DNA Phages". Cold Spring Harbor, NY: Cold Spring Harbor Laboratory, pp 31–49.

Wickner RB, Wright M, Wickner S, Hurwitz J (1972): Conversion of φX174 and fd single-stranded DNA to replicative forms in extracts of Escherichia coli. Proc Natl Acad Sci USA 69:3233–3237.

Wickner W (1983): M13 coat protein as a model of membrane assembly. Trends Biochem Sci 8:90–94.

Wickner W, Brutlag D, Schekman R, Kornberg A (1973): RNA synthesis initiates in vitro conversion of M13 DNA to its replicative form. Proc Natl Acad Sci USA 69:965–969.

Wilson DR, Finlay BB (1998): Phage display: applications, innovations, and issues in phage and host biology. Can J Microbiol 44:313–329.

Wollin R, Bruse GW, Jansson PE, Lindberg AA (1989): Definition of the phage G13 receptor as structural domains of trisaccharides in Salmonella and Escherichia coli core oligosaccharide. J Mol Recognit 2:37–43.

Yen TSB, Webster RE (1981): Bacteriophage fl gene II and X proteins. Isolation and characterization of the products of two overlapping genes. J Biol Chem 256:11259–11265.

Yen TSB, Webster RE (1982): Translational control of bacteriophage fl gene II and gene X proteins by gene V protein. Cell 29:337–345.

Zagursky RJ, Berman ML (1984): Cloning vectors that yield high levels of single-stranded DNA for rapid DNA sequencing. Gene 27:183–191.

Zaman GJR, Kaan AM, Schoenmakers JGG, Konings RNH (1992): Gene V protein-mediated translational regulation of the synthesis of gene II protein of the filamentous bacteriophage M13: A dispensable function of the filamentous-phage genome. J Bacteriol 174:595–600.

Zinder ND (1975): "RNA Phages". Cold Spring Harbor, NY: Cold Spring Harbor Laboratory.

Zinder ND, Boeke J (1982): The filamentous phage (Ff) as vectors for recombinant DNA. Gene 19:1–10.

Zinder ND, Horiuchi K (1985): Multiregulatory element of filamentous bacteriophages. Microbiol Rev 49:101–106.

Zinder ND, Valentine RC, Roger M, Stoeckenius W (1963): fl, a rod shaped male-specific bacteriophage that contains DNA. Virol 20:638–640.

Zoller MJ, Smith M (1983): Oligonucleotide-directed mutagenesis of DNA fragments cloned into M13 vectors. Methods Enzymol 100:468–500.

7

Restriction-Modification Systems

ROBERT M. BLUMENTHAL AND XIAODONG CHENG

Department of Microbiology and Immunology, Medical College of Ohio, Toledo, Ohio 43614–5806;
Biochemistry Department, Emory University, Atlanta, Georgia 30322–4218

Modern Microbial Genetics, Second Edition. Edited by
Uldis N. Streips and Ronald E. Yasbin. ISBN 0–471–38665–0
Copyright © 2002 Wiley-Liss, Inc.

I. A CUT ABOVE (AND ANOTHER BELOW)

In 1978 the Nobel Prize for Physiology or Medicine was awarded to Werner Arber, Daniel Nathans, and Hamilton Smith "for the discovery of restriction enzymes and their application to problems of molecular genetics" (for more information, visit *http://www.nobel.se/laureates/medicine-1978.html*). While most excellent science is never recognized by a Nobel Prize, and though *many* scientists have contributed to our understanding of restriction-modification systems

(not least the coworkers of the prize winners), the Nobels do indicate the importance of this field of study as well as identifying some of its key pioneers.

Werner Arber discovered restriction enzymes during the 1960s, nicely illustrating the point that discoveries of major importance can be made as completely unanticipated benefits of pursuing basic research problems. Dr. Arber was trying to understand a phenomenon of bacteriophage biology that had been seen in the previous decade. The term "restriction" comes from the observation that some strains of *E. coli* greatly reduced the ability of bacteriophages to form plaques (Bertani and Weigle, 1953; Luria and Human, 1952) (one of the original observers of this phenomenon, Salvador Luria, later won a Nobel for "discoveries concerning the replication mechanism and the genetic structure of viruses"). As illustrated in Figure 1, a stock of bacteriophage that has not previously been grown in a strain possessing a given restriction-modification system forms plaques with low efficiency. However, the few bacteriophage that have escaped restriction and formed plaques have been "modified" and are now resistant to restriction if replated on the same host. This resistance is specific to the particular restriction-modification system and is lost when the bacteriophage is grown in a different host strain. Arber found that the phenomenon had two opposing aspects, restriction and modification, and proposed that each was catalyzed by an enzyme that would in one case cut the DNA and in the other protect it. He demonstrated that protection involved transfer of a methyl group from S-adenosyl-L-methionine (AdoMet)(Kuhnlein and Arber, 1972), and suggested what seems obvious only in retrospect—that the two activities competed for the same segments of DNA as defined by the nucleotide sequence.

Hamilton Smith confirmed Arber's hypothesis in 1970 (Smith and Wilcox, 1970). He purified a restriction enzyme, and showed that it cut DNA into large fragments but that it didn't cut DNA from the host bacterium. His key observation was that all DNA fragments generated by this enzyme had the same nucleotides at their ends, indicating cleavage within a short symmetrical sequence (Kelly and Smith, 1970). As discussed below, not all restriction endonucleases cleave within the recognized DNA sequence, and not all recognize symmetrical sequences, but Dr. Smith fortunately focused on one of the enzymes that show this behavior (later called *type II*).

Dan Nathans realized that the sequence specificity of restriction endonucleases such as the one isolated by Dr. Smith provided a way to characterize DNA molecules, by repeatably breaking them into defined segments. Again, what in retrospect seems obvious was actually a very clever inductive leap. He generated a restriction map for the small animal virus SV40, and also suggested several other important uses for restriction enzymes (Danna and Nathans, 1971). (Drs. Smith and Nathans also suggested the now-standard nomenclature with a three-letter designation indicating the Genus and SPecies in which the enzyme was discovered, followed by any strain designations and a roman numeral to indicate the order in which an enzyme was discovered in a given host; e.g., *Bam*H I was the first restriction enzyme found in *Bacillus amyyloliquefaciens* strain H. (See Smith and Nathans, 1973).

Two other Nobel laureates played important roles in the development of restriction endonucleases as powerful tools, though the prize they later shared was for discovery of RNA splicing. The use of restriction endonucleases together with agarose gel electrophoresis was first reported by Phillip Sharp (Sharp et al., 1973). The first systematic attempt to isolate type II endonucleases from the full panoply of bacterial sources was made by Richard Roberts—by the early 1980s nearly three-quarters of the known restriction enzymes had been characterized in his lab. Dr. Roberts still maintains the most comprehensive database on restriction-modification systems (*http://rebase.neb.com/*)(Roberts and Macelis, 2000). Now, nearly a half-century after their discovery, it is astonishing how important these enzymes have become to science, medicine, and biotechnology, and even more astonishing how much we have yet to learn about them.

II. LIFESTYLES OF THE SMALL AND PROKARYOTIC

A. Where Are Restriction-Modification Systems Found?

This chapter describes the roles, mechanisms of action, regulation, and evolution of restriction-modification systems. At the time of this writing, the turn of the millenium, we still lack a surprising amount of basic information on these topics. This is a reflection of our relatively poor understanding of the microbial world in general, which of necessity has focused on those species that we can easily grow in vitro. It is estimated that under 1% of soil microflora corresponds to characterized species (Roszak and Colwell, 1987). Even the microflora associated with our own bodies, which as a result of medically oriented research have received far more scrutiny than most bacterial communities, are poorly understood. For example, whole-population PCR amplification of rRNA sequences revealed that fully three-quarters of the human intestinal microflora did not correspond to any known species (Suau et al., 1999).

Over 3000 restriction-modification systems have been discovered so far, and they have been found across the full spectrum of known bacterial species. This includes both eubacteria and archaea. Even the bacteria with the smallest known genomes (the mycoplasmas, with ~800,000 bp – about a fifth the amount of DNA in *E. coli*) make room for restriction-modification systems (Himmelreich et al., 1997). The extreme thermophile *Pyrococcus* produces an extremely thermostable restriction-modification system (Morgan et al., 1998). Of the over 30 bacterial genomes now fully sequenced (as of spring 2001), only those from the obligate intracellular parasites *Buchnera*, *Chlamydia* and *Rickettsia* appear to lack candidate restriction-modification systems. *Buchnera* only grow inside a eukaryotic host cell and are passed only vertically (Wernegreen, 2000), but *Chlamydia* has an extracellular phase and bacteriophages are known to attack it (Hsia et al., 2000; Liu et al., 2000; Storey et al., 1989). No restriction-modification system is native to a eukaryote, though oddly enough a family of viruses that grow on the unicellular eukaryotic alga *Chlorella* do produce restriction-modification systems (whose roles are unclear) (Van Etten and Meints, 1999).

Purely as an aside, one bacterial restriction-modification system was introduced into mammalian (mouse) cells, to see if the system could provide protection against DNA viruses such as adenovirus (Kwoh et al., 1988). The methyltransferase gene was introduced first, to protect endogenous DNA, and the restriction endonuclease gene was then introduced in a second step. Both genes were expressed and the mouse cells remained viable, yet no protective effect was achieved. While it is hard to draw conclusions from a single result, and a negative one at that, further experiments of this kind may reveal basic reasons for which restriction systems haven't appeared in eukaryotic cells; these may include the nature of the DNA packaging in the nucleus, the extensive compartmentalization of the cell, or even simply the huge amount of DNA (several hundredfold more than in bacteria).

On the one hand, if we know relatively little about the microbial world then it is hard to justify any proposed general role for the restriction-modification systems. On the other hand, while our understanding may be skewed by the small subset of bacteria with which we have worked to date, we know enough to make tentative hypotheses. This first section of the chapter will provide an overview of the basic features of bacterial life, from a genetic perspective.

B. Abundance of Bacteria and Bacteriophages in Natural Environments

Bacteria (which, for the purposes of this chapter, include both eubacteria and archaea) collectively form the bulk of the earth's biomass. This is illustrated by the observation that bacterial peptidoglycan (cell wall) fragments are the major source of dissolved organic nitrogen in seawater (McCarthy et al., 1998). The density of bacteria in natural settings can be quite high, even in environments one might suppose were not particularly congenial, such as 460 million cells per cm^3 in sediments under open ocean and 3.5×10^{11} cells per cm^2 in desert scrubland (Whitman et al., 1998). These populations almost always include multiple species, so a diversity of genetic information is often separated into subsets by only the thickness of two bacterial cell walls. In aquatic environments, even bacteria that are physically separated may be continuously exposed to one another's DNA.

This tremendous bacterial biomass represents an irresistable opportunity for exploitation by farmers, predators, and parasites. The farmers include multicellular organisms that provide havens and nutrients for bacteria in return for bacterial services that range from photosynthesis in fungi and sponges (Bewley et al., 1996; Flowers et al., 1998; Gehrig et al., 1996), through nitrogen fixation (rhizobia in the root nodules of leguminous plants) (Gualtieri and Bisseling, 2000), to cellulose degradation in the midguts of termites and in the stomachs of ruminant mammals (Ohkuma and Kudo, 1996; Varga and Kolver, 1997), and even light production (vibrios in flashlight fish and squid) (Ruby and McFall-Ngai, 1999). The predators include organisms that consume bacteria such as other bacteria (*Myxococcus* preying upon *E. coli*), unicellular protozoa (amoeba predation appears to limit bacterial populations in soils), and filter-feeding animals such as bivalve mollusks (Earampamoorthy and Koff, 1975; McBride and Zusman, 1996; Rodriguez-Zaragoza, 1994). The parasites are for the most part viruses called bacteriophages, though there are also bacteria (Bdellovibrio) that invade and grow within other bacteria (McCann et al., 1998).

Where there are high numbers of bacteria, there are generally high numbers of bacteriophage (see Hendrix, this volume). For example, measurements in the open ocean give bacteriophage particle counts ranging from 70,000 to 15 million per milliliter (Bergh et al., 1989; Wommack and Colwell, 2000)! There are few groups of bacteria not known to be the targets of bacteriophages, and there are almost certainly a huge variety of bacteriophages yet to be discovered. The bacteriophages have diverse morphologies, as described elsewhere in this text, and their genomes can be made of dsDNA, ssDNA, dsRNA, or ssRNA. The tailed dsDNA bacteriophages are a particularly large group that continue to exchange genes with one another (Hendrix et al., 1999). It is also worth restating that a subset of the DNA bacteriophages can enter a state known as *lysogeny*, in which their DNA is integrated into a host bacterium's genomic DNA (a "prophage"). Prophages are widespread, and to take *E. coli* as an example, they (or their inactivated genetic remains) can constitute several percent of the genome (Campbell, 1996). If prophage DNA enters a cell that is not already lysogenized by that bacteriophage, the repressor(s) maintaining the lysogenic state are not present and a lytic infection can ensue (a phenomenon called *zygotic induction*) (Feinstein and Low, 1982). Thus even large extracellular DNA molecules may, from the viewpoint of bacteria, represent Trojan horses carrying a hidden bacteriophage up to the cells' walls.

I. Disadvantage faced by asexual, clonal organisms

In evolutionary terms, bacteria have the advantage of short generation times (in at least some environments) and haploidy (so that potentially beneficial mutations are immediately expressed). However, bacteria also suffer the tremendous disadvantage of clonal propagation. The fact that both daughter cells resulting from cell division are genetically identical to the mother cell means that each cell lineage must continually "reinvent the wheel." Consider cells adapting to a warmer niche, for example—if one lineage develops a more thermostable version of protein X while another lineage develops a more thermostable protein Y, the only way for a strictly clonal bacterium to possess both traits is to independently evolve both. This is why bacterial resistance to multiple antibiotics was not, in the 1940s and 1950s, expected to be a serious problem even though resistance to single antibiotics was a well-established phenomenon (Davies, 1997).

The limitations of strict clonal propagation may explain why, despite the potential risks of importing foreign DNA into the cell (of which possible prophages is just one), bacteria don't simply exclude all immigrant genes and live in xenophobic isolation. The evidence suggests that few of the bacteria we know about have cut themselves off in this way, though individual species range widely on the continuum between being strictly clonal and being panmictic (Smith et al., 1993). Either such isolationism is difficult to achieve, or its consequences are too severe, since most bacteria either encourage immigration by producing complex multienzyme systems to import and incorporate exogenous DNA, or permit it to occur under the control of plasmids and certain types of bacteriophage (Lawrence, 1999).

2. Mechanisms of gene transfer and genomic evidence for horizontal gene exchange

We of course know, to our sorrow, that bacteria can acquire resistance to multiple antibiotics with extreme rapidity (measured in years, not millennia), and that the same patterns of resistances can spread between bacteria that are only distantly related to one another (Davies, 1997). As described in other chapters of this text, we now know that this intercellular DNA transfer occurs at substantial rates via three basic mechanisms (Davison, 1999):

- Transformation. The uptake of naked DNA from the extracellular environment.
- Transduction. The introduction of bacterial DNA into a new cell by a bacteriophage.
- Conjugation. The direct transfer of DNA between physically associated bacteria.

With entire bacterial genomes now being sequenced routinely, we can see the extent to which gene exchange has occurred (Ochman et al., 2000). In comparing the genomes of related bacteria, we have learned that large contiguous segments of apparently imported DNA can be found in many locations. The foreign nature of these DNA segments is inferred from their absence in other, closely related genomes, as well as by their having total %GC and codon biases that are atypical for the (new) host bacterium. For example, nearly a fifth of *Escherichia coli* genes have apparently been imported just since the divergence of *E. coli* and its close relative *Salmonella* (Lawrence and Ochman, 1998; Martin, 1999). Even populations that appear to be clonal may simply have undergone recent selective sweeps (Elena et al., 1996a; Elena et al., 1996b; Guttman, 1997; Guttman and Dykhuizen, 1994).

III. PROTECTOR, PARASITE, OR PERMUTER? ROLES OF RESTRICTION-MODIFICATION SYSTEMS

As Arber proposed and Smith first demonstrated, restriction-modification systems consist of two components that, respectively, restrict and modify DNA. Restriction involves an endonuclease breaking DNA by hydrolyzing the phosphodiester backbone on both strands, while modification involves a methyltransferase adding a chemical group to a DNA base at a position that blocks the paired restriction activity. Both the restriction activity and the modification activity are specific for the same DNA sequence. The obvious rationale for this pairing is that while restriction might carry out any (or all) of the roles described below, the cell must protect its own DNA from being attacked. It is still unclear what roles restriction-modification systems play, and what original role(s) may have provided the selectable advantage that explains the extraordinarily widespread distribution of these systems in the bacterial world. Three models stand out, and it is important to note that they are not mutually exclusive.

A. Defense against Bacteriophage Infection

This is the most obvious explanation for restriction-modification systems, and like other "obvious" explanations may seem to be self-evident and not worth exploring. However, given the evidence in support of alternative roles, it is important to review the evidence regarding the bacteriophage defense hypothesis. This issue may come to have clinical significance as well, with attempts to develop bacteriophages as therapeutic antibacterial agents (Merril et al., 1996).

There are numerous examples of restriction-modification systems that have been directly proven to protect the bacterial host from bacteriophage infection. This protection often results in a decrease of as much as 10^5-fold in the plaquing efficiency of unmodified bacteriophage; though the actual extent of restriction depends on several factors (not all of them well defined), such as the level of expression of the restriction-modification system, the relative activities of methyltransferase and endonuclease, the number of recognition sequences in the bacteriophage DNA, the presence of unusual bases in the bacteriophage DNA, and the kinetics of bacteriophage replication. While it is important to note that only a tiny fraction of the > 3000 known systems have been tested in this manner, there is no question that defense against bacteriophage infection is a role played by many (if not all) restriction-modification systems.

A second line of evidence that strongly supports a defensive role for restriction-modification systems is the fact that bacteriophages themselves have evolved numerous countermeasures (Bickle and Kruger, 1993; Kruger and Bickle, 1983). These range from the simple expedient of eliminating substrate sequences from the bacteriophage genome up to the production of proteins that specifically inhibit restriction. For example, bacteriophage ϕ1 grows on *Bacillus subtilis*, where it would encounter a restriction-modification system that recognizes the sequence CGCG. In the roughly 10^5 base pairs of the ϕ1 genome, one would expect to find about 400 occurrences of the four-nucleotide CGCG sequence ($10^5/4^4$), but there is in fact only one occurrence. Regarding bacterio-

phage proteins that inhibit restriction, there are a variety of approaches including direct inhibition of the endonuclease and the opposite strategy of stimulating the protective methyltransferase, for example (Bandyopadhyay et al., 1985; Loenen and Murray, 1986; Spoerel et al., 1979). Some bacteriophages have even acquired their own DNA methyltransferases, which are unique in being able to recognize and protect multiple distinct DNA sequences (Trautner et al., 1996). As restriction-modification systems represent a barrier against foreign DNA in general and not just bacteriophage DNA (Matic et al., 1996), an interesting addendum to this line of evidence about the protective role is that some plasmids have also developed antirestriction proteins produced early in conjugative transfer, such as Ard (Althorpe et al., 1999; Belogurov et al., 1992; Chilley and Wilkins, 1995).

A third line of evidence comes from a prediction of the bacteriophage defense hypothesis that selection should operate, to an extent, on a population and not only on individual cells (Raleigh and Brooks, 1998). This is because, as indicated in Figure 1, there is a low but measurable escape rate where the bacteriophage DNA becomes methylated before any endonuclease cleavage can occur. The bacteriophages released following such escape contain fully methylated DNA. If a population of cells has only one restriction-modification system, then not only is every cell at risk of infection from escapees, but the methylation of the bacteriophage will be maintained since every cell is producing the methyltransferase. In this way, an entire population could be eliminated. Two ways to prevent this catastrophe are production of several restriction-modification systems (so the net escape rate is made vanishingly small) and ensuring that populations will be heterogeneous in terms of the restriction-modification systems they are actually producing even if they all have the same restriction-modification genes (so that escapees from one cell can only productively infect a subset of the other cells).

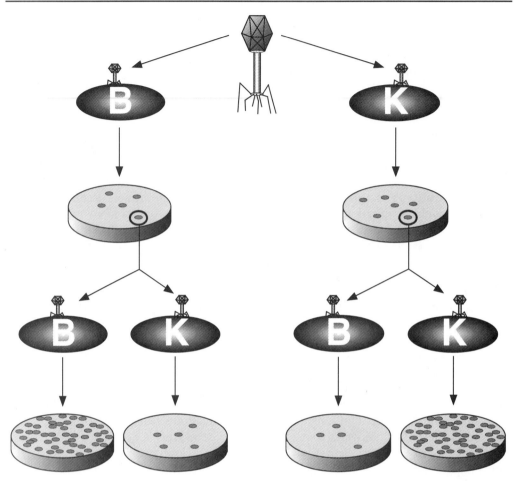

Fig. 1. Evidence for restriction-modification systems from bacteriophage plating experiments. In this hypothetical experiment, a bacteriophage with a genome of unmodified double-stranded DNA is plated on two bacterial host strains carrying different restriction-modification systems. The letters refer to the *E. coli* type I systems named *Eco*B and *Eco*K. The unmodified bacteriophage is restricted, but a few infecting genomes are modified before restriction can occur and these form plaques containing bacteriophage with fully modified genomes. When these bacteriophage are subsequently plated on the two host strains, the B-grown bacteriophage plate efficiently on the B strain and the K-grown bacteriophage plate efficiently on the K strain. This shows that the modification is specific to a particular restriction-modification system. The next set of platings (not shown) would reveal that the B-grown bacteriophage that was plated on the K strain had lost its B modification, and would plate efficiently on the K but not the B strain.

The latter of these two strategies represents a population-level phenotype, and there is quite a bit of evidence that this strategy is used. First, the genome sequence for *Neisseria meningitidis* (Tettelin et al., 2000) reveals the presence of genes for over 20 restriction-modification systems. Furthermore, repeat sequences associated with six of these systems suggest that they are subject to phase variation—a process in which "slipped strand" mispairing between adjacent repeats on the two DNA strands leads to high-frequency insertion or deletion that abolishes or restores expression of the associated genes. This would ensure that a population of cells is heterogeneous in their repertoires of active restriction-modification systems even though the population is (in essence) genetically homogeneous. Assuming that these six systems are turned on or off independently, a population of these bacteria could contain cells with 64 (2^6) distinct restriction phenotypes. A number of restriction-modification genes in other bacteria are associated with homopolymeric repeats or even inversion elements that can vary the production or specificity of the system. *Haemophilus* provides another example of turning a restriction system on or off by slipped-strand mispairing of a repeated sequence (De Bolle et al., 2000), while *Mycoplasma* provides an example of control of synthesis and specificity using a remarkable set of inversion elements (Dybvig et al., 1998).

A second example that appears to support population-level selection involves *Dpn*I and *Dpn*II in *Streptococcus pneumoniae*. *Dpn*II is a classic restriction-modification system that cleaves the sequence GATC when it is unmethylated, and marks "self" DNA by methylating the adenine in that sequence. *Dpn*I, in contrast, has no modification methyltransferase, and the restriction endonuclease cleaves GATC only when the adenines of both strands *are* methylated. Both restriction-modification systems are specified by gene casettes that can replace one another at a defined location in the *S. pneumoniae* chromosome (Lacks et al., 1986). These two systems are mutually incompatible. They cannot be co-expressed in the same cell, but if both are present in a population, then escapees from one system are specifically targeted by the second system. As an aside, it is not clear how both systems are maintained when populations go through the inevitable bottlenecks associated with host-to-host transmission (Bergstrom et al., 1999). *S. pneumoniae* is a naturally transformed species, and the two *Dpn* cassettes can replace one another following transformation, but in the absence of unremitting bacteriophage selection (and defense provided by the mixed population), it would be hard to see how this pairing could be maintained.

There is no evidence that directly contradicts the bacteriophage defense hypothesis, but it is clear that restriction-modification systems have limits as defensive systems. For example, they have the problem of modified escapees referred to above. There are also classes of bacteriophage that have RNA genomes, or single-stranded DNA genomes (that do, however, form double-stranded replication intermediates), against which the known restriction-modification systems would not operate.

It should be noted that there are approaches bacteria take to defend themselves from bacteriophage infection that do not involve restriction-modification systems, and bacteriophage countermeasures are also seen in these cases. For example, growth in a biofilm protects bacteria from many agents, as the bacteria are embedded in a polysaccharide matrix. However, some bacteriophages produce specific hydrolases that digest the biofilm matrix allowing access to the bacteria (Hughes et al., 1998). Figure 2 shows two other bacteriophage defense systems that do not rely on DNA hydrolysis. The Lit system is itself specified by a prophage and protects the lysogenized population by a programmed cell death when another bacteriophage infects the cell. Capsid morphogenesis of bacteriophage T4 appears to involve translation elongation factor EF-Tu binding the major coat protein, and this complex is recognized and destroyed by the *lit* product which is a specific protease (Bingham et al., 2000). Prr illustrates both another defensive system not based on DNA hydrolysis, and the interplay between bacterium and bacteriophage. In this case the bacteriophage produced a protein inhibitor of the host cell's restriction-modification system, and the host responded by putting a suicide enzyme onto the restriction-modification protein complex. When the bacteriophage-produced inhibitor binds, it activates the PrrC suicide enzyme (which is a ribonuclease specific for tRNA^{Lys}) (Penner et al., 1995). The bacteriophage, not too surprisingly, has responded to this by producing tRNA repair enzymes.

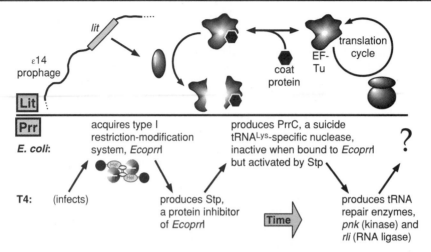

Fig. 2. Examples of alternatives to restriction-modification systems in defending cells against bacteriophage infection. Both systems are found in *E. coli*, and both are suicide systems triggered by bacteriophage T4. Lit is a protease specific for the complex between EF-Tu and the major T4 coat protein. Destruction of EF-Tu kills the cell and aborts the infection. PrrC is a ribonuclease specific for tRNA^{Lys}. PrrC remains in an inactive complex with the restriction-modification protein unless the T4 Stp product binds. In this case bacteriophage T4 has taken the next step in producing enzymes that repair the damaged tRNA.

In summary, the abundance of bacteriophages in the environments occupied by bacteria, and the available direct evidence, strongly support the bacteriophage defense hypothesis. There is evidence that plasmids conferring even weak resistance to bacteriophage infection are by this means maintained in natural populations (Feldgarden et al., 1995). This does not rule out other additional roles for restriction-modification systems, and it does not mean that bacteriophage defense was necessarily the original role of these systems.

B. Acting As an Addiction Module ("Selfish" Behavior)

Restriction-modification systems possess some of the properties that characterize addiction modules, which can be described as having selfish behavior (Dawkins, 1989). Addiction modules function to maintain a genetic element such as a plasmid by killing any segregants that have lost that element (Engelberg-Kulka and Glaser, 1999). Typically addiction systems contain two components; a toxin and an antitoxin. The antitoxin is less stable than the toxin, so segregants will

lose protection before they lose the toxin. As illustrated in Figure 3, there are two major classes of addiction modules. Both have protein toxins, but they differ in that the antitoxin in one class is a toxin-binding protein (Rawlings, 1999) and in the other class is an antisense RNA that blocks translation of the toxin mRNA (Gerdes et al., 1997).

It occurred to one group of scientists that restriction-modification systems had many features of addiction modules. They are often specified by mobile genetic elements such as plasmids. They have a potentially toxic component (the restriction endonuclease) and a second protein that, while not inhibiting the "toxin," protects the cell against its activity (the modification methyltransferase). In fact, when plasmids carrying some restriction-modification systems are lost, the plasmid-free segregants are killed (Kobayashi, 1998; Naito et al., 1995; Nakayama and Kobayashi, 1998). This is not because the methyltransferase is less stable than the endo-nuclease; probably it is because as each enzyme is diluted by cell growth, the DNA becomes partially unprotected while endonuclease is still present (Fig. 3, bottom); there is direct evidence that

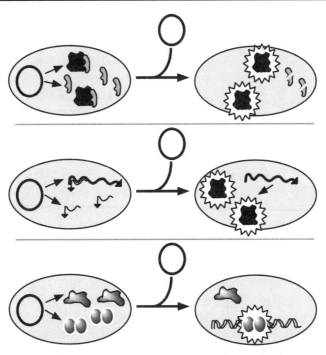

Fig. 3. Addiction modules. These systems exhibit selfish behavior, in that cells losing one of these modules are selectively killed. The top two panels illustrate the two major classes of addiction modules. The top panel represents the class in which a polypeptide antitoxin of low stability is produced along with the toxin; if the genes are lost, the antitoxin will disappear before the toxin does. The middle panel represents the class in which the "antitoxin" is actually an antisense RNA that prevents translation of the toxin gene; if the genes are lost, the antisense RNA disappears before the toxin mRNA and toxin is produced. The bottom panel represents a type II restriction-modification system, which exhibits the behavior of an addiction module. In this case the toxin is the restriction endonuclease, and the antitoxin is the protective methyltransferase; if the genes are lost, both enzymes may disappear in parallel but the result will be partially unprotected DNA in the presence of the endonuclease.

cell death in these cases is associated with DNA cleavage (Handa et al., 2000). Another phenomenon that could be explained by the addiction module hypothesis is that some restriction-modification systems have very rarely occurring substrate sites. An endonuclease with a cleavage site that is expected to occur only once in several tens of thousands of bp of DNA will probably not provide protection against too many bacteriophages. However, to act as an addiction module, the target is the entire bacterial genome so these restriction-modification systems would be no less "fit" than a system with a more frequently occurring substrate sequence.

The restriction-modification systems showing selfish behavior belong to a class in which the restriction endonuclease and modification methyltransferase are separate, independently active proteins (type II; see below). When these studies were repeated using a restriction-modification system in which the endonuclease and methyltransferase are part of one large protein complex, no selfish behavior was seen (O'Neill et al., 1997). Furthermore an enzyme like *Dpn*I (described earlier) has no partner methyltransferase and cannot easily be explained as part of an addiction module.

In summary, there is clear evidence that some restriction-modification systems (type II) can function as addiction modules, and attempts to understand the evolution and regulation of these systems must take this into consideration. However, some classes of restriction-modification system fail to act as addiction modules, and their evolution may have been driven by other selectable phenotypes.

C. Stimulating Recombination with Incoming Foreign DNA

Not everything that cuts DNA does so solely to degrade it. Some host-(Ban and Yang, 1998) and bacteriophage-coded (Carlson and Kosturko, 1998) endonucleases play (or may play) recombinational roles. This may also be true of restriction-modification systems (Barcus and Murray, 1995; Price and Bickle, 1986).

1. Immediate effects of cleaving DNA in vivo

Particularly in the bacteriophage defense hypothesis, restriction endonucleases are thought of as purely destructive agents. The DNA is broken on both strands and then is degraded or disappears. The reality is a bit more complex.

First, the restriction endonucleases do not all do the same amount of damage. If we ignore for the moment the varied di-, tri-, and tetranucleotide frequencies in specific genomes, it is clear that (on average) an endonuclease with a four-nucleotide specificity will have more closely spaced substrate sequences than an endonuclease with an eight-nucleotide specificity. Given equal chances of having the four possible nucleotides at each position, the probability of having a particular sequence at a given point on the DNA is $1/4^n$, where n is the number of nucleotides in the substrate sequence. Thus the AluI endonuclease, recognizing AGCT, should on average cleave every 256 bp in a DNA of random sequence ($1/4^4$), while the PvuII endonuclease (CAGCTG, $1/4^6$) should cleave every 4096 bp and SgfI (GCGATCGC, $1/4^8$) should cleave every 65,536 bp. The eight-nucleotide sites could thus easily fail to occur in a genome the size of that belonging to bacteriophage λ (under 50,000 bp). Furthermore the fragment ends generated by cleavage differ. Some enzymes

cut off-center with respect to their substrate sequences, yielding a complementary pair of 5' single-strand extensions that can readily hybridize and be religated. Others generate 3' single-strand extensions, also religatable. Still others generate blunt ends, on which some DNA ligases (e.g., the E. coli NAD-dependent enzyme) do not act efficiently. Finally, some restriction endonucleases (type I; see below) generate DNA ends that are apparently unligatable—one strand is degraded to yield a 70 to 100-nucleotide 3' extension, and while the nature of the 5' ends is not yet clear, they cannot be labeled with polynucleotide kinase even following phosphatase treatment (Endlich and Linn, 1985).

The major players in the immediate fate of DNA molecules cleaved in vivo by restriction endonucleases are (in E. coli) DNA ligase and the RecBCD complex. When the endonuclease EcoR I is expressed in the absence of protective methylation, yielding double-strand breaks in the chromosomal DNA, cell survival is highly dependent on the cell's NAD-dependent DNA ligase (Heitman et al., 1989). Interestingly, when either EcoR I or EcoR V endonuclease generates nicks in the DNA in vivo (cleaving one strand only), repair requires not only ligase but also RecA and RecB (Heitman et al., 1999). This suggests that, at least under some conditions, nicking is more damaging than a double-strand break. The proposed explanation for this surprising result is that nicks in the template strand can lead to collapse of a replication fork, yielding a single DNA end that can only be rescued by recombinational repair with a sister duplex. For our purposes the important points are that nicking is not irrelevant and that limited numbers of nicks and double-strand breaks can be repaired.

2. Actions of the RecBCD complex

Repair would seem to be counter to the bacteriophage defense or addiction module roles proposed for restriction-modification systems. It fits nicely, however, with the hypothesis that restriction promotes recombination of certain types of incoming DNA. If the cleaved DNA is not quickly ligated again, the free ends will be bound by the RecBCD complex. As described elsewhere in this text, the RecBCD complex has helicase and nuclease activity and rapidly degrades both separated strands. This would fit the selfish or defensive goals of restriction-modification systems quite well, but it would be suicidal to a cell with a break in its chromosome. As illustrated in Figure 4, the fate of the DNA turns out to depend on whether or not χ (chi, for crossover hotspot instigator) sequences are present (Kuzminov et al., 1994). In E. coli, χ sequences have eight nucleotides (GCTGGTGG) that appear to be tightly bound when encountered by RecBCD (Wang et al., 2000). The χ sequences are highly over-represented in E. coli DNA, and other bacterial species appear to have distinct but functionally equivalent χ sequences (Lao and Forsdyke, 2000). If no χ sequence is encountered, as would be the case for a cleaved bacteriophage λ chromosome, the DNA is degraded from one end to the other. In contrast, if a χ sequence is encountered, the RecBCD complex radically changes its behavior—the $3' \rightarrow 5'$ nucleolysis ceases, while the $5' \rightarrow 3'$ nucleolysis con-

tinues for a time, generating a long $3'$ single-strand extension. Furthermore, RecBCD actively loads RecA onto the extension, making it competent for homologous recombination (Anderson et al., 1999; Churchill et al., 1999)—if there is any homologous DNA in the cell with which it can recombine. It would be interesting to know if the type I restriction-modification systems, which like RecBCD generate long $3'$ extensions, have the ability to facilitate RecA loading—if so it would certainly change our view of these restriction complexes. In this regard it is interesting that a type I restriction-modification system has been shown (in particular genetic backgrounds, at least) to stimulate a certain type of illegitimate recombination (Kusano et al., 1997).

In this view, restriction-modification systems simply ensure that incoming DNA is rapidly scanned by RecBCD, and the presence or absence of χ sequences (rather than the methylation status of the DNA) determines whether or not the DNA is destroyed. Looked at another way, for incoming DNA that does have χ sequences the restriction endonuclease facilitates its recombination with endogenous DNA by providing multiple entry points for RecBCD.

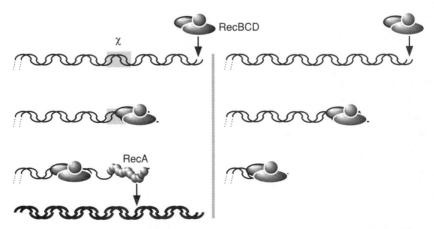

Fig. 4. Action of RecBCD. The RecBCD complex binds to free DNA ends and processively degrades both strands. On the right, the DNA contains no chi (χ) sequence, and RecBCD continues its nucleolytic activity. On the left, the RecBCD encounters a χ sequence, generates a single-strand extension, and facilitates the loading of RecA onto the single-stranded DNA. The DNA and proteins are not drawn to the same scale.

The strongest evidence in support of this view is that the length of incorporated DNA, when homologous DNA is introduced into a cell, depends on the presence of restriction-modification systems and on the methylation status of the incoming DNA (Arber, 2000; Milkman et al., 1999). In addition to making the incoming DNA more highly recombinogenic, the fragmentation would also separate a potentially useful segment of DNA from flanking segments that might be deleterious to the cell were they to be incorporated into the genome; this would increase the chances of the useful segment being retained (Milkman, 1999).

D. Summary—What Is the Role of Restriction-Modification Systems?

An analogy may be helpful. A bacterium "attempting" to expand its range into a new ecological niche is like a business attempting to expand its market into a new country. The bacteria/businesses already occupying the new area represent competitors, but they also represent an extremely rich and useful treasury of information—based on decades (or millions of years) of trial-and-error experience—on exactly how to thrive in the new niche. On the one hand, it is hard to overstate the value of this information pool to the newcomer, and there is clear and compelling evidence for a torrential flow of DNA between bacterial species. On the other hand, the new business can't hire employees away from competitors without screening them, or they could end up bringing in corporate spies or other malefactors. Similarly the new bacterium could pay a heavy price for indiscriminate importation of genetic information that could include parasitic genetic elements. Restriction-modification systems (RMSs) can be thought of as providing screening without forming an insuperable barrier to genetic imports. A subset of restriction-modification systems were also adapted by mobile genetic elements to serve as addiction modules (or originally evolved to serve that role and happened to serve as defensive systems). The integra-

tion of phage defense and recombination roles would not be unique to the restriction-modification systems. We've already seen that an enzyme thought of as purely recombinational, RecBCD, participates in the restriction process, and it should not surprise anyone that there are bacteriophage products such as that of the γgam gene that inhibit RecBCD (Murphy, 1991; Salaj-Smic et al., 1997).

Bacteria face a difficult balancing act. If they are too promiscuous in their exchange of DNA, they risk some dread and lethal disease such as bacteriophage (or prophage) infection. If they refuse to engage in DNA exchange altogether, their descendants will never have the advantages available to other bacteria of taking useful new genes from the worldwide genetic library, and the line could well die out or be relegated to isolated or harsh environments. Put in less anthropomorphic terms, the cells that most efficiently survive and spread to new niches will be those that have a balance, appropriate to their environment, between acquisition of new DNA sequences and rejection of potentially dangerous ones (Ochman et al., 2000). Restriction-modification systems, together with the recombination and repair machinery, play an important role in achieving this balance. When these systems cut up incoming DNA into roughly gene-sized pieces, they appear to be processing the DNA into more recombinogenic fragments, allowing useful segments to recombine independently of potentially deleterious flanking segments, *and* inactivating bacteriophage DNA all at once.

IV. TYPES OF RESTRICTION-MODIFICATION SYSTEMS

The restriction-modification systems fall into several classes defined by subunit composition and cofactor requirements. It is highly unlikely that we have discovered all of the classes, and some simplified classification scheme may be needed. At present, however, the groupings shown in Figure 5, and described below, are broadly (if unenthusiastically) accepted. The

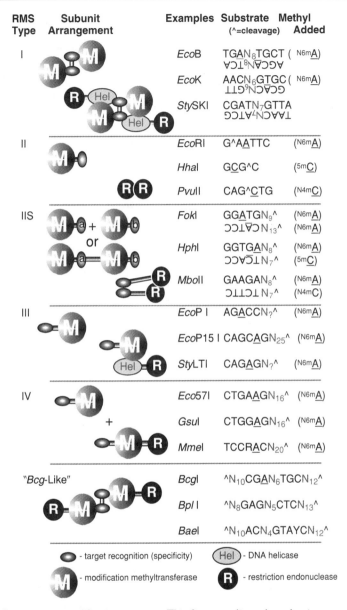

RMS Type	Subunit Arrangement	Examples	Substrate (^=cleavage)	Methyl Added
I		*Eco*B	TGAN$_8$TGCT (N6m$\underline{A}$) ACGA$\underline{N}_8$NTCA	
		*Eco*K	AACN$_6$GTGC (N6m$\underline{A}$) GCACN_9TT	
		*Sty*SKI	CGATN$_7$GTTA TAAC$\underline{N}_7$ATCG	
II		*Eco*RI	G^A$\underline{A}$TTC	(N6m$\underline{A}$)
		*Hha*I	GC$\underline{G}$^C	(5m$\underline{C}$)
		*Pvu*II	CAG^$\underline{C}$TG	(N4m$\underline{C}$)
IIS		*Fok*I	GG$\underline{A}$TGN$_9$^ CC$\underline{T}$AC N$_{13}$^	(N6m$\underline{A}$) (N6m$\underline{A}$)
		*Hph*I	GGTG$\underline{A}$N$_8$^ CC$\underline{T}$C N$_7$^	(N6m$\underline{A}$) (5m$\underline{C}$)
		*Mbo*II	GA$\underline{A}$GAN$_8$^ CTTC N$_7$^	(N6m$\underline{A}$) (N4mC)
III		*Eco*P I	AG$\underline{A}$CCN$_?$^	(N6m$\underline{A}$)
		*Eco*P15 I	CAGC$\underline{A}$GN$_{25}$^	(N6m$\underline{A}$)
		*Sty*LTI	CAG$\underline{A}$GN$_?$^	(N6m$\underline{A}$)
IV		*Eco*57I	CTGA$\underline{A}$GN$_{16}$^	(N6m$\underline{A}$)
		*Gsu*I	CTGG$\underline{A}$GN$_{16}$^	(N6m$\underline{A}$)
		*Mme*I	TCCR$\underline{A}$CN$_{20}$^	(N6m$\underline{A}$)
"*Bcg*-Like"		*Bcg*I	^N$_{10}$CG$\underline{A}$N$_6$TGCN$_{12}$^	
		Bpl I	^N$_8$GAGN$_5$CTCN$_{13}$^	
		*Bae*I	^N$_{10}$ACN$_4$GTAYCN$_{12}$^	

- target recognition (specificity)   (Hel) - DNA helicase

(M) - modification methyltransferase   (R) - restriction endonuclease

Fig. 5. Types of restriction-modification systems. This figure outlines the subunit composition and recognition sequences for major groups of restriction-modification systems. The key at the bottom indicates the meaning of the various shapes. The structures are highly schematic and not meant to reflect actual three-dimensional structures (which, in most cases, have not been determined). The point of cleavage is indicated by "^" and the underlined base is targeted by the methyltransferase. In type I and III systems a subset of the subunits can function independently as a methyltransferase; in type IV there is both an independent methyltransferase and a second one fused to the endonuclease. For some systems with asymmetric substrates both DNA strands are shown; go to *http://rebase.neb.com* for additional information on recognition sequences and cleavage points.

huge majority of restriction-modification systems that have been characterized belong to type II so it is worth noting that, for the non-type II systems described below, the properties described come from study of a very small number of enzymes. It is also worth noting that however big the gaps in our understanding of the biochemical, kinetic, and structural features of the various types of restriction-modification systems, there are even bigger gaps in our understanding of how the various types might differ in roles, functional advantages or disadvantages, and evolutionary histories. We have seen that type II, but not type I, systems can function as addiction modules (at least for the few systems tested), and that some systems can play recombinational roles. Is there an advantage to a type III system over a type IIS in phage defense, or do the differences between them merely reflect their different evolutionary histories? Do type IIE systems have more effects on recombination than type IIQ systems? Are there advantages to having a mix of types in addition to the advantage provided by having a mix of specificities? These are the sorts of questions that we have not even begun to answer.

A. Type I Restriction-Modification Systems

The type I systems are the most complex of the known systems (Davies et al., 1999; Rao et al., 2000; Redaschi and Bickle, 1996a). They are heteropentamers, with two modification subunits (HsdM, where "Hsd" stands for host specificity determinant), two subunits that have both endonucleolytic and helicase activities (HsdR), and one subunit that confers DNA sequence specificity on the complex (HsdS). The M_2S complex is active as a protective methyltransferase, but the intact M_2SR_2 heteropentamer is required for restriction, and restriction only occurs in the presence of divalent cation (usually Mg^{2+} is used), AdoMet, and ATP. The full complex recognizes an asymmetric bipartite sequence that, if it is unmethylated, is bound tightly and activates the ATP-driven helicase

activity. DNA is pulled through the bound complex from both sides, creating ever-larger loops (Studier and Bandyopadhyay, 1988); on closed circular DNAs the complex generates positive supercoils ahead and negative supercoils behind (Janscak and Bickle, 2000). When the translocating complex reaches an impassable barrier (a second translocating type I complex, even a Holliday junction) then cleavage occurs (Janscak et al., 1999); a bound repressor does not provide enough of a barrier to trigger cleavage (Dreier et al., 1996), and linear DNA molecules containing just one binding site are poorly cleaved (Rosamond et al., 1979). The cleavage can thus occur thousands of bp away from the recognized sequence. The nature of the cleavage, referred to above, is also unique among restriction-modification systems in that a long 3' single-strand extension is generated by making numerous closely-spaced cleavages on the complementary strand. There are several additional features of the type I systems that are not understood. These complexes act stoichiometrically (i.e., they don't turn over), and in vitro the ATPase activity continues long after cleavage activity has ceased (Eskin and Linn, 1972a; Eskin and Linn, 1972b; Yuan et al., 1972); there is even some evidence for association of type I systems with the cytoplasmic membrane (Holubova et al., 2000). Clearly, much more basic research is needed on this family of multienzyme complexes.

Standard screening assays for restriction endonucleases are designed to detect the more biotechnologically useful type II enzymes, so it should come as no surprise that few type I systems have been found. Based on sequence similarity and subunit complementation, four families have been identified (types IA–ID) (Fuller-Pace et al., 1985). These systems were once thought to occur only in the Enterobacteriaceae (the family that includes *E. coli*), but genomic sequencing has revealed them to occur in a variety of other bacterial genera including *Campylobacter*, *Haemophilus*, *Helicobacter*,

Lactobacillus, Mycoplasma, Pasteurella, Streptococcus, and *Ureaplasma* (though they have not yet been found in archaea). More accurately, genomic sequencing has revealed open reading frames that closely resemble known type I restriction-modification genes; whether or not they are active systems remains to be proved.

B. Type II Restriction-Modification Systems

The type II systems are at the opposite extreme from the type I systems, being the most structurally simple of the restriction-modification systems. The modification methyltransferase is a free, asymmetric monomer, and the restriction endonuclease is a free homodimer (Fig. 6). Some exceptions exist in which the methyltransferase (Karreman and de Waard, 1990; Lee et al., 1996) or endonuclease (Hsieh et al., 2000) are expressed as two polypeptides that associate to form the active enzyme. Some type II methyltransferases dimerize at high concentration, though the functional significance of this is unclear (Dubey et al., 1992), and some type II endonucleases are active as tetramers that presumably bind two sites at a time (Siksnys et al., 1999).

The key defining features of type II systems are the completely independent activities of the methyltransferase and endonuclease,

Fig. 6. Member proteins of the *Pvu* II type II restriction-modification system. The restriction endonuclease (left) is a homodimer that encircles the DNA; one subunit is shown in yellow, and the other in white. The modification methyltransferase (orange, right) is a monomer. These two proteins recognize the same sequence (CAGCTG), though they bind it in very different ways.

and the simplicity of their substrate requirements. The methyltransferase requires only AdoMet for activity. The endonuclease activity requires a divalent cation (usually Mg^{2+} is used), and neither ATP nor AdoMet has any effect. The cleavage occurs at fixed positions that are symmetrically disposed relative to the symmetrical recognition sequence, and within or adjacent to that sequence. For this reason the type II endonucleases are extremely useful in gene cloning, genetic engineering, and various diagnostic tests (phylogenetic, medical, forensic, etc.), and this usefulness explains why several thousand of these systems have been identified and why a couple hundred of them are produced commercially.

The simplicity of type II restriction-modification systems is only relative to systems such as those of type I. Some type II systems recognize partially degenerate sequences (e.g., *Hinc*II, GTYRAC, where Y is a pYrimidine and R is a puRine) or gapped sequences (e.g., *Bgl*I, GCCN₅GGC, where "N" is aNy nucleotide). Some recognize asymmetrical sequences (more accurately quasisymmetrical, so some call them *type IIQ enzymes*), for example, *Btr*I recognizes 5'-CAC ∧ GTC on one strand ("∧" indicates the point of cleavage), where the central four bp are symmetrical but the outer bp are not (Degtyarev et al., 2000).

The type II endonucleases can be split into two subgroups based on their ability to cleave single isolated recognition sequences, with certain enzymes (called by some *type IIE*) requiring two simultaneously bound recognition sequences in order to give efficient cleavage. The type IIE enzymes are not characterized by having unusual recognition sequences, and their substrate requirements and cleavage positions are typical of type II enzymes. At least one of the gapped-sequence endonucleases (*Sfi*I; GGCCN₅ GGCC) is type IIE, though others (*Bgl*I) cleave single sites efficiently (Gormley et al., 2000). Some of the type II systems with ungapped sequences also show the need for second binding sites (*Eco*RII, CCWGG where W is A or

T [Weak pairing]), and the second site can contain uncleavable phosphorothioate linkages and still activate cleavage at the first site (Petrauskene et al., 1998; Reuter et al., 1998). It isn't clear if the type IIE enzymes form an evolutionary subgrouping among the broader group of type II endonucleases because the overall sequence conservation among endonucleases as a group is remarkably low (see Section VI). More of the structural and kinetic information that has been gleaned from type II enzymes is described below, and the take-home lesson is that we still have much to learn even for this "simple" group.

C. Type IIS Restriction-Modification Systems

Early in the characterization of type II systems, some very strange enzymes were found. The first sign that all was not normal was the decidedly asymmetric nature of the recognized sequences (Szybalski et al., 1991). When the sites of cleavage were determined, it was found that both DNA strands were cleaved off to one side of the recognized sequence (Sugisaki, 1978; Sugisaki and Kanazawa, 1981)—hence the name S (for Shifted cleavage). One of the most puzzling aspects of these systems was how their recognition sequences could be maintained in a protected state. If we take *Fok*I as an example, one strand is 5'-GGATG while the other is 5'-CATCC. An adenine methyltransferase specific for GGATG would protect one strand, and the hemimethylated duplex (methylated on just one strand) would be resistant to endonuclease cleavage, but every time a replication fork passes, one of the daughter duplexes will be completely unmethylated and subject to cleavage. There are several ways this problem might be solved, and the type IIS systems chose the most obvious route—making two methyltransferases (sometimes expressed as a single fusion protein) that recognize the two strands. In the case of *Fok*I, as shown in Figure 5, one methyltransferase activity recognizes GGATG, methylating the A, while

the other activity methylates the A in CATCC (Leismann et al., 1998).

Nevertheless, these enzymes have been described as a subset of type II because the methyltransferase requires only AdoMet, and the endonuclease requires only Mg^{2+}. It was at one point thought that type IIS endonucleases differed from standard type II enzymes in being active as monomers. However, while the IIS endonucleases are monomeric in solution and do have weak cleavage activity in that form, the homodimers (which may form on the DNA) are far more active (Bitinaite et al., 1998).

D. Type III Restriction-Modification Systems

The best-characterized type III restriction-modification systems are carried by lysogenic bacteriophages. As with the type I systems, before the advent of genomic sequencing only a handful of type III systems had been identified, and all were in Gram-negative bacteria. A search through the available bacterial genomes (finished and unfinished, at *http://www.ncbi.nlm.nih.gov/ Microb_blast/unfin_databases.html*), using the Res subunit of *Eco*PI as the subject, identified many matches having expect scores between 10^{-15} and 10^{-86}. (In simplified terms the expect score gives the probability of finding an equivalent match between random amino acid sequences; Karlin and Altschul, 1993). These matches were across the spectrum of gram-negative bacteria including *Neisseria, Actinobacillus, Dehalococcoides, Chlorobium, Bordetella*, and *Helicobacter*. No matches were found in archaea, and the matches to gram-positive bacteria (*Corynebacterium* and *Clostridium*, as of spring 2001) had expect scores in the more questionable 10^{-1} to 10^{-2} range.

DNA sequence specificity of type III systems is provided by the methyltransferase subunit, which is different from the case in type I systems, and the type III recognition sequences are asymmetric but not bipartite. However, the type III restriction-modification systems resemble type I systems in some key respects. The specificity/methyl-transferase subunit (called "Mod") is independently active, while the subunit reponsible for endonuclease activity ("Res") is only active as a complex with Mod and restriction activity is stimulated by AdoMet. Another basic similarity to type I systems includes the helicase domain and requirement for ATP associated with Res. A type III enzyme bound to a recognition site does begin ATP-powered translocation, though with two differences from type I translocation—a bound repressor is enough to block the translocation, and cleavage is neither at the point of the blockage nor even triggered by the blockage (Meisel et al., 1995). Efficient cleavage by type III restriction-modification systems requires two complexes bound to unmethylated sites. This is more like the type IIE than like type I enzymes; the type I enzymes do cleave where two translocating complexes collide, but the second type I complex can be replaced by other translocation barriers. However, in the case of type III systems, the two sites must be in opposite orientation with respect to one another; this is not true of the type IIE systems. The requirement for paired, oppositely oriented sites represents another solution to the problem that type IIS systems solved with multiple methyltransferase activities—how to ensure that there are never completely unmethylated daughter duplexes. In the case of the type III systems, there is only one methyltransferase activity (Meisel et al., 1991). If only one strand is methylated, but restriction requires a second site in the opposite orientation, then one member of each pair of sites will always be methylated in daughter duplexes (Meisel et al., 1992). Interestingly the chromosome of bacteriophage T7 contains 36 recognition sequences for the type III restriction-modification system *Eco*P15I, but all 36 sites are in the same orientation and T7 DNA is completely refractory to *Eco*P15I cleavage (Kruger et al., 1995); in the absence of selection, the chance that 36 sites would all appear in the same orientation is $1/2^{36}$ (about 1 in 7×10^{10}).

One of the many remaining puzzles about type III restriction-modification systems is how cleavage is triggered. Unlike the type I systems, type III systems will still bind to and begin apparently wasteful translocation at methylated sites (at least in vitro), and so, when two translocating complexes meet, they must still be able to sense when one of the two sites is methylated. Another type III puzzle is how the system functions given that a bound repressor was shown to block translocation. Thanks to textbook figures and the conditions employed for most in vitro experiments, we tend to think of DNA as a naked double helix sporting the occasional bound protein, resembling a palm tree with a monkey hanging on halfway up. The truth is quite different, and DNA is no more naked in bacteria than it is in eukaryotes—in *E. coli* a variety of so-called histonelike proteins of high abundance and limited specificity (HU, IHF, H-NS, etc.) cover the DNA, binding as close together as every 60 bp (Azam and Ishihama, 1999; Blumenthal et al., 1996; Segall et al., 1994). If *lac* repressor blocks type III complex translocation in vivo, how do site pairs separated by several hundred bp ever get cleaved?

E. Type IV Restriction-Modification Systems

The type IV restriction-modification systems are a proposed grouping (Janulaitis et al., 1992a; Janulaitis et al., 1992b). The enzymes have asymmetrical recognition sequences and shifted cleavage positions like the type IIS enzymes (Fig. 5), and are often categorized as type IIS for that reason. However, the enzymes belonging to type IV differ from standard type IIS enzymes in two key respects. First, the endonuclease is fused to a methyltransferase, and second, endonuclease activity is stimulated by AdoMet. As these properties conflict with the defining properties of type II enzymes, many are reluctant to classify this group as type IIS. If nothing else, this illustrates the problems in categorizing a diverse group of enzymes that very probably represents a mix of evolutionary convergence and divergence. There are proposals to classify restriction-modification systems strictly by the type of cleavage generated, instead of the mix of structure, specificity, and substrate requirements; this would simplify things but would group together systems that differ substantially in structure, evolutionary history, and possibly even roles.

In the few cases studied, the methyltransferase activity intrinsic to the endonuclease protein methylates one strand of the recognition sequence, but does so in a place that does not block cleavage by the endonuclease. Unless this intrinsic methylation blocks or substantially slows cleavage under in vivo conditions, one wonders what selective pressure has maintained this methylating activity. The systems also include a second methyltransferase that, like some of the type IIS enzymes, protectively methylates both strands of the recognized sequence.

F. "Bcg-Like" Restriction-Modification Systems

The "Bcg-like" restriction-modification systems are named for their archetype, *Bcg* I from *Bacillus coagulans* (Kong et al., 1993, 1994). So far nobody has suggested calling them "type V," though in structural and functional terms this might be justified. Like the type IV restriction-modification systems, the Bcg-like systems have methyltransferase and endonuclease fused into a single polypeptide (the A or α subunit); as with other dual function systems AdoMet is both methyl donor for the methyltransferase and stimulator of endonuclease activity. There is no associated helicase activity, and restriction does not require ATP. Unlike the type IV systems, the Bcg-like systems have a second B (or β) subunit that provides specificity (similar to the HsdS subunit in the type I systems); the solution complex is A$_2$B (Kong, 1998; Piekarowicz et al., 1999).

The Bcg-like systems, in a second similarity to HsdS of type I systems, recognize gapped asymmetrical sequences (Fig. 5). However, the cleavage pattern of Bcg-like systems is unique—they make a *pair* of double-strand breaks, one pair to each side of the recognition sequence, thus removing the recognition sequence as a short double-stranded oligonucleotide. The advantage of this double cleavage is not known, though one possibility is that the pair of double-strand breaks would be more difficult to repair so that this mode of cleavage might provide better defense against bacteriophages (or more effective selfish behavior). The Bcg-like enzymes resemble the type IIE endonucleases in requiring a pair of recognition sequences for efficient cleavage; in fact *Bcg* I can dimerize (forming a heterohexamer) (Kong and Smith, 1998).

V. MODIFICATION

A. Use of AdoMet and the Conserved Methyltransferase Tertiary Structure

AdoMet is, after ATP (and not counting things like water), the second most frequently used enzyme substrate (Cantoni, 1975). A large family of methyltransferase enzymes use AdoMet, including all known DNA methyltransferases (go to *http://www.expasy. ch/enzyme/enzyme-search-ec.html* and list enzymes having EC numbers that begin 2.1.1.; only about 10% of methyltransferases use other methyl donors such as folate). The AdoMet-dependent methyltransferases appear to be an ancient family, in that almost all of them share a highly conserved core structure even though the identity between amino acid sequences can be 10% or lower due (presumably) to divergent evolution (Fauman et al., 1999). Where sequence conservation is seen, it takes the form of isolated motifs that correspond for the most part to loops at the ends of beta strands (Chandrasegaran and Smith, 1988; Klimasauskas et al., 1989; Malone et al., 1995; Posfai et al., 1988).

It is striking that the DNA methyltransferases belonging to restriction-modification systems have the same core structures as the methyltransferases that act on small molecules or proteins. This structure is a seven-stranded β-sheet made up of a series of α/β

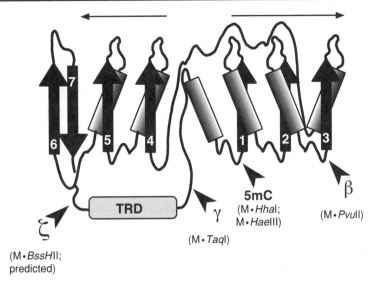

Fig. 7. Consensus structure and circular permutation of DNA methyltransferases. The core of DNA methyltransferases is a seven-stranded sheet (numbered arrows) flanked by ?-helices (cylinders). TRD stands for target-recognizing domain, a region with primary responsibility for DNA sequence specificity. The arrowheads indicate the points at which various methyltransferase families have their amino and carboxyl termini; these determinations come from sequence comparisons, but have been confirmed by X-ray crystallography for the enzymes listed in parentheses. The α family (not shown) has the TRD inserted between strands 3 and 4 rather than being a circular permutant of the other families, and this has been confirmed by crystallography of one of these enzymes (*Dpn* II).

motifs (an alpha helix followed by a beta strand; see Fig. 7). In the standard configuration, strands 1 to 3 of the sheet go from the middle to one end, then the protein loops back to the center and strands 4 to 7 go from the middle to the other end (6 ↑ 7 ↓ 5 ↑ 4 ↑ 1 ↑ 2 ↑ 3 ↑, where the arrow points to the carboxyl end of the strand). This generates what is called a "topological switch point" between strands 1 and 4 in the center of the sheet, where the loops diverge to create a deep cleft; AdoMet is bound in this cleft. Another feature is the reversed β-hairpin formed by strands 6 and 7—its role is unclear, but it provides another signature of the conserved structural core of the AdoMet-dependent methyltransferases.

B. Three Types of DNA Methylation (5mC, N4mC, N6mA)

If chemical alkylating agents such as dimethyl sulfate are added to DNA, methylation can occur at a variety of positions.

However, cells producing a DNA methyltransferase would be selected against if the enzyme added methyl groups at positions that destabilized the DNA or interfered with base pairing. Among restriction-modification systems, the methyltransferases have only been found to act at one of three positions: a ring carbon of cytosine (generating 5-methylcytosine, 5mC) or the exocyclic amino groups of either cytosine or adenine (generating N4-methylcytosine, N4mC, or N6-methyladenine, N6mA). None of these methylations interferes with base pairing (Fig. 8), and all three result in methyl groups exposed in the major groove (Fig. 9).

The only known damage that results from enzymatic DNA methylation has to do with oxidative deamination of cytosine. As described by Yasbin, this text, cytosine undergoes spontaneous deamination to form uracil, but this lesion can be repaired following the action of an enzyme that

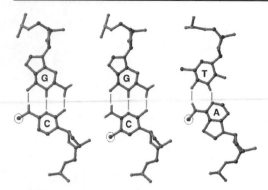

Fig. 8. Products of cellular DNA methylation. Only three DNA methylations (dotted circles) have been found thus far to be carried out by cells—N4-methylcytosine (left), C5-methylcytosine (middle), and N6-methyladenine (right). None of these three interferes with base pairing (dotted lines) or nucleotide stability. All three have been found in association with restriction-modification systems.

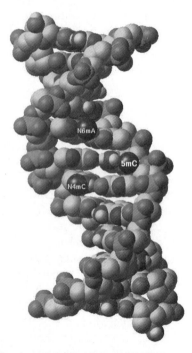

Fig. 9. Location of added methyl groups on DNA double helix. All three methylations lead to exposure of methyl groups in the major groove.

scans DNA and removes all uracils (uracil-D-glycosylase). N4mC has a reduced rate of deamination, and it has been suggested that N4mC methyltransferases might be rela-

tively prevalent among thermophiles for this reason (Ehrlich et al., 1985). In contrast, 5mC deaminates as readily as unmethylated cytosine, but the product is not uracil but the normal DNA base thymine (Duncan and Miller, 1980). In fact there is abundant evidence that 5mC sites are hotspots for C → T mutation (Cooper and Krawczak, 1989). Cells produce enzymes (Vsp, for very short patch repair) designed to repair the resultant G-T mismatches (Gabbara et al., 1994; Hennecke et al., 1991; Lieb et al., 1986). Some restriction-modification systems with 5mC-generating methyltransferases even include a *vsp* gene to repair the damage (Kulakauskas et al., 1994).

C. Fitting DNA into the Consensus Catalytic Pocket: Base Flipping

It should seem surprising that the same core structure works for two AdoMet-dependent methyltransferases that act on substrates differing in mass by over an order of magnitude. Yet catechol-*O*-methyltransferase, which acts on catechols, has the same core structure as *Hha* I DNA methyltransferase (Schluckebier et al., 1995); the catechol dopamine has a MW of about 150 while a 12 mer duplex oligonucleotide substrate has a MW of nearly 8000. In fact the rare exceptions to the rule that AdoMet-dependent methyltransferases have a common core structure comes from enzymes that act on corrins, the large flat multi-ring structures such as cobalamin (Dixon et al., 1999).

The solution to this apparent conundrum is that the DNA methyltransferases don't methylate DNA per se, they methylate a specific purine or pyrimidine base within a DNA molecule. An individual nucleotide averages roughly 330 in MW. The elegant solution of the DNA methyltransferases is shown in Figure 10, and was originally discovered for the *Hha* I 5mC methyltransferase (Klimasauskas et al., 1994), which is the modification enzyme from a type II restriction-modification system. In the protein-DNA complex the target cytosine is no longer buried within the double helix; it has

A

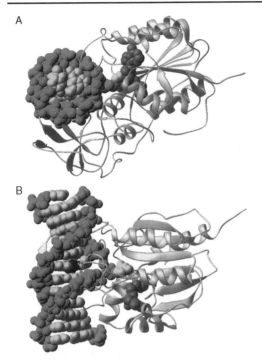

B

Fig. 10. Base flipping by DNA methyltransferases. Structure of M•*Hha* I (PDB code 1MHT) complexed with a substrate duplex oligonucleotide and the methyl donor AdoMet (red). The cytosine to be methylated, generating (in this case) C5-methylcytosine, has been "flipped" out of the double helix by rotation on the flanking sugar-phosphate bonds, and is next to the AdoMet. This flipped complex was trapped by using the suicide substrate 5-fluorocytosine. **A:** End view, with DNA helical axis projecting out of the page. This is also an end view of the seven-stranded sheet that forms the core of the catalytic portion of the methyltransferase (upper right). **B:** Side view; a 90° rotation of the structure shown in (A).

been rotated 180° on its flanking sugar-phosphate bonds so that it projects out into a typically concave catalytic pocket. No covalent bonds were broken to carry out this process, which is called *base flipping*—the base-pairing hydrogen bonds were broken, and the stacking interactions with adjacent base pairs was lost. Base flipping has since been found in a variety of other enzymes that act on DNA bases (Blumenthal and Cheng, 2001; Cheng and Blumenthal, 1996; Goedecke et al., 2001; Roberts and Cheng, 1998).

There are still a number of things we don't understand about base flipping, including how it is initiated and how (or if) it is related to recognition of the substrate sequence. Two interesting features are known, however. First, in cocrystals of *Hha* I methyltransferase with a DNA substrate having an abasic site at the position of the target cytosine, the enzyme still moves the sugar-phosphate backbone to the "flipped-out" position (O'Gara et al., 1998). Second, base flipping will work with most any base at the target position; this lack of specificity has been used to measure base flipping because 2-aminopurine fluorescence increases dramatically when it is removed from the stacking environment in double helical DNA (Allan et al., 1998; Holz et al., 1998). Though the methyltransfer reaction is generally more sensitive to the base at the target position than is the base-flipping step, even here it is worth noting that at least some methyltransferases that generate N6mA can generate N4mC if a cytosine is in the flipped position (Jeltsch et al., 1999b).

D. A Brief Excursion: Independent DNA Methyltransferases

It is important to note that not all methyltransferases belong to restriction-modification systems. Independent methyltransferases can play a variety of roles in cells ranging from bacterial through mammalian. The Dam methyltransferase found among the Enterobacteriaceae controls expression of some genes, influences the initiation of chromosome replication, and identifies the parental DNA strand for the mismatch correction system (Barras and Marinus, 1989; Messer and Noyer-Weidner, 1998). In *Salmonella*, Dam plays a critical role in controlling pathogenesis, so much so that Dam⁻ strains may make good vaccine strains (Garcia-Del Portillo et al., 1999). Dam generates N6mA in the sequence GATC, and a related enzyme that generates N6mA in the sequence GANTC (CcrM) plays critical roles in controlling gene expression for pathogenesis or the cell division/differentiation cycle in *Brucella*, *Caulobacter*, and *Rhizobium* (Reisenauer et al., 1999; Robertson et al., 2000; Wright et al., 1997). Many more independent regulatory DNA methyltransferases probably exist in bacteria and archaea, waiting only for genomic sequences to reveal them as candidates.

The regulatory methyltransferases found in bacteria have so far been limited to N6mA-generating enzymes. In eukaryotes the only DNA methyltransferases found to date generate 5mC. This methylation is rare among organisms with a genome size $< 10^8$ bp but nearly universal among organisms with larger genomes, including most plants and animals (Bestor, 1990). The mammalian methyltransferase Dnmt1 (DNA methyltransferase 1) is a large protein, though the carboxyl-terminal third has the typical structure and conserved sequence motifs of the bacterial 5mC methyltransferases (Margot et al., 2000). This protein is a maintenance methyltransferase, which means that it has low activity on unmethylated DNA and much higher activity on hemimethylated DNA, and this behavior is conferred by the amino-proximal two-thirds of the protein (Bestor, 1992). One or more separate enzymes (de novo methyltransferases) initiate methylation at a given position (Okano et al., 1999). Among mammalian cells, methylated DNA is preferentially bound by proteins, at least one of which is part of a histone deacetylating complex; this puts the adjacent chromatin into a transcriptionally inactive state (Ng et al., 2000). Thus DNA methylation in mammals is associated with transcriptional silencing, but it also plays other roles (Bird and Wolffe, 1999; Robertson and Jones, 2000).

There is some support for a model in which the original role of DNA methylation in eukaryotes was a defensive one (reminiscent of the phage defense hypothesis for restriction-modification systems) (Wolffe and Matzke, 1999)—methylation would silence invading selfish genetic elements such as the numerous Alu elements found in mammalian DNA (Bestor, 1996; Walsh and Bestor,

1999). This is consistent with the roles played by DNA methylation in silencing introduced transgenes in plants (Dieguez et al., 1998; Jakowitsch et al., 1999), and in marking transposons and other duplicated regions for hypermutation in Neurospora (Irelan and Selker, 1997; Margolin et al., 1998; Windhofer et al., 2000).

With regard to the objectives of this chapter, it is not yet clear how the independent methyltransferases are related to those belonging to restriction-modification systems (Dryden, 1999). Did these independent methyltransferases appear first, selected for their defensive and regulatory roles and creating a permissive background for the development of nucleases with overlapping specificity? Or, alternatively, did simpler methyltransferases belonging to restriction-modification systems (simpler in the sense of methylating every available substrate sequence) become associated with regulatory domains that ensured that only appropriate occurrences of the substrate would be methylated?

E. Permuted Families of DNA Methyltransferases

There have been two major impediments to the phylogenetic analysis of DNA methyltransferases. One is the aforementioned low overall sequence conservation, with such conservation as exists limited to unevenly spaced conserved motifs. In fact a structure-guided sequence alignment of all of the (then-available) structures for AdoMet dependent methyltransferases revealed only two well-conserved positions in the entire protein (Fauman et al., 1999). The second impediment is the fact that the DNA methyltransferases have apparently undergone circular permutation to generate three families, with a fourth family resulting from an insertion into the core structure (Fig. 7) (Jeltsch, 1999; Malone et al., 1995). The 5mC and γ families differ only in the placement of one α-helix and its associated conserved motif (respectively at the carboxyl and amino ends of the protein) (Schluckebier et al.,

1995). The β family amino terminus is just upstream of strand 3, which means that several of the conserved motifs are in a permuted order relative to the other methyltransferases (Gong et al., 1997). The α family (not shown) has an insertion between strands 3 and 4 (Tran et al., 1998). Other families probably exist; one (ζ) is indicated in Figure 7 based on supporting sequence information but no structural confirmation (Sethmann et al., 1999). To date, such permutation has only been seen among the AdoMet-dependent methyltransferases that act on DNA. It is not clear why this is the case, though it can be difficult to identify permutations based only on the pattern of conserved sequence motifs. There is one report of a candidate family of RNA methyltransferases with permuted arrangement, though this needs structural confirmation (Bujnicki et al., 2002). Additional DNA methyltransferase structures will help in defining phylogenetic relationships, by guiding the sequence alignments, but initial attempts have been made to map the evolutionary history of these enzymes (Bujnicki, 1999; Bujnicki and Radlinska, 1999a,b).

VI. RESTRICTION

A. Conserved Structure of the Catalytic Core May Represent Convergent Evolution

Unlike the DNA methyltransferases, among which the highly conserved core structure and conserved sequence motifs suggest divergence from a common ancestor, the endonuclease sequences are so dissimilar as to suggest nothing at all, while the structures suggest a mix of convergent and divergent evolution with no single ancestor. Among the type II endonucleases, where the enzyme is an independent homodimer, there are three functional regions: the dimerization interface, the DNA sequence-recognition region, and the phosphodiesterase or catalytic center. The catalytic centers of these endonucleases resemble those of other nucleases, such as the exonuclease of bacteriophage λ and the mismatch

repair endonuclease MutH (Fig. 11). The nuclease catalytic center in all of these structurally characterized nucleases comprises a five-stranded β-sheet with a ↑↓↑↑↓ relative strand orientation, flanked by a pair of α-helices (Kovall and Matthews, 1998, 1999). In all cases there is a requirement for Mg^{2+} or some other suitable divalent cation. The acidic sidechains coordinating the cation provide the only hint of a conserved sequence motif among these enzymes, for example, Glu^{85}-Asp^{119}-Glu^{129}-Lys^{131} in λ exonuclease corresponds to Glu^{55}-Asp^{58}-Glu^{68}-Lys^{70} in *Pvu* II endonuclease—clearly this is not the sort of conserved motif that can be used to identify nucleases from the primary sequence alone. It is reminiscent of the spatially conserved catalytic amino acids found among unrelated proteases (Fischer et al., 1994).

With a structurally conserved catalytic center, one might suppose that the endonucleases diverged from a common ancestor. There is some evidence to support this view (Bujnicki, 2000). However several points mitigate against any model based exclusively on divergence from a single ancestor. First, the dimerization interfaces and DNA sequence-recognition regions of various restriction endonucleases look utterly unlike

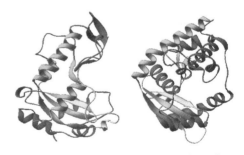

Fig. 11. Structures of Two Deoxyribonucleases. Single subunits from the *E. coli* MutH mismatch nuclease (left) and the exonuclease of bacteriophage λ (right) are shown. In this orientation the DNA helical axis, if shown, would project out of the page. The λ Exo (PDB 1AVQ) is active as a trimer that encircles the DNA, while MutH (PDB 1AZO; note "O" not zero) is active as a dimer. Regions of catalysis of phosphadiester cleavage (light) and DNA binding (dark) are indicated.

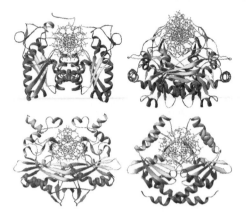

Fig. 12. Structures of four type II restriction endonucleases. All four homodimers are orientated such that the DNA helical axis projects out of the page. The endonucleases shown are (clockwise from upper left, with PDB codes in parentheses) *Bam*H I (1BHM), *Bgl* I (1DMU), *Pvu* II (1PVI), and *Eco*R V (1AZ0; note zero not "O"). Regions of catalysis of phosphodiester cleavage (light) and DNA binding (dark) are indicated.

one another (Fig. 12). This would still leave open the possibility of a conserved nuclease module becoming associated with various other elements. However, a second point mitigating against divergence is that the catalytic centers are not specified by a con- tiguous region of the gene. In other words, if the protein was stretched out one would see that structural elements comprised by the catalytic center are interspersed among elements of the dimerization interface and/or DNA sequence-recognition region. This is illustrated for the *Pvu* II endonuclease in Figure 13.

A third argument against the model that endonucleases diverged from a single ancestor is based on similarities of individual endonucleases to members of other families. The most striking example of this involves the type IIE restriction endonuclease *Nae* I. This endonuclease is just one missense mutation from being a topoisomerase-recombinase (Jo and Topal, 1995). Alignment to the sequence of a DNA ligase revealed that one region of *Nae* I had all of the conserved catalytic amino acids except one. When Leu[43] was changed to Lys (L43K), to match the ligase alignment, *Nae* I relaxed supercoiled DNA giving DNA topoisomers and recombined DNA to give dimers. The L43K mutation also led to other behaviors typical of topoisomerases but not of restriction endonucleases (including wild-type *Nae* I), such as binding to single-stranded

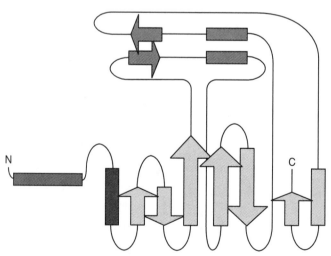

Fig. 13. Linear order of structural elements in the *Pvu* II restriction endonuclease. The color scheme is the same as in Figures 11 and 12. Arrows represent strands and rectangles represent helices. Note that the catalytic sheet-based region is not formed by a single contiguous segment of the protein.

DNA, sensitivity to intercalating agents and formation of a covalent bond to the newly exposed 5' DNA end (Jo and Topal, 1996a,b). Thus *Nae* I endonuclease, unlike other known restriction endonucleases, belongs to the topoisomerase-recombinase family of proteins. Interestingly one of the methyltransferases (*Sss* I) has topoisomerase activity in the presence of Mg^{2+} (Matsuo et al., 1994); full-length alignment of *Sss* I with other 5mC methyltransferases indicates that this is not a fusion of two distinct functional domains, and its evolutionary and functional significance are unclear.

B. Roles of the Divalent Cation

All of the endonucleases require a divalent metal cation such as Mg^{2+} in order to catalyze phosphodiester bond cleavage. Divalent metal cations can play a variety of roles in phosphoryltransfer reactions, some of which are activating the attacking nucleophile (water), stabilizing the intermediate pentavalent state of the phosphorus, and helping in removal of the leaving group (Gerlt, 1993; Kyte, 1995). The endonucleases may be heterogeneous even in the uses to which the Mg^{2+} is put. For example, type II endonucleases such as *Bam*H I use a two-metal mechanism of cleavage, while *Bgl* II has been proposed to use a single-metal mechanism (Galburt and Stoddard, 2000; Lukacs et al., 2000). *Bgl* I and *Bam*H I each have two Mg^{2+} per subunit, one of them near the leaving group; the two Mg^{2+} in *Eco*R V subunits are in a different orientation relative to the DNA, and neither one is near the leaving group (Baldwin et al., 1999; Sam and Perona, 1999; Stanford et al., 1999). One important caveat is that neither the one-metal *Bgl* II complex nor the *Eco*R V complex just described have been shown to be catalytically competent. Thus it is possible, though not yet proved, that even among the type II endonucleases there are striking differences in the architecture of the catalytic centers.

Another difference in Mg^{2+} utilization among restriction endonucleases has to do with recognition of the specific DNA substrate sequence. Many type II endonucleases specifically bind their substrate sequences whether or not divalent cation is present, while others such as *Eco*R V will only bind specifically if divalent cation is present (Erskine and Halford, 1998; Martin et al., 1999).

C. Another Brief Excursion: Independent Endonucleases

Just as there are DNA methyltransferases that don't belong to a restriction-modification system, there are sequence-specific DNA endonucleases that act independently. We will discuss two groups of independent endonucleases.

I. Methylation-dependent endonucleases

If the role of restriction endonucleases is to cut foreign DNA, then any clear indication that a DNA is foreign can be used to control the activity of the restricting enzyme. The nucleases in restriction-modification systems recognize DNA as being foreign when it lacks protective methylation. It works just as well to detect non-self-patterns of DNA methylation, but in this case one would have a single-component system consisting of a nuclease that cleaves in response to a particular sequence only when it *is* methylated. Several such systems are known, and they can work synergistically with restriction-modification systems. The first such enzyme to be recognized as such was *Dpn* I (de la Campa et al., 1988), which was discussed earlier in the context of the *Dpn* I–*Dpn* II cassette system in *Streptococcus pneumoniae*. *Dpn* I recognizes GATC, and only cleaves if the adenines on both strands have been methylated. It may be possible for many restriction endonucleases to change to a form that requires methylation for cleavage; in one line of experiments the *Bam*H I endonuclease was altered to a form that only cleaves its GGATCC sequence if the A is methylated (Whitaker et al., 1999).

McrA, McrBC, and Mrr were all characterized in *E. coli* on the basis of their ability to restrict methylated DNA (Modified Cytosine Restriction, Methyl puRine Restriction). The Mcr system was actually first identified in the 1950s as leading to restriction of T-even bacteriophage mutants (Luria and Human, 1952). These mutants failed to attach glucose to the cytosine in their DNA, so the genes responsible for the restriction were originally named Rgl (restricts glucose-less) A and B (Revel, 1967; Revel and Georgopoulos, 1969). It has been suggested that the Mcr/Rgl systems arose in an offensive-defensive cycle between *E. coli* and its bacteriophages (Revel, 1983). The T-even bacteriophages, in this model, gained resistance to many restriction endonucleases by replacing the cytosine in their DNA with 5-hydroxymethylcytosine (5hmC), the host responded with methylation-dependent restriction endonucleases (Mcr/Rgl), and the bacteriophages responded yet again by linking glucose to the hydroxyl on the 5hmC such that the DNA is resistant to Mcr/Rgl.

In the 1980s several groups were cloning restriction-modification system genes, and found that some of them could only be expressed in a few *E. coli* strains. This was originally suspected to be due to misregulation, with restriction occurring before the DNA of the host was protected by the incoming restriction-modification system. Actually the problem was that expression of the methyltransferase itself was lethal to many *E. coli* strains; even cloning DNA that didn't code for a methyltransferase but that was itself methylated led to problems, including the biasing of genomic libraries from plant and mammalian sources (Blumenthal, 1986; Blumenthal et al., 1985; Heitman and Model, 1987; Noyer-Weidner et al., 1986; Woodcock et al., 1988). These phenomena led to the rediscovery of the RglA and RglB (McrA and McrBC) systems (Raleigh, 1987; Raleigh et al., 1988, 1989; Raleigh and Wilson, 1986; Ross et al., 1989; Ross and Braymer, 1987), and to the discovery of Mrr (methyl puRine

restriction) (Kelleher and Raleigh, 1991; Kretz et al., 1991; Waite-Rees et al., 1991).

McrA is specified by a defective prophage called ε14, and it digests DNA that has been modified by the 5mC methyltransferases *Hpa*II (C<u>C</u>GG) or *Sss* I (<u>C</u>G), though other evidence rules out a simple <u>C</u>G substrate specificity. Mrr is chromosomally coded and digests DNA containing either 5mC or N6mA in particular contexts; DNA modified by *Sss* I or *Hha* I (GC<u>G</u>C) is a target, as is DNA modified by any of eight N6mA methyltransferases such as *Hpa* I (GTTA<u>A</u>C) or *Pst* I (CTGC<u>A</u>G). It is difficult to define a consensus recognition pattern for either McrA or Mrr. Not a great deal is known about the biochemistry of these two systems, but quite a lot has been learned about the third.

McrBC does not restrict in response to N6mA; it recognizes three forms of modified cytosine: 5mC, 5hmC, and N4mC. The consensus pattern is RmC (R = puRine), and there must be two such sites at least 20 to 30 bp and at most 2 to 3 kbp apart (Stewart and Raleigh, 1998). McrBC is unique among endonucleases in that its movement on the DNA is powered by GTP rather than by ATP (Sutherland et al., 1992), and it has some features that distinguish it from other G proteins (Pieper et al., 1999). It is a multi-subunit complex, with one McrC subunit associated with several (perhaps four) McrB subunits to form an active enzyme (Panne et al., 1998). For regulatory purposes the chromosomal *mcrB* gene produces two versions of McrB due to an alternative internal translation initiator, and the enzymatically inactive smaller product (McrBs) competes with full-length McrBL for binding to McrC (Panne et al., 1998). Another interesting feature of *mcrBC* is its genetic location. In *E. coli* K-12, these genes lie in a hypervariable cluster (Barcus et al., 1995) that also includes the EcoK type I restriction-modification system and the *mrr* methylation-dependent endonuclease (Fig. 14); this cluster has been called the "immigration control region," and ironically, based on its nucleotide

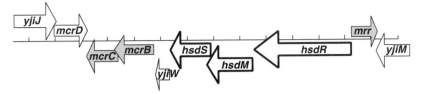

Fig. 14. The immigration control region (ICR) of *E. coli*. This highly polymorphic region includes the genes for a type I restriction-modification system (bold arrows) and for two restriction systems that target methylated DNA (gray arrows). Salmonella has a similar region (including a close homologue to the uncharacterized *yjiW* gene). The genes in this region are tightly clustered—only 134 bp separates the starts of the oppositely oriented *hsdR* and *mrr* coding regions, while *yjiW* is separated from the preceding *hsdS* gene by 171 bp and from the following *mcrB* gene by 162 bp.

composition, it is itself an immigrant (Raleigh, 1992; Raleigh et al., 1989).

Sequences similar to McrA, McrBC, and Mrr occur in other bacteria besides *E. coli*, though activity has not been confirmed. The distribution of these restriction systems may be tentatively gauged by looking for extremely close matches in the microbial genomes database, as described above. Mrr matches having TBLASTN expect scores $< 10^{-15}$ were found (as of spring 2000) in *Salmonella*, *Deinococcus*, *Porphyromonas*, *Mycobacterium*, *Methanobacterium*, and *Thiobacillus*. For McrA, such matches were found in *Rhodobacter* and *Vibrio*. Only one organism had such matches to both McrB and McrC—*Staphylococcus aureus*; however, several organisms had such matches to McrB alone, including *Clostridium*, *Streptococcus*, *Campylobacter*, *Yersinia*, *Porphyromonas*, and *Helicobacter*.

2. Homing endonucleases

These enzymes have several interesting biochemical features, but their most striking property is that some are within mobile single-gene elements (Gimble, 2000; Jurica and Stoddard, 1999). This has to be the ultimate in selfish genes (Edgell et al., 1996), though there are suggestions that these elements might provide some benefit to their hosts (Dalgaard, 1994). Homing endonuclease genes (often abbreviated HEGs) are found throughout the biological world—bacteria, archaea, plants and animals, and

even mitochondria and chloroplasts contain them. To ensure that integration into a gene doesn't reduce the fitness of their new host, and thus reduce their spread, HEGs make themselves phenotypically neutral via one of two strategies. Some HEGs lie within group I or group II introns, and so are spliced out of the mRNA following transcription (Gorbalenya, 1994, Quirk et al., 1989). Other HEGs lie within inteins, and so are spliced out of the protein following translation (Derbyshire et al., 1997; Pietrokovski, 1994, 1998). Since they have no phenotypic cost, the most widespread HEGs are associated with the most highly conserved genes, such as the genes for DNA polymerase, gyrase, or RecA, often in the most highly conserved regions of those highly conserved genes.

The mechanism for initial HEG integration at a given site varies with the type of element, but it is not known in detail for many of them. Transfer from one allele to a sister allele in the same cell (homing) is somewhat better understood. In both cases the initiating step is believed to be generation of a double-strand break in the target DNA by the sequence-specific homing endonuclease. In the case of homing, repair of this break involves homologous recombination with the HEG-containing (uncleaved) allele. For our purposes the relevant part of this process is generation of the double-strand break. In fact an artificial HEG was made that used the *Eco*R I type II restriction endonuclease, so generating the double-strand break appears to be the only significant action of the authentic homing endonucleases (Eddy and Gold, 1992). Homing endonucleases recognize much longer sequences than restriction endonucleases (14–31 bp *vs.* 4–8 bp); for example, the intron-carried homing endonuclease I-*Ceu* I (from an rRNA gene in *Chlamydomonas*) recognizes the sequence TAACTA TAACGGTCCTAA∧GGTAGCGA, cleaving at the ∧ (and at the appropriate position on the opposite strand) to generate a four nucleotide 3′ extension (Gauthier et al., 1991; Marshall and Lemieux, 1992). However, for homing endonucleases some of the nucleotides in the recognized sequence are more important than others (for I-*Ceu* I these are underlined), while with restriction endonucleases all positions tend to contribute substantially to the interaction (e.g., see Alves et al., 1995).

The homing endonucleases fall into three families (LADLIDADG and GIY-YIG, named for conserved sequence motifs, and ββα-Me, named for a conserved structural motif). The ββα-Me family also includes a nonspecific nuclease from *Serratia*, the colicin E7 and E9 DNase domains, and endonuclease VII of bacteriophage T4 (Kuhlmann et al., 1999). There is no correspondence between endonuclease family and intein versus intron association (Gimble, 2000; Jurica and Stoddard, 1999). None of the homing endonucleases appear to be related to the known restriction endonucleases, reinforcing the idea that nucleases arose convergently from many ancestors.

VII. ACHIEVING AND VARYING SPECIFICITY

Whether acting defensively, selfishly, recombinationally, or all three, there would be advantages for restriction-modification systems in being able to develop new sequence specificities. In this section we will see that this ability has been designed into a number of restriction-modification systems.

A. Recombination of the Specificity Subunit in Type I Systems

The architecture of type I restriction-modification systems is suited in several ways to the diversification of sequence specificities. Recall that in type I systems the sequence specificity is determined by the HsdS subunit. The type I systems fall into several subfamilies (to date A–C), defined by the ability of their subunits to cross-complement. Providing an additional specificity can be as simple as providing a new HsdS subunit, making use of the HsdM and HsdR subunits already produced by a given bacterium. For example, in a *Lactococcus* strain producing a type I system, plasmids were found that carried *hsdS* genes but not *hsdR* or *hsdM*, and the plasmid-coded HsdS subunits resulted in active type I restriction-modification systems with new specificities (Schouler et al., 1998). The Bcg-like systems resemble the type I systems in having a separate specificity subunit (called B or β), and it is possible that a similar phenomenon might occur with plasmids bringing additional B subunits into cells that already produce a Bcg-like system. The *bcgIB* product shows intriguing similarity to the HsdS subunits of several type I systems in ClustalW alignments (not shown). Along the same lines, a new type III Mod subunit might be able to join with some endogenous Res subunits, though this has not yet been seen.

A second feature of HsdS that promotes variation in specificity is its modular architecture. The bipartite recognition sequences are each recognized by separated regions of the HsdS protein (Fig. 15). For example, *Eco*A I recognizes GAGN$_7$GTCA, with one region of the HsdS subunit recognizing GAG and the other recognizing GTCA (Cowan et al., 1989). HsdS specificity can be altered in three ways. First, two different *hsdS* genes can recombine in the conserved spacer region that joins the two recognition regions. For example, the *Salmonella hsdS* genes for the *Sty*SP system (recognizing

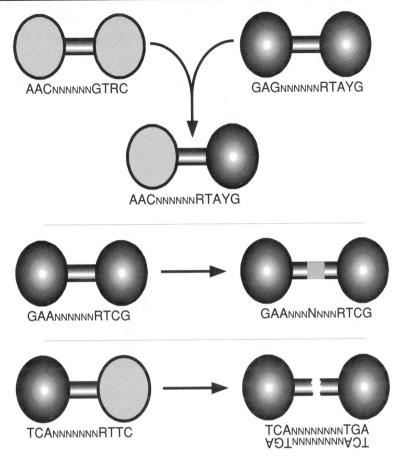

Fig. 15. Modular structure of HsdS allows generation of new specificities in type I restriction-modification systems. In the top panel, two *hsdS* genes have recombined to produce a hybrid HsdS with a new, hybrid specificity. In the example shown here, *hsdS* genes from *Sty*SP (*left*) and *Sty*SB have recombined to yield *Sty*SQ (see text). In the middle panel, the *hsdS* gene from *Eco*R124 I has undergone slipped-strand replication or unequal crossing over, so a repeated 4 aa sequence in the linker region appears three times rather than two. This results in an HsdS with the same sequence specificity but with an extra bp in the spacer between the two recognized sequences (*Eco*R124/3 I). The bottom panel illustrates that some HsdS polypeptides that are truncated in the linker region can still form a functional HsdS by dimerization of the amino-terminal half that is still made. In the example shown, *hsdS* from *Eco*DXX I was interrupted by a transposon, and the dimerized HsdS fragment leads to a functional restriction-modification complex that has a symmetrical sequence specificity.

AACN$_6$GTRC) and the *Sty*SB system (GAGN$_6$RTAYG) recombine to give a new *hsdS* (called *Sty*SQ, AACN$_6$RTAYG) (Bullas et al., 1976; Gann et al., 1987). Second, the length of the spacer region can change due to slipped-strand mispairing of a repeated nucleotide sequence during replication, and this changes the distance between the two recognized sequence portions. For example *Eco*R124 I recognizes GAAN$_6$RTCG and *Eco*R124/3 I recognizes the same sequence but with an N$_7$ spacer; their respective HsdS subunits have identical sequences except that they differ in the length of the region joining the two specificity portions, with *Eco*R124 I having two copies of a 4 amino acid sequence and *Eco*R124/3 I having three copies (Price et al., 1989).

Third, due to the modular architecture of HsdS, even deletions or polar insertions that truncate the subunit can leave a functional restriction-modification system with a new specificity (Abadijieva et al., 1993; MacWilliams et al., 1994; Meister et al., 1993). For example, EcoDXX I recognizes $TCAN_7 RTTC$, and transposon insertion into the middle of its hsdS gene leads to production of just the portion recognizing the TCA—this dimerizes to form a functional restriction-modification complex that recognizes the gapped palindrome $TCAN_8 TGA$.

B. Separate Domains for Specificity and Cleavage in Type IIS Endonucleases

In the type IIS endonucleases, as noted above, there are separate domains that respectively bind a specific sequence and catalyze strand cleavage (Fig. 5). This modular architecture allows cleavage to occur a turn or more of the double helix from where the recognition sequence lies. The structure of the IIS restriction endonuclease Fok I is exactly what you would expect from this description (Wah et al., 1997, 1998). The DNA specificity portion of FokI includes a helix-turn-helix motif. In theory, new specificities of type IIS systems might arise by recombination that replaces the specificity portion with another sequence-specific DNA-binding protein. The problem is that this would change the specificity of the endonuclease but not of the separate protective methyltransferase. Accordingly this type of specificity change remains a potentially useful laboratory tool that is unlikely to occur at a substantial rate in nature. Hybrid type IIS endonucleases have in fact been made, and they do show cleavage to one side of the expected new recognition sequence in each case (Chandrasegaran and Smith, 1999; Kim and Chandrasegaran, 1994; Kim et al., 1996, 1998; Smith et al., 1999). As an aside, a similar strategy has been used to target a DNA methyltransferase to a region adjacent to specific sequences (Xu and Bestor, 1997).

C. Changing the Specificity of Type II Restriction-Modification Systems

The preceding discussion of type IIS specificity alteration points out a problem faced by the type II systems in general—any change in restriction specificity must be accompanied by a change in the specificity of the separate methyltransferase. In general, none of the type II endonucleases is particularly easy to alter in terms of specificity, even when this is attempted in the laboratory (Dorner et al., 1999; Flores et al., 1995; Grabowski and Alves, 1995; Ivanenko et al., 1998; Lukacs et al., 2000). The restriction endonucleases have defined regions primarily responsible for sequence specificity (see Fig. 12), but recognition is not confined exclusively to those regions; rather, recognition appears to involve a complex interface between DNA and protein (Galburt and Stoddard, 2000; Winkler, 1994), as one might expect for a protein that can easily kill the cell if it gets careless.

Notwithstanding the difficulties of changing type II endonucleases, they clearly *have* diversified their specificities over time as several hundred distinct specificities have already been identified. In theory the complex recognition process could work to their advantage in this process: most changes in specificity might take several steps, with the intermediates having very low catalytic activity. This model would yield the counterintuitive result that the restriction-modification systems most successful in generating new specificities would be those having complex interfaces to the DNA that could *only* change through low-activity intermediates. A low-activity endonuclease with a new specificity would create an environment selective for alteration in the methyltransferase (or for association with a different methyltransferase) to protect the host's DNA, which would in turn allow for recovery of catalytic activity by the altered endonuclease. This is speculative, of course, but meant to show that there are alternatives to models requiring the methyltransferase to change first. We simply know very little about how this process of change occurs.

D. Modular Target Recognition Domains in Multispecific Type II Methyltransferases

Sequence specificity by the type II methyltransferases is similar to that of the type II endonucleases in the sense that there is a defined region with primary responsibility for recognition (called the *target-recognizing domain*, or TRD)(Lauster et al., 1989), but that in many methyltransferases other parts of the protein contribute to the process. Aside from this, recognition by the two groups of enzymes is very different. The endonucleases are generally symmetrical homodimers that wrap around the DNA, while the methyltransferases are generally asymmetrical monomers that interact predominantly with one strand of the DNA (Fig. 6). Furthermore the methyltransferases bind DNA-containing mismatches of the target (methylatable) base and flip that incorrect base into the catalytic pocket with high efficiency; even in the methyltransfer step, some enzymes that normally generate N6mA will generate N4mC if the target A is replaced by C (Jeltsch et al., 1999a). Both methyltransferases and endonucleases can exhibit sloppiness (cleavage of a site one base off from being a true site is called *star* activity, as in *Eco*R I*), though this is seen more in vitro with inappropriate buffer conditions than in vivo where such activity could be lethal (Woodbury Jr. et al., 1980).

Our question here is how do the type II methyltransferases change their specificity? To answer this by extrapolation, we can look at the adaptations made by a group of methyltransferases for which rapid acquisition of new specificities is an essential property. This group consists of the methyltransferases coded for by bacteriophages that attack *Bacillus*. These bacteriophage-coded enzymes are multispecific, and protect the bacteriophage DNA from a variety of restriction-modification systems carried by their host bacteria. From sequence analysis, these are typical 5mC-generating methyltransferases, with all of the conserved sequence motifs (though none has yet been structurally characterized). The only difference from monospecific methyltransferases is that the TRD region (the area between conserved motifs VIII and IX) is unusually large. Even here, there are some monospecific methyltransferases with even larger TRD regions for reasons that are not yet clear (Master and Blumenthal, 1997; Zhang et al., 1993). What makes the multispecific methyltransferases stand out is that their TRD regions are organized into modular, adjacent, nonoverlapping TRDs that are *entirely* responsible for sequence specificity. That is, in these multispecific enzymes, all responsibility for sequence recognition has been concentrated into the TRD. One can excise a particular TRD from one enzyme, and that enzyme loses one of its specificities; if that TRD is introduced into another multispecific enzyme, that enzyme gains the expected new specificity (Trautner et al., 1988; Trautner et al., 1996; Walter et al., 1992). It has even been possible to make hybrid TRDs with novel specificities (Lange et al., 1995). One multispecific enzyme has one of its TRDs at the amino terminus of the protein, so there is apparently some flexibility in this design (Sethmann et al., 1999). This type of TRD exchange is much more difficult to achieve with monospecific methyltransferases, but the limited results nevertheless confirm the role of the TRD regions in these enzymes as well (Klimasauskas et al., 1991; Mi and Roberts, 1992). It would be interesting to know if the multispecific enzymes pay a price in extent of specificity for this concentration of recognition into the TRD, or if they pay a kinetic price for having multiple TRDs competing for access to the DNA. However, either price is not high enough to prevent the enzyme from filling its protective role. With regard to our focus on evolution of new specificities among restriction-modification systems, might the multispecific enzymes be an important source of new specificities, with methyltransferases that become paired with endonucleases then losing all unneeded TRDs to improve kinetic efficiency?

VIII. REGULATION

A. Mobility of Restriction-Modification System Genes

The key issue for regulation of the genes for restriction-modification systems is the fact that they will periodically enter new host cells having completely unprotected DNA. Many restriction-modification systems are specified by plasmids (Roberts and Halford, 1993). Some of these plasmids are conjugative and can direct their own transfer to new cells, while others have *mob* genes that allow the plasmids to hitchike—using the transfer systems of other, conjugative, plasmids present in the same cell (Derbyshire et al., 1987). Other restriction-modification systems are specified by lysogenic bacteriophages and can move via transduction (Kita et al., 1999). Still others are chromosomally coded, with no special sequences conferring genetic mobility—even in these cases transfer can occur via Hfr-type conjugation, generalized transduction, or transformation. The bottom line is that for a system to be mobile, it has to have some means of ensuring that the new host's DNA is modified before endonuclease activity appears. Furthermore, in order to have any effect (whether defensive, selfish, or recombinational), at some point the restriction-modification system must shift its pattern of expression more in favor of the endonuclease activity. This switching pattern poses a problem for all restriction-modification systems, but it is a particularly acute problem for type II systems in which the endonuclease is independent of the methyltransferase. Ironically, given the tremendous ecological and biotechnological importance of these systems, relatively little is known about their regulation. Nevertheless, we do have at least a list of basic strategies employed by various systems.

B. Generation of Readily Repaired Breaks in the DNA

As noted earlier, the restriction endonucleases vary in the types of DNA ends generated by their cleavage reactions. Type I systems appear to generate breaks that could only be repaired via recombination, as do type II endonucleases that generate blunt ends. Other type II endonucleases generate breaks that have 4-nucleotide complementary single-strand extensions (5′ or 3′ depending on the enzyme); these are good substrates for DNA ligases. Cells do have limited capacity to repair damage due to expression of endonuclease activity before the new host's DNA is fully protected, as indicated by the fact that cells lacking the protective methyltransferase can remain viable despite low-level expression of certain endonucleases (even type II; e.g., see Gingeras and Brooks, 1983). Generating readily repaired DNA breaks might make the regulatory problem somewhat less critical, in that cells carrying these systems could better tolerate a short period of DNA cleavage before full protection was established.

C. Subunit Architectures and Differences in Processivity

In all systems other than type II or type IIS, the endonuclease activity is physically linked to and dependent on the methyltransferase protein (Fig. 5). This linkage and the associated stimulatory effect of AdoMet on cleavage activity, help ensure that the endonuclease is active only when the methyltransferase is also active. This may provide an important level of control in recovery from starvation or other stresses that could lead to undermodification of the DNA, but the physical linkage is not sufficient by itself to allow mobility of these systems. For example, the type III system *Sty*LT I cannot be moved to new cells unless the *mod* gene is first moved by itself (De Backer and Colson, 1991a; De Backer and Colson, 1991b), though some type III systems can be moved.

The relative subunit affinities can apparently be tuned to aid in establishment. In at least one type I system, the binding of the first HsdR subunit to form RM_2S occurs with high affinity, but restriction requires a second HsdR subunit to bind and this occurs with much lower affinity (Janscak et al., 1998); thus restriction activity wouldn't appear until high intracellular levels of HsdR are reached. In another type I system, a similar but distinct mechanism appears to delay restriction—the MS complex binds HsdR as efficiently as does the M_2S complex, but only the latter gives rise to restriction activity. So in this system restriction-competent complexes won't appear until high intracellular levels of HsdM are reached (Dryden et al., 1997).

In the case of type II restriction-modification systems, some protection might result from the fact that the methyltransferases are active as monomers, while the homodimeric endonucleases need to accumulate to high enough levels to promote dimerization (Greene et al., 1981). Furthermore there is some evidence that endonuclease subunit multimerization can be inhibited by a regulatory peptide in type II (Adams and Blumenthal, 1995) and possibly type I systems (Belogurov and Delver, 1995). Another possible difference between methyltransferase and endonuclease in type II systems might favor methylation during the period of establishment in a new host—the extent of processivity. A highly processive enzyme would remain associated with the DNA, and act at all recognition sites it encountered in scanning that DNA molecule, while a highly distributive enzyme would dissociate after each reaction. Protection of DNA in a new host, or remethylation of newly replicated DNA, would occur most efficiently if the methyltransferase was processive. This has been studied for only a few enzymes, but in some type II systems it appears that the methyltransferase is more processive than the endonuclease (Jeltsch and Pingoud, 1998; Surby and Reich, 1996; Wright et al., 1999).

D. Controlling Endonucleases with Proteases

The preceding regulatory strategies are mostly passive, in that they rely on intrinsic properties of the proteins with no outside intervention. There are also active mechanisms for regulating restriction-modification systems so as to enhance their mobility. One example involves proteolytic turnover of endonuclease subunits. A relatively straightforward example of this may be provided by the type III restriction-modification system *Eco*P1 I (Redaschi and Bickle, 1996b). Infection of *E. coli* by the carrier, bacteriophage P1, rapidly leads to methyltransferase activity, but restriction activity appears only after a substantial lag. This lag doesn't appear to be due to transcription-level regulation, and the investigators suggest that it is due to proteolysis of free Res subunits. In this model both Mod and Res accumulate in parallel, but Mod is immediately active, while Res is degraded until both subunits reach concentrations favoring heterodimerization.

When type I (A or B) systems are transferred into naïve *E. coli* cells, there is a lag of roughly 15 generations before restriction activity appears (Prakash-Cheng and Ryu, 1993), and yet there is no evidence that this lag results from control of *hsdR* transcription (Makovets et al., 1998; O'Neill et al., 1997; Prakash-Cheng et al., 1993). The exact basis for this lag is not yet clear, but it depends on the ClpXP protease (Makovets et al., 1999). The evidence to date suggests that ClpXP degrades HsdR when it is in an actively translocating complex on chromosomal DNA. There are many conditions under which unmethylated recognition sequences could be generated in Hsd$^+$ cells, following DNA recombination and repair, for example, and there would be a strong selection to eliminate type I complexes that had been activated by such "normal" unmodified sequences. There is in fact a phenomenon called *restriction alleviation* that follows DNA damage and some other insults, and such cells transiently become nonrestricting (or very poorly restricting) with respect to restriction-modification systems other than type II (Day, 1977; Hiom and Sedgwick, 1992; Kelleher and Raleigh, 1994). The effect of ClpXP on type I systems is believed to be part of this restriction alleviation, though it is not yet clear how restriction alleviation is made to result when unmodified recognition sequences appear on the chromosome and yet not when the unmodified sites are on incoming bacteriophage DNA (Makovets et al., 1999). It should be noted that type IC systems do not appear to use the ClpXP regulatory system, and the basis for their regulation is still unclear (Kulik and Bickle, 1996).

E. Regulating Translation and RNA Stability

The step preceding assembly and proteolytic control (if any) of restriction-modification proteins is translation. The relative amounts of different restriction-modification proteins can be controlled passively by having mRNA stabilities or translation initiator strengths of differing fixed values (see e.g., Lacks and Greenberg, 1993). Some active control is needed, however, to get the switching behavior needed during establishment. There are several known examples of suspicious structures or phenomena suggestive of translation-or stabilty-level control, but in no case is there a clear understanding of the mechanisms involved.

One possible way to link endonuclease to methyltransferase expression is to have the two genes translationally coupled, with the methyltransferase gene upstream. Translational coupling means that translation of the downstream gene depends on ribosomes that have translated the upstream gene; if the coupling is complete, then no free ribosomes can bind and initiate translation of the downstream gene (Adhin and van Duin, 1990; Andre et al., 2000; Rex et al., 1994). The *Cfr*9 I type II restriction-modification system provides an example in which translation of the downstream endonuclease gene is coupled to

that of the upstream methyltransferase gene, with an RNA hairpin structure contributing to the coupling (Lubys et al., 1994).

Another possible approach to actively controlling gene expression is to actively control mRNA stability. An example of this is provided by the type II *Lla* I system (O'Sullivan and Klaenhammer, 1998). This is a three-gene operon, with the methyltransferase and endonuclease genes preceded by a small gene that plays a regulatory role. This gene, *llaIC* (see Szybalski et al., 1988, for nomenclature rules), has no effect on transcription, but is associated with increased stability of the mRNA. Interestingly C•*Lla*I shows sequence similarity to the plasmid copy-number control protein Rop, which is an RNA-binding and stabilizing protein (Predki et al., 1995).

F. Regulating Transcription

To achieve a switch in the methyltransferase/endonuclease ratio following establishment, one could either repress transcription of the methyltransferase gene or activate that of the endonuclease gene. Various type II restriction-modification systems show evidence of each strategy. For example, the *Kpn*2 I system produces a small protein that, after it accumulates, represses transcription of the methyltransferase gene (Lubys et al., 1999). In some cases this repression is mediated by the methyltransferase itself. In the *Eco*R II, *Sso* I, and *Msp* I systems the methyltransferases have an amino-terminal extension relative to other DNA methyltransferases, and this extension has a sequence-specific DNA-binding helix-turn-helix motif—when the methyltransferases accumulate, they bind to their own promoter regions and repress transcription (Karyagina et al., 1997; Som and Friedman, 1994; Som and Friedman, 1997). This mode of regulation has the nice theoretical feature that any increase in DNA content (as follows a nutritional shift up) or in the number of unmethylated sites will tend to compete for the "repressor" and transiently increase production of the methyltransferase.

The second strategy is to boost endonuclease production. Most systems that do this contain members of a family of regulatory proteins. These family members were the original C proteins (Brooks et al., 1991; Tao et al., 1991), though, as indicated above, that term is now used for any regulatory protein specified by a restriction-modification system. The C proteins we are considering here appear to be a family of transcription activators that include a helix-turn-helix motif. Type II restriction-modification systems that include C proteins can occur in several relative gene orientations (Anton et al., 1997); the one constant is that the C gene is upstream of, and in the same orientation as, the endonuclease gene—in fact the two often overlap (Tao et al., 1991). The reason for this conserved orientation is that the C proteins are autogenous activators—the promoter they activate is their own. Transcription of the downstream endonuclease gene is thus stimulated as C protein accumulates in a positive feedback loop, and in theory this should result in a very sharp transition—in essence a two-state system. The logic is that when one of these systems enters a new host cell, the C and endonuclease genes are poorly expressed until the required activator accumulates, while the methyltransferase gene is immediately expressed at the maximal level. Consistent with this, pre-expressing a C gene in a cell makes it impossible to introduce the intact restriction-modification system, presumably because the endonuclease is expressed prematurely (Nakayama and Kobayashi, 1998; Vijesurier et al., 2000).

Most of the C proteins bind to closely related sequences called C-boxes that usually occur in pairs upstream of the C genes (Rimseliene et al., 1995; Vijesurier et al., 2000), and sometimes occur upstream of the methyltransferase genes where they may be responsible for C-mediated repression (Bart et al., 1999). As an aside, the C proteins possess two intriguing properties. First, C proteins from cells as different as the "Gram-positive" soil bacterium Bacillus and

the "Gram-negative" enteric bacterium Proteus can cross-complement (Ives et al., 1995); this has been proposed to play a role in selfish behavior since a restriction-modification system entering a cell with a resident C+ system will prematurely (and lethally) express the endonuclease (Nakayama and Kobayashi, 1998). Second, the C boxes occur in a region of the target promoter at which binding is usually associated with repression instead of activation (Vijesurier et al., 2000). C genes of this activator family have been found in type II systems from a range of bacteria, but to date, none have been reported from one of the archaea; this may reflect differences in the transcription machinery (Baumann et al., 1995) with which C proteins might be incompatible.

The most straightforward and safe means one could imagine for regulating transcription of a restriction-modification system would be to tie expression of the endonuclease gene to DNA methylation. Surprisingly it took a long time to find such a system, suggesting that this is a relatively rare mechanism. Nevertheless, the CfrBI system contains a substrate site for its own methyltransferase that overlaps the −35 promoter hexamer of the methyltransferase gene and is adjacent to the −10 hexamer of the oppositely-oriented endonuclease gene. Both in vivo and in vitro, native methylation of this site boosts transcription of the endonuclease gene while depressing that of the methyltransferase gene (Beletskaya et al., 2000).

IX. WHAT NEXT?

There are a variety of very basic questions left to answer about restriction-modification systems. A list of the most important questions would include (but not be limited to) the following:

• Theories and models aside, what roles do these systems actually play in modulating gene flow in the biosphere?
• How do the abundant type II systems achieve changes in specificity?

• How are the various regulatory features of restriction-modification systems integrated to allow mobility?

• How is this regulation achieved for systems with no known regulators, such as *Eco*R I or *Sal* I (Alvarez et al., 1993; O'Connor and Humphreys, 1982)?

• How is regulation of restriction-modification systems integrated with the physiology of the host cell in terms of responses to starvation, DNA damage, quorum sensing, and the like? There is at least one methyltransferase that is expressed in response to induction of transformation competence in its host (Lacks et al., 2000).

• What is the basis for restriction alleviation and how does it distinguish between unmethylated sites on the chromosome and on incoming DNA?

• What actually happens following restriction by a type I system in vivo? Is the activated system and its continuing ATPase activity destroyed, or if not, does this harm the cell?

• Why aren't functional restriction-modification systems found in eukaryotes? Are they made by the largest prokaryotic cells (some of which rival eukaryotes in size, see e.g., Angert et al., 1996; Guerrero et al., 1999; Schulz et al., 1999)?

Addressing these questions would profitably fill the time of many bacterial geneticists. It's no clearer in this field than in any other how many (more) Nobel prizes await, but it's obvious that discoveries with fundamental importance to microbial genetic ecology and bacterial physiology remain to be made by studying restriction-modification systems.

REFERENCES

Abadijieva A, Patel J, Webb M, Zinkevich V, Firman K (1993): A deletion mutant of the type IC restriction endonuclease *Eco*R124I expressing a novel DNA specificity. Nucleic Acids Res 21:4435–4443.

Adams GM, Blumenthal RM (1995): Gene *pvuIIW*: A possible modulator of *Pvu*II endonuclease subunit association. Gene 157:193–199.

Adhin MR, van Duin J (1990): Scanning model for translational reinitiation in eubacteria. J Mol Biol 213:811–818.

Allan BW, Beechem JM, Lindstrom WM, Reich NO (1998): Direct real time observation of base flipping by the *Eco*RI DNA methyltransferase. J Biol Chem 273:2368–2373.

Althorpe NJ, Chilley PM, Thomas AT, Brammar WJ, Wilkins BM (1999): Transient transcriptional activation of the IncI1 plasmid anti-restriction gene (*ardA*) and SOS inhibition gene (*psiB*) early in conjugating recipient bacteria. Mol Microbiol 31:133–42.

Alvarez MA, Chater KF, Rosario MR (1993): Complex transcription of an operon encoding the *Sal*1 restriction-modification system of *Streptomyces albus* G. Mol Microbiol 8:243–252.

Alves J, Selent U, Wolfes H (1995): Accuracy of the *Eco*RV restriction endonuclease: binding and cleavage studies with oligodeoxynucleotide substrates containing degenerate recognition sequences. Biochem 34:11191–11197.

Anderson DG, Churchill JJ, Kowalczykowski SC (1999): A single mutation, RecB(D1080A) eliminates RecA protein loading but not Chi recognition by RecBCD enzyme. J Biol Chem 274:27139–27144.

Andre A, Puca A, Sansone F, Brandi A, Antico G, Calogero RA (2000): Reinitiation of protein synthesis in *Escherichia coli* can be induced by mRNA cis-elements unrelated to canonical translation initiation signals. FEBS Lett 468:73–78.

Angert ER, Brooks AE, Pace NR (1996): Phylogenetic analysis of *Metabacterium polyspora*: clues to the evolutionary origin of daughter cell production in *Epulopiscium species*, the largest bacteria. J Bacteriol 178:1451–1456.

Anton BP, Heiter DF, Benner JS, Hess EJ, Greenough L, Moran LS, Slatko BE, Brooks JE (1997): Cloning and characterization of the *Bgl*II restriction-modification system reveal a possible evolutionary footprint. Gene 187:19–27.

Arber W (2000): Genetic variation: molecular mechanisms and impact on microbial evolution. FEMS Microbiol Rev 24:1–7.

Azam TA, Ishihama A (1999): Twelve species of the nucleoid-associated protein from *Escherichia coli*. Sequence recognition specificity and DNA binding affinity. J Biol Chem 274:33105–33113.

Baldwin GS, Sessions RB, Erskine SG, Halford SE (1999): DNA cleavage by the *Eco*RV restriction endonuclease: Roles of divalent metal ions in specificity and catalysis. J Mol Biol 288:87–103.

Ban C, Yang W (1998): Structural basis for MutH activation in *E. coli* mismatch repair and relationship of MutH to restriction endonucleases. EMBO J 17:1526–1534.

Bandyopadhyay PK, Studier FW, Hamilton DL, Yuan R (1985): Inhibition of the type I restriction-modification enzymes *Eco*B and *Eco*K by the gene 0.3 protein of bacteriophage T7. J Mol Biol 182:567–578.

Barcus VA, Murray NE (1995): Barriers to recombination: Restriction. Cambridge, Cambridge University Press.

Barcus VA, Titheradge AJB, Murray NE (1995): The diversity of alleles at the hsd locus in natural populations of *Escherichia coli*. Genetics 140:1187–1197.

Barras F, Marinus MG (1989): The great GATC: DNA methylation in *E. coli*. Trends Genet 5:139–143.

Bart A, Dankert J, van der Ende A (1999): Operator sequences for the regulatory proteins of restriction modification systems. Mol Microbiol 31:1277–1278.

Baumann P, Qureshi SA, Jackson SP (1995): Transcription: New insights from studies on Archaea. Trends Genet 11:279–283.

Beletskaya IV, Zakharova MV, Shlyapnikov MG, Semenova LM, Solonin AS (2000): DNA methylation at the *Cfr*BI site is involved in expression control in the *Cfr*BI restriction-modification system. Nucleic Acids Res 28:3817–3822.

Belogurov AA, Delver EP (1995): A motif conserved among the type I restriction-modification enzymes and antirestriction proteins: A possible basis for mechanism of action of plasmid-encoded antirestriction functions. Nucleic Acids Res 23:785–787.

Belogurov AA, Delver EP, Rodzevich OV (1992): IncN plasmid pKM101 and IncI1 Plasmid ColIb-P9 encode homologous antirestriction proteins in their leading regions. J Bacteriol 174:5079–5085.

Bergh O, Borsheim KY, Bratbak G, Heldal M (1989): High abundance of viruses found in aquatic environments. Nature 340:467–468.

Bergstrom CT, McElhany P, Real LA (1999): Transmission bottlenecks as determinants of virulence in rapidly evolving pathogens. Proc Natl Acad Sci USA 96:5095–5100.

Bertani G, Weigle JJ (1953): Host controlled variation in bacterial viruses. J Bacteriol 65:113–121.

Bestor TH (1990): DNA methylation: evolution of a bacterial immune function into a regulator of gene expression and genome structure in higher eukaryotes. Phil Trans R Soc Lond B 326:179–187.

Bestor TH (1992): Activation of mammalian DNA methyltransferase by cleavage of a Zn binding regulatory domain. EMBO J 11:2611–2617.

Bestor TH (1996): DNA methyltransferases in mammalian development and genome defense. In Russo VEA, Martienssen RA, Riggs AD (eds): "Epigenetic Mechanisms of Gene Regulation". Cold Spring Harbor, NY: Cold Spring Harbor Press, pp 61–76.

Bewley CA, Holland ND, Faulkner DJ (1996): Two classes of metabolites from *Theonella swinhoei* are localized in distinct populations of bacterial symbionts. Experientia 52:716–722.

Bickle TA, Kruger DH (1993): Biology of DNA restriction. Microbiol Rev 57:434–450.

Bingham R, Ekunwe SI, Falk S, Snyder L, Kleanthous C (2000): The major head protein of bacteriophage T4

binds specifically to elongation factor Tu. J Biol Chem 275:23219–23226.

Bird AP, Wolffe AP (1999): Methylation-induced repression—Belts, braces, and chromatin. Cell 99: 451–454.

Bitinaite J, Wah DA, Aggarwal AK, Schildkraut I (1998): *Fok*I dimerization is required for DNA cleavage. Proc Natl Acad Sci USA 95:10570–10575.

Blumenthal RM (1986): *E. coli* can restrict methylated DNA and may skew genomic libraries. Trends Biotechnol 4:302–305.

Blumenthal RM, Borst DW, Matthews RG (1996): Experimental analysis of global gene regulation in *Escherichia coli*. Prog Nucleic Acid Res Mol Biol 55:1–86.

Blumenthal RM, Cheng X (2001): A Taq attack displaces bases. Nat Struct Biol 8:101–103.

Blumenthal RM, Gregory SA, Cooperider JS (1985): Cloning of a restriction-modification system from *Proteus vulgaris* and its use in analyzing a methylase-sensitive phenotype in *Escherichia coli*. J Bacteriol 164:501–509.

Brooks JE, Nathan PD, Landry D, Sznyter LA, Waite-Rees P, Ives CL, Moran LS, Slatko BE, Benner JS (1991): Characterization of the cloned *Bam*HI restriction modification system: its nucleotide sequence, properties of the methylase, and expression in heterologous hosts. Nucleic Acids Res 19:841–850.

Bujnicki JM (1999): Comparison of protein structures reveals monophyletic origin of AdoMet-dependent methyltransferase family and mechanistic convergence rather than recent differentiation of N4-cytosine and N6-adenine DNA methylation. In Silico Biology 1:e0016.

Bujnicki JM (2000): Phylogeny of the restriction endonuclease-like superfamily inferred from comparison of protein structures. J Mol Evol 50:39–44.

Bujnicki JM, Feder M, Radlinska M, Rychlewski L, Blumenthal RM (2002): Structure prediction and phylogenetic analysis of a family of proteins homologous to the MT-A70 subunit of the human mRNA: m^6A methyltransferase. J Mol Evol (submitted).

Bujnicki JM, Radlinska M (1999a): Molecular evolution of DNA-(cytosine-N4) methyltransferases: evidence for their polyphyletic origin. Nucleic Acids Res 27:4501–4509.

Bujnicki JM, Radlinska M (1999b): Molecular phylogenetics of DNA 5mC-methyltransferases. Acta Microbiol Pol 48:19–30.

Bullas LR, Colson C, van Pel A (1976): DNA restriction and modification systems in Salmonella. SQ, a new system derived by recombination between the SB system of *Salmonella typhimurium* and the SP system of *Salmonella potsdam*. J Gen Microbiol 95:166–172.

Campbell AM (1996): Cryptic Prophages. In Neidhardt FC, Curtiss III R, Ingraham JL, Lin ECC, Low KB, Magasanik B, Reznikoff WS, Riley M, Schaechter M, Umbarger HE (eds): "*Escherichia coli* and *Salmonella*:

Cellular and Molecular Biology." Washington, DC: ASM Press, pp 2041–2046.

Cantoni GL (1975): Biological methylation: selected aspects. Annu Rev Biochem 44:435–451.

Carlson K, Kosturko LD (1998): Endonuclease II of coliphage T4: A recombinase disguised as a restriction endonuclease? Mol Microbiol 27:671–676.

Chandrasegaran S, Smith HO (1988): "Amino Acid Sequence Homologies among Twenty-five Restriction Endonucleases and Methylases", Vol 1. New York, Adenine Press.

Chandrasegaran S, Smith J (1999): Chimeric restriction enzymes: What is next? Biol Chem 380:841–848.

Cheng X, Blumenthal RM (1996): Finding a basis for flipping bases. Structure 4:639–645.

Chilley PM, Wilkins BM (1995): Distribution of the *ardA* family of antirestriction genes on conjugative plasmids. Microbiology 141:2157–2164.

Churchill JJ, Anderson DG, Kowalczykowski SC (1999): The RecBC enzyme loads RecA protein onto ssDNA asymmetrically and independently of chi, resulting in constitutive recombination activation. Genes Dev 13:901–911.

Cooper DN, Krawczak M (1989): Cytosine methylation and the fate of CpG dinucleotides in vertebrate genomes. Hum Genet 83:181–188.

Cowan GM, Gann AAF, Murray NE (1989): Conservation of complex DNA recognition domains between families of restriction enzymes. Cell 56:103–109.

Dalgaard JZ (1994): Mobile introns and inteins: Friend or foe? Trends Genet 10:306–307.

Danna K, Nathans D (1971): Specific cleavage of simian virus 40 DNA by restriction endonuclease of *Hemophilus influenzae*. Proc Natl Acad Sci USA 68: 2913–2917.

Davies GP, Martin I, Sturrock SS, Cronshaw A, Murray NE, Dryden DTF (1999): On the structure and operation of type I DNA restriction enzymes. J Mol Biol 290:565–579.

Davies JE (1997): Origins, acquisition and dissemination of antibiotic resistance determinants. Ciba Found Symp 207:15–27.

Davison J (1999): Genetic exchange between bacteria in the environment. Plasmid 42:73–91.

Dawkins R (1989): "The Selfish Gene," 2d ed. Oxford, Oxford University Press.

Day RS (1977): UV-induced alleviation of K-specific restriction of bacteriophage lambda. J Virol 21: 1249–1251.

De Backer O, Colson C (1991a): Transfer of the genes for the *Sty*LTI restriction-modification system of *Salmonella typhimurium* to strains lacking modification ability results in death of the recipient cells and degradation of their DNA. J Bacteriol 173:1328–1330.

De Backer O, Colson C (1991b): Two-step cloning and expression in *Escherichia coli* of the DNA restriction-

modification system *Sty*LTI of *Salmonella typhimurium*. J Bacteriol 173:1321–1327.

De Bolle X, Bayliss CD, Field D, van de Ven T, Saunders NJ, Hood DW, Moxon ER (2000): The length of a tetranucleotide repeat tract in *Haemophilus influenzae* determines the phase variation rate of a gene with homology to type III DNA methyltransferases. Mol Microbiol 35:211–222.

de la Campa AG, Springhorn SS, Kale P, Lacks SA (1988): Proteins encoded by the *Dpn*I restriction gene cassette. Hyperproduction and characterization of the *Dpn*I endonuclease. J Biol Chem 263: 14696–14702.

Degtyarev SK, Belichenko OA, Lebedeva NA, Dedkov VS, Abdurashitov MA (2000): *Btr*I, a novel restriction endonuclease, recognises the nonpalindromic sequence 5′-CACGTC(-3/-3)-3′. Nucleic Acids Res 28:e56.

Derbyshire KM, Hatfull G, Willetts N (1987): Mobilization of the nonconjugative plasmid RSF1010: A genetic and DNA sequence analysis of the mobilization region. Mol Gen Genet 206:161–168. Published erratum in Mol Gen Genet 209:411.

Derbyshire V, Wood DW, Wu W, Dansereau JT, Dalgaard JZ, Belfort M (1997): Genetic definition of a protein-splicing domain: functional mini-inteins support structure predictions and a model for intein evolution. Proc Natl Acad Sci USA 94:11466–11471. Published erratum in Proc Natl Acad Sci USA 95:762.

Dieguez MJ, Vaucheret H, Paszkowski J, Mittelsten Scheid O (1998): Cytosine methylation at CG and CNG sites is not a prerequisite for the initiation of transcriptional gene silencing in plants, but it is required for its maintenance. Mol Gen Genet 259: 207–215.

Dixon MM, Fauman EB, Ludwig ML (1999): The black sheep of the family: AdoMet-dependent methyltransferases that do not fit the consensus structural fold. In Cheng X, Blumenthal RM, (eds): "*S*-Adenosylmethionine-Dependent Methyltransferases: Structures and Functions." Singapore: World Scientific Publishing, pp 39–54.

Dorner LF, Bitinaite J, Whitaker RD, Schildkraut I (1999): Genetic analysis of the base-specific contacts of *Bam*HI restriction endonuclease. J Mol Biol 285:1515–1523.

Dreier J, MacWilliams P, Bickle TA (1996): DNA cleavage by the type IC restriction-modification enzyme *Eco*R124II. J Mol Biol 264:722–733.

Dryden DTF (1999): Bacterial DNA methyltransferases. In Cheng X, Blumenthal RM (eds): "*S*-Adenosylmethionine-dependent Methyltransferases: Structures and Functions." Singapore, World Scientific, pp 283–340.

Dryden DTF, Cooper LP, Thorpe PH, Byron O (1997): The in vitro assembly of the *Eco*KI type I DNA restriction/modification enzyme and its in vivo implications. Biochem 36:1065–1076.

Dubey AK, Mollet B, Roberts RJ (1992): Purification and characterization of the *Msp*I DNA methyltransferase cloned and overexpressed in *E. coli*. Nucleic Acids Res 20:1579–1585.

Duncan BK, Miller JH (1980): Mutagenic deamination of cytosine residues in DNA. Nature 287:560–561.

Dybvig K, Sitaraman R, French CT (1998): A family of phase-variable restriction enzymes with differing specificities generated by high-frequency gene rearrangements. Proc Natl Acad Sci USA 95:13923–13928.

Earampamoorthy S, Koff RS (1975): Health hazards of bivalve-mollusk ingestion. Ann Intern Med 83:107–110.

Eddy SR, Gold L (1992): Artificial mobile DNA element constructed from the *Eco*RI endonuclease gene. Proc Natl Acad Sci USA 89:1544–1547.

Edgell DR, Fast NM, Doolittle WF (1996): Selfish DNA: The best defense is a good offense. Curr Biol 6:385–388.

Ehrlich M, Gama-Sosa MA, Carreira LH, Ljungdahl LG, Kuo KC, Gehrke CW (1985): DNA methylation in thermophilic bacteria: N4-methylcytosine, 5-methylcytosine, and N6-methyladenine. Nucleic Acids Res 13:1399–1412.

Elena SF, Cooper VS, Lenski RE (1996a): Punctuated evolution caused by selection of rare beneficial mutations. Science 272:1797–1802.

Elena SF, Cooper VS, Lenski RE (1996b): Punctuated evolution caused by selection of rare beneficial mutations. Science 272:1802–1804.

Endlich B, Linn S (1985): The DNA restriction endonuclease of *Escherichia coli* B. II. Further studies of the structure of DNA intermediates and products. J Biol Chem 260:5729–5738.

Engelberg-Kulka H, Glaser G (1999): Addiction modules and programmed cell death and antideath in bacterial cultures. Annu Rev Microbiol 53:43–70.

Erskine SG, Halford SE (1998): Reactions of the *Eco*RV restriction endonuclease with fluorescent oligodeoxynucleotides: Identical equilibrium constants for binding to specific and non-specific DNA. J Mol Biol 275:759–772.

Eskin B, Linn S (1972a): The deoxyribonucleic acid modification and restriction enzymes of *Escherichia coli* B. J Biol Chem 247:6192–6196.

Eskin B, Linn S (1972b): The deoxyribonucleic acid modification and restriction enzymes of *Escherichia coli* B. II. Purification, subunit structure, and catalytic properties of the restriction endonuclease. J Biol Chem 247:6183–6191.

Fauman EB, Blumenthal RM, Cheng X (1999): Structure and evolution of AdoMet-dependent methyltransferases. In Cheng X, Blumenthal RM (eds): "*S*-Adenosylmethionine-Dependent Methyltransferases: Structures and Functions." Singapore: World Scientific, pp 3–38.

Feinstein SI, Low KB (1982): Zygotic induction of the *rac* locus can cause cell death in *E. coli*. Mol Gen Genet 187:231–235.

Feldgarden M, Golden S, Wilson H, Riley MA (1995): Can phage defence maintain colicin plasmids in *Escherichia coli*? Microbiology 141:2977–2984.

Fischer D, Wolfson H, Lin SL, Nussinov R (1994): Three-dimensional, sequence order-independent structural comparison of a serine protease against the crystallographic database reveals active site similarities: Potential implications to evolution and to protein folding. Protein Sci 3:769–778.

Flores H, Osuna J, Heitman J, Soberon X (1995): Saturation mutagenesis of His114 of *Eco*RI reveals relaxed-specificity mutants. Gene 157:295–301.

Flowers AE, Garson MJ, Webb RI, Dumdei EJ, Charan RD (1998): Cellular origin of chlorinated diketopiperazines in the dictyoceratid sponge *Dysidea herbacea* (Keller). Cell Tissue Res 292:597–607.

Fuller-Pace FV, Cowan GM, Murray NE (1985): *Eco*A and *Eco*E: Alternatives to the *Eco*K family of type I restriction and modification systems of *Escherichia coli*. J Mol Biol 186:65–75.

Gabbara S, Wyszynski M, Bhagwat AS (1994): A DNA repair process in *Escherichia coli* corrects U : G and T : G mismatches to C : G at sites of cytosine methylation. Mol Gen Genet 243:244–248.

Galburt EA, Stoddard BL (2000): Restriction endonucleases: One of these things is not like the others. Nat Struct Biol 7:89–91.

Gann AAF, Campbell AJB, Collins JF, Coulson AFW, Murray NE (1987): Reassortment of DNA recognition domains and the evolution of new specificities. Mol Microbiol 1:13–22.

Garcia-Del Portillo F, Pucciarelli MG, Casadesus J (1999): DNA adenine methylase mutants of *Salmonella typhimurium* show defects in protein secretion, cell invasion, and M cell cytotoxicity. Proc Natl Acad Sci USA 96:11578–11583.

Gauthier A, Turmel M, Lemieux C (1991): A group I intron in the chloroplast large subunit rRNA gene of *Chlamydomonas eugametos* encodes a double-strand endonuclease that cleaves the homing site of this intron. Curr Genet 19:43–47.

Gehrig H, Schussler A, Kluge M (1996): *Geosiphon pyriforme*, a fungus forming endocytobiosis with Nostoc (cyanobacteria), is an ancestral member of the Glomales: Evidence by SSU rRNA analysis. J Mol Evol 43:71–81.

Gerdes K, Gultyaev AP, Franch T, Pedersen K, Mikkelsen ND (1997): Antisense RNA-regulated programmed cell death. Annu Rev Genet 31:1–31.

Gerlt JA (1993): Mechanistic principles of enzyme-catalyzed cleavage of phosphodiester bonds. In Linn SM, Lloyd RS, Roberts RJ (eds): "Nucleases." Cold Spring Harbor, NY: Cold Spring Harbor Laboratory Press, pp 1–34.

Gimble FS (2000): Invasion of a multitude of genetic niches by mobile endonuclease genes. FEMS Microbiol Lett 185:99–107.

Gingeras TR, Brooks JE (1983): Cloned restriction/modification system from *Pseudomonas aeruginosa*. Proc Natl Acad Sci USA 80:402–406.

Goedecke K, Pignot M, Goody RS, Scheidig AJ, Weinhold E (2001): Structure of the N6-adenine DNA methyltransferase M•TaqI in complex with DNA and a cofactor analog. Nat Struct Biol 8:121–125.

Gong W, O'Gara M, Blumenthal RM, Cheng X (1997): Structure of *Pvu* II DNA-(cytosine N4) methyltransferase, an example of domain permutation and protein fold assignment. Nucleic Acids Res 25:2702–2715.

Gorbalenya AE (1994): Self-splicing group I and group II introns encode homologous (putative) DNA endonucleases of a new family. Protein Sci 3:1117–1120.

Gormley NA, Bath AJ, Halford SE (2000): Reactions of *Bgl*I and other Type II restriction endonucleases with discontinuous recognition sites. J Biol Chem 275:6928–6936.

Grabowski G, Alves J (1995): Transformation of the *Eco*RI restriction endonuclease to an enzyme with altered specificity: Development of a positive in vivo selection system. Biol Chem Hoppe-Seyler 376:S102.

Greene PJ, Gupta M, Boyer HW, Brown WE, Rosenberg JM (1981): Sequence Analysis of the DNA Encoding the *Eco* RI Endonuclease and Methylase. J Biol Chem 256:2143–2153.

Gualtieri G, Bisseling T (2000): The evolution of nodulation. Plant Mol Biol 42:181–194.

Guerrero R, Haselton A, Sole M, Wier A, Margulis L (1999): *Titanospirillum velox*: A huge, speedy, sulfur-storing spirillum from Ebro Delta microbial mats. Proc Natl Acad Sci USA 96:11584–11588.

Guttman DS (1997): Recombination and clonality in natural populations of *Escherichia coli*. Trends Ecol Evol 12:16–22.

Guttman DS, Dykhuizen DE (1994): Detecting selective sweeps in naturally occurring *Escherichia coli*. Genetics 138:993–1003.

Handa N, Ichige A, Kusano K, Kobayashi I (2000): Cellular responses to postsegregational killing by restriction-modification genes. J Bacteriol 182:2218–2229.

Heitman J, Ivaneko T, Kiss A (1999): DNA nicks inflicted by restriction endonucleases are repaired by a RecA- and RecB-dependent pathway in *Escherichia coli*. Mol Microbiol 33:1141–1151.

Heitman J, Model P (1987): Site-specific methylases induce the SOS DNA repair response in *Escherichia coli*. J Bacteriol 169:3243–3250.

Heitman J, Zinder ND, Model P (1989): Repair of the *Escherichia coli* chromosome after in vivo scission by the *Eco*RI endonuclease. Proc Natl Acad Sci USA 86:2281–2285.

Hendrix RW, Smith MCM, Burns RN, Ford ME, Hatfull GF (1999): Evolutionary relationships among diverse bacteriophages and prophages: All the world's a phage. Proc Natl Acad Sci USA 96:2192–2197.

Hennecke F, Kolmar H, Brundl K, Fritz HJ (1991): The *vsr* gene product of *E. coli* K-12 is a strand- and sequence-specific DNA mismatch endonuclease. Nature 353:776–778.

Himmelreich R, Plagens H, Hilbert H, Reiner B, Herrmann R (1997): Comparative analysis of the genomes of the bacteria *Mycoplasma pneumoniae* and *Mycoplasma genitalium*. Nucleic Acids Res 25:701–712.

Hiom KJ, Sedgwick SG (1992): Alleviation of *Eco*K DNA restriction in *Escherichia coli* and involvement of UmuDC activity. Mol Gen Genet 231:265–275.

Holubova I, Vejsadova S, Weiserova M, Firman K (2000): Localization of the type I restriction-modification enzyme *Eco*KI in the bacterial cell. Biochem Biophys Res Commun 270:46–51.

Holz B, Klimasauskas S, Serva S, Weinhold E (1998): 2-Aminopurine as a fluorescent probe for DNA base flipping by methyltransferases. Nucleic Acids Res 26:1076–1083.

Hsia RC, Ting LM, Bavoil PM (2000): Microvirus of *Chlamydia psittaci* strain guinea pig inclusion conjunctivitis: isolation and molecular characterization. Microbiology 146:1651–1660.

Hsieh P-C, Xiao J-P, O'Loane D, Xu S-Y (2000): Cloning, expression, and purification of a thermostable nonhomodimeric restriction enzyme, *Bsl*I. J Bacteriol 182:949–955.

Hughes KA, Sutherland IW, Jones MV (1998): Biofilm susceptibility to bacteriophage attack: the role of phage-borne polysaccharide depolymerase. Microbiology 144:3039–3047.

Irelan JT, Selker EU (1997): Cytosine methylation associated with repeat-induced point mutation causes epigenetic gene silencing in *Neurospora crassa*. Genetics 146:509–523.

Ivanenko T, Heitman J, Kiss A (1998): Mutational analysis of the function of Met137 and Ile197, two amino acids implicated in sequence-specific DNA recognition by the *Eco*RI endonuclease. Biol Chem 379:459–465.

Ives CL, Sohail A, Brooks JE (1995): The regulatory C proteins from different restriction-modification systems can cross-complement. J Bacteriol 177:6313–6315.

Jakowitsch J, Papp I, Moscone EA, van der Winden J, Matzke M, Matzke AJ (1999): Molecular and cytogenetic characterization of a transgene locus that induces silencing and methylation of homologous promoters in trans. Plant J 17:131–140.

Janscak P, Bickle TA (2000): DNA supercoiling during ATP-dependent DNA translocation by the type I restriction enzyme *Eco*AI. J Mol Biol 295:1089–1099.

Janscak P, Dryden DTF, Firman K (1998): Analysis of the subunit assembly of the type IC restriction-modification enzyme *Eco*R124I. Nucleic Acids Res 26:4439–4445.

Janscak P, MacWilliams MP, Sandmeier U, Nagaraja V, Bickle TA (1999): DNA translocation blockage, a general mechanism of cleavage site selection by type I restriction enzymes. EMBO J 18:2638–2647.

Janulaitis A, Petrusyte M, Maneliene Z, Klimasauskas S, Butkus V (1992a): Purification and properties of the *Eco*57I restriction endonuclease and methylase—Prototypes of a new class (type IV). Nucleic Acids Res 20:6043–6049.

Janulaitis A, Vaisvila R, Timinskas A, Klimasauskas S, Butkus V (1992b): Cloning and sequence analysis of the genes coding for *Eco*57I type IV restriction-modification enzymes. Nucleic Acids Res 20:6051–6056.

Jeltsch A (1999): Circular permutations in the molecular evolution of DNA methyltransferase. J Mol Evol 49:161–164.

Jeltsch A, Christ F, Fatemi M, Roth M (1999a): On the substrate specificity of DNA methyltransferases. J Biol Chem 274:19538–19544.

Jeltsch A, Christ F, Fatemi M, Roth M (1999b): On the substrate specificity of DNA methyltransferases: Adenine-N6 DNA methyltransferases also modify cytosine residues at position N4. J Biol Chem 274:19538–19544.

Jeltsch A, Pingoud A (1998): Kinetic characterization of linear diffusion of the restriction endonuclease *Eco*RV on DNA. Biochemistry 37:2160–2169.

Jo K, Topal MD (1995): DNA topoisomerase and recombinase activities in *Nae* I restriction endonuclease. Science 267:1817–1820.

Jo K, Topal MD (1996a): Changing a leucine to a lysine residue makes *Nae*I endonuclease hypersensitive to DNA intercalative drugs. Biochemistry 35:10014–10018.

Jo K, Topal MD (1996b): Effects on *Nae*I-DNA recognition of the leucine to lysine substitution that transforms restriction endonuclease NaeI to a topoisomerase: a model for restriction endonuclease evolution. Nucleic Acids Res 24:4171–4175.

Jurica MS, Stoddard BL (1999): Homing endonucleases: structure, function and evolution. Cell Mol Life Sci 55:1304–1326.

Karlin S, Altschul SF (1993): Applications and statistics for multiple high-scoring segments in molecular sequences. Proc Natl Acad Sci USA 90:5873–5877.

Karreman C, de Waard A (1990): *Agmenellum quadruplicatum* M.*Aqu*I, a novel modification methylase. J Bacteriol 172:266–272.

Karyagina A, Shilov I, Tashlitskii V, Khodoun M, Vasil'ev S, Lau PCK, Nikolskaya I (1997): Specific binding of *Sso*II DNA methyltransferase to its promoter region provides the regulation of *Sso*II restriction-modification gene expression. Nucleic Acids Res 25:2114–2120.

Kelleher JE, Raleigh EA (1991): A novel activity in *Escherichia coli* K-12 that directs restriction of DNA modified at CG dinucleotides. J Bacteriol 173:5220–5223.

Kelleher JE, Raleigh EA (1994): Response to UV damage by four *Escherichia coli* K-12 restriction systems. J Bacteriol 176:5888–5896.

Kelly TJJ, Smith HO (1970): A restriction enzyme from *Hemophilus influenzae* II: Base sequence of the recognition site. J Mol Biol 51:393–409.

Kim Y, Chandrasegaran S (1994): Chimeric restriction endonuclease. Proc Natl Acad Sci USA 91:883–887.

Kim Y-G, Cha J, Chandrasegaran S (1996): Hybrid restriction enzymes: Zinc finger fusions to *Fok*I cleavage domain. Proc Natl Acad Sci USA 93:1156–1160.

Kim Y-G, Smith J, Durgesha M, Chandrasegaran S (1998): Chimeric restriction enzyme: Gal4 fusion to *Fok*I cleavage domain. Biol Chem 379:489–495.

Kita K, Tsuda J, Kato T, Okamoto K, Yanase H, Tanaka M (1999): Evidence of horizontal transfer of the EcoO109I restriction-modification gene to *Escherichia coli* chromosomal DNA. J Bacteriol 181:6822–6827.

Klimasauskas S, Kumar S, Roberts RJ, Cheng X (1994): *Hha*I methyltransferase flips its target base out of the DNA helix. Cell 76:357–369.

Klimasauskas S, Nelson JL, Roberts RJ (1991): The sequence specificity domain of cytosine-C5 methylases. Nucleic Acids Res 19:6183–6190.

Klimasauskas S, Timinskas A, Menkevicius S, Butkiene D, Butkus V, Janulaitis A (1989): Sequence motifs characteristic of DNA [cytosine-N4] methylases: Similarity to adenine and cytosine-C5 DNA-methylases. Nucleic Acids Res 17:9823–9832.

Kobayashi I (1998): Selfishness and death: *Raison d'être* of restriction, recombination and mitochondria. Trends Genet 14:368–374.

Kong H (1998): Analyzing the functional organization of a novel restriction modification system, the *Bcg*I system. J Mol Biol 279:823–832.

Kong H, Morgan RD, Maunus RE, Schildkraut I (1993): A unique restriction endonuclease, *Bcg*I, from *Bacillus coagulans*. Nucleic Acids Res 21:987–991.

Kong H, Roemer SE, Waite-Rees PA, Benner JS, Wilson GG, Nwankwo DO (1994): Characterization of *Bcg*I, a new kind of rectriction-modification system. J Biol Chem 269:683–690.

Kong H, Smith CL (1998): Does *Bcg*I, a unique restriction endonuclease, require two recognition sites for cleavage? Biol Chem 379:605–609.

Kovall RA, Matthews BW (1998): Structural, functional, and evolutionary relationships between lambda-exonuclease and the type II restriction endonucleases. Proc Natl Acad Sci USA 95:7893–7897.

Kovall RA, Matthews BW (1999): Type II restriction endonucleases: structural, functional and evolutionary relationships. Curr Opin Chem Biol 3:578–583.

Kretz PL, Kohler SW, Short JM (1991): Identification and characterization of a gene responsible for inhibiting propagation of methylated DNA sequences in mcrA mcrB1 Escherichia coli strains. J Bacteriol 173:4707–4716.

Kruger DH, Bickle TA (1983): Bacteriophage survival: multiple mechanisms for avoiding the deoxyribonucleic acid restriction systems of their hosts. Microbiol Rev 47:345–360.

Kruger DH, Kupper D, Meisel A, Reuter M, Schroeder C (1995): The significance of distance and orientation of restriction endonuclease recognition sites in viral DNA genomes. FEMS Microbiol Rev 17:177–184.

Kuhlmann UC, Moore GR, James R, Kleanthous C, Hemmings AM (1999): Structural parsimony in endonuclease active sites: Should the number of homing endonuclease families be redefined? FEBS Lett 463:1–2.

Kuhnlein U, Arber W (1972): Host specificity of DNA produced by Escherichia coli. XV. The role of nucleotide methylation in vitro B-specific modification. J Mol Biol 63:9–19.

Kulakauskas S, Barsomian JM, Lubys A, Roberts RJ, Wilson GG (1994): Organization and sequence of the HpaII restriction-modification system and adjacent genes. Gene 142:9–15.

Kulik EM, Bickle TA (1996): Regulation of the activity of the type IC EcoR124I restriction enzyme. J Mol Biol 264:891–906.

Kusano K, Sakagami K, Yokochi T, Naito T, Tokinaga Y, Ueda E, Kobayashi I (1997): A new type of illegitimate recombination is dependent on restriction and homologous interaction. J Bacteriol 179:5380–5390.

Kuzminov A, Schabtach E, Stahl FW (1994): Chi sites in combination with RecA protein increase the survival of linear DNA in Escherichia coli by inactivating ExoV activity of RecBCD nuclease. EMBO J 13:2764–2776.

Kwoh TJ, Obermiller PS, McCue AW, Kwoh DY, Sullivan SA, Gingeras TR (1988): Introduction and expression of the bacterial PaeR7 restriction endonuclease gene in mouse cells containing the PaeR7 methylase. Nucleic Acids Res 16:11489–11506.

Kyte J (1995): "Mechanism in Protein Chemistry." New York: Garland.

Lacks SA, Ayalew S, de la Campa AG, Greenberg B (2000): Regulation of competence for genetic transformation in Streptococcus pneumoniae: Expression of dpnA, a late competence gene encoding a DNA methyltransferase of the DpnII restriction system. Mol Microbiol 35:1089–1098.

Lacks SA, Greenberg B (1993): Atypical ribosome binding sites and regulation of gene expression in the DpnII restriction enzyme system of S. pneumoniae. FASEB J 7:A1082.

Lacks SA, Mannarelli BM, Springhorn SS, Greenberg B (1986): Genetic basis of the complementary DpnI and DpnII restriction systems of S. pneumoniae: An intercellular cassette mechanism. Cell 46:993–1000.

Lange C, Wild C, Trautner TA (1995): Altered sequence recognition specificity of a C5-DNA methyltransferase carrying a chimeric "target recognizing domain." Gene 157:127–128.

Lao PJ, Forsdyke DR (2000): Crossover hot-spot instigator (Chi) sequences in Escherichia coli occupy distinct recombination/transcription islands. Gene 243:47–57.

Lauster R, Trautner TA, Noyer-Weidner M (1989): Cytosine-specific type II DNA methyltransferases: A conserved enzyme core with variable target-recognizing domains. J Mol Biol 206:305–312.

Lawrence JG (1999): Gene transfer, speciation, and the evolution of bacterial genomes. Curr Opin Microbiol 2:519–523.

Lawrence JG, Ochman H (1998): Molecular archaeology of the Escherichia coli genome. Proc Natl Acad Sci USA 95:9413–9417.

Lee K-F, Liaw Y-C, Shaw P-C (1996): Overproduction, purification and characterization of M.EcoHK31I, a bacterial methyltransferase with two polypeptides. Biochem J 314:321–326.

Leismann O, Roth M, Friedrich T, Wende W, Jeltsch A (1998): The Flavobacterium okeanokoites adenine-N6-specific DNA-methyltransferase M.FokI is a tandem enzyme of two independent domains with very different kinetic properties. Eur J Biochem 251:899–906.

Lieb M, Allen E, Read D (1986): Very short patch mismatch repair in phage lambda: Repair sites and length of repair tracts. Genetics 114:1041–1060.

Liu BL, Everson JS, Fane B, Giannikopoulou P, Vretou E, Lambden PR, Clarke IN (2000): Molecular characterization of a bacteriophage (Chp2) from Chlamydia psittaci. J Virol 74:3464–3469.

Loenen WA, Murray NE (1986): Modification enhancement by the restriction alleviation protein (Ral) of bacteriophage lambda. J Mol Biol 190:11–22.

Lubys A, Jurenaite S, Janulaitis A (1999): Structural organization and regulation of the plasmid-borne type II restriction-modification system Kpn2I from Klebsiella pneumoniae RFL2. Nucleic Acids Res 27:4228–4234.

Lubys A, Menkevicius S, Timinskas A, Butkus V, Janulaitis A (1994): Cloning and analysis of translational control for genes encoding the Cfr9I restriction-modification system. Gene 141:85–89.

Lukacs CM, Kucera R, Schildkraut I, Aggarwal AK (2000): Understanding the immutability of restriction enzymes: crystal structure of BglII and its DNA substrate at 1.5 Å resolution. Nat Struct Biol 7:134–140.

Luria SE, Human ML (1952): A nonhereditary, host-induced variation of bacterial viruses. J Bacteriol 64:557–569.

MacWilliams M, Meister J, Jutte H, Bickle T (1994): Generation of a new type-I restriction-modification specificity by transposition. J Cell Biochem S18C:136.

Makovets S, Doronina VA, Murray NE (1999): Regulation of endonuclease activity by proteolysis prevents breakage of unmodified bacterial chromosomes by type I restriction enzymes. Proc Natl Acad Sci USA 96:9757–9762.

Makovets S, Titheradge AJB, Murray NE (1998): ClpX and ClpP are essential for the efficient acquisition of genes specifying type IA and IB restriction systems. Mol Microbiol 28:25–35.

Malone T, Blumenthal RM, Cheng X (1995): Structure-guided analysis reveals nine sequence motifs conserved among DNA amino-methyl-transferases, and suggests a catalytic mechanism for these enzymes. J Mol Biol 253:618–632.

Margolin BS, Garrett-Engele PW, Stevens JN, Fritz DY, Garrett-Engele C, Metzenberg RL, Selker EU (1998): A methylated Neurospora 5S rRNA pseudogene contains a transposable element inactivated by repeat-induced point mutation. Genetics 149:1787–1797.

Margot JB, Aguirre-Arteta AM, Di Giacco BV, Pradhan S, Roberts RJ, Cardoso MC, Leonhardt H (2000): Structure and function of the mouse DNA methyltransferase gene: Dnmt1 shows a tripartite structure. J Mol Biol 297:293–300.

Marshall P, Lemieux C (1992): The I-CeuI endonuclease recognizes a sequence of 19 base pairs and preferentially cleaves the coding strand of the *Chlamydomonas moewusii* chloroplast large subunit rRNA gene. Nucleic Acids Res 20:6401–6407.

Martin AM, Horton NC, Luseti S, Reich NO, Perona JJ (1999): Divalent metal dependence of site-specific DNA binding by *Eco*RV endonuclease. Biochemistry 38:8430–8439.

Martin W (1999): Mosaic bacterial chromosomes: a challenge en route to a tree of genomes. Bioessays 21:99–104.

Master SS, Blumenthal RM (1997): A genetic and functional analysis of the unusually large variable region in the M.*Alu*I DNA–(cytosine C5)-methyltransferase. Mol Gen Genet 257:14–22.

Matic I, Taddei F, Radman M (1996): Genetic barriers among bacteria. Trends Microbio 4:69–73.

Matsuo K, Silke J, Gramatikoff K, Schaffner W (1994): The CpG-specific methylase SssI has topoisomerase activity in the presence of Mg^{2+}. Nucleic Acids Res 22:5354–5359.

McBride MJ, Zusman DR (1996): Behavioral analysis of single cells of *Myxococcus xanthus* in response to prey cells of *Escherichia coli*. FEMS Microbiol Lett 137:227–231.

McCann MP, Solimeo HT, Cusick F, Jr, Panunti B, McCullen C (1998): Developmentally regulated protein synthesis during intraperiplasmic growth of *Bdellovibrio bacteriovorus* 109J. Can J Microbiol 44:50–55.

McCarthy MD, Hedges JI, Benner R (1998): Major bacterial contribution to marine dissolved organic nitrogen. Science 281:231–234.

Meisel A, Bickle TA, Kruger DH, Schroeder C (1992): Type III restriction enzymes need two inversely oriented recognition sites for DNA cleavage. Nature 355:467–469.

Meisel A, Kruger DH, Bickle TA (1991): M.*Eco*P15 methylates the second adenine in its recognition sequence. Nucleic Acids Res 19:3997.

Meisel A, Mackeldanz P, Bickle TA, Kruger DH, Schroeder C (1995): Type III restriction endonucleases translocate DNA in a reaction driven by recognition site-specific ATP hydrolysis. EMBO J 14:2958–2966.

Meister J, MacWilliams M, Hubner P, Jutte H, Skrzypek E, Piekarowicz A, Bickle TA (1993): Macroevolution by transposition: Drastic modification of DNA recognition by the type I restriction enzyme following Tn5 transposition. EMBO J 12:4585–4591.

Merril CR, Biswas B, Carlton R, Jensen NC, Creed GJ, Zullo S, Adhya S (1996): Long-circulating bacteriophage as antibacterial agents. Proc Natl Acad Sci USA 93:3188–3192.

Messer W, Noyer-Weidner M (1998): Timing and targeting: the biological functions of Dam methylation in *E. coli*. Cell 54:735–737.

Mi S, Roberts RJ (1992): How M.*Msp*I and M.*Hpa*II decide which base to methylate. Nucleic Acids Res 20:4811–4816.

Milkman R (1999): Gene transfer in *Escherichia coli*. In Charlebois RL (ed): "Organization of the Prokaryotic Genome." Washington, DC: ASM Press, pp 291–309.

Milkman R, Raleigh EA, McKane M, Cryderman D, Bilodeau P, McWeeny K (1999): Molecular evolution of the *Escherichia coli* chromosome: V. Recombination patterns among strains of diverse origin. Genetics 153:539–554.

Morgan R, Xiao J-P, Xu S-Y (1998): Characterization of an extremely thermostable restriction enzyme, *Psp*GI, from a *Pyrococcus* strain and cloning of the *Psp*GI restriction-modification system in *Escherichia coli*. Appl Environ Microbiol 64:3669–3673.

Murphy KC (1991): Lambda Gam protein inhibits the helicase and chi-stimulated recombination activities of *Escherichia coli* RecBCD enzyme. J Bacteriol 173:5808–5821.

Naito T, Kusano K, Kobayashi I (1995): Selfish behavior of restriction-modification systems. Science 267:897–899.

Nakayama Y, Kobayashi I (1998): Restriction-modification gene complexes as selfish gene entities: roles of a regulatory system in their establishment, maintenance, and apoptotic mutual exclusion. Proc Natl Acad Sci USA 95:6442–6447.

Ng HH, Jeppesen P, Bird A (2000): Active repression of methylated genes by the chromosomal protein MBD1. Mol Cell Biol 20:1394–1406.

Noyer-Weidner M, Diaz R, Reiners L (1986): Cytosine-specific DNA modification interferes with plasmid establishment in *Escherichia coli* K12: Involvement of *rglB*. Mol Gen Genet 205:469–475.

O'Connor CD, Humphreys GO (1982): Expression of the *Eco*RI restriction-modification system and the construction of positive-selection cloning vectors. Gene 20:219–229.

O'Gara M, Horton JR, Roberts RJ, Cheng X (1998): Structures of *Hha*I methyltransferase complexed with substrates containing mismatches at the target base. Nat Struct Biol 5:872–877.

O'Neill M, Chen A, Murray NE (1997): The restriction-modification genes of *Escherichia coli* K-12 may not be selfish: They do not resist loss and are readily replaced by alleles conferring different specificities. Proc Natl Acad Sci USA 94:14596–14601.

O'Sullivan DJ, Klaenhammer TR (1998): Control of expression of *Lla*I restriction in *Lactococcus lactis*. Mol Microbiol 27:1009–1020.

Ochman H, Lawrence JG, Groisman EA (2000): Lateral gene transfer and the nature of bacterial innovation. Nature 405:299–304.

Ohkuma M, Kudo T (1996): Phylogenetic diversity of the intestinal bacterial community in the termite *Reticulitermes speratus*. Appl Environ Microbiol 62:461–468.

Okano M, Bell DW, Haber DA, Li E (1999): DNA methyltransferases Dnmt3a and Dnmt3b are essential for de novo methylation and mammalian development. Cell 99:247–257.

Panne D, Raleigh EA, Bickle TA (1998): McrBS, a modulator peptide for McrBC activity. EMBO J 17:5477–5483.

Penner M, Morad I, Snyder L, Kaufmann G (1995): Phage T4-coded Stp: Double-edged effector of coupled DNA and tRNA-restriction systems. J Mol Biol 249:857–868.

Petrauskene OV, Babkina OV, Tashlitsky VN, Kazankov GM, Gromova ES (1998): *Eco*RII endonuclease has two identical DNA-binding sites and cleaves one of two co-ordinated recognition sites in one catalytic event. FEBS Lett 425:29–34.

Piekarowicz A, Golaszewska M, Sunday AO, Siwinska M, Stein DC (1999): The *Hae*IV restriction modification system of *Haemophilus aegyptius* is encoded by a single polypeptide. J Mol Biol 293:1055–1065.

Pieper U, Schweitzer T, Groll DH, Gast F-U, Pingoud A (1999): The GTP-binding domain of McrB: More than just a variation on common theme? J Mol Biol 292:547–556.

Pietrokovski S (1994): Conserved sequence features of inteins (protein introns) and their use in identifying new inteins and related proteins. Protein Sci 3:2340–2350.

Pietrokovski S (1998): Modular organization of inteins and C-terminal autocatalytic domains. Protein Sci 7:64–71.

Posfai J, Bhagwat AS, Roberts RJ (1988): Sequence motifs specific for cytosine methyltransferases. Gene 74:261–265.

Prakash-Cheng A, Chung SS, Ryu J-I (1993): The expression and regulation of hsdK genes after conjugative transfer. Mol Gen Genet 241:491–496.

Prakash-Cheng A, Ryu J (1993): Delayed expression of in vivo restriction activity following conjugal transfer of *Escherichia coli hsd*^{K} (restriction-modification) genes. J Bacteriol 175:4905–4906.

Predki PF, Nayak LM, Gottlieb MB, Regan L (1995): Dissecting RNA-protein interactions: RNA-RNA recognition by Rop. Cell 80:41–50.

Price C, Bickle TA (1986): A possible role for DNA restriction in bacterial evolution. Microbiol Sci 3:296–299.

Price C, Lingner J, Bickle TA, Firman K, Glover SW (1989): Basis for changes in DNA recognition by the *Eco*R124 and *Eco*R124/3 Type I DNA restriction and modification enzymes. J Mol Biol 205:115–125.

Quirk SM, Bell-Pedersen D, Belfort M (1989): Intron mobility in the T-even phages: High frequency inheritance of group I introns promoted by intron open reading frames. Cell 56:455–465.

Raleigh EA (1987): Restriction and modification in vivo by *Escherichia coli* K12. Methods Enzymol 152:130–141.

Raleigh EA (1992): Organization and function of the *mcrBC* genes of *Escherichia coli* K-12. Mol Microbiol 6:1079–1086.

Raleigh EA, Brooks JE (1998): Restriction modification systems: where they are and what they do. In De Bruijn FJ, Lupski JR, Weinstock GM (eds): "Bacterial Genomes." New York: Chapman and Hall, pp 78–92.

Raleigh EA, Murray NE, Revel H, Blumenthal RM, Westaway D, Reith AD, Rigby PWJ, Elhai J, Hanahan D (1988): McrA and McrB restriction phenotypes of some *E. coli* strains and implications for gene cloning. Nucleic Acids Res 15:1563–1575.

Raleigh EA, Trimarchi R, Revel H (1989): Genetic and physical mapping of the *mcrA (rglA)* and *mcrB (rglB)* loci of *Escherichia coli* K-12. Genetics 122:279–296.

Raleigh EA, Wilson G (1986): *Escherichia coli* K-12 restricts DNA containing 5-methylcytosine. Proc Natl Acad Sci USA 83:9070–9074.

Rao DN, Saha S, Krishnamurthy V (2000): ATP-dependent restriction enzymes. Prog Nucleic Acid Res Mol Biol 64:1–63.

Rawlings DE (1999): Proteic toxin-antitoxin, bacterial plasmid addiction systems and their evolution with special reference to the pas system of pTF-FC2. FEMS Microbiol Lett 176:269–277.

Redaschi N, Bickle TA (1996a): DNA restriction and modification systems. In Neidhardt FC, Curtiss III R, Ingraham JL, Lin ECC, Low KB, Magasanik B, Reznikoff WS, Riley M, Schaechter M, Umbarger HE (eds): "*Escherichia coli* and *Salmonella*: Cellular and Molecular Biology." Washington, DC: ASM Press, pp 773–781.

Redaschi N, Bickle TA (1996b): Posttranscriptional regulation of *Eco*P1I and *Eco*P15I restriction activity. J Mol Biol 257:790–803.

Reisenauer A, Kahng LS, McCollum S, Shapiro L (1999): Bacterial DNA methylation: A cell cycle regulator? J Bacteriol 181:5135–5139.

Reuter M, Kupper D, Meisel A, Schroeder C, Krueger DH (1998): Cooperative binding properties of restriction endonuclease *Eco*RII with DNA recognition sites. J Biol Chem 273:8294–8300.

Revel H (1967): Restriction of nonglycosylated T-even bacteriophage: Properties of permissive mutants of *E. coli* B and K-12. Virol 31:688–701.

Revel HR (1983): DNA modification: Glucosylation. In Mathews CK, Kutter EM, Mosig G, Berget P (eds): "Bacteriophage T4." Washington, DC: ASM Press, pp 156–165.

Revel HR, Georgopoulos CP (1969): Restriction of nonglucosylated T-even bacteriophages by prophage P1. Virol 39:1–17.

Rex G, Surin B, Besse G, Schneppe B, McCarthy JE (1994): The mechanism of translational coupling in *Escherichia coli*. Higher order structure in the *atpHA* mRNA acts as a conformational switch regulating the access of *de novo* initiating ribosomes. J Biol Chem 269:18118–18127.

Rimseliene R, Vaisvila R, Janulaitis A (1995): The *eco*72IC gene specifies a *trans*-acting factor which influences expression of both DNA methyltransferase and endonuclease from the *Eco*72I restriction-modification system. Gene 157:217–219.

Roberts RJ, Cheng X (1998): Base flipping. Annu Rev Biochem 67:181–198.

Roberts RJ, Halford SE (1993): Type II restriction enzymes. In Linn SM, Lloyd RS, Roberts RJ (eds): "Nucleases." Cold Spring Harbor, NY: Cold Spring Harbor Laboratory Press, pp 35–88.

Roberts RJ, Macelis D (2000): REBASE-Restriction enzymes and methylases. Nucleic Acids Res 28: 306–307.

Robertson GT, Reisenauer A, Wright R, Jensen RB, Jensen A, Shapiro L, Roop II RM (2000): The *Brucella abortus* CcrM DNA methyltransferase is essential for viability, and its overexpression attenuates intracellular replication in murine macrophages. J Bacteriol 182:3482–3489.

Robertson KD, Jones PA (2000): DNA methylation: Past, present and future directions. Carcinogenesis 21:461–467.

Rodriguez-Zaragoza S (1994): Ecology of free-living amoebae. Crit Rev Microbiol 20:225–241.

Rosamond J, Endlich B, Linn S (1979): Electron microscopic studies of the mechanism of action of the restriction endonuclease of *Escherichia coli* B. J Mol Biol 129:619–635.

Ross TK, Achberger EC, Braymer HD (1989): Identification of a second polypeptide required for McrB restriction of 5-methylcytosine-containing DNA in *Escherichia coli* K12. Mol Gen Genet 216:402–407.

Ross TK, Braymer HD (1987): Localization of a genetic region involved in McrB restriction by *Escherichia coli* K-12. J Bacteriol 169:1757–1759.

Roszak DB, Colwell RR (1987): Survival strategies of bacteria in the natural environment. Microbiol Rev 51:365–379.

Ruby EG, McFall-Ngai MJ (1999): Oxygen-utilizing reactions and symbiotic colonization of the squid light organ by *Vibrio fischeri*. Trends Microbiol 7:414–420.

Salaj-Smic E, Marsic N, Trgovcevic Z, Lloyd RG (1997): Modulation of *Eco*KI restriction in vivo: role of the lambda Gam protein and plasmid metabolism. J Bacteriol 179:1852–1856.

Sam MD, Perona JJ (1999): Catalytic roles of divalent metal ions in phosphoryl transfer by *Eco*RV endonuclease. Biochemistry 38:6576–6586.

Schluckebier G, O'Gara M, Saenger W, Cheng X (1995): Universal catalytic domain structure of Ado-Met-dependent methyltransferases. J Mol Biol 247:16–20.

Schouler C, Gautier M, Ehrlich SD, Chopin M-C (1998): Combinational variation of restriction modification specificities in *Lactococcus lactis*. Mol Microbiol 28:169–178.

Schulz HN, Brinkhoff T, Ferdelman TG, Marine MH, Teske A, Jorgensen BB (1999): Dense populations of a giant sulfur bacterium in Namibian shelf sediments. Science 284:493–495.

Segall AM, Goodman SD, Nash HA (1994): Architectural elements in nucleoprotein complexes: interchangeability of specific and non-specific DNA binding proteins. EMBO J 13:4536–4548.

Sethmann S, Ceglowski P, Willert J, Iwanicka-Nowicka R, Trautner TA, Walter J (1999): M.Φ*Bss*HII, a novel cytosine-C5-DNA-methyltransferase with target-recognizing domains at separated locations of the enzyme. EMBO J 18:3502–3508.

Sharp PA, Sugden B, Sambrook J (1973): Detection of two restriction endonuclease activities in *Haemophilus parainfluenzae* using analytical agarose-ethidium bromide electrophoresis. Biochem 12:3055–3063.

Siksnys V, Skirgaila R, Sasnauskas G, Urbanke C, Cherny D, Grazulis S, Huber R (1999): The *Cfr*10I restriction enzyme is functional as a tetramer. J Mol Biol 291:1105–1118.

Smith HO, Nathans D (1973): A suggested nomenclature for bacterial host modification and restriction systems and their enzymes. J Mol Biol 81:419–423.

Smith HO, Wilcox KW (1970): A restriction enzyme from *Hemophilus influenzae*: I. Purification and general properties. J Mol Biol 51:379–391.

Smith J, Berg JM, Chandrasegaran S (1999): A detailed study of the substrate specificity of a chimeric restriction enzyme. Nucleic Acids Res 27:674–681.

Smith JM, Smith NH, O'Rourke M, Spratt BG (1993): How clonal are bacteria? Proc Natl Acad Sci USA 90:4384–4388.

Som S, Friedman S (1994): Regulation of EcoRII methyltransferase: effect of mutations on gene expression and in vitro binding to the promoter region. Nucleic Acids Res 22:5347–5353.

Som S, Friedman S (1997): Characterization of the intergenic region which regulates the MspI restriction-modification system. J Bacteriol 179:964–967.

Spoerel N, Herrlich P, Bickle TA (1979): A novel bacteriophage defence mechanism: the anti-restriction protein. Nature 278:30–34.

Stanford NP, Halford SE, Baldwin GS (1999): DNA cleavage by the EcoRV restriction endonuclease: pH dependence and proton transfers in catalysis. J Mol Biol 288:105–116.

Stewart FJ, Raleigh EA (1998): Dependence of McrBC cleavage on distance between recognition elements. Biol Chem 379:611–616.

Storey CC, Lusher M, Richmond SJ, Bacon J (1989): Further characterization of a bacteriophage recovered from an avian strain of Chlamydia psittaci. J Gen Virol 70:1321–1327.

Studier FW, Bandyopadhyay PK (1988): Model for how type I restriction enzymes select cleavage sites in DNA. Proc Natl Acad Sci USA 85:4677–4681.

Suau A, Bonnet R, Sutren M, Godon JJ, Gibson GR, Collins MD, Dore J (1999): Direct analysis of genes encoding 16S rRNA from complex communities reveals many novel molecular species within the human gut. Appl Environ Microbiol 65:4799–4807.

Sugisaki H (1978): Recognition sequence of a restriction endonuclease from Haemophilus gallinarum. Gene 3:17–28.

Sugisaki H, Kanazawa S (1981): New restriction endonucleases from Flavobacterium okeanokoites (FokI) and Micrococcus luteus (MluI). Gene 16:73–78.

Surby MA, Reich NO (1996): Contribution of facilitated diffusion and processive catalysis to enzyme efficiency: Implications for the EcoRI restriction-modification system. Biochem 35:2201–2208.

Sutherland E, Coe L, Raleigh EA (1992): McrBC: a multisubunit GTP-dependent restriction endonuclease. J Mol Biol 225:327–358.

Szybalski W, Blumenthal RM, Brooks JE, Hattman S, Raleigh EA (1988): Nomenclature for bacterial genes coding for class-II restriction endonucleases and modification methyltransferases. Gene 74:279–280.

Szybalski W, Kim SC, Hasan N, Podhajska AJ (1991): Class-IIS restriction enzymes—A review. Gene 100:13–26.

Tao T, Bourne JC, Blumenthal RM (1991): A family of regulatory genes associated with type II restriction-modification systems. J Bacteriol 173:1367–1375.

Tettelin H, Saunders NJ, Heidelberg J, Jeffries AC, Nelson KE, Eisen JA, Ketchum KA, Hood DW, Peden JF, Dodson RJ, et al. (2000): Complete genome sequence of Neisseria meningitidis serogroup B strain MC58. Science 287:1809–1815.

Tran PH, Korszun ZR, Cerritelli S, Springhorn SS, Lacks SA (1998): Crystal structure of the DpnM DNA adenine methyltransferase from the DpnII restriction system of Streptococcus pneumoniae bound to S-adenosylmethionine. Structure 6:1563–1575.

Trautner TA, Balganesh TS, Pawlek B (1988): Chimeric multispecific DNA methyltransferases with novel combinations of target recognition. Nucleic Acids Res 16:6649–6658.

Trautner TA, Pawlek B, Behrens B, Willert J (1996): Exact size and organization of DNA target-recognizing domains of multispecific DNA-(cytosine-C5)-methyltransferases. EMBO J 15:1434–1442.

Van Etten JL, Meints RH (1999): Giant viruses infecting algae. Annu Rev Microbiol 53:447–494.

Varga GA, Kolver ES (1997): Microbial and animal limitations to fiber digestion and utilization. J Nutr 127:819S–823S.

Vijesurier RM, Carlock L, Blumenthal RM, Dunbar JC (2000): Role and mechanism of action of C.PvuII, a regulatory protein conserved among restriction-modification systems. J Bacteriol 182:477–487.

Wah DA, Bitinaite J, Schildkraut I, Aggarwal AK (1998): Structure of FokI has implications for DNA cleavage. Proc Natl Acad Sci USA 95:10564–10569.

Wah DA, Hirsch JA, Dorner LF, Schildkraut I, Aggarwal AK (1997): Structure of the multimodular endonuclease FokI bound to DNA. Nature 388:97–100.

Waite-Rees PA, Keating CJ, Moran LS, Slatko BE, Hornstra LJ, Benner JS (1991): Characterization and expression of the Escherichia coli Mrr restriction system. J Bacteriol 173:5207–5219.

Walsh CP, Bestor TH (1999): Cytosine methylation and mammalian development. Genes Dev 13:26–34.

Walter J, Trautner TA, Noyer-Weidner M (1992): High plasticity of multispecific DNA methyltransferases in the region carrying DNA target recognizing enzyme modules. EMBO J 11:4445–4450.

Wang J, Chen R, Julin DA (2000): A single nuclease active site of the Escherichia coli RecBCD enzyme catalyzes single-stranded DNA degradation in both directions. J Biol Chem 275:507–513.

Wernegreen JJ (2000): Decoupling of genome size and sequence divergence in a symbiotic bacterium. J Bacteriol 182:3867–3869.

Whitaker RD, Dorner LF, Schildkraut I (1999): A mutant of BamHI restriction endonuclease which requires N6-methyladenine for cleavage. J Mol Biol 285:1525–1536.

Whitman WB, Coleman DC, Wiebe WJ (1998): Prokaryotes: The unseen majority. Proc Natl Acad Sci USA 95:6578–6583.

Windhofer F, Catcheside DE, Kempken F (2000): Methylation of the foreign transposon Restless in vegetative mycelia of *Neurospora crassa*. Curr Genet 37:194–199.

Winkler FK (1994): Restriction endonucleases, the ultimate in sequence specific DNA recognition. J Mol Recog 6:9.

Wolffe AP, Matzke MA (1999): Epigenetics: Regulation through repression. Science 286:481–486.

Wommack KE, Colwell RR (2000): Virioplankton: Viruses in aquatic ecosystems. Microbiol Mol Biol Rev 64:69–114.

Woodbury Jr. CPJ, Downey RL, von Hippel PH (1980): DNA site recognition and overmethylation by the *Eco*RI methylase. J Biol Chem 255:11526–11533.

Woodbury Jr. CP, Hagenbuchle O, von Hippel PH (1980): DNA site recognition and reduced specificity of the *Eco*RI endonuclease. J Biol Chem 255:11534–11546.

Woodcock DM, Crowther PJ, Diver WP, Graham M, Bateman C, Baker DJ, Smith SS (1988): RglB facilitated cloning of highly methylated eukaryotic DNA: The human L1 transposon, plant DNA, and DNA methylated in vitro with human DNA methyltransferase. Nucleic Acids Res 16:4465–4482.

Wright DJ, Jack WE, Modrich P (1999): The kinetic mechanism of *Eco*RI endonuclease. J Biol Chem 274:31896–31902.

Wright R, Stephens C, Shapiro L (1997): The CcrM DNA methyltransferase is widespread in the alpha subdivision of proteobacteria, and its essential functions are conserved in *Rhizobium meliloti* and *Caulobacter crescentus*. J Bacteriol 179:5869–5877.

Xu G-L, Bestor TH (1997): Cytosine methylation targetted to pre-determined sequences. Nature Genet 17:376–378.

Yuan R, Heywood J, Meselson M (1972): ATP hydrolysis by restriction endonuclease from *E. coli* K. Nat New Biol 240:42–43.

Zhang B, Tao T, Wilson GG, Blumenthal RM (1993): The M*Alu*I DNA-(cytosine C5)-methyltransferase has an unusually large, partially dispensable, variable region. Nucleic Acids Res 21:905–911.

8

Recombination

STEPHEN D. LEVENE AND KENNETH E. HUFFMAN

Department of Molecular and Cell Biology, University of Texas at Dallas, Richardson, Texas 75083–0688

I. INTRODUCTION

A. Biological Significance of Mobile DNA Elements

The development of and interest in the idea of mobile genetic elements derived from Barbara McClintock's seminal papers describing the transposition of what she called "controlling elements" in maize during the 1950s (McClintock, 1955, 1956). McClintock demonstrated that movement of specific genetic elements to new chromosomal locations affected the expression of nearby genes and caused chromosomal breakages in a developmentally regulated manner. Since McClintock's remarkable discovery, the dynamic nature of mobile genetic elements has been observed in almost every prokaryotic and eukaryotic organism investigated to date.

The genome of any organism must posses two key features. First, it must be stable

Modern Microbial Genetics, Second Edition. Edited by
Uldis N. Streips and Ronald E. Yasbin. ISBN 0–471–38665–0
Copyright © 2002 Wiley-Liss, Inc.

enough to pass accurate information through inheritance, ensuring the survival of progeny. However, the genome must also be dynamic in order to respond to selective environmental pressures. Therefore any successful biological system must maintain a delicate balance between genome integrity and flexibility. In both prokaryotic and eukaryotic systems, recombination is one of the key mechanisms that regulates genome integrity. Chromosomal breakages and mutations, stemming from problems in DNA replication or environmental stress, can be repaired through recombination pathways (Evans and Alani, 2000; Foaini et al., 2000; Haber, 2000; Kreuzer, 2000). In some cases organisms contain multiple recombination pathways by which damage can be repaired, underscoring the importance of this process.

The mobility of DNA sequence elements is also a driving force in evolutionary patterns. Recombination provides a mechanism by which DNA can be moved, deleted and amplified to effect these changes. These movements are often tightly regulated, such as those involved in gene rearrangements, DNA amplification and deletion, and genome integration events. In other examples, the movements are rare and only become visible when selective pressures are imposed on large populations. One of the most startling examples of recombination-driven evolution involves the inheritance of antibiotic resistance genes among certain populations of bacteria (Davies, 1994). The biological consequences of recombination are ubiquitous.

In recent years recombination has become an invaluable tool in the biological laboratory for both genome manipulation and genetic analysis. Understanding mechanisms of recombination is therefore central to any discussion of the functional properties of any particular genome. Moreover recombination systems offer the opportunity to harness the power of mobile DNA elements for use in genetic therapies and other medical applications.

B. Genetic Recombination: Background and Perspective

The modern field of recombination, by most accounts, began in 1964 with Robin Holliday's hypothetical four-stranded DNA structure (Holliday, 1964) (Fig. 1). This structure, later to be named the Holliday junction, consists of two DNA duplexes associated by a single-stranded crossover and was proposed to explain gene conversions previously observed in fungi. Holliday possessed one advantage over McClintock in her earlier work: knowledge of the structure of DNA. Knowing that DNA was double stranded, he proposed a mechanism by which two chromosomes in close proximity could exchange DNA strands, thereby effecting genomic alterations. He also noted

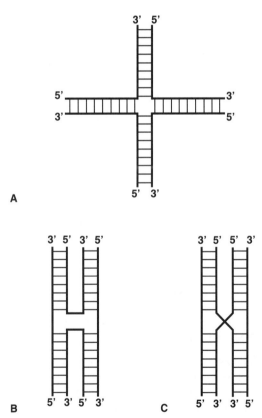

Fig. 1. Plane projections of three different structures of a four-way DNA junction, or Holliday junction. **A:** Symmetric planar cross. **B:** Antiparallel, open junction; **C:** Crossed, parallel junction.

the necessity for homology between the exchanging strands. Historically the Holliday junction and the necessity for extensive DNA homology have defined classical recombination, but it is important to note that there are many different recombination reaction mechanisms that proceed through a variety of intermediate structures.

II. RECOMBINATION SYSTEMS

A. General or Homologous Recombination

The most familiar recombination systems are those that are involved in general or homologous recombination. This class of recombination events is defined by the exchange of homologous sequences between double-stranded DNA molecules and results in recombinant product molecules that contain genetic information originally present in each of the parental molecules. There is generally no limit to how much DNA can be exchanged in this process, ranging from tens to thousands of base pairs, so long as homology is maintained between recombining duplexes. The basis of recognition in homologous systems is pairing of the exchanging DNA sequences. Protein components of these systems have other roles in the recombination reaction such as DNA strand juxtaposition, recruitment of cofactors such as ATP or accessory proteins, catalysis of strand cleavage and rejoining reactions, and heteroduplex extension or branch migration (Kowalczykowski et al., 1994).

A key intermediate in general recombination pathways is the four-stranded Holliday junction (Fig. 1). One of the hallmarks of this four-way junction is its ability to undergo branch migration, which can extend the heteroduplex region for thousands of base pairs (see Fig. 2). Branch migration can occur spontaneously by thermal fluctuations, resulting in unidimensional diffusion of the junction's branch point, or can be driven by helicase-dependent ATP hydrolysis (Yu et al., 1997). The final step of homologous recombination involves resolution

of the four-way junction into two distinct duplex molecules. Resolution in homologous systems can occur in a variety of ways, depending on the specificity of resolving enzymes and the particular conformation of the DNA intermediate. The locations of strand cleavages that resolve the junction direct the formation of particular products. Homologous recombination systems are involved in numerous cellular functions, including the repair of genomic damage caused by mismatched base pairs, chromosomal breaks, or deletions, mating-type conversion, antigenic variation, DNA replication, and meiosis.

B. Homologous Recombination in *Escherichia coli*

The paradigm for general recombination is abstracted from the set of RecA-dependent homologous recombination systems of *E. coli*. Although our understanding of other homologous recombination systems in both prokaryotic and eukaryotic organisms is rapidly improving, none of these approach the extent to which the RecA-dependent pathway has been characterized. At least 25 proteins have been shown to play some role in all types of homologous recombination in *E. coli* (Bianco et al., 1998); many of these proteins have functional homologs in other organisms though there is often little structural homology.

The strand-exchange protein RecA plays a central role in nearly all *E. coli* homologous recombination pathways. The conservation of functional RecA homologs underscores the biological importance of DNA strand-exchange proteins: all free-living organisms examined to date possess a RecA-like protein (Kowalczykowski and Eggleston, 1994). RecA-dependent homologous recombination involves at least four distinct stages (Fig. 2):

1. Initiation

Initiation encompasses the processing of DNA at a double-stranded break to generate a single-stranded DNA segment required for strand invasion of a duplex DNA homolog

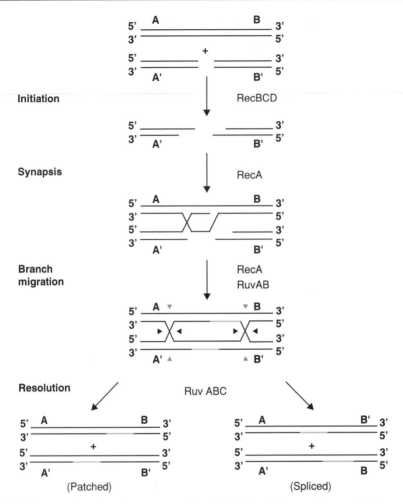

Fig. 2. Model for general or homologous recombination in *E. coli*. Stages of recombination are shown in bold; enzyme systems associated with each step are also given. Letters indicating the locations of arbitrary DNA sequence elements are given for reference. Initiation occurs at a double-stranded break and involves resection of 3′ single-stranded overhangs by RecBCD. Synapsis and strand invasion by one of the processed ends into the intact duplex is mediated by RecA protein to form a D-loop structure. DNA synthesis then extends the free 3′ ends to form a double Holliday-junction structure; newly synthesized DNA is shown in gray. Both Holliday junctions undergo branch migration promoted by RecA and RuvAB. Resolution of the junctions by the RuvABC complex involves a second set of strand-exchange steps and can take place in either of two planes through each Holliday junction. Resolution that occurs via strand exchange at both sets of dark arrowheads or both sets of light arrowheads generates recombinant products shown at bottom left (patched products). In contrast, resolution at opposing combinations of orientations, one set of dark arrowheads and one set of light arrowheads, yields recombinant products shown at bottom right (spliced products). (Adapted from Kowalczykowski, 2000, with permission from the publisher.)

by a RecA-ssDNA complex. The DNA processing that occurs during initiation in *E. coli* is carried out by the recombination-specific helicases RecBCD and RecQ, probably in conjunction with some exogenous exonu-

clease activity, such as that of the protein RecJ. In addition to helicase activity, the RecBCD complex also has an intrinsic nuclease activity; the specificity of this nuclease activity is modulated dramatically by the presence of an 8 bp DNA sequence element, χ, which functions as a hotspot for homologous recombination (Eggleston and West, 1997; Kowalczykowski, 2000). Upon encountering a χ site in the appropriate orientation, RecBCD switches from a $3' \rightarrow 5'$ to a $5' \rightarrow 3'$ exonuclease activity. This strand-polarity switch in the exonuclease activity of RecBCD leads to the preferential formation of DNA molecules bearing a $3'$-terminal single-stranded overhang, which are ideal substrates for formation of a RecA-ssDNA filament known as the presynaptic filament.

2. Synapsis

This stage involves steps that lead to homologous pairing and strand exchange. In the presence of ATP or nonhydrolyzable ATP analogues, RecA protein binds in a highly cooperative manner to single-stranded DNA to produce a nucleoprotein filament in which the ssDNA is stretched to nearly 1.5 times its original length. Both in vitro and in vivo, binding of RecA to ssDNA is assisted by *E. coli* single-stranded binding protein (SSB), which facilitates RecA binding through the destabilization of internal DNA secondary structure. Although ATP binding stabilizes the ssDNA-specific form of RecA, ATP hydrolysis is not required either for formation of the presynaptic filament or for homologous pairing or strand exchange.

Synapsis occurs between the presynaptic filament, which contains the invading segment of single-stranded DNA, and a double-stranded homologous target sequence. Binding of the presynaptic filament to dsDNA is initially random, but the RecA-ssDNA filament carries out a rapid and efficient search for sequence homology by a mechanism not fully understood at present. Upon locating the homologous target sequence, the RecA-DNA filament generates what is known as a joint molecule, in which the invading single strand displaces the complementary strand of the duplex to form a D loop. Current structural and biochemical data overwhelmingly support a model for a joint molecule-RecA complex that contains

three DNA strands (Cox, 1995); this view is in contrast to previously proposed models that invoke a four-stranded DNA intermediate (Howard-Flanders et al., 1984). Thus it seems likely that the homology-search mechanism involves weak and transient binding of the target dsDNA to a secondary binding site on the presynaptic filament. Upon locating the homologous region, the RecA-ssDNA filament and/or double-stranded target sequence likely undergo conformational rearrangements that transfer the displaced complementary strand to the secondary DNA-binding site. However, the joint molecule remains relatively unstable subject to additional processing events.

Joint-molecule recombination intermediates are stabilized by the formation of the Holliday junction, which requires both DNA synthesis and strand-joining activities. The mechanistic details of this process remain largely unknown, although both polymerase I and topoisomerase I activities have been implicated in this step.

3. Branch migration

Assembly of RecA filaments on ssDNA occurs exclusively in a $5' \rightarrow 3'$ direction. This activity probably continues on joint molecules, advancing the branch point of the Holliday junction in the same direction with respect to the incoming DNA strand at a rate of about 6 nt s^{-1} (Bedale and Cox, 1996). However, because branch-migration activity seems to be largely bidirectional, the RuvAB helicase complex is thought to be the principal factor involved in this phase of homologous recombination. RuvA protein binds to the Holliday junction and recruits RuvB, the latter assembling into a typical ringlike, hexameric helicase structure that surrounds each of two duplex branches of the Holliday intermediate (Yu et al., 1997). These two helicase structures translocate the duplex DNA in opposing directions, causing DNA to be "pumped out" of the center of the junction, thereby facilitating branch migration. Although branch-migration proceeds at a rate comparable to that

promoted by RecA-binding activity (Tsaneva et al., 1992), RuvAB catalyzes branch
migration bidirectionally depending on
which pair of duplex Holliday-junction
arms are bound by the RuvB hexamers.

4. Resolution

The Holliday junction is specifically cleaved
by the endonuclease activity of RuvC (Connolly et al., 1991). The RuvC endonuclease is
highly specific for Holliday junctions, and its
cleavage activity occurs in concert with the
branch-migration activity of RuvAB, presumably to locate RuvC at preferred cleavage
sites. RuvC is capable of cleaving the Holliday junction in either of two ways leading to
two potential sets of resolution products
(Fig. 2). However, protein-DNA interactions
probably distort the structure of the junction
and thereby generate a preference for one of
these (van Gool et al., 1999).

C. Site-Specific Recombination

Unlike general recombination, site-specific
recombination events involve the interaction
of defined DNA sequence elements. These
sequences are highly specialized, carry specific binding sites for the recombination proteins as well as the point of genetic exchange,
and are usually present in extremely low
copy number in the genome. Often these sites
are present in pairs. However, they are sometimes present only as a single copy as in
the case of the bacteriophage λ integration
site in the *E. coli* genome. This extraordinary
degree of specificity leads to precisely defined genetic rearrangements. In the examples considered here, the rearrangements
that occur are essentially uniquely defined.

Another important attribute of a site-
specific recombination locus is the polarity
of the recombination site. These loci are frequently nonpalindromic and therefore have
an intrinsic polarity. Recombination normally occurs only when a pair of recombination sites has been juxtaposed in a particular
spatial alignment, thereby imparting both
positional and orientational specificity to
these systems (Gellert and Nash, 1987;

Nash, 1996). This specificity has important
biological consequences; moreover the site-
orientation specificity leads to the formation
of specific DNA topologies in the recombination products. The topological specificity of
site-specific recombination systems has been
exploited to great effect in unraveling the
mechanisms of many site-specific recombinases.

DNA homology normally plays a very
limited role in site-specific recombination,
more a feature of specific recombinase-DNA
interactions than a necessity for homologous
pairing or strand exchange. Unlike virtually
all other modes of recombination, site-
specific recombination is conservative in that
no DNA is gained or lost during the recombination reaction. This aspect of site-specific
recombination applies both at the level of
genetic information (recombination products
are merely permutations of the original parental DNA) and at the level of actual DNA
nucleotides (no DNA synthesis or nucleolytic
degradation is involved). In contrast, significant levels of DNA synthesis activity are required both for homologous recombination
(see above) and transposition (see below).

Initiation of site-specific recombination
begins with the binding of the recombination
proteins to their respective recognition
sequences within recombining loci. Upon
binding to the target sites, protein-protein
interactions among the recombination proteins facilitate the synapsis of recombination
sites. Well-defined protein-DNA contacts
allow site-specific recombinases to cleave
their DNA targets with the specificity of
restriction endonucleases, whereas protein-
protein interactions direct strand exchange.
An early step in virtually all site-specific recombination pathways is the formation of a
covalently linked protein-DNA intermediate
during the initial strand-cleavage reaction.

All site-specific recombination systems
that have been investigated to date fall into
two superfamilies: the integrase and resol-
vase/invertase families (Table 1). Particular
examples from both of these families are
discussed below. Products of reactions

TABLE 1. Site-Specific Recombination Systems

Function	Element or Recombinase	Host or Context	Recombinase Superfamily
Diversity/gene expression	*hin*	*S. typhimurium*	Resolvase/invertase
	gin	Bacteriophage Mu	Resolvase/invertase
	pin	*E. coli*	Resolvase/invertase
	SpoIV *cisA*	*Bacillus*	Resolvase/invertase
	flm	*E. coli*	Integrase
Dimer reduction	*res*	Transposon Tn*3*	Resolvase/invertase
	res	Transposon Tn*21*	Resolvase/invertase
	cre	Bacteriophage P1	Integrase
	xer	*E. coli*	Integrase
	flp	*S. cerevisiae*	Integrase
Integration/excision	λ *int*	Bacteriophage λ	Integrase
	Tn*916*	*Enterococcus*	Integrase
	Tn*1545*	*Streptococcus*	Integrase

Source: Excerpted from Nash (1996).

carried out by the integrase superfamily vary widely depending on the orientation and disposition of recombination sites; this variability permits systems such as λ-integrase to participate in both integrative and excisive recombination in a highly regulated fashion. The resolvase/invertase mechanisms are characterized by a well-defined DNA geometry in the synaptic intermediate and, as a consequence, tightly controlled product topologies. The two superfamilies are also distinct in terms of the intermediate structure of the DNA segments undergoing recombination; whereas λ-integrase-type mechanisms proceed through a Holliday intermediate, the resolvase/invertase mechanisms do not. Site-specific recombination systems participate in a wide range of biological processes in both prokaryotes and eukaryotes: viral integration, antigenic variation, gene duplication and copy-number control, and the integration of antibiotic resistance cassettes.

1. Integrative and excisive recombination in the λ-integrase system

The λ-integrase (λ-int) system is vital to the lysogenic stage of the life cycle of bacteriophage λ and is one of the most intensively studied site-specific recombination systems. A notable feature of this system is the nonsym-

metrical nature of the integrative and excisive recombination reactions: although strand exchange activities are identical both for integration and excision of the phage-λ genome, each reaction has distinct requirements for specific DNA sequences at the recombining loci and subsets of protein cofactors involved in recombination (see Hendrix, this volume).

Integration of phage λ occurs at a unique 25 bp site, termed *attB*, on the 4.6 Mbp *E. coli* chromosome. The catalytic activity for strand exchange resides in the λ-encoded integrase protein (int), which functions in concert with a number of DNA-binding accessory proteins: the integration host factor (IHF) and factor for inversion stimulation (FIS) proteins of *E. coli*, and the λ-excisionase (Xis), which is phage-encoded. In contrast to the *attB* site, which by itself has negligible affinity for the recombination proteins, the recombination locus on the phage genome, *attP*, is about 250 bp in size and has multiple binding sites for int and the accessory factors (Fig. 3). Integrative recombination most likely involves assembly of int and IHF proteins to form an organized nucleoprotein structure called the *intosome* (Better et al., 1982), which subsequently captures a protein-free *attB* site during synapsis (Richet et al., 1988). Products of the integrative recombination reaction

P1   H1   P2

```
ACAGGTCACT AATACCATCT AAGTAGTTGA TTCATAGTGA CTGCATATGT TGTGTTTTAC
TGTCCAGTGA TTATGGTAGA TTCATCAACT AAGTATCACT GACGTATACA ACACAAAATG
```

                                       H2

```
AGTATTATGT AGTCTGTTTT TTATGCAAAA TCTAATTTAA TATATTGATA TTTATATCAT
TCATAATACA TCAGACAAAA AATACGTTTT AGATTAAATT ATATAACTAT AAATATAGTA
```
                                                             attP

                       C

```
TTTACGTTTC TCGTTCAGCT TTTTTATACT AAGTTGGCAT TATAAAAAAG CATTGCTTAT
AAATGCAAAG AGCAAGTCGA AAAAATATGA TTCAACCGTA ATATTTTTTC GTAACGAATA
```
                               C'   H'

```
CAATTTGTTG CAACGAACAG GTCACTATCA GTCAAAATAA AATCATTATT
GTTAAACAAC GTTGCTTGTC CAGTGATAGT CAGTTTTATT TTAGTAATAA
```
       H'   P'1   P'3
                                        P'2

   B   B'

```
CTGCTTTTTT ATACTAACTT G
GACGAAAAAA TATGATTGAA C
```
                attB
           O

P1   H1   P2

```
ACAGGTCACT AATACCATCT AAGTAGTTGA TTCATAGTGA CTGCATATGT TGTGTTTTAC
TGTCCAGTGA TTATGGTAGA TTCATCAACT AAGTATCACT GACGTATACA ACACAAAATG
```

                                       H2

```
AGTATTATGT AGTCTGTTTT TTATGCAAAA TCTAATTTAA TATATTGATA TTTATATCAT
TCATAATACA TCAGACAAAA AATACGTTTT AGATTAAATT ATATAACTAT AAATATAGTA
```
                                                            attL

                       C   B'

```
TTTACGTTTC TCGTTCAGCT TTTTTATACT AACTTG
AAATGCAAAG AGCAAGTCGA AAAAATATGA TTGAAC
```
                               O

   B   C'   H'

```
CTGCTTTTTT ATACTAAGTT GGCATTATAA AAAAGCATTG CTTATCAATT TGTTGCAACG
GACGAAAAAA TATGATTCAA CCGTAATATT TTTTCGTAAC GAATAGTTAA ACAACGTTGC
```
              O

```
AACAGGTCAC TATCAGTCAA AATAAAATCA TTATT
TTGTCCAGTG ATAGTCAGTT TTATTTTAGT AATAA
```
                                                     attR

       P'1   P'3
                   P'2

Fig. 3. DNA-sequence organization of target sites involved in integrative and excisive recombination mediated by the λ-int system. Sites occupied by int protein on *att*P are of two types: "core" binding sites, designated C and C'; and "arm-type" binding sites, P1, P2, and P' I − −P'3. IHF binding occurs at sites H1, H2, and H'. Binding sites for proteins involved in excisive recombination, Xis and FIS, are not shown. The sequence of *att*B shows sites B and B', which are occupied by catalytically active int monomers during recombination, and the overlap region, O, which is the sequence element involved in strand exchange during recombination.

are a functionally distinct pair of new recombination sites, called *attL* and *attR*, that are no longer competent to participate in subsequent rounds of integrative recombination (Fig. 3). Instead, these sites are substrates for excisive recombination, a reaction that requires FIS and Xis in addition to int and IHF. By coupling recombination to intracellular levels of specific protein factors, tight regulation of the phage-λ life cycle can be achieved in vivo.

2. Structural and topological consequences of λ-int recombination

The natural role of the λ-int system is to generate a circular DNA fusion product from recombination sites residing on two circular DNA substrates, the wild-type chromosome and a circularized λ genome, and to excise the integrated λ prophage via recombination of two sites present on a single DNA circle. Useful model systems for investigating the mechanism of int recombination (and other site-specific recombination systems) are stripped-down versions of the natural substrates, generally plasmid DNAs that contain a copy of one or both recombination sites (either *attP/attB* or *attL/attR*). An overview of the possible reactions involving pairs of recombination sites on circular substrates is shown in Figure 4. When two sites are present on separate circles, only the fusion reaction is possible. However, the case where two loci are present on the same circle leads to two possible outcomes depending on the relative orientations of the two sites. If the sites are directly repeated, then recombination results in a deletion reaction that is the opposite of

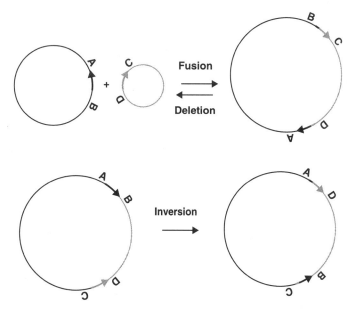

Fig. 4. Prototype site-specific recombination reactions involving target sites on circular DNA molecules. Arrows denote target sites and letters indicating the locations of arbitrary DNA sequence elements are given for reference. All site-specific recombination reactions involving sites of defined polarity conserve polarity during recombination, thus recombination entails the exchange of respective head and tail portions of the arrows that indicate target sites. Fusion reactions are intermolecular recombination events that result in circular products with directly oriented sites; the reverse of this pathway is an intramolecular deletion reaction, resulting in the formation of two distinct circular products. Circular deletion products may be unlinked, or linked one or more times to form a catenane, depending on the topology of the substrate and the mechanism of recombination. Inversion occurs on circular DNA substrates with inversely oriented target sites. Products of inversion reactions may be either unknotted or knotted circles, again depending on substrate topology and recombinase mechanism.

the fusion reaction shown in Figure 4. If the sites are inversely repeated, then the product of recombination is a single circle that has undergone inversion; that is, the relative orientation of segments of DNA between the recombination sites has been inverted.

When supercoiled DNA substrates are used in reconstituted in vitro recombination reactions, it is possible to examine the topological changes that take place during recombination. For intramolecular recombination reactions, supercoiled plasmid substrates bearing inversely oriented sites generate knotted recombination products, whereas supercoiled substrates containing directly repeated sites generate topologically linked circles called catenanes (Fig. 5). Knots and catenanes, being particular topologies of a circle in three-dimensional space, are classified according to the number and arrangement of irreducible or minimal crossings in a two-dimensional projection of the figure's axis. The knots and catenanes that are formed during recombination are never random, but instead a highly restricted subset of

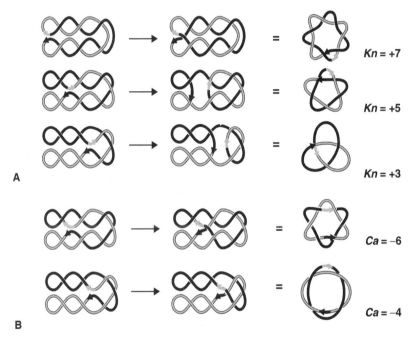

Fig. 5. Topology of products generated by λ-integrative recombination on circular DNA molecules. Diagrams show planar projections of negatively supercoiled DNA substrates undergoing intramolecular recombination. Recombination sites, indicated by arrows, divide the DNA contour into two domains, shown as black and outlined gray curves. Random Brownian motion of recombination sites (left column) leads to site synapsis in DNA conformations that involve varying numbers of interdomainal supercoils (supercoils involving separate DNA domains). Only interdomainal supercoils are trapped in the form of knot or catenane crossings by strand-exchange steps in recombination (*middle column*). The resulting topologies are shown in the form of diagrams (*right column*) that depict only the number and topological sign of irreducible crossings in each knotted or catenated product, which are given to the right below each figure. These diagrams correspond to actual products in which extraneous supercoils have been removed by nicking of one DNA strand. **A:** Inversely oriented sites. Inversion generates knotted products that are separated by intervals of +2 knot crossings; these knots belong to the so-called torus class because of the property that these knots can all be inscribed on the surface of a torus. Only three examples of knotted products are shown. **B:** Directly oriented sites. In addition to unlinked circles, deletion reactions generate (−) torus catenanes that also differ by steps of two crossings. Only two examples of catenated products are shown.

all the possible knotted or catenated structures that can be formed. For example, all of the knots with up to 13 irreducible crossings are known—there are over 12,000 topologically distinct knots. Integrative recombination on a circular substrate with inverted sites yields only seven of the possible knots containing up to 13 irreducible crossings, each containing an odd number of crossings. Among all possible recombination mechanisms that can lead to the formation of a knotted DNA product, the formation of this particular set of observed products can be ascribed uniquely to a particular mechanism (Fig. 5).

3. Regulation of flagellin gene expression in S. typhimurium

Site-specific recombination accounts for the phenomenon of phase variation first observed in the 1920s and linked to a genetic rearrangement in the 1950s (Lederberg Iino, 1956). Sites for the Hin recombination system flank a promoter region that regulates two genes; these encode the flagellin H1 protein and a repressor, rh2, of an alternate flagellin protein, H2. Hin mediates a DNA-inversion event that orients the promoter either toward the H1 and rh2 genes, an orientation appropriate for expression of H1 and rh2, or away from these genes to express the alternate pair of genes H2 and rh1 (Fig. 6A). Thus the alternate expression of two flagellin proteins is modulated by Hin recombination, which generates both orientations of a common regulatory region.

A great deal has been learned about the mechanism of this system by examining the topology of Hin recombination (Heichman et al., 1991) and that of a homologous recombination system, Gin, in bacteriophage Mu (Kanaar et al., 1990; Kanaar et al., 1988). Hin recombination requires in addition to the Hin recombinase and a pair of recombination sites, called *hixL* and *hixR*, the accessory protein FIS, which binds to an enhancer DNA sequence (Heichman Johnson, 1990). Besides the juxtaposition of the Hin-bound recombination sites, the syn-

apsis in the Hin system also requires the FIS-bound enhancer to be present. Although the enhancer sequence is not explicitly involved in any of the cleavage or strand-exchange reactions, its required participation in synapsis leads to a particular set of recombination-product topologies (Fig. 6B).

D. Transposition

Transposable elements are mobile segments of DNA that can insert into nonhomologous target sites. The first prokaryotic element was discovered by Taylor in bacteriophage Mu, which was so named because of the mutations it caused in its *E.coli* host. Taylor observed that mu did not have a specific attachment site like phage λ, but was inserted almost randomly, causing mutations in genes and regulatory regions that had been disrupted (Taylor, 1963). Transposable elements may carry noncoding segments of DNA, the genome of a virus or phage, or antibiotic resistance elements. However, essential features of these elements are (1) they encode at least one protein factor that is involved in insertion, the transposase, and (2) the presence of terminal sequences that are recognized by the transposase and function as donor recombination sites. The transposase binds specifically to the end sequences, and either alone or in conjunction with accessory proteins, it is generally responsible for target-site selectivity. Transposon-end sequences are specific to each element and are frequently identical or consensus sequences that are arranged as a pair of terminal inverted repeats (see Whittle and Salyery, this volume).

Target specificity varies widely among transposition systems and is characterized by avoidance of particular loci in the targeted genome as much as expressed preferences for particular sites of integration. Target-site specificity ranges from very weak, as in the case of phage mu, to moderate (e.g., Tn10, IS10), to high (e.g. Tn7) (Craig, 1997); for some classes of transposable elements there is a weak consensus that corresponds to preferential insertion in

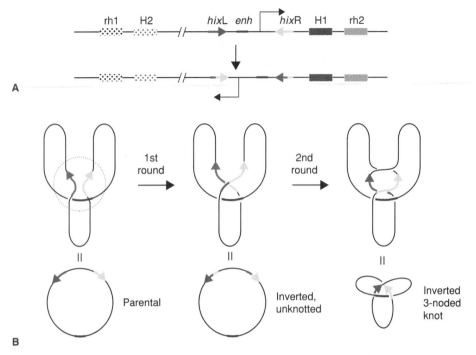

Fig. 6. DNA inversion carried out by the Hin recombinase of *S. typhimurium*. **A:** Expression of alternate sets of flagellin genes regulated by Hin site-specific recombination. Target sequences recognized by the Hin recombinase, *hix*L and *hix*R, flank a promoter (right arrow) that governs expression of the HI flagellin gene and a repressor of flagellin H2 expression (rh2). An enhancer sequence element (*enh*), which is a binding site for the DNA-binding protein FIS, is also required for Hin-mediated inversion. Hin recombination inverts the promoter-containing segment of DNA relative to the HI–rh2 segment, thereby inactivating HI and rh2 expression, and activating expression of the alternate set of genes H2 and rh1. **B:** Topological effects of Hin recombination on a circular DNA substrate. Synapsis of the sequence elements participating in Hin recombination, *hix*L, *hix*R, and *enh*, is depicted using a circular DNA bearing all three sites; a dashed circle in the figure at left denotes the synaptosome or complex of proteins and DNA involved in site pairing and strand exchange. An initial round of Hin recombination generates an inverted, unknotted DNA circle whereas a second round of recombination produces a 3-crossing knotted DNA with the primary sequence of the parental substrate. Topological diagrams are based on those of Kanaar et al. (1988).

A + T-rich regions, but this is not a universal characteristic. In some cases a sequence-dependent structure such as an intrinsic DNA bend is targeted rather than a particular sequence, per se (Craig, 1997). A low degree of target specificity suggests that transposition reactions must be tightly regulated in order to prevent the accumulation of excessive mutation by the host genome.

Transposable elements are divided into two general classes: transposons, which use a DNA intermediate for direct insertion, and retrotransposons, which proceed through an RNA intermediate. Within these classes

mechanistic details vary widely. However, a universal feature is that transposons are directly inserted into the target site by a series of DNA cleavage and strand-transfer reactions mediated by the transposase and any necessary accessory factors. Recombination in these cases may simply involve induction of a double-stranded break and strand transfer of a nearly intact duplex segment of DNA from donor to target (conservative transposition, Fig. 7A) or may involve fusion of the donor and target via duplication of the transposon (replicative transposition, Fig. 7B). The latter pathway is characterized by

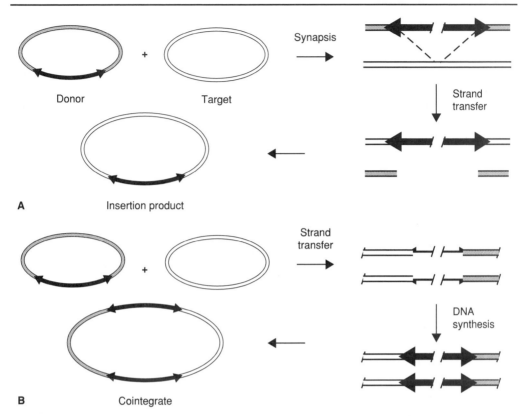

Fig. 7. Conservative and replicative transposition pathways. **A:** Conservative transposition. Strand transfer takes place between a nearly intact double-stranded DNA segment flanked by transposon ends. Integration into the target molecule leaves a double-stranded break in the donor molecule, which is lost during conservative transposition. **B:** Replicative transposition. Nicking occurs on the upper and lower DNA strands at opposite boundaries of the transposon; the free 3' ends are transferred to the target DNA with the resulting single-stranded gaps being filled by DNA synthesis. Replicative transposition thereby generates a cointegrate product containing two copies of the transposable element.

nicking of the upper and lower DNA strands at opposite boundaries of the transposon and transfer of both of the free 3' ends to the target DNA. This leads to duplicate copies of the transposon because the transposon/donor boundary remains intact on opposing ends of the element. The resulting single-stranded gap between duplex donor and target sequences, corresponding to the complementary strand of each copy of the transposon, is subsequently filled by DNA synthesis to generate a circular DNA molecule called a cointegrate. General aspects of these mechanisms apply to retroviral integration as well as mobile DNA elements; in the

former case a complete DNA copy of the retroviral RNA is made by an element-encoded reverse transcriptase and used as the donor in the transposition reaction.

In contrast, retrotransposition pathways involve participation of a reverse transcriptase activity directly in the recombination reaction. In these cases a DNA target site is cleaved and an exposed 3'-OH group in the cleaved DNA is used as a primer for reverse transcriptase, which uses a transposable-element RNA as the template. A complementary DNA segment is thereby copied directly from an RNA donor into the target site.

I. Phage Mu, a model transposable element

Upon infection of its *E. coli* host, phage Mu integrates by conservative transposition into the bacterial genome (see Fig. 7). The resulting prophage can be induced by a variety of environmental factors, resulting in multiple rounds of replicative transposition. In the Mu system, transposition is obligatory for replication and thus the phage genome is replicated only as a cointegrate structure. Insertions occur at many sites in the genome, frequently within several kb of one another. In the final stage of the lytic cycle, packaging is initiated at one end of the Mu genome and continues until approximately 39 kb of DNA has been incorporated into the phage head. The wild-type Mu genome is 37.5 kb and thus about a 1.5 kb of host DNA adjacent to Mu is incorporated into the phage head. Deletion derivatives of Mu, called *mini-Mu*, can package still larger amounts of host DNA and are more useful as transducing phages.

The biochemistry of Mu transposition has been studied extensively (Craigie and Mizuuchi, 1985; Mizuuchi, 1992; Mizuuchi and Craigie, 1986). The transposase is a protein called MuA. MuA has both DNA binding and endonuclease activities needed to form the strand-transfer intermediate. However, MuA normally acts in a multiprotein complex with several accessory proteins (MuB, IHF, and HU). MuB is an activator of MuA and provides some degree of target-site selectivity, whereas HU is an *E. coli* non-sequence-specific DNA-binding protein. An extremely interesting feature of Mu is target immunity; copies of Mu do not insert into DNA molecules that already contain a copy of the transposon (Adzuma and Mizuuchi, 1988). This effect is achieved at the level of single DNA molecules and not through overall inhibition of transposition activity. The molecular basis for target immunity in the Mu system is the sequestration of MuB protein by copies of MuA that remain bound to MuA binding sites on the integrated transposon.

E. Illegitimate Recombination

Illegitimate recombination occurs at DNA sequences that share little or no homology with their exchange partners. These recombinases most likely comprise the most primitive of recombination systems because of their inherent lack of recognition specificity. Illegitimate recombination is divided into two classes; end-joining and strand-slippage. The end-joining reaction in eukaryotes is very efficient and has been shown to allow broken chromatids to undergo replication by fusing their ends together, presumably to prevent them from being recognized by DNA damage checkpoints and degraded. One of the hallmarks of end-joining in eukaryotes is that it readily occurs in the absence of homology. In prokaryotes, however, end-joining reactions do require short regions of micro-homology. Strand-slippage occurs most often in regions containing tri- or tetranucleotide repeats. These strand-slippage reactions can delete or amplify these repetitive sequences often precipitating deleterious genetic defects.

Other illegitimate recombination reactions can be seen in type I and II topoisomerase reactions and in transposition or site-specific recombination events involving aberrant substrate molecules. Illegitimate recombination may also be an important player in the extensive genetic rearrangements that occur in cancerous cells. The loss of normal cell cycle checkpoints may allow damaged chromosomes to enter S phase and become subject to end-joining reactions or cause the accumulation of amplified palindromic sequences. Illegitimate recombination can mediate a number of chromosomal rearrangements, some of which may be deleterious to the organism, whereas others may lead to an evolutionarily favorable reorganization of the genome.

REFERENCES

Adzuma K, Mizuuchi K (1988): Target immunity of Mu transposition reflects a differential distribution of MuB protein. Cell 53:257–266.

Bedale WA, Cox M (1996): Evidence for the coupling of ATP hydrolysis to the final (extension) phase of RecA protein-mediated DNA strand exchange. J Biol Chem 271:5725–5732.

Better M, Lu C, Williams RC, Echols H (1982): Site-specific DNA condensation and pairing mediated by the int protein of bacteriophage lambda. Proc Natl Acad Sci USA 79:5837–5841.

Bianco PR, Tracy RB, Kowalczykowski SC (1998): DNA strand exchange proteins: A biochemical and physical comparison. Front Biosci 3:D570–603.

Connolly B, Parsons CA, Benson FE, Dunderdale HJ, Sharples GJ, Lloyd RG, West SC (1991): Resolution of Holliday junctions in vitro requires Escherichia coli ruvC gene product. Proc Natl Acad Sci USA 88:6063–6067.

Cox MM (1995): Alignment of 3 (but not 4) DNA strands within a RecA protein filament. J Biol Chem 270:26021–26024.

Craig NL (1997): Target site selection in transposition. Annu Rev Biochem 66:437–474.

Craigie R, Mizuuchi K (1985): Mechanism of transposition of bacteriophage Mu: Structure of a transposition intermediate. Cell 41:867–876.

Davies J (1994): Inactivation of antibiotics and the dissemination of resistance genes. Science 264:375–382.

Eggleston AK, West SC (1997): Recombination initiation: Easy as A, B, C, D...chi? Curr Biol 7:R745–749.

Evans E, Alani E (2000): Roles for mismatch repair factors in regulating genetic recombination. Mol. Cell Biol. 20:7839–7844.

Foaini M, Pellicioli A, Lopes M, Lucca C, Ferrari M, Liberi G, Muzi Falconi M, Plevanil P (2000): DNA damage checkpoints and DNA replication controls in Saccharomyces cerevisiae. Mutat Res 451:187–196.

Gellert M, Nash H (1987): Communication between segments of DNA during site-specific recombination. Nature 325:401–404.

Haber JE (2000): Partners and pathways repairing a double-strand break. Trends Genet 16:259–264.

Heichman KA, Johnson RC (1990): The Hin invertasome: Protein-mediated joining of distant recombination sites at the enhancer. Science 249:511–517.

Heichman KA, Moskowitz IP, Johnson RC (1991): Configuration of DNA strands and mechanism of strand exchange in the Hin invertasome as revealed by analysis of recombinant knots. Genes Dev 5:1622–1634.

Holliday R (1964): A mechanism for gene conversion in fungi. Genet Res 5:282–304.

Howard-Flanders P, West SC, Stasiak A (1984): Role of RecA protein spiral filaments in genetic recombination. Nature 309:215–219.

Kanaar R, Klippel A, Shekhtman E, Dungan JM, Kahmann R, Cozzarelli NR (1990): Processive recombi

nation by the phage Mu Gin system: Implications for the mechanisms of DNA strand exchange, DNA site alignment, and enhancer action. Cell 62:353–366.

Kanaar R, van de Putte P, Cozzarelli NR (1988): Gin-mediated DNA inversion: Product structure and the mechanism of strand exchange. Proc Natl Acad Sci USA 85:752–756.

Kowalczykowski SC (2000): Initiation of genetic recombination and recombination-dependent replication. Trends Biochem Sci 25:156–165.

Kowalczykowski SC, Dixon DA, Eggleston AK, Lauder SD, Rehrauer WM (1994): Biochemistry of homologous recombination in Escherichia coli. Microbiol Rev 58:401–465.

Kowalczykowski SC, Eggleston AK (1994): Homologous pairing and DNA strand-exchange proteins. Annu Rev Biochem 63:991–1043.

Kreuzer KN (2000): Recombination-dependent DNA replication in phage T4. Trends Biochem. Sci. 4:165–173.

Lederberg J, Iino T (1956): Phase variation in salmonella. Genetics 41:743–757.

McClintock B (1955): Intranuclear systems controlling gene action and mutation. Brookhaven Symp Biol 8:58–74.

McClintock B (1956): Controlling elements and the gene. Cold Spring Harbor Symp Quant Biol 21:197–216.

Mizuuchi K (1992): Transpositional recombination: Mechanistic insights from studies of mu and other elements. Annu Rev Biochem 61:1011–1051.

Mizuuchi K, Craigie R (1986): Mechanism of bacteriophage mu transposition. Annu Rev Genet 20:385–429.

Nash HA (1996): Site-specific recombination: Integration, excision, resolution, and inversion of defined DNA segments. In Neidhardt FC (ed): "Escherichia coli and Salmonella: Cellular and Molecular Biology," Vol 2. Washington, DC: ASM Press, pp 2363–2376.

Richet E, Abcarian P, Nash HA (1988): Synapsis of attachment sites during lambda integrative recombination involves capture of a naked DNA by a protein-DNA complex. Cell 52:9–17.

Taylor AL (1963): Bacteriophage-induced mutation in E. coli. Proc Natl Acad Sci USA 50:1043–1051.

Tsaneva IR, Muller B, West SC (1992): ATP-dependent branch migration of Holliday junctions promoted by the RuvA and RuvB proteins of E. coli. Cell 69:1171–1180.

van Gool AJ, Hajibagheri NM, Stasiak A, West SC (1999): Assembly of the Escherichia coli Ruv ABC resolvasome directs the orientation of Holliday junction resolution. Genes Dev 13:1861–1870.

Yu X, West SC, Egelman EH (1997): Structure and subunit composition of the RuvAB-Holliday junction complex. J Mol Biol 266:217–222.

9

Molecular Applications

Department of Biochemistry and Molecular Biology, University of Louisville School of Medicine,
Louisville, Kentucky 40292

I. Introduction 243
II. Why Clone?................................ 244
III. Tools for Molecular Cloning 244
    A. Restriction Enzymes 244
    B. Vectors 246
    C. DNA Libraries............................ 249
IV. Screening Strategies 249
    A. Screening by Functional Activity 250
    B. Screening with Homologous Genes............. 250
    C. Using Proteomics.......................... 250
    D. Screening by Linkage Using BAC
       (Bacterial Artificial Chromosome) Libraries 251
    E. cDNA Cloning............................ 252
V. Special Considerations......................... 252
    A. Transposons as Cloning Tools................. 252
    B. Phage Display 254
VI. Yeast Two-Hybrid Systems...................... 255

I. INTRODUCTION

Molecular cloning is the isolation of a unique piece of DNA, usually representing a gene or gene fragment, from an organism. It is in some sense a misnomer since molecules cannot really be cloned. However, organisms can be. And by cloning a host organism containing an exogenous gene or other DNA fragment of some particular interest, the DNA molecule itself can be cloned. Molecular cloning relies on the ability to express genes from one genetic background in a totally different genetic background. In its 30-year history this simple laboratory technique has revolutionized the biological sciences, given rise to the entire new discipline of molecular biology, and changed the way society views biology, ethics, and biological scientists. Despite its overwhelming impact, molecular cloning had rather meager beginnings. In the early 1970s Dan Nathans, studying the genetics of bacteriophage/host interactions, and more specifically the phenomenon of host resistance to phage infection, uncovered the restriction-modification systems (Nathans and Smith, 1975). He and others immediately recognized the potential of these enzymes, since the restriction enzyme component of some of the RM systems was able to cleave DNA at unique sequences,

Modern Microbial Genetics, Second Edition. Edited by
Uldis N. Streips and Ronald E. Yasbin. ISBN 0–471–38665–0
Copyright © 2002 Wiley-Liss, Inc.

generating specific fragments of any DNA molecule. Prior to that time, using the available DNA degrading enzymes, DNA could only be cleaved into random fragments. The discovery of restriction enzymes, and their ability to cleave DNA into defined fragments, was the key event that led directly to the techniques of molecular cloning.

II. WHY CLONE?

This is an important question particularly in view of the vast amount of already available DNA sequence information and the use of polymerase chain reaction, which is technically a much simpler way of generating unique pieces of DNA. Is molecular cloning a thing of the past, a technology that is no longer useful? The answer is a resounding no! Cloning is still a primary tool for generating unique pieces of DNA and one of the most important tools in the arsenal of the molecular biologist.

The primary reason for molecular cloning is to isolate a DNA of particular interest. Unlike proteins, whose amino acid sequence imparts unique physical properties as well as informational content, the sequence of bases in DNA does not impart any unique biochemical or biophysical property that can be used to differentiate one piece of DNA representing a particular gene from another piece of DNA representing a different gene. Molecular cloning obviates the need for biochemical isolation. By introducing unique fragments of DNA joined to a plasmid, bacteriophage, or other vector, and then selecting host cells containing those DNAs, the DNA itself can in essence be isolated. Cloned DNA fragments are fundamental to DNA sequence analysis, development of probes for Southern and northern blot analysis, site-directed mutagenesis, and a host of other molecular applications (for a thorough discussion of these methods, see *Current Protocols in Molecular Biology*, Ausubel et al., or *Molecular Cloning*, Sambrook et al.). There are commercial applications to cloning as well. Bacteria containing a particular DNA or gene could be, with a little additional manipulation, used to produce a protein product of the cloned gene in what has come to be known as genetic engineering. The list of pharmaceutical products produced by genetic engineering—from antibodies to insulin to tissue plasminogen activator (TPA)—is now quite long. Cloned genes can also be used to transform a bacterium (see the chapter on transformation by Streips, this volume) to provide new and useful phenotypes. For example, "bioremediation" approaches have been developed to clean up toxic chemicals resulting from environmental accidents (U.S. Environmental Protection Agency, 2000). Plants and animals can also be modified by introduction of cloned genes. A significant percentage of crops used for human consumption are now modified by recombinant DNA procedures to provide useful properties like drought or pest resistance. These so-called GMOs (genetically modified organisms) have created considerable controversy in European and Asian agricultural markets (hearings before the Subcommittee on Basic Research of the Committee on Science, U.S. GPO 2000). Transgenic animals, particularly mice, in which mutated genes have been introduced to generate gene "knockouts" are widely used in biomedical research to mimic human diseases (Miesfeld, 1999). And finally, gene therapy using cloned genes, while still in its infancy and not without its problems, may ultimately be used to treat diseases and improve human health (Templeton and Lasic, 2000).

III. TOOLS FOR MOLECULAR CLONING

A. Restriction Enzymes

The primary tools for molecular cloning are a set of restriction endonucleases and other enzymes that allow the researcher to manipulate DNA in a test tube. Molecular cloning is in essence the ability to alter DNA in defined ways so that it can be introduced into vectors (see below) for cloning. Restriction endonucleases, or restriction enzymes in the vernacular, are endonucleases that cleave DNA at specific base sequences. According to information from New England Biolabs, over 10,000 bacteria and archebacteria have been screened for restriction enzymes, and more than 3000 different enzymes containing more than 200 different sequence recognition specificites have been found (New England Bio-labs Catalog, 2001). Restriction enzymes fall into one of four classes, types I, II, IIS, and III, based primarily on the type of recognition/cleavage specificity or the co-factors required for their activity (see the chapter on restriction-modification by Blumenthal and Cheng, this volume). Most characterized restriction enzymes, and certainly the most useful ones for molecular cloning, belong to type II or IIS. Type II restriction enzymes recognize symmetric DNA sequences or palindromes, and cleave within the recognition site generating 3'OH and 5'P ends. The recognition specificities are generally four, six, or eight base pairs, with six being the most common (Table 1). There are a few enzymes with five or seven base pair recognition sequences where a central nonasymmetric base pair is surrounded by four or six asymmetric base pairs. Type IIS enzymes also recognize specific base sequences, but cleave the DNA some distance away (up to 20 bp) from the recognition sequence.

Restriction endonucleases cleave DNA at specific recognition sites by making either symmetric or asymmetric cleavages (Fig. 1). Any particular restriction enzyme will make one or the other type of cut. The resulting DNA ends differ. Asymmetrically cleaved DNA from a staggered cut produces DNA ends with short single-strand overhangs (Fig. 1). Because of the palindromic nature of the cleavage/recognition sites, the single-stranded overhangs are complementary and capable of base pairing. These DNAs are sometimes said to have "sticky ends." The sticky ends allow two different DNAs cleaved with the same restriction enzyme to anneal and be easily joined by DNA ligase to form a chimeric or recombinant molecule (Fig. 2). Two different restriction enzymes cleaving at different recognition sites, will allow two cleaved DNAs to be joined in a specific orientation. This is an important feature for directional cloning that inserts one DNA into another in a particular orientation. There are also restriction enzymes that cleave symmetrically, leaving blunt ends that contain no single-strand overhang. These molecules can also be ligated to give a recombinant DNA, although ligation is somewhat more difficult, requiring specific ligases, and directional cloning is not possible. Still there are circumstances where blunt ends are required in a cloning experiment. For example, if the two DNAs to be joined were, out of necessity, cleaved with different restriction enzymes, they would generate different, noncomplementary single-stranded ends. The DNA ends could be made blunt by filling in nucleotides using a DNA polymerase or cleaving off the single-stranded overhangs with a nuclease. In practice, DNA polymerase from the

TABLE 1. Restriction Enzymes and Their Recognition Sequences

Restriction Enzyme	Recognition Sequence
*Eco*RI	G'AATTC
Hind III	A'AGCTT
Hha I	GCG'C
Stu I	AGG'CCT
Fse I	GGCCGG'CC

Note: The (') represents the site of cleavage. Only one strand of DNA is shown, written in the 5' to 3' direction.

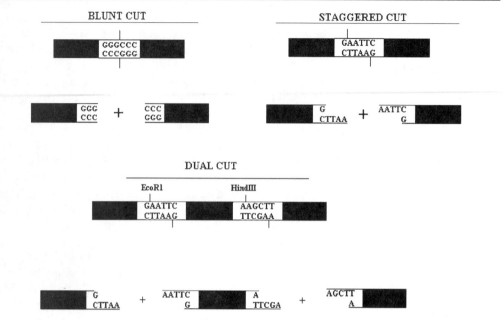

Fig. 1. Types of restriction enzyme cleavages. Staggered cuts leave single-stranded complementary or "sticky ends," blunt cuts leave blunt ends, and dual cuts cleaved with two different restriction enzymes leave single-stranded overhangs with different sequences; dual cuts are useful for directional cloning.

bacteriophage T4 can do either reaction, filling in a 3′ recessed end with its 5′ to 3′ polymerase or clipping off a 3′ overhang with its 3′ to 5′ proofreading exonuclease. The "Klenow" fragment of DNA polymerase I can also be used to fill in an overhanging site. These enzymes generate blunt ends capable of blunt end ligation to make a recombinant DNA.

B. Vectors

One of the DNAs used to construct a recombinant DNA molecule for a molecular cloning experiment is a vector. Vectors are capable of carrying some other "foreign" DNA into a bacterial cell where it can be replicated, and in some cases expressed. The primary properties of a vector are (1), that it can be replicated in a suitable host, (2) that it is able to carry a sufficient amounts of "foreign" or "stuffer" DNA to be useful, and (3) that it contains some type of selectable marker to facilitate selection of bacteria containing the recombinant DNA. In addition, most modern vectors have "multiple-cloning sites" containing restriction enzyme recognition sequences for a number of commercially available restriction enzymes (Table 1). This provides flexibility for the molecular biologist in choosing restriction enzymes to cleave the DNA they are trying to clone. Many vectors also contain observable or selectable markers that allow a recombinant vector to be distinguished from a nonrecombinant vector. This property facilitates molecular cloning by minimizing background transformation resulting from undigested or religated vector, and is of great practical importance. Such markers include things like interruption of a lacZ gene by introduction of a recombinant DNA, or interruption of a bacterial suicide gene by introduction of a recombinant DNA. The former allows recombinants to be identified by blue-white screening, that is, growing transformants in the presence of substrate for β-galactosidase that turns a colony blue when a functional lacZ is produced but does not if the lacZ gene has been interrupted by a recombinant DNA. The latter

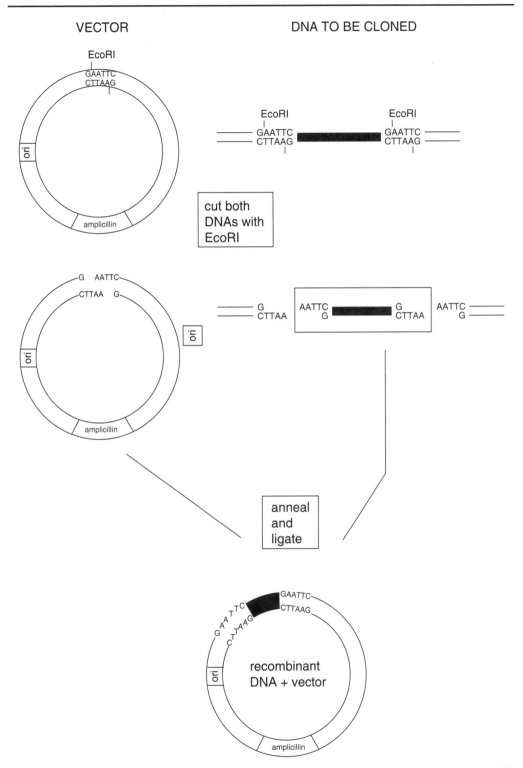

Fig. 2. Generation of a recombinant DNA. The vector DNA carries a unique EcoRI site ampiclillin resistance gene and an origin of replication.

allow recombinants to be selected since interruption of a suicide gene such as *ccdB* (control of cell death), which allows bacteria transformed with a recombinant vector to survive, while any transformants with an uninterrupted, nonrecombinant vector would express *ccdB* and be killed (see *Current Protocols in Molecular Biology*, Ausubel et al., Vol. 1, 2001).

Plasmids are undoubtedly the most widely used of the bacterial cloning vectors. They are relatively small and easy to isolate and manipulate in vitro. They contain autonomous bacterial origins of replication and natural or genetically engineered genes for resistance to antibiotics (usually ampicillin, tetracycline or kanamycin). There is a plethora of plasmid vectors for specialized applications including plasmids with T7 or T3 bacteriophage promoters for making RNA transcripts of cloned genes (pGEM vectors); vectors with specialized translation sites for efficient translation in either bacterial or eukaryotic systems (e.g., pCITE vectors); vectors containing reporter genes like luciferase or β-galactosidase, which are used to clone and examine promoters. All of these are properties that can be incorporated into a cloning strategy.

Bacteriophage can also be used as molecular cloning vectors. Among the most popular of these is the temperate phage λ, in part because its biology has been so widely studied and well understood (see chapter by Hendrix, this volume). DNAs to be cloned can be inserted into regions of the λ genome that are not needed for lytic growth and can therefore be deleted (the middle third of the genome). There are convenient restriction enzyme sites in λ phage vectors for introducing DNAs to be cloned into these regions. Once a linear recombinant λ DNA is made, it is packaged in vitro into phage heads using packaging extracts. The packaging extracts contain empty phage heads, unattached phage tails, and the phage encoded proteins required for DNA packaging. These are made by complementation, combining two separate extracts of bacteria each infected with phage mutants defective in different steps of phage DNA packaging. The final extract is then capable of packaging phage DNA, including recombinant phage DNA, in vitro. One of the advantages of phage are the large number of recombinant clones that can be obtained. A typical packaging extract can generate 2×10^9 plaque forming particles per ug of DNA, compared with 10^8 transformants per μg DNA for plasmid DNA transformation. In addition phage plaques can be generated at high density on a plate, making it technically easier to screen large numbers of recombinants to find a gene of interest (see Section IV below). A second advantage over plasmids is the ability to clone larger pieces of DNA. Most commonly used plasmids are size restricted and in general can accommodate less than 9 to 10 kb of insert DNA. The λ phage are also size restricted, but they can generally accommodate up to 20 kb of DNA Clearly, any effort to clone a large gene as a single piece of DNA requires the use of vectors that can accomodate such DNAs. There are other bacteriophage vectors that are also used in molecular cloning, often for very specific applications. For example, M13 phage is used for cloning DNAs to be sequenced. Historically this was because the M13 is a single-stranded DNA phage, and single-stranded DNAs were easier to sequence (see the chapter on ssDNA phages by LeClerc, this volume). The phage DNA does go through a transient replicative form that allows in vitro manipulation, namely insertion of DNAs to be cloned, but the final form packaged in the virus is single stranded.

Cosmids combine some of the useful features of both bacteriophage and plasmid vectors. They were developed to allow cloning of even larger pieces of DNA (up to 40–50 kb), and in essence represent a delivery system for large recombinant molecules. These vectors contain antibiotic resistance genes and origins of replication (like plasmids), but they also contain *cos* sites that allow them to be packaged in vitro into phage heads for infection. When the cosmid-containing phage infects a bacterium, it injects its linear DNA into the cell where, because of its "sticky ends," it circularizes and behaves as a very large plasmid, there by conferring antibiotic resistance and replicating autonomously (*Current Protocols in Molecular Biology*, Ausubel et al. Vol. 1, 2001).

Bacterial artificial chromosomes (BACs) are vectors capable of carrying very large fragments of DNA, up to 500 kb. They are plasmids that contain elements of the low copy-number F factor replicator (Shizuya et. al., 1992). The F factor replicator has several essential genes, *parA, parB* and *parC oriS, and repE*. The *parABC* genes maintain a low copy number, while *oriS* and *repE* are required for replication of the plasmid DNA. An example of such a BAC vector is pBelo-BAC11. In addition to genes for the required F factor replicator, this plasmid contains a selectable marker for chloramphenicol and a cloning site with unique HindIII, BamH1 restriction enzyme sites within the *lacZ* gene for blue-white screening. These sites are flanked by a Not1 restriction enzyme site that allows easy removal of the inserted DNA. BACs are commonly used as cloning vectors for large genome sequencing projects, like the Human Genome Project.

C. DNA Libraries

It is important to understand that generating recombinant DNAs in vitro is only the first step in molecular cloning. When DNA from an entire organism or cell is cleaved with a restriction enzyme and ligated to a vector in a test tube, what results is a collection of recombinant DNA molecules and not a single molecular species. Such a collection is called a DNA library. To illustrate, let us suppose genomic DNA from *Mycobacterium tuberculosis* was cleaved with the restriction enzyme EcoRI, which recognizes the palindromic hexanucleotide sequence GAATTC. One would expect to generate some 1100 DNA fragments from such a digest, based on the expected random frequency of finding GAATTC in a genome the size of DNA

from *M. tuberculosis*. To clone a particular fragment, the DNA digest would be mixed with an EcoRI cut vector and ligated in a single reaction to generate 1100 or so different vector/insert DNA molecules. This represents the DNA library. In order to sustain the library, the DNAs would be transformed in *E. coli* such that each transformant would contain a single recombinant plasmid. This would represent an *M. tuberculosis* genomic library in *E. coli*. Generating a DNA library is relatively straightforward. Screening the library to identify the clone of interest is not and usually represents the real work of molecular cloning. Screening requires unique cloning strategies designed for the particular gene or DNA of interest.

IV. SCREENING STRATEGIES

The key to effective screening is to devise a unique strategy that will be highly selective for the target DNA. The strategies that have been used are diverse, but they can generally be divided into functional screening and screening based on DNA sequence similarities.

An important question to consider before embarking on a cloning project, is how many clones need to be screened to identify the one of interest. Such a value cannot be calculated precisely. However, a good approximation can be made by considering the average size of the DNA fragments in the library being screened and the overall size of the genome from the source of the DNA being cloned. If we return to our example of *M. tuberculosis*, its genome size is 4.1×10^6 bp of DNA. If the fragments of DNA in the library were on average 4 kb (4000 bp), then in theory 1025 clones ($4.1 \times 10^6/4000$) would be needed to ensure that all fragments of DNA are in the library. Unfortunately, this is an underestimate of the number of clones actually needed. Clarke and Carbon (1976) have devised a statistical formula to estimate the actual number of clones needed to give a specific probability of finding a clone of interest. For a probability (P) of having a DNA fragment in a library, the number of clones that need to be screened (N) is given by the formula:

$$N = \frac{\ln(1 - P)}{\ln(1 - f)},$$

where f is the average size of the cloned DNA fragments divided by the total genome size in base pairs. Thus, for our example of the *M. tuberculosis* genome of 4.1×10^6 bp and an average size of DNA fragments in the library of 4000 bp, we would have to construct a library containing 4700 clones to have a 99% probability that all DNA fragments will be represented in the library.

A. Screening by Functional Activity

Functional screening is undeniably rapid, and it is the simplest approach to identify clones for a gene of interest. Unfortunately, it is limited to those cases where the target gene imparts some selectable or observable phenotype. Cloning antibiotic resistance genes provides a simple example. When a DNA library containing an antibiotic resistance gene of interest is transformed into a bacterial strain sensitive to the antibiotic, transformants with the resistance gene are easily identified by growth on media containing the specific antibiotic, since bacterial clones surviving the antibiotic treatment must carry the antibiotic resistance gene. There are other such selectable markers. For example, genes involved in a biosynthetic pathway for an essential nutrient could be selected by transformation into bacterial auxotrophs for the nutrient. There are also genes coding for various hydrolytic enzymes for which there are available substrates containing chromophores that change or produce a color when hydrolyzed by the enzyme. This would provide an easily observable if not selectable phenotype to clone the specific hydrolytic gene.

B. Screening with Homologous Genes

In most cases there are not easily observable phenotypes associated with a gene of interest, and other strategies are needed to identify specific clones. One approach relies on DNA sequence similarities for functionally homologous genes that had previously been cloned. This approach is based on the ability of a DNA segment from one gene to anneal or hybridize to DNA from a functionally related, and thus very probably structurally similar, gene. For example, if the goal is to clone a gene from *Mycobacterium bovis* where the orthologous gene from *Mycobacterium tuberculosis* has already been cloned, one could generate a DNA hybridization probe from the *M. tuberculosis* gene that would hybridize to clones made from a *M. bovis* gene library. Assuming that the genes are functionally homologous, there is reasonable chance that they would also be structurally related and have enough sequence similarity that a radiolabeled probe for one gene would hybridize with the other. This does not have to be left solely to chance. Prior to screening the library, a Southern blot of *M. bovis* DNA using the probe from *M. tuberculosis* should be able to determine if there were sufficient structural similarity for the heterologous probe to be used to identify the gene of interest. Once cross-hybridization has been established, a colony hybridization screening approach is used to identify the gene of interest (Fig. 3). *E. coli* transformed with the *M. bovis* gene library is grown on a plate and transferred to nylon filters. The bacteria is lysed on the filter (in situ) with detergent, the DNA denatured with alkali and annealed with a radiolabeled *M. tuberculosis* probe. After washing the filter to remove any nonspecifically bound probe, the filters are exposed to X-ray film or a phosphorimager to identify clones carrying the *M. bovis* gene homologous to the *M. tuberculosis* probe. This approach is widely used to clone genes from related organisms, where functional or structural information about the genes is known.

C. Using Proteomics

There are circumstances where homologous genes have not been cloned and approaches such as those described above are not possible. In such cases it is still possible to clone a specific gene using structural information about its corresponding protein. This requires that a part of the amino acid

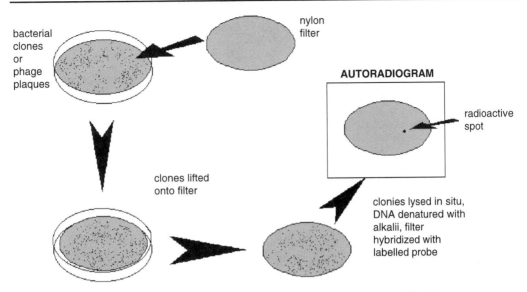

Fig. 3. Screening recombinant DNA libraries by colony or plaque hybridization. The library is grown on agar plates and lifted onto nylon filters. Bacteria or phage on the filters are lysed in situ and the DNA denatured with alkali. The filters are hybridized with radiolabeled probes, and after washing to remove nonspecific hybridization, the filters are exposed to X-ray film or a phosphoimager.

sequence of the protein be known, and with rapid advances in proteomics, obtaining partial amino acid sequences is straightforward. In general, proteins (the proteome) can be separated by two-dimensional electrophoresis. The spots associated with a particular protein are excised from the gel and subjected to partial protein sequence analysis by mass spectrometry (MALDI-TOF). The partial amino acid sequences can then be reverse-translated into oligonucleotide sequences using the genetic code. For example, the amino acid sequence [met-phe-asn-cys-trp] could be reverse-translated into the DNA coding sequence [ATGTT(T/C)AA (T/C)TG(T/C)TGG]. There is ambiguity (the bases in parentheses) associated with this because of degeneracy in the genetic code. However, it is not difficult to make a set of degenerate oligonucleotides from the amino acid sequence information. Although this generates a mixture of oligonucleotides, presumably the correct one is contained within that mixture. Oligonucleotides can be used directly as probes if sufficient sequence infor-

mation is available. In theory, an oligonucleotide of 14 to 15 residues, representing five amino acids in the contiguous protein sequence, will be unique in a genome as complex as the human genome. Unfortunately, this estimate presupposes that DNA sequences are random, which they are not, and in practice, oligonucleotides as short as 15 are usually not gene specific. An added level of specificity can be obtained by using degenerate oligonucleotides derived from protein sequence information to generate a PCR product from either genomic or cDNA (see below). The PCR product can also be used to probe a DNA library and can be sequenced directly to provide structural information verifying the PCR product as an authentic gene probe.

D. Screening by Linkage Using BAC (Bacterial Artificial Chromosome) Libraries

Linkage analysis has been a standard genetic tool for many years and physical linkage maps of many genomes are now available.

This information can be used to clone genes. If one gene or piece of DNA (i.e., a DNA marker) is closely linked physically to another that has already been cloned, then screening a DNA library with the known gene can identify a clone that contains both the known and unknown genes (Fig. 4). This approach generally requires that the DNA library contain very large fragments of DNA. One would encounter such large fragments in cosmid or BAC libraries. In fact BAC libraries have been widely used in such approaches, and the recent announcement that the human genome has been sequenced is in large part due to the identification of overlapping BAC clones that cover virtually all of the human genome. Computer analysis of sequence data from overlapping BAC clones allows long pieces of DNA sequence to be assembled into contiguous segments (or contigs). Individual clones with DNAs of interest can be subsequently identified and subcloned into smaller vectors.

Another very useful cloning approach, which was used extensively to solve the human genome, is the generation of sequence tagged sites (STSs). This shotgun approach uses rapid high-throughput sequencing to solve genome structures. The approach relies on having randomly overlapping pieces of DNA in a library. Clones are not identified functionally or by any sort of screening but simply randomly selected and sequenced. Any such sequenced piece of

DNA can be mapped to a chromosomal site, generating a new STS. The STS then becomes a DNA marker that can be used to identify and clone new closely linked genes. The more DNA markers there are along a chromosome, the easier it is to find and clone new genes.

E. cDNA Cloning

cDNAs are DNA copies of RNA constructed in vitro by using the retroviral enzyme, reverse transcriptase. They differ from genomic DNA because they represent only expressed genes. In addition, because they are copies of RNA transcripts, they do not contain promoter sequences or other transcriptional regulatory sites not transcribed into RNA. Nevertheless, there are advantages to using cDNAs for cloning. Notably it reduces the number of clones that need to be screened to identify a target gene of interest. This is because most cells do not express all genes at all times, so in theory, fewer clones need to be screened to identify a gene known to be expressed in the cell used as a source for cDNA. This is more important in eukaryotic systems, where the amount of nontranscribed genomic DNA is much greater than in prokaryotic systems.

Just as STSs are randomly sequenced pieces of genomic DNA, libraries of cDNA can also be randomly sequenced to generate ESTs (expressed sequence tags). These are similar to STSs, but they represent only expressed genes. ESTs can be localized to a chromosomal site pinpointing an expressed gene to a particular locus. This is particularly advantageous in complex eukaryotes where much of the genomic DNA does not encode genes.

V. SPECIAL CONSIDERATIONS

A. Transposons as Cloning Tools

Transposons are mobile genetic elements that have the ability to insert randomly into a genome. They have been best characterized in bacteria (see the chapter on transposons by Whittle and Salyers; this volume), yeast, and *Drosophila*, but they occur in virtually

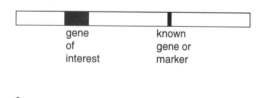

gene   known
of   gene or
interest   marker

size of cloned DNA fragment

Fig. 4. Cloning by linkage. If a known gene or DNA segment is linked on the same piece of a DNA clone as another gene of interest, screening for the known gene will identify the clones containing the new gene of interest.

all organisms. Transposons can be of tremendous value in mapping and cloning new genes. The insertion of a transposon into a gene generally inactivates the gene and provides a fixed reference point for its cloning. The inactivated gene is identified by loss of a phenotype and can be cloned by using the transposon DNA as a probe to screen a library. Alternatively, because many transposons encode antibiotic resistance genes, a transposon inserted into a genome can be excised with restriction enzymes, circularized either directly or in a vector, and transformed into antibiotic-sensitive bacteria. Selecting for antibiotic resistance then identifies transformants carrying the transposon and any flanking DNA that it picked up during the restriction enzyme digestion.

When the restriction map of a transposon is known, it is possible to generate a map of the genomic regions flanking the transposon (Fig. 5). A particular restriction enzyme may cut only once within the transposon. Digesting genomic DNA containing the inserted transposon will yield two fragments carrying portions of the transposon linked to host genomic DNA. These fragments would be the proximal portions of the transposon and genomic DNA up to the first location of the same restriction enzyme site and the distal portion of the transposon linked to genomic DNA up to the next genomic site for the enzyme. Different restriction enzymes can be used to construct a detailed map of the chromosomal region flanking the transposon. This information can be used in

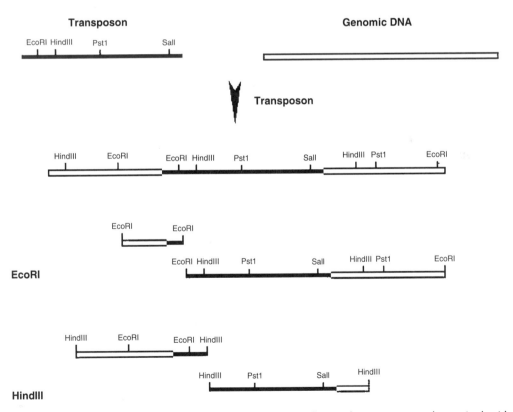

Fig. 5. Transposon mediated cloning. A transposon inserted into the genome can be excised with restriction enzymes that will cut the DNA both inside the transposon and outside of it so that the resulting DNA fragments contains DNA sequences that flank the inserted transposon.

designing strategies to clone DNA flanking the inserted transposon.

B. Phage Display

Phage display and phage-based interaction cloning offer powerful tools to identify interacting proteins and clone their corresponding cDNAs. It was initially designed and used to identify receptor protein kinases (Skolnik et al., 1991). However, phage display has been adapted to screen a large variety of interacting proteins from transcription factors (Blanar et al., 1992) to protein kinase C substrates (Chapline et al., 1993) to proteins that interact with tumor supressors (Kaelin et al., 1994). Phage display starts with a known protein called the *bait*, which is suspected of interacting with one or more target proteins. A cDNA phage expression library can be screened using a radiaolabled bait protein to identify those phage particles expressing cDNA for proteins capable of interacting with the bait. The phage library is constructed such that target protein cDNAs are expressed as fusion proteins with a phage capsid protein (e.g., T7 gene10 major capsid protein). This directs target proteins to the surface of the phage particle allowing several options for screening. For traditional plaque screening (as in Fig. 3), the phage plaques are transferred to a nylon

membrane, and since the proteins of interest are expressed on the surface of the phage particle, the phage need not be lysed. The radiolabeled bait protein will bind to the immobilized phage expressing an appropriate fusion protein. Purification of the plaque results in cloning the cDNA for the target protein. Phage expressing an interacting protein can also be enriched by using a biopanning approach (Fig. 6). Here the bait protein is immobilized on a solid support (e.g., an ELISA plate) and an amplified phage library is allowed to adhere to the immobilized bait. After washing to remove phage nonspecifically associated with the surface, the specifically bound phage can be eluted with a mildly chaotropic agent (guanidinuim salts or urea), amplified through another round of infection and subjected to a second round of selection. In practice, three to four rounds of selection are sufficient to ensure that all the selected phage are expressing proteins able to interact with the bait. This powerful methodology can be adapted in any number of ways to study interacting protein systems. For example, random mutagenesis of a target protein can be used to study how mutations affect interacting proteins. In this case the phage library could be constructed to contain of a collection of randomly mutated

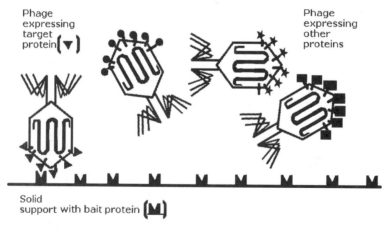

Fig. 6. Bio-panning a phage display library. Phage expressing a target protein can be isolated by binding to a bait protein immobilized on a solid support like an ELISA plate.

target proteins that could be rapidly screened for their ability or inability to interact with the bait protein. This approach can define portions of the target protein critical for interacting with the bait. Phage display has also been suggested as an alternative to monoclonal antibody production (for review, see Winters et al. 1994). In theory, a phage library expressing fragments of antibody molecules capable of recognizing antigens could be screened with an antigen. The appropriate phage clone can then be used for production of the antibody fragment.

VI. YEAST TWO-HYBRID SYSTEMS

The yeast two-hybrid system is a second method to study interacting proteins, using the yeast *Saccromyces cerevisiae*. Yeast two-hybrid systems take advantage of the need for interacting protein molecules to drive yeast transcription. To understand how this method works requires a brief look at eukaryotic transcriptional regulation. In most eukaryotes, transcription is regulated by sets of gene-specific regulatory factors that bind unique *cis*-acting DNA elements and then recruit more general factors that help assemble an active transcription complex. This recruitment requires protein-protein interactions. In many cases transcriptional regulatory proteins contain distinct modules, protein structural domains that carry out different functions. For example, the proteins might contain a DNA-binding domain needed to bind a specific DNA sequence and a separate activation domain involved in recruiting other protein factors to activate transcription at the particular locus. Often such domains are modular and can be removed and replaced with domains from other proteins. This was first demonstrated by Brent and Ptashne (1985) who fused the DNA binding domain of LexA (see the chapter on DNA repair by Yasbin, this volume) to the activation domain of the yeast GAL4 transcription factor and showed functional transcriptional activation from a gene containing the LexA DNA-binding site. The key

point is that protein-protein interactions are necessary for recruitment of the factors needed for transcription. The general strategy in a yeast two-hybrid systems is to use this requirement for protein-protein interactions to activate expression of a reporter gene. Figure 7 illustrates this approach. If one links the DNA binding domain of a specific transcription factor to a bait protein, then activation of a reporter gene containing the appropriate DNA recognition sequence would occur only if a target protein capable of interacting with the bait were fused to an activation domain for another transcription factor (Fig. 7A). Note that the bait protein itself need not be a transcription factor; it is simply fused to the DNA-binding domain of one. The only requirements for the bait protein in fact are that it not be actively excluded from the yeast nucleus and that it not be capable of activating transcription by itself. In one version of this method (Fig. 7B), a plasmid is constructed to code for a fusion protein between the bait and the DNA-binding domain of Lex A (vectors to do this are commercially available). This fusion protein would bind to a reporter gene containing the LexA operator in front of a LacZ reporter gene. Activation of LacZ would then be dependent on the bait protein interacting with a second fusion protein containing a transcriptional activation domain. Since these domains are modular, such a protein could be made by cloning cDNAs in frame with the activation domain of the GAL1 transcription factor (vectors for making such a fusion library are also available commercially). Interaction between the bait fusion protein and an interacting protein fused to the GAL1-activation-domain would recruit transcription initiation factors to the site and activate transcription of the reporter gene. Yeast cells expressing cDNA for such an interacting protein would turn blue when grown in the presence of X-gal. In practice, there is significant background associated with spurious activation of lacZ. The chances of identifying specific cDNAs for an interacting proteins are improved if

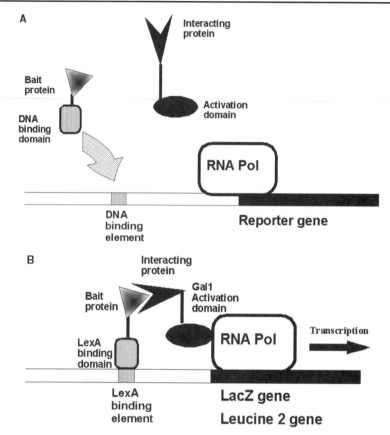

Fig. 7. Yeast two-hybrid screening. **A:** General scheme showing a bait protein fused to a DNA-binding domain. As the bait binds to its DNA recognition sequence, it recruits an interacting partner fused to an activation domain able to activate transcription of a reporter gene. **B:** Example of the lexA DNA-binding domain, gal1 activation domain, and a lacZ reporter gene. The leucine 2 gene is used to increase specificity.

the yeast also contain a selectable marker activated by the protein-protein interaction. Typically a chromosomal copy of the gene for leu2, which has been modified to contain the lexA site in place of its normal activation sites, is used. Such a yeast strain is auxotrophic for leucine in the absence of lexA binding gene-activating complex. By selecting for growth on leucine minus plates, and for colonies that turn blue in the presence of Xgal, one can improve the chances of identifying interacting proteins. Such yeast clones would contain a plasmid-bourne copy of a cDNA for the protein that interacts with the bait. There are many modifications of these methods, using different DNA-binding domains and different activation

domains, but all have in common the requirement that protein-protein interaction is required to activate expression of a reporter gene. It is important to realize that any type of interaction cloning, be it phage display, yeast two-hybrid, or other modifications, is qualitative and not a quantitative representation of protein-interacting systems. There are limits to these systems based on both the affinity of the two proteins for each other and the kinetics of their interaction.

REFERENCES

Ausbel FM, Brent R, Kingston RE, Moore DD, Seidman JG, Smith JA, Struhl K (eds) (2001): "Current Protocols in Molecular Biology" Vol 1. New York: Wiley.

Brent R, Ptashne M (1985): A eukaryotic transcriptional activator bearing the DNA specificity of a prokaryotic repressor. Cell 43:729–736.

Clark L, Carbon J (1976): A colony bank containing synthetic Col E1 hybrid plasmids representative of the entire *E. coli* genome Cell 9:91–99.

Miesfeld RL (1999): "Applied Molecular Genetics" New York: Wiley.

Nathans D, Smith HO (1975): Restriction endonucleases in the analysis and restructuring of DNA molecules. *Ann Rev Biochem* 44:273–293.

Smyth-Templeton N, Lasic DD (eds) (2000): "Gene therapy: Therapeutic Mechanisms and Strategies". New York: Dekker.

US Environmental Protection Agency. Office of Solid Waste and Emergency Response, Technology Innovation Office (2000): "Engineered Approaches to in Situ Bioremediation of Chlorinated Solvents: Fundamentals and Field Applications". Washington, DC: GPO.

US Congress. House, Committee on Science, Subcommittee on Basic Research (2000): Plant genome science : From the lab to the field to the market, parts I-III. Hearings before the Subcommittee on Basic Research of the Committee on Science, House of Representatives, One Hundred Sixth Congress, first session, August 3, October 5, and October 19, 1999. Washington, DC: GPO.

Section 2: GENETIC RESPONSE

10

Genetics of Quorum Sensing Circuitry in *Pseudomonas aeruginosa*: Implications for Control of Pathogenesis, Biofilm Formation, and Antibiotic/Biocide Resistance

DANIEL J. HASSETT, URS A. OCHSNER, TERESA DE KIEVIT, BARBARA H. IGLEWSKI, LUCIANO PASSADOR, THOMAS S. LIVINGHOUSE, TIMOTHY R. MCDERMOTT, JOHN J. ROWE, AND JEFFREY A. WHITSETT

Department of Molecular Genetics, Biochemistry, and Microbiology, University of Cincinnati, College of Medicine, Cincinnati, Ohio 45267–0524; Department of Microbiology, University of Colorado Health Sciences Center, Denver, Colorado 80262; Department of Microbiology and Immunology, University of Rochester School of Medicine, Rochester, New York 14642; Department of Chemistry and Biochemistry, Montana State University, Bozeman, Montana 59717; Department of Land Resources and Environmental Sciences, Montana State University, Bozeman, Montana 59717; Department of Biology, University of Dayton, Dayton, Ohio 45469; Division of Pulmonary Biology, Children's Hospital Medical Center, Cincinnati, Ohio 45229–3039

I. INTRODUCTION

Bacterial quorum sensing (QS) is a cell density–dependent form of cell-to-cell communication, utilized by a number of gram-negative and gram-positive bacteria, whereby low molecular weight diffusible molecules synthesized by one organism trigger gene activation in others (Greenberg, 1997). In Gram-negative bacteria, the major signaling molecules, or autoinducers, are acylated homoserine lactones (HSL). However, alternative signaling molecules exist. For example, diketopiperazines (DKPs)- (26), 3-hydroxy-palmitic acid methyl ester (Flavier et al., 1997), butyrolactones (Gamard et al., 1997),

Modern Microbial Genetics, Second Edition. Edited by Uldis N. Streips and Ronald E. Yasbin. ISBN 0–471–38665–0
Copyright © 2002 Wiley-Liss, Inc.

and 2-heptyl-3-hydroxy-4-quinolone (Pesci et al., 1999) have been identified as signal molecules in Gram-negative QS systems. In contrast, Gram-positive bacteria typically make use of small peptide signals that interact with a two-component histidine kinase signal transduction system. It is interesting to note that various *Streptomyces* species utilize butyrolactones in their QS process (Horinouchi and Beppu, 1994).

Pseudomonas aeruginosa is ubiquitous in nature. It is an important opportunistic pathogen of humans, especially in chronic lung infection of individuals afflicted with cystic fibrosis (CF). The organism is capable of forming biofilms on multiple surfaces including CF and urinary tract epithelial cells, catheters, and industrial piping systems. Typically such biofilms have proved to be highly refractory to antibiotic/biocide treatment. Multiple *P. aeruginosa* genes involved in animal, plant, and nematode virulence, biofilm architecture and resistance to hydrogen peroxide (H_2O_2), are controlled by QS. Given the myriad of genes regulated by QS in *P. aeruginosa*, it is an excellent model organism for the study of these communication/ regulatory systems. The goal of this chapter is to reflect on new discoveries in QS circuitry and to map a course for future evolution of the field, especially in the design of QS structural analogues that could inhibit such processes.

II. HSL-BASED SIGNALING

Because most of the information about QS genetics in *P. aeruginosa* concerns HSL-based autoinducers, we will focus our efforts in this chapter on the machinery of these systems. HSL-based QS in *P. aeruginosa* is a multi-tiered process governed by two gene tandems, *lasRlasI* and *rhlRrhlI* (Passador et al., 1993; Pearson et al., 1994, 1995). When cell populations of *P. aeruginosa* approach the stationary growth phase (i.e., when cell densities are high), transcription of QS-regulated genes is activated. The *las* system is composed of LasR, a transcriptional activator, and LasI, an autoinducer

synthase that produces one of two primary *Pseudomonas* HSL's (*N*-(3-oxododecanoyl)-L-homoserine lactone; 3O-C_{12}-HSL). The second tier consists of RhlR, which, like LasR, is a transcriptional activator, and RhlI, an autoinducer synthase that catalyzes the synthesis of the second predominant HSL (*N*-butyryl-L-homoserine lactone; C_4-HSL). The *lasRlasI* system has been shown to activate expression of *lasI*, *lasB*, *lasA*, *apr*, and *toxA* (Pesci et al., 1997). Similarly the *rhlRrhlI* system is known to activate *rhlI*, *rhlAB*, *lasB*, and *rpoS* and loci involved in chitinase, casein protease, staphylolytic activity and pyocyanin production (Brint and Ohman, 1995; Latifi et al., 1996; Pesci et al., 1997; Winson et al., 1995). A recent comprehensive work by Greenberg and colleagues using Tn5-B22 mutagenesis has identified approximately 25 additional QS-regulated genes (Whiteley et al., 1999). A synopsis of the hierarchy of genes under QS control are depicted in Figure. 1 and does not include those found by Greenberg and colleagues due to complexity.

Vfr (Albus et al., 1997), a cyclic AMP repressor protein (CRP) homolog of the *Escherichia coli* global transcriptional regulator CAP (West et al., 1994) appears to positively control expression of *lasRlasI*, and thus *rhlRrhlI*. Organisms devoid of Vfr are unable to produce the virulence determinants exotoxin A and protease (West et al., 1994). More recently the global response regulator GacA, which along with GacS forms a two-component sensor-kinase regulator, was shown to positively control expression of *lasR* and *rhlR*, as well as the production of C4-HSL and several other virulence determinants including pyocyanin, cyanide, and lipase (Reimmann et al., 1997). A recently described negative regulator, RsaL, appears to inhibit the expression of *lasI* resulting in a decrease in production of 3O-C_{12}-HSL and a concomitant decrease in the production of various virulence factors (De Kiewit and Iglewshi, 1999). Thus there is a mounting body of literature of not only genes controlled by QS but of the regulators

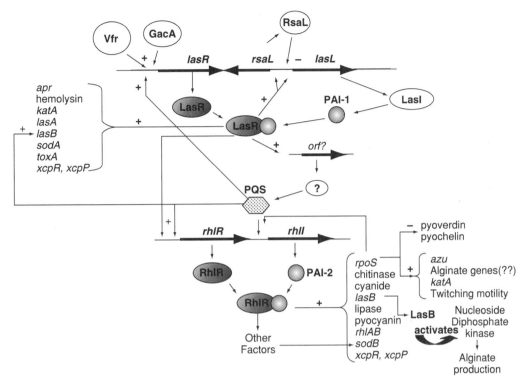

Fig. I. Schematic diagram outlining the tier of QS circuitry in *P. aeruginosa* and the genes under *las* and *rhl* control. Gene products; *apr*, alkaline protease, *azu*, azurin; *katA*, the major, constitutive catalase, KatA; *lasA*, LasA protease; *lasB*, elastase; *lasI*, 3O-C$_{12}$-HSL autoinducer synthase; *lasR*, LasR transcriptional activator; PQS, *Pseudomonas* quinolone signal; *rhlAB*; rhamnolipid biosynthesis; *rhlI*, C$_4$-HSL autoinducer synthase; *rhlR*, RhlR transcriptional activator; *rpoS*, stationary phase sigma factor; *rsaL*, repressor of *lasI* transcription, RsaL; *sodA*, Mn-superoxide dismutase; *sodB*, Fe-superoxide dismutase; *toxA*, exotoxin A; *xcpRP*; Xcp translocation machinery. The LasB protease is required for intracellular cleavage and activation of nucleoside diphosphate kinase, which is required for alginate production.

controlling such processes. Finally, QS has gained much of its notoriety because functional systems have been documented in over 35 different bacteria. Not surprisingly, there is now a Web site available that attempts to provide the most current information (*www.nottingham.ac.uk/quorum*).

III. USE OF MULTIPLE HSL MOLECULES BY LASR AND RHLR

LasR and RhlR require their cognate HSL's, 3O-C$_{12}$-HSL and C$_4$-HSL, for optimal transcriptional activation of target genes. All of the autoinducer synthase molecules thus far

examined have been found to use *S*-adenosyl-methionine (SAM) and the appropriate fatty acid conjugated to acyl carrier protein as substrates (Hoang et al., 1999; More et al., 1996; Parsek et al., 1999; Schaefer et al., 1996). Studies using purified RhlI have shown that in addition to generating *N*-butyryl-ʟ-homoserine lactone from butyryl-ACP and SAM, it can also generate *N*-hexanoyl-ʟ-homoserine lactone and *N*-octanoyl-ʟ-homoserine lactone when hexanoyl-and octanoyl-ACL are provided as substrates, together with SAM (Hoang et al., 1999; Parsek et al., 1999). Thus RhlI can generate HSL's with acyl side chains of 4, 6, or 8 carbons, although the rate of synthesis of the C$_6$-and C$_8$-HSL molecules is lower than the rate of C_4-HSL (Parsek et al., 1999). Such findings are consistent with the fact that in addition to $3O$-C_{12}-HSL and C_4-HSL, minor HSL products can be detected in *P. aeruginosa* culture supernatants (Jones et al., 1993; Pearson et al., 1994). At present the exact purpose of these minor HSL's is unclear. One possible role for these compounds may be to "fine-tune" the QS circuitry. Noncognate HSLs can frequently activate a given R-protein, but at lower induction levels than the cognate HSL. In this manner, the minor HSL products found in *P. aeruginosa* supernatants may function as negative competitors of autoinduction. This phenomenon has been observed in *P. aeruginosa*, where the *las* signal molecule $3O$-C_{12}-HSL is able to effectively compete with C_4-HSL for RhlR binding (Pesci et al., 1997). Similarly in *V. fischeri*, a second HSL synthase, AinS, directs the synthesis of C_8-HSL, which appears to function as a competitive inhibitor of *V. fischeri* bioluminescence (Kuo et al., 1996).

A. A Third Autoinducer Is a 4-Quinolone

More recently, a third structurally-distinct signaling molecule (2-heptyl-3-hydroxy-4-quinolone) was described and called PQS (*Pseudomonas* quinolone signal) (Pesci et al., 1999). Preliminary studies have shown that PQS regulates *lasB* expression to some degree by a mechanism that remains poorly understood. Production of PQS is dependent on the *las* system, yet RhlR is required for PQS activity. PQS was most recently found to be essential for optimal *rhlI* activation (McKnight et al., 2000). Currently the identity of the putative R-protein with which PQS interacts and its precise role in *P. aeruginosa* QS is unknown.

B. Biofilm Formation and Resistance to Oxidizing Biocides Are Dependent on QS

P. aeruginosa forms highly recalcitrant biofilms on multiple surfaces of clinical, environmental and industrial importance. QS has been implicated in biofilm architecture and resistance to the biocide sodium dodecyl sulfate (SDS) (Davies et al., 1998). Wild-type bacteria grown in the presence of ample and continuous nutrient flow form multilayered biofilms consisting of apparently highly ordered microcolony, mushroom, pillar, and streamerlike structures. In contrast, a mutant (PAO-JP1) deficient in $3O$-C_{12}-HSL production forms a thin, tightly packed biofilm with a thickness of only two to three cell layers, which were easily detached by treatment with SDS. Strains deficient in C_4-HSL production formed biofilms similar to that of the parent strain, indicating that it is the *las* and not the *rhl* system that is important for *P. aeruginosa* biofilm architecture and integrity under the conditions described in that work.

Recently Hassett et al. (1999b) have shown that genes important in the response to oxidative stress are controlled by QS. Oxidizing biocides such as H_2O_2, HOCl, monochloramine and chlorine dioxide are used in the treatment of biofilms (Hassett et al., 1999a). Specifically, *katA*, encoding the constitutively expressed catalase, KatA, requires both the *las* and *rhl* systems for optimal activity. Furthermore the double HSL mutant, PAO-JP2, was exquisitely susceptible to H_2O_2 in biofilms relative to wild-type bacteria. Another gene important in the oxidative stress response is *sodA*, which encodes a manganese-cofactored superoxide dismutase that is important for optimal resistance to superoxide (O_2^-) (Hassett et al., 1995). In contrast to *katA*, QS mediated activation of *sodA* requires only the *las* system (21). However, catalase, at least in *E. coli*, appears to be a very important enzyme for resistance to H_2O_2, hypochlorite, and monochloramine (Dukan and Touati, 1996).

IV. QS AND *P. AERUGINOSA* VIRULENCE

The pivotal role played by QS in *P. aeruginosa* virulence has been established using different animal models. In the neonatal mouse model of pneumonia, *P. aeruginosa lasR, lasI, rhlI,* and *lasIrhlI* mutants were significantly decreased in virulence when compared to the parent strain (Pearson et al., 2000; Tang et al., 1996). Similarly, in a burned mouse model, the above-mentioned mutant strains were found to be less virulent than the parental strain in vivo (Rumbaugh et al., 1999a,b). The total number of bacteria recovered from the spleens, livers, and skin of mice infected with these mutants was also significantly lower, indicating that QS plays an important role in *P. aeruginosa* dissemination throughout the body of burned mice. In another study employing three different infection models, QS was shown to be important for cross-phylum virulence of *P. aeruginosa*. Researchers found that a *lasR* mutant exhibited greatly reduced virulence in *Caenorhabditis elegans* (nematode), *Arabidopsis thaliana* (plant), and a burned mouse model of infection (Tan, 1999). It is interesting to note that *gacA* and *toxA* mutants possessed decreased virulence properties in the three infection models (Rahme et al., 1995, 1997; Tan et al., 1999). GacA, a global activator in *P. aeruginosa*, was previously shown to regulate expression of *lasR*, *rhlR*, and production of C_4-HSL (Reimmann et al., 1997), as well as *toxA* encoding exotoxin A (Gambello et al., 1993). Finally, *P. aeruginosa* QS has also been shown to play a role in the pathogenesis of this organism during human infections. A correlation was observed between *lasR*, *lasA*, *lasB*, and *toxA* transcript accumulation in sputa from the lungs of cystic fibrosis (CF) patients infected with *P. aeruginosa* (Storey et al., 1998). Thus it appears that during chronic lung infection, the *las* quorum-sensing system actively regulates gene expression. Taken together, these findings suggest that strategies designed to interfere with *P. aeruginosa* QS may be an effective means of treating or preventing infections caused by this organism.

V. AUTOINDUCER STRUCTURAL ANALOGUES AS INHIBITORS OR ACTIVATORS OF QS CIRCUITRY

The above-mentioned findings suggest that inhibition of QS in *P. aeruginosa* would decrease or inhibit production of multiple virulence factors, and potentially prevent complex biofilm formation and ultimately resistance to a variety of biocides. One mechanism for blocking QS circuitry that is being aggressively pursued by various researchers in academia and industry makes use of structural analogs of $3O\text{-}C_{12}\text{-}HSL$ or $C_4\text{-}HSL$ that could bind to LasR or RhlR, preventing binding of wild-type autoinducer, thereby inhibiting activation of *las* and *rhl*-controlled genes (Fig. 2B). The identification of HSL analogues that inhibit QS processes is not without precedent. Such studies have been conducted in *Vibrio fischeri* (Eberhard et al., 1986; Schaefer et al., 1996), *Erwinia carotovora* (Chhabra et al., 1993), and *Agrobacterium tumefaciens* (Zhang et al., 1993; Zhu et al., 1998). Except for the potent antagonism reported in the *A. tumefaciens* system (Zhu et al., 1998), no strong antagonists have been identified. In *P. aeruginosa* (Passador et al., 1996) it has been demonstrated that LasR can bind to and be activated by a variety of structural analogues of $3O\text{-}C_{12}\text{-}HSL$. Chain length appeared to be a critical factor in the ability of these compounds to function as HSLs, since molecules having chain lengths of six carbons or less were ineffective at both binding and activating LasR. Nonetheless, LasR was able to bind to and be activated by a number of compounds that differed in structure from $3O\text{-}C_{12}\text{-}HSL$ in chain length, chain substituents, or lactone-ring heteroatom composition. Such findings are in agreement with others mentioned above that suggest this family of transcriptional activator proteins can recognize a variety of compounds. Most important, these studies suggest that the approach to identify antagonists is feasible.

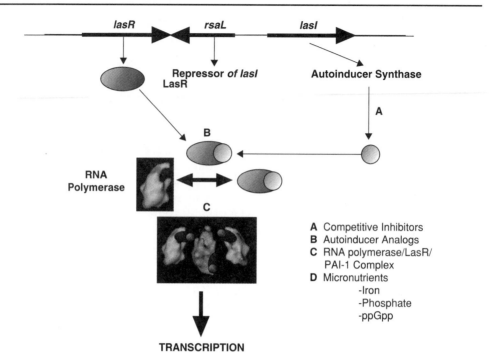

Fig. 2. Diagram showing four possible ways to inhibit the process of QS in *P. aeruginosa* using the *las* system as an example. **A:** Compounds could be designed to act as competitive inhibitors of LasI, the autoinducer synthase that is required for 3O-C$_{12}$-HSL synthesis. **B:** Autoinducer structural analogues (for examples, see Fig. 3) could be designed to inhibit LasR/3O-C$_{12}$-HSL interactions, with the ultimate goal of paralyzing transcription of genes under *las* control. **C:** An uncharacterized interaction between LasR/3O-C$_{12}$-HSL and RNA polymerase is essential for transcriptional activation of *las*-controlled genes. The three-dimensional structure of *E. coli* RNA polymerase holoenzyme was used as a model. For this and other RNA models, see *http://www.rockefeller.edu/labheads/darst/structures.htm*). Because the nature of such interactions is poorly understood on both a mechanistic and molecular level, they offer yet another means to cripple QS. **D:** Many of the genes controlled by QS are under the control of micronutrients (e.g., iron, phosphate). The role of such micronutrients in the process of QS is poorly understood.

LasR null mutants are typically markedly reduced in both elastase and rhamnolipid production. Recent work by van Delden et al. (1998) has demonstrated that the *lasR* mutant, PAO-R1, is capable of reverting to a proteolytically proficient state when grown under high-stress conditions. Attempts to isolate suppressor mutations in *P. aeruginosa* strains deficient in both the *las* and *rhl* systems were unsuccessful. These data indicate that successful antagonist analogues will need to inhibit both the *las* and *rhl* circuits to completely abrogate production of virulence factors and biofilm-related proteins.

Traditionally the R-protein/HSL interaction has been the target for analog studies (Figs. 2A, B). Other areas of the QS mechanism may also prove fruitful as targets. Studies of C$_4$-HSL synthesis by RhlI (Parsek et al., 1999) indicate that analogues of various intermediates in the synthetic pathway can function as strong antagonists. Furthermore, little is known of the molecular basis for interaction between R-protein-HSL complexes and RNA polymerase (Fig. 2C). This final interaction represents yet another feature of QS circuitry that could be targeted for novel drug therapies.

VI. RATIONALE FOR THE DESIGN OF MORE NOVEL AUTOINDUCER ANALOGUES

The structural elucidation of 3O-C$_{12}$-HSL in 1994 (Pearson et al., 1994) stimulated the search for molecular analogues that would possess alternative, or enhanced, bioactivity profiles (Fig. 3). We have focused on two approaches toward fulfilling this objective. In the first of these, electronic perturbation of the C-3 carbonyl function was achieved by substitution of the individual pairs of flanking

hydrogens by fluorine atoms. Although in a dissimilar context, analogous structural modifications were shown to induce significant activity modulations resulting in inhibitory activity in serine protease substrates by Abeles (Gelb et al., 1985; Imperiali and Abeles, 1986). Specific molecular examples of autoinducer analogues in this category include the homoserine lactone derived amides 2a and 2b. In addition the perfluorinated analogue 3b of the natural autoinducer 3a has been synthesized, as has the trifluoroacetyl derivative of homoserine lactone 4. As part of a completely

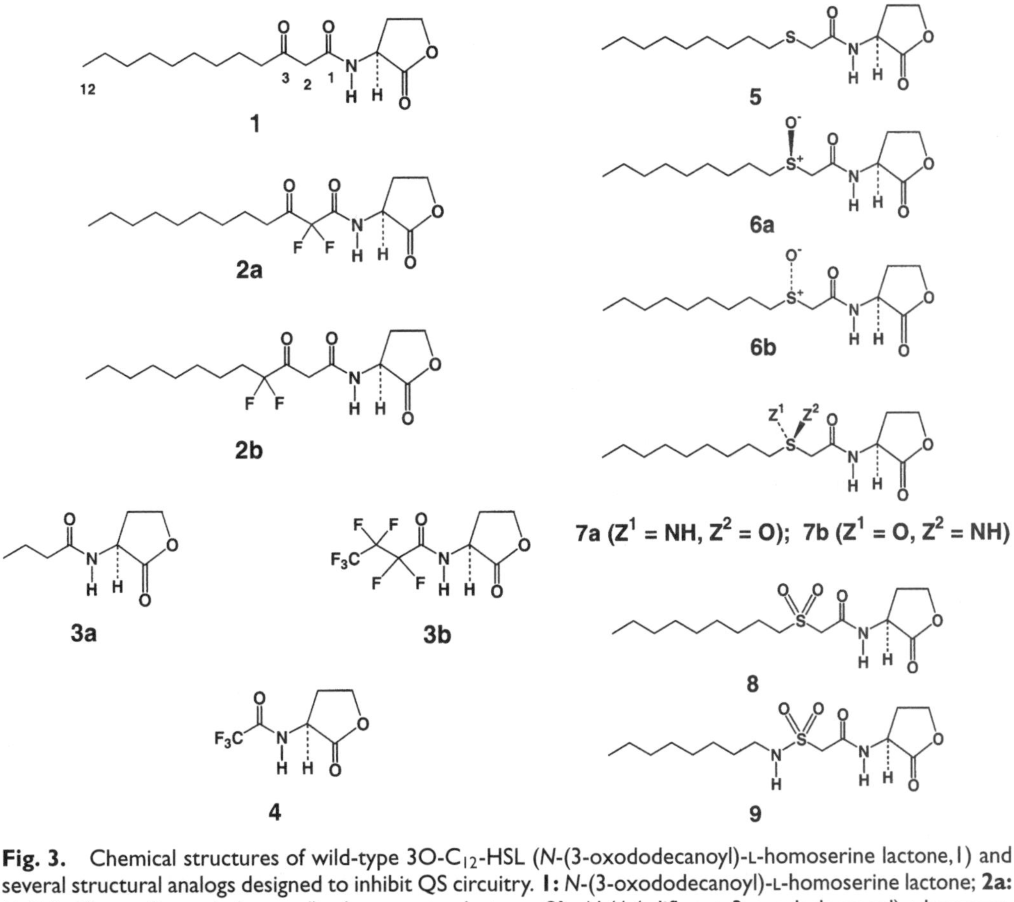

Fig. 3. Chemical structures of wild-type 3O-C$_{12}$-HSL (*N*-(3-oxododecanoyl)-L-homoserine lactone, 1) and several structural analogs designed to inhibit QS circuitry. **1:** *N*-(3-oxododecanoyl)-L-homoserine lactone; **2a:** *N*-(2,2-difluoro-3-oxododecanoyl)-L-homoserine lactone; **2b:** *N*-(4,4-difluoro-3-oxododecanoyl)-L-homoserine lactone; **3a:** *N*-(butanoyl)-L-homoserine lactone; **3b:** *N*-(2,2,3,3,4,4,4-heptafluorobutanoyl)-L-homoserine lactone; **4:** *N*-(2,2,2-trifluoroethanoyl)-L-homoserine lactone; **5:** *N*-[2-(n-nonanylthio)ethanoyl]-L-homoserine lactone; **6a:** *N*-(R)-[2-(n-nonanylsulfinyl)ethanoyl]-L-homoserine lactone; **6b:** *N*-(S)-[2-(n-nonanylsulfinyl)ethanoyl]-L-homoserine lactone; **7a:** *N*-(R)-[(S-nonyl-S-(2-ethanoyl)]-L-homoserine lactone]sulfoximine; **7b:** *N*-(S)-[(S-nonyl-S-(2-ethanoyl)]-L-homoserine lactone]sulfoximine; **8:** *N*-[(nonylsulfonyl)ethanoyl]-L-homoserine lactone; **9:** *N*-[[(nonylamino)sulfonyl]ethanoyl]-L-homoserine lactone.

unrelated approach to analogue design, the C-3 carbonyl has been deleted entirely and replaced by heteroatomic structural surrogates based on sulfur linkages. This strategy was suggested by the known polarity differences that exist between sulfoxide moieties (and their derivatives) and the carbonyl function. In addition, unlike a simple carbonyl linkage, the sulfur atom in these compounds can be an asymmetric center, thereby allowing an additional conformational parameter that can be manipulated in SAR studies. The series selected for initial evaluation as autoinducer analogues in this category include the chiral sulfoxides 6a,b and sulfoximines 7a,b as well as the achiral sulfide 5, the corresponding sulfone 8 and sulfonamide 9 as examples.

VII. NUTRITIONAL OVERRIDE OF QS

It is clear that HSL's are critical for activation of QS regulatory circuitry in *P. aeruginosa*. However, prior to the discovery of QS in this organism, the production of several gene products now known to be QS-regulated were first found to be up-regulated by nutritional factors. To date, iron has been implicated most often. For example, elastase synthesis occurs maximally when iron is restricted from the cell, specifically, at micromolae (presumably ample) levels (Sokol et al., 1982). Another example is *sodA*, encoding a manganese-cofactored superoxide dismutase (Mn-SOD) (Hassett et al., 1992, 1993). Mn-SOD expression was originally found to occur in mucoid, alginate-overproducing bacteria, in nonmucoid organisms that had been iron restricted, and in *fur* (ferric uptake regulator) mutants (Hassett et al., 1992, 1993, 1996). More recent work has demonstrated that *sodA* expression is dependent on LasR-$3O$-C_{12}-HSL; that is, no Mn-SOD was found in *lasI* or *lasIrhlI* mutants (Hassett et al., 1999b). Interestingly *sodA* transcription could be observed in an iron-starved *lasI* mutant (Bollinger et al., submitted for publication), an event that could only be forced in stationary phase cells. This suggested that an accessory factor that is only present at high

cell densities is necessary for non-QS regulated expression of *sodA*. Another example of nutritional overlapping control of QS-regulated activity in *P. aeruginosa* includes pyocyanin production. This exoproduct is typically observed in stationary phase cells, is controlled positively by RhlR-C_4-HSL (Brint and Ohman, 1995) and negatively by RpoS (Suh et al., 1999), but can also be induced by phosphate starvation (Hassett et al., 1992). Thus, supplementation of phosphate inhibits pyocyanin production and its is possible that it could paralyze *rhl* circuitry.

Integration of QS with nutrient availability should not be unexpected, and it is likely no coincidence that QS-based regulation is observed in stationary phase cultures when cell densities approach levels that represent a significant nutrient sink; indeed, growth rates are declining due to limitation of some nutrient(s). Lazazzera (2000) has recently reviewed a similar concept that appears to be operable in *Bacillus subtilis*, and there are several reports that suggest QS and nutrient sensing in Gram-negative bacteria are integrated. Viewing the growing list of QS-controlled genes from a physiologic perspective shows that several essential cellular functions are affected (catalase, superoxide dismutase, stationary phase sigma factors [RpoS], etc.), and it would seem a mistake for a bacterium to place such important functions under the control of a single regulatory system. In particular, QS relies on the presence of sufficient quantities of specific signaling molecules that are diffusible but may not approach sufficient levels under open, flowing conditions (e.g., open ocean planktonic marine organisms). Such conditions are entirely unlike fixed volume culture flasks often used to study this phenomenon. However, conditions conducive to QS regulation could prevail in biofilms, where cell densities can approach 10^{10} to 10^{12} per cm^3, and could lead to highly localized nutrient demand as well as restricted metabolite diffusion in some biofilm zones. Both conditions could lead to nutritional stress and to the accumulation of signal molecules.

It is more likely that bacteria sense their environment (e.g., adequate or limiting nutrients, cell density), using multiple metabolic indicators or signaling molecules. In a similar way cellular responses could then be mediated by ratios of two or more metabolites that serve regulatory roles or that participate in regulatory overlap. One potential metabolite that has been linked to cell-to-cell communication is ppGpp, best known for its role in the stringent response (reviewed in Cashel et al., 1996). In a series of studies with a marine *Vibrio* sp., strain S14, Kjelleberg and colleagues made a number of interesting observations that connect carbon starvation, QS, and ppGpp levels. The carbon starvation response in this bacterium induces expression of at least 123 proteins (Ostling et al., 1996; Srinivasan et al., 1998). Thirty-one of these carbon starvation inducible proteins could also be induced in logarithmically growing cells simply by the addition of small volumes of supernatant fluids from stationary phase cultures (Srinivasan et al., 1998), implying the role of some type of signaling metabolite in their regulation. Induction of nearly 30% of these proteins was abolished by the addition of a furanone that had been previously shown to interfere with HSL-based cell signaling (Givskov et al., 1996; Gram et al., 1996). Furthermore synthesis of most of the up-regulated carbon stimulon proteins were also found to be controlled by the *Vibrio* RelA and SpoT homologues (Ostling et al., 1995, 1996); these latter proteins mediate the stringent response by regulating levels of (p)ppGpp (Cashel et al., 1996). Another example that integrates ppGpp-sensitive gene regulation to cell-to-cell signaling is fruiting body formation in *Myxococcus xanthus* (Harris et al., 1998). When experiencing carbon, nitrogen, or phosphorus deprivation, myxobacteria enter into a developmental cycle whereby cells aggregate into multicellular fruiting bodies (reviewed in Shimkets, 1999); the latter is arguably the most elegant example of a biofilm structure. Coordination of this communitywide re-sponse is initiated via extracellular signals, with the earliest signal being the "A-signal" (Kuspa et al., 1986, 1992). A-signal production and fruiting structure differentiation are both dependent upon RelA-dependent ppGpp synthesis (Harris et al., 1998; Singer and Kaiser, 1995).

The examples above linking cell–cell signaling with nutrient starvation and intracellular metabolites demonstrate the complexity of QS systems. When considering possible strategies for manipulating QS for controlling bacterial infections and/or biofilms, the affects of multiple control mechanisms or regulatory override must be recognized. The onset of QS-regulated gene expression relative to the accumulation of autoinducers or to metabolites such as (p)ppGpp, and the timing of their accumulation relative to cell nutritional changes represents a critical and under-represented area of research.

ACKNOWLEDGMENTS

This work was supported by Public Health Service Grant AI-40541 (D.J.H.), AI-33713 (B.H.I.), Cystic Fibrosis Foundation Pilot Grants (D.J.H.) and PASSAD9510 (L.P.), National Science Foundation Center for Biofilm Engineering Cooperative Agreement EEC-8907039 (T.R.M.), and the Canadian Cystic Fibrosis Foundation (T.R.D.).

REFERENCES

Albus AM, Pesci EC, Runyen-Janecky L, West SEH, Iglewski BH (1997): Vfr controls quorum sensing in *Pseudomonas aeruginosa*. J Bacteriol 179:3928–3935.

Bollinger N, Hasset DJ, Iglewski BH, Custerton JW, McDermott TR (2001): Gene expression in *Pseudomonas aeruginosa*: evidence for iron override effects on quorum sensing and biofilm-specific gene regulation. J Bacteriol 183:1990–1996.

Brint JM, Ohman DE (1995): Synthesis of multiple exoproducts in *Pseudomonas aeruginosa* is under the control of RhlR-RhlI, another set of regulators in strain PAO1 with homology to the autoinducer-responsive LuxR-LuxI family. J Bacteriol 177:7155–7163.

Cashel M, Gentry DR, Hernandez VJ, Vinella D (1996): The stringent response. In Neidhardt FC, Curtiss R, Ingraham JL, Lin ECC, Brooks K, Magasanik B, Resnikoff WS, Riley M, Schaechter M, Umbarger HE (ed) *Escherichia coli* and *Salmonella*. Cellular and Molecular Biology. Washington, DC: ASM Press, 1458–1496.

Chhabra SR, Stead P, Bainton NJ, Salmond GP, Stewart GS, Williams P, Bycroft BW (1993): Autoregulation of carbapenem biosynthesis in *Erwinia carotovora* by analogues of *N*-(3-oxohexanoyl)-L-homoserine lactone. J Antibiot (Tokyo) 46:441–454.

Davies DG, Parsek MR, Pearson JP, Iglewski BH, Costerton JW, Greenberg EP (1998): The involvement of cell-to-cell signals in the development of a bacterial biofilm. Science 280:295–298.

De Kievit TR, Iglewski BH (1999): Quorum sensing, gene expression, and *Pseudomonas* biofilms. Methods Enzymol 310:117–128.

Dukan S, Touati D (1996): Hypochlorous acid stress in *Escherichia coli*: Resistance, DNA damage, and comparison with hydrogen peroxide stress. J Bacteriol 178:6145–6150.

Eberhard A, Widrig CA, McBath P, Schineller JB (1986): Analogs of the autoinducer of bioluminescence in *Vibrio fischeri*. Arch Microbiol 146:35–40.

Flavier AB, Clough SJ, Schell MA, Denny TP (1997): Identification of 3-hydroxypalmitic acid methyl ester as a novel autoregulator controlling virulence in *Ralstonia solanacearum*. Mol Microbiol 26:251–259.

Gamard P, Sauriol F, Benhamou N, Belanger RR, Paulitz TC (1997): Novel butyrolactones with antifungal activity produced by *Pseudomonas aureofaciens* strain 63–28. J Antibiot (Tokyo) 50:742–749.

Gambello MJ, Kaye S, Iglewski BH (1993): LasR of *Pseudomonas aeruginosa* is a transcriptional activator of the alkaline protease gene (*apr*) and an enhancer of exotoxin A expression. Infect Immun 61:1180–1184.

Gelb MH, Svaren JP, Abeles RH (1985): Fluoro ketone inhibitors of hydrolytic enzymes. Biochem 24:1813–1817.

Givskov M, de Nys R, Manefield M, Gram L, Maximilien R, Eberl L, Molin S, Steinberg PD, Kjelleberg S (1996): Eukaryotic interference with homoserine lactone-mediated prokaryotic signalling. J Bacteriol 178:6618–6622.

Gram L, de Nys R, Maximillen R, Givskov M, Steinberg PD, Kjelleberg S (1996): Inhibitory effects of secondary metabolites from the red alga *Delisea pulchra* on swarming mobility of *Proteus mirabilis*. Appl Environ Microbiol 62:4284–4287.

Greenberg EP (1997): Quorum sensing in Gram-negative bacteria. ASM News 63:371–377.

Harris BZ, Kaiser D, Singer M (1998): The guanosine nucleotide (p)ppGpp initiates development and A-factor production in *Myxococcus xanthus*. Gen Dev 12:1022–1035.

Hassett DJ, Charniga L, Bean KA, Ohman DE, Cohen MS (1992): Antioxidant defense mechanisms in *Pseudomonas aeruginosa*: Resistance to the redox-active antibiotic pyocyanin and demonstration of a manganese-cofactored superoxide dismutase. Infect Immun 60:328–336.

Hassett DJ, Elkins JG, Ma J-F, McDermott TR (1999a): *Pseudomonas aeruginosa* biofilm sensitivity to biocides: Use of hydrogen peroxide as model antimicrobial agent for examining resistance mechanisms. Meth Enzymol 310:599–608.

Hassett DJ, Howell ML, Ochsner U, Johnson Z, Vasil M, Dean GE (1997): An operon containing *fumC* and *sodA* encoding fumarase C and manganese superoxide dismutase is controlled by the ferric uptake regulator (Fur) in *Pseudomonas aeruginosa*: *fur* mutants produce elevated alginate levels. J Bacteriol 179:1452–1459.

Hassett DJ, Howell ML, Sokol PA, Vasil M, Dean GE (1997): Fumarase C activity is elevated in response to iron deprivation and in mucoid, alginate-producing *Pseudomonas aeruginosa*: Cloning and characterization of *fumC* and purification of native FumC. J Bacteriol 179:1442–1451.

Hassett DJ, Ma J-F, Elkins JG, McDermott TR, Ochsner UA, West SEH, Huang C-T, Fredericks J, Burnett S, Stewart PS, McPheters G, Passador L, Iglewski BH (1999b): Quorum sensing in *Pseudomonas aeruginosa* controls expression of catalase and superoxide dismutase genes and mediates biofilm susceptibility to hydrogen peroxide. Mol Microbiol 34:1082–1093.

Hassett DJ, Schweizer HP, Ohman DE (1995): *Pseudomonas aeruginosa sodA* and *sodB* mutants defective in manganese- and iron-cofactored superoxide dismutase activity demonstrate the importance of the iron-cofactored form in aerobic metabolism. J Bacteriol 177:6330–6337.

Hassett DJ, Sokol P, Howell ML, Ma J-F, Schweizer HP, Ochsner U, Vasil ML (1996): Ferric uptake regulator (Fur) mutants of *Pseudomonas aeruginosa* demonstrate defective siderophore-mediated iron uptake and altered aerobic metabolism. J Bacteriol 178:3996–4003.

Hassett DJ, Woodruff WA, Wozniak DJ, Vasil ML, Cohen MS, Ohman DE (1993): Cloning of the *sodA* and *sodB* genes encoding manganese and iron superoxide dismutase in *Pseudomonas aeruginosa*: Demonstration of increased manganese superoxide dismutase activity in alginate-producing bacteria. J Bacteriol 175:7658–7665.

Hoang TT, Ma Y, Stern RJ, McNeil MR, Schweizer HP (1999): Construction and use of low-copy number T7 expression vectors for purification of problem proteins: Purification *of Mycobacterium tuberculosis* RmlD and *Pseudomonas aeruginosa* LasI and RhlI proteins, and functional analysis of purified RhlI. Gene 237:361–371.

Holden MT, Ram Chhabra S, de Nys R, Stead P, Bainton NJ, Hill PJ, Manefield M, Kumar N, Labatte M, England D, Rice S, Givskov M, Salmond GP, Stewart GS, Bycroft BW, Kjelleberg S, Williams P (1999): Quorum-sensing cross talk: Isolation and chemical characterization of cyclic dipeptides from *Pseudomonas aeruginosa* and other Gram-negative bacteria. Mol Microbiol 33:1254–1266.

Horinouchi S, Beppu T (1994): A-factor as a microbial hormone that controls cellular differentiation and secondary metabolism in *Streptomyces griseus*. Mol Microbiol 12:859–864.

Imperiali B, Abeles RH (1986): Inhibition of serine proteases by peptidyl fluoromethyl ketones. Biochem 25:3760–3767.

Irani VR, Darzins A, Rowe JJ (1997): Snr, new genetic loci common to the nitrate reduction systems of *Pseudomonas aeruginosa* PAO1. Curr Microbiol 35:9–13.

Jones S, Yu B, Bainton NJ, Birdsall M, Bycroft BW, Chhabra SR, Cox AJ, Golby P, Reeves PJ, Stephens S et al. (1993): The lux autoinducer regulates the production of exoenzyme virulence determinants in *Erwinia carotovora and Pseudomonas aeruginosa*. EMBO J 12:2477–2482.

Kuo A, Callahan SM, Dunlap PV (1996): Modulation of luminescence operon expression by *N*-octanoyl-L-homoserine lactone in ainS mutants of *Vibrio fischeri*. J Bacteriol 178:971–976.

Kuspa A, Kroos L, Kaiser D (1986): Intercellular signaling is required for developmental gene expression in *Myxococcus xanthus*. Dev Biol 117:267–276.

Kuspa A, Plamann L, Kaiser D (1992): Identification of heat-stable A-factor from *Myxococcus xanthus*. J Bacteriol 174:3319–3326.

Latifi A, Foglino M, Tanaka K, Williams P, Lazdunski A (1996): A hierarchical quorum-sensing cascade in *Pseudomonas aeruginosa* links the transcriptional activators LasR and RhlR (VsmR) to expression of the stationary phase sigma factor RpoS. Mol Microbiol 21:1137–1146.

Lazazzera BA (2000): Quorum sensing and starvation: signals for entry into stationary phase. Curr Opin Microbiol 3:177–182.

McKnight SL, Iglewski BH, Pesci EC (2000): The pseudomonas quinolone signal regulates *rhl* quorum sensing in *Pseudomonas aeruginosa*. J Bacteriol 182:2702–2708.

More MI, Finger LD, Stryker JL, Fuqua C, Eberhard A, Winans SC (1996): Enzymatic synthesis of a quorum-sensing autoinducer through use of defined substrates. Science 272:1655–1658.

Ostling J, Flardh K, Kjelleberg S (1995): Isolation of a carbon starvation regulatory mutant in a marine *Vibrio* strain. J Bacteriol 177:6978–6982.

Ostling J, Holmquist L, Kjelleberg S (1996): Global analysis of the carbon starvation response of a marine *Vibrio* species with disruptions in genes homologous to *relA* and *spoT*. J Bacteriol 178:4901–4908.

Parsek MR, Val DL, Hanzelka BL, Cronan Jr JE, Greenberg EP (1999): Acyl homoserine-lactone quorum-sensing signal generation. Proc Natl Acad Sci USA 96:4360–4365.

Passador L, Cook JM, Gambello MJ, Rust L, Iglewski BH (1993): Expression of *Pseudomonas aeruginosa* virulence genes requires cell-to-cell communication. Science 260:1127–1130.

Passador L, Tucker KD, Guertin KR, Journet MP, Kende AS, Iglewski BH (1996): Functional analysis of the *Pseudomonas aeruginosa* autoinducer PAI. J Bacteriol 178:5995–6000.

Pearson JP, Feldman M, Iglewski BH, Prince A (2000): *Pseudomonas aeruginosa* cell-to-cell signaling is required for virulence in a model of acute pulmonary infection. Infect Immun 68:4331–4334.

Pearson JP, Gray KM, Passador L, Tucker KD, Eberhard A, Igkewski BH, Greenberg EP (1994): Structure of the autoinducer required for expression of *Pseudomonas aeruginosa* virulence genes. Proc Natl Acad Sci 91:197–201.

Pearson JP, Passador L, Iglewski BH, Greenberg EP (1995): A second N-acylhomoserine lactone produced by *Pseudomonas aeruginosa*. Proc Natl Acad Sci 92:1490–1494.

Pesci EC, Milbank JB, Pearson JP, McKnight S, Kende AS, Greenberg EP, Iglewski BH (1999): Quinolone signaling in the cell-to-cell communication system of *Pseudomonas aeruginosa*. Proc Natl Acad Sci 96:11229–11234.

Pesci EC, Pearson JP, Seed PC, Iglewski BH (1997): Regulation of *las* and *rhl* quorum sensing in *Pseudomonas aeruginosa*. J Bacteriol 179:3127–3132.

Rahme LG, Stevens EJ, Wolfort SF, Shao J, Tompkins RG, Ausubel FM (1995): Common virulence factors for bacterial pathogenicity in plants and animals. Science 268:1899–1902.

Rahme LG, Tan MW, Le L, Wong SM, Tompkins RG, Calderwood SB, Ausubel FM (1997): Use of model plant hosts to identify *Pseudomonas aeruginosa* virulence factors. Proc Natl Acad Sci USA 94:13245–13250.

Reimmann C, Beyeler M, Latifi A, Winteler H, Foglino M, Lazdunski A, Haas D (1997): The global activator GacA of *Pseudomonas aeruginosa* PAO positively controls the production of the autoinducer *N*-butyryl-homoserine lactone and the formation of the virulence factors pyocyanin, cyanide, and lipase. Mol Microbiol 24:309–319.

Rumbaugh KP, Griswold JA, Hamood AN (1999a): Contribution of the regulatory gene *lasR* to the pathogenesis of *Pseudomonas aeruginosa* infection of burned mice. J Burn Care Rehabil 20:42–49.

Rumbaugh KP, Griswold JA, Iglewski BH, Hamood AN (1999b): Contribution of quorum sensing to the virulence of *Pseudomonas aeruginosa* in burn wound infections. Infect Immun 67:5854–5862.

Schaefer AL, Val DL, Hanzelka BL, Cronan Jr JE, Greenberg EP (1996): Generation of cell-to-cell signals in quorum sensing: acyl homoserine lactone synthase activity of a purified *Vibrio fischeri* LuxI protein. Proc Natl Acad Sci USA 93:9505–9509.

Shimkets LJ (1999): Intercellular signaling during fruiting-body development of *Myxococcus xanthus*. Annu Rev Microbiol 53:525–549.

Singer M, Kaiser D (1995): Ectopic production of guanosine penta- and tetraphosphate can initiate early developmental gene expression in *Myxococcus xanthus*. Genes Dev 9:1633–1644.

Sokol PA, Cox CD, Iglewski BH (1982): *Pseudomonas aeruginosa* mutants altered in their sensitivity to the effect of iron on toxin A or elastase yields. J Bacteriol 151:783–787.

Srinivasan S, Ostling J, Charlton T, de Nys R, Takayama K, Kjellberg S (1998): Extracellular signal molecule(s) involved in the carbon starvation response of marine *Vibrio* sp. strain S14. J Bacteriol 180:201–209.

Storey DG, Ujack EE, Rabin HR, Mitchell I (1998): *Pseudomonas aeruginosa lasR* transcription correlates with the transcription of *lasA*, *lasB*, and *toxA* in chronic lung infections associated with cystic fibrosis. Infect Immun 66:2521–2528.

Suh SJ, Silo-Suh L, Woods DE, Hassett DJ, West SEH, Ohman DE (1999): Effect of *rpoS* mutation on the stress response and expression of virulence factors in *Pseudomonas aeruginosa*. J Bacteriol 181:3890–3897.

Tan MW, Rahme LG, Sternberg JA, Tompkins RG, Ausubel FM (1999): *Pseudomonas aeruginosa* killing of *Caenorhabditis elegans* used to identify *P. aeruginosa* virulence factors. Proc Natl Acad Sci USA 96:2408–2413.

Tang HB, DiMango E, Bryan R, Gambello M, Iglewski BH, Goldberg JB, Prince A (1996): Contribution of specific *Pseudomonas aeruginosa* virulence factors to pathogenesis of pneumonia in a neonatal mouse model of infection. Infect Immun 64:37–43.

Van Delden C, Pesci EC, Pearson JP, Iglewski BH (1998): Starvation selection restores elastase and rhamnolipid production in a *Pseudomonas aeruginosa* quorum-sensing mutant. Infect Immun 66:4499–4502.

West SE, Sample AK, Runyen-Janecky LJ (1994): The *vfr* gene product, required for *Pseudomonas aeruginosa* exotoxin A and protease production, belongs to the cyclic AMP receptor protein family. J Bacteriol 176:7532–7542.

Whiteley M, Lee KM, Greenberg EP (1999): Identification of genes controlled by quorum sensing in *Pseudomonas aeruginosa*. Proc Natl Acad Sci 96:13904–13909.

Winson MK, Camara M, Latifi A, Foglino M, Chhabra SR, Daykin M, Bally M, Chapon V, Salmond GP, Bycroft BW, Lazdunski A, Stewart GSAB, Williams P (1995): Multiple *N*-acyl-L-homoserine lactone signal molecules regulate production of virulence determinants and secondary metabolites in *Pseudomonas aeruginosa*. Proc Natl Acad Sci 92:9427–9431.

Zhang L, Murphy PJ, Kerr A, Tate ME (1993): Agrobacterium conjugation and gene regulation by *N*-acyl-L-homoserine lactones. Nature 362:446–448.

Zhu J, Beaber JW, More MI, Fuqua C, Eberhard A, Winans SC (1998): Analogs of the autoinducer 3-oxooctanoyl-homoserine lactone strongly inhibit activity of the TraR protein of *Agrobacterium tumefaciens*. J Bacteriol 180:5398–5405.

11

Endospore Formation in *Bacillus subtilis*: An Example of Cell Differentiation by a Bacterium

CHARLES P. MORAN JR.

Department of Microbiology and Immunology, Emory University School of Medicine, Atlanta, Georgia 30322

I. INTRODUCTION

A. Morphological Stages of Development

Many species of bacteria are capable of undergoing remarkably complex cellular differentiations. In each case the process involves the activation and sequential expression of a large number of genes. One type of cellular differentiation is the formation of endospores by *Clostridium* and *Bacillus* species. In no example of bacterial differentiation is it completely understood how gene expression is regulated, but the most exten-

sively characterized system, and probably the one most ameanable to genetic analysis, is that of endospore formation in *Bacillus subtilis* (for reviews on sporulation, see Errington, 1993; Stragier and Losick, 1996).

Bacillus subtilis is a rod-shaped Gram-positive bacterium, which like most bacteria, multiplies by binary fission. However, in response to nutrient depletion, the bacterium differentiates in a complex developmental process that culminates in the production of

Modern Microbial Genetics, Second Edition. Edited by Uldis N. Streips and Ronald E. Yasbin. ISBN 0–471–38665–0

a new cell type, the endospore. The endo-spore is a dormant cell that can survive numerous environmental insults (e.g., dessi-cation, heat, chemical stresses, and nutrient depletion). Although the endospore is a dor-mant cell, it responds to specific nutrient signals by germinating to produce a vegeta-tive cell that resumes multiplication by binary fission.

Endospore development is a morphologic-ally complex process that proceeds through a well-defined series of morphological stages (Fig. 1). Multiplication of vegetative cells by binary fission involves a cell division that produces two identical daughter cells. However, at the onset of sporulation the cell divides asymetrically to produce a small cell and a large cell. These two cells have different developmental fates. The smaller cell, forespore, develops into the mature endospore. The larger cell, known as the mother cell, develops into a terminally differentiated cell that nurtures the develop-ing endospore, providing to it a large number of products. Ultimately the mother cell dies when the endospore is mature.

In the next most conspicuous morpho-logical event after the asymmetric cell div-ision the peptidoglycan layer between the forespore and mother cells is hydrolyzed, and the mother cell engulfs the forespore in a phagocytic-like process in which the mother cell cytoplasmic membrane sur-rounds the membrane of the forespore. The engulfed forespore protoplast floats within the mother cell cytoplasm where it is sur-rounded by both its own original membrane and a second membrane that originated from the mother cell cytoplasmic membrane. This cell-within-a-cell form of development is why this type of bacterial spore is called an endospore. At this stage both cells, forespore and mother cell, are metabolically active. Therefore the developing endospore is built from the inside and outside simultaneously, with some products synthesized within the forespore where they accumulate (e.g., proteins that bind and protect the spore DNA), while other products are synthesized in the mother cell cytoplasm from where they are donated to the endospore. In the later stages of development a thick layer of peptidoglycan accumulates between the two membranes of the forespore, and a thick layer of proteins forms a coat that surrounds and protects the peptidoglycan layer.

B. Developmentally Regulated Gene Expression

Endospore formation requires the expres-sion of over 100 genes that are not required

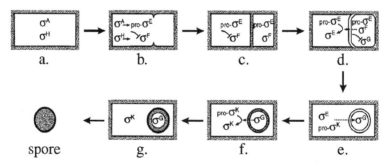

Fig. 1. Morphological stages of endospore formation in *B. subtilis*. The most conspicuous morphological changes include an asymmetric cell division (**c**), engulfment of the forespore by the mother cell membrane (**d** and **e**), synthesis of the cortex (**f**), assembly of the coat surrounding the forespore (**g**), and lysis of the mother cell to release the dormant spore (**h**). Also shown are the sigma factors (σ) active during sporulation The dashed arrows represent signals from one cell that signal activation of a σ in the other cell. The symbol (⊥) indicates the anti-sigma factor SpoIIAB that represses the activity of σF and σG.

for growth or survival of vegetative cells. These genes are regulated primarily at the level of transcription, and are transcribed in a specific temporal sequence. Transcription of sporulation genes is also regulated in a cell-type specific manner; some genes are transcribed exclusively in the forespore, whereas other genes are transcribed exclusively in the mother cell.

C. The Cascade of RNA Polymerase Sigma Factors

Regulation of both the temporal pattern of transcription and the cell-type specific transcription of genes during endospore development is primarily governed by a family of transcription factors known as RNA polymerase sigma factors (reviewed in Losick and Stragier, 1992; Kroos et al., 1999). RNA polymerase in bacteria is made of several protein subunits. The core enzyme is composed of two alpha, one beta, and one beta' subunits. The sigma subunit, or sigma factor, associates with the core RNA polymerase to form the holoenzyme. Association of the sigma with RNA polymerase confers on the polymerase its ability to bind to and utilize specific promoters. All bacteria have a primary sigma factor (called sigma A, σ^A, in *B. subtilis*) that directs transcription of most housekeeping genes. RNA polymerase containing σ^A ($E\sigma^A$) recognizes specific promoters because the sigma interacts with two specific sequences of DNA located 35 and 10 base pairs upstream from the start point of transcription, the -35 and -10 regions, respectively, of the promoter. Most bacteria contain secondary sigma factors that enable RNA polymerase to utilize promoters that are defined by different sequences of DNA. Five secondary sigma factors, each confering a different specificity for promoter recognition on RNA polymerase, direct transcription during endospore formation in *B. subtilis* (see Helmann, this volume).

One of the secondary sigma factors required for endospore development, σ^H, is present in vegetative cells. However, the other secondary sigma factors required for endospore development are synthesized after the onset of sporulation in the order σ^F, σ^E, σ^G, and σ^K. The sequential appearance of these sigma factors produces the temporally regulated pattern of gene transcription. Therefore genes whose transcription is directed by σ^F are transcribed before genes whose transcription is dependent on σ^E, σ^G, or σ^K. Although these secondary sigma factors define the major temporal classes of gene transcription during sporulation, each class is subdivided by ancillary transcription factors that work with the sigma factors. These are DNA-binding proteins that repress or activate transcription of specfic genes. For example, σ^K is active during the final stages of development. When σ^K becomes active, it directs the transcription of several genes, one of which encodes a DNA-binding protein called GerE. When GerE accumulates, it binds to and represses transcription from a subset of σ^K dependent promoters. GerE also binds to and activates a different subset of σ^K-dependent promoters that are only active when both σ^K and GerE are present. Therefore GerE subdivides the σ^K regulon into genes that are transcribed throughout the period of σ^K activity, genes that are transcribed early but are shut off when GerE accumulates, and genes whose transcription is activated only in the final stage of σ^K activity when GerE is present.

The secondary sigma factors also play the central role in the cell-type specific transcription of genes. σ^F and σ^G are active exclusively in the forespore, whereas, σ^E and σ^K are active exclusively in the mother cell. Therefore genes whose transcription is dependent on one of these sigma factors can only occur in one cell type, either the forespore or mother cell. Transcription of genes is coordinated between the two cell types. The mechanisms that coordinate the gene expression in the two cells types involve regulation of sigma factor activity. (These mechanisms are discussed below in Section IV.)

II. SIGNAL TRANSDUCTION AND THE INITIATION OF SPORULATION

A. Phosphorylation of the Transcriptional Activator Spo0A

Most of the developmental gene expression required for endspore formation is the result of the production of RNA polymerase sigma factors. But how is the synthesis of the first sporulation-induced sigma factors regulated? σ^F and σ^E are the first two sigma factors produced after the initiation of sporulation. The structural genes encoding these sigma factors are part of the *spoIIA* and *spoIIG* operons, respectively. Transcription of both operons is activated within minutes after the initiation of sporulation. The promoter for the *spoIIG* operon is used by RNA polymerase containing σ^A, the primary sigma factor in vegetative cells. The promoter for the *spoIIA* operon is used by RNA polymerase containing σ^H. σ^H, like σ^A, is present in vegetative cells; therefore these sigma factors cannot account for the activation of the *spoIIA* and *spoIIG* promoters at the onset of sporulation. Activation of these promoters requires an additional factor, a DNA-binding protein known as Spo0A. Spo0A is similar to a large family of proteins that form the response regulator class of two-component regulatory systems. Like other regulators of this class, Spo0A is active only when phosphorylated on a specific aspartate residue. Once phosphorylated, it binds to specific sites on the *spoIIA* and *spoIIG* promoters where it interacts with RNA polymerase to activate transcription. Binding of phosphorylated Spo0A to other sites on the chromosome activates, or some cases represses, transcription of other genes.

B. The Phosphorelay System, Phosphatases, and Integration of Multiple Systems

Spo0A is the key to the initiation of sporulation. Spo0A is present in growing cells, and its phosphorylation triggers the initiation of sporulation. Phosphorylation of Spo0A is controlled by a phosphorelay system that is more complex than most other two-component systems (see the chapter by Bayles and Fujimoto, this volume) (Fig. 2) (reviewed in Fabret et al., 1999; Hoch, 2000; Perego, 1998). Unlike simple two-component systems in which a kinase phosphorylates the response regulator, the phosphoryl group that is added to Spo0A is first passed from one of two kinases to another response regulator-like protein called Spo0F. The phosphoryl group is then tranfered from Spo0F~P to Spo0A by a phosphotransferase called Spo0B. The concentration of Spo0A~P is the key determinant of whether the cell will sporulate or continue to divide. The concentration of Spo0A~P depends on the rate of flow of phosphoryl groups through the phosphorelay. The rate of flow of phosphoryl groups to Spo0A is modulated by the control of kinase activity, and by phosphatases that remove phosphoryl groups from the stream (Fig. 2).

The complexity of the network of regulators that control of Spo0A~P concentration allows the cell to integrate several signals, a

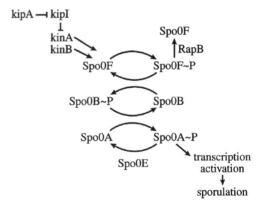

Fig. 2. The phosphorelay system that controls the initiation of sporulation. Kinases KinA and KinB feed phosphate into the phosphorelay. The flow of phosphoryl groups to Spo0A is contolled by KipI, which contols KinA activity, and by phosphatases RapB and Spo0E, which remove phosphoryl groups from Spo0F and Spo0A, respectively. Phosphorylated Spo0A (Spo0A~P) acts as a DNA-binding protein to directly activate some promoters and repress transcription from other promoters.

way of considering other responses to nutrient depletion, before committing to the most drastic response of producing a dormant cell. Other responses, such as chemotaxis and the development of the competent state for DNA uptake, are discussed in other chapters. Nutrient depletion may affect phosphate flow to Spo0A by modulating kinase activity. KipI, an inhibitor of KinA, the major kinase of the phosphorelay, and KipA, an anti-kinase inhibitor, are encoded in an operon that is responsive to glucose and nitrogen levels. Recently, A. L. Sonenshein and his colleagues found that CodY, a DNA-binding protein, is also a GTP-binding protein, and CodY acts as an intracellular sensor of GTP levels (Ratnayake-Lecamwasam et al., 2001). Under conditions of nutrient excess the intracellular level of GTP is high so CodY binds GTP and acts as a transcriptional repressor of several genes. Under conditions of nutrient depletion the intracellular level of GTP decreases. In the absence of bound GTP CodY fails to acts as a transcriptional repressor. CodY may regulate the initiation of sporulation in response to nutrient conditions by regulating the transcription of one of more structural genes for components of the phosphorelay.

Other responses to nutrient depletion and other physiological signals compete with the sporulation response by affecting the flow of phosphate through the phosphorelay. The development of competence affects the activity of a phosphatase, RapA, which dephosphorylates Spo0F. A second Spo0F~P phosphatase, RapB, is expressed in growing cells. Another phosphatase Spo0E dephosphorylates Spo0A~P directly, but it is unknown how its activity is regulated. Presumably Spo0E phosphatase activity is used to integrate another, as yet unknown, input into the decision to sporulate. Another level of complexity was revealed by the discovery of peptides that regulate phosphatase activity. PhrA is a 44 amino acid peptide that inhibits RapA activity. This peptide is secreted and then a peptide consisting of at least the five amino acyl residues at the car-

boxy terminal end PhrA is internalized where it may act directly on the phosphatase. It is not known whether the main role of PhrA is to serve as a timing device or as an intercellular signal. However, it is evident that PhrA is essential for inhibition of the RapA phosphatase and thereby for the initiation of sporulation.

III. DETERMINATION OF A CELL'S DEVELOPMENTAL FATE BY A PHOSPHATASE AND ANTI-SIGMA FACTOR

Early during endospore formation the asymetric cell division produces two cells that have different developmental fates. Different sets of genes are transcribed in each of these two cells because the activities of specific sigma factors are restricted to specific cells (for review on the control of sigma activity during sporulation, see Kroos et al., 1999; Losick, 1994; Losick and Stragier, 1992). Specific genes are transcribed exclusively in the forespore within minutes after the asymmetric division. Transcription of these genes in the forespore is directed by σ^F, one of the sigma factors synthesized early after the initiation of sporulation. Although σ^F is produced in the predivisional cell within minutes after the onset of sporulation, σ^F is not active in the predivisional cell. After the asymmetric cell division, σ^F remains inactive in the mother cell but becomes active in the forespore. It is the activity of σ^F in the forespore that leads to expression of forespore specific genes, and as is discussed in the next section, σ^F activity in the forespore also signals the activation of mother cell specific gene expression. Since activation of σ^F in the forespore determines its developmental fate, it is important to understand how σ^F activity is restricted to the forespore.

In order to discuss the regulation of σ^F activity three additional proteins must be introduced (Fig. 3). The first is SpoIIAB, an anti-sigma factor that is capable of binding to σ^F and keeping it inactive. SpoIIAB is produced in the predivisional cell at the same

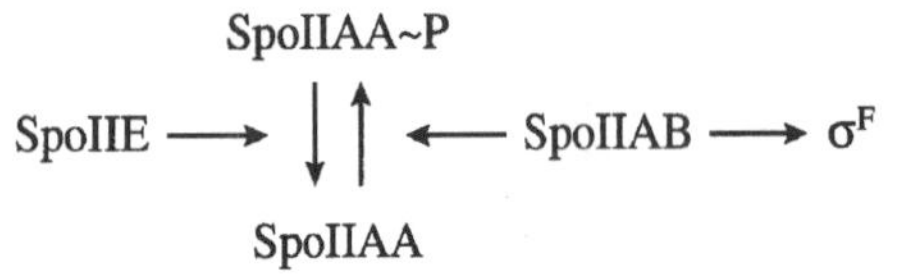

$$\text{SpoIIE} \longrightarrow \underset{\text{SpoIIAA}}{\overset{\text{SpoIIAA} \sim \text{P}}{\updownarrow}} \longleftarrow \text{SpoIIAB} \longrightarrow \sigma^F$$

Fig. 3. Regulation of σ^F activity by the anti-sigma factor SpoIIAB. SpoIIAB binds to σ^F inhibiting its function. SpoIIAB can also bind to SpoIIAA but cannot stably bind σ^F and SpoIIAA simultaneously. SpoIIAB also is a kinase that phosphorylates SpoIIAA. However, SpoIIAA~P does not bind SpoIIAB. In the predivisional cell (Fig. I, panel b) almost all of SpoIIAA is found in its phosphorlated form (SpoIIAA~P). Therefore SpoIIAB is free to bind and inhibit σ^F. After the asymmetric septation (Fig. I, panel c) SpoIIAA~P is dephosphorylated by the phosphatase SpoIIE exclusively in the forespore compartment. The dephosphorylated SpoIIAA binds SpoIIAB releasing σ^F, which is therefore active exclusively in the forespore.

time that σ^F is produced. SpoIIAB binds to σ^F in the predivisional cell antagonizing its activity. SpoIIAB also binds and antagonizes σ^F in the mother cell after cell division.

SpoIIAB releases σ^F in the forespore because of the action of an anti-anti-sigma, SpoIIAA. SpoIIAA binds to SpoIIAB forcing it to release σ^F. The SpoIIAA protein is present in the predivisional cell, and after the asymmetric cell division it is present in both the mother cell and forespore. However, SpoIIAA disrupts the SpoIIAB-σ^F complex exclusively in the forespore because in the predivisional cell and mother cell SpoIIAA remains phosphorylated. SpoIIAB is the kinase that phosphorylates SpoIIAA. The phosphorylated form of SpoIIAA (SpoIIAA~P) does not bind to SpoIIAB. Therefore SpoIIAA~P leaves SpoIIAB free to bind and inhibit σ^F. SpoIIAA~P is dephosphorylated in the forespore by a membrane bound phosphatase called SpoIIE. SpoIIE is synthesized in the predivisional cell, but after the assymmetric cell division SpoIIE is located in the septum that divides the forespore and mother cell. The reason that SpoIIE is active exclusively in the forespore is a topic that is currently being investigated. One hypothesis states that SpoIIE may be located primarily on the forespore side of the septum. However, even if SpoIIE is located on both

sides of the septum, its concentration would be much higher in the forespore because the volume of the forespore is much smaller than the mother cell. One hypothesis suggests that this difference in volume may be sufficient to account for the forespore-specific activation of σ^F through the action of the SpoIIE phosphatase.

In summary, σ^F activation in the forespore determines the developmental fate of this cell, and as described below, dictates the fate of the mother cell. σ^F activity is regulated by an anti-sigma factor (SpoIIAB) and an anti-anti-sigma factor (SpoIIAA), whose phosphorylation state is controlled by a phosphatase (SpoIIE) whose activity is restricted to the forespore.

IV. COORDINATION OF TWO CELLS' DEVELOPMENTAL PROGRAMS BY CRISSCROSS REGULATION OF SIGMA FACTOR ACTIVITY

A. Regulation of σ^E Activity

Like σ^F, σ^E is synthesized in the predivisional cell but is not active until after the asymmetric cell division during endspore development. However, the mechanism controlling σ^E activity is very different from that controlling σ^F. σ^E, is initially produced as an inactive precursor (pro-σ^E) that contains 27 amino acids at its amino terminus,- which must be removed to produce the active form of σ^E. The protease (SpoIIGA) that processes pro-σ^E to its mature form is membrane bound and located in the septum that separates the mother cell and forespore. Activation of the protease requires the product of the *spoIIR* locus, which is transcribed in the forespore under the direction of σ^F. The *spoIIR* product is a small peptide that is secreted through the forespore membrane so that it can interact with the membrane bound SpoIIGA protease to activate processing of pro-σ^E to form active σ^E. σ^E is active exclusively in the mother cell. It is clear that its activation depends on processing by SpoIIGA, but it is not known whether this is the mechanism of control is responsible for restricting σ^E activity to the mother cell. Some evidence supports a model in which σ^E is degraded in the forespore by an unidentified protease.

Activation of σ^E requires *spoIIR* expression that is dependent on σ^F; therefore σ^E activation is dependent on σ^F activation, which in turn is coupled to the asymmetric cell division. In this way both σ^F and σ^E activation is coupled to completion of the division septum. Coupling of sigma activation to a morphological landmark (in this case, the completion of the division septum) appears to be a reoccurring theme throughout development of the endospore (see below). A second theme is illustrated by the fact that activation of σ^E in the mother cell is regulated by events in the forespore (the σ^F-dependent expression of *spoIIR*). Gene expression in the two cells may be coordinated to ensure the proper assembly of spore components from the inside and outside of the forespore during development. The coordination of gene expression in the two cells is regulated by a crisscross pattern of regulation of sigma factor activity (Fig. 1). σ^F in the forespore is an essential prerequisite for activation of σ^E in the mother cell. σ^E activity in the mother cell is required for activation of σ^G in the forespore. Finally, σ^G activity in the forespore is required for the activation of σ^K in the mother cell. The mechanisms controlling these sigma factors are described below.

B. Regulation of σ^K Activity

We will examine regulation of σ^K activity first because its mechanism seems superficially similar to that regulating σ^E activity. The structural gene encoding σ^K is transcribed in the mother cell by RNA polymerase containing σ^E. The primary product pro-σ^K contains 22 amino acids at its amino terminal end, which must be proteolytically removed to produce the active σ^K. However, here the similarity ends, since the mechanisms of processing for these two mother cell sigma factors are not homologous. The pro-

tease that activates σ^E appears to be an aspartate protease, whereas the protease that activates σ^K is a member of a subset of zinc metalloproteases. Processing of σ^K in the mother cell is dependent on events in the forespore, but the mechanism that transduces the signal from forespore to the protease is not homologous to the SpoIIR pathway that signals processing of σ^E.

In its default state the SpoIIGA protease that processes pro-σ^E is inactive. It is activated by the *spoIIR* product, which is produced in the forespore. In its default state the protease that processes pro-σ^K (SpoIVFB) is active. It resides in the forespore membrane where it is held inactive by two other proteins. The forespore-expressed gene product that signals pro-σ^K processing (*spoIVB*) is not homologous to SpoIIR, and it acts by antagonizing the inhibitors of the SpoIVFB protease, thus activating the proteolytic activation of σ^K. Transcription of the *spoIVB* gene in the forespore, which signals pro-σ^K processing, is directed by σ^G. Therefore σ^K activation in the mother cell is tied to the activation of σ^G activation in the forespore, which, as discussed below, may be coupled to another morphological landmark.

C. Regulation of σ^G Activity

σ^G activity also is regulated. σ^F directs transcription of *spoIIIG*, the structural gene for σ^G. However, σ^G does not become active until after the forespore protoplast is engulfed by the mother cell (Fig. 1). Activation of σ^G in the forespore is dependent on the activity of σ^E in the mother cell. It is not known how σ^G activity is regulated in the forespore, or how σ^E dependent events in the mother cell affect this regulation. According to one hypothesis, the anti-sigma factor SpoIIAB may regulate σ^G activity. However, this model raises a number of unanswered questions. For example, in order for SpoIIAB to sequentially regulate both σ^F and σ^G, the SpoIIAA-dependent inhibition of SpoIIAB's anti-σ activity must be temporary so that, after σ^F directs expression of σ^G, SpoIIAB will be active to inhibit the newly

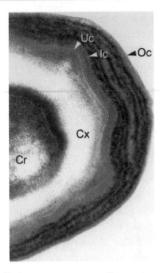

Fig. 4. Endospore structure. Shown is an electron micrograph of a thin cross section of a *B. subtilis* endospore. The proteinaeous coat forms three layers: an striated electron-dense outer coat (Oc), a lamellar inner coat (Ic), and diffuse undercoat (Uc). A thick layer of peptidoglycan known as the cortex (Cx) lies beneath the coat. The core of the spore is also indicated (Cr). The radius of this spore is approximately 0.5 μm.

synthesized σ^G. It is known that the unphosphorylated form of SpoIIAA disappears after σ^F is activated; however, it is not known how SpoIIAB is inactivated during this later stage to release σ^G activity.

V. MORE UNANSWERED QUESTIONS

The assembly of structures in bacteria requires that proteins be targeted to specific subcellular locations. For example, proteins are targeted to the division septum during bacterial cell division. Morphogenesis of the endospore is a complex process in which many proteins are targeted to specific locations. For example, the mature spore is surrounded by coat composed of over two dozen proteins. These proteins are arranged into at least three layers around the forespore membrane (Fig. 4). Most of the spore coat proteins are synthesized in the mother cell. However, it is not known how they are targeted to their final locations. Future studies of spore coat assembly may lead to important insights into how bacteria build complex cellular structures (for reviews on the the spore coat, see Driks, 1999; Henriques and Moran, 2000).

REFERENCES

Driks A (1999): *Bacillus subtilis* spore coat. Microbiol Mol Biol Rev 63:1–20.

Errington J (1993): *Bacillus subtilis* sporulation: Regulation of gene expression and control of morphogenesis. Microbiol Rev 57:1–33.

Fabret C, Feher VA, Hoch JA (1999): Two-component signal transduction in *Bacillus subtilis*: how one organism sees its world. J Bacteriol 181:1975–1983.

Henriques AO, Moran CP, Jr (2000): Structure and assembly of the bacterial endospore coat. Methods 20:95–110.

Hoch JA (2000): Two-component and phosphorelay signal transduction. Curr Opin Microbiol 3:165–170.

Kroos L, Zhang B, Ichikawa H, Yu YT (1999): Control of sigma factor activity during *Bacillus subtilis* sporulation. Mol Microbiol 31:1285–1294.

Losick R (1994): RNA polymerase sigma factors and cell-specific gene transcription in a simple developing organism. Harvey Lect 90:1–17.

Losick R, Stragier P (1992): Crisscross regulation of cell-type-specific gene expression during development in *B. subtilis*. Nature 355:601–604.

Perego M (1998): Kinase-phosphatase competition regulates *Bacillus subtilis* development. Trends Microbiol 6:366–370.

Ratnayake-Lecamwasam M, Serror P, Wong KW, Sonenshein AL (2001): *Bacillus subtilis* CodY represses early-stationary-phase genes by sensing GTP levels. Genes Dev 15:1093–1003.

Stragier P, Losick R (1996): Molecular genetics of sporulation in *Bacillus subtilis*. Annu Rev Genet 30:297–341.

12

Stress Shock

ULDIS N. STREIPS

Department of Microbiology and Immunology, School of Medicine, University of Louisville, Louisville, Kentucky 40292

I. INTRODUCTION

The molecular study on the effects of heat shock in living organisms was initiated in the 1970s by Tissieres in *Drosophila melanogaster*, where chromosome puffing resulted from exposure of the fruit flies to elevated temperatures (Tissieres et al., 1974). Since that time it has been found that the response to environmental stress is universal and results in the synthesis of a subset of proteins called the *stress shock proteins*. All manner of stress conditions elicit a response, and some of the conditions overlap in the proteins that are produced. For instance, the heat shock protein GroEL is produced prominently not only after heat shock but also under several other stress conditions. Just to mention a few, bacteria react to stresses of heat, cold, anoxia, chemicals, growth conditions and aging, oxidative conditions, and many more. As an example, this chapter will describe the responses of *Escherichia coli* and *Bacillus subtilis* to environmental stresses, though mostly concentrating on heat stress. Both of these organisms respond physiologically to stress by organizing several relevant genes and operons under coordinate control by specific regulatory molecules. These sets of genes are defined as regulons.

II. ESCHERICHIA COLI RESPONSE

Escherichia coli is a Gram-negative, nonsporeforming, motile bacterium, approximately 3.0 to 0.5 μm in size. This organism has regulons for oxidative responses, stringent response, nitrogen, carbon, and phosphate utilization, and others (see Neidhardt, 1987). Chemotaxis and flagellar assembly genes are coordinately controlled (Arnosti and Chamberlin, 1989). SOS repair and heat shock are especially large regulons. In *E. coli* heat shock can be assayed by two dimensional gel electrophoresis and proteomics (see Gross, 1996). In Figure 1 is a representative pair of gels demonstrating the upshift in synthesis of specific proteins, following temperature elevation.

Control of Response

A major sigma factor for the heat shock response in *E. coli* is sigma 32 and is one of three sigma factors that control stress in *E. coli*. This sigma factor, associated to the RNA polymerase core unit, recognizes specific heat shock promoters. Sigma 32 is synthesized from the *rpoH* gene and transcribes

Modern Microbial Genetics, Second Edition. Edited by Uldis N. Streips and Ronald E. Yasbin. ISBN 0–471–38665–0
Copyright © 2002 Wiley-Liss, Inc.

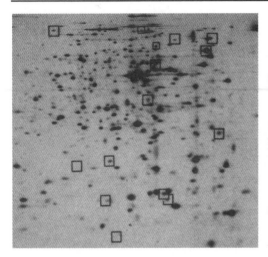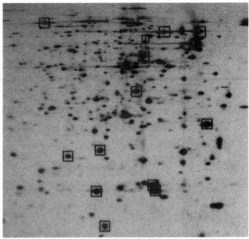

Fig. 1. Heat shock response in *Escherichia coli* strain W3110. Cells were pulse-labeled for five minutes with
[35]S-methionine at 28 C (left panel) and following a 3 minute shift to 42 C (right panel). The heat shock
proteins or the location of these proteins is indicated by a box. The autoradiograms are oriented such that
the acid side is to the right and the high molecular weight proteins are at the top. The protein at the top right
side which is well represented at 28 C and is highly induced at 42 C is GroEL. In the box immediately above it
and slightly to the right is DnaK. (Courtesy of Dr. L. F. Peruski, Ph.D. thesis, University of Michigan, 1988.)

at least 20 or more genes following heat shock, including the most studied molecular chaperones, the DnaK synthesizing operon and the GroEL expressing operon, as well as the proteases Lon, HflB, HsIUV, and the Clp proteins (Gross, 1996). The small heat shock proteins IpbA and IpbB are also transcribed by the sigma 32 holoenzyme and are involved in both heat shock and oxidative stress responses (Kitigawa et al., 2000). The sigma 32 gene is itself transcribed by the *rpoD* gene product, sigma 70 (sigma D), the housekeeping sigma factor in *E. coli*. In fact the *rpoH* gene has four promoters, two recognized by sigma D and one by sigma E (*rpoE* gene product, below) (Wang and Kaguni, 1989). Heat shock leads to an increase in the intracellular pools of sigma 32. Deletion of the *rpoH* gene results in cells unable to grow above 20°C (Zhou et al., 1988). This gene is also related to genes involved in controlling cell division (Crickmore and Salmond, 1986).

Sigma E is in a regulon that is activated by heat shock and other extracellular stresses, and it transcribes about 10 genes. Part of the sigma E operon is the gene for RseA, which

is an anti-sigma factor. Additionally RseB, also synthesized from this operon, mitigates the effect of RseA on sigma E (Missiakas and Raina, 1997). As transcription by sigma E proceeds, the pool of RseA also rises, and this regulates the level of transcription from the sigma E controlled regulon. Another gene transcribed by sigma E is *htrA*, which is vital for high-temperature survival (below) (Lipinska et al., 1989). Transcription by sigma E is controlled by a two-component signaling pathway, CpxR-CpxA (Danese et al., 1995).

A third sigma factor, sigma S, controls a global stress regulon of over 50 genes that are involved in responses to growth conditions and aging, as well as hyperosmotic and acid stresses (Loewen et al, 1998). This sigma is the product of the *rpoS* gene, and the expression of this gene and stability of its mRNA are regulated by members of the cold shock protein family CspC and CspE (Phadtare and Inouye, 2001). CspC and CspE are produced at normal growth temperatures, and they also participate in chromosome segregation and control of RNA synthesis for CspA—the major cold shock protein in *E. coli* (Bae et al., 1999). Studies with hyperosmotic stress have revealed that sigma S cooperates with both sigma 32 and sigma E in inducing genes to deal with a variety of stress conditions (Bianchi and Baneyx, 1999). In *Salmonella* sigma S activates virulence genes and in *E. coli* invasiveness (Wang and Kim, 2000).

III. *BACILLUS SUBTILIS* RESPONSE

Bacillus subtilis is a Gram-positive, motile, spore-forming, naturally competent, rod-shaped organism, approximately $4.0 \times 1.0\,\mu$m in size. The lifestyle of this organism is also regulated by a series of regulons. The *B. subtilis* heat shock response has been intensively investigated (see Hecker et al., 1996). It was first described in bacilli by Wachlin and Hecker (1984) and Streips and Polio (1985). This heat shock response, also assayed by two-dimensional gel electrophoresis, proved to be far more extensive than the one reported in *E. coli*, with at least 67 heat shock proteins induced (Miller et al., 1991). In addition the response to cold shock induced at least 53 polypeptides, some common to heat shock response (Lottering and Streips, 1995).

Control of Response

The careful work of Schumann, Hecker, and coworkers has delineated this extensive response into four classes of heat shock proteins (Hecker et al., 1996; Derre et al., 1999). Class I genes, which include the often studied molecular chaperones, the GroEL producing operon, and the DnaK synthesizing operon, are regulated by a repressor, Hcr, which recognizes the conserved, inverted repeat, CIRCE element in the promoter/operator region. This element has a consensus sequence of TTAGCACTC-N_9-GAGTGC-TAA. It has been found in over 50 species, always in front of the *groEL* and *dnaK* operons. It is postulated that the repressor binds to the CIRCE element during normal growth temperatures, allowing a minimal level of transcription. However, after heat

and other stress shocks, the repressor dissociates, allowing full transcription. It is known that the response decays after downshift to normal growth temperature. The repressor may then reassociate to the CIRCE element to inactivate the operons again. It is also known that the presence of the CIRCE element at the 5′ end of mRNA and close to the Shine-Dalgarno sequence will destabilize the mRNA and shorten its half-life. This may be a secondary mechanism for ensuring appropriate levels of heat shock proteins at all times (Homuth et al., 1999).

Class II heat shock genes are regulated by one of the sigma factors in bacilli, sigma B (see Haldenwang, 1995) (see the chapter by Helmann, this volume). These genes belong to the extensive stress shock regulon, which includes more than 100 proteins induced by heat shock and other stress conditions and growth conditions (Volker et al., 1999). A mutant for sigma B failed to induce any of the proteins under stress conditions, and the cells don't survive high heat nor several other stress conditions well, but sporulate normally (Duncan et al., 1987; Volker et al., 1999). Sigma B is encoded from an operon and is preceded by two promoters, one recognized by vegetative sigma A and a second by sigma B itself (Wise and Price, 1995). Since the gene is autoregulated, there is the opportunity for sigma B to mediate the acceleration of the stress response. The regulator for sigma B activity is RsbW, an anti-sigma factor, which is a member of the same operon in which the sigma B gene resides (Brown and Hughes, 1995). RsbW complexes with sigma B inactivating it. Another protein from the operon, RsbV can short-circuit this regulation by binding RsbW and sequestering it from reacting with sigma B (Dufour and Haldenwang, 1994). This reaction is dependent on the phosphorylation state of RsbV. Unphosphorylated RsbV is active, whereas phosphorylated is not. RsbW is the kinase which phosphorylates RsbV. This process is ATP dependent. Since under good growth conditions there is plentiful ATP, and RsbV is phosphorylated and inactive, then sigma B would be inactive and the stress regulon would be silent. However, upon stress a protein coded by the third gene from this operon, RsbU, a RsbV-P phosphatase is activated, thereby activating sigma B. The sensing of stress to activate this process may come from ribosomes (Zhang et al., 2001). The sigma B regulon appears to be multifaceted and important to a soil organism, such as *B. subtilis*, to ensure survival under a plethora of conditions.

Class III heat shock genes are not regulated by the CIRCE element nor by sigma B. The main genes in this class belong to the Clp protease family (*clpC*, *clpE*, *clpP*) (analogous genes produce heat shock proteins in *E. coli*). These enzymes are postulated to remove denatured proteins. In fact a mutation in MecB, a member of the ClpC group, is very sensitive to heat shock and osmotic stress and also affects the competence regulon (Msadek et al., 1994) (see Streips transformation chapter, this volume) All data to date suggest that the major sigma A is used to transcribe these genes. The class III genes are negatively regulated by the CtsR protein, which interacts in the operator region with a direct repeat heptanucleotide (A/GGTCAAA NAN A/GGTCAAA) (Derre et al., 1999b). In ClpE production, CtsR binds to the promoter, where there are five CtsR binding sites.

Class IV heat shock genes also are controlled independently of the CIRCE element and sigma B, and furthermore they are not regulated by CtsR. These include ClpX, FtsH, Lon, HtpG, and others (Derre et al., 1999a). The details on the regulation of these genes still need to be elucidated.

IV. GENERAL FACTS

In most cells several of the proteins induced at high temperatures are also present at normal growth temperatures (i.e., GroEL and DnaK) and must have promoter recognition for both the vegetative and heat shock sigma factors. These proteins also must be vital to normal growth functions and stress conditions. Other proteins appear to be un-

detectable during normal growth or are not vital for cellular survival (i.e., LysU and Lon).

The heat shock response is immensely conserved, with every living organism capable of inducing heat shock. Many of the proteins have sequence identity across species lines, either at the DNA or protein level (see Lindquist and Craig, 1988). The Hsp70 heat shock protein of humans and *Drosophila melanogaster* shares approximately 50% amino acid homology with the DnaK protein of bacteria. The *E. coli* GroEL protein is highly homologous with plant and animal heat shock and regulatory proteins. Proteins related to GroEL are displayed on the surface of some bacteria during infection (Haregewoin et al., 1989).

It is known that the heat shock proteins provide transient heat resistance to the cells (Kusukawea and Yura, 1988). For instance, the DnaK protein constitutes about 4.3% of the protein found in heat-shocked cells. Mutants for this protein are heat sensitive and grow aberrantly in *B. subtilis* (Staples et al., 1992a). *Escherichia coli* mutants with a deleted *dnaK* gene also exhibit poor gowth, poor viability, cell division aberrancy, and cold sensitivity at non-heat shock temperatures (Bukau and Walker, 1989). It is evident that this and most likely many of the other stress proteins also have a role in the normal growth of the cell.

The purpose of generating these proteins during stress conditions is still being actively investigated, though it is quite likely that one major purpose is to remove proteins that have been misfolded or denatured due to heat or other stresses (see Lindquist and Craig, 1988). As mentioned above, several of the heat shock proteins are proteases. In addition it is known that DnaK cooperates with ClpB to solubilize aggregates of thermolabile proteins (Mogk et al., 1999). The HtrA protein is a periplasmic, rather than cytoplasmic, protease whose activity is critical for high heat survival of *E. coli*, as mentioned above. This molecule forms a ring structure and destroys denatured proteins that arrive into the periplasm, thus providing a necessary function in a specific compartment other than the cytoplasm (Kim et al., 1999). It also known that the GroEL and GroES chaperones and the DnaK, DnaJ, and GrpE group participate in protein folding (Thomas and Baneyx, 1996). Nonnative proteins are bound to the GroEL ring structure and encapsulated by GroES. Then, following ATP hydrolysis, the polypeptide can achieve a native state, or perhaps repeat the process if not complete (Sigler et al., 1998). Furthermore the level of GroEL within the cell is critical. When this level is lowered, some heat shock proteins are induced due to stabilization of sigma 32, but also some proteins involved in intermediary metabolism are affected (Kanemori et al., 1994). So GroEL may control the proper expression of proteins other than just stress shock related ones. It is clear that as research in this area continues, these functions will be even more clearly elucidated and other, still unknown, purposes for stress-induced proteins will be revealed.

Neidhardt's laboratory did extensive work in documenting the overlap of the heat shock regulon with regulons controlling other cellular functions (Neidhardt and VanBogelen, 1987). In addition it is known that sporulation interfaces with heat shock in *B. subtilis* (Khoury et al., 1990). Bacteriophage interface with the heat shock proteins for their own morphogenetic purposes (Bahl et al., 1987; Sand et al., 1995; Staples et al., 1992b). Heat shock proteins alter the lysis capability of bacteriophage φX174 (Young et al., 1989). Topoisomerase activity is influenced by heat shock (Camacho-Carranza et al., 1995). In sum, many of the proteins induced in response to heat and other stresses perform vital and far-ranging functions in cells both during stress and normal growth.

This brief summary and update on stress shock research is designed to provide the reader with ample references and specific ideas where further research and opportunities for discovery in this area may lie. Good hunting!

REFERENCES

Arnosti DN, Chamberlin MJ (1989): Secondary σ factor controls transcription of flagellar and chemotaxis genes in *Escherichia coli*. Proc Natl Acad Sci USA 86:830–834

Bae W, Phadtare S, Severinov K, Inouye M (1999): Characterization of *Escherichia coli cspE*, whose product negatively regulates transcription of *cpsA*, the gene for the major cold shock protein. Mol Micobiol 31:1429–1441.

Bahl H, Echols H, Straus DB, Court D, Crowl R, Georgopoulos CP (1987): Induction of the heat shock response of *E. coli* through stabilization of the σ^{32} by the phage lambda cIII protein. Genes Dev 1:57–64.

Bianchi AA, Baneyx F (1999): Hyperosmotic shock induces the σ^{32} and σ^E stress regulons of *Escherichia coli*. Mol Microbiol 34:1029–1038.

Brown KL, Hughes KT (1995): The role of anti-sigma factors in gene regulation. Mol Microbiol 16:397–404.

Bukau B, Walker G (1989): Cellular defects caused by deletion of the *Escherichia coli dnaK* gene indicate roles for heat shock protein in normal metabolism. J Bacteriol 171:2337–2346.

Camacho-Carranza R, Membrillo-Hernandez J, Ramirez-Santos J, Castro-Dorantes J, de Sanchez V, Gomez-Eichelmann MC (1995): Topoisomerase activity during the heat shock response in *Escherichia coli* K-12. J Bacteriol: 3619–3622.

Crickmore N, Salmond GPC (1986): The *Escherichia coli* heat shock regulatory gene is immediately downstream of a cell division operon: The *fam* mutation is allelic with *rpoH*. Mol Gen Genet 205:535–539.

Danese PN, Snyder WB, Cosma C, Davis LJ, Silhavy TJ (1995): The Cpx two-component signal transduction pathway of *Escherichia coli* regulates transcription of the gene specifying the stress-inducible periplasmic protease DegP. Genes Dev 9:387–398.

Derre I, Rapaport G, Devine K, Rose M, Msadek T (1999a): ClpE, a novel type of HSP100 ATPase, is part of the CtsR heat shock regulon of *Bacillus subtilis*. Mol Microbiol 32:581–593.

Derre I, Rapaport G, Msadek T (1999b): CtsR, a novel regulator of stress and heat shock response, controls *clp* and molecular chaperone gene expression in Gram-positive bacteria. Mol Microbiol 31:117–131.

Dufour A, Haldenwang WG (1994): Interactions between a *Bacillus subtilis* anti-σ factor (RsbW) and its antagonist (RsbV). J Bacteriol 176:1813–1820.

Duncan ML, Kalman SS, Thomas M, Price CW (1987): Gene encoding the 37,000 dalton minor sigma factor of *Bacillus subtilis* RNA polymerase: Isolation, nucleotide sequence, chromosomal locus, and cryptic function. J Bacteriol 169:771–778.

Gross CA (1996): Function and regulation of the heat shock proteins.In Neidhardt FC, Curtiss III R, Ingraham, JL, Lin ECC, Low KB, Magasanik B (eds): "*Escherichia coli* and *Salmonella typhimurium*:

Cellular and Molecular Biology." Washington, DC: American Society for Microbiology, pp 1382–1399.

Haldenwang WG (1995): The sigma factors of *Bacillus subtilis*. Microbiol Rev 59:1–30.

Haregewoin A, Soman G, Hom RC, Finberg RW (1989): Human γδ+ T cells respond to mycobacterial heat-shock protein. Nature 340:309–312.

Hecker M, Schumann, Volker U (1996): Heat shock and general stress response in *Bacillus subtilis*. Mol Microbiol 19:417–428.

Homuth G, Mogk A, Schumann W (1999): Post-transcriptional regulation of the *Bacillus subtilis dnaK* operon. Mol Microbiol 32:1183–1197.

Kanemori M, Mori H, Yura T (1994): Effects of reduced levels of GroE chaperones on protein metabolism: Enhanced synthesis of heat shock proteins during steady-state growth of *Escherichia coli*. J Bacteriol 176:4235–4242.

Khoury PH, Qoronfleh MW, Streips UN, Slepecky RA (1990): Altered heat resistance in spores and vegetative cells of a mutant from *Bacillus subtilis*. Curr Microbiol 21:249–253.

Kim KI, Park S-C, Kang SH, Cheong G-W, Chung CH (1999): Selective degradation of unfolded proteins by the self-compartmentalizing HtrA protease, a periplasmic heat shock protein in *Escherichia coli*. J Mol Biol 294:1363–1374.

Kitigawa M, Mtsumura Y, Tsuchido T (2000): Small heat shock proteins, IbpA and IbpB, are involved in resistances to heat and superoxide stresses in *Escherichia coli*. FEMS Microbiol Lett 184:165–171.

Kusukawa N, Yura T (1988): Heat shock protein GroE of *Escherichia coli*: Key protective roles against thermal stress. Genes Dev 2:874–882.

Lindquist S, Craig EA (1988): The heat shock proteins. Annu Rev Genet 22:631–677.

Lipinska B, Fayet O, Baird L, Georgopoulos C (1989): Identification, characterization, and mapping of the *Escherichia coli htrA* gene, whose product is essential for bacterial growth only at elevated temperatures. J Bacteriol 171:1574–1584.

Lottering EA, Streips UN (1995): Induction of cold shock in *Bacillus subtilis*. Curr Microbiol 30:193–199.

Miller BS, Kennedy TE, Streips UN (1991): Molecular characterization of specific heat shock proteins in *Bacillus subtilis*. Curr Microbiol 22:231–236.

Missiakas D, Raina S (1997): Signal transduction pathways in response to protein misfolding in the extracytoplasmic compartments of *E. coli*: Role of two new phosphoprotein phosphatases PrpA and PrpB. EMBO J16:1670–1685.

Mogk A, Tomoyasu T, Goloubinoff P, Rudinger S, Roder D, Langen H, Bukau B (1999): Identification of thermolabile *Escherichia coli* proteins: Prevention and reversion of aggregation by DnaK and ClpB. EMBO J 24:6934–6949.

Msadek T, Kunst F, Rapaport G (1994): MecB of *Bacillus subtilis*, a member of the ClpC ATPase family, is a pleiotropic regulator controlling competence gene expression and growth at high temperature. Proc Natl Acad Sci USA 91:5788–5792.

Neidhardt FC (1987): Multigene systems and regulons. In Neidhardt FC, Ingraham JL, Low KB, Magasanik B, Schaecter M, Umbarger HE (eds): "*Escherichia coli* and *Salmonella typhimurium*: Cellular and Molecular Biology," Vol 2. Washington, DC: American Society for Microbiology, pp 1313–1317.

Neidhardt FC, Van Bogelen RA (1987): Heat shock response. In Neidhardt FC, Ingraham JL, Low KB, Magasanik B, Schaecter M, Umbarger HE (eds): "*Escherichia coli* and *Salmonella typhimurium*: Cellular and Molecular Biology," Vol 2. Washington, DC: American Society for Microbiology, pp 1134–1345.

Phadtare S, Inouye M (2001): Role of the CspC and CspE in regulation of expression of RpoS and UspA, the stress response proteins in *Escherichia coli*. J Bacteriol 183:1205–1214.

Sand O, Desmet L, Toussaint A, Pato M (1995): The *Escherichia coli* DnaK chaperone machine and bacteriophage Mu late transcription. Mol Microbiol 15:977–984.

Sigler PB, Xu Z, Rye HS, Burston SG, Fenton WA, Horwich AL (1998): Structure and function in GroEL-mediated protein folding. Annu Rev Biochem 67:581–608.

Staples RR, Miller BS, Hoover ML, Chou Q, Streips UN (1992a): Initial studies on a *Bacillus subtilis* mutant lacking the DnaK-homologue protein. Curr Microbiol: 24:143–149.

Staples RR, Miller BS, Streips UN (1992b): Bacteriophage φ105 induces the GroEL-homologue protein in *Bacillus subtilis*. Antonie van Leeuwehoek J Microbiol 61:339–342.

Streips UN, Polio FW (1985): Heat shock proteins in bacilli. J Bacteriol 162:434–437.

Thomas JG, Baneyx F (1996): Protein folding in the cytoplasm of *Escherichia coli*: requirements for the DnaK-DnaJ-GrpE and GroEL-GroES molecular chaperone machines. Mol Microbiol 21:1185–1196.

Tissieres, H, Mitchell HK, Tracy J (1974): Protein synthesis in salivary glands of *Drosophila melanogaster*: relation to chromosome puffs. J Mol Biol 84:389–398.

Volker U, Maul B, Hecker M (1999): Expression of the σB-dependent general stress regulon confers multiple stress resistance in *Bacillus subtilis*. J Bacteriol 181:3942–3948.

Wachlin G, Hecker M (1984): Proteinbiosynthesen nach hitzeschock in *Bacillus subtilis*. Z Allg Mikrobiol 24:397–398.

Wang Q, Kaguni JM (1989): A novel sigma factor is involved in expression of the *rpoH* gene of *Escherichia coli*. J Bacteriol 171:4248–4253.

Wang Y, Kim KS (2000): Effect of *rpoS* mutations on stress-resistance and invasion of brain microvascular endothelial cells in *Escherichia coli* K1. FEMS Microbiol Lett 182:241–247.

Wise AA, Price CW (1995): Four additional genes in the *sigB* operon of *Bacillus subtilis* that control activity of the general stress factor σ^B in response to environmental signals. J Bacteriol 177:123–133.

Young KD, Anderson RJ, Hafner RJ (1989): Lysis of *Escherichia coli* by the bacteriophage ϕX174 E protein: Inhibition of lysis by heat shock proteins. J Bacteriol 171:4334–4341.

Zhang S, Scott, JM, Haldenwang WG (2001): Loss of ribosomal protein L11 blocks stress activation of the *Bacillus subtilis* transcription factor σ^B. J Bacteriol 183:2316–2321.

Zhou Y-N, Kusukawa J, Erickson JW, Gross CA, Yura T (1988): Isolation and characterization of *Escherichia coli* mutants that lack the heat shock sigma factor σ^{32}. J Bacteriol 170:3640–3649.

13

Genetic Tools for Dissecting Motility and Development of *Myxococcus xanthus*

PATRICIA L. HARTZELL

Department of Microbiology, Molecular Biology, and Biochemistry, University of Idaho, Moscow, Idaho 83844–3052

**Modern Microbial Genetics, Second Edition. Edited by
Uldis N. Streips and Ronald E. Yasbin. ISBN 0–471–38665–0
Copyright © 2002 Wiley-Liss, Inc.**

I. WHAT THE MYXOBACTERIA HAVE TO OFFER US

A. History

Studies of the Gram-negative myxobacteria have broadened our understanding of prokaryotic biology since they were first described by Roland Thaxter, a Harvard botanist, over a century ago (Pfister, 1993; Thaxter, 1892). The myxobacteria were initially thought to be eukaryotes because they produce complex three-dimensional structures that resemble the fruiting bodies produced by slime molds. Thaxter's astute observations of the cell morphology, motility, and fruiting body formation of the myxobacteria led him to conclude that these organisms were bacteria. While others disputed this claim, Thaxter went on to publish descriptions of myxospore formation and germination that supported his ideas. His intricate sketches of the multicellular structures produced by these remarkable organisms can still be seen at the Farlow Reference Library at Harvard University.

The work of Helen Krzemieniewska and Seweryn Krzemieniewski in Poland renewed interest in the myxobacteria during the 1920s and 1930s. However, the studies of myxobacteria that persist today began with the work of H. Kühlwein in the 1950s and continued to develop as H. Reichenbach, E. Rosenberg, R. Burchard, M. Dworkin, D. Kaiser, D. Zusman, S. Inouye, H. Schairer, M. Inouye, D. White, and others, entered the field. The work that has emerged from these labs has shown us the potential for myxobacteria as antibiotic producers, environmental scavengers, and models for social interactions and development. The capacity to carry out such diverse processes is consistent with the fact that these organisms have one of the largest genomes among the prokaryotes, with twice as many genes as *E. coli*. Many of these genes are required for specialized functions, including adventurous (independent, or single-cell) gliding, social (group-dependent) gliding, and the complex, starvation-induced developmental cycle, which is unique to the myxobacteria. Genetic analysis of gliding and development has led to the discovery of novel extracellular signaling systems, a large family of serine-threonine kinases, a small Ras-like protein, regulatory timers, and a pheromone. Efforts to understand these remarkable, genetically tractable bacteria will continue to yield new surprises.

B. A Social Network Helps the Myxobacteria to Survive in Nature

Many of the rich organic odors that permeate the forest air are generated by *Streptomyces sps.* and the myxobacteria. The myxobacteria group includes organisms such as *Myxococcus xanthus, Stigmatella aurantiaca, Nannocystis exedens, Corallococcus coralloides, Chondromyces apiculatus*, and *Archangium sp.* that can be isolated from plants, soil, and decaying matter (Burchard, 1984; Reichen-

bach and Dworkin, 1992). Isolates belonging to different genera can be distinguished in part by the shape of their vegetative cells and their pattern(s) of gliding, a form of surface translocation that does not involve flagella (Burchard, 1984). Gliders can move over many types of relatively dry surfaces, so they are able to inhabit niches that are normally inhospitable to bacteria that depend on flagellar-mediated swimming movement.

The success of myxobacteria in nature is due to their capacity to degrade whole organisms and complex macromolecules. These aerobic chemoorganotrophs release hydrolytic enzymes, including nucleases, proteases, polysaccharidases, and lipases, which allow the myxobacteria to lyse, and then consume, other microbial cells (Zahavi and Ralt, 1984; Sudo and Dworkin, 1972). Myxobacteria seem to thrive best as dense swarms of cells—a multicellular existence reminiscent of social behavior. For example, secretion of hydrolytic enzymes results in a burst of nutrients that must quickly be consumed by the foraging cells (Zahavi and Ralt, 1984). Because gliding, the only means of directed movement, is slower than the rate of diffusion of hydrolyzed materials, groups of cells can feed more efficiently than an individual cell. Hence social behavior may compensate for the slow speed of gliding because a group of cells can consume material before it has a chance to diffuse.

C. *M. xanthus* Undergoes a Complex Developmental Program in Response to Starvation

1. Fruiting body morphogenesis

Myxobacteria are the only gliding bacteria that have the capacity to form a multicellular structure, called a fruiting body, as part of their life cycle. As shown in Figure 1, fruiting bodies come in a variety of colors, shapes, and sizes. Typical fruiting bodies of *Myxococcus xanthus* have a soft mound shape, whereas *Stigmatella* and *Archangium* produce fruiting bodies with sporangioles that have a treelike appearance (Reichenbach,

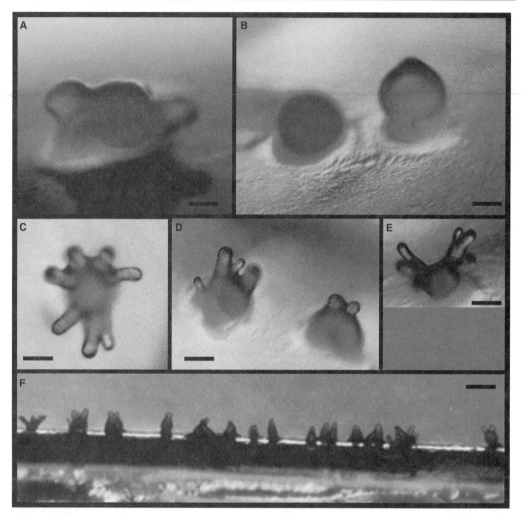

Fig. 1. Fruiting bodies of myxobacteria. **A, B:** Fruiting bodies of *M. xanthus* after five days on starvation medium; bar = 50 μm. **C, D, E:** Fruiting bodies of a myxobacterium isolated from rabbit dung in Idaho; bar = 100 μm. **F:** Side view of fruiting bodies of *M. xanthus*; bar = 200 μm. Samples were viewed with a Nikon FXA microscope equipped with diffraction interference contrast (DIC).

1984). The number of cells that congregate to form the fruiting body can range as high as 500,000, making these structures large enough to be seen with the naked eye. The production of carotenoids yields fruiting bodies that are red, yellow, or orange in color (Reichenbach and Kleinig, 1984).

Fruiting bodies are produced in response to starvation (Wireman and Dworkin, 1975). When the supply of nutrients is exhausted, the cells cease to divide and begin to express genes that are required for fruiting body for-

mation and production of quiescent, heat-resistant spores. Macroscopically, changes in cell behavior can be seen within the first hour as the frequency of cell reversal increases and the rate of colony expansion decreases (Fig. 2) (Jelsbak and Sogaard-Anderson, 1999). Hundreds of thousands of starving cells glide toward one another and construct the fruiting body that soon will be filled with differentiated spores. Cells begin to coordinate their movements to align tightly and to aggregate to a high cell

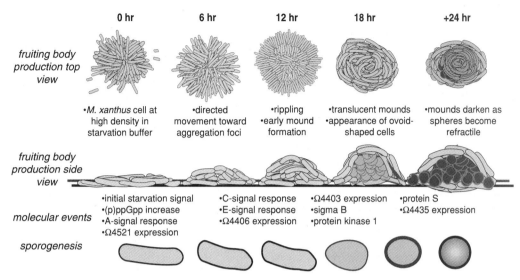

	0 hr	6 hr	12 hr	18 hr	+24 hr

fruiting body production top view

•*M. xanthus* cell at high density in starvation buffer

•directed movement toward aggregation foci

•rippling
•early mound formation

•translucent mounds
•appearance of ovoid-shaped cells

•mounds darken as spheres become refractile

fruiting body production side view

molecular events

•initial starvation signal
•(p)ppGpp increase
•A-signal response
•Ω4521 expression

•C-signal response
•E-signal response
•Ω4406 expression

•Ω4403 expression
•sigma B
•protein kinase 1

•protein S
•Ω4435 expression

sporogenesis

Fig. 2. Stages of fruiting body formation and sporogenesis. Fruiting body development and sporulation can be induced by placing an aliquot (20 μl) of a dense culture of *M. xanthus* (5 × 10^9 cells/ml) on buffered agar. Within six hours, cells within a small range begin to congregate and produce ripples, which are coordinated, wavelike movements (top view). As more cells enter the mound, the elevation increases (side view). A few of the molecular events that coincide with morphological changes are listed. Within the raised mound, cells begin to restructure their cell wall and ovoid-shaped cells can be seen by 18 hours. After 24 hours, fruiting bodies appear dark and grainy due to the presence of refractile, heat-resistant spores, which are coated with specific proteins, such as the calcium-binding protein S.

density. Cell alignment and aggregation are critical for signaling (Kim and Kaiser, 1990b). As cells align, they undergo a process called *rippling*, which generates dramatic wavelike patterns that are reminiscent of the cAMP-generated patterns produced during development of the slime mold *Dictyostelium discoideum* (Shimkets and Kaiser, 1982). Within 12 and 24 hours after the onset of development, aggregates of cells coalesce into mounds that increase in size to form a compact fruiting body. The ability to switch from the vegetative (growth) cycle to the developmental cycle ensures the survival of these organisms.

2. Development culminates in production of spores

Mature fruiting bodies are filled with spores—differentiated cells that have increased resistance to extremes of temperature, ultraviolet light, and desiccation (White, 1984). Sporulation is a property shared by many organisms, including many fungi, the myxobacteria, and bacteria of the genera *Bacillus*, *Clostridium*, and *Streptomyces*, whose survival in nature depends on the ability to endure cold winters, hot, dry summers, and periods of starvation (see Moran, this volume).

Sporogenesis in the myxobacteria is fundamentally different than in *Bacillus*, which produces endospores. During development, the endospore forms within a compartment of the *Bacillus* cell and acquires features that distinguish it from the terminally differentiated mother cell. Organisms such as *Streptomyces* and *Aspergillus* form multiple spores by septation of a filamentous cell. In contrast, each myxospore arises in toto from a rod-shaped myxobacterial cell that undergoes transformation. The cells of some myxobacterial strains undergo a dramatic shape change during spore formation. During the early stages of development, *M. xanthus* cells retain the long, thin rod shape that is characteristic of vegetative cells, but after 20 hours on starvation agar, they appear ovoid.

After another 10 to 12 hours the dull ovoid-shaped cells ripen into optically refractile spheres that are resistant to 50°C heat and sonication (Sudo and Dworkin, 1969).

3. Germination

When the environmental conditions become suitable for growth, spores can recover from the quiescent state and germinate to become vegetative cells. Very shortly after addition of amino acids the spores become sensitive to heat and detergent (SDS) treatment (Elías and Murillo, 1991) and refractility, which is one of the last visible features acquired during spore formation, disappears. A rod-shaped structure emerges through the outer wall of the spore and within about 5 hours, the spores have differentiated into rod-shaped cells. An empty spore shell remains behind (Voelz and Dworkin, 1962). During germination there is a burst of new RNA synthesis and new protein synthesis (Otani et al., 1995). Both calcium and protease activity are critical for germination because spores incubated with nutrients plus protease inhibitors, such as phenylmethylsulfonyl fluoride or the calcium chelator, EGTA, fail to germinate (Otani et al., 1995). Consistent with this, spores provided with only casitone and $MgSO_4$, but lacking calcium, are unable to germinate.

D. Vegetative Growth and Starvation-Induced Development

1. Growth of M. xanthus in rich and defined media

Amino acids or hydrolyzed protein, such as casitone, provide the carbon and energy source needed to support growth of *M. xanthus* and other myxobacteria in the laboratory (Dworkin, 1962). When aerated at 32°C in 1% casitone medium at pH7.6, *M. xanthus* will grow with a minimum doubling time of about 3.5 hours (Dworkin, 1962). Mono and disaccharides are not used as sources of energy, although pyruvate and acetate will support growth. A subset of myxobacteria can use cellulose and starch as a carbon and energy source (Reichenbach

and Dworkin, 1992). *M. xanthus* can be grown with a generation time of 6.5 hours in M1, a chemically defined medium containing 17 amino acids and minerals (Zusman and Rosenberg, 1971), or with a generation time of 22 to 36 hours in A1 medium, a defined medium containing seven amino acids, pyruvate, $(NH_4)_2SO_4$, and vitamins (Bretscher and Kaiser, 1978).

As cells enter the stationary phase of growth, secondary metabolites are produced (Reichenbach and Höfle, 1993). Myxobacteria are prolific producers of antibiotics and secondary metabolites such as myxovirescin (antibiotic TA)(Gerth et al., 1982), which adheres avidly to surfaces, and cytotoxic epothilons, which mimic the biological effect of taxol (Tang et al., 2000). Myxalin, a glycopeptide produced by *M. xanthus*, possesses antithrombic effects (Akoum et al., 1992). *Stigmatella aurantiaca*, a close relative of *M. xanthus*, produces myxothiazol, a compound that inhibits electron transport by targeting the cytochrome bc_1-complex. Myxothiazol is synthesized by a unique combination of polyketide synthases and nonribosomal peptide synthetases, which are activated by a 4′-phosphopantetheinyl transferase (Silakowski et al., 1999).

2. Induction of the development cycle

Myxobacteria can be induced to form fruiting bodies by starving the cells for nutrients (Dworkin, 1963). Three methods are commonly used to induce fruiting: rapid nutrient depletion, gradual nutrient depletion, and submerged culture.

The rapid nutrient depletion method provides a reproducible, synchronized way to induce development. Myxobacterial cells are grown in rich medium to a density of about 5×10^8 and harvested by centrifugation. The cell pellet is suspended in a nutrient-free buffer to about 5×10^9 to achieve a high cell density, and 20 μl aliquots are placed on 2% agar (Wireman and Dworkin, 1975) or buffered agar (1.5% agar with 10 mM Tris, 1 mM potassium phosphate, and 5 mM $MgSO_4$) (Bretscher and Kaiser,

1978; Kroos et al., 1986). Under these conditions roughly 200 to 400 fruiting bodies will form in about 24 hours within a spot that initially contained 10^8 cells.

Clone fruiting medium is used to induce fruiting body formation as a result of gradual nutrient depletion (Hagen et al., 1978). This medium more closely mimics the conditions that myxobacteria probably encounter in nature because it contains a limited supply of nutrients. Sufficient nutrients are available for *M. xanthus* cells to form colonies, but by the time cells reach a high density, the supply of nutrients is exhausted at which point fruiting bodies begin to form. Although fruiting body formation on clone fruiting medium may take up to two weeks, it is relatively easy to screen for developmental mutants using this medium.

The wild-type strain and a subset of mutants are able to develop in tissue culture wells. If about 2×10^9 cells suspended in buffer are placed in a polystyrene tissue culture well (24-well Falcon 3047), wild-type cells will form fruiting bodies on the bottom of the well under the buffer (Kuner and Kaiser, 1982). This technique, which is called *submerged culture*, makes it relatively easy to label developing cells with radioisotopes, monitor the release of molecules such as CsgA (C-signal, described below), and determine the effects of specific inhibitors of development. Submerged cultures of a *csg* mutant were used as a bioassay during purification of C-signal (Kim and Kaiser, 1990c). The *csg* mutant fails to develop in submerged culture because it cannot produce the C-signal. However, when column fractions containing C-signal purified from developing wild-type cells were added to the tissue culture well, the mutant was able produce fruiting bodies and mature spores.

II. GENETIC TOOLS

A. The 9.4 Mb Genome of *M. xanthus*

1. Physical map

Myxobacteria form a phylogenetically coherent group within the δ division of the purple bacteria. The circular *M. xanthus* genome has a G + C content of >67% (Mesbah et al., 1989). The genome of *M. xanthus* strains FB, DZ1 and DK101 is 9.45 Mbp, while that of strain DK1622, a derivative of DK101, is 9.2 Mb (Chen et al., 1990, 1991, He et al., 1994). The genome of *Stigmatella aurantiaca*, a close relative of *M. xanthus*, has been estimated at 9.35 Mbp (Neumann et al., 1993). DK1622 carries only a single copy of a cryptic phage genome that exists in multiple copies in its parent, DK101 (Chen et al., 1990, 1991; Starich et al., 1985). Cosmid (Hager et al., 2001), λ-ZAP (White and Hartzell, 2000), plasmid (Kimsey, H. Kaiser, D. unpublished results), and yeast artificial chromosome (YAC) libraries (Kuspa et al., 1989) have been constructed from *M. xanthus* DNA, and the sequence of the *M. xanthus* genome has recently been completed by the Cereon Microbial Genome group.

Because *M. xanthus* is near the theoretical limit of G + C content (Woese and Bleyman, 1972), the third codon (wobble) positions are rich in G or C. For example, the G + C content in the third position of a codon is typically 88–92% within an open reading frame. This helps to maintain the overall high G + C content of the chromosome and compensates for the need to produce proteins with amino acids derived from AT-rich codons, such as Asn, Lys, Phe, Ser, Tyr, and Ile (Bibb et al., 1984). Translations of non-coding regions typically are disproportionately rich in amino acids derived from GC-rich codons, including Pro, Gly, Arg, and Ala.

2. Functional map

Digestion of wild-type genomic DNA with *Ase*I generates 16 fragments that can be separated on CHEF gels (Chen et al., 1990). The chromosomal location of genes with known function was determined by hybridization between a transposon marker linked with the gene of interest and *Ase*I-digested chromosomal DNA (Chen et al., 1991). Because the transposons Tn5-*lac*, Tn*phoA*-132, Tn5-132, and Tn5-*lac* each have an internal *Ase*I site,[1] digestion of DNA from strains that carry one of these transposons with *Ase*I will generate 17 fragments. Hence one fragment that is present in the wild-type will be missing from the transposon mutant, and two new, smaller *Ase*I fragments will appear. The size of the lost fragment and the sizes of the new fragments make it possible to determine the position of the transposon to a unique *Ase*I fragment on the physical map (Chen et al., 1991). This technique has been used to locate insertions of Tn*5-lac* in A and S motility genes on the physical map of the *M. xanthus* genome (MacNeil et al., 1994a).

The *mariner* transposon, *magellan4* (Rubin et al., 1999), is being used to disrupt genes required for gliding motility and starvation-induced fruiting body formation and sporulation. Because these genes encode functions that are unique to gliding bacteria and myxobacteria, information obtained from their sequence should provide new insights into our understanding of these processes.

3. Resistance determinants, plasmids, and retroelements

The expression of β-lactamase by wild-type *M. xanthus* confers natural resistance to ampicillin, but cells are sensitive to kanamycin, tetracycline, and spectinomycin/streptomycin. Transposons and derivatives of pUC vectors that confer resistance to $40 \mu g/ml$ kanamycin or $12 \mu g/ml$ tetracycline are routinely used (Kaiser, 1984). Plasmids carrying the *aadA* gene from plasmid R100 confer resistance to 0.8 mg/ml spectinomycin and 1 mg/ml streptomycin (Magrini et al., 1998).

Plasmid pBGS18, a pUC18 derivative that carries the kanamycin-resistance determinant (Spratt et al., 1986), cannot replicate in *M. xanthus*, and it gives rise to Km[R] colonies only if it carries the phage attachment site or a region of homology with the chromosome (described below). Plasmids that integrate at

the phage attachment site are stable, but plasmids that integrate by homologous recombination appear to exist in equilibrium with the circular form of the plasmid. Intact plasmid can be recovered from *M. xanthus* after many generations and used to transform an *E. coli* host (Hartzell, 1997).

Although naturally occurring plasmids have not been isolated from any strain of myxobacteria, plasmids pMx-1 (O'Connor and Zusman, 1983) and pGC3, a kanamycin-resistant derivative of pMx-1 (Crawford and Shimkets, 2000a), have been engineered to replicate in *M. xanthus*. Transformation efficiencies show that the number of pGC3 plasmids that can be reisolated from *M. xanthus* is about 10^4-fold greater than the number of plasmids that arises following an excision event. This shows that pGC3 can replicate in *M. xanthus* (Crawford and Shimkets, 2000a).

Several myxobacteria, including *M. xanthus* and *Stigmatella aurantiaca*, carry different forms of satellite nucleic acid, called msDNA (multicopy single-stranded DNA), shown in Figure 3, and mrDNA (Lampson et al., 1989). These multicopy retroelements form a DNA-RNA hybrid joined by a phosphodiester linkage (Dhundale et al., 1987; Inouye and Inouye, 1991). The genes for msDNA are cotranscribed with the *rev* gene (reverse transcriptase) and deletion of *rev* abolishes production of msDNA. msDNA is synthesized from a precursor RNA that, after cleavage, serves as both primer and template for reverse transcriptase. There are two inverted repeats that are required for folding of the RNA into a stem-loop structure from which synthesis of DNA occurs (Hsu et al., 1989). Because mutations in the msDNA retron, which includes the genes for *msr* (msRNA), *msd* (msDNA), and *rev*, do not affect growth, motility, or development of *M. xanthus* cells, the function of the retroelement is unclear (Dhundale et al., 1988).

---

[1] Tn*5-lac* has two *Ase*I restiction sites.

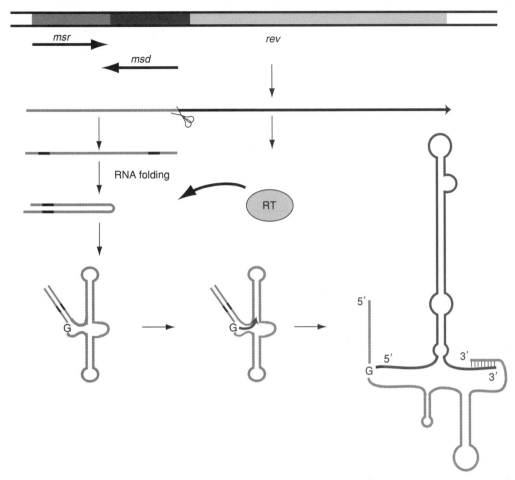

Fig. 3. Production of msDNA, a retron from *M. xanthus*. The single stranded msDNA molecule Mx162 is a 77 base RNA joined to a 162 base DNA. A single mRNA that includes the code for msRNA (*msr*), msDNA (*msd*) and reverse transcriptase, is cleaved before the *rev* coding region. Inverted repeats (black boxes) are required for folding of the RNA, which then serves as primer and template for DNA synthesis.

B. Transduction

Several bacteriophage have been characterized for use in the myxobacteria. The *Escherichia coli* phage P1 can act as a donor for the transfer of DNA from *E. coli* to *M. xanthus* (Kaiser and Dworkin, 1975). Myxophages Mx4 (Campos et al., 1978) and Mx8 (Martin et al., 1978) are generalized transducing phages that are used routinely for strain construction and analysis.

1. Coliphage P1

Phage P1 is able to inject DNA into *M. xanthus*, but is unable to multiply or estab-

lish lysogeny in this host (Kaiser and Dworkin, 1975; O'Connor and Zusman, 1983). As a result derivatives of P1 that carry a transposon, such as Tn5, will infect most strains of *M. xanthus* and yield stable Km^R transductants that carry Tn5 insertions but lack phage DNA. P1 also can mediate specialized transduction of plasmids from *E. coli* to *M. xanthus*. Plasmid pREG411, which was developed by Ron Gill (Shimkets et al., 1983), is a derivative of pBR322 that carries the P1 *inc* fragment encoding P1-specific incompatibility. pREG411 forms a co-integrate with infecting P1 *cam clr-100* by recombination between the homologous P1-incompatibility DNA sequences. Upon thermal induction to lytic growth, P1 will package within the P1 region of the co-integrate, yielding phage particles that carry the entire pREG11 plasmid. When these P1 stocks are used to infect *M. xanthus*, homologous recombination between P1-*inc* regions will regenerate the original plasmid, which can integrate onto the *M. xanthus* chromosome if homologous sequences are present.

2. Myxophage Mx4

The morphology of the lytic phage Mx4 is similar to coliphage P1 (Geisselsoder et al., 1978). Unlike the lytic coliphages, that lyse their starving host, growth of the lytic myxophage arrests if the host cell is starved for nutrients, and the phage remains in a dormant state until nutrients become available. Mx4 *ts-27htf-1 hrm-1* is a temperature-sensitive derivative that transduces at high frequency in *M. xanthus* (Geisselsoder et al., 1978). This provides a powerful tool for transduction because phage lysates prepared at 25°C fail to replicate when introduced into an *M. xanthus* host at 32°C. Use of this phage may be limited somewhat by its low efficiency of plating when grown on fully motile strains of *M. xanthus*.

3. Myxophage Mx8

Mx8 *clp2* is a clear plaque mutant of the phage Mx8 that is used to make high titer lysates for generalized transduction (Martin et al., 1978). Mx8 is temperate phage whose cycle resembles that of phage P22 from *Salmonella*. Single plaques of Mx8 grown on *M. xanthus* strain DZ1 contain about 10^7 to 10^8 plaque-forming particles that carry terminally repetitious DNA molecules that are circularly permutated. Phage particles can be purified by high-speed centrifugation for 30 minutes and are stable in buffer at 4°C.

Although the sequence of the 49.5 kb Mx8 genome has been completed (Accession AF396866, P. Youderian), it is unclear how Mx8 makes the decision between lysogenic and lytic growth because the regulatory

factors identified in other phage paradigms, such as alternative sigma factors and antiterminators have not been detected. Remarkably the Mx8 prophage is maintained even when *M. xanthus* cells undergo starvation-induced development. The function of particular Mx8 genes has been studied by integration of plasmids that carry these genes plus Mx8 *attP-int* at the *attB* locus, or by integration of plasmids that carry these genes plus *mglBA* at the *mgl* locus (Magrini et al., 1999; Salmi et al., 1998). When integrated at the *mgl* locus of a nonmotile *M. xanthus* Δ*mgl* strain, motile recombinants express the Mx8 genes constitutively. For example, these studies show that Mx8 encodes a nonessential DNA adenine methylase, Mox, which modifies adenine residues in occurrences of *Xho*I and *Pst*I recognition sites, CTCGAG and CTGCAG, respectively, on both phage DNA and the host chromosome (Magrini et al., 1997). Mox may protect phage DNA against restriction upon infection of particular strains of *Myxococcus*.

4. Integration of plasmid DNA at the Mx8 attachment site

Mx8 integrates as a prophage by site-specific recombination between a preferred site, *attP*, on the Mx8 genome and a preferred site, *attB*, on the *M. xanthus* chromosome (Stellwag et al., 1985). Plasmids that carry the *attP* and *int* (integrase) phage genes can integrate into the *attB* locus provide the best tool for complementation studies because the efficiency of site-specific recombination and integration at *attB* is much greater than homologous recombination. For example, if plasmid pPLH335, which carries a 2 kb chromosomal *mgl* fragment and the *attP-int* fragment, is introduced into a wild-type *M. xanthus* strain, 97% of Km^R colonies result from integration at *attB* while only 3% result from integration at the *mgl* locus (Hartzell, unpublished).

A minimal 2.2 kb fragment of the Mx8 genome that contains the *attP-int* genes required for integration at *attB* has been defined. The integrase is the only product required in *trans* for site-specific integration between *attP* and the 3′ ends of either of two tandem tRNA^{Asp} genes, *trnD1* and *trnD2*, that are located within *attB* (Magrini et al., 1999). Recombination between *attP* and the *attB1* site within *trnD1* is highly favored and often is accompanied by a deletion of the region between *trnD1* and *trnD2*. The *cis*-acting *attP* site lies within the *int* coding sequence. Upon integration, the 3′ end of the *int* gene is modified to make a less active form of Int (Tojo et al., 1996; Magrini et al., 1999). The lower specific activity may limit excisive recombination in lysogens that carry a single copy of the prophage.

5. Mapping by cotransduction

Mx4 and Mx8 phage can package about 50 kb fragments of DNA, permitting fine-structure genetic mapping in *M. xanthus* by cotransduction. The cotransduction frequency (C) can be used to approximate the physical distance (d) between markers in *M. xanthus* using the Wu equation (Kaiser, 1984; Wu, 1966).[2] The Wu method assumes that the entire length of the transduced fragment is homologous to a chromosomal equivalent, that these fragments are packaged at random, and that the likelihood for recombination between donor and recipient is equal over the length of the fragment (Sanderson and Hurley, 1987). Because the majority of cotransductions in *M. xanthus* involve a transposon or a plasmid insertion in the donor that is not present in the recipient, a segment of the incoming DNA is not homologous to the host chromosome. To estimate physical distance in these cases, it is necessary to reduce L (length in minutes where 1 min = 50 kb) by the amount of nonhomologous DNA (Sanderson and Hurley, 1987). The transposons Tn5 and Tn5-*lac* have 5.8 and 12 kb of nonhomologous DNA, respectively. For transduction of a Tn5-linked[3] marker, the equation becomes $d = 0.89(1 - c^{1/3})$.

The 5.8 kb transposon Tn5::1901 cotransduces with a frequency of about 80% with the *mglA8* allele (Sodergren and Kaiser, 1983; Stephens et al., 1989). If the Wu equation is modified to account for the size of the transposon, the distance between markers is estimated to be about 3.6 kb (Table 1). Sequence analysis shows that the actual distance between Tn5::1901 and the *mglA8* mutation is 1 kb, which is about a third of the distance predicted by the Wu equation. Additional crosses in which a Tn5 insertion is present on the donor-transducing fragment but not in the recipient consistently yield cotransduction frequencies that are lower than predicted by the Wu equation, given the known distances between sequenced insertions and point mutations. These results show that the Wu formula yields overestimates of molecular distances in *M. xanthus*. If Tn5 and other transposons contain high-efficiency packaging start signals (*pac* sites) (Margolin, 1987), the intact transposon would be packaged at a reduced frequency, which might account for disparities between actual and calculated distances.

6. Electroporation

Plasmid and phage DNA can be introduced into *M. xanthus* and its relatives, including *S. aurantiaca*, by electroporation (Kashefi and

---

[2] $C = (1 - d/L)^3$ [or $d = L - LC^{1/3}$, where d is the distance in minutes (based on conjugation in *E. coli* and *Salmonella*) and C is the cotransduction frequency when L is 1]. Solving the Wu equation for d gives:

$(c)L = ((1 - d/L)^3)L$
$(c)L = (L - d)^3$
$(cL)^{1/3} = ((L - d)^3)^{1/3}$
$(cL)^{1/3} = L - d$
$(-1)(cL)^{1/3} = (L - d)(-1)$
$-1(cL)^{1/3} = -L + d$
$+L - 1(cL)^{1/3} = (-L + d) + L$
$L - (cL)^{1/3} = d$

[3] $L = 1 = 50$ kb for Mx8 packaged wild-type DNA. If the donor has the 5.8 kb Tn5 marker, then $L \approx 0.89$.

TABLE 1. Comparison of Actual versus Predicted Distances between Genetic Markers

%CT	$(1 - C^{1/3})$	Predicted Distance in kb = dx50		Tn5 in donor $(L = 0.884)^a$		Actual Distance[b]
		Homologous DNA $L = 1$				
90	0.036	$d = 0.036$	1.8 kb	0.0318	1.59 kb	0.5 kb[c]
80	0.073	$d = 0.073$	3.65 kb	0.0645	3.23 kb	1.0 kb[d]
70	0.115	$d = 0.115$	5.75 kb	0.102	5.1 kb	NA
60	0.16	$d = 0.160$	8 kb	0.141	7.07 kb	NA
50	0.21	$d = 0.210$	10.5 kb	0.186	9.28 kb	NA
40	0.268	$d = 0.268$	13.4 kb	0.237	11.8 kb	NA
30	0.336	$d = 0.363$	18.2 kb	0.297	14.85	5.5 kb

[a] Tn5 is 5.8 kb (50 kb − 5.8 = 44.2; 44.2/50 = 0.884).
[b] Actual distance is based on sequence analysis of *cglB*, *mglA*, and *upsC*
[c] Based on cotransduction frequency and sequence analysis of *cglB*.
[d] Based on cotransduction frequency and sequence analysis of *mglA*.

Hartzell, 1995; Stamm et al., 1999). To prepare the DNA and the cells for electroporation, trace amounts of salt must be removed. Log-phase cells ($\approx 8 \times 10^8$ cells) are washed several times with water and mixed in a 0.1 cm cuvette with DNA that has been dialyzed against water. Cells pulsed with 0.65 kV [25 µF, 400 Ω] typically yield a time constant of about 9.4 ms and must be transferred immediately to 1 ml of rich medium (without antibiotic) and aerated at 32°C for two to four hours. The entire sample is then added to 4 ml of rich medium with 0.7% agar and spread on rich medium containing antibiotic (White and Hartzell, 2000).

The overall efficiency of electroporation depends on the size of the plasmid and whether integration occurs by homologous recombination or site-specific recombination at the prophage attachment site. Vectors in the 3 to 12 kb range that lack the *attP* fragment need at least 200 bp of homology with the chromosome in order for homologous recombination to be observed. If a plasmid carries a fragment with 1 to 2 kb of homology, about 500 recombinants per µg of DNA can be obtained from a wild-type strain and > 2000 recombinants per µg can be obtained from strains such as DK101[4] (Wall et al., 1999). The recovery of electro-porants increases about 50-fold if vectors that carry the *int-attP* fragment are used.

Homologous recombination with a single crossover between a circular piece of DNA (plasmid DNA) and the chromosome generates a merodiploid. As shown in Figure 4, if the incoming plasmid carries an *internal* fragment of a gene, integration will generate a disruption and may yield a mutant phenotype (unless the 3′ end of the gene is dispensable). Although this is a convenient way to construct a mutant, there are potential problems inherent with merodiploid strains. As described earlier, an integrated plasmid may be in equilibrium with the circular form, a merodiploid strain may undergo gene conversion (Stephens and Kaiser, 1987), and integration of plasmid DNA may have a negative or positive affect on expression on genes downstream of the insertion. These problems can be avoided by introducing linear DNA, which favors allele replacement by double recombination (Magrini et al., 1998). Prior to electroporation, plasmid DNA is cleaved at a site adjacent to a region of homology between the *M. xanthus* DNA on the plasmid and the genomic DNA target. This technique has been used to cross the *aadA* gene conferring resistance to Sp[R] Sm[R], flanked on both sides by *M. xanthus* DNA, onto the chromosome

[4] DK101 carries a mutation in the *pilQ* gene (sglA), which encodes a secretin for type IV pili biogenesis.

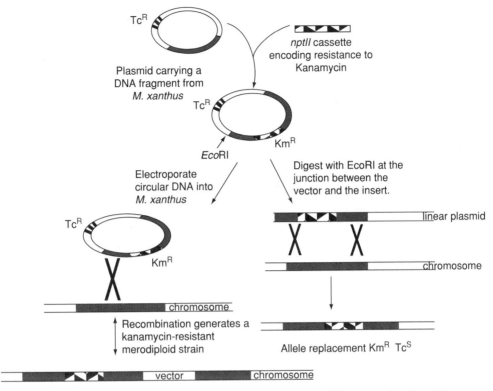

Fig. 4. Disruption of a nonessential gene. Replacement of an internal fragment of a plasmid-borne gene with an antibiotic resistance determinant, such as *npt*II, may be used to generate a null mutant. Integration of the circular form of the plasmid by homologous recombination with a single crossover gives rise to a merodiploid that interrupts the chromosomal gene. Integration of a linear plasmid by homologous recombination requires double crossover events and gives rise to Km^R gene disruptions that cannot excise.

(Magrini et al., 1998). It was possible to obtain recombinants that were resistant to spectinomycin and streptomycin but sensitive to kanamycin because the *nptII* gene encoding resistance to kanamycin resided on the plasmid outside of the region of homology and was lost after the double recombination event (see Fig. 4).

C. Transposons

1. Tn5-*lac*

The transposon Tn5-*lac*, shown in Figure 5, was developed as a tool to monitor the expression of genes during growth and development in *M. xanthus* (Kroos and Kaiser, 1984). To introduce the transposon into *M. xanthus*, a derivative of phage P1 that carries Tn5-*lac* was constructed and P1 : : Tn5-*lac*

lysates were prepared by thermal induction (Kroos and Kaiser, 1984). Upon infection, transposition of Tn5-*lac* into a nonessential region of the chromosome will yield colonies within about five days after plating on rich medium with kanamycin. If the promoterless *lacZ* gene is in the same orientation as the operon in which it has inserted, expression of β-galactosidase will reflect the activity of the *M. xanthus* promoter. Wild-type *M. xanthus* does not produce β-galactosidase. Transposition-generated fusions between *lacZ* and promoters that are induced during development have yielded a set of temporal markers that demarcate stages of development and identify genes whose expression is regulated in response to cell-cell signaling (see Section VI D). Tn5-*lac* also has been used

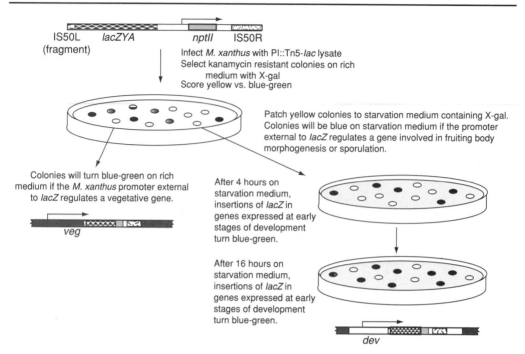

Fig. 5. Transposon Tn5-*lac* fuses expression of *lacZ* to an external promoter. Tn5-*lac* can transpose from P1 into *M. xanthus* cells giving rise to Km[R] colonies. If the insertion occurs in the proper orientation within a vegetatively expressed transcription unit, colonies will turn blue-green on rich medium supplemented with X-gal. Insertions that occur in transcription units that are expressed during development will yield blue-green colonies only on starvation medium containing X-gal.

to determine how mutations that affect motility alter gene expression during growth and development (Kroos et al., 1988; Shi and Zusman, 1993).

Identification of DNA flanking transposons such Tn5-*lac* (or Tn5), which lack an origin of replication, can be accomplished by in situ cloning (Laue and Gill, 1994), inverse PCR (Hartl and Ochman, 1994), or cloning the transposon and flanking DNA into a vector that carries an origin of replication. To clone in situ, as shown in Figure 6, a plasmid carrying a region of homology with the transposon is introduced by electroporation or P1 transduction into the strain with the transposon insertion (Laue and Gill, 1994). When plated on medium containing two antibiotics, such as kanamycin and tetracycline, recombinants that carry both the original (Km[R]) transposon and the incoming (Tc[R]) plasmid are selected. Stable

Km[R] Tc[R] derivatives can arise only if the plasmid has undergone homologous recombination with the transposon. Chromosomal DNA is then digested with a restriction enzyme that cuts at a unique site in the vector and nearby in chromosomal DNA. After ligation, the plasmid and flanking *M. xanthus* DNA can be recovered in *E. coli*.

2. magellan4

The transposon *Himar1* is a mobile genetic element of the *mariner* superfamily in eukaryotes (Hartl et al., 1997). The *mariner* transposon can move from donor to recipient DNA by a "cut-and-paste" mechanism that requires transposase but not host-encoded proteins (Hartl et al., 1997). The plasmid pMycoMar is a version of the mini-*Himar1* element *magellan4*, which was constructed for use as a suicide donor in prokaryotes (Akerley et al., 1998; Rubin et al.,

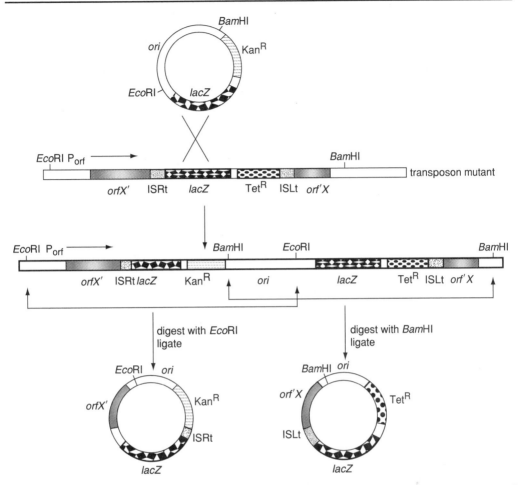

Fig. 6. In situ cloning. To identify genes disrupted by an insertion element that lacks an *E. coli* origin of replication a plasmid carrying a functional *ori* and a region of homology with the transposon is introduced by PI transduction or electroporation. Digestion of the resulting merodiploid with *Eco*RI liberates the Km^R marker and *M. xanthus* DNA located upstream of the insertion. Digestion with *Bam*HI liberates the Tc^R marker and *M. xanthus* DNA located downstream of the insertion.

1999). Unlike Tn*5* and its derivatives, which show marked site and regional specificities of insertion (hot spots), transposition of *magellan4* occurs more randomly at TA dinucleotide target positions (Rubin et al., 1999). When electroporated into *M. xanthus* pMycoMar, transposes at random locations with high efficiency (P. Youderian et al., unpublished results; White and Hartzell, 2000). The 2.2 kb transposon and flanking DNA can be recovered from *M. xanthus* chromosomal DNA by digesting with a restriction enzyme that does not cut within

the transposon. The *magellan4* transposon carries the R6Kδ origin of replication, and recovery of kanamycin-resistant colonies in *E. coli* requires the *pir*-encoded π protein.

III. MUTAGENESIS: CONSTRUCTING DELETION MUTANTS

The analysis of nonessential genes ideally involves construction of a strain with a mutation that deletes a large amount of the target gene without affecting the expression of genes downstream. Markerless deletion mutants can be constructed by integrating a gene carrying an internal deletion, and then enriching for loss of the integrated plasmid plus the wild-type copy of the gene, leaving the engineered copy on the chromosome. For example, *M. xanthus* strain DK6204 lacks a 850 bp region of the *mglBA* operon, rendering the mutant nonmotile and unable to develop (Hartzell and Kaiser, 1991). To construct DK6204, a 2 kb chromosomal fragment containing *mglBA* was digested with *Bal*I to remove about 75% of the coding region, and the resulting plasmid was introduced into wild-type *M. xanthus*. Motile, Km^R colonies that carried the full-length *mglBA* and the Δ*mglBA* were passed daily in medium lacking antibiotic to allow for excision of the vector. After 20 days, about 10% of the Km^s colonies with a nonmotile phenotype were found to have lost the wild-type copy of *mglBA*.

More expedient methods for constructing markerless deletion mutations employ genes such as *sacB* or *galK* that provide a counterselection. The *sacB* gene confers sucrose sen-

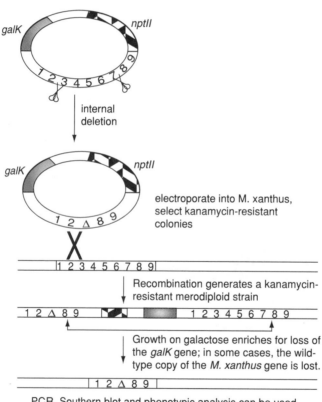

Fig. 7. Generating a null mutation using the *E. coli galK* gene as a counterselection. A plasmid with the gene of interest with an internal fragment removed, *npt*II, and *galK*, is introduced into *M. xanthus* and Km^R colonies are selected. Because GalK converts galactose into galactose-phosphate, which is toxic to *M. xanthus*, growth of this mutant in rich medium with 1–2% galactose but lacking kanamycin enriches for excision of the integrated plasmid. If the gene is not essential, some of the Gal^R Km^S derivatives will have undergone excision of the wild-type copy of the gene.

sitivity on its *M. xanthus* host when it is in the same orientation as adjacent *M. xanthus* genes (Ried and Collmer, 1987; Wu and Kaiser, 1996). A plasmid carrying *sacB* and in-frame deletions of *pilA* and *pilS* were used to construct null mutants in these *pil* genes (Wu and Kaiser, 1996). The *galK* gene, which encodes galactokinase, also has been used as a counterselection for constructing deletion mutants in *M. xanthus* (Ueki et al., 1996). As shown in Figure 7, a plasmid carrying *galK*, *nptII*, and the 5′ and 3′ ends of the *aglU* gene (Δ*aglU*), was introduced into *M. xanthus* strain DK1622 to generate a Km^R merodiploid (White and Hartzell, 2000). After growth on 2% galactose, which is toxic in the presence of GalK, excision of the integrated plasmid carrying *galK* and allele exchange yielded Km^S, galactose-resistant strains missing the wild-type copy of *aglU*, which is required for A-motility.

IV. REGULATION OF MOTILITY AND DEVELOPMENT

A. Regulatory Components

1. Sigma factors

M. xanthus uses at least seven sigma factors, shown in Table 2, to regulate expression of genes during vegetative growth and development. The major sigma factor, *sigA* (Inouye, 1990), has been purified and shown in vitro to transcribe the *vegA* gene (Biran and Kroos, 1997). A vegetative promoter consensus, TAGACA-17-TAAGGG, was derived by alignment of the *vegA* promoter and the *aphII* promoter with the *E. coli* σ^{70} consensus sequence. Alternative sigma factors CarQ and RpoEI are involved in regulation of extracytoplasmic factors (ECF) for production of carotenoids and control of motility, respectively.

Like *B. subtilis*, *M. xanthus* uses different sigma factors to regulate expression of genes

TABLE 2. Major Sigma Factors of *Myxococcus xanthus* and *Stigmatella aurantiaca*

Gene	Protein	Function	Reference
sigA (*rpoD, asgC*)	Sigma A; M$_r$ 80.4 kDa in *M. xanthus* by Inouye and 105 kDa by Biran and Kroos; M$_r$ 79.9 in *S. aurantiaca*	Major vegetative sigma factor	Biran and Kroos, 1997; Inouye, 1990
sigB	Sigma B; M$_r$ 33 kDa (original estimate 21.5 kDa)	Sporulation maturation	Apelian and Inouye, 1990, 1993
sigC	Sigma C; M$_r$ 33 kDa	Expression of genes involved in negative regulation of initiation of development	Apelian and Inouye, 1993
sigD	Sigma D	Stationary phase and development	Ueki and Inouye, 1998
carQ	ECF (extracytoplasmic factor)	Expression of carotenoids	Martinez-Argudo et al., 1998
rpoEI	ECF (extracytoplasmic factor)	Motility behavior; may interact with FrzZ	Ward et al., 1998a
rpoN	Sigma 54	Essential; required for expression of *csgA*, *pilA*, and *sdeK*	Keseler and Kaiser, 1997

during specific stages of development. Expression of genes early in development is controlled by the SigC protein while at least one sigma factor, SigB, acts late in development to regulate spore maturation (Apelian and Inouye, 1990, 1993). The *rpoN* gene, encoding σ^{54}, is essential for growth of *M. xanthus* (Keseler and Kaiser, 1997). σ^{54} regulates *pilA*, which is required for social motility, and several developmental genes, including myxobacterial hemagglutinin (MBHA) (Romeo and Zusman, 1991), *sdeK*, (Garza et al., 1998) and Tn*5-lac* :: 4521(Keseler and Kaiser, 1995). Act1, a σ^{54} activator protein similar to NtrC, is required for full expression of the *csgA* gene during development (Gorski et al., 2000).

2. Analysis of regulatory elements and genetic units

Modified pUC and Litmus℗ vectors that carry a minimal 2.2 kb Mx8 *attP int* region allow for the analysis of genes and their regulatory regions in single copy upon site-specific integration at the *attB* locus (Magrini et al., 1999). Although integration of genes at the *attB* locus is useful for complementation studies, the level of expression for a gene at *attB* is lower than the level of expression for the same gene at its chromosomal site (Bradner and Kroos, 1998). To overcome this problem, vectors that fuse a regulatory region to a reporter gene, such as *lacZ*, can be integrated at the native locus to delimit *cis*-acting elements. As shown in Figure 8, homologous recombination generates two versions of the same regulatory region. Only Lac$^+$ colonies are chosen to ensure that recombination has fused *lac* expression with the regulatory elements on the recipient chromosome. Hence expression of *orfX* will depend on the regulatory element carried on the incoming (donor) fragment. This method has been used to define the regulatory region required for expression of

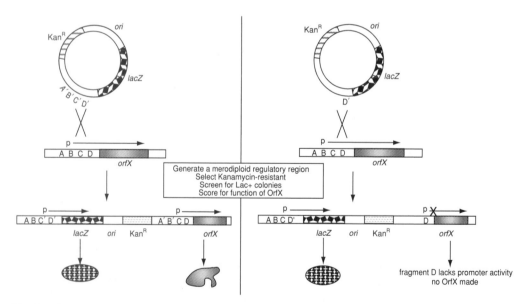

Fig. 8. Analysis of regulatory elements in *M. xanthus* A plasmid carrying the *lacZ* gene under control of the *orfX* regulatory region (A • B • C • D) is introduced into *M. xanthus* and blue-green, Km^R colonies are selected on rich medium with X-gal. If the regulatory region is complete, normal amounts of OrfX will be produced and the phenotype of the Km^R derivative will be wild-type (right panel). If essential elements of the regulatory region have been deleted, no OrfX will be produced and the strain will exhibit the *orfX$^-$* phenotype *(left panel)*.

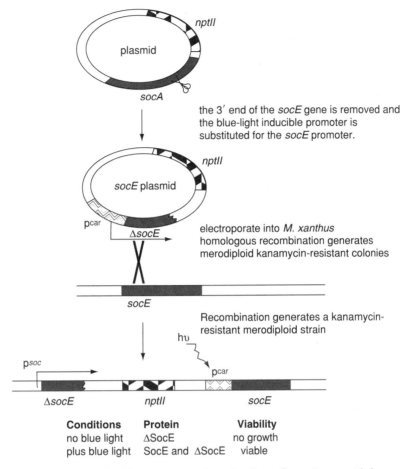

Fig. 9. Use of the blue-light inducible promoter to show that the *socE* gene is essential. A truncated form of the *socE* gene is placed under control of the *car* promoter (P$_{hv}$), which responds to blue light. Integration of the plasmid generates strain in which the full-length SocE protein is made only if cells are grown in blue light. If SocE is essential, no growth will occur when cells are grown in the dark.

ypp, the yellow-pigment production gene (Laue and Gill, 1994). A similar technique has been used to examine the regulatory elements upstream of *4521*, which carries an insertion of Tn*5-lac* in a gene that is induced during development (Gulati et al., 1995).

3. Analysis of essential genes with the blue-light inducible promoter

M. xanthus responds to blue light by producing carotenoid pigments. Carotenoid synthesis in *M. xanthus* is regulated by the *car* genes (Balsalobre et al., 1987), include *carQ*, which encodes an extracytoplasmic sigma factor (ECF), and *carR*, which encodes an anti-sigma factor located in the inner membrane (Hodgson, 1993). In the dark, CarR sequesters CarQ to the membrane, but when cells are exposed to light, degradation of CarR liberates the CarQ sigma factor to increase transcription of the *carQRS* regulon. Increased production of CarS relieves CarA-mediated repression of the *crtEBDC* operon.

The blue-light responsive *carQRS* regulatory domain, p$_{hv}$, has been used to study genes that are essential for growth and for studying the effect of ectopic expression. As shown in Figure 9, p$_{hv}$ was used to show that *socE* is essential for growth in a *csgA*$^+$ strain and that depletion of SocE arrests growth and induces sporulation under nutrient-rich conditions (Crawford and Shimkets, 2000b). The *hv* promoter also was used to determine the role of (p)ppGpp in initiation of development. The *E. coli relA* gene encoding (p)pp Gpp synthetase I was placed under the control of the *hv* promoter and integrated at the chromosomal Mx8 prophage attachment site in *M. xanthus* (Singer and Kaiser, 1995). Blue-light dependent expression of RelA resulted in an increase in (p)ppGpp, which was sufficient to activate development specific gene expression.

V. GLIDING MOTILITY

A. Genetic Analysis of Gliding Motility

I. Social motility

Social motility requires type IV pili (Tfp: type four pili) (Wu and Kaiser, 1995, Wall and Kaiser, 1999), fibrils (or the perception of fibrils) (Behmlander and Dworkin, 1991; Dworkin, 1999), exopolysaccharide (Weimer et al., 1998), and lipopolysaccharide (LPS) (Bowden and Kaplan, 1998). The *pil* cluster in *M. xanthus* contains 17 genes, 14 of which encode products that are about 30% identical to type IV pilin and pilus assembly proteins of *Pseudomonas aeruginosa* (Wall and Kaiser, 1999). A single *pilA* gene encodes the major pilin subunit that forms a coiled-coil structure with other PilA monomers to form the pilus fiber. Components that form the base of the pilus are related to the type II secretion system. Assembly of the pilus in *M. xanthus* also requires the function of an ABC exporter complex containing PilG, PilH, and PilI, proteins that are not found in *Ps. aeruginosa* Tfp (Wu et al., 1998). S-motility also requires the function of a tetratricopeptide repeat (TPR) protein called *Tgl* (Rodriguez-soto and Kaiser, 1997) and SglK, a DnaK homologue (Weimer et al., 1998). The Tgl protein, like pili, is located at the poles of the cell. Although *tgl⁻* mutants lack pili, they can be stimulated transiently to glide if physically associated with a *tgl⁺* strain (Hodgkin and Kaiser, 1977). The *sglK* mutant produces pili but lacks extracellular

fibrils that may be needed for cell-cell interactions during S-motility (Weimer et al., 1998).

The requirement that cells be within one cell length of one another in order to move by S-motility suggests that *M. xanthus* can use pili and/or fibrils to communicate or attach to one another. S-motility is very similar to Tfp-mediated twitching motility in organisms such as *Pseudomonas* and *Neisseria* (Semmler et al., 1999). One model suggests that S-motility and twitching motility function through extension and retraction of pili (Merz et al., 2000). According to this model, when pili anchor to a surface, the cell can be pulled toward the site of attachment as pili retract. The molecular components, PilF and PilT, are predicted to be critical for extension (PilT) and retraction (PilF). Both PilF and PilT contain ATP-binding sites consistent with the need to hydrolyze ATP during extension or retraction.

2. Adventurous motility

Cells lacking S motility can still move by the A-motility system, producing lacelike flares of individual cells at the edge of a colony. A-motility requires CglB, a 44 kDa cysteine-rich lipoprotein, which has been shown by immunofluorescence to localize to the cell surface (Rodriguez and Spormann, 1999). This is consistent with the finding that addition of either *cglB⁺* cells or purified CglB protein is sufficient to restore motility to *cglB⁻* cells.

A-motility also requires proteins that share identity with the *E. coli* Tol proteins. The *aglU* gene, which encodes a homologue of TolB (White and Hartzell, 2000) and at least five other genes encoding homologues of TolQ, TolR, and a second TolB, are essential for A-motility. In *E. coli*, the Tol proteins form heterooligomeric membrane-associated macromolecular transport complexes which are required for both biopolymer transport and outer membrane stabilization in Gram-negative bacteria (Lazzaroni et al., 1999). AglU, like its TolB counterpart, contains WD-repeat motifs that likely form multiple β-propellar platforms involved in protein-protein interactions. The Tol complexes in *M. xanthus* may play a role in transport of gliding components. Although mutations in the *tol* gene homologues do not affect the ability of cells to glide during development, the production of heat-resistant spores is abolished in these mutants (MacNeil et al., 1994b). This hints that during development, these transport complexes play an auxiliary role.

3. The GTPase is required for both motility systems

The *mglA* gene encodes a 22 kDa Ras-like GTPase that is essential for gliding and fruiting body development but is not essential for growth. A single mutation in *mglA* affects both A-and S-motility and *mglA* mutants produce colonies that are indistinguishable from A$^-$S$^-$ double mutants. The yeast GTPase, Sar1p, can complement an *mglA* mutant (Hartzell, 1997). *mglA* is transcribed as part of an operon with *mglB*, which encodes a protein with a Ras guanyl nucleotide release factor (GNRF) consensus motif. In the absence of GNRF, the activity of eukaryotic Ras is reduced. Consistent with this, MglB mutants have reduced rates for A$^-$ and S-motility (Hartzell and Kaiser, 1991; Spormann and Kaiser, 1999), presumably due to reduced activity of MglA. Time-lapse studies of individual cells show that *mglA* mutants are capable of gliding but that they make no net movement because they reverse direction 17 times more often than cells of the wild-type strain (2.9/min for Δ*mgl* versus 0.17/min for wild-type) (Spormann and Kaiser, 1999). This suggests that one function of MglA is to regulate the reversal frequency, perhaps by interacting directly or indirectly with the Frz proteins. Indeed, a mutation in the C-terminal domain of the *frzCD* gene, which results in a truncated FrzCD MCP, can partially suppress a mutation in *mglA* (Spormann and Kaiser, 1999). Because MglA affects both A and S motility, it may act to coordinate the direction of movement for these two gliding systems.

4. Gliding

Gliding is controlled by multiple sets of chemotaxis proteins. The most detailed studies of gliding in myxobacteria have focused on *M. xanthus*, which produces flat, spreading colonies. The edges of wild-type colonies develop a delicate, lacelike appearance (flares) as groups of cells and isolated cells glide outward from the initial point of growth. Isolated cells of *M. xanthus* move at about 3.8 μm/min on a 1.5% agar surface, while cells in close proximity (< 1 cell length) move at about 5 μm/min (Spormann, 1999), reversing their direction once every 5 to 7 minutes (Blackhart and Zusman, 1985). The reversal frequency of gliding is controlled by homologues of the enteric chemotaxis (Che) proteins. The first chemotaxis genes identified in *M. xanthus* were the *frz* genes (McBride et al., 1989), whose products coordinate the movements of cells to form a fruiting body during development. A second set of chemotaxis genes, called *dif*, is required for social gliding motility (Yang et al., 1998).

Environmental signals that affect methylation of FrzCD, an MCP (methyl-accepting chemotaxis protein) include chemorepellants, such as isoamylalcohol, and chemoattractants, such as phosphatidylethanolamine. Isoamylalcohol causes demethylation of FrzCD and increases the frequency with which cells reverse their direction of movement (McBride et al., 1992) while phosphatidylethanolamine causes a decrease in cell reversal frequency (Kearns and Shimkets, 1998). Although a poorly soluble fatty acid molecule is unlikely to serve as a chemical stimulant for fast moving cells, such as swimming bacteria, the slow rate of diffusion for PE and related compounds is compatible with the slow rate of gliding.

Although a wide variety of environmentally important organisms move by gliding, it remains a poorly understood process. Genetic studies identify >60 loci that comprise two independent sets of gliding genes—those that control adventurous or A-motility and those that control social or S-motility

(Hodgkin and Kaiser, 1979a, 1979b; Mac-Neil et al., 1994a). Mutations in the A-motility genes (*agl* or *cgl*) abolish motility of isolated cells, but mutants are able to glide using S motility if cells are within one cell length of each other. S⁻ mutants (*dsp*, *pil*, *sgl*, *tgl*) can glide as single cells using A-motility, yet show reduced cohesiveness and swarming and produce less exopolysaccharide (EPS) (Arnold and Shimkets, 1988). Hence *M. xanthus* has two systems that enable gliding. A⁻ and S⁻ single mutants (A − S+ and A + S−) have reduced motility compared with a wild-type A⁺S⁺ strain. Double mutants having an A⁻S⁻ genotype produce small colonies with smooth edges (Hodgkin and Kaiser, 1977) and are nonmotile when viewed by time-lapse under the microscope (Spormann and Kaiser, 1995).

B. Gliding Motors

I. Adaptation to different conditions

The A- and S-motility systems are adapted for optimal movement in different environments (Shi and Zusman, 1993). Mutants using only the A system (A⁺S⁻), produce colonies that resemble the wild-type strain on 1.5% agar but move poorly on 0.3% agar. The opposite is true for A⁻S⁺ strains, which move poorly on 1.5% agar but behave more like the wild-type strain on 0.3% agar. This is consistent with the behavior of cells by videomicroscopy—A⁺S⁻ cells can move as single cells and A⁻S⁺ cells need to be near other cells to move. *M. xanthus* cells may preferentially use A motility on firm, dry surfaces, and S-motility on soft, wet surfaces (Shi and Zusman, 1993).

2. Lateral A-motility motors and polar S-motility motors

Clues to the location and distribution of gliding motors in *M. xanthus* have come from analysis of wild-type and mutant strains treated with cephalexin (Spormann, 1999; Sun et al., 1999). Cephalexin blocks cell separation during growth, causing cells to form long filaments (Segall et al., 1985). Videomicroscopic analysis shows that filaments formed by wild-type (A⁺S⁺) cells move at about the same rate as untreated cells, suggesting that gliding motors are distributed over the length of the cell and increase in proportion to the cell length. Although the rate of gliding of cephalexin-treated A⁺S⁻ cells is normal, the rate of gliding of cephalexin-treated A⁻S⁺ cells is severely reduced (Sun et al., 1999). These data suggest that A-motility is powered by components that are distributed over the length of the cell, whereas S-motility is powered by components at the poles of the cells, which become limiting as the cell length increases.

C. The Search for New Gliding Genes

I. Screening for nonmotile mutants

The early work of Hodgkin and Kaiser showed that a single mutation in an A- or S-motility gene reduces gliding, and that mutants that carry two mutations—one in an A gene and one in an S gene, are nonmotile (Hodgkin and Kaiser, 1979a,b). Transposons Tn*5*, Tn*5-lac*, and *magellan4*, have been used to identify new A- and S-motility genes (P. Youderian et al, unpublished; MacNeil et al., 1994a; White and Hartzell, 2000). As shown in Figure 10, insertions in A-motility genes (or S-motility genes) can be obtained by introducing a transposon into a strain that is S⁻, then screening among the Km^R colonies for nonmotile (A⁻S⁻) mutants. To confirm that the transposon has disrupted an A-motility gene, Mx4 or Mx8 phage lysate of the nonmotile strain is used to transduce a fully motile wild-type strain to Km^R. If the original transposon inserted in an A-motility gene, each Km^R transductant will lack A-motility (Fig. 11).

Strains that lack S-motility produce colonies of a size intermediate between wild-type and nonmotile mutants, and they move poorly on 0.3% agar relative to 1.5% agar. Many Tn*5-lac* insertions affecting motility yield transcriptional fusions that express β-galactosidase at high levels during growth and development and give a

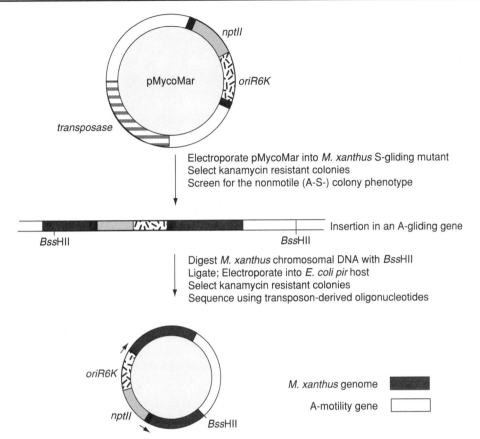

Fig. 10. Targeting of motility genes by transposon mutagenesis. Introduction of transposons such as Tn*5-lac* and *magellan4* into a strain with a defect in an S-motility gene (or A-motility gene) will result in colonies with a nonmotile phenotype in some cases due to disruption of a gene in the A-motility system (or S-motility if the original mutant is A⁻). The plasmid pMycoMar contains a transposon derived from *Himar1*, a *mariner* family element isolated from the horn fly *Haematobia irritans*, which was modified to carry the conditional R6Kγ origin of replication, and the kanamycin resistance gene from Tn*5*. When pMycoMar is electroporated into *M. xanthus*, transposition occurs at a high frequency. Plasmids that carry the insertion and flanking DNA can be recovered after digestion of chromosomal DNA with an enzyme that does not cut within the transposon. Sequence of the disrupted gene is obtained using primers that anneal with the ends of the transposon.

Lac⁺ phenotype. These strains can be used to identify mutations in genes that regulate motility and to identify environmental factors that affect expression of motility genes.

2. Protein-protein interactions lead to identification of new genes

New insights into gliding and control of gliding have come also from experiments that use the yeast two-hybrid system. One of the two cheW-like domains of the FrzZ protein interacts with the C-terminal domain of a protein, AbcA, that is related to ATP-binding cassette (ABC) tranporter proteins (Ward et al., 1998). AbcA is similar to transporters that are involved in export of small molecules and its role in chemotaxis may be to export a self-generated, "autochemotactic" signal that controls social gliding (for review, see Ward and Zusman, 1999).

1. Generate mutants

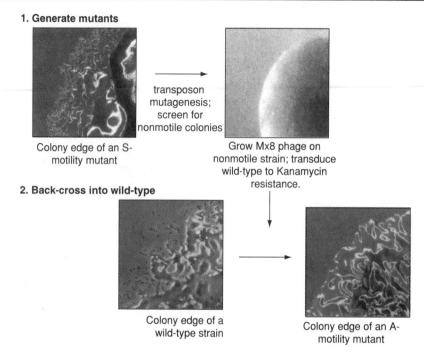

Colony edge of an S-motility mutant

transposon mutagenesis; screen for nonmotile colonies

Grow Mx8 phage on nonmotile strain; transduce wild-type to Kanamycin resistance.

2. Back-cross into wild-type

Colony edge of a wild-type strain

Colony edge of an A-motility mutant

Fig. 11. Identification of genes involved in A- and S-motility using a genetic test. *M. xanthus* strains that carry two mutations—one in any A gliding gene and one in any S gliding gene—are nonmotile. 1: To enrich for transposon mutations in A gliding genes, an S⁻ mutant is mutagenized with the transposon and Km^R mutants are screened for the A⁻S⁻ (nonmotile) phenotype. 2: To confirm that the transposon is in an A-motility gene, phage lysates are used to transduce the insertion into a naïve host to generate a mutant whose colony phenotype has the characteristic A⁻ edge.

VI. STARVATION-INDUCED FRUITING BODY FORMATION AND SPORULATION

A. Initiation of Development

1. Regulation of the stringent response

The biochemical and morphological events that produce the mature fruiting body have been studied in some detail for *M. xanthus*. The RelA-dependent stringent response is activated in response to starvation for amino acids, the primary source for carbon and energy. Ectopic expression of the *E. coli relA* gene from the *car* promoter (p*hv*, described earlier) in *M. xanthus* activates transcription of developmental-specific genes in the absence of a starvation signal (Singer and Kaiser, 1995). As ribosomes stall in the absence of charged tRNAs, the ribosome-associated protein RelA produces the effectors guanosine-5′-diphosphate-3′ diphosphate (pp Gpp) and guanosine-5′-triphosphate-3′ diphosphate (pppGpp), which are predicted to shift certain enzymatic activities in the cell, including redirecting the function of RNA polymerase. After abrupt starvation, the level of ppGpp increases 10- to 15-fold within 30 minutes, then decreases slightly and persists for about 8 hours at a level that is about 5-fold higher than vegetative levels (Crawford and Shimkets, 2000b; Manoil and Kaiser, 1980). When nutrients are available, the activity of RelA in *M. xanthus* is inhibited by the SocE protein, but when cells are starved for nutrients, *socE* transcription declines (Crawford and Shimkets, 2000a), making it possible for the level of CsgA to increase (Crawford and Shimkets, 2000b; Shimkets, 1999).

2. Bsg protease

The activity of a 90 kDa ATP-dependent protease called BsgA also is required early in development (Gill et al., 1993). The *bsgA* mutants appear to be unable to initiate a required cell-cell interaction and consequently fail to transcribe normal levels of many developmentally induced genes (Gill et al., 1993). The role of Bsg protease may be to degrade regulatory proteins that repress the expression of developmental-specific genes during vegetative growth. The mechanism by which Bsg is activated during development is unknown.

B. Extracellular Signals Are Required for Morphogenesis and Sporulation

I. Isolation of signaling mutants

The multicellular nature of fruiting body morphogenesis and sporulation in *M. xanthus* requires cell-cell communication and exchange of signals, which predicts that mutants unable to produce critical extracellular signals needed for sporulation should be able to develop if provided with the missing signal molecule. Mutants defective in production of extracellular signals were initially identified based on the ability of a wild-type strain to complement the sporulation phenotype of the mutant when the two cell types are allowed to develop together. As a result synergistic or extracellular complementation groups were identified by McVittie (McVittie et al., 1962) and Hagen (Hagen et al., 1978). A simplified version of the Hagen experiment is shown in Figure 12. Wild-type cells (yellow in color) were treated with chemical mutagens or ultraviolet, and mutants that exhibited normal vegetative growth, but failed to produce fruiting bodies or spores (Spo$^-$) on starvation medium, were identified (Hagen et al., 1978). To identify mutants whose sporulation defect was caused by the inability to produce a critical extracellular signal molecule, a red-colored Spo$^+$ strain was mixed with each yellow spo$^-$ mutant and allowed to develop. After several days, the vegetative cells were killed by heating, and the heat-resistant spores were quantified on nutrient-rich germination plates. While only "red" cells survived from the majority of mixtures, in some cases, both red (wild-type) and yellow (mutant) cells germinated. When the yellow cells were subjected to another starvation test, in the absence of the wild-type strain, they retained the Spo$^-$ defect, indicating that they had not undergone exchange of genetic information with the wild-type cells in the original mixture. These experiments revealed that the wild-type strain produced signal molecules during development that could be shared with the Spo$^-$ mutants. Sufficient amounts of these "shared" signals were available to rescue, or complement extracellularly, the sporulation defect of the mutant transiently. Initially the mutants were classified into four complementation groups: A, B, C, and D. A mutant from group A could provide a missing factor for a mutant from group B, C, or D, but not another mutant from group A. A fifth group, E, was identified later. The genetic and biochemical analysis of three extracellular signal molecules, Asg (A-signal), Csg (C-signal), and Esg (E-signal), are described below.

2. The Asg (A-signal) pathway

The A-signal mutants, *asgA, asgB*, and *asgC*, arrest during the first few hours of development and fail to produce heat-resistant spores (LaRossa et al., 1983) because they are unable to produce or release the extracellular A-signal molecule(s) (Hagen et al., 1978). The A-signal was identified biochemically as material from developing wild-type cells that could rescue the sporulation and gene expression defects of the *asg* mutants (Kuspa et al., 1992a; Plamann et al., 1992). Both heat-stable (amino acids, including Pro, Tyr, Phe, Trp, Leu, and Ile,) and heat-labile (protease) forms of A-signal could be isolated from developing cells. The *asg* mutants show a strong response to amino acids in the 100 μM range, far below the concentration of amino acids (about 10 mM) required for vegetative growth (Kuspa et al., 1992b).

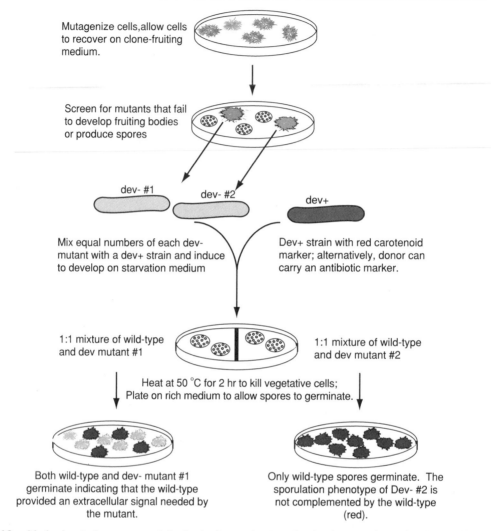

Fig. 12. Method to isolate mutants defective in the production of a developmental signal molecule. Cells that have acquired a mutation in a gene required for development will be able to grow on clone-fruiting agar but will be unable to form fruiting bodies when the nutrients are depleted. To determine if a mutant is unable to produce an essential signal molecule, it is mixed with a marked development-proficient strain and allowed to develop. If the sporulation defect of the mutant can be rescued with a signal provided by the proficient strain, then both strains will germinate on rich medium (mutant 1). A mutant whose developmental defect does not block production of a signal will not be rescued by the proficient strain and will not produce heat-resistant spores.

To understand how A-signaling is controlled, the *asg* genes and their products were characterized. The *asgA* gene encodes a member of the two-component family of regulatory proteins. AsgA is unusual because it contains an N-terminal receiver response regulator domain with a conserved aspartate residue, and a C-terminal transmitter kinase domain with a conserved histidine that can be autophosphorylated (Plamann et al., 1995; Li and Plamann, 1996). *asgB* encodes a transcription factor having a sigma-factor-like domain (Plamann et al., 1994) and the *asgC* 767 mutation is in *sigA*, the major sigma factor in *M. xanthus* (Davis et al., 1995). Recently the requirement for AsgD, another two-component regulatory protein in A-signaling has been identified (Cho and Zusman, 1999). The Asg proteins may control general secretory pathways because in addition to lacking the extracellular A-signal, they secrete less protein (Plamann et al., 1992) and fail to secrete endoglucanase during growth (Bensmail et al., 1998).

To understand how cells respond to the A-signal, mutations that restore expression of A-signal dependent genes in the absence of the A-signal were isolated. Expression of ß-galactosidase from Tn5-*lac*::4521 during development is very low in an *asg* mutant background. An *asgB480* Tn5-*lac*::4521 strain was mutagenized and second-site mutations that restore expression of β-galactosidase were obtained (Kaplan et al., 1991). Because these mutations bypass the requirement for A-signaling, they were predicted to identify genes encoding A-signal response elements. One of these second-site suppressors, *rfbA*, (originally named *sasA*), is a component of the RfbABC complex, an ABC transporter involved in O-antigen production (Guo et al., 1996). Several of the suppressor mutations mapped to the *sasB* locus (Kaplan et al., 1991) which codes for the regulatory proteins SasS, SasR, and SasN. Although a point mutation in *sasS*, which encodes a putative histidine kinase sensor, restores expression of Tn5-*lac*::4521 in an *asgB* mutant, disruption of *sasS* abolishes expression of Tn5-*lac*::4521 (Yang and Kaplan, 1997). These data suggest that SasS is a positive regulator of gene expression. The membrane-associated SasS is predicted to activate SasR, a putative response regulator, and may act downstream of SasP, another positive regulator (Guo et al., 2000; Plamann and Kaplan, 1999). In a strain carrying a disruption of *sasN*, repression of Tn5-*lac*::4521 during growth is abolished, suggesting that SasN is a negative regulatory element (Xu et al., 1998).

3. The C-signal pathway

Like the *asg* mutants, the *csg* mutant is unable to aggregate or form fruiting bodies and produces fewer than 0.01% the wild-type complement of heat-resistant spores. The *csg* mutation does not affect vegetative growth or motility, which is consistent with its role as a developmental-specific signal.

The *csgA* gene encodes a 24.6 kDa NAD$^+$-dependent alcohol dehydrogenase (Lee et al., 1995) that appears to function at two different stages in development. CsgA is required during the early stages of development to regulate production of (p)ppGpp (Crawford and Shimkets, 2000b). The enzymatic function of CsgA protein is critical for signaling because mutations that destroy the NAD$^+$-binding motif affect the ability of CsgA to function as a signaling molecule. Although the requirement of NAD$^+$ for an extracellular enzyme is unusual, group A streptococci produce an extracellular dehydrogenase that is important for its interaction with its mammalian host (Pancholi and Fischetti, 1997).

Bypass mutations that suppress the sporulation defect of a *csgA$^-$* mutant are due to enhanced expression of a second member of the short chain alcohol dehydrogenase family, called SocA (suppressor of csgA) (Lee and Shimkets, 1994). The *socA* mutation rescues the sporulation defect of a *csgA* mutant.[5] SocA is essential for vegetative growth but expression of *socA* normally is decreased during development. Because SocA shares only 30% identity with CsgA, this result argues that CsgA functions as a dehydrogenase during early stages of development.

At later stages in development, CsgA is involved in rippling, aggregation and sporulation. A 17 kDa form of C-signal protein has been purified from wild-type cells about 6 to 10 hours after the onset of development using a bioassay described below (Kim and Kaiser, 1990c). Extraction of active CsgA requires detergent which suggests that the

---

[5] Spores of *csgA$^-$socA$^-$* strain germinate immediately within the fruiting body, even in the absence of nutrients.

protein is associated with the extracellular matrix or membrane (Kim and Kaiser, 1990c). This is consistent with the finding that cell alignment and end-to-end contact are important for transmission of the signal (Kim and Kaiser, 1990a; Kaiser, 1999; Sogaard-Andersen et al., 1996). The 17 kDa form of CsgA is able to rescue the aggregation, gene expression, and sporulation defects of a csgA⁻ mutant. The 17 kDa CsgA is derived from the 24.6 kDa CsgA protein which suggests that CsgA is processed during development.

Clues to the mechanism by which CsgA affect both sporulation and aggregation have come from identification of events downstream of C-signaling. C-signaling is required for expression of the *devTRS* genes, which are involved in sporulation (Sogaard-Andersen and Kaiser, 1996), for its own expression (Kim and Kaiser, 1991) and for activation of the Frz system, which controls the frequency of cell reversals (Li, S.-F. et al., 1992; Sogaard-Andersen and Kaiser, 1996). Another link in the Csg transduction pathway is FruA, a putative DNA-binding protein (Ogawa et al., 1996), that acts downstream of C-signal to regulate expression of the *devTRS* (Ellehauge et al., 1998).

4. The E-signal

During the early stages of development, a third extracellular signal, Esg, is produced. The *esg* locus encodes the E1α and E1β subunits of the E1 decarboxylase responsible for oxidative decarboxylation of the branched-chain keto acids 2-ketoisovalerate, 2-keto-3-methylvalerate, and 2-ketoisocaproate to the short branched-chain fatty acyl-CoA esters, isobutyryl-CoA, 2-methylbutyryl-CoA, and isovaleryl-CoA (Downard et al., 1993). Addition of fatty acyl-CoA esters to *esg* mutants, which have reduced levels of long branched-chain fatty acids, rescues their growth defect, restores production of yellow pigment, and increases the yield of heat-resistant spores by 1400-fold (Toal et al., 1995).

C. Phase Variation Is Important for Production of Heat-Resistant Spores

Cultures of *M. xanthus* typically undergo phase variation on nutrient medium between a bright yellow variant that forms rough, swarming colonies and an unpigmented variant that forms tan, shiny colonies (Burchard and Dworkin, 1966). A predominantly yellow colony contains about 1–5% tan variants, whereas a predominantly tan colony contains about 25% yellow variants. The yellow-to-tan switch in *M. xanthus* affects the color and texture of colonies, consistent with a change in outer membrane or cell surface components. At low cell densities or low nutrient concentrations, yellow variants grow faster, but at high cell densities or high nutrient concentrations, tan variants grow faster (Laue and Gill, 1994). During the stationary phase of growth there is an increase in the ratio of tan to yellow variants.

A number of *M. xanthus* sporulation mutants show altered phase variation phenotypes. For example, certain *asg* mutants are predominantly tan variants and rarely produce yellow phase variant colonies. To determine if phase variation plays a role in development, a mutant that cannot phase-vary (phase-locked tan mutant) was isolated (Laue and Gill, 1995). When starved for nutrients, this mutant fails to form fruiting bodies or make heat-resistant spores. Remarkably, when cells of the tan phase-locked mutant are allowed to develop with cells from a predominantly yellow culture, heat-resistant myxospores, *derived primarily from the tan cells*, are produced. This result hints that tan cells within a normal (phase-variation proficient) culture of *M. xanthus* may be the progenitors of spores and that yellow and tan variants within the *M. xanthus* colony have different fates during development (Laue and Gill, 1995).

There is evidence that physiological differences between tan and yellow variants may contribute to their different fates. For example, when *M. xanthus* cultures are treated with glucosamine, > 50% of the cells lyse and

10% form spores (Mueller and Dworkin, 1991). A disproportionate number of the spores are derived from the tan cells in the mixture because tan variants are more resistant to glucosamine-induced lysis (Mueller and Dworkin, 1991). The ability of tan cells to resist lysis may account for their enhanced survival during fruiting body development.

D. Gene Expression during Development

Reporter genes have been used to illustrate patterns of gene expression in starving cells. As described earlier, the transposon Tn5-*lac* has been used to generate mutants of *M. xanthus* in which expression of *lacZ* is regulated by an exogenous promoter (Kroos et al., 1988; Kroos and Kaiser, 1984, 1987). A subset of the mutants carrying a Tn5-*lac* insertion was found to express β-galactosidase only during starvation-induced development (Kroos et al., 1986). These insertions are in genes that are expressed at a low level during growth but increase at least three fold when starved for nutrients. Because the timing and level of *lacZ* expression is unique for each insertion mutant, a pattern of developmental events can be associated with each insertion. Many of these developmental markers have been used to establish the relationship between particular events during development. For example, expression of *lacZ* from Tn5-*lac* :: 4445 (Ω4445) peaks four hours after the initial starvation event and produces 1700 units of β-galactosidase/ min·mg in the wild-type genetic background (Kroos and Kaiser, 1987). When expression of *lacZ* from Tn5-*lac*: :4445 is measured in a strain that carries an *csgA* mutation,[6] 1200 units of β-galactosidase are produced (Kroos and Kaiser, 1987). Hence the C-signaling process that requires active *csgA* is not required for expression of Ω4445. In contrast, when expression of *lacZ* from Tn5-*lac* :: 4445 is measured in a strain that carries a *bsg* mutation, only 620 units of

β-galactosidase are produced. This indicates that the Bsg protein is needed for full expression of the Ω4445 gene (Kroos and Kaiser, 1987). Taken together, these studies show that the function of Bsg is required at an earlier stage of development than the function of CsgA. A comparison of *lacZ* expression from different insertions of Tn5-*lac* in wild-type and mutant backgrounds reveals that expression of genes during the first 4 to 6 hours of development is *independent* of C-signaling, but genes that are expressed >6 hours are dependent on C-signaling. Analysis of the regulatory regions upstream of Csg-dependent genes has identified a consensus sequence—CAYYCCY where Y = pyrimidine—near the promoter (Bradner and Kroos, 1998). This sequence has been named the *C-box* because it may represent a binding site for a regulatory protein that is activated in response to C-signaling.

REFERENCES

Akerley BJ, Rubin EJ, Cailli A, Lampe DJ, Robertson HM, Mekalanos JJ (1998): Systematic identification of essential genes by in vitro *mariner* mutagenesis. Proc Natl Acad Sci USA 95:8927–8932.

Akoum A, Guidoin R, King MW, Marois Y, Sigot M, Sigot-Luizard MF (1992): A new bioactive molecule for improving vascular graft patency: Exploratory trials in dogs. Clin Invest Med 15:318–330.

Apelian D, Inouye S (1990): Development-specific sigma-factor essential for late-stage differentiation of *Myxococcus xanthus*. Genes Dev 4:1396–1403.

Apelian D, Inouye S (1993): A new putative sigma factor of *Myxococcus xanthus*. J Bacteriol 175:3335–3342.

Arnold JW, Shimkets LJ (1988): Cell surface properties correlated with cohesion in *Myxococcus xanthus*. J Bacteriol 170:5771–5777.

Balsalobre JM, Ruiz-Vasquez RM, Murillo FJ (1987): Light induction of gene expression in *Myxococcus xanthus*. Proc Natl Acad Sci USA 84:2359–2362.

Behmlander RM, Dworkin M (1991): Extracellular fibrils and contact-mediated cell interactions in *Myxococcus xanthus* J Bacteriol 173:7810–7821.

Bensmail L, Quillet L, Petit F, Barray S, Guespin-Michel JF (1998): Regulation of the expression of a gene encoding beta-endoglucanase secreted by *Myxococcus xanthus* during growth: Role of genes involved in developmental regulation. Res Microbiol 149:319–326.

---

[6] When the transposon from the donor is transferred to a recipient, the original site of the insertion and its corresponding regulation is preserved because recombination is more frequent than transposition (Kaiser, 1984).

Bibb MJ, Findlay PR, Johnson MW (1984): The relationship between base composition and condon usage in bacterial genes and its use for the simple and reliable identification of protein coding sequences. Gene 30:157–166.

Biran D, Kroos L (1997): In vitro transcription of Myxococcus xanthus genes with RNA polymerase containing sigma A, the major sigma factor in growing cells. Mol Microbiol 25:463–472.

Blackhart BD, Zusman DR (1985): "Frizzy" genes of Myxococcus xanthus are involved in control of frequency of reversal of gliding motility. Proc Natl Acad Sci USA 82:8767–8770.

Bowden MG, Kaplan HB (1998): The Myxococcus xanthus lipopolysaccharide O-antigen is required for social motility and multicellular development. Mol Microbiol 30:275–284.

Bradner JP, Kroos L (1998): Identification of the Ω4400 regulatory region, a developmental promoter of Myxococcus xanthus. J Bacteriol 180:1995–2004.

Bretscher AP, Kaiser D (1978): Nutrition of Myxococcus xanthus, a fruiting myxobacterium. J Bacteriol 133:763–768.

Burchard RP (1984): Gliding motility and taxes. In Rosenberg E (ed): "Myxobacteria: Development". New York: Springer, pp 139–164.

Burchard RP, Dworkin M (1966): Light-induced lysis and carotinogenesis in Myxococcus xanthus. J Bacteriol 91:535–545.

Campos JM, Geisselsoder J, Zusman DR (1978): Isolation of bacteriophage MX4, a generalized transducing phage for Myxococcus xanthus. J Mol Biol 119:167–178.

Chen H, Keseler IM, Shimkets LJ (1990): Genome size of Myxococcus xanthus determined by pulsed-field gel electrophoresis. J Bacteriol 172:4206–4213.

Chen HW, Kuspa A, Keseler IM, Shimkets LJ (1991): Physical map of the Myxococcus xanthus chromosome. J Bacteriol 173:2109–2115.

Cho K, Zusman DR (1999): AsgD, a new two-component regulator required for A-signalling and nutrient sensing during early development of Myxococcus xanthus. Mol Microbiol 34:268–281.

Crawford EW, Shimkets LJ (2000a): The Myxococcus xanthus socE and csgA genes are regulated by the stringent response. Mol Microbiol 37:788–799.

Crawford EW, Jr, Shimkets LJ (2000b): The stringent response in Myxococcus xanthus is regulated by SocE and the CsgA C-signaling protein. Genes Dev 14:483–492.

Davis JM, Mayor J, Plamann L (1995): A missense mutation in rpoD results in an A-signalling defect in Myxococcus xanthus. Mol Microbiol 18:943–952.

Dhundale A, Furuichi T, Inouye M, Inouye S (1988): Mutations that affect production of branched RNA-linked msDNA in Myxococcus xanthus. J Bacteriol 170:5620–5624.

Dhundale A, Lampson B, Furuichi T, Inouye M, Inouye S (1987): Structure of msDNA from Myxococcus xanthus: Evidence for a long, self-annealing RNA precursor for the covalently linked, branched RNA. Cell 51:1105–1112.

Downard J, Ramaswamy SV, Kil KS (1993): Identification of esg, a genetic locus involved in cell-cell signaling during Myxococcus xanthus development. J Bacteriol 175:7762–7770.

Dworkin M (1962): Nutritional requirements for vegetative growth of Myxococcus xanthus. J Bacteriol 85:250–257.

Dworkin M (1963): Nutritional regulation of morphogenesis in Myxococcus xanthus. J Bacteriol 86:67–72.

Dworkin M (1999): Fibrils as extracellular appendages of bacteria: Their role in contact-mediated cell-cell interactions in Myxococcus xanthus. BioEssays 21:590–595.

Elías M, Murillo FJ (1991): Induction of germination in Myxococcus xanthus fruiting body spores. J Gen Microbiol 137:381–388.

Ellehauge E, Norregaard-Madsen M, Sogaard-Andersen L (1998): The FruA signal transduction protein provides a checkpoint for the temporal co-ordination of intercellular signals in Myxococcus xanthus development. Mol Microbiol 30:807–817.

Garza AG, Pollack JS, Harris BZ, Lee A, Keseler IM, Licking EF, Singer M (1998): SdeK is required for early fruiting body development in Myxococcus xanthus. J Bacteriol 180:4628–4637.

Geisselsoder J, Campos JM, Zusman DR (1978): Physical characterization of bacteriophage Mx4, a generalized transducing phage for Myxococcus xanthus. J Mol Biol 119:179–189.

Gerth K, Irschik H, Reichenbach H, Trowitzsch W (1982): The myxovirescins, a family of antibiotics from Myxococcus virescens (Myxobacterales). J Antibiot 35:1454–1459.

Gill RE, Karlok M, Benton D (1993): Myxococcus xanthus encodes an ATP-dependent protease which is required for developmental gene transcription and intercellular signaling. J Bacteriol 175:4538–4544.

Gorski L, Gronewald T, Kaiser D (2000): A sigma(54) activator protein necessary for spore differentiation within the fruiting body of Myxococcus xanthus. J Bacteriol 182:2438–2444.

Gulati P, Xu D, Kaplan HB (1995): Identification of the minimum regulatory region of a Myxococcus xanthus A-signal-dependent developmental gene. J Bacteriol 177:4645–4651.

Guo D, Bowden MG, Pershad R, Kaplan HB (1996): The Myxococcus xanthus rfbABC operon encodes an ATP-binding cassette transporter homolog required for O-antigen biosynthesis and multicellular development. J Bacteriol 178:1631–1639.

Guo D, Wu Y, Kaplan HB (2000): Identification and characterization of genes required for early Myxococ-

cus xanthus developmental gene expression. J Bacteriol 182:4564–4571.

Hagen DC, Bretscher AP, Kaiser D (1978): Synergism between morphogenetic mutants of *Myxococcus xanthus*. Dev Biol 64:284–296.

Hager E, Tse H, Gill R (2001): Identification and characterization of SpdR mutations that bypass the BsgA protease-dependent regulation of developmental gene expression in *Myxococcus xanthus*. Mol Microbiol 39:765–780.

Hartl DL, Lohe AR, Lozovskaya ER (1997): Modern thoughts on an ancyent marinere: Function, evolution, regulation. Annu Rev Genet 31:337–358.

Hartl DL, Ochman H (1994): Inverse polymerase chain reaction. Methods Mol Biol 31:187–196.

Hartzell PL (1997): Complementation of *Myxococcus xanthus* sporulation and motility defects by a eukaryotic RAS homolog. Proc Natl Acad Sci USA 97:9881–9886.

Hartzell PL, Kaiser D (1991): Upstream gene of the *mgl* operon controls the level of MglA protein in *M. xanthus*. J Bacteriol 172:7625–7535.

He Q, Chen H, Kuspa A, Cheng Y, Kaiser D, Shimkets LJ (1994): A physical map of the *Myxococcus xanthus* chromosome. Proc Natl Acad Sci USA 91:9584–9587.

Hodgkin J, Kaiser D (1977): Cell-to-cell stimulation of movements in non-motile mutants of *Myxococcus xanthus*. Proc Natl Acad Sci USA 74:2938–2942.

Hodgkin J, Kaiser D (1979a): Genetics of gliding motility an *Myxococcus xanthus* (Myxobacterales): genes controlling movement of single cells. Mol Gen Genet 171:167–171.

Hodgkin J, Kaiser D (1979b): Genetics of gliding motility in *Myxococcus xanthus* (Myxobactererales): Two gene systems control movement. Mol Gen Genet 171:177–191.

Hodgson DA (1993): Light-induced carotenogenesis in *Myxococcus xanthus*: Genetic analysis of the *carR* region. Mol Microbiol 7:471–488.

Hsu MY, Inouye S, Inouye M (1989): Structural requirements of the RNA precursor for the biosynthesis of the branched RNA-linked multicopy single-stranded DNA of *Myxococcus xanthus*. J Biol Chem 264:6214–6219.

Inouye M, Inouye S (1991): Retroelements in bacteria. Trends Biochem Sci 16:18–21.

Inouye S (1990): Cloning and DNA sequence of the gene coding for the major sigma factor from *Myxococcus xanthus*. J Bacteriol 172:80–85.

Jelsbak L, Sogaard-Anderson L (1999): The cell surface-associated intercellular C-signal induces behavioral changes in individual *Myxococcus xanthus* cells during fruiting body morphogenesis. Proc Natl Acad Sci USA 96:5031–5036.

Kaiser D (1984): Genetics of Myxobacteria. In Rosenberg E (ed): "Myxobacteria: Development and Cell Interactions". New York: Springer.

Kaiser D (1999): Cell fate and organogenesis in bacteria. Trends Gen 15:273–277.

Kaiser D, Dworkin M (1975): Gene transfer to myxobacterium by *Escherichia coli* phage P1. Science 187:653–654.

Kaplan HB, Kuspa A, Kaiser D (1991): Suppressors that permit A-signal-independent developmental gene expression in *Myxococcus xanthus*. J Bacteriol 173:1460–1470.

Kashefi K, Hartzell PL (1995): Genetic suppression and phenotypic masking of a *Myxococcus xanthus frz*F-defect. Mol Microbiol 15:483–494.

Kearns DB, Shimkets LJ (1998): Chemotaxis in a gliding bacterium. Proc Natl Acad Sci USA 95:11957–11962.

Keseler I, Kaiser D (1995): An early A-signal-dependent gene in *Myxococcus xanthus* has a sigma 54-like promoter. J Bacteriol 177:4638–4644.

Keseler IM, Kaiser D (1997): Sigma54, a vital protein for *Myxococcus xanthus*. Proc Natl Acad Sci USA 94:1979–1984.

Kim SK, Kaiser D (1990a): Cell alignment required in differentiation of *Myxococcus xanthus*. Science 249:926–928.

Kim SK, Kaiser D (1990b): Cell motility is required for the transmission of C-factor, an intercellular signal that coordinates fruiting body morphogenesis of *Myxococcus xanthus*. Genes Dev 4:896–904.

Kim SK, Kaiser D (1990c): Purification and properties of *Myxococcus xanthus* C-factor, an intercellular signaling protein. Proc Natl Acad Sci USA 87:3635–3639.

Kim SK, Kaiser D (1991): C-factor has distinct aggregation and sporulation thresholds during Myxococcus development. J Bacteriol 173:1722–1728.

Kroos L, Hartzell P, Stephens K, Kaiser D (1988): A link between cell movement and gene expression argues that motility is required for cell-cell signaling during fruiting body development. Genes Dev 2:1677–1685.

Kroos L, Kaiser D (1984): Construction of Tn*5-lac*, a transposon that fuses *lacZ* expression to exogenous promoters, and its introduction into *Myxococcus xanthus*. Proc Natl Acad Sci USA 81:5816–5820.

Kroos L, Kaiser D (1987): Expression of many developmentally regulated genes in *Myxococcus* depends on a sequence of cell interactions. Genes Dev 1:840–854.

Kroos L, Kuspa A, Kaiser D (1986): A global analysis of developmentally regulated genes in *Myxococcus xanthus* Dev Biol 117:252–266.

Kuner JM, Kaiser D (1982): Fruiting body morphogenesis in submerged cultures of *Myxococcus xanthus*. J Bacteriol 151:458–461.

Kuspa A, Plamann L, Kaiser D (1992a): Identification of heat-stable A-factor from *Myxococcus xanthus*. J Bacteriol 174:3319–3326.

Kuspa A, Plamann L, Kaiser D (1992b): A-signalling and the cell density requirement for *Myxococcus xanthus* development. J Bacteriol 174:7360–7369.

Kuspa A, Vollrath D, Cheng Y, Kaiser D (1989): Physical mapping of the *Myxococcus xanthus* genome by random cloning in yeast artificial chromosomes. Proc Natl Acad Sci USA 86:8917–8921.

Lampson BC, Inouye M, Inouye S (1989): Reverse transcriptase with concomitant ribonuclease H activity in the cell-free synthesis of branched RNA-linked msDNA of *Myxococcus xanthus*. Cell 56:701–707.

LaRossa R, Kuner J, Hagen D, Manoil C, Kaiser D (1983): Developmental cell interactions of *Myxococcus xanthus*: Analysis of mutants. J Bacteriol 153:1394–1404.

Laue BE, Gill RE (1994): Use of a phase variation-specific promoter of *Myxococcus xanthus* in a strategy for isolation a phase-locked mutant. J Bacteriol 176:5341–5349.

Laue BE, Gill RE (1995): Using a phase-locked mutant of *Myxococcus xanthus* to study the role of phase variation in development. J Bacteriol 177:4089–4096.

Lazzaroni JC, Germon P, Ray MC, Vianney A (1999): The Tol proteins of *Escherichia coli* and their involvement in the uptake of biomolecules and outer membrane stability. FEMS Microbiol Lett 15:191–197.

Lee BU, Lee K, Mendez J, Shimkets LJ (1995): A tactile sensory system of *Myxococcus xanthus* involves an extracellular NAD(P)(+)-containing protein. Genes Dev 9:2964–2973.

Lee K, Shimkets LJ (1994): Cloning and characterization of the *socA* locus which restores development to *Myxococcus xanthus* C-signaling mutants. J Bacteriol 176:2200–2209.

Li S-F, Lee B-U, Shimkets LJ (1992): *csgA* expression entrains *Myxococcus xanthus* development. Genes Dev 6:401–410.

Li Y, Plamann L (1996): Purification and in vitro phosphorylation of *Myxococcus xanthus* AsgA protein. J Bacteriol 178:289–292.

MacNeil SD, Calara F, Hartzell PL (1994a): New clusters of genes required for gliding motility in *Myxococcus xanthus*. Mol Microbiol 14:61–71.

MacNeil SD, Mouzeyan A, Hartzell PL (1994b): Genes required for both gliding motility and development in *Myxococcus xanthus*. Mol Microbiol 14:785–795.

Magrini V, Creighton C, White D, Hartzell PL, Youderian P (1998): The *aadA* gene of plasmid R100 confers resistance to spectinomycin and streptomycin in *Myxococcus xanthus*. J Bacteriol 180:6757–6760.

Magrini V, Creighton C, Youderian P (1999): Site-specific recombination of temperate *Myxococcus xanthus* phage Mx8: Genetic elements required for integration. J Bacteriol 181:4050–4061

Magrini V, Salmi D, Thomas D, Herbert SK, Hartzell PL, Youderian P (1997): Temperate *Myxococcus*

xanthus phage Mx8 encodes a DNA adenine methylase, Mox. J Bacteriol 179:4254–4263.

Magrini V, Storms ML, Youderian P (1999): Site-specific recombination of temperate *Myxococcus xanthus* phage Mx8: Regulation of integrase activity by reversible, covalent modification. J Bacteriol 181:4062–4070.

Manoil C, Kaiser D (1980): Accumulation of guanosine tetraphosphate and guanosine pentaphosphate in *Myxococcus xanthus* during starvation and myxospore formation. J Bacteriol 252:297–304.

Margolin P (1987): Generalized transduction. In Neidhardt FC (ed): "*Escherichia coli* and *Salmonella typhimurium*". Washington, DC: ASM Press, pp 1154–1168.

Martin S, Sodergren E, Masuda T, Kaiser D (1978): Systematic isolation of transducing phages for *Myxococcus xanthus*. Virol 88:44–53.

McBride MJ, Kohler T, Zusman DR (1992): Methylation of FrzCD, a methyl-accepting taxis protein of *Myxococcus xanthus*, is correlated with factors affecting cell behavior. J Bacteriol 174:4246–4257.

McBride MJ, Weinberg RA, Zusman DR (1989): "Frizzy" aggregation genes of the gliding bacterium *Myxococcus xanthus* show sequence similarities to the chemotaxis genes of enteric bacteria. Proc Natl Acad Sci USA 86:424–428.

McVittie A, Messik F, Zahler SA (1962): Developmental biology of *Myxococcus*. J Bacteriol 84:546–551.

Merz AJ, So M, Scheetz MP (2000): Pilus retraction powers bacterial twitching motility. Nature 407:98–102.

Mesbah M, Premachandran U, Whitman WB (1989): Precise measurement of the G + C content of deoxyribonucleic acid by high-performance liquid chromatography. Int J Syst Bacteriol 39:159–167.

Mueller C, Dworkin M (1991): Effects of glucosamine on lysis, glycerol formation, and sporulation in *Myxococcus xanthus*. J Bacteriol 173:7164–7175.

Neumann B, Pospiech A, Schairer HU (1993): A physical and genetic map of the *Stigmatella aurantiaca* DW4/3.1chromosome. Mol Microbiol 10:1087–1099.

O'Connor KA, Zusman DR (1983): Coliphage P1–mediated transduction of cloned DNA from *Escherichia coli* to *Myxococcus xanthus*: use for complementation and recombinational analyses. J Bacteriol 155:317–329.

Ogawa M, Fujitani S, Mao X, Inouye S, Komano T (1996): FruA, a putative transcription factor essential for the development of *Myxococcus xanthus*. Mol Microbiol 22:757–767.

Otani M, Inouye M, Inouye S (1995): Germination of myxospores from the fruiting bodies of *Myxococcus xanthus*. J Bacteriol 1177:4261–4265.

Pancholi V, Fischetti VA (1997): Regulation of the phosphorylation of human pharyngeal cell proteins by group A streptococcal surface dehydrogenase:

signal transduction between streptococci and pharyngeal cells. J Exp Med 186:1633–1643.

Pfister DH (1993): Roland Thaxter and the Myxobacteria. In Dworkin M, Kaiser D (eds): "*Myxobacteria II*". Washington, DC: ASM Press, pp 1–11.

Plamann L, Davis JM, Cantwell B, Mayor J (1994): Evidence that *asgB* encodes a DNA-binding protein essential for growth and development of *Myxococcus xanthus*. J Bacteriol 176:2013–2020.

Plamann L, Kaplan HB (1999): Cell-density sensing during early development in *Myxococcus xanthus*. In Dunny GM, Winans SC (eds): "Cell-Cell Sigaling in Bacteria". Washington DC: ASM Press, pp 67–82.

Plamann L, Kuspa A, Kaiser D (1992): Proteins that rescue A-signal-defective mutants of *Myxococcus xanthus*. J Bacteriol 174:3311–3318.

Plamann L, Li Y, Cantwell B, Mayor J (1995): The *Myxococcus xanthus asgA* gene encodes a novel signal transduction protein required for multicellular development. J Bacteriol 177:2014–2020.

Reichenbach H (1984): Myxobacteria: A most peculiar group of social prokaryotes. In Rosenberg E (ed): "Myxobacteria: Development". New York: Springer, pp 1–50.

Reichenbach H, Dworkin M (1992): The myxobacteria. In Balows A, Truper HG, Dworkin M, Harder W, Schleifer KH (eds): "The Prokaryotes". New York: Springer, pp 3416–3487.

Reichenbach H, Höfle G (1993): Production of bioactive secondary metabolites. In Dworkin M, Kaiser D (eds): "Myxobacteria II". Washington, DC: ASM Press, pp 347–397.

Reichenbach H, Kleinig H (1984): Pigments of myxobacteria. In Rosenberg E (ed): "Myxobacteria: Development". New York: Springer, pp 127–137.

Ried JL, Collmer A (1987): An *nptI-sacB-sacR* cartridge for constructing directed, unmarked mutations in gram-negative bacteria by marker exchange-eviction mutagenesis. Gene 57:239–246.

Rodriguez AM, Spormann AM (1999): Genetic and molecular analysis of *cglB*, a gene essential for single-cell gliding in *Myxococcus xanthus*. J Bacteriol 181:4381–4390.

Rodriguez-soto JP, Kaiser D (1997): The *tgl* gene: Social motility and stimulation in *Myxococcus xanthus*. J Bacteriol 179:4361–4371.

Romeo JM, Zusman DR (1991): Transcription of the myxobacterial hemagglutinin gene is mediated by a sigma 54–like promoter and a cis-acting upstream regulatory region of DNA. J Bacteriol 173:2969–2976.

Rubin EJ, Akerley BJ, Novik VN, Lampe DJ, Husson RN, Mekalanos JJ (1999): In vivo transposition of *mariner*-based elements in enteric bacteria and mycobacteria. Proc Natl Acad Sci USA 96:1645–1650.

Salmi D, Magrini V, Hartzell PL, Youderian P (1998): Genetic determinants of immunity and integration of temperate *Myxococcus xanthus* phage Mx8. J Bacteriol 180:614–621.

Sanderson KE, Hurley JA (1987): Linkage map of *Salmonella typhimurium*. In Neidhardt FC (ed): "*Escherichia coli* and *Salmonella typhimurium*." Washington, DC: ASM Press, pp 877–918.

Segall JE, Ishihara A, Berg HC (1985): Chemotactic signaling in filamentous cells of *Escherichia coli*. J Bacteriol 161:51–59.

Semmler AB, Whitchurch CB, Mattick JS (1999): A re-examination of twitching motility in *Pseudomonas aeruginosa*. Microbiol 145:2863–2873.

Shi W, Zusman DR (1993): The two motility systems of *Myxococcus xanthus* show different selective advantages on various surfaces. Proc Natl Acad Sci USA 90:3378–3382.

Shimkets LJ (1999): Intercellular signaling during fruiting-body development of *Myxococcus xanthus*. Ann Rev Microbiol 53:525–549.

Shimkets LJ, Gill RE, Kaiser D (1983): Developmental cell interactions in *Myxococcus xanthus* and the *spoC* locus. Proc Natl Acad Sci 80:1406–1410.

Shimkets LJ, Kaiser D (1982): Induction of coordinated cell movement in *Myxococcus xanthus*. J Bacteriol 152:451–461.

Silakowski B, Schairer HU, Ehret H, Kunze B, Weinig S, Nordsiek G, Brandt P, Blocker H, Hofle G, Beyer S, Muller R (1999): New lessons for combinatorial biosynthesis from myxobacteria: The myxothiazol biosynthetic gene cluster of *Stigmatella aurantiaca* DW4/3-1. J Biol Chem 274:37391–37399.

Singer M, Kaiser D (1995): Ectopic production of guanosine penta-and tetraphosphate can initiate early developmental gene expression in *Myxococcus xanthus*. Genes Dev 9:1633–1644.

Sodergren E, Kaiser D (1983): Insertions of Tn*5* near genes that govern stimulatable cell motility in *Myxococcus*. J Mol Biol 167:295–310.

Sogaard-Andersen L, Kaiser D (1996): C factor, a cell-surface-associated intercellular signaling protein, stimulates the cytoplasmic Frz signal transduction system in *Myxococcus xanthus*. Proc Natl Acad Sci USA 2:2675–2679.

Sogaard-Andersen L, Slack FJ, Kimsey H, Kaiser D (1996): Intercellular C-signaling in *Myxococcus xanthus* involves a branched signal transduction pathway. Genes Dev 10:740–754.

Spormann AM (1999): Gliding motility in bacteria: Insights from studies of *Myxococcus xanthus* Microbiol. Mol Biol Rev 63:621–641.

Spormann AM, Kaiser D (1995): Gliding movements of *Myxococcus xanthus*. J Bacteriol 177:5846–5852.

Spormann AM, Kaiser D (1999): Gliding mutants of *Myxococcus xanthus* with high reversal frequencies and small displacements. J Bacteriol 181:2593–2601.

Spratt BG, Hedge PJ, Heesey ST, Edelman A, Broome-Smith JK (1986): Kanamycin-resistant vectors that are analogues of pUC8, pUC9, pEMBL8 and pEMBL9. Gene 41:337–342.

Stamm I, Leclerque A, Plaga W (1999): Purification of cold-shock-like proteins from *Stigmatella aurantiaca*—Molecular cloning and characterization of the cspA gene. Arch Microbiol 172:175–181.

Starich T, Cordes P, Zissler J (1985): Transposon tagging to detect a latent virus in *Myxococcus xanthus*. Science 230:541–513.

Stellwag E, Fink JM, Zissler J (1985): Physical characterization of the genome of the *Myxococcus xanthus* bacteriophage Mx8. Mol Gen Genet 199:123–132.

Stephens K, Hartzell PL, Kaiser D (1989): Gliding motility in *Myxococcus xanthus*: *mgl* locus, RNA, and predicted protein products. J Bacteriol 171:819–830.

Stephens K, Kaiser D (1987): Genetics of gliding in *Myxococcus xanthus*: Molecular cloning of the *mgl* locus. Mol Gen Genetics 207:256–266.

Sudo D, Dworkin M (1972): Bacteriolytic enzymes produced by *Myxococcus xanthus*. J Bacteriol 110:236–245.

Sudo SZ, Dworkin M (1969): Resistance of vegetative cells and microcysts of *Myxococcus xanthus*. J Bacteriol 98:883–887.

Sun H, Yang Z, Shi W (1999): Effect of cellular filamentation on adventurous and social gliding motility of *Myxococcus xanthus*. Proc Natl Acad Sci USA 96:15178–15183.

Tang L, Shah S, Chung L, Carney J, Katz L, Khosla C, Julien B (2000): Cloning and heterologous expression of the epothilone gene cluster. Science 287:640–642.

Thaxter R (1892): On the Myxobacteriaceae, a new order of Schizomycetes. Botan Gazette 17:389–406.

Toal DR, Clifton SW, Roe BA, Downard J (1995): The *esg* locus of *Myxococcus xanthus* encodes the E1 alpha and E1 beta subunits of a branched-chain keto acid dehydrogenase. Mol Microbiol 16:177–189.

Tojo N, Sanmiya K, Sugawara H, Inouye S, Komano T (1996): Integration of bacteriophage Mx8 into the *Myxococcus xanthus* chromosome causes a structural alteration at the C-terminal region of the IntP protein. J Bacteriol 178:4004–4011.

Ueki T, Inouye S, Inouye M (1996): Positive-negative KG cassettes for construction of multigene deletions using a single drug marker. Gene 183:153–157.

Wall D, Kaiser D (1999): Type IV pili and cell motility. Mol Microbiol 32:1–10.

Wall D, Kolenbrander PE, Kaiser D (1999): The *Myxococcus xanthus pilQ* (*sglA*) gene encodes a secretin homolog required for type IV pilus biogenesis, social motility and development. J Bacteriol 181:24–33.

Ward MJ, Mok KC, Astling DP, Lew H, Zusman DR (1998): An ABC transporter plays a developmental aggregation role in *Myxococcus xanthus*. J Bacteriol 180:5697–5703.

Ward MJ, Zusman DR (1999): Motility in *Myxococcus xanthus* and its role in developmental aggregation. Curr Opin Microbiol 2:624–629.

Weimer RM, Creighton C, Stassinopoulos A, Youderian P, Hartzell PL (1998): A chaperone in the HSP70 family controls production of extracellular fibrils in *Myxococcus xanthus*. J Bacteriol 180:5357–5368.

White D (1984): Structure and function of myxobacteria cells and fruiting bodies. In Rosenberg E (ed): "Myxobacteria: Development." New York: Springer, pp 52–67.

White DJ, Hartzell PL (2000): AglU, a protein required for gliding motility and spore maturation of *Myxococcus xanthus*, is related to WD-repeat proteins. Mol Microbiol 36:662–678.

Wireman JW, Dworkin M (1975): Morphogenesis and developmental interactions in *Myxococcus xanthus*. Science 189:516–522.

Woese CR, Bleyman MA (1972): Genetic code limit organisms—Do they exist? J Mol Evol 1:223–229.

Wu SS, Kaiser D (1995): Genetic and functional evidence that type IV pili are required for social gliding motility in *Myxococcus xanthus*. Mol Microbiol 18:547–558.

Wu SS, Kaiser D (1996): Markerless deletions of *pil* genes in *Myxococcus xanthus* generated by counterselection with the *Bacillus subtilis sacB* gene. J Bacteriol 178:5817–5821.

Wu SS, Wu J, Cheng YL, Kaiser D (1998): The *pilH* gene encodes an ABC transporter homologue required for type IV pilus biogenesis and social gliding motility in *Myxococcus xanthus*. Mol Microbiol 29:1249–1261.

Wu TT (1966): A model for three-point analysis of random general transduction. Genetics 54:405–410.

Xu D, Yang C, Kaplan HB (1998): *Myxococcus xanthus sasN* encodes a regulator that prevents developmental gene expression during growth. J Bacteriol 180:6215–6223.

Yang C, Kaplan HB (1997): *Myxococcus xanthus sasS* encodes a sensor histidine kinase required for early developmental gene expression. J Bacteriol 179:7759–7767.

Yang Z, Geng Y, Xu D, Kaplan HB, Shi W (1998): A new set of chemotaxis homologues is essential for *Myxococcus xanthus* social motility. Mol Microbiol 30:1123–1130.

Zahavi A, Ralt D (1984): Social adaptations in myxobacteria. In Rosenberg, E (ed): "Myxobacteria: Development." New York: Springer, pp 215–220.

Zusman DR, Rosenberg E (1971): Division cycle of *Myxococcus xanthus*: II. Kinetics of stable and unstable ribonucleic acid synthesis. J Bacteriol 105:801–810.

14

Agrobacterium Genetics

Department of Microbiology, Oregon State University, Corvallis, Oregon 97331

I. INTRODUCTION

This chapter will summarize our current understanding of the central feature of the *Agrobacterium tumefaciens* story: interkingdom genetic exchange. *Agrobacterium*-mediated gene transfer made possible modern plant molecular genetics and genetic engineering. In addition recent discoveries have uncovered remarkable similarities among virulence protein secretion systems employed by *A. tumefaciens* and by several bacterial species that afflict humans. Students will find *A. tumefaciens* an excellent model system to study topics as diverse as conjugation, nuclear targeting, protein secretion, microbial ecology, chemical signaling, and a variety of gene regulation strategies. Many important aspects of crown gall disease, for example, oncogenes, chemotaxis, host range, opines, conjugation, and *Agrobacterium* ecology will not be discussed in detail. The final section will discuss the role of *Agrobacterium* in creating genetically engineered plants and the controversy surrounding this technology.

II. *AGROBACTERIUM*: NATURE'S GENETIC ENGINEER

Agrobacterium tumefaciens, a Gram-negative soil bacterium, belongs to the family

**Modern Microbial Genetics, Second Edition. Edited by
Uldis N. Streips and Ronald E. Yasbin. ISBN 0–471–38665–0**

Rhizobiaceae and shares many similarities with members of the closely related genus *Rhizobium*. *Agrobacterium* and the *Rickettsia*, which also live in association with eukaryotic host cells, share a very close phylogenetic relationship too (Weisburg et al., 1985). Indeed, a suite of genes discovered recently in the genome of *Rickettsia prowazekii* were named

Fig. 1. **A:** Wild-type and shooty galls on stems of *Nicotiana tabacum* (tobacco) inoculated with wild-type or auxin biosynthesis mutant (*iaaM* or *iaaH* mutants) *Agrobacterium tumefaciens* strains. **B, C:** Wild-type (B) or rooty galls (C) on *Kalanchoe daigremontiana* stems inoculated with wild-type or cytokinin biosynthesis mutant (*ipt* mutant) strains of *A. tumefaciens*. **D:** Wild-type and large mutant galls on *K. daigremontiana* leaves inoculated with wild-type or tumor morphology large (*tml*) mutant strains of *A. tumefaciens*.

after their homologues in *A. tumefaciens* (Andersson et al., 1998). Crown gall tumors (Fig. 1) form on most dicotyledonous plants (De-Cleene and DeLey, 1976) when virulent strains of *A. tumefaciens*, containing a 200 kilobase-pair (kb) tumor-inducing (Ti) plasmid (Fig. 2, from Ream, 1989), infect wounded plant tissue. A specific segment of the Ti plasmid, the T-DNA (Fig. 3), enters plant cells and stably integrates into plant nuclear DNA (Chilton et al., 1977, 1980). The T-DNA encodes enzymes for biosynthesis of plant growth hormones indole acetic acid (IAA, an auxin) and isopentenyl adenosine monophosphate (ipA, a cytokinin) (Fig. 4), thereby causing transformed cells to grow as crown gall tumors (Nester et al., 1984; Binns and Thomashow, 1988; Winans, 1992; Ream, 1989; Christie, 1997; Zhu et al., 2000; Sheng and Citovsky, 1996). However, T-DNA transfer and integration does not require tumorigenesis or T-DNA encoded proteins (Hoekema et al., 1983; Ream et al., 1983). This fact has allowed genetic engineers to use *A. tumefaciens* to transfer beneficial genes into plants in place of the T-DNA oncogenes (Klee et al., 1987).

Agrobacterium tumefaciens is nature's genetic engineer. The first genetically modified plant cells were not produced by humans. Instead, plants were first engineered by *A. tumefaciens*. These bacteria genetically transform host cells with genes that cause rapid growth and production of large quantities of opines, which are used as nutrients by the tumor-inducing bacteria (Guyon et al., 1980; Petit et al., 1983). Many opines are derived from sugars and amino acids, which provide both carbon and nitrogen to the bacteria (Winans, 1992). Transformed plant cells synthesize and secrete significant quantities of specific opines, and the tumor-inducing bacteria carry genes (outside the T-DNA and usually on the Ti plasmid) required to

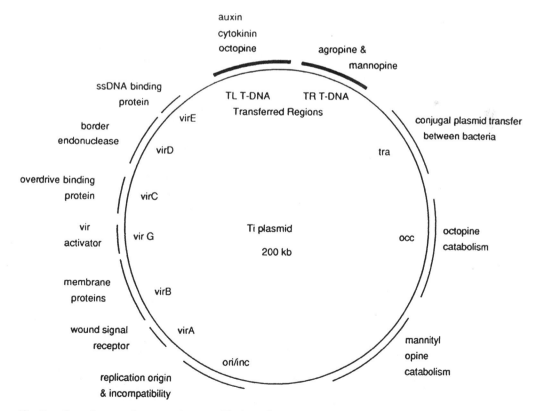

Fig. 2.   Genetic map of an octopine-type Ti plasmid.

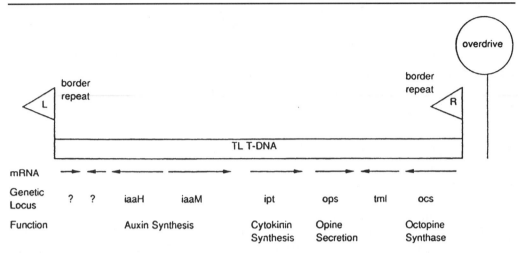

Fig. 3. Genetic map of the TL T-DNA of an octopine-type Ti plasmid. The gene symbols indicate indole acetamide hydrolase (*iaaH*), tryptophan monooxygenase (*iaaM*), isopentenyl transferase (*ipt*), opine secretion (*ops*), tumor morphology large (*tml*; mutants in this gene produce abnormally large tumors), and octopine synthase (*ocs*).

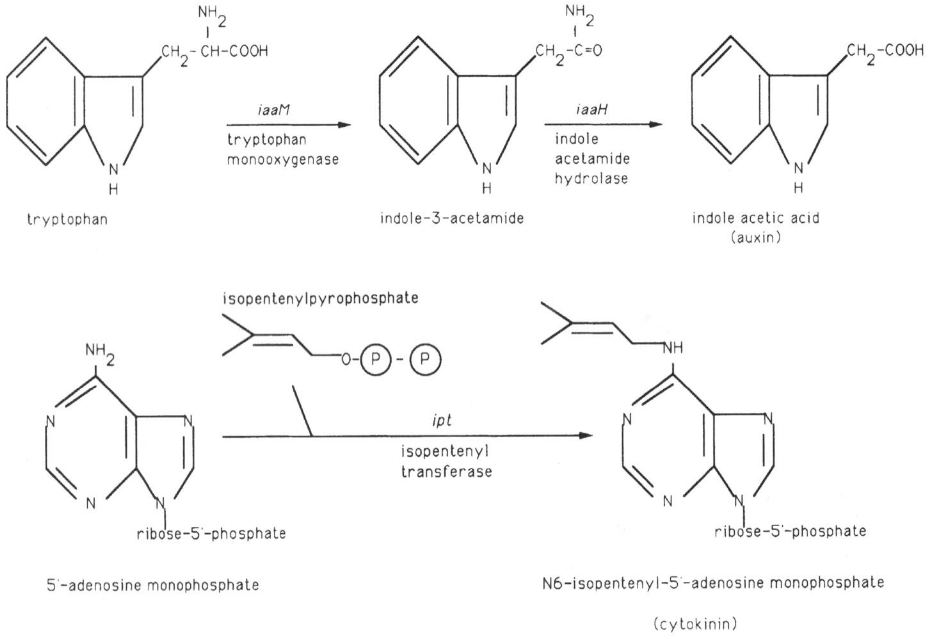

Fig. 4. Pathways for biosynthesis of auxin (top) and cytokinin (bottom) by enzymes encoded in the *Agrobacterium tumefaciens* T-DNA.

catabolize the same opines synthesized by the tumor. More than 20 different opines exist, and each strain induces and catabolizes a specific set of opines. Generally, each *A. tumefaciens* strain catabolizes only the

opines synthesized by tumors it induces. In addition some opines induce conjugal transfer of self-transmissible Ti plasmids between strains of *Agrobacterium* (Petit et al., 1978; Ellis et al., 1982), thereby conferring on other strains the ability to catabolize extant opines. Apparently *A. tumefaciens* strains create a niche (a crown gall tumor synthesizing particular opines) that offers an environment favorable for growth of the inducing strain.

III. INTERKINGDOM GENE TRANSFER

Agrobacterium tumefaciens transfers the T-DNA portion of its Ti plasmid and virulence proteins VirD2, VirE2, and VirF into host cells during crown gall tumorigenesis (Figs. 2–3, from Ream, 1989); for reviews see Sheng and Citovsky (1996), Christie (1997), and Zhu et al. (2000). VirD2 nicks border sequences at the T-DNA ends and attaches covalently to the 5′ end of the nicked strand (Wang et al., 1987; Yanofsky et al., 1986; Ward and Barnes, 1988; Young and Nester, 1988; Herrera-Estrella et al., 1988; Howard et al., 1989; Durrenberger et al., 1989). A type IV secretion system encoded by the *virB* operon and *virD4* mediates export to plant cells of the VirD2-T-DNA complex (T-complex) as well as VirE2 single-stranded DNA (ssDNA) binding protein (SSB) (Citovsky et al., 1988, 1989; Gietl et al., 1987; Das, 1988; Christie et al., 1988b; Sen et al., 1989) and VirF protein, which is required for tumorigenesis in some hosts (e.g., tomato and *Nicotiana glauca*) but not others (Melchers et al., 1990). VirE2 may coat single-stranded T-DNA copies (T-strands) as they are displaced from the nicked Ti plasmid or later, inside plant cells. However, VirE2 and VirF function in plant cells; transgenic plants that express either of these proteins produce tumors when inoculated with *A. tumefaciens* mutants that lack intact copies of the corresponding *vir* gene (Citovsky et al., 1992; Regensburg-Tuink and Hooykaas, 1993). Nuclear localization signals (NLS) in VirD2 and VirE2 target the T-complex into the nucleus where T-DNA integrates into the genome (Citovsky et al., 1994; Sheng and Citovsky, 1996).

A. Key Early Experiments

A series of key observations led to the discovery of interkingdom gene transfer. Crown gall tumor cells continue to proliferate and produce opines even after the tumor-inducing bacteria are killed with antibiotics (Braun, 1958). These observations suggested that *A. tumefaciens* transmits genes for tumor maintenance and opine synthesis to plant cells and that, once established, these genes encode all the functions necessary to confer the transformed phenotype. Because virulence depends on the presence of a Ti plasmid (Watson et al., 1975; Van Larebeke et al., 1974), this extrachromosomal element seemed likely to carry the oncogenes. Hybridization between specific Ti plasmid sequences and DNA isolated from axenic (bacteria-free) tumor cells proved this hypothesis; DNA from nontransformed plant cells did not hybridize (Chilton et al., 1977). Subsequent work established that tumor cells often contain a specific portion of the Ti plasmid, called the T-DNA, integrated into the nuclear DNA of the host (Thomashow et al., 1980; Chilton et al., 1980; Willmitzer et al., 1980), and tumor cells express genes responsible for the transformed phenotype (Garfinkel and Nester, 1980; Garfinkel et al., 1981; Willmitzer et al., 1982). We now have a reasonably detailed understanding of this gene transfer system, and recently the complete DNA sequence of a Ti plasmid was compiled (Zhu et al., 2000). The remainder of this chapter will cover more recent work on T-DNA transfer: genetic analysis of *cis*-acting T-DNA border (origin of transfer) sequences, biochemical characterization of proteins that interact with border sequences and T-DNA, and studies of membrane proteins that facilitate export of T-DNA and virulence proteins into host cells.

B. Regulation of Virulence Genes: The *vir* Regulon

T-DNA transfer requires, in *trans*, virulence (*vir*) genes located on the Ti plasmid outside of the T-DNA (Stachel and Nester, 1986;

Garfinkel and Nester, 1980). Wounded plant cells produce phenolic defense compounds (e.g., acetosyringone) and release sugars that induce expression of Ti plasmid *vir* genes (Stachel et al., 1985). A constitutive gene, *virA*, encodes an inner membrane signal receptor/kinase protein that responds to *vir*-inducing compounds by phosphorylating itself and VirG (Stachel et al., 1985; for reviews, see Winans, 1992; Hooykaas and Beijersbergen, 1994). Activated VirG protein stimulates transcription of its own gene and other *vir* operons (Winans, 1992). Promoters for these operons contain similar sequence motifs that facilitate their coordinate regulation by a common transcription factor (Das et al., 1986; Tate, 1987; Winans, 1992).

The *virC* and *virD* operons also respond to regulation by the chromosomal locus *ros* (Close et al., 1987; Tait and Kado, 1988). A mutation in *ros* elevates expression of both of these operons, independent of *virA*, *virG*, and *vir*-inducing phenolic compounds. Thus factors within the bacterial cell as well as signal molecules produced by host plant cells regulate the *vir* loci.

C. Protein Secretion Apparatus

Export of the T-DNA-VirD2 complex and other virulence proteins (VirE2 and VirF) requires at least 12 membrane-associated proteins: 11 encoded by the *virB* operon and another encoded by *virD4* (for reviews, see Winans et al., 1996; Segal et al., 1999b; Lessl and Lanka, 1994; Christie, 1997). The VirB proteins and VirD4 belong to a family of type IV secretion systems, which includes the *Bordetella pertussis* toxin liberation (Ptl) proteins (Covacci and Rappuoli, 1993; Weiss et al., 1993), *Legionella pneumophila* vir homologues (Lvh) and some Icm/Dot proteins (Segal et al., 1999a,b; Segal and Shuman, 1998): *Helicobacter pylori* Cag proteins (Tummuru et al., 1995; Censini et al., 1996), *Rickettsia prowazekii* VirB proteins (Andersson et al., 1998), and conjugation proteins from IncPα plasmid RP4 (Trb; Lessl et al., 1992; Brahn et al., 2000), IncN plasmid

pKM101 (Tra; Pohlman et al., 1994), and IncW plasmid R388 (Trw; Kado, 1994). Thus type IV secretion systems facilitate two important processes: (1) secretion of virulence factors from pathogen to host, and (2) promiscuous (broad-host-range) conjugation of plasmid DNA. The *A. tumefaciens* VirB/VirD4 transporter is the most versatile of these systems and mediates both promiscuous gene transfer and export of virulence proteins.

The genes that encode type IV secretion systems of *A. tumefaciens*, *L. pneumophila*, *H. pylori*, *B. pertussis*, *R. prowazekii*, and plasmids RP4, pKM101, and R388 share sequence homologies and similar arrangements within operons (Segal et al., 1999b). Functional similarities also exist. For example, both the *A. tumefaciens* VirB/VirD4 and *L. pneumophila* Icm/Dot systems mediate conjugation of IncQ plasmid RSF1010 between bacteria, and the presence of RSF1010 abolishes the virulence of both pathogens (Binns et al., 1995; Stahl et al., 1998; Vogel et al., 1998; Segal et al., 1999a). In addition, VirD4 from *A. tumefaciens* can substitute for TraG from pTiC58 in plasmid conjugation (Hamilton et al., 2000). The involvement of closely related type IV secretion systems in both conjugation and protein export suggests that conjugation may be a specialized form of protein export in which the exported protein—the DNA-nicking protein VirD2 in this case—is covalently attached to DNA.

D. The Conjugation Model of T-DNA Transfer

I. Promiscuous conjugation

In many ways T-DNA transfer from *A. tumefaciens* to plant cells resembles broad-host-range plasmid conjugation between bacteria. In each case, a multi-subunit endonuclease binds an origin of transfer (*oriT*) sequence forming a relaxosome, and one endonuclease subunit (relaxase) nicks the DNA and covalently attaches to the 5' end (Ward and Barnes, 1988; Young and Nester, 1988;

Herrera-Estrella et al., 1988; Howard et al., 1989; Panesgrau et al., 1993; Jasper et al., 1994; Lessl and Lanka, 1994). The donor transfers a single DNA strand, together with the bound protein, to the recipient via a type IV secretion system.

The conjugation model of T-DNA transfer gained strong support from an unexpected quarter. The *A. tumefaciens* VirB/VirD4 system can transfer a broad-host-range mobilizable bacterial plasmid (RSF1010), which does not contain a T-DNA border sequence, into plant cells where the plasmid DNA integrates into the nuclear genome (Buchanan-Wollaston et al., 1987). In addition to the VirB/VirD4 secretion system, interkingdom transfer of RSF1010 requires the *oriT* sequence and mobilization (*mob*) proteins, which create a site-specific nick within *oriT* (see Porter, this volume). The ability of *A. tumefaciens* to mobilize a broad-host-range plasmid into plant cells supports the conjugation model and opens the possibility that plants potentially receive a great variety of information from many species of gram-negative bacteria.

The *A. tumefaciens* VirB/VirD4 system also promotes conjugation of plasmid DNA into other bacteria (Beijersbergen et al., 1992; Steck and Kado, 1990; Gelvin and Habeck, 1990; Fullner et al., 1996a; Fullner and Nester, 1996b), plants (Buchanan-Wollaston et al., 1987), fungi (Bundock et al., 1995; de Groot et al., 1998), or human cells (Kunik et al., 2001) indicating that type IV secretion systems can export a variety of proteins and protein-DNA complexes into a broad range of recipient cells.

At least one *A. tumefaciens vir* protein can function as part of a different conjugation system. An essential conjugation protein, known as the *coupling protein*, appears to link the relaxosome to the transmembrane DNA/protein secretion apparatus, which is also called the *mating pair formation system*. Several coupling proteins, for example, TraG of plasmid RP4 and TraD of F, show limited sequence similarity to *A. tumefaciens* VirD4 (Lessl and Lanka, 1994). In fact

VirD4 can substitute for its pTiC58 homologue, TraG, during conjugation of RSF1010 via the pTiC58 *trb*-encoded mating bridge, thereby proving that VirD4 is a conjugation protein (Hamilton et al., 2000).

2. Border sequences

T-DNAs from several different *Agrobacterium* strains have very similar 23 base-pair (bp) border sequences at each T-DNA end. T-DNA transfer requires the right-hand border in its wild-type orientation (Peralta and Ream, 1985; Wang et al., 1984; Shaw et al., 1984). Inversion of the right border reduces virulence drastically (Peralta and Ream, 1985; Wang et al., 1984), and the rare tumors that develop contain most or all of the 200 kb Ti plasmid (Miranda et al., 1992). Deletion of the right border abolishes tumorigenesis (Hepburn and White, 1985; Christie et al., 1988a), whereas removal of the left border does not affect virulence (Joos et al., 1983), indicating that T-DNA transfer begins at the right border, moves leftward through the T-DNA, and terminates at the left border (Fig. 5). T-DNA borders share both sequence and functional similarities with the *oriT* of broad-host-range conjugative plasmid RP4 (Lessl and Lanka, 1994). Thus T-DNA transfer from *A. tumefaciens* into plants strongly resembles plasmid conjugation between bacteria.

Another *cis*-acting sequence, called *overdrive*, flanks right-hand (but not left-hand) border sequences and stimulates T-DNA transfer several hundredfold (Peralta et al., 1986). Unlike the border sequence, *overdrive* functions in either orientation and at considerable distances on either side of the right-hand border sequence (Ji et al., 1988). Efficient T-DNA transfer requires only two *cis*-acting sequences: the right-hand border sequence and *overdrive*. The following paragraphs will explore the interaction of these DNA sequences with virulence proteins.

3. The relaxosome

Several *vir* operons encode proteins that participate in DNA-protein interactions

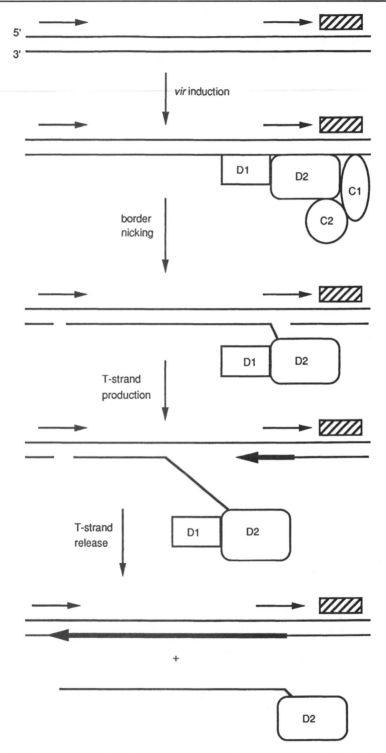

Fig. 5. Model for T-strand production. Short arrows indicate border sequences; hatched box is overdrive. Bold arrow indicates strand displacement (helicase) activity and possible DNA systhesis.

necessary for T-DNA transfer. The first two genes of the *virD* operon (*virD1* and *virD2*) encode a site-specific nicking enzyme that nicks the bottom strand of T-DNA border sequences between the third and fourth base (Wang et al., 1987; Yanofsky et al., 1986). (For *vir* operons containing more than one gene, the number that follows the gene name indicates the position of the gene in the operon rather than an allele number for a specific mutation.) VirD2 protein attaches covalently to the $5'$ end of the nicked DNA (Ward and Barnes, 1988; Young and Nester, 1988; Herrera-Estrella et al., 1988; Howard et al., 1989; Durrenberger et al., 1989) via a phosphodiester bond with a specific tyrosine residue (Vogel and Das, 1992). The nick within the right border sequence initiates production of T-strands, which are full-length single-stranded copies of the bottom strand of the T-DNA (Stachel et al., 1986a; Albright et al., 1987; Jayaswal et al., 1987) that the bacteria export into plant cells (Stachel and Zambryski, 1986b; Yusibov et al., 1994; Tinland et al., 1994). Thus, early events in T-DNA transfer resemble those in bacterial conjugation: a multiple-subunit nicking enzyme binds an *oriT* sequence forming a DNA-protein complex called the relaxosome, which creates a site-specific nick in one strand of the *oriT* sequence (see Porter, this volume). During this process, one endonuclease subunit covalently binds the nicked DNA.

The *overdrive* sequence lies near the right-hand T-DNA border and, together with the VirC1 and VirC2 proteins, stimulates tumorigenesis several hundredfold (Peralta et al., 1986). VirC1 binds *overdrive*, and the VirD2 nicking protein also interacts with this sequence (Toro et al., 1989). Although the precise role of *overdrive* and the *virC*-encoded proteins in the relaxosome remain unknown, they appear to distinguish the right-and left-hand border sequences (the origin and terminus of T-DNA transfer). Because T-DNA transfer is unidirectional (Miranda et al., 1992), plant cells will receive the oncogenes and opine synthesis genes only if transfer

begins at the right border. In order to avoid unproductive transfer events that begin at the left border, the relaxosome must distinguish between right-and left-hand border sequences, which are functionally equivalent in their interaction with the VirD1/VirD2 nicking enzyme (Yanofsky et al., 1986; Albright et al., 1987). Apparently *overdrive* allows the transfer apparatus to recognize the right-hand border as the origin of transfer, perhaps by helping to tether the relaxosome to the VirB/VirD4 secretion/mating bridge apparatus.

4. T-strands

Induction of *vir* expression and border nicking leads to formation of intermediates in T-DNA transfer (T-strands), which are full-length, linear, single-stranded DNA molecules comprised of the bottom strand of the T-DNA (Fig. 5) (Stachel et al., 1986a, 1987; Veluthambi et al., 1988). Following a proteinase digestion, DNA isolated from *vir*-induced *A. tumefaciens* cells and subjected to agarose gel electrophoresis can be transferred to nitrocellulose filters by blotting *without denaturation*. These conditions permit hybridization of only DNAs with single-stranded regions to complementary labeled DNA or RNA probes. Only probes corresponding to the *top* strand of the T-DNA anneal to T-strands, proving that T-strands are derived from the bottom strand. Treatment with an endonuclease (S1) or exonuclease (*E. coli* exonuclease VII or T4 DNA polymerase) specific for single-stranded DNA destroys T-strands and demonstrates the presence of a free $3'$ end (exo VII or T4 DNA polymerase) and possibly a free $5'$ end (exo VII).

Mutations in *virA*, *virG*, *virD1*, and *virD2* abolish T-strand production (Stachel et al., 1987; Veluthambi et al., 1988); *virA* and *virG* are required because these genes control expression of the *virD* operon. T-strand displacement, which occurs $5'$ to $3'$, likely requires helicase activity and may be accompanied by synthesis of a new copy of the bottom strand, although this has not been shown. The genes that encode other proteins that are probably involved in T-strand production, for example, helicase, DNA polymerase, and topoisomerase, have not been identified and may lie on one of the two *A. tumefaciens* chromosomes.

5. Secreted single-stranded DNA-binding protein: VirE2

The *virE* operon encodes two proteins: VirE1 (65 amino acids) and VirE2 (533 amino acids; (Winans et al., 1987). The single-stranded DNA-binding (SSB) activity of VirE2 does not depend on VirE1 (Citovsky et al., 1988, 1989; Gietl et al., 1987; Das, 1988; Christie et al., 1988b; Sen et al., 1989). In *E. coli* that contain the Lon protease, VirE1 protein stabilizes VirE2 (McBride and Knauf, 1988), suggesting that these proteins interact physically. Indeed, protein interaction cloning (yeast two hybrid; see Geoghegan, this volume) studies showed that VirE2 contains two separable domains that bind VirE1 (Sundberg et al., 1996; Sundberg and Ream, 1999). In *A. tumefaciens*, VirE2 is equally stable with or without VirE1 (Sundberg et al., 1996). VirE1, a secretory chaperone, promotes transport of VirE2 protein into plant cells via the VirB/VirD4 secretion system (Deng et al., 1999; Zhou and Christie, 1999; Sundberg and Ream, 1999). Thus both proteins are essential for tumorigenesis (Sundberg et al., 1996).

VirE2 likely has multiple roles in T-DNA transfer. VirE2 binding protects single-stranded DNA from nuclease attack in vitro (Citovsky et al., 1989; Sen et al., 1989) and inside plant cells (Yusibov et al., 1994; Rossi et al., 1996); however, absence of VirE2 does not diminish T-strand accumulation in *A. tumefaciens* (Stachel et al., 1987; Veluthambi et al., 1988). The presence of nuclear localization signals (NLS) in VirE2 (Citovsky et al., 1992) suggests that it enters plant nuclei during infection. The two NLSs of VirE2 continue to function when VirE2 is bound to ssDNA (Zupan et al., 1996), even though the NLS domains overlap regions of VirE2 involved in cooperativity and ssDNA

binding (Citovsky et al., 1992). Fluorescein-labeled ssDNA bound by VirE2 enters the nucleus of plant cells injected with the complex, whereas in the absence of VirE2, the DNA remains cytoplasmic (Zupan et al., 1996). Thus, inside plant cells, VirE2 shields T-strands from nuclease attack and targets them to the nucleus.

VirE2 is exported from bacterial cells and functions inside plant cells. VirE2 may coat T-strands as they are displaced from the nicked Ti plasmid (the T-complex model) (Sheng and Citovsky, 1996; Zupan and Zambryski, 1997) (Fig. 6) or later, inside plant cells (the separate export model) (Sundberg et al., 1996) (Fig. 7). However, VirE2 is necessary only in plant cells; transgenic plant cells that express VirE2 produce tumors

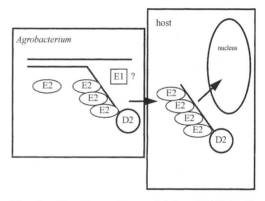

Fig. 6. The T-complex model for T-DNA/VirE2 export.

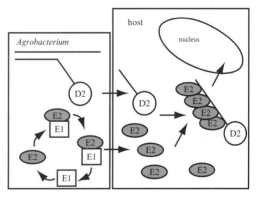

Fig. 7. The separate export model for VirE2 & T-DNA.

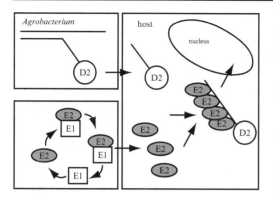

Fig. 8. "Complementation" by mixed infection. Infection of the same host cell by mutant *A. tumefaciens* lacking *virE2* (top) and T-DNA (bottom) induces tumors.

when inoculated with *virE2*-mutant *A. tumefaciens* (Citovsky et al., 1992). Coinoculation of a *virE2* mutant and a *virE+* strain lacking T-DNA results in tumor formation, even though each strain alone is avirulent and unable to exchange genes by conjugation (Otten et al., 1984). This "complementation" by mixed infection (Fig. 8) requires the VirB/VirD4 transporter (Otten et al., 1984; Christie et al., 1988a), and both strains must bind to plant cells (Christie et al., 1988a). T-strands accumulate to normal levels in bacterial cells without VirE2 (Stachel et al., 1987; Veluthambi et al., 1988), and *A. tumefaciens* can transfer these uncoated T-strands into plant cells (Citovsky et al., 1992; Sundberg et al., 1996). T-strands were detected inside wild-type plant cells infected by a *virE2* mutant (Yusibov et al., 1994). Export of VirE2, but not of T-strands, from *A. tumefaciens* requires VirE1 (Sundberg et al., 1996). Thus *A. tumefaciens* appears able to export into plant cells, independently, either VirE2 protein or uncoated T-strand DNA; one does not depend on the other for transfer (Sundberg et al., 1996; Sundberg and Ream, 1999).

Interaction cloning experiments identified protein contacts between VirE2 and VirE1, its secretory chaperone (Sundberg et al., 1996; Deng et al., 1999; Zhou and Christie, 1999; Sundberg and Ream, 1999). VirE1

binds VirE2 domains involved in binding ssDNA and self-association, and VirE1 may facilitate VirE2 export by preventing VirE2 aggregation and premature binding of VirE2 to ssDNA (Deng et al., 1999; Sundberg and Ream, 1999).

The T-complex model rests largely on the observation that antibodies specific for VirE2 can precipitate T-strands from lysates of *A. tumefaciens* cultured in the presence of acetosyringone, a *vir*-inducing compound (Christie et al., 1988a). Although this may indicate an interaction between VirE2 and T-strands in living *A. tumefaciens* during infection (Sheng and Citovsky, 1996; Zupan and Zambryski, 1997), another interpretation is possible. Because at least 30 minutes elapsed between cell lysis and the addition of VirE2-specific antibodies (Christie et al., 1988a), VirE2 may have bound T-strand DNA after disruption of the cells. Also induction of *vir* genes in the absence of recipient plant cells is an artificial situation that provides no outlet for nascent T-strands. T-strand transfer resembles plasmid conjugation in many ways (Firth et al., 1996; Winans et al., 1996; Baron and Zambryski, 1996; Kado, 1994; Christie, 1997; Sheng and Citovsky, 1996; Zupan and Zambryski, 1997; Lessl and Lanka, 1994); conjugal DNA metabolism appears to occur at the transmembrane export channel, with ssDNA transferred directly to the recipient (Firth et al., 1996). Indeed, proper contact between recipient and donor cells triggers conjugal DNA processing (Firth et al., 1996; Ou and Reim, 1978; Kingsman and Willetts, 1978). T-strand production likely occurs in a similar manner, at the VirB/VirD4 macromolecule export apparatus. Thus, during a normal infection, T-strands may leave bacterial cells as they are displaced from the Ti plasmid, without significant exposure to the bacterial cytoplasm. T-strands may accumulate in acetosyringone-treated *A. tumefaciens* only because recipient cells are absent.

The separate export model seeks to explain several observations: (1) VirE2-producing plant cells restore pathogenicity to *virE*-mutant *A. tumefaciens* (Citovsky et al., 1992); (2) VirE2 made in one *A. tumefaciens* strain can interact productively with T-strands generated in another (during mixed infections) (Christie et al., 1988a; Otten et al., 1984); (3) export of VirE2 requires VirE1 (Sundberg et al., 1996) and VirB1 (K.J. Fullner, personal communication), whereas T-strand transfer does not; (4) the presence of plasmid RSF1010 in *A. tumefaciens* blocks VirE2 export but merely reduces T-strand transfer (Binns et al., 1995); and (5) expression of the Osa (oncogenesis suppressing activity) protein in *A. tumefaciens* prevents export of VirE2 but not of T-strands (Lee et al., 1999). From these studies we know *A. tumefaciens* can export VirE2 and T-strand DNA separately under special circumstances; however, they do not indicate whether wild-type *A. tumefaciens* transfers VirE2 and T-strand DNA into plant cells separately or as a complex.

6. A pilot protein: VirD2

VirD2 endonuclease attaches covalently to T-strand DNA through a phosphodiester bond between the 5′ end of the T strand and a conserved tyrosine residue in VirD2 (reviewed in Sheng and Citovsky, 1996). T strands apparently enter plant cells via the VirB/VirD4 secretion system by virtue of this attachment. T-DNA border-specific endonuclease activity lies entirely within the N-terminal half of VirD2 (Ward and Barnes, 1988). Conserved residues at the C-terminus of VirD2 are important for tumorigenesis (Shurvinton et al., 1992; Rossi et al., 1993), although they are not needed for T-strand production. Instead, this domain contains a bipartite NLS that helps target T strands to plant nuclei (Howard et al., 1992; Rossi et al., 1993; Tinland et al., 1992; Shurvinton et al., 1992). Thus VirD2 pilots T strands into plant cells, and VirD2 may also lead T strands through the nuclear pore.

VirD2 participates in T-DNA integration, as indicated by several observations. First, the left and right ends of T-DNAs are joined to plant DNA via different mechanisms. Sequences of plant-to-T-DNA junctions indicate that right-hand ends of integrated T-DNAs very often correspond exactly to the base at which T strands attach to VirD2; in contrast, the left end of the T-DNA varies by hundreds of bases (Gheysen et al., 1991; Matsumoto et al., 1990; Mayerhofer et al., 1991). This preservation of right-hand T-DNA ends suggests that VirD2 protects them from nuclease attack (Jasper et al., 1994) and ligates the 5′ ends of T strands to plant DNA (Pansegrau et al., 1993; Jasper et al., 1994). Second, specific mutations in *virD2* either reduce integration (but not nuclear entry) of T-DNA (Narasimhulu et al., 1996) or result in T-DNAs with aberrant right-hand ends (Tinland et al., 1995), indicating that proper joining of the VirD2-bound end of a T strand to plant DNA requires wild-type VirD2. Third, VirD2 has ligase activity. Purified VirD2 cleaves ssDNAs containing the bottom strand of the T-DNA border sequence (Jasper et al., 1994; Pansegrau et al., 1993), and VirD2 can ligate these cut ssDNA molecules to reform the original substrate (Jasper et al., 1994) or join the VirD2-bound portion to another oligonucleotide in a sequence-specific manner (Pansegrau et al., 1993). This same ligase activity may join the 5′ end of T strands to plant DNA.

7. The VirB pilus

Pathogenic bacteria use pili to export virulence-associated macromolecules (usually proteins) into plant and animal cells (Fullner et al., 1996a; Roine et al., 1997; Ginocchio et al., 1994). *A. tumefaciens* produces a VirB/VirD4-dependent pilus upon induction of the *vir* regulon (Fullner et al., 1996a). VirB2 is probably the major component of the Vir pilus, and it shares sequence similarity to TraA, the pilin encoded by plasmid F (Lai and Kado, 1998). Both VirB2 and TraA are processed to smaller forms and exported (Firth et al., 1996; Shirasu and Kado, 1993). F pili contain only TraA, but at least 13 other Tra proteins act during pilus assembly (Firth et al., 1996). Four VirB proteins have sequences and subcellular locations similar to Tra counterparts: VirB3 (TraL), VirB4 (TraC), VirB5 (TraE), and VirB10 (TraB) (Christie, 1997).

VirB4, an inner-membrane protein with ATPase activity (Shirasu et al., 1994; Fullner et al., 1994; Berger and Christie, 1993), is required for transport of another pilus-assembly protein, VirB3, to its location in the outer membrane (Jones et al., 1994). The protein translocation activity of VirB4 is consistent with its similarity to TraL, an ATPase needed for F pilus assembly (Firth et al., 1996).

VirB5, found in the inner membrane and periplasm (Thorstenson et al., 1993), shares sequence similarity with an inner-membrane protein (TraE) involved in F pilus assembly (Firth et al., 1996) and with pKM101 TraC, which acts extracellularly (Pohlman et al., 1994). The role of VirB5 in pilus assembly remains unknown.

VirB10 functions in concert with VirB9 and VirB11; elevated levels of all three proteins are required to overcome RSF1010-mediated inhibition of T-DNA transfer (Ward et al., 1991; Binns et al., 1995). The similarity between TraB, an inner-membrane protein required for F pilus assembly (Firth et al., 1996), and VirB10, located in the inner and outer membranes (Thorstenson et al., 1993; Finberg et al., 1995), suggests that VirB9, VirB10, and VirB11 may participate in VirB pilus assembly. VirB11 has both ATPase and protein kinase activities (Christie et al., 1989; Rashkova et al., 1997); this ATPase activity may drive assembly of a structure containing VirB9 and VirB10, or it may provide the force to move macromolecules through the pore after assembly.

Interactions occur between several VirB proteins. VirB7, a lipoprotein located in the outer membrane (Fernandez et al., 1996), forms disulfide bonds with and stabilizes VirB9, which is associated with both the outer and inner membranes (Spudich et al., 1996; Anderson et al., 1996; Das et al., 1997). VirB8 may also affect accumulation of VirB9 (Berger and Christie, 1994). Assembly of

VirB10 into high molecular weight aggregates requires VirB9, although VirB9 is not a component of the VirB10-containing aggregates (Beaupre et al., 1997; Ward et al., 1990; Finberg et al., 1995). Instead, VirB9 forms high molecular weight complexes that do not include VirB10 (Beaupre et al., 1997). These aggregates may form a transmembrane macromolecule transporter.

VirB1 influences the efficiency of protein export from *A. tumefaciens* to plant cells; however, loss of VirB1 reduces but does not abolish tumorigenesis (Berger and Christie, 1994), indicating that it cannot be a crucial component of the pilus. During extracellular complementation, VirB1 is required for export of VirE2 but not the VirD2-T strand complex (K. J. Fullner, personal communication). Because approximately 600 VirE2 molecules are needed to coat one T strand, reduced transport efficiency should affect the ability of the bacteria to export a useful amount of VirE2 more profoundly than VirD2-T strand transfer. VirB1 contains sequences similar to lysozyme and may have glycosidase activity (Mushegian et al., 1996; Baron et al., 1997); it occurs in both inner and outer membranes, and a processed form of VirB1 is exported (Baron et al., 1997). These properties may allow VirB1 to create openings in the bacterial cell wall (Mushegian et al., 1996; Baron et al., 1997).

8. Gateway to the pore: VirD4 coupling protein

VirD4 resembles TraG of RP4 and TraD of F plasmid (Firth et al., 1996; Lessl and Lanka, 1994); these proteins appear to connect relaxosomes with membrane-associated secretion/mating bridge systems (Lessl and Lanka, 1994; Cabezon et al., 1997; Firth et al., 1996). VirD4 is similar enough to pTi-encoded TraG that it can substitute for TraG, allowing conjugal transfer of RSF1010 through the pTi Trb pilus into recipient bacteria (Hamilton et al., 2000). Unlike most VirB proteins, a VirD4 analogue has not been reported in the *B. pertussis* Ptl toxin export system. However, other type IV secretion systems, including those in

H. pylori, L. pneumophila, and *R. prowazekii*, contain a protein similar to VirD4 (Segal et al., 1999). Thus VirD4 and most VirB proteins have homologues not only among conjugation proteins but also among proteins devoted solely to toxin secretion.

The function of VirD4 appears more elaborate than an interface between relaxosome and transmembrane pore. Export of VirE2 (in the absence of T-strand DNA) requires VirD4, establishing its involvement in protein transport. Formation of the VirB pilus requires VirD4 (Fullner et al., 1996a), indicating that it participates in translocation of pilus proteins as well. In this regard, VirD4 differs significantly from TraD, which is not needed for F pilus production (Firth et al., 1996).

E. Integration Products

Structures of integrated T-DNAs contributed to our understanding of the transformation process. However, the results require cautious interpretation because T-DNAs in established tumor lines may have rearranged subsequent to the initial integration events. Thus the structures examined may not resemble the initial integration products, although numerous T-DNAs remained stable after lengthy propagation of transformed tissue (Van Lijsebettens et al., 1986). Indeed, nononcogenic T-DNAs in transgenic plants remain stable through meiosis and mitosis (Barton et al., 1983). Many integrated T-DNAs do not undergo obvious rearrangements and remain colinear with the corresponding portion of the Ti plasmid. T-DNAs reside at a variety of locations in the plant genome, often as single copies or short tandem arrays (direct or inverted repeats) (Thomashow et al., 1980; Zambryski et al., 1980; Lemmers et al., 1980; Peerbolte et al., 1987), and separate T-DNAs can integrate independently at different locations in the genome of a single plant cell (Chyi et al., 1986).

Alterations of the host target site that accompany T-DNA insertion suggest that this process follows a complex pathway. In one case, T-DNA integration produced a 158 base-pair direct repeat of host sequences as well as a base change, a small deletion, and "filler" DNA of unknown origin that resembles nearby host sequences (Gheysen et al., 1987, 1991). Other transformed plant cells contained a particular truncated T-DNA integrated at several different locations in the genome. Apparently the plant cell copied an aberrant T-DNA before integrating the copies at multiple sites. Thus T-DNA integration occurs via a complex mechanism with several steps, including replication events and ligation of T-DNA to plant DNA.

IV. PLANT GENETIC ENGINEERING

Agrobacterium tumefaciens provides a convenient and effective means to produce transgenic plants. Gene transfer from *A. tumefaciens* to plants requires the right border and *overdrive* sequences in *cis*, but the transferred DNA need not reside on the Ti plasmid, nor does transfer and integration depend on specific genes within the T-DNA (Klee et al., 1987). Thus we can substitute genes of our choice for those normally found in the T-DNA, and plant cells that inherit innocuous foreign genes can, under appropriate tissue culture conditions, regenerate into morphologically normal plants that express novel genes (Klee et al., 1987). Generally, foreign genes in transgenic plants remain stable through meiosis and behave as Mendelian traits (Klee et al., 1987). *A. tumefaciens* strains used to engineer plants harbor a disarmed Ti plasmid that lacks oncogenes but retains the virulence genes. Transgenes flanked by T-DNA borders reside either in a separate broad-host-range plasmid (a "binary" system) or become cointegrated into the disarmed Ti plasmid (a "*cis*" system) by homologous recombination between sequences found both on the Ti plasmid and a shuttle plasmid bearing the engineered T-DNA (Klee et al., 1987; Fraley et al., 1985). Usually the T-DNA will include a drug resistance gene designed to render transformed plant cells resistant to a compound toxic to plant cells, for example, kanamycin, hygromycin, glyphosate, and other herbicides. These markers provide a simple selection for plant cells that contain the engineered T-DNA.

Although *A. tumefaciens* does not induce tumors on monocotyledonous plants, it can transmit T-DNA into monocots, including important crop species such as rice and corn. (Monocots and dicots differ significantly in phytohormone physiology, which explains why monocots do not respond to *A. tumefaciens* oncogenes and why 2,4-dichlorophenoxyacetic acid (2,4-D), a synthetic auxin, kills the dandelions but not the grass in lawns.) The *A. tumefaciens* VirB/VirD4 conjugation system transforms dicotyledonous plants under natural circumstances, and in the laboratory it can also transfer DNA into bacteria, yeast, filamentous fungi, and human cells. Given the promiscuity of this gene transfer system, it is not surprising that *A. tumefaciens* can deliver T-DNA to monocotyledonous plants.

A. tumefaciens-mediated gene transfer is the preferred method to create transgenic plants. The proteins associated with T-DNA, VirD2 and VirE2, help maintain the integrity of the integrated genes and reduce the frequency of duplications and rearrangements, which often affect transgene expression. For example, transcription through inverted repeat copies of a transgene will produce double-stranded RNA. This aberrant RNA may trigger post-transcriptional silencing of the transgene through systemic sequence-specific degradation of the RNA (Fire et al., 1998; Timmons and Fire, 1998; Montgomery et al., 1998; Montgomery and Fire, 1998; Angell and Baulcombe, 1997; Hamilton and Baulcombe, 1999; Ratcliff et al., 1997). Electroporation, transformation, or microprojectile bombardment can introduce "naked" DNA into plant cells, but integration is inefficient and almost always results in multiple tandem copies that suffer frequent rearrangements.

A novel approach combines the "biolistic" or microprojectile bombardment method with the integration-promoting activity of VirD2 protein. "Agrolistic" transformation uses bombardment with tungsten particles coated with double-stranded T-DNA to introduce DNA into plant cells (Hansen et al., 1997; Hansen and Chilton, 1996). In contrast to standard biolistic transformation, the T-DNA includes border sequences as well as *virD1* and *virD2* fused to plant promoters. Plant cells that receive the T-DNA via particle bombardment produce VirD1 and VirD2 prior to integration of the incoming DNA. The VirD1/VirD2 enzyme nicks the border sequences, which yields an integrated T-DNA with a precise right end (Hansen et al., 1997; Hansen and Chilton, 1996). This approach is important for species recalcitrant to regeneration from tissue culture, since cells within embryos can be transformed.

A. tumefaciens-mediated leaf disk transformation is widely used for species amenable to regeneration from unorganized callus (Horsch et al., 1985). In this method aseptic leaf disks are infected with disarmed *A. tumefaciens* harboring the transformation vector. After a period of cocultivation, antibiotics are added to kill the bacteria. Eventually the leaf disks are transferred to agar containing an antibiotic or herbicide toxic to nontransformed plant cells, which permits only transgenic cells to form callus. The callus is transferred first to cytokinin-rich shoot-inducing medium and then to root-inducing medium, resulting in transgenic plants. For some species, such as *Arabidopsis thaliana*, a floral dip method yields transgenic seeds and obviates the need for tissue culture (Clough and Bent, 1998). Flowers are dipped in a culture of disarmed *A. tumefaciens* and allowed to set seed. In a typical procedure, approximately 0.1% of the seeds will contain the transgenes; these seeds are identified by their ability to germinate on medium containing the appropriate herbicide.

All plant transformation methods, including *A. tumefaciens*-mediated transfer, suffer from one serious limitation: the inability to target transgenes to specific chromosomal locations at a useful frequency. In bacteria, fungi, and mammalian cells, transgenes flanked on both sides with host chromosomal DNA can be introduced into specific locations via homologous recombination. Although such homologous recombination events can occur during plant transformation, they constitute a very small fraction of the total number of integration events. The chromosomal location of a transgene can affect expression of both the transgene and chromosomal genes at the site of insertion. On average, plant scientists must examine 100 transformed plants to find one that exhibits appropriate transgene expression without affecting other important agronomic traits. An efficient integration system based on homologous recombination would allow engineers to place transgenes at specific chromosomal locations that allow good transgene expression without affecting host genes. For these reasons control over transgene insertion site is an important tool that plant genetic engineers currently lack.

A. tumefaciens–mediated transformation has been used to create a number of highly successful transgenic crop plants sown extensively in the United States and elsewhere during the late 1990s. The first transgenic plant marketed was a tomato variety that contains an antisense transgene that blocks expression of polygalacturonase. Because this enzyme contributes to fruit softening, the transgenic tomatoes remain firm even when they are harvested ripe and shipped. In 1999, soybean, corn, cotton, and canola occupied the most acreage, with 21.6, 11.1, 3.7, and 3.4 million hectares planted worldwide (Ferber, 1999).

Herbicide tolerance and insect resistance are the most common traits introduced into transgenic crops. For example, "Bt" corn and cotton produce *Bacillus thuringiensis* insecticidal protein, which kills the larvae of lepidopteran insects. This highly specific insecticidal protein affects the digestive tracts of larvae that ingest plant material contain-

ing the protein. These varieties have prevented billions of dollars of damage by the European corn borer and the boll weevil. U.S. farmers planted 1 million hectares of Bt cotton in 1998, which reduced application of chemical insecticides by 450,000 kilograms, increased yields by 39 million kilograms, and raised profits by $92 million (Ferber, 1999). Although Bt-resistant insects may (or may not) emerge as the result of extensive use of Bt-producing plants, to date this has not occurred.

Another potential problem with Bt-producing plants is harm to nontarget species. A laboratory study demonstrated that pollen from a Bt corn line could kill monarch butterfly larvae that ingested a sufficient amount of the pollen along with milkweed, their food source (Losey et al., 1999). Field studies suggest that pollen from Bt corn causes little or no harm to monarch caterpillars (Ferber, 1999; Wraight et al., 2000; Niiler, 1999; Hodgson, 1999). The most widely sown varieties of Bt corn produce less insecticidal protein than the variety tested initially, and milkweed within 1 meter of a cornfield are unlikely to become coated with enough pollen to cause harm (Ferber, 1999). In addition the times of corn pollen shed and monarch larvae feeding do not overlap in most of the corn belt during a typical summer, which greatly reduces or eliminates exposure of the larvae to the insecticidal protein. Loss of habitat to cities, suburbs, and any type of agriculture likely represents the greatest human threat to the monarch butterfly.

Herbicides play an important role in modern agriculture. They eliminate competition between weeds and crops, thereby increasing yields, and herbicides preserve precious topsoil by reducing the need to cultivate. Despite these benefits, some herbicides cause environmental damage by contaminating soil and water with toxic residues. When evaluating herbicide-tolerant transgenic crop plants, engineers must consider more than the efficacy of the compound on weeds and the level of resistance in the engineered plants. The toxicity of the compound, its longevity and mobility in soil, and the likelihood of herbicide resistance genes crossing into weedy species should also influence decisions to introduce herbicide resistance into crop plants.

Tolerance to a widely used broad-spectrum herbicide, glyphosate, has been introduced into several important crops, including corn, soybeans, and canola. Glyphosate is environmentally safe due to its extremely low toxicity and short life (typically days) in soil (Grossbard and Atkinson, 1985). Glyphosate inhibits 5-*enol*pyruvylshikimic acid-3-phosphate synthase (EPSP synthase), an enzyme in the shikimic acid pathway that catalyzes production of 5-*enol*pyruvylshikimate-3-phosphate from shikimate-3-phosphate (Grossbard and Atkinson, 1985). The shikimic acid pathway, which occurs only in plants and microbes, generates the aromatic amino acids phenylalanine, tyrosine, and tryptophan. Because this pathway does not occur in animals, glyphosate has very low animal toxicity. For example, sodium chloride ($LD_{50} = 3.75$ g/kg) is more toxic to rats than glyphosate ($LD_{50} = 4.8$ g/kg) (Windholz et al., 1983). Thus glyphosate-tolerant soybeans, canola, and corn permit environmentally safe weed control and reduce the need for cultivation, thereby decreasing soil erosion.

Resistance to viruses has been introduced into crops by incorporating transgenes that encode the virus coat protein. Coat protein is normally made late in the viral life cycle for good reason: premature expression of coat protein disrupts virus production (Bendahmane and Beachy, 1999). Another strategy, production of double-stranded viral RNAs, has proved effective against viruses with RNA genomes. In plants, nematodes, fungi, and flies, double-stranded RNA molecules trigger systemic, sequence-specific destruction of all RNA molecules having sufficient sequence identity (Dougherty and Parks, 1995; Smith et al., 1994; Lindbo and Dougherty, 1992a, b; Dougherty et al., 1994; Waterhouse et al., 1998; Jorgensen et al.,

1998; Montgomery et al., 1998; Ruiz et al., 1998; Chuang and Meyerowitz, 2000; Fire et al., 1998; Timmons and Fire, 1998; Montgomery et al., 1998; Montgomery and Fire, 1998; Angell and Baulcombe, 1997; Hamilton and Baulcombe, 1999; Ratcliff et al., 1997). Either method can lead to durable virus-resistant crop plants that require no additional input from the farmer. One example of commercial success are transgenic papaya trees resistant to papaya ringspot virus. These transgenic trees saved the papaya industry in Hawaii, where ringspot virus had destroyed groves of conventional trees (Ferber, 1999).

The most widely acclaimed and complex genetically modified crop, "golden" rice, was engineered to improve its nutritional value (Guerinot, 2000; Ye et al., 2000). Three transgenes were introduced that allow rice plants to produce β-carotene (provitamin A) in rice endosperm (Ye et al., 2000). Vitamin A deficiency is a serious public health problem in many areas of Asia, Africa, and Latin America, where rice is a staple food. Worldwide, vitamin A deficiency contributes to at least 1 million deaths and 350,000 cases of blindness annually (Ye et al., 2000; Nash, 2000). Wild-type rice endosperm synthesizes geranylgeranyl diphosphate, a β-carotene precursor, but lacks the enzymes phytoene synthase, carotene desaturase, and lycopene-β-cyclase, which are required to convert this precursor into β-carotene. Genes designed to express these enzymes in rice endosperm were constructed and introduced (via *A. tumefaciens*-mediated gene transfer) into the rice genome on two separate T-DNAs, one of which also contained a selectable marker, a gene (*aphIV*) encoding resistance to hygromycin (Ye et al., 2000). The genes for phytoene synthase (*psy*) and lycopene-β-cyclase (*lcy*) were both obtained from daffodil, and each was fused to the endosperm-specific rice glutelin (*gtl*) promoter. The gene encoding carotene desaturase (*crtI*) was obtained from a bacterial source, *Erwinia uredovora*, and fused to the cauliflower mosaic virus 35S (CaMV 35S) promoter (Ye et al., 2000). The resulting engineered rice plants produce grain containing nutritionally useful quantities of provitamin A, which give the grains a yellow color.

V. SUMMARY

In the past 20 years the study of *Agrobacterium tumefaciens* has passed from its infancy to maturity. During that time the promise of interkingdom gene transfer has been fulfilled with the introduction of the the first genetically engineered crop plants, which thus far have proved highly successful. Increased understanding of pathogen-to-host gene transfer and protein secretion has accompanied this technological development, and recent genome sequencing projects have uncovered the close relationship of type IV secretion systems in *A. tumefaciens* and other pathogens. *Agrobacterium*-mediated gene transfer has allowed crop scientists to use any gene with potential benefits, regardless of its source. This technology has also increased the speed with which new cultivars can be produced. Despite the proven successes and enormous potential of plant biotechnology, "genetically modified organisms" (or GMOs, as transgenic plants are known in the popular press) have been vehemently rejected by several European nations and met vocal opposition from small groups in the United States. With food processors using "GMO-free" as a marketing tool, some farmers, who are otherwise pleased with transgenic varieties, have decided to return to nonengineered crops (Enserink, 1999; Kilman, 2000). Despite protests in Europe and North America, where food is plentiful today, the role of plant biotechnology in food production will likely increase in the future. Nations with large populations to feed are enthusiastic proponents of plant biotechnology (Nash, 2000; Tepfer et al., 2000), and these countries need not rely on the United States and Europe to carry this technology forward. Intelligent and highly motivated scientists from India and China, in particular, have trained in the best European and North American laboratories, and these individuals have brought this inexpensive and simple technology to their homelands (Tepfer et al., 2000). Progress will continue, although leadership in this area may change. The potential of plant biotechnology to make relatively rapid and dramatic improvements in crop plants is extremely important for the future as human populations grow, arable land decreases, climates change, and pathogens evolve. The average crop yield is just 20% of that obtained in a bumper crop year, which means that environmental and pathogen stresses presently claim 80% of our food (Boyer, 1982). As population growth and crop losses increase demand for food in the future, plant biotechnology will become increasingly important.

REFERENCES

Albright LM, Yanofsky MF, Leroux B, Ma D, Nester EW (1987): Processing of the T-DNA of *Agrobacterium tumefaciens* generates border nicks and linear, single-stranded T-DNA. J Bacteriol 169:1046–1055.

Anderson LB, Hertzel AV, Das A (1996): *Agrobacterium tumefaciens* VirB7 and VirB9 form a disulfide-linked protein complex. Proc Natl Acad Sci USA 93:8889–8894.

Andersson SGE, Zomorodipour A, Andersson JO, Sicheritz-Ponten T, Alsmark, UCM, Podowski RM, Naslund AK, Eriksson AS, Winkler HH, Kurland CG (1998): The genome sequence of *Rickettsia prowazekii* and the origin of mitochondria. Nature 396:133–140.

Angell SM, Baulcombe DC (1997): Consistent gene silencing in transgenic plants expressing a replicating potato virus X RNA. EMBO J 16:3675–3684.

Baron C, Llosa M, Zhou S, Zambryski PC (1997): VirB1, a component of the T-complex transfer machinery of *Agrobacterium tumefaciens*, is processed to a C-terminal secreted product, VirB1*. J Bacteriol 179:1203–1210.

Baron C, Zambryski PC (1996): Plant transformation: A pilus in *Agrobacterium* T-DNA transfer. Curr Biol 6:1567–1569.

Barton KA, Binns AN, Matzke A, Chilton MD (1983): Regeneration of intact tobacco plants containing full length copies of genetically engineered T-DNA to R1 progeny. Cell 32:1039–1043.

Beaupre CE, Bohne J, Dale EM, Binns AN (1997): Interactions between VirB9 and VirB10 membrane proteins involved in movement of DNA from *Agrobacterium tumefaciens* into plant cells. J Bacteriol 179:78–89.

Beijersbergen A, Dulk-Ras AD, Schilperoort RA, Hooykaas PJJ (1992): Conjugative transfer by the virulence system of *Agrobacterium tumefaciens*. Science 256:1324–1327.

Bendahmane M, Beachy RN (1999): Control of tobamovirus infections via pathogen-derived resistance. Adv Virus Res 53:369–386.

Berger BR, Christie PJ (1993): The *Agrobacterium tumefaciens virB4* gene product is an essential virulence protein requiring an intact nucleoside triphosphate-binding domain. J Bacteriol 175:1723–1734.

Berger BR, Christie PJ (1994): Genetic complementation analysis of the *Agrobacterium tumefaciens virB* operon: *virB2* through *virB11* are essential virulence genes. J Bacteriol 176:3646–3660.

Binns AN, Beaupre CE, Dale EM (1995): Inhibition of VirB-mediated transfer of diverse substrates from *Agrobacterium tumefaciens* by the IncQ plasmid RSF1010. J Bacteriol 177:4890–4899.

Binns AN, Thomashow MF (1988): Cell biology of *Agrobacterium* infection and transformation of plants. Annu Rev Microbiol 42: 575–606.

Braun AC (1958): A physiological basis for the autonomous growth of the crown gall tumor cell. Proc Natl Acad Sci USA 44:344–349.

Boyer JS (1982): Plant productivity and environment. Science 218:443–448.

Buchanan-Wollaston V, Passiatore JE, Cannon F (1987): The *mob* and *oriT* mobilization functions of a bacterial plasmid promote its transfer to plants. Nature 328:172–175.

Bundock P, den Dulk-Ras A, Beijersbergen A, Hooykaas PJJ (1995): Trans-kingdom T-DNA transfer from *Agrobacterium tumefaciens* to *Saccharomyces cerevisiae*. EMBO J 14:3206–3214.

Cabezon E, Sastre JI, de la Cruz F (1997): Genetic evidence of a coupling role for the TraG protein family in bacterial conjugation. Mol Gen Genet 254:400–406.

Censini S, Lange C, Xiang Z, Crabtree JE, Ghiara P, Bordovsky M, Rappuoli R, Covacci A (1996): *cag*, a pathogenicity island of *Helicobacter pylori*, encodes type I-specific and disease-associated virulence factors. Proc Natl Acad Sci USA 93:14648–14653.

Chilton MD, Drummond MH, Merlo DJ, Sciaky D, Montoya AL, Gordon MP, Nester EW (1977): Stable incorporation of plasmid DNA into higher plant cells: The molecular basis of crown gall tumorigenesis. Cell 11:263–271.

Chilton MD, Saiki RK, Yadav N, Gordon MP, Quetier F (1980): T-DNA from *Agrobacterium* Ti plasmid is in the nuclear DNA fraction of crown gall tumor cells. Proc Natl Acad Sci USA 77:4060–4064.

Christie PJ, Ward JE, Winans SC, Nester EW (1988a): The *Agrobacterium tumefaciens virE2* gene product is a single-stranded-DNA-binding protein that associates with T-DNA. J Bacteriol 170:2659–2667.

Christie PJ, Ward JE, Winans SC, Nester EW (1988b): The *Agrobacterium tumefaciens virE2* gene product is a single-stranded-DNA-binding protein that associates with T-DNA. J Bacteriol 170:2659–2667.

Christie PJ, Ward JE, Gordon MP, Nester EW (1989): A gene required for transfer of T-DNA to plants encodes an ATPase with autophosphorylating activity. Proc Natl Acad Sci USA 86:9677–9681.

Christie PJ (1997): *Agrobacterium tumefaciens* T-complex transport apparatus: A paradigm for a new family of multifunctional transporters in eubacteria. J Bacteriol 179:3085–3094.

Chuang CF, Meyerowitz EM (2000): Specific and heritable genetic interference by double-stranded RNA in *Arabidopsis thaliana*. Proc Natl Acad Sci USA 97:4985–4990.

Chyi YS, Jorgensen RA, Goldstein D, Tanksley SD, Loaiza-Figueroa F (1986): Locations and stability of *Agrobacterium*-mediated T-DNA insertions in the *Lycopersicon* genome. Mol Gen Genet 204:64–69.

Citovsky V, De Vos G, Zambryski P (1988): Single-stranded DNA binding protein encoded by the *virE* locus of *Agrobacterium tumefaciens*. Science 240:501–504.

Citovsky V, Wong ML, Zambryski P (1989): Cooperative interaction of *Agrobacterium* VirE2 protein with single-stranded DNA: Implications for the T-DNA transfer process. Proc Natl Acad Sci USA 86:1193–1197.

Citovsky V, Zupan J, Warnick D, Zambryski P (1992): Nuclear localization of *Agrobacterium* VirE2 protein in plant cells. Science 256:1802–1805.

Citovsky V, Warnick D, Zambryski P (1994): Nuclear import of *Agrobacterium* VirD2 and VirE2 proteins in maize and tobacco. Proc Natl Acad Sci USA 91:3210–3214.

Close TJ, Rogowsky PM, Kado CI, Winans SC, Yanofsky MF, Nester EW (1987): Dual control of *Agrobacterium tumefaciens* Ti plasmid virulence genes. J Bacteriol 169:5113–5118.

Clough SJ, Bent AF (1998): Floral dip: A simplified method for *Agrobacterium*-mediated transformation of *Arabidopsis thaliana*. Plant J 16:735–743.

Covacci A, Rappuoli R (1993): Pertussis toxin export requires accessory genes located downstream from the pertussis toxin operon. Mol Microbiol 8:429–434.

Das A, Stachel SE, Ebert P, Allenza P, Montoya A, Nester EW (1986): Promoters of *Agrobacterium tumefaciens* Ti plasmid virulence genes. Nucleic Acids Res 14:1355–1364.

Das A (1988): *Agrobacterium tumefaciens virE* operon encodes a single-stranded DNA-binding protein. Proc Natl Acad Sci USA 85:2909–2913.

Das A, Anderson LB, Xie YH (1997): Delineation of the interaction domains of *Agrobacterium tumefaciens* VirB7 and VirB9 by use of the yeast two-hybrid assay. J Bacteriol 179:3404–3409.

de Groot MJ, Bundock P, Hooykaas PJJ, Beijersbergen AG (1998): *Agrobacterium tumefaciens*-mediated transformation of filamentous fungi. Nature Biotech 16:839–842.

DeCleene M, DeLey J (1976): The host range of crown gall. Bot Rev 42, 389–466.

Deng W, Chen L, Peng WT, Liang X, Sekiguchi S, Gordon MP, Comai L, Nester EW (1999): VirE1 is a specific molecular chaperone for the exported single-stranded-DNA-binding protein VirE2 in *Agrobacterium*. Mol Microbiol 31:1795–1807.

Dougherty WG, Lindbo JA, Smith HA, Parks TD, Swaney S, Proebsting WM (1994): RNA-mediated virus resistance in transgenic plants: Exploitation of a cellular pathway possibly involved in RNA degradation. Mol Plant-Microbe Interact 7:544–552.

Dougherty WG, Parks TD (1995): Transgenes and gene suppression: Telling us something new? Curr Opin Cell Biol 7:399–405.

Durrenberger F, Crameri A, Hohn B, Koukolikova-Nicola Z (1989): Covalently bound VirD2 protein of *Agrobacterium tumefaciens* protects the T-DNA from exonucleolytic degradation. Proc Natl Acad Sci USA 86:9154–9158.

Ellis JG, Kerr A, Petit A, Tempe J (1982): Conjugal transfer of nopaline and agropine Ti plasmids: The role of agrocinopines. Mol Gen Genet 186:269–274.

Enserink M (1999): Ag biotech moves to mollify its critics. Science 286:1666–1668.

Ferber D (1999): GM crops in the cross hairs. Science 286:1662–1666.

Fernandez D, Dang TAT, Spudich GM, Zhou XR, Berger BR, Christie PJ (1996): The *Agrobacterium tumefaciens virB7* gene product, a proposed component of the T-complex transport apparatus, is a membrane-associated lipoprotein exposed at the periplasmic surface. J Bacteriol 178:3156–3167.

Finberg KE, Muth TR, Young SP, Maken JB, Heitritter SM, Binns AN, Banta LM (1995): Interactions of VirB9, -10, and -11 with the membrane fraction of *Agrobacterium tumefaciens*: solubility studies provide evidence for tight associations. J Bacteriol 177:4881–4889.

Fire A, Xu SQ, Montgomery MK, Kostas SA, Driver SE, Mello CC (1998): Potent and specific genetic interference by double-stranded RNA in *Caenorhabditis elegans*. Nature 391:806–811.

Firth N, Ippen-Ihler K, Skurray RA (1996): Structure and function of the F factor and mechanism of conjugation. In Neidhardt FC, Curtiss III R, Ingraham JL, Lin ECC, Low KB, Magasanik B, Rsznikoff WS, Riley M, Schaechter M, and Umbarger HC, (eds): "*Escherichia coli* and *Salmonella*: Cellular and molecular biology." Washington, DC: ASM Press), pp 2377–2401.

Fraley RT, Rogers SG, Horsch RB, Eichholtz DA, Flick JS, Fink CL, Hoffmann NL, Sanders PR (1985): The SEV system: A new disarmed Ti plasmid vector system for plant transformation. Bio/Technol 3:629–635.

Fullner KJ, Stephens KM, Nester EW (1994): An essential virulence protein of *Agrobacterium tumefaciens*, VirB4, requires an intact mononucleotide binding domain to function in transfer of T-DNA. Mol Gen Genet 245:704–715.

Fullner KJ, Lara JC, Nester EW (1996a): Pilus assembly by *Agrobacterium* T-DNA transfer genes. Science 273:1107–1109.

Fullner KJ, Nester EW (1996b): Temperature affects the T-DNA transfer machinery of *Agrobacterium tumefaciens*. J Bacteriol 178:1498–1504.

Garfinkel DJ, Nester EW (1980): *Agrobacterium tumefaciens* mutants affected in crown gall tumorigenesis and octopine catabolism. J Bacteriol 144:732–743.

Garfinkel DJ, Simpson RB, Ream LW, White FF, Gordon MP, Nester EW (1981): Genetic analysis of crown gall: Fine structure map of the T-DNA by site-directed mutagenesis. Cell 27:143–153.

Gelvin SB, Habeck LL (1990): *vir* genes influence conjugal transfer of the Ti plasmid of *Agrobacterium tumefaciens*. J Bacteriol 172:1600–1608.

Gheysen G, Van Montagu M, Zambryski P (1987): Integration of *Agrobacterium tumefaciens* transfer DNA (T-DNA) involves rearrangements of target plant DNA sequences. Proc Natl Acad Sci USA 84:6169–6173.

Gheysen G, Villarroel R, Van Montagu M (1991): Illegitimate recombination in plants: A model for T-DNA integration. Genes Devel 5:287–297.

Gietl C, Koukolikova-Nicola Z, Hohn B (1987): Mobilization of T-DNA from *Agrobacterium* to plant cells involves a protein that binds single-stranded DNA. Proc Natl Acad Sci USA 84:9006–9010.

Ginocchio CC, Olmsted SB, Wells CL, Galan JE (1994): Contact with epithelial cells induces the formation of surface appendages on *Salmonella typhimurium*. Cell 76:717–724.

Grahn AM, Haase J, Bamford DH, Lanka E (2000). Components, of the RP4 conjugative apparatus form an envelope structure bridging inner and oulter membranes of donor cells: implications for related macromolecule transport systems, J Bacteriol 182:1564–1574.

Grossbard E, Atkinson D (1985): "The Herbicide Glyphosate." London: Butterworths:

Guerinot, ML (2000): The Green Revolution strikes gold. Science 287:241–243.

Guyon P, Chilton MD, Petit A, Tempe J (1980): Agropine in "null-type" crown gall tumors: Evidence for generality of the opine concept. Proc Natl Acad Sci USA 65:2693–2697.

Hamilton AJ, Baulcombe DC (1999): A species of small antisense RNA in posttranscriptional gene silencing in plants. Science 286:950–952.

Hamilton CM, Lee H, Li PL, Cook DM, Piper KR, Beck von Bodman S, Lanka E, Ream W, Farrand SK (2000): TraG from RP4 and VirD4 from Ti plasmids confer relaxosome specificity to the conjugal transfer system of pTiC58. J Bacteriol 182:1541–1548.

Hansen G, Chilton MD (1996): "Agrolistic" transformation of plant cells: Integration of T-strands generated in planta. Proc Natl Acad Sci USA 93:14978–14983.

Hansen G, Shillito R, Chilton MD (1997): T-strand integration in maize protoplasts after codelivery of a T-DNA substrate and virulence genes. Proc Natl Acad Sci USA 94:11726–11730.

Hepburn A, White J (1985): The effect of right terminal repeat deletion on the oncogenicity of the T-region of pTiT37. Plant Mol Biol 5:3–11.

Herrera-Estrella A, Chen Z, Van Montagu M, Wang K (1988): VirD proteins of *Agrobacterium tumefaciens* are required for the formation of a covalent DNA-protein complex at the 5′ terminus of T-strand molecules. EMBO J 7:4055–4062.

Hodgson J (1999): Monarch Bt-corn paper questioned. Nature Biotech 17:627.

Hoekema A, Hirsch PR, Hooykaas PJJ, Schilperoort RA (1983): A binary plant vector strategy based on separation of *vir* and T-region of the *Agrobacterium tumefaciens* Ti plasmid. Nature 303:179–180.

Hooykaas PJJ, Beijersbergen GM (1994): The virulence system of *Agrobacterium tumefaciens*. Ann Rev Phytopathol 32:157–179.

Horsch RB, Fry JE, Hoffmann NL, Eichholtz D, Rogers SG (1985): A simple and general method of transferring genes into plants. Science 227:1229–1231.

Howard EA, Winsor BA, De Vos G, Zambryski P (1989): Activation of the T-DNA transfer process in *Agrobacterium* results in the generation of a T-strand-protein complex: Tight association of VirD2 with the 5′ ends of T-strands. Proc Natl Acad Sci USA 86:4017–4021.

Howard EA, Zupan JR, Citovsky V, Zambryski PC (1992): The VirD2 protein of *Agrobacterium tumefaciens* contains a C-terminal bipartite nuclear localization signal: Implications for nuclear uptake of DNA in plant cells. Cell 68:109–118.

Jasper F, Koncz C, Schell J, Steinbiss HH (1994): *Agrobacterium* T-strand production *in vitro*: Sequence-specific cleavage and 5′ protection of single-stranded DNA templates by purified VirD2 protein. Proc Natl Acad Sci USA 91:694–698.

Jayaswal RK, Veluthambi K, Gelvin SB, Slightom JL (1987): Double-stranded T-DNA cleavage and the generation of single-stranded T-DNA molecules in *E. coli* by a *virD* encoded border specific endonuclease from *Agrobacterium tumefaciens*. J Bacteriol 169:5035–5045.

Ji JM, Martinez A, Dabrowski M, Veluthambi K, Gelvin SB, Ream W (1988): The *overdrive* enhancer sequence stimulates production of T-strands from the *Agrobacterium tumefaciens* tumor-inducing plasmid.

In "Molecular Biology of Plant-Pathogen Interactions"; UCLA Symposia on Molecular and Cellular Biology, Staskawicz B, Ahlquist P, Yoder O, eds: New York: Alan R. Liss.

Jones AL, Shirasu K, Kado CI (1994): The product of the *virB4* gene of *Agrobacterium tumefaciens* promotes accumulation of VirB3 protein. J Bacteriol 176:5255–5261.

Joos H, Timmerman B, Van Montagu M, Schell J (1983): Genetic analysis of transfer and stabilisation of *Agrobacterium* DNA in plant cells. EMBO J 2:2151–2160.

Jorgensen RA, Atkinson RG, Forster RLS, Lucas WJ (1998): An RNA-based information superhighway in plants. Science 279:1486–1487.

Kado CI (1994): Promiscuous DNA transfer system of *Agrobacterium tumefaciens*: Role of the *virB* operon in sex pilus assembly and synthesis. Mol Microbiol 12:17–22.

Kilman S (2000): U.S. farmers cutting back on crops that have been genetically modified. Wall Street J April 3:A34.

Kingsman A, Willetts N (1978): The requirements for conjugal DNA synthesis in the donor strain during F*lac* transfer. J Mol Biol 122:287–300.

Klee HJ, Horsch RB, Rogers SG (1987): *Agrobacterium*-mediated plant transformation and its further applications to plant biology. Annu Rev Plant Physiol 38:467–486.

Kunik T, Tzfira T, Kapulni K Y, Gafni Y, Dingwall C, Citovsky V (2001): Genetic transformation of HeLa cells by *Agrobacterium*. Proc Natl Acad Sci USA 98:1871–1876.

Lai EM, Kado CI (1998): Processed VirB2 is the major subunit of the promiscuous pilus of *Agrobacterium tumefaciens*. J Bacteriol 180:2711–2717.

Lee LY, Gelvin SB, Kado CI (1999): pSa causes oncogenic suppression of *Agrobacterium* by inhibiting VirE2 protein export. J Bacteriol 181:186–196.

Lemmers M, Debeuckeleer M, Holsters M, Zambryski P, Depicker A, Hernalsteens JP, Van Montagu M, Schell J (1980): Internal organization and integration of Ti plasmid DNA in nopaline crown gall tumors. J Mol Biol 144:355–378.

Lessl M, Balzer D, Pansegrau W, Lanka E (1992): Sequence similarities between the RP4 Tra2 and the Ti VirB region strongly support the conjugation model for T-DNA transfer. J Biol Chem 267:20471–20480.

Lessl M, Lanka E (1994): Common mechanisms in bacterial conjugation and Ti-mediated T-DNA transfer to plant cells. Cell 77:321–324.

Lindbo JA, Dougherty WG (1992a): Pathogen-derived resistance to a potyvirus: Immune and resistant phenotypes in transgenic tobacco expressing altered forms of a potyvirus coat protein nucleotide sequence. Mol Plant-Microbe Interact 5:144–153.

Lindbo JA, Dougherty WG (1992b): Untranslatable transcripts of the tobacco etch virus coat protein gene sequence can interfere with tobacco etch virus replication in transgenic plants and protoplasts. Virol 189:725–733.

Losey JE, Rayor LS, Carter ME (1999): Transgenic pollen harms monarch larvae. Nature 399:214.

Matsumoto S, Ito Y, Hosoi T, Takahashi Y, Machida Y (1990): Integration of *Agrobacterium* T-DNA into a tobacco chromosome: Possible involvement of DNA homology between T-DNA and plant DNA. Mol Gen Genet 224:309–316.

Mayerhofer R, Koncz-Kalman Z, Nawrath C, Bakkeren G, Crameri A, Angelis K, Redei GP, Schell J, Hohn B, Koncz C (1991): T-DNA integration: A mode of illegitimate recombination in plants. EMBO J 10:697–704.

McBride KE, Knauf VC (1988): Genetic analysis of the *virE* operon of the *Agrobacterium* Ti plasmid pTiA6. J Bacteriol 170:1430–1437.

Melchers LS, Maroney MJ, Dulk-Ras AD, Thompson DV, van Vuuren HAJ, Schilperoort RA, Hooykaas PJJ (1990): Octopine and nopaline strains of *Agrobacterium tumefaciens* differ in virulence; molecular characterization of the *virF* locus. Plant Mol Biol 14:249–259.

Miranda A, Janssen G, Hodges L, Peralta EG, Ream W (1992): *Agrobacterium tumefaciens* transfers extremely long T-DNAs by a unidirectional mechanism. J Bacteriol 174:2288–2297.

Montgomery MK, Fire A (1998): Double-stranded RNA as a mediator in sequence-specific genetic silencing and co-suppression. Trends Genet 14:255–258.

Montgomery MK, Xu SQ, Fire A (1998): RNA as a target of double-stranded RNA-mediated genetic interference in *Caenorhabditis elegans*. Proc Natl Acad Sci USA 95:15502–15507.

Mushegian AR, Fullner KJ, Koonin EV, Nester EW (1996): A family of lysozyme-like virulence factors in bacterial pathogens of plants and animals. Proc Natl Acad Sci USA 93:7321–7326.

Narasimhulu SB, Deng X, Sarria R, Gelvin SB (1996): Early transcription of *Agrobacterium* T-DNA genes in tobacco and maize. Plant Cell 8:873–886.

Nash JM (2000): Grains of hope. Time 156:38–46.

Nester EW, Gordon MP, Amasino RM, Yanofsky MF (1984): Crown gall: A molecular and physiological analysis. Ann Rev Plant Physiol 35:387–413.

Niiler E (1999): GM corn poses little threat to monarch. Nature Biotech 17:1154.

Otten L, DeGreve H, Leemans J, Hain R, Hooykaas PJJ, and Schell J (1984): Restoration of virulence of *vir* region mutants of *Agrobacterium tumefaciens* strain B653 by coinfection with normal and mutant *Agrobacterium* strains. Mol Gen Genet 175:159–163.

Ou JT, Reim RL (1978): F-mating materials able to generate a mating signal in mating with HfrH *dnaB*(Ts) cells. J Bacteriol 133:442–445.

Panesgrau W, Schoumacher F, Hohn B, Lanka E (1993): Site-specific cleavage and joining of single-stranded DNA by VirD2 protein of *Agrobacterium tumefaciens* Ti plasmids: Analogy to bacterial conjugation. Proc Natl Acad Sci USA 90:11538–11542.

Pansegrau W, Schoumacher F, Hohn B, Lanka E (1993): Site-specific cleavage and joining of single-stranded DNA by VirD2 protein of *Agrobacterium tumefaciens* Ti plasmids: Analogy to bacterial conjugation. Proc Natl Acad Sci USA 90:11538–11542.

Peerbolte R, te Lintel-Hekkert W, Barfield DG, Hoge JHC, Wullems GJ, Schilperoort RA (1987): Structure, organization and expression of transferred DNA in *Nicotiana plumbaginifolia* crown gall tissues. Planta 171:393–405.

Peralta EG, Ream LW (1985): T-DNA border sequences required for crown gall tumorigenesis. Proc Natl Acad Sci USA 82:5112–5116.

Peralta EG, Hellmiss R, Ream W (1986): *Overdrive*, a T-DNA transmission enhancer on the *A. tumefaciens* tumour-inducing plasmid. EMBO J 5:1137–1142.

Petit A, Tempe J, Kerr A, Holsters M, Van Montagu M, Schell J (1978): Substrate induction of conjugative activity of *Agrobacterium tumefaciens* Ti plasmids. Nature 271:570–571.

Petit A, David C, Dahl GA, Ellis JG, Guyon P, Casse-Delbert F, Tempe J (1983): Further extension of the opine concept: Plasmids in *Agrobacterium rhizogenes* cooperate for opine degradation. Mol Gen Genet 190:204–214.

Pohlman RF, Genetti HD, Winans SC (1994): Common ancestry between IncN conjugal transfer genes and macromolecular export systems of plant and animal pathogens. Mol Microbiol 14:655–668.

Rashkova S, Spudich GM, Christie PJ (1997): Characterization of membrane and protein interaction determinants of the *Agrobacterium tumefaciens* VirB11 ATPase. J Bacteriol 179:583–591.

Ratcliff F, Harrison BD, Baulcombe DC (1997): A similarity between viral defense and gene silencing in plants. Science 276:1558–1560.

Ream LW, Gordon MP, Nester EW (1983): Multiple mutations in the T-region of the *Agrobacterium tumefaciens* tumor-inducing plasmid. Proc Natl Acad Sci USA 80:1660–1664.

Ream W (1989): *Agrobacterium tumefaciens* and interkingdom genetic exchange. Ann Rev Phytopathol 27:583–618.

Regensburg-Tuink AJG, Hooykaas PJJ (1993): Transgenic *N glauca* plants expressing bacterial virulence gene *virF* are converted into hosts for nopaline strains of *A. tumefaciens*. Nature 363:69–71.

Roine E, Wei W, Yuan J, Nurmiaho-Lassila EL, Kalkkinen N, Romantschuk M, He SY (1997): Hrp pilus: An *hrp*-dependent bacterial surface appendage produced by *Pseudomonas syringae* pv. *tomato* DC3000. Proc Natl Acad Sci USA 94:3459–3464.

Rossi L, Hohn B, Tinland B (1993): The VirD2 protein of *Agrobacterium tumefaciens* carries nuclear localization signals important for transfer of T-DNA to plants. Mol Gen Genet 239:345–353.

Rossi L, Hohn B, Tinland B (1996): Integration of complete transferred DNA units is dependent on the activity of virulence E2 protein of *Agrobacterium tumefaciens*. Proc Natl Acad Sci USA 93:126–130.

Ruiz MT, Voinnet O, Baulcombe DC (1998): Initiation and maintenance of virus-induced gene silencing. Plant Cell 10:937–946.

Segal G, Shuman HA (1998): Intracellular multiplication and human macrophage killing by *Legionella pneumophila* are inhibited by conjugal components of IncQ plasmid RSF1010. Mol Microbiol 29: 197–208.

Segal G, Purcell M, Shuman HA (1999a): Host cell killing and bacterial conjugation require overlapping sets of genes within a 22-kb region of the *Legionella pneumophila* genome. Proc Natl Acad Sci USA 95:1669–1674.

Segal G, Russo JJ, Shuman HA (1999b): Relationships between a new type IV secretion system and the *icm/dot* virulence system of *Legionella pneumophila*. Mol Microbiol 34:799–809.

Sen P, Pazour GJ, Anderson D, Das A (1989): Cooperative binding of the VirE2 protein to single-stranded DNA. J Bacteriol 171:2573–2580.

Shaw CH, Watson M, Carter G (1984): The right hand copy of the nopaline Ti plasmid 25 bp repeat is required for tumour formation. Nucleic Acids Res 12:6031–6041.

Sheng J, Citovsky V (1996): *Agrobacterium*-plant cell DNA transport: Have virulence proteins, will travel. Plant Cell 8:1699–1710.

Shirasu K, Kado CI (1993): Membrane location of the Ti plasmid VirB proteins involved in the biosynthesis of a pilin-like conjugative structure on *Agrobacterium tumefaciens*. FEMS Microbiol Lett 111:287–294.

Shirasu K, Koukolikova-Nicola Z, Hohn B, Kado CI (1994): An inner-membrane-associated virulence protein essential for T-DNA transfer from *Agrobacterium tumefaciens* to plants exhibits ATPase activity and similarities to conjugative transfer genes. Mol Microbiol 11:581–588.

Shurvinton CE, Hodges L, Ream W (1992): A nuclear localization signal and the C-terminal omega sequence in the *Agrobacterium tumefaciens* VirD2 endonuclease are important for tumor formation. Proc Natl Acad Sci USA 89:11837–11841.

Smith HA, Swaney SL, Parks TD, Wernsman EA, Dougherty WG (1994): Transgenic plant virus resistance mediated by untranslatable sense RNAs: Expression, regulation, and fate of nonessential RNAs. Plant Cell 6:1441–1453.

Spudich GM, Fernandez D, Zhou XR, Christie PJ (1996): Intermolecular disulfide bonds stabilize VirB7 homodimers and VirB7/VirB9 heterodimers during biogenesis of the *Agrobacterium tumefaciens* T-complex transport apparatus. Proc Natl Acad Sci USA 93:7512–7517.

Stachel SE, Messens E, Van Montagu M, Zambryski P (1985): Identification of the signal molecules produced by wounded plant cells that activate T-DNA transfer in *Agrobacterium tumefaciens*. Nature 318:624–629.

Stachel SE, Nester EW (1986): The genetic and transcriptional organization of the *vir* region of the A6 Ti plasmid of *Agrobacterium tumefaciens*. EMBO J 5:1445–1454.

Stachel SE, Timmerman B, Zambryski P (1986a): Generation of single-stranded T-DNA molecules during the initial stages of T-DNA transfer from *Agrobacterium tumefaciens* to plant cells. Nature 322:706–712.

Stachel SE, Zambryski P (1986b): *Agrobacterium tumefaciens* and the susceptible plant cell: A novel adaptation of extracellular recognition and DNA conjugation. Cell 47:155–157.

Stachel SE, Timmerman B, Zambryski P (1987): Activation of *Agrobacterium tumefaciens* vir gene expression generates multiple single-stranded T-strand molecules from the pTiA6 T-region: Requirement of 5' *virD* gene products. EMBO J 6:857–863.

Stahl LE, Jacobs A, Binns AN (1998): The conjugal intermediate of plasmid RSF1010 inhibits *Agrobacterium tumefaciens* virulence and VirB-dependent export of VirE2. J Bacteriol 180:3933–3939.

Steck TR, Kado CI (1990): Virulence genes promote conjugative transfer of the Ti plasmid between *Agrobacterium* strains. J Bacteriol 172:2191–2193.

Sundberg C, Meek L, Carroll K, Das A, Ream W (1996): VirE1 protein mediates export of the single-stranded DNA-binding protein VirE2 from *Agrobacterium tumefaciens* into plant cells. J Bacteriol 178:1207–1212.

Sundberg CD, Ream W (1999): The *Agrobacterium tumefaciens* chaperone-like protein, VirE1, interacts with VirE2 at domains required for single-stranded DNA binding and cooperative interaction. J Bacteriol 181:6850–6855.

Tait RC, Kado CI (1988): Regulation of the *virC* and *virD* promoters of pTiC58 by the *ros* chromosomal mutation of *Agrobacterium tumefaciens*. Mol Microbiol 2:385–392.

Tate ME (1987): *A. tumefaciens* pTiA6 and C58 *virC* and *D* promoter alignment. Nucleic Acids Res 15:6739.

Tepfer D, Smeekens S, Lepri O (2000): Plant biotechnology in the real world. Intl Herald Tribune, February 15.

Thomashow MF, Nutter R, Postle K, Chilton MD, Blattner FR, Powell A, Gordon MP, Nester EW (1980): Recombination between higher plant DNA

and the Ti plasmid of *Agrobacterium tumefaciens*.
Proc Natl Acad Sci USA 77:6448–6452.

Thorstenson YR, Kuldau GA, Zambryski PC (1993):
Subcellular localization of seven VirB proteins of
Agrobacterium tumefaciens: Implications for the formation of a T-DNA transport structure. J Bacteriol
175:5233–5241.

Timmons L, Fire A (1998): Specific interference by
ingested dsRNA. Nature 395:854–854.

Tinland B, Koukolikova-Nicola Z, Hall MN, Hohn B
(1992): The T-DNA-linked VirD2 protein contains
two distinct functional nuclear localization signals.
Proc Natl Acad Sci USA 89:7442–7446.

Tinland B, Hohn B, Puchta H (1994): *Agrobacterium
tumefaciens* transfers single-stranded transferred
DNA (T-DNA) into the plant cell nucleus. Proc
Natl Acad Sci USA 91:8000–8004.

Tinland B, Schoumacher F, Gloeckler V, Bravo-Angel
A, Hohn B (1995): The *Agrobacterium tumefaciens*
virulence D2 protein is responsible for precise integration of T-DNA into the plant genome. EMBO J
14:3585–3595.

Toro N, Datta A, Carmi OA, Young C, Prusti RK,
Nester EW (1989): The *Agrobacterium tumefaciens*
virC1 gene product binds to overdrive, a T-DNA
transfer enhancer. J Bacteriol 171:6845–6849.

Tummuru MKR, Sharma SA, Blaser MJ (1995): *Helicobacter pylori picB*, a homologue of the *Bordetella
pertussis* toxin secretion protein, is required for induction of IL-8 in gastric epithelial cells. Mol Microbiol
18:867–876.

Van Larebeke N, Engler G, Holsters M, Van den El-
sacker S, Zaenen I, Schilperoort RA, Schell J (1974):
Large plasmid in *Agrobacterium tumefaciens* essential
for crown gall inducing activity. Nature 252:169–170.

Van Lijsebettens M, Inze D, Schell J, Van Montagu M
(1986): Transformed cell clones as a tool to study T-
DNA integration mediated by *Agrobacterium tumefaciens*. J Mol Biol 188:129–145.

Veluthambi K, Ream W, Gelvin SB (1988): Virulence
genes, borders, and overdrive generate single-stranded
T-DNA molecules from the A6 Ti plasmid of *Agrobacterium tumefaciens*. J Bacteriol 170:1523–1532.

Vogel AM, Das A (1992): Mutational analysis of *Agrobacterium tumefaciens virD2*: Tyrosine 29 is essential
for endonuclease activity. J Bacteriol 174:303–308.

Vogel JP, Andrews HL, Wong SK, Isberg RR (1998):
Conjugative transfer by the virulence system of *Legionella pneumophila*. Science 279:873–876.

Wang K, Herrera-Estrella L, Van Montagu M, Zambryski P (1984): Right 25 bp terminus sequence of the
nopaline T-DNA is essential for and determines direction of DNA transfer from *Agrobacterium* to the
plant genome. Cell 38:455–462.

Wang K, Stachel SE, Timmerman B, Van Montagu M,
Zambryski P (1987): Site-specific nick in the T-DNA
border sequence as a result of *Agrobacterium vir* gene
expression. Science 235:587–591.

Ward ER, Barnes WM (1988): VirD2 protein of *Agrobacterium tumefaciens* very tightly linked to the 5′ end
of T-strand DNA. Science 242:927–930.

Ward JE, Dale EM, Nester EW, Binns AN (1990):
Identification of a VirB10 protein aggregate in the
inner membrane of *Agrobacterium tumefaciens*. J Bacteriol 172:5200–5210.

Ward JE, Dale EM, Binns AN (1991): Activity of the
Agrobacterium T-DNA transfer machinery is affected
by *virB* gene products. Proc Natl Acad Sci USA
88:9350–9354.

Waterhouse PM, Graham MW, Wang MB (1998):
Virus resistance and gene silencing in plants can be
induced by simultaneous expression of sense and antisense RNA. Proc Natl Acad Sci USA 95:13959–
13964.

Watson B, Currier TC, Gordon MP, Chilton MD, Nester EW (1975): Plasmid required for virulence of
Agrobacterium tumefaciens. J Bacteriol 123:255–264.

Weisburg W, Woese C, Dobson M, Weiss E (1985): A
common origin of rickettsiae and certain plant pathogens. Science 230:556–558.

Weiss AA, Johnson FD, Burns DL (1993): Molecular
characterization of an operon required for pertussis
toxin secretion. Proc Natl Acad Sci USA 90:2970–
2974.

Willmitzer L, Debeuckeleer M, Lemmers M, Van Montagu M, and Schell J (1980): DNA from Ti plasmid
present in nucleus and absent from plastids of crown
gall plant cells. Nature 287:359–361.

Willmitzer L, Simons G, Schell J (1982): The TL-DNA
in octopine crown gall tumors codes for seven well-
defined polyadenylated transcripts. EMBO J 1:139–
146.

Winans SC, Allenza P, Stachel SE, McBride KE, Nester
EW (1987): Characterization of the *virE* operon of the
Agrobacterium Ti plasmid pTiA6. Nucleic Acids Res
15:825–838.

Winans SC (1992): Two-way chemical signaling in *Agrobacterium*-plant interactions. Micro Revs 56:12–31.

Winans SC, Burns DL, Christie PJ (1996): Adaptation
of a conjugal transfer system for the export of pathogenic macromolecules. Trends Microbiol 4:64–68.

Windholz M, Budavari S, Blumetti RF, Otterbein ES
(1983): "The Merck Index" Rahway: Merck.

Wraight, CL, Zangerl AR, Carroll MT, Berenbaum MR
(2000): Absence of toxicity of *Bacillus thuringiensis*
pollen to black swallowtails under field conditions.
Proc Natl Acad Sci USA 97:7700–7703.

Yanofsky MF, Porter SG, Young C, Albright LM,
Gordon MP, Nester EW (1986): The *virD* operon of
Agrobacterium tumefaciens encodes a site-specific en-
donuclease. Cell 47:471–477.

Ye X, Al-Babili S, Kloti A, Zhang J, Lucca P, Beyer P,
Potrykus I (2000): Engineering the provitamin A
(beta-carotene) biosynthetic pathway into (caroten-
oid-free) rice endosperm. Science 287: 303–305.

Young C, Nester EW (1988): Association of the VirD2 protein with the 5′ end of T strands in *Agrobacterium tumefaciens*. J Bacteriol 170:3367–3374.

Yusibov VM, Steck TR, Gupta V, Gelvin SB (1994): Association of single-stranded transferred DNA from *Agrobacterium tumefaciens* with tobacco cells. Proc Natl Acad Sci USA 91:2994–2998.

Zambryski P, Holsters M, Kruger K, Depicker A, Schell J, Van Montagu M, Goodman H (1980): Tumor DNA structure in plant cells transformed by *A. tumefaciens*. Science 209:1385–1391.

Zhou XR, Christie PJ (1999): Mutagenesis of the *Agrobacterium* VirE2 single-stranded DNA-binding protein identifies regions required for self-association and interaction with VirE1 and a permissive site for hybrid protein construction. J Bacteriol 181:4342–4352.

Zhu J, Oger PM, Schrammeijer B, Hooykaas PJJ, Farrand SK, Winans SC (2000): The bases of crown gall tumorigenesis. J Bacteriol 182:3885–3895.

Zupan J, Zambryski P (1997): The *Agrobacterium* DNA transfer complex. Crit Rev Plant Sci 16:279–295.

Zupan JR, Citovsky V, Zambryski P (1996): *Agrobacterium* VirE2 protein mediates nuclear uptake of single-stranded DNA in plant cells. Proc Natl Acad Sci USA 93:2392–2397.

15

Two-Component Regulation

KENNETH W. BAYLES AND DAVID F. FUJIMOTO

Department of Microbiology, Molecular, and Biochemistry, The College of Agriculture, University of Idaho, Moscow, Idaho 83844–3052; Biology Department LS-416, San Diego State University, San Diego, California 92182

I. INTRODUCTION

Bacteria are constantly exposed to changing environmental conditions. Imagine *Escherichia coli* living in the human intestine being bombarded with an abundance of nutrients after a meal or *Staphylococcus aureus* going from the desolate environment of the skin surface to the bloodstream via a fresh wound. Depending on the environment, bacteria may experience changes in osmolarity, carbon or nitrogen sources, and oxygen levels, to name a few. To survive, bacteria must sense their environment and quickly adapt to these types of changes. This is efficiently achieved using sensory proteins that comprise what are known as two-component regulatory systems. Remarkably this protein family has the ability to respond to a nearly unlimited diversity of environmental signals.

Put into simplistic terms, the function of all two-component regulatory systems, indeed all signal transduction systems, is to convert an extracellular stimulus into a chemical signal that the cell can recognize. In a sense bacterial two-component regulatory systems function as translators to interpret environmental signals and translate them into bacterial language. As will be seen in the following sections, the variable regions of individual members of the two-component regulatory system family reflect the diverse nature of the environmental "dialects" present, while the conserved regions correspond to the language of the bacterial cell. Although we have learned a great deal about the molecular mechanisms involved in signal transduction via two-component regulatory systems, much remains to be discovered.

In this chapter we illustrate the structural and functional characteristics of typical two-

Modern Microbial Genetics, Second Edition. Edited by Uldis N. Streips and Ronald E. Yasbin. ISBN 0–471–38665–0 Copyright © 2002 Wiley-Liss, Inc.

component regulatory systems and describe how the components of these systems interact during signal transduction. The information presented is intended to provide a general review of the mechanics of a prototypical two-component regulatory system and to give examples of those systems that vary from this central theme.

II. THE PROTOTYPICAL TWO-COMPONENT REGULATORY SYSTEM

As the name implies, the most common two-component regulatory system consists of a cytoplasmic membrane-associated protein referred to as a sensor protein (or histidine kinase) and a cytoplasmic protein known as a response regulator (see Fig. 1). Both of

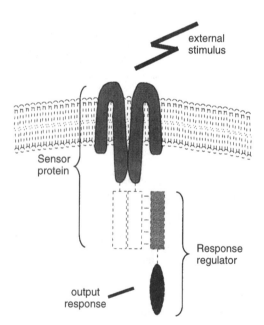

Fig. 1. Schematic representation of a membrane-bound sensor protein dimer and its associated response regulator. Signal transduction is initiated by an external stimulus that triggers a cascade of events leading to an output response. Specific contact points (short lines) between sensor protein and response regulator pairs mediate the specificity exhibited between these two classes of proteins. The output response is typically altered gene expression but can also be manifested as a reversal in flagellar rotation or changes in enzymatic activity.

these proteins contain separable domains, connected by flexible linker regions, that have defined roles in signal transduction. Prototypical sensor proteins have a membrane-associated input (or sensor) domain that is optimally positioned to sense extracellular signals. Once this occurs, the cytoplasmic transmitter domain undergoes autophosphorylation at a conserved histidine residue, activating the sensor protein. This phosphorylation event is mediated by an autokinase activity, one of two enzymatic activities that sensor proteins exhibit. Next the phosphorylated histidine residue of the sensor protein serves as a substrate for the transfer of a phosphoryl group to a conserved aspartic acid residue located within the receiver domain of the sensor proteins cognate response regulator. This phosphorylation event then stimulates the output domain of the response regulator, triggering an output response (e.g., activation of a set of genes) usually related to the original environmental stimulus. Thus signal transduction begins with the stimulation of the input domains of sensor proteins and ends with a molecular response mediated by the output domains of response regulators.

Given the existence of a diversity of two-component regulatory systems in bacteria (at least 30 in *E. coli* and 34 in *Bacillus subtilis*), there must be exquisite specificity between sensor protein/response regulator pairs so that "wires" are not inappropriately crossed. This specificity is undoubtedly mediated by specific molecular interactions between cognate sensor proteins and response regulators, probably located in the nonconserved regions of these proteins. On the other hand, molecular cross talk between different members of two-component pairs has also been shown to occur (Fisher et al., 1995). This likely represents the overlapping roles of some genes under the control of different two-component regulatory systems.

A. Sensor Proteins

As mentioned above, the input domains of sensor proteins are typically membrane-associated, reflecting their roles in sensing extra-cellular signals. The lack of conservation within the input domains of sensor proteins is consistent with the variety of extracellular signals that this family of proteins senses (Fabret et al., 1999). Most of these domains have two membrane-spanning segments that loop part of the domain out into the peri-plasmic space (or extracellular environment in gram-positive bacteria), presumably so that it can interact with soluble signaling molecules. Other sensor proteins contain six or more membrane-spanning segments with little being exposed to the outside, suggesting their interaction with membrane-associated signals. The transmitter domains of all sensor proteins are thought to protrude into the cytoplasm of the cell (Fig. 1) where they can interact with the receiver domains of their cognate response regulators. Despite the large number of sensor proteins that have been characterized, and our growing knowledge of how these systems work, rela-tively little is known about the initial inter-action of the signals with the input domains.

Although sensor proteins exhibit an aver-age of only 25% amino acid sequence iden-tity with each other (Stock et al., 1995), there are several unique signature sequences, des-ignated H, N, G1, F, and G2 blocks (see Fig. 2), found within the approximately 240 amino acids that make up the transmitter domains. These sequences are abbreviated by using the letter of the most conserved amino acid in each block. The H block, which is found in the N terminus of the transmitter domain, contains the conserved histidine residue that serves as the site for autophosphorylation. This sequence is quite variable compared to the other five con-served sequence blocks and is located within a separate subdomain, DHp (Dimerization histidine phosphotransfer), that is involved in the dimerization and autophosphoryla-tion of these proteins (Dutta et al., 1999). Many studies have shown that mutations within the H block eliminate autophosphor-ylation activity by eliminating the ability to generate the phosphohistidine intermediate

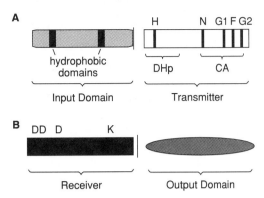

Fig. 2. Domain organization of a prototypical sensor protein (**A**) and a response regulator (**B**). The domains of sensor proteins include an input domain, which is usually membrane associated and responsible for the detection of external signals, and a transmitter domain that transmits the signal to its associated response regulator. Transmitter domains contain subdomains, DHp and CA, that contain con-served sequence motifs (H, N, G1, F, and G2) common to most sensor proteins. The domains of response regulators include a receiver domain that receives the signal from its associated sensor protein and an output domain that is typically a transcrip-tional regulator. As illustrated, receiver domains also contain conserved sequence motifs (DD, D, and K).

during signal transduction (Parkinson and Kofoid, 1992).

Within the C terminus of transmitter domains are the remaining four signature sequences of these proteins that are required for autokinase activity. Presumably these se-quences are arranged in the tertiary struc-ture, called the CA (catalytic ATP-binding) subdomain, forming a nucleotide binding region (Dutta et al., 1999). Indeed, the two glycine-rich G blocks are similar to motifs found in most nucleotide binding proteins (stocke et al., 1995). Mutagenesis studies have revealed that the G1, G2, and N blocks are involved in autokinase activity (Patkin-son and Kofoid, 1992). The conserved F block is located between the G blocks, but its function is not well characterized. Al-though these blocks exist in the prototypical sensor protein, it is not uncommon to find sensor proteins that are missing one or more of these sequences. Interestingly biochemical and genetic studies indicate that sensor proteins with defective DHp subdomains can complement mutant sensor proteins containing defective CA subdomains. Thus a CA subdomain within a sensor protein binds ATP and transautophosphorylates the conserved histidine residue on the DHp subdomain of the partner subunit within a dimer (Dutta et al., 1999).

Once the sensor protein is phosphorylated, the phosphoryl group is then immediately transferred to its cognate response regulator. The rate-limiting step in signal transduction appears to be the initial phosphorylation of the transmitter domain. How then is the autokinase activity regulated? Most studies indicate that the input domain of sensor proteins is a negative regulator of transmitter activity. In several instances deletion of the input domain results in constitutive activity of the transmitter (Parkinson, 1992). Thus signal recognition by the input domain leads to the relief of transmitter repression most likely as a result conformational changes within the sensor protein dimer. A more detailed picture of the conformational changes that occur during signal transduction awaits a better overall understanding of the domain organization and overall molecular structure of sensor proteins.

B. Response Regulators

As described above, the receiver domain of a given response regulator accepts a phosphoryl group from the transmitter domain of its cognate sensor protein. The result of this is an output response such as the transcriptional activation of specific genes. As with the transmitter domains of the sensor protein family, the 120 amino acid receiver domains also contain signature sequences (Fig. 2) that are common among members of the response regulator family. The site of phosphorylation occurs near the middle of the receiver domain on an invariant aspartic acid residue designated "D." Near the N and C termini are "DD" and "K" sequences, respectively, that are also found in nearly all response regulator proteins. It is believed

that the invariant aspartate residues produce an acid pocket involved in the phosphorylation at the "D" site, while the lysine residue plays a role in the phosphorylation-induced conformational changes (Parkinson, 1995). Sequences that are not conserved within the receiver domains (as well as within the transmitter domains of sensor proteins) are thought to contain motifs involved in the specific interactions between cognate sensor proteins and response regulators. Unfortunately, little is known about the molecular details that govern the specificity between these proteins.

Since sensor proteins share sequences in common with kinases and exhibit kinase activity, it was somewhat surprising to find that transfer of the phosphoryl group during signal transduction appears to be catalyzed by their cognate response regulators. One of the most striking pieces of evidence demonstrating this was the observation that these proteins can be phosphorylated by any one of a variety of small-molecule phosphodonors (Lukat et al., 1992). Thus the enzymatic activity necessary for the phosphorylation of the conserved aspartic acid residues within response regulators must reside within these proteins. The biological significance of phosphorylation of response regulators using phosphodonors is unclear although it may be involved in the fine-tuning of the response to external stimulation.

Similar to the regulatory control that input domains possess over transmitter domains, receiver domains also seem to play an important role in the regulation of output function. Unlike sensor proteins, however, receivers convey both positive and negative effects on output activity. In some cases, receiver domains control their output domains by inhibiting their activity, while in others, the receiver domains seem to activate output activity.

C. Regulation of Signal Transduction

One important aspect of bacterial two-component regulatory systems is that their activities are controlled at multiple levels.

We have already discussed the phospho-transfer events that lead to activation of an output response. How then does the system return to the unstimulated state after the stimulus has been removed? Once the bacterial cell has elicited an appropriate response to an extracellular signal and adapted accordingly, it must then quickly return to the unstimulated state when the signal is no longer present. If the phosphorylation events generated by two-component regulatory systems were irreversible, the output response would continue regardless of the presence or absence of input. Two-component regulatory systems overcome this by reducing the stability of the phosphoryl linkages to histidine residues of sensor proteins and/or aspartic acid residues of response regulators. How is this done? Studies demonstrate that sensor and response regulator proteins maintain a delicate balance between kinase and phosphatase activities. The phosphatase activity, which removes the phosphoryl group from the aspartic acid residue, is primarily contained within the receiver domains of response regulators. Nonphosphorylated transmitters modulate this activity to control receiver phosphorylation levels via an ATP-dependent mechanism, possibly involving conformational interactions. Sensor proteins are also thought to contain phosphatase activity, which can modulate phosphorylation levels of the transmitter domain. Due to variability in the levels of kinase and phosphatase activities, the half-lives of the phosphorylated receivers vary widely, ranging from a few seconds to several minutes. Besides having a major impact on the extent of the output response, the half-life of a phosphorylated receiver domain influences the ability of these regulatory systems to quickly respond to changing environmental conditions.

III. A SPECTRUM OF FUNCTIONS

Two-component regulatory systems provide a variety of regulatory functions critical for the bacterial cell to adapt to its environment.

The following sections are designed to give the reader a better understanding of the scope of regulatory functions that these systems carry out, while at the same time revealing the many variations from the prototype.

A. Osmolarity Changes and Porin Regulation

In their simplest form, two-component regulatory systems convey information across the cytoplasmic membrane from a sensor protein to a response regulator, which then affects the expression of specific genes. In this first example we will discuss a typical two-component regulatory system, the *E. coli* EnvZ and OmpR proteins involved in sensing osmolarity changes in the environment. In response to changing osmolarity, the response regulator of this system, OmpR, regulates two genes (*ompF* and *ompC*) encoding the porin proteins, OmpF and OmpC. Both of these porins form trimers in the outer membrane, with the main difference between these being the pore size that they form. OmpF multimerizes in the outer membrane to form a pore with diameter of 1.16 nm, while OmpC proteins form a 1.08 nm diameter pore. Intuitively, the difference between these pore sizes may seem trivial; however, the rates of passive diffusion are dramatically different when comparing one to the other. Thus, to maintain similar diffusion rates across the outer membrane, the OmpC porin predominates in high-osmolarity conditions, while the OmpF porin predominates in low-osmolarity conditions. Although the total amount of OmpF and OmpC proteins remains constant, the ratio of these proteins varies depending on the osmolarity of the environment.

So how does EnvZ and OmpR control the OmpF/OmpC ratio? Not surprisingly the level of phosphorylated OmpR protein plays a major role in the regulation of *ompF* and *ompC* transcription in response to osmolarity (i.e., mutants lacking EnvZ and OmpR produce constitutively low levels of OmpF and OmpC, regardless of the external osmolarity). Under high-osmolarity conditions, *ompC* expression is activated and *ompF* expression is repressed resulting in the predominance of smaller pores. When low-osmolarity conditions are encountered, only the *ompF* gene is expressed resulting in the predominance of larger pores. Intermediate conditions adjust the OmpF/OmpC ratio accordingly, resulting in an average pore size that provides an optimal diffusion rate. Some key clues that led to a better understanding of this regulatory mechanism were the observations that the level of OmpR-P increases with increasing osmolarity (Forst et al., 1990) and that OmpR-P levels are directly proportional to *ompC* expression. In contrast, OmpR-P concentrations were found to be inversely related to *ompF* expression. How could this reciprocal regulation correlate to the amount of OmpR-P present? One possibility is that OmpR-P has both positive and negative effects on *ompF* transcription, because of the differential affinities of OmpR-P for various regions upstream of the *ompF* promoter. At low OmpR-P concentrations (under low-osmolarity conditions), the sequences spanning −46 to −96 relative to the transcription start site (the high-affinity binding site) are occupied by OmpR-P, activating *ompF* transcription. When the concentration of OmpR-P rises (under high-osmolarity conditions), binding sites spanning −350 to −380 relative to the transcription start site (the low-affinity binding site) also become occupied by OmpR-P resulting in repression of *ompF* transcription. The presence of intrinsic DNA curvature and an IHF (a histonelike protein known to bend DNA) binding site between the two OmpR-P binding sites within the *ompF* promoter region strongly suggests a looping model in the repression of *ompF* transcription. Thus, under high-osmolarity conditions, OmpR-P could form a stable repression loop (with the help of IHF) by binding simultaneously to the high-and low-affinity binding sites. On the other hand, under low-osmolarity conditions, OmpR-P, which is present at low levels, would bind only to the high-affinity binding site activating transcription of *ompF*. As for *ompC*, its promoter region contains a single low-affinity binding site that would only become occupied to activate gene expression in the presence of high OmpR-P concentrations. Finally, the fine-tuning of this system in response to intermediate-osmolarity conditions could be achieved by adjusting the ratio of kinase and phosphatase activities associated with the EnvZ and OmpR proteins. This would control the levels of phosphorylated OmpR protein present in the cell and, in turn, the relative expression levels of the *ompC* and *ompF* genes.

Although we know a great deal about the EnvZ/OmpR system, little is known about how osmolarity is sensed. The problem of sensing changes in osmolarity turns out to be a tricky one, since factors contributing to osmolarity are so complex. Is the signal a single molecule that represents the overall osmolarity of the environment? Or is it a physical consequence of osmolarity changes? The quick answer is that we don't know for sure, but at least one study suggests that the latter possibility is correct. Based on the effects of membrane perturbants on *ompC/ompF* expression, it has been proposed that the EnvZ sensor protein functions to monitor membrane integrity (Rampersaud and Inouye, 1991). EnvZ has two transmembrane spanning domains and, like most sensor proteins, is located in the cytoplasmic membrane. The N and C termini are positioned in the cytoplasm, with a section of the protein between the N and C terminal in the periplasm. Interestingly mutations within the input domain of EnvZ result in a constitutive high-osmolarity phenotype indicating that the default state (in the absence of signal) is one that confers a high-osmolarity phenotype. Based on these results, it has been hypothesized that the signal must be present in a low-osmolarity environment (Pratt and Silhavy, 1995).

B. Quorum Sensing and Staphylococcal Virulence

The *agr* locus of *S. aureus* encodes a transcription regulatory system that differentially regulates expression of exoprotein and cell wall protein genes. This locus is comprised of a four-gene operon and a divergently transcribed regulatory RNA molecule, termed RNAIII (Fig. 3). The *agrA* and *agrC* genes of this operon encode a response regulator and a sensor protein, while the *agrD* and *agrB* genes encode an octapeptide signaling molecule and a protein that is believed to be involved in the processing and/or transport of the octapeptide, respectively (Kornblum et al, 1990). Studies indicate that high concentrations of octapeptide within the extracellular fluid are recognized by the AgrC sensor protein, leading to the activation of the AgrA response regulator. Activated AgrA then promotes the transcription of RNAIII, which in turn affects the expression of Agr-regulated genes via a mechanism that is not well understood.

The differential effect of Agr on extracellular and cell wall–associated protein production is best illustrated by analysis of in vitro grown cultures (Fig. 3). During the exponential growth, cell wall–associated proteins are produced while exoprotein production is repressed. As the cells enter stationary phase, a transition occurs in which cell wall protein expression diminishes and exoprotein production commences. The molecular mechanism of this growth phase regulation can be attributed to the accumulation of AgrD octapeptide in the culture medium. Once a "threshold" level of octapeptide has accumulated, the AgrC sensor protein presumably undergoes autophosphorylation, triggering the stimulation of the Agr regulatory cascade. By measuring the octapeptide concentration within the surrounding medium, the Agr regulatory system senses, indirectly, bacterial cell density.

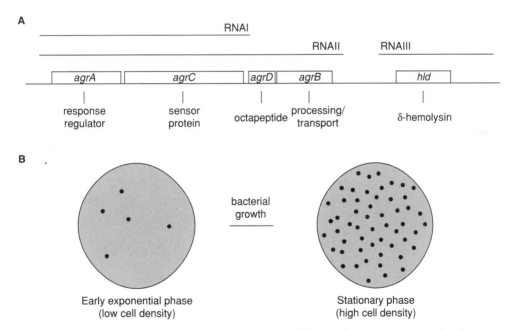

Fig. 3.   The *agr* regulatory system of S. *aureus*. The *agr* locus (**A**) encodes a sensor protein (*agrC*), response regulator (*agrA*), and proteins required for the production (*agrD*) and processing or transport (*agrB*) of the AgrD octapeptide. Activation of the Agr two-component regulatory system (by the accumulation of AgrD) results in increased transcription of the regulatory RNAIII molecule (which also encodes δ-hemolysin) and, ultimately, altered virulence factor expression. (**B**) As the density of the S. *aureus* cells (filled circles) increases during the transition from exponential growth to stationary phase, AgrD molecules (small dots) accumulate until they reach a threshold concentration that triggers the Agr signal transduction cascade.

The role of the Agr regulatory system during pathogenesis has been hypothesized to enable the bacteria to differentially express various genes during the course of infection. For example, genes whose products are involved in establishing an infection (e.g. adherence and resistance to host immune responses) are expressed initially, while those genes whose products are involved in maintaining or expanding the infection (e.g., toxins and proteases) are turned on later. Thus an encounter with *S. aureus* would initially involve bacteria (which are relatively few in number) that are in a state that maximizes adherence to host tissue (perhaps involving matrix-binding proteins), resistance to opsinophagocytosis (using antibody-binding proteins, e.g., protein A) and development of an abscess (involving blood coagulating factors, e.g., coagulase). Once the infection is established and the bacterial cell density has reached high levels, a shift in gene expression occurs resulting in the production of exoproteins involved in the tissue destruction and inflammation that is associated with many staphylococcal diseases.

C. The Phosphorelay and Sporulation Initiation in *B. subtilis*

We have focused so far on two-component regulatory systems that contain a single sensor protein and a single response regulator. In many cases more complex situations occur in which more than two components are involved in signal transduction. Several two-component regulatory systems exist that contain additional receiver domains cova-

lently linked to a sensor protein, presumably to increase the level of regulation that could be imposed on these signal transduction systems. In an extreme case multiple sensor and response regulator proteins are linked together into a so-called phosphorelay, to process complex signaling events during the initiation of sporulation in *B. subtilis* (see Moran, this volume).

Induction of the phosphorelay in *B. subtilis* is thought to be the initial trigger of a complex cascade of regulatory events necessary for the expression of genes involved in spore development. The triggering of the phosphorelay is known to be induced by a broad range of environmental stimuli, including metabolic, cell cycle and cell density information. Clearly, the molecular events that lead to the initiation of sporulation are very complex and the phosphorelay appears to be at the center of it all. As shown in Figure 4, the components of the phosphorelay include KinA, KinB, Spo0F, Spo0B, and Spo0A. The KinA and KinB components contain homology to sensor proteins, while Spo0F and Spo0A are response regulator proteins. Spo0B is a protein that is unique to the phosphorelay and functions as a phosphotransferase. The flow of the phosphorelay begins with KinA and KinB, which input phosphate into the system. The phosphoryl groups of either KinA or KinB are then transferred to Spo0F, which then serves as a phosphodonor for the formation of phosphorylated Spo0A, a reaction that is catalyzed by Spo0B. Once Spo0A is phosphorylated, it then functions to regulate a variety of genes that are involved in the initial stages of sporulation. Thus, the con-

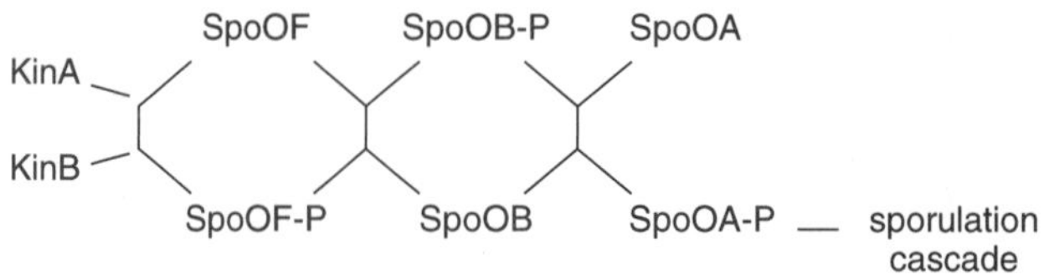

Fig. 4. The phosphorelay of *B. subtilis*. Phosphorylation of Spo0F by either KinA or KinB can initiate the phosphorelay, which is comprised of a series of phosphotransfer reactions between transmitter and receiver domains of the protein components of this system. Input at any point along the phosphorelay can alter the signaling necessary to initiate the sporulation cascade.

version of nonphosphorylated Spo0A to its phosphorylated form is the key determinant in the molecular decision as to whether the bacterial cell will sporulate or will remain in a vegetative state. Clearly, this phosphorelay is much more complicated than most two-component regulatory systems. Why should this be?

If one considers the many environmental factors that contribute to the initiation of sporulation, along with the consequences of this event, it becomes apparent why such a system has evolved. The initiation of sporulation is not as simple as sensing a single environmental factor that impacts only a small fraction of the overall physiology of the bacterium. The consequences of inducing sporulation would be quite severe if it was done unnecessarily. Thus the decision to induce sporulation is based more on the overall state of the bacterial cell, which obviously is indicated by a combination of factors. If the decision to induce sporulation is made in haste, the cell is faced with the consequences of undergoing an essentially irreversible developmental process while other cells continue to proliferate at a potentially exponential rate! Obviously, in a feast or famine world, bacteria that make this decision incorrectly will be at a nearly hopeless disadvantage.

Since the discovery of the phosphorelay that controls the initiation of sporulation, several other phosphorelay systems that lie at the heart of complex developmental pathways have been identified. These include sporulation development and timing in *Myxococcus xanthus* (Cho and Zusman, 1999), fruiting body development in *Dictyostelium* (Thomason et al., 1999), and hyphal development in *Candida albicans* (Calera and Calderone, 1999). Additional phosphorelays involved in developmental processes in other organisms are sure to be discovered.

D. Chemotaxis and Atypical Output Responses

The output responses that we have discussed until now entail transcriptional regulation of genes under the response regulator's control. In this section, atypical output responses will be discussed. The best characterized of these are those mediated by the Che proteins, which are important determinants in modulating chemotactic responses.

Bacterial motility is commonly mediated by the coordinated action of flagella whose spinning action propels the organisms through their environment. Directed movement, or chemotaxis, is accomplished by altering the rotation of the flagella either clockwise (CW) or counterclockwise (CCW). CCW rotation causes the bundling and coordinated rotation of the flagella at one pole of the cell, resulting in the propulsion of the cell forward in a "smooth swimming" movement. CW rotation disperses the bundle and causes a sporadic "tumbling" movement. Thus, when the cell is traveling toward an attractant (i.e., certain sugars and amino acids) or away from a repellant (i.e., fatty acids and alcohols), the flagella rotate in a CCW manner to continue its movement in that direction. On the other hand, if the cell goes off course (i.e., away from an attractant or toward a repellant), the flagella rotate in a CW manner causing a tumble and reorientation of the direction of movement, until the cell is pointed in the optimal direction again. In this manner the bacterial cell accomplishes net movement toward an attractant and away from a repellant.

Obviously the key event in this chemotaxis strategy is to control the rotation of the flagella in response to the attractants and repellants that are present in the surrounding medium. Once again, members of the two-component regulatory system family of proteins lie at the heart of this mechanism. The sensor protein of this system is termed CheA, which is associated with another protein, CheW, that aids in CheA autophosphorylation. CheA is unique in that rather than being stimulated directly by extracellular signals, CheW-mediated interactions with transmembrane receptor proteins known as MCPs (methyl-accepting chemotaxis proteins) trigger autophosphorylation. Once this occurs the phosphoryl group from CheA is transferred to the response regulator of this system, CheY. Rather than functioning as a transcriptional regulator, CheY-P then interacts with the flagellar motor and increases the likelihood of CW rotation and tumbling. Thus MCP-mediated signaling ultimately triggers tumbling.

In a homogeneous environment, the bacteria are known to tumble every two to four seconds resulting in what is referred to as a "random walk," ideal for taking random samples of the environment. Once a gradient of attractant or repellant is encountered, the frequency of tumbling changes dramatically. If the concentration of an attractant is increasing or of a repellant is decreasing as the cell moves, tumble frequency is reduced and the cells swim for longer periods. If the concentration of an attractant is decreasing or of a repellant is increasing as the cell moves, tumble frequency is increased and the cells tumble about. This approach results in a net migration up an attractant gradient and down a gradient of repellant. Remarkably many of the molecular details of how the bacteria sense attractant and repellant gradients are well characterized.

Studies demonstrate that an attractant signal from the MCPs decreases CheA autophosphorylation leading to decreased CheY-P levels and a CCW rotation of the flagellum (smooth swimming). In contrast, a repellant signal from the MCPs increases CheA autophosphorylation, leading to increased CheY-P levels and a CW rotation of the flagellum (tumbling). Gradients of attractants and repellants are detected as a function of changes in the methylation state of the MCPs, which is controlled by a complex balance between methylases (enzymes, e.g., CheR, that add methyl groups) and methylesterases (enzymes, e.g., CheB, that remove methyl groups). The binding of attractants makes MCPs more accessible to methylation, while the binding of repellants makes MCPs less accessible. As a result of methylation-induced conformational changes in these proteins, the methylated MCPs trigger CheA dephosphor-

ylation and CCW flagellar rotation while unmethylated MCPs lead to CheA autophosphorylation and CW rotation. Adaptation in a gradient occurs as a result of methylation-induced changes in affinities of the MCPs for the attractant signal. As the MCPs become increasingly methylated, their affinities for attractants are reduced so that the signaling state returns to prestimulus levels. Thus, by requiring a constant signal to induce methylation, this mechanism allows for the bacterium to continually sample the environment and respond to a gradient of attractant. Again, two-component signaling is an important part of these processes. In another example of an atypical output response, the methylesterase activity of CheB, which controls the methylation levels, is stimulated by CheA-mediated phosphorylation!

IV. CONCLUSIONS

Although a great deal of progress has been made in the area of bacterial signal transduction since the discovery of two-component regulatory systems, much remains to be done. For example, despite the identification of numerous sensor proteins from a variety of bacterial species, relatively little is known about the initial signaling event that triggers autophosphorylation and signal transduction. What part of the sensor proteins are necessary for interacting with their specific signals, and how do these interactions trigger the intermolecular autophosphorylation event? Also, what is the nature of the sequences that provide the specificity of sensor proteins for their cognate response regulators? Finally, how do changes in the phosphorylation state of response regulators lead to modification of the output response? Answers to all of these questions will surely be forthcoming as more structural information on these fascinating and multifunctional proteins is obtained.

REFERENCES

Calera JA, Calderone R (1999): Flocculation of hyphae is associated with a deletion in the putative CaHK1 two-component histidine kinase gene from *Candida albicans*. Microbiol 145:1431–1442.

Cho K, Zusman DR (1999): Sporulation timing in *Myxococcus xanthus* is controlled by the *espAB* locus. Mol Microbiol 34(4):714–725.

Dutta R, Qin L, Inouye M (1999): Histidine kinases: Diversity of domain organization. Mol Microbiol 34(4):633–640.

Fabret C, Feher VA, Hoch JA (1999): Two-component signal transduction in *Bacillus subtilis*: How one organism sees its world. J Bacteriol 181(7):1975–1983.

Fisher SL, Jiang W, Wanner BL, Walsh CT (1995): Cross-talk between the histidine protein kinase VanS and the response regulator PhoB. Characterization and identification of a VanS domain that inhibits activation of PhoB. J Biol Chem 270:23143–23149.

Forst S, Delgado J, Rampersaud A, Inouye M (1990): In vivo phosphorylation of OmpR, the transcription activator of the *ompF* and *ompC* genes in *Escherichia coli*. J Bacteriol 172(6):3473–3477.

Kornblum J, Kreiswirth BN, Projan SJ, Ross H, Novick RP (1990): Agr: A polycistronic locus regulating exoprotein synthesis in *Staphylococcus aureus*. In Novick RP (ed): "Molecular Biology of the Staphylococci." New York: VCH Publishers, pp. 373–402.

Lukat GS, McCleary WR, Stock AM, Stock JB (1992): Phosphorylation of bacterial response regulator proteins by low molecular weight phospho-donors. Proc Natl Acad Sci USA 89:718–722.

Parkinson JS (1995): Genetic approaches for signaling pathways and proteins. In Hoch JA, Silhavy TJ (eds): "Two-Component Signal Transduction." Washington, DC: ASM Press, pp 9–23.

Parkinson JS, Kofoid EC (1992): Communication modules in bacterial signalling proteins. Ann Rev Genet 26:71–112.

Pratt LA, Silhavy TJ (1995): Porin regulon of *Escherichia coli*. In Hoch JA, Silhavy TJ, (eds): "Two-Component Signal Transduction." Washington, DC: ASM Press, pp 105–127.

Rampersaud A, Inouye M (1991): Procaine, a local anesthetic, signals through the EnvZ receptor to change the DNA binding affinity of the transcriptional activator protein OmpR. J Bacteriol 173(21):6882–6888.

Stock JB, Surette MG, Levit M, Park P (1995): Two-component signal transduction systems:structure-function relationships and mechanisms of catalysis. In Hoch JA, Silhavy TJ (eds), "Two-Component Signal Transduction." Washington, DC: ASM Press, pp 25–51.

Thomason PA, Traynor D, Stock JB, Kay RR (1999): The RdeA-RegA system, a eukaryotic phospho-relay controlling cAMP breakdown. J Biol Chem 274(39):27379–27384.

16

Molecular Mechanisms of Quorum Sensing

CLAY FUQUA AND MATTHEW R. PARSEK

Department of Biology, Indiana University, Bloomington, Indiana 47405; Department of Civil Engineering,
Northwestern University, Evanston, Illinois 60208

**Modern Microbial Genetics, Second Edition. Edited by
Uldis N. Streips and Ronald E. Yasbin. ISBN 0–471–38665–0**
Copyright © 2002 Wiley-Liss, Inc.

I. INTRODUCTION

The ability of bacteria to monitor their own population density and modulate gene expression accordingly has been termed quorum sensing. Although there are a number of different mechanisms by which quorum sensing can function, perhaps the best-characterized quorum sensing mechanism is the acyl-homoserine lactone (acyl-HSL) based system used by a growing number of Gram-negative bacteria (Bassler, 1999; Dunny and Leonard, 1997; Fuqua and Greenberg, 1998; Fuqua et al., 1996). This type of quorum sensing has thus far exclusively been found among members of the Proteobacteria. The physiological processes regulated by quorum sensing in these different species vary from conjugal plasmid transfer to bioluminescence. However, a common theme is that many of the physiological processes involve some type of interaction with a eukaryotic host (Parsek and Greenberg, 2000).

Acyl-HSL based quorum sensing was first described in the marine luminescent bacterium, *Vibrio fischeri*. *V. fischeri* is capable of colonizing the light organs of a variety of marine fish and squids, where it grows to very high cell densities (10^{10} cells/ml) and produces light (Nealson and Hastings, 1979; Ruby, 1996). In 1970 Nealson, Platt, and Hastings reported that *V. fischeri* produced an extracellular factor, which they called an "autoinducer," that regulated production of the light-producing enzyme, luciferase (Nealson et al., 1970). We now know that this "autoinducer" is an acyl-HSL, a freely diffusible signaling molecule. The following 20 years saw much research focused on the genetics of quorum sensing; however, this regulatory strategy was thought to be unique to certain marine species of *Vibrio*. In the early to mid-1990s acyl-HSL-based quorum sensing systems were identified in the plant pathogens, *Erwinia* spp. and *Agrobacterium tumefaciens*, as well as the opportunistic human pathogen *Pseudomonas aeruginosa* (Bainton et al., 1992; Beck von Bodman and Farrand, 1995; Gambello and Iglewski, 1991; Passador et al., 1993; Piper et al., 1993; Zhang et al., 1993). Currently over 50 species have been shown to produce acyl-HSLs. With the knowledge derived from sequenced bacterial genomes, the number of bacterial species found to utilize quorum sensing will continue to grow.

The basic molecular scheme of acyl-HSL based quorum sensing is rather simple. The bacterium produces an acyl-HSL synthase (LuxI-type protein) that synthesizes acyl-HSLs at a low basal level. The acyl-HSL diffuses out of the cell, down its concentration gradient and is lost to the environment. However, at a critical cell density, the local concentration of acyl-HSL builds to a threshold level at which it interacts with a transcriptional regulator (LuxR-type protein). This acyl-HSL-transcriptional regulator complex modulates expression of quorum sensing regulated genes. In many cases this involves positive autoregulation of the quorum sensing *luxI* and *luxR* homologues. Although this illustrates the general molecular components of quorum sensing, studies of different systems has revealed that there are interesting variations of this general scheme. This chapter will review what is currently known about the different aspects of the molecular biology of quorum sensing with acyl-HSLs.

II. THE ECOLOGY OF QUORUM SENSING

The environmental context of quorum sensing is an interesting point to consider. Although a variety of Gram-negative bacterial species use quorum sensing to regulate different physiological processes, the environments in which these species encounter high population densities greatly varies. Certain systems, such as the colonization of the light organs of marine animals by *Vibrio fischeri*, represent well-defined niches in which a quorum would be achieved (McFall-Ngai and Ruby, 1991; Ruby, 1996; Ruby and McFall-Ngai, 1992). Previous studies have demonstrated that the concentration of acyl-HSLs in such environments exceeds concentrations required to induce quorum sensing regulated genes in laboratory experiments. The cepaloid squid, *Euprymna scolopes*, feeds branched chain amino acids to the bacterial populations within the light organ, to

provide a nutritional environment capable of supporting high numbers of *V. fischeri* and bioluminescence, which is an energy-intensive process (Graf and Ruby, 1998). This is in contrast to free seawater, an oligotrophic environment, where *V. fischeri* does not luminesce and is found at very low cell densities (a few cells/ml). *E. scolopes* has also been shown to expel most of the bacteria in the light organ in a circadian pattern every 24 hours. This has been suggested to explain why *V. fischeri* populations are higher in ambient seawater located within the squid habitat ranges.

Most species that utilize acyl-HSL based quorum sensing do not have a clearly identified physical location where high cell densities are achieved. For many plant and animal pathogens, quorum sensing is thought to provide the bacterium a means to delay the onset of virulence factor production and the ensuing immunological response by the host, until sufficient bacteria have been amassed. Pathogenic interactions may also involve the formation of surface-associated communities of microorganisms, called biofilms. Bacterial biofilms can form on the surface of in-dwelling medical devices, artificial joints, and dead tissue. Recent studies have estimated that at least 67% of human infections involve bacterial biofilms (Costerton et al., 1999). *P. aeruginosa* is one example of a pathogen that that uses quorum sensing to regulate expression of virulence factors and also has been shown to form biofilms of clinical importance. The high cell densities achieved in biofilms can result in inducing concentrations of acyl-HSLs (see Hassett et al., this volume).

Many symbiotic and pathogenic interactions of bacteria with a eukaryotic host involve pure cultures of bacteria. However, outside the host, these species coexist in the environment with a variety of different bacterial species. One must assume that at least a small percentage of this population consists of organisms harboring acyl-HSL-based quorum sensing. The question of interspecies sensing of acyl-HSLs is crucial for understanding the dynamics of environmental microbial communities. Such an interaction could potentially be either synergistic or competitive. McKenney et al. observed that cell-free extracts of *P. aeruginosa* induced the formation of *Burkolderia cepacia* quorum sensing-regulated exoproducts (McKenney et al., 1995). Cell-free extract from a *P. aeruginosa* strain with reduced acyl-HSL synthesis did not stimulate production of *B. cepacia* exoproducts, suggesting that *P. aeruginosa* acyl-HSLs were modulating expression of the exoproducts. Recently Anderson et al. (1999) demonstrated in liquid and biofilm culture that acyl-HSL sensitive transcriptional reporters in quorum sensing mutant backgrounds of either *P. aeruginosa* or *Serratia liquefaciens* were capable of perceiving acyl-HSLs produced by wild-type strains of *S. liquefaciens* or *P. aeruginosa* (M. Givskov, personal communication). Perhaps it is not surprising that a bacterial species possessing a quorum sensing system is capable of perceiving acyl-HSLs produced by another species. Another possibility is that a bacterial species may specifically respond to acyl-HSLs produced by another species in an R-protein independent manner. This may make sense in an environment where two species commonly encounter one another. The ability to sense the presence of an acyl-HSL producing species that was either a competitor or a syntroph would provide an advantage.

III. ACYL-HSL STRUCTURE AND SYNTHESIS

A. A Comparison of Known acyl-HSL Structures

In 1981 Anatol Eberhard and colleagues determined the first acyl-HSL structure (Eberhard et al., 1981). This study was conducted on the primary acyl-HSL of *Vibrio fischeri*, 3-oxo-hexanoyl-HSL. Since 1981 over 10 acyl-HSL structures have been determined, usually using a combination of high-performance liquid chromatography (HPLC) purification and nuclear magnetic resonance imaging (NMR) (Schaefer et al., 2000). The

identified acyl-HSL structures all have in common a homoserine lactone ring moiety, which imparts a hydrophilic character on the molecule. However, the acyl side chain of different acyl-HSLs varies in length, degree of substitution, and saturation. These characteristics modulate the hydrophobicity of the acyl-HSL. The amphipathic nature of acyl-HSLs presumably allows them to freely navigate the hydrophobic portion of the cell membrane as well as the aqueous environment inside and outside the cell. The nature of the acyl side chain also provides specificity to the quorum sensing system. This is of particular importance in species that have multiple quorum sensing systems, such as *Pseudomonas aeruginosa*. The two quorum sensing systems of *P. aeruginosa* use structurally distinct acyl-HSLs (butyryl-HSL and 3-oxo-dodecanoyl-HSL), minimizing interactions between the two systems. Acyl-HSL with the longest acyl side chain identified to date are tetradecanoyl-HSLs produced by *Rhizobium leguminosarum* and *Rhodobacter sphaeroides*, while the acyl-HSL with the shortest acyl side chain is butyryl-HSL (Puskas et al., 1997). Since the acyl side chain is derived from fatty acid biosynthesis (see Section C below), the length of the side chain could conceivably be longer.

B. The Substrates for acyl-HSL Synthesis and Their Relationship to Cellular Physiology

Initial studies conducted by Engebrecht et al. (1983) determined that the *V. fischeri luxI* gene was the only gene required for *Escherichia coli* to direct the synthesis of 3-oxo-hexanoyl-HSL. This was later found to be true for *traI* of *Agrobacterium tumefaciens* and *lasI* of *P. aeruginosa*. Studies conducted with crude extracts of *V. fischeri* showed that the synthesis of 3-oxo-hexanoyl-HSL (or VAI for *v.* bro Fischesi auto induce) occurred after the addition of *S*-adenosylmethionine (SAM) and 3-oxo-hexanoyl-CoA, implying that these common cellular compounds may be substrates for synthesis of VAI (Eberhard et al., 1991). Experiments conducted with amino acid auxotrophs of *E. coli* suggested homoserine lactone itself might serve as the source of the homoserine lactone ring moiety of VAI, although this was later disproved (Hanzelka and Greenberg, 1996). A breakthrough occurred when two groups were able to purify and demonstrate acyl-HSL synthesis in vitro. TraI and LuxI fusion proteins were shown in vitro to synthesize acyl-HSLs from SAM and acylated-acyl carrier protein (ACP) (More et al., 1996; Schaefer et al., 1996). In vitro studies with RhlI, as well as the LuxI and TraI fusion proteins, were hampered by the fact that these enzymes had very low specific activities (Jiang et al., 1998; More et al., 1996; Schaefer et al., 1996). A report using active, purified native RhlI suggested that the in vivo substrates for acyl-HSL synthesis are probably SAM, an acyl-ACP based on kinetic data (Parsek et al., 1999).

SAM and acyl-ACP are common metabolic intermediates, central to several cellular processes. In *E. coli*, methionine is converted to SAM by the *metK* gene product, SAM synthetase. SAM is required for a variety of nucleic acid methylation reactions, the synthesis of polyamines such as spermidine, and a variety of steps in single carbon metabolism. Acyl-ACPs are intermediates of fatty acid biosynthesis and can also be used as a substrate in lipid A biosynthesis (Magnuson et al., 1993). The available pools of both substrates reflect the general metabolic state of the cell. Presumably actively growing cells tend to have larger pools of these substrates than starved cells. This may have a significant impact on acyl-HSL synthesis, especially for acyl-ACP where there are very small amounts found free in the cell and kinetic competition for it is fierce (Magnuson et al., 1993). Ultimately understanding how the metabolic state of the cell modulates acyl-HSL synthesis will be important for understanding the quorum sensing mechanism of populations in complex environments.

C. The Enzymatic Reaction Mechanism and Inhibitors of acyl-HSL Synthesis

One of the first suggestions put forth about a potential enzymatic reaction mechanism involved the formation of an acylated-enzyme intermediate, characteristic of a ping-pong reaction mechanism (Sitnikov et al., 1995). The finding that acyl-ACP is a substrate of acyl-HSL synthesis supported the proposed ping-pong reaction mechanism, since the enzymes involved in fatty acid biosynthesis, which also use acyl-ACP as a substrate, utilize a similar mechanism (DAgnolo et al., 1973; Garwin et al., 1980). A ping-pong reaction mechanism would initially involve a transfer of the acyl group from acyl-ACP to an active site residue on the acyl-HSL synthase. This acyl-enzyme intermediate would then bind to SAM, resulting in amide bond formation prior to, after, or simultaneously with lactonization of SAM to yield the homoserine lactone ring. However, attempts to isolate an acyl-enzyme intermediate in vitro were unsuccessful and mutagenesis of acyl-HSL synthases failed to identify conserved amino acids that could act as an acceptor site for acyl group transfer (Hanzelka et al., 1997; Parsek et al., 1997).

Inhibitors of acyl-HSL synthesis and the enzymatic reaction mechanism were determined for RhlI. The reaction products, 5′-methylthioadenosine and holo-ACP were found to inhibit the reaction. Various analogues of SAM, such as *S*-adenosylhomocysteine and sinefungin, were also found to be potent inhibitors of acyl-HSL synthesis. A kinetic analysis using the substrates and inhibitors suggested that RhlI uses a sequential ordered reaction mechanism to synthesize butyryl-HSL (Parsek et al., 1999). The reaction proceeds initially through the binding of SAM to RhlI (see Fig. 1). Subsequent binding of butyryl-ACP is followed by amide bond formation, and then holo-ACP is released. Butyryl-HSL is released after lactonization of the homoserine lactone ring. The final step of the enzymatic mechanism is release of 5′-methylthioadenosine. Whether

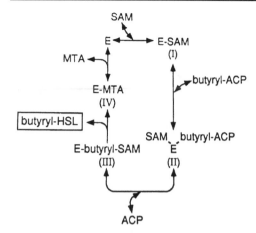

Fig. 1. Enzymatic mechanism of acyl-HSL synthesis by RhlI. This diagram shows the steps predicted to occur in acyl-HSL synthesis by RhlI. The reaction mechanism is sequential ordered. E represents the enzyme, RhlI. The initial step is binding of SAM followed by binding of acyl-ACP. ACP is released followed by amide bond formation between SAM and the acyl group. Lactonization of the homoserine lactone ring occurs and then release of butyryl-HSL. The final step is release of the product, MTA. (Reproduced from Parsek et al., 1999, with permission of the publisher.)

other LuxI-type proteins share this reaction mechanism remains to be determined.

IV. LUXI-TYPE PROTEINS

A. Structure-Function Studies of LuxI-type acyl-HSL Synthases

Over 20 LuxI family members have been identified and sequenced. While all acyl-HSL synthases presumably catalyze the formation of an amide bond between the acyl side chain and the amino group of SAM, many synthesize acyl-HSLs with structurally different acyl side chains. Comparisons of their primary amino acid sequences reveal proteins that vary in length from 194 to 226 amino acids. Amino acid sequence alignments have identified 10 completely conserved amino acids within the LuxI family, with most of them lying within the amino-terminal half of the proteins. Seven out of the 10 amino acids are charged residues (Fig. 2). The amino-terminal portions of LuxI-type proteins display the highest homology,

with little homology seen toward the carboxy terminus. This led to the suggestion that carboxy-terminus was involved in acyl side chain specificity (Hanzelka et al., 1997).

Random mutagenesis of LuxI demonstrated that 13 separate point mutations in LuxI had a negative effect on acyl-HSL synthase activity (Fig. 2; and Hanzelka et al., 1997). Eleven of the mutations yielded a peptide with no detectable activity, while two of the mutations exhibited reduced activity. Seven of these 11 mutations clustered within a region spanning amino acids 25 to 70. This lead the authors to propose that this region may constitute the active site of the protein involved in amide bond formation. Many of these point mutations corresponded to the completely conserved residues R25, E44, D46, D49, and R70 (wild-type amino acid position on LuxI). Four of these mutations (at positions R104, A133, E150, and G164) were located toward the carboxy terminus of the protein, within the region spanning residues 104 to 164. These mutations were proposed to lie in amino acids critical for the binding of the acyl side chain of the acyl-HSL. Site-directed mutagenesis demonstrated that none of the cysteine residues in LuxI were critical for activity, casting doubt on the role of a cysteine as a binding site for acyl side chain transfer from acyl-ACP (Hanzelka et al., 1997).

A subsequent report on the *P. aeruginosa* acyl-HSL synthase, RhlI revealed similar trends as seen in the LuxI study. Eight critical residues required for acyl-HSL synthesis were found using random mutagenesis (Parsek et al., 1997). Seven of these residues corresponded to similar mutations found in LuxI (see Fig. 2). The one mutation unique to RhlI, E101 (wild-type amino acid position on LuxI), corresponded to a residue that is conserved among LuxI family members. All the mutations fell within a region spanning residues 24–104 of RhlI. This finding reinforced the idea of an amino terminal active site encompassing this region of conservation among LuxI family members. No residues in the carboxy-terminal region of

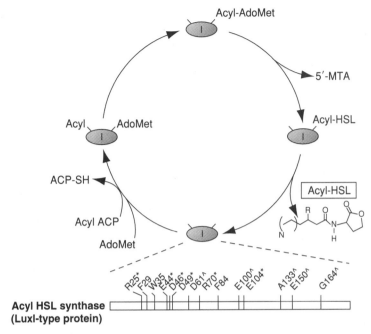

Fig. 2. Mutational map of acyl-HSL synthases. The LuxI-type protein is represented as a single bar. Amino acid residues conserved throughout the LuxI family of proteins are in bold. Residues at which mutations are known to abolish activity for RhlI and LuxI are indicated by an asterisk, while those that affect only LuxI have a caret.

the protein were found to be critical for activity of RhlI. Although at three of the positions (A133, E150, and G164 of LuxI) RhlI had the same amino acid residues as LuxI, site-directed mutagenesis of RhlI at these residues confirmed that they were not critical for activity, suggesting that RhlI and LuxI have different requirements for residues in the carboxy-terminus. Site-directed mutagenesis of cysteines in RhlI confirmed that no cysteine was absolutely required for acyl-HSL synthesis by acting as an acceptor site for acyl-HSL transfer (Parsek et al., 1997). This finding also holds true for the *A. tumefaciens* acyl-HSL synthase TraI (C. Fuqua, unpublished results).

Future studies of acyl-HSL synthases will involve elucidation of the substrate interaction components of the enzyme, as well as a more careful mapping of the active sites. Combining substrate binding assays and mutagenesis could potentially derive this information. Ultimately X-ray crystallography

of an acyl-HSL synthase, both free and complexed with its substrates would yield the most detailed information concerning the structure and function of LuxI-type proteins.

B. The AinS Family of acyl-HSL Synthases

While most genes encoding acyl-HSL synthases show homology to the LuxI family, two genes not homologous to *luxI; ainS* of *V. fischeri* and *luxM* of *V. harveyi* have also been implicated in acyl-HSL synthesis (Bassler et al., 1993; Gilson et al., 1995). These genes encode 46-kDa and 25-kDa proteins, respectively. The C-terminal 218 amino acids of AinS share similarity with the 216 amino acid LuxM, showing 34% identity. The *ainS* gene has been shown to direct production of octanoyl-HSL in *E. coli*, although expression of *luxM* in *E. coli* does not direct *E. coli* to make the *V. harveyi* acyl-HSL, 3-hydroxybutanoyl-HSL. Chromosomal mutations in either *ainS* of *V. fischeri* or *luxM* of *V. harveyi* result in their inability to make their respective acyl-HSLs. The *ainS* gene has DNA sequence elements called *lux* boxes (see Section VIC) upstream of its promoter, suggesting that expression of *ainS* is under control of *luxI/luxR*. The *luxM* gene positively regulates bioluminescence in *V. harveyi*. Although the genes regulated discretely by *ainS* remain to be identified, Kuo et al. (1996) proposed that the octanoyl-HSL modulates the *lux* quorum sensing system by competing with the *lux* system acyl-HSL, 3-oxohexanoyl-HSL, for binding to LuxR.

The substrates for acyl-HSL synthesis by AinS appear to be similar to those used by LuxI-type proteins. Studies with a purified maltose binding protein-AinS fusion indicated that the homoserine lactone ring is derived from *S*-adenosylmethionine (Hanzelka et al., 1999). These studies also indicated that the acyl side chain of octanoyl-HSL could be derived from either octanoyl-CoA or octanoyl-ACP as a substrate. The in vitro kinetics of acyl-HSL synthesis suggest that both of these compounds may serve as a substrate in vivo for AinS, unlike RhlI which uses primarily acyl-ACP substrates. Whether this represents a characteristic feature of AinS family members remains to be determined. The use of both acyl-ACP and -CoA substrates may represent a strategy the cell uses to ensure that it has substrates for acyl-HSL synthesis during the different physiological states represented by fatty acid biosynthesis (source of acyl-ACP) and fatty acid β-oxidation (source of acyl-CoA). Compounds that inhibit AinS activity such as *S*-adenosylhomocysteine and holo-ACP, are similar to those found to inhibit RhlI, possibly indicating that AinS has a similar reaction mechanism as LuxI family members (Hanzelka et al., 1999).

V. ACYL-HSL RELEASE AND ACCUMULATION

A. Diffusion and Membrane Interactions of acyl-HSLs

In *V. fischeri* the distribution of 3-oxo-hexanoyl-HSL is dictated largely by its concentration gradient across the bacterial envelope. In 1985 Kaplan and Greenberg demonstrated that radiolabeled 3-oxo-hexanoyl-HSL rapidly diffused into and out of bacterial cells (Kaplan and Greenberg, 1985). By analogy to *V. fischeri*, acyl-HSLs from other bacteria are thought to freely diffuse across the bacterial envelope. In addition the general observation that exogenous addition of acyl-HSLs in heterologous hosts suggests that their transmembrane traffic relies on passive diffusion. However, acyl-HSLs with side chains of up to 14 carbons in length have been identified, and it is questionable whether these more hydrophobic molecules would readily cross membranes or remain soluble in the aqueous cytoplasm (Fuqua et al., 1996). Genetic analysis of quorum sensing in *P. aeruginosa* has demonstrated the MexAB-OprD efflux system, a general transporter of hydrophobic molecules from phospholipid bilayers, plays a role in the release of 3-oxo-dodecanoyl-HSL from producing cells (Evans et al., 1998; Pearson et al., 1999). This observation suggests that 3-oxo-dodecanoyl-HSL is a substrate for this efflux system, and this provides indirect evidence that a substantial amount of the larger signal molecule partitions into the lipid bilayer. It remains to be determined how general this phenomenom may be, and which other acyl-HSLs are assisted by membrane transporters.

B. Environmental Conditions Resulting in Elevated acyl-HSL Concentration

Accumulation of acyl-HSLs due to increased population density is thought to be the primary mechanism dictating interaction with the LuxR-type receptor and ultimately alteration of target gene expression. The light organs of symbiotic animals provide such an environment for *V. fischeri*, in which the bacterial colonizers are supported to a high-density fostering concomitant increases in 3-oxo-hexanoyl-HSL concentrations (Ruby and Asato, 1993). In other bacteria that lack such a specialized inducing habitat, the composition of the appropriate environment is less well characterized, but it probably involves formation of aggregates or surface-associated biofilms (see Section II above). The demonstration of acyl-HSL production in monospecies and multispecies biofilms from different environments in fact supports such an idea (McLean et al., 1997; Stickler et al., 1998). Although a dense population appears to be a common way in which to elevate acyl-HSL concentrations, theoretically many other factors can influence this dynamic process. Mathematical modeling of LuxR interactions with acyl-HSLs suggests that there are multiple pathways to induction (James et al., 2000). The flow characteristics of any given environment should have profound effects on signal accumulation, as high flow rate will tend to accelerate removal of the acyl-HSLs while a drop to zero flow will foster accumulation. Likewise the dimensions and diffusion characteristics of the environment should affect induction. The chemical nature of the environment will also greatly influence the level of bioactive acyl-HSL. For example, under alkaline conditions most acyl-HSLs are chemically unstable, affecting their active concentration. Likewise chemical interactions of acyl-HSLs with extracellular compounds of biotic or abiotic origin may act to sequester acyl-HSLs that would otherwise be active. Finally, the regulation of bacterial physiology can significantly influence the buildup of the pheromones. Environmental factors often regulate the expression of the acyl-HSL synthase genes including the positive feedback of their cognate LuxR-type protein, additional diffusible signals, and other regulatory proteins (see below). Also, as mentioned above, the concentration of acyl-HSLs may be influenced by cellular efflux systems. All of these features can affect the inducing character of a given environment.

VI. LUXR-TYPE PROTEINS

LuxR-type proteins from different bacterial genera share end-to-end sequence identity of 18–23%. There are two clusters of stronger conservation between residues 66–138 and 183–229 (relative to LuxR), roughly defining an acyl-HSL interaction domain and a DNA binding motif, respectively (Fig. 3). The carboxy-terminal regions of LuxR-type proteins share homology with the larger FixJ-NarL superfamily of prokaryotic transcription factors many of which are two-component type response regulators (Kahn and Ditta, 1991). The DNA-binding activity of these response regulators is differentially controlled by phosphorylation of a conserved aspartate residue in the amino-terminal half of these proteins. In contrast, the amino-terminal halves of LuxR-type proteins interact with acyl-HSLs and do not share detectable sequence homology with two-component response regulators (Fuqua et al., 1994). However, as detailed below, LuxR-type proteins and other members of the FixJ-NarL superfamily may share several mechanistic features.

The role of LuxR-type proteins in responding to acyl-HSLs can be divided into several key steps: (1) recognition and binding of acyl-HSLs, (2) conformational changes and alterations in multimerization in response to binding of the pheromone, (3) binding or release of specific cis-acting sites proximal to target genes, and (4) for most LuxR-type proteins, transcriptional activation (Fig. 3). We will discuss each of these events in the approximate order in which they occur.

A. Perception of the Signal

I. LuxR-type proteins are acyl-HSL receptors

Based on genetic and biochemical experiments, the LuxR protein of V. fischeri, and several other LuxR homologues, are known to be acyl-HSL receptors. E. coli cells expressing either luxR or lasR of P. aeruginosa (aided by co-expression with the GroESL chaperone complex) bind significant amounts of their cognate pheromone (Adar and Ulitzur, 1993; Hanzelka and Greenberg, 1995; Pearson et al., 1997). Stable association of acyl-HSLs with cells expressing the genes for these proteins is thought to reflect binding to the receptor proteins. In support

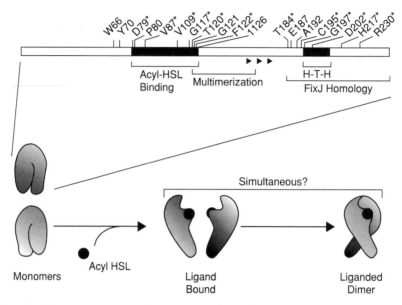

Fig. 3. Functional domain structure and activation pathway of LuxR-type proteins. A general LuxR-type protein is depicted by the open bar. Functional modules defined by mutational analyses or sequence homology are indicated by black fill or brackets. Amino acid residues conserved throughout the LuxR family of proteins are indicated in bold, and positions where mutations affect the proteins activity of LuxR itself are labeled with an asterisk (nonconserved residues implicated by mutation are not in bold). The three triangles indicate positions where deletion of the amino-terminal module results in acyl-HSL independence for LuxR and LasR. A tentative model for acyl-HSL interaction and formation of active multimers is indicated. The bracket labeled "simultaneous" indicates that the act of binding acyl-HSL and multimerization may not be separable events.

of this, mutations in *luxR* identified by a lack of response to 3-oxo-hexanoyl-HSL also result in reduced or abolished cell-association of the compound when expressed in *E. coli*. Expression of the N-terminus of LuxR (Res. 10–194) is sufficient to impart binding in this whole cell assay (Hanzelka and Greenberg, 1995). The most direct evidence for acyl-HSL binding to LuxR-type proteins is that purified preparations of TraR from *A. tumefaciens* and CarR from *Erwinia carotovora* bind their cognate acyl-HSLs (3-oxo-octanoyl-HSL and 3-oxo-hexanoyl-HSL) in an equimolar ratio of acyl-HSL to protein (Welch et al., 2000; Zhu and Winans, 1999).

2. Defining the acyl-HSL binding site

The *V. fischeri* LuxR protein has been functionally divided into two discrete regions (Fig. 3; and Stevens and Greenberg, 1999, for a detailed review). Roughly the amino-

terminal 70% of the protein is involved in interactions with 3-oxo-hexanoyl-HSL, while the carboxy-terminal region contains a helix-turn-helix motif (HTH) and is required for DNA binding. Mutations in *luxR* that abolish the cellular response to acyl-HSLs map between residues 79–127 and 184–230 (Shadel et al., 1990; Slock et al., 1990). Several mutations within the amino-terminal cluster (V82I and H127Y) that result in loss of response to the pheromone can be partially rescued by the addition of higher concentrations of 3-oxo-hexanoyl-HSL, suggesting that these reduce LuxR-acyl-HSL interaction (Shadel et al., 1990; Slock et al., 1990). A LuxR protein truncated by 55 codons at its carboxy-terminus (Res. 195–250) is sufficient for binding to 3-oxo-hexanoyl-HSL (Hanzelka and Greenberg, 1995). Taken together, these results suggest that the amino terminal region defines an acyl-HSL interaction site. As mentioned above, although the overall level of identity among LuxR-type proteins from different bacterial genera is low (18–25%), it is significantly stronger within the region corresponding to the LuxR acyl-HSL interaction site (Fig. 3). It seems likely that this region of other LuxR-type proteins is responsible for acyl-HSL interactions, but this has yet to be proved for any protein other than LuxR itself.

3. Interactions of the acyl-HSLs and LuxR-type proteins

The conformation of acyl-HSLs that are associated with LuxR-type proteins and the chemical interaction between the pheromone and its binding site are not well understood. Organic extraction of purified LuxR-type protein bound to acyl-HSLs and whole cells that have sequestered acyl-HSLs through interactions with LuxR-type proteins releases biologically active signal molecules, suggesting that binding to the receptor does not irreversibly alter the acyl-HSL chemistry (Adar et al., 1992; Hanzelka and Greenberg, 1995; Zhu and Winans, 1999).

Several studies have explored the interaction between acyl-HSLs and LuxR-type receptors by examining acyl-HSL analogues for activity (Chhabra et al., 1993; Eberhard et al., 1986; Passador et al., 1996; Schaefer et al., 1996; Zhu et al., 1998). In general, LuxR homologues recognize features of both the homoserine lactone moiety and the acyl chain. In all cases disruption of the homoserine lactone ring structure dramatically reduced the activity of the pheromone. Acyl-HSL analogues, which conserve the basic ring configuration (e.g., thiolactone and lactam derivatives) and have acyl side chains of similar length to the cognate acyl-HSL, retain strong inducing activity. However, alterations to the size and orientation of the ring completely abolish activity, and side chain lengths that differ greatly from the cognate acyl-HSL show dramatically reduced activity. Modifications at the β-position reduce, but do not abolish activity. Binding assays with the acyl-HSL analogues

reveal much the same pattern of activity (Passador et al., 1996; Schaefer et al., 1996). Although several potential competitive inhibitors have been identified that strongly bind LuxR-type proteins but do not induce well, the concentrations required for productive inhibition are often quite high. Elevated expression of the TraR protein of *A. tumefaciens* results in a reduction in specificity that is paralleled by a general recalcitrance to competitive inhibition, as most inhibitors themselves resulted in activation (Zhu et al., 1998). However, TraR expressed at normal levels was much more specific for its cognate acyl-HSL, and was effectively inhibited by acyl-HSL analogues. This observation suggests that analogue studies where the LuxR-type protein is strongly expressed may lead to erroneously poor inhibition results.

4. Subcellular localization of LuxR-type proteins

Analogous to steroid receptors LuxR-type acyl-HSL receptors are cytoplasmically localized. Neither LuxR nor any of its homologues possesses membrane-spanning sequences. However, it remains unclear whether the interaction with acyl-HSLs occurs free in the aqueous cytoplasm or is associated in some way with the membrane. LuxR itself is avidly associated with membranes in *V. fischeri* (Kolibachuk and Greenberg, 1993). As such, it has been proposed that LuxR is amphipathic and in intimate contact with the interior leaflet of the cytoplasmic membrane bilayer. Truncations of LuxR that remove the amino-terminal 156 residues are fully soluble and localized in the *E. coli* cytoplasm (Stevens et al., 1994). However, full-length protein or truncations retaining the amino-terminal domain are highly insoluble (Hanzelka and Greenberg, 1995). A plausible model is that the acyl-HSL interaction site on LuxR is associated with the membrane and interacts with acyl-HSLs as they traverse the membrane. This model is difficult to test, and it remains unclear whether other LuxR-type proteins share this property. However, there is precedent for such interactions, as perception of hydrophobic, plant-released flavonoid Nod factors by the NodD protein of *Rhizobium leguminosarum* has been proposed to occur via contact of this regulatory protein with the ligand in the membrane bilayer (Schlaman et al., 1989).

B. Conformational Changes in Response to acyl-HSL Binding

1. An inhibitory role for the amino-terminal region of LuxR-type proteins

It is clear that acyl-HSL interaction radically alters the activity of LuxR-type proteins, usually resulting in stimulation of DNA-binding activity. For LuxR itself and LasR from *P. aeruginosa*, truncated proteins that lack the acyl-HSL interaction region (LuxRΔN and LasRΔN, respectively) are constitutive transcriptional activators (Anderson et al., 1999; Choi and Greenberg, 1991; Pesci et al., 1999). The carboxy-terminal regions comprise constitutive activation modules, which can function independently when liberated from the amino-terminal portion of the protein. A model has been proposed in which the DNA-binding domains of LuxR and LasR, and by extension other LuxR-type proteins, are held in check by interactions with the amino-terminal region of each protein (Fig. 1). Interaction with the respective acyl-HSLs presumably unmasks the carboxy-terminal domain of the protein relieving inhibition. This is consistent with several other members of the NarL-FixJ superfamily. Similar amino-terminal truncations of the FixJ two-component response regulator from *Rhizobium meliloti* result in constitutive DNA binding and transcriptional activation (Da Re et al., 1994). Furthermore the three-dimensional structure of the NarL protein of *E. coli* suggests that the unphosphorylated amino-terminal domain would occlude the DNA binding segment of the protein, thereby preventing association with target promoters (Baikalov et al., 1996). These observations lead to a model

for LuxR-type proteins and similar members of the FixJ-NarL superfamily that interaction with the ligand (whether an acyl-HSL or a phosphoryl group) relieves inhibition by the amino-terminal half of the protein, and thereby affects the activity of the carboxy-terminal DNA-binding segment.

2. Multimerization of LuxR-type proteins

Several lines of evidence suggest that members of the LuxR family act as multimers. Truncated LuxR proteins lacking from 15–89 of the carboxy-terminal amino acid residues, act as dominant negative inhibitors of the wild type protein (Choi and Greenberg, 1992). These mutant deletion proteins lack some or all of the residues required for DNA binding, but retain the ability to interact with full length LuxR, presumably through formation of inactive heterodimers. Likewise several point mutations in *luxR* that result in nonfunctional LuxR proteins have a dominant inhibitory effect over the wild-type protein (Choi and Greenberg, 1992). In *A. tumefaciens* a homologue of TraR, called TraS, lacks the carboxy terminal 52 residues including the HTH sequence of wild-type TraR (Oger et al., 1998; Zhu and Winans, 1998). This natural deletion derivative acts in a dominant negative fashion over wild-type TraR (Zhu and Winans, 1998). Although these genetic studies suggest that LuxR and TraR form multimers, they do not address whether multimerization is affected by acyl-HSL interaction, nor do they reveal a stoichiometry for the complex. Biochemical experiments with purified TraR and CarR from *E. carotovora* indicate that protein dimers are proficient for DNA binding (Zhu and Winans, 2001; Welch, et al., 2000). This biochemical evidence is consistent with the dyad symmetry of the DNA binding sites for many LuxR-type proteins (*lux*-type boxes, see below) (Fuqua and Winans, 1996).

A common mode of receptor activation is ligand-induced multimerization, and it is plausible that acyl-HSLs stimulate multimerization of LuxR-type proteins. In vitro

studies of TraR suggest that acyl-HSL interaction promotes dimerization, consistent
with such a model (S. C. Winans, personal
communication). Interestingly the CarR protein of *E. carotovora* exists as a preformed
dimer in the absence of acyl-HSL that is
converted to a higher order multimer(s) in
response to acyl-HSL addition (Welch et al.,
2000). In both cases protein dimers are proficient for DNA binding, and acyl-HSL
interaction enhances multimer formation.
However, these observations are not consistent with the finding that the LuxRΔN constitutive activator, consisting of only the
carboxy-terminal DNA-binding module, is
monomeric in solution (Stevens et al.,
1994). If the constitutive activity of LuxRΔN
truly reflects the normal activation mechanism of full-length LuxR, multimerization is
not required or induced by acyl-HSL binding. However, interpretation of the LuxRΔN
studies is complicated by the observation
that LuxR and LuxRΔN require substantially different contacts with RNA polymerase during transcriptional activation (A.
Stevens, personal communication). Furthermore it is not clear whether other members
of the LuxR family share the identical mechanism of activation as LuxR. The integration
between ligand-induced activation and multimerization remains a subject of intense investigation.

C. Control of Target Gene Expression

1. cis-Acting sequences for acyl-HSL regulated genes

As with other transcriptional regulators,
thus far all members of the LuxR family
require DNA sequences associated with their
target genes to activate transcription. The
DNA sequence element required for activation of *lux* genes was historically called
the *lux* operator, and more recently the *lux*
box (Devine et al., 1989; Gray et al., 1994).
The *lux* box is a 20 base-pair inverted repeat
centered at pos. −42.5 relative to the *lux*
operon transcriptional start site (Egland
and Greenberg, 1999). Mutational analysis

of the *lux* box suggests that transcriptional
activation by LuxR is strictly dependent on
its position relative to the *lux* promoter and
that both arms of the dyad repeat are required (Devine et al., 1989; Egland and
Greenberg, 1999).

Inspection of several promoters regulated
by LuxR-type proteins from different bacteria has revealed the presence of inverted
repeat sequences ranging from 18 to 22 bp
in length, with primary sequence similarity
with the *V. fischeri lux* box (Fuqua et al.,
1996). These *lux*-type boxes are often
centered just upstream of the −35 promoter
element. Full expression of the *P. aeruginosa
lasB* gene requires two 20 bp *lux*-type boxes,
one centered at −42 (proximal to the −35
element), and a second further upstream at
pos. −102 (Anderson, et al., 1999; Rust et al.,
1996). The mechanism by which the upstream site contributes is not yet known.
Examination of the upstream regions of
quorum sensing controlled (*qsc*) genes in *P.
aeruginosa* has revealed 14/39 genes with
likely *lux*-type boxes, adding to previously
identified *las*-regulated genes (Pesci and
Iglewski, 1999; Whiteley et al., 1999). In *A.
tumefaciens* there are several 18 bp *lux*-type
boxes required to activate expression of at
least three different target operons by TraR
(Fuqua and Winans, 1996). One of these
elements is positioned in a 16 bp gap between
the −35 elements of a pair of divergent *tra*
promoters, overlapping each −35 element by
one base pair, and is required to activate
expression from either promoter. A similar
lux-type box that overlaps the −35 element
of a third TraR-regulated promoter is also
required for activation of downstream genes
(Fuqua and Winans, 1996).

Although *lux*-type boxes appear to be
common elements upstream of target
operons, there are several examples of promoters for which these elements are not dis-
cernable but that are clearly regulated by
LuxR-type regulators. In *A. tumefaciens*,
the *traM* and *traR* genes lack canonical *lux*-
type boxes but are activated by TraR (Fuqua
et al., 1995; Fuqua and Winans, 1996). Both

of these genes are activated less strongly than TraR-regulated genes with the canonical *lux*-type boxes (C. Fuqua, manuscript in preparation). In *P. aeruginosa* the *lasI* gene is strongly activated by LasR, but has a *lux*-type box that is only a weak match with other LasR-dependent elements, and is located unusually close to the *lasI* transcriptional start site (Pesci and Iglewski, 1999; Seed et al., 1995). These examples point out the conservation of the *lux*-type sequence elements while also exemplifying some of the diversity of promoter architecture of quorum sensing-regulated promoters within the same bacterium and among different microbes.

2. Binding of DNA by LuxR-type proteins

The LuxRΔN deletion derivative mentioned above lacks the amino-terminal 156 amino acid residues, exhibits acyl-HSL-independent transcriptional activation, and is soluble when purified from *E. coli* (Choi and Greenberg, 1991; Stevens et al., 1994). Specific in vitro binding of the *lux* box by LuxRΔN requires the presence of RNA Polymerase (RNAP) (Stevens et al., 1994). Neither protein binds the *lux* box independently, although LuxRΔN alone binds weakly to DNA further upstream. It is not yet possible to assess the contribution of each protein to DNA binding, nor is it clear whether the requirement for RNAP is shared by full-length LuxR.

Several other LuxR-type proteins have now been examined in vitro. In all cases the acyl-HSL increases or modifies the DNA binding interaction of the protein. At least two LuxR-type proteins, LasR from *P. aeruginosa* and CarR from *E. carotovora*, are reported to bind to promoter elements in the absence of acyl-HSL (Welch et al., 2000; You et al., 1996). In both cases addition of acyl-HSL altered the interactions with the DNA element, increasing affinity and resulting in higher-order complexes that were likely to be multimers. Likewise TraR from *A. tumefaciens*, purified as a complex with its cognate acyl-HSL, binds quite

specifically to a single genetically defined *lux*-type box (Zhu and Winans, 1999). Subsequent removal of the acyl-HSL by detergent treatment reduces the DNA-binding activity of TraR, and this can be partially restored by addition of acyl-HSL to this preparation.

Consistent with those LuxR-type proteins described above, the ExpR$_{Ech}$ protein of *Erwinia chrysanthemi* binds to a number of potential target promoters in the absence of its cognate acyl-HSL (Nasser et al., 1998). Addition of the acyl-HSL alters the contacts of ExpR$_{Ech}$ with these promoter sequences. However, several LuxR proteins, including ExpR$_{Ech}$ (but excluding CarR) from species of *Erwinia*, are likely to be repressors rather than transcriptional activators (Andersson et al., 2000; Beck von Bodman et al., 1998). LuxR-type proteins that act as repressors probably have significant differences with LuxR-type transcriptional activators and mechanistic features may not be consistent (see below).

3. Transcriptional control

Most LuxR-type proteins activate expression of their target genes. Null mutations in genes encoding LuxR-type proteins generally result in loss of target gene expression. Likewise LuxRΔN and TraR have been shown to be sufficient to activate target gene transcription in vitro, with purified DNA templates and RNAP (Stevens and Greenberg, 1997; Zhu and Winans, 1999). Deletion analysis of LuxR suggested that mutations within the region at the very carboxy-terminus of the protein resulted in LuxR proteins that remained proficient for DNA binding but could not activate transcription of the *luxI* gene (i.e., positive control or PC mutants) (Choi and Greenberg, 1992). However, much of this work assessed DNA binding proficiency by testing for autorepression of *luxR* expression, which occurs at noninducing concentrations of 3-oxo-C6-HSL (Shadel and Baldwin, 1992). More recent work has called into question the role of the LuxR C-terminus in activation, but the actual sequences required for positive control have not been identified (Egland and Greenberg, 2000). Mutant TraR proteins with the PC phenotype have in fact mutations that map in the amino-terminus of the protein (Luo and Farrand, 1999). It is possible that the discrepancies between LuxR and TraR reflect true mechanistic differences.

The position of most *lux*-type boxes relative to their promoter sequences suggests that LuxR-type proteins interact directly with RNAP (Fuqua et al., 1996). As with other regulated promoters, transcriptional activation likely occurs through contacts with RNAP, and the location of these binding sites implicates the carboxy terminal domain of the RNAP alpha subunit (alphaNTD) or the sigma subunit, or both (Rhodius and Busby, 1998). The positional dependence of the *lux* box itself is consistent with an ambidextrous activator that contacts both of these control sites simultaneously (Egland and Greenberg, 1999).

4. LuxR-type proteins that function as repressors

In different species and subspecies of *Erwinia*, mounting evidence suggests that several LuxR-type proteins act as transcriptional repressors (Andersson et al., 2000; Beck von Bodman and Farrand, 1995; Beck von Bodman et al., 1998). Perhaps the most compelling evidence exists for the EsaR protein of *E. stewartii*, which clearly exerts negative control on its target genes (Beck von Bodman et al., 1998). Addition of acyl-HSL relieves repression of the target genes. A model is suggested where EsaR, and other repressors in the LuxR family, bind to DNA target sequences in the absence of ligand, thereby sterically blocking access of RNAP to the promoter. Interaction with the acyl-HSL abolishes or alters DNA binding, and the repressed genes are expressed. Clearly, while acyl-HSL binding may be similar to LuxR-type activators, the molecular consequences of ligand interaction must be dramatically different for the repressors. Sequence comparisons of all LuxR-type proteins do not reveal any striking differences between members of the LuxR family that function as activators and those that are repressors (Stevens and Greenberg, 1999).

VII. CONTROL OF *luxR* AND *luxI* HOMOLOGUE GENE EXPRESSION

Most, if not all, acyl-HSL quorum sensors are embedded into additional regulatory circuitry, expanding the range of environmental signals that affect target gene expression beyond population density. This is most commonly through regulation of the gene encoding the LuxR-type regulator, but it also occurs by regulating expression of the *luxI*-type acyl-HSL synthase gene (Fig. 4).

A. Expression and Stability of *luxR* Homologues

Quorum sensing is often integrated into other regulatory pathways through control of the expression or stability of the LuxR-type protein. The expression of the *V. fischeri luxR* gene itself is regulated by the cyclic AMP receptor protein (CRP), thereby placing the Lux quorum sensor under the influence of the CRP pathway (Dunlap and Greenberg, 1988). Another clear example of such upstream regulation is control of the gene encoding TraR, where expression is strictly dependent on opines, compounds produced by plants infected with *A. tumefaciens* (Fuqua and Winans, 1996; Piper et al., 1999). Interestingly expression of the *lasR* gene in *P. aeruginosa* is affected by a CRP homologue, called Vfr (Albus et al., 1997). Expression of *lasR* is also influenced by the GacA two-component response regulator, a common regulator of virulence in *P. aeruginosa* (Reimmann et al., 1997). The combined effect of these two pathways on *lasR* expression is not yet understood. An intriguing variation on the regulation of *luxR*-type gene expression is the control of *rhlR*, a second *P. aeruginosa* LuxR-type protein responsive to butyryl-HSL, by the LasR

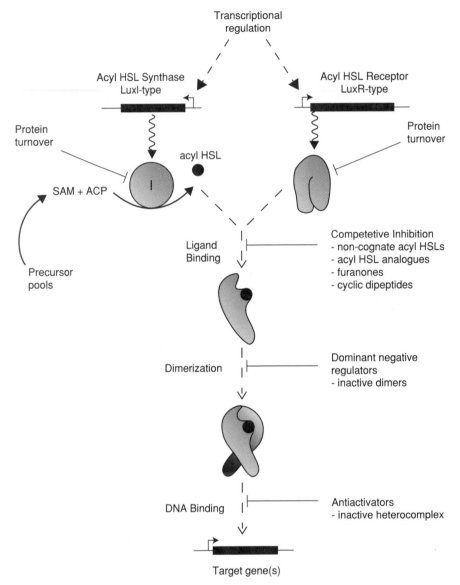

Transcriptional
regulation

Acyl HSL Synthase
LuxI-type

Acyl HSL Receptor
LuxR-type

Protein
turnover

Protein
turnover

acyl HSL

SAM + ACP

Ligand
Binding

Competetive Inhibition
- non-cognate acyl HSLs
- acyl HSL analogues
- furanones
- cyclic dipeptides

Precursor
pools

Dimerization

Dominant negative
regulators
- inactive dimers

DNA Binding

Antiactivators
- inactive heterocomplex

Target gene(s)

Fig. 4. Modulation of quorum sensing. Dashed arrows trace the primary path to transcriptional control by LuxR-type proteins and acyl-HSL. The different points in the signal transduction pathway where other factors can influence quorum sensing are indicated with text. Squiggles extending from *luxR*- and *luxI*-type genes represent transcription and translation. Stalked arrows indicate promoters. Lines with end bars are inhibitory or negative for activation of target genes.

protein and 3-oxo-dodecanoyl-HSL (Latifi et al., 1996; Pesci et al., 1997). Therefore the Las and Rhl quorum sensors, while in many ways discrete systems, compose a hierarchical signal cascade. Multiple quorum sensing systems in the same microbe are becoming more common as these regulatory systems are deliberately sought after as well as through whole genome sequencing projects.

The half-lives of mRNA transcripts encoding LuxR-type proteins, and the stability of the proteins themselves, will also affect the cellular pools of these receptor proteins and thus responsiveness to acyl-HSLs (Fig. 4). While there is currently no information addressing the stability of transcripts encoding *luxR*-type proteins, there is evidence that interaction with the acyl-HSL ligand alters the proteolytic degradation of the CarR and TraR proteins (Welch et al., 2000; Zhu and Winans, 1999). Intracellular pools of TraR are stabilized by binding to the acyl-HSL. Therefore changes in the overall proteolytic activity within the cell can affect the half-life of these LuxR-type proteins, and perhaps modulate the response to increasing levels of acyl-HSL.

B. Expression of LuxI-type acyl-HSL Synthase Genes

Acyl-HSL synthase genes can also be directly controlled by upstream regulatory systems that affect expression of the genes or the half-lives of the mRNA transcript (Fig. 4). As described above for *P. aeruginosa*, the GacAS two-component regulatory systems is required for quorum sensing control of phenazine antibiotic synthesis in the plant biocontrol agent *Pseudomonas aureofaciens* (Chansey et al., 1999). However, in this case GacA activates the *phzI* gene, elevating synthesis of hexanoyl-HSL. In *P. aeruginosa* the acyl-HSL synthase *lasI* is regulated by an inhibitory protein called RsaL that specifically inhibits transcription of the *lasI* gene (De Kievit et al., 1999). Recent studies have also revealed a negative role for the *P. aeruginosa* stationary phase sigma factor RpoS in regulating *rhlI* expression (Whiteley et al., 2000). In addition to these specific examples, LuxR-type proteins often activate expression of their cognate acyl-HSL synthase counterpart, establishing a positive feedback loop (Fuqua and Greenberg, 1998). Although this positive feedback is not required for cell-density responsiveness, it is likely involved in buffering the expression of the target genes from fluctuations in acyl-HSL concentration.

As with the LuxR-type proteins, changes in turnover rates of acyl-HSL synthase mRNA transcripts and proteins can conceivably alter quorum sensing, by altering the amount of acyl-HSL synthesized. In *Erwinia carotovora* subsp. *carotovora* the *rsmA* gene product destabilizes the mRNA encoding the HslI protein responsible for synthesizing 3-oxo-hexanoyl-HSL, and thereby reduces production of the pheromone (Cui et al., 1995). Homologues of RsmA are widespread in bacteria, and it is possible that regulated stability of acyl-HSL synthase transcripts is a common feature (White et al., 1996).

VIII. MODULATION OF THE QUORUM SENSING MECHANISM BY ADDITIONAL FACTORS

A number of diffusible antagonists as well as other regulatory proteins can impinge upon the function of acyl-HSL quorum sensors through interactions with LuxR-type proteins. Several such factors, produced by the bacteria themselves or host organisms, have now been identified.

A. Modulatory Signals of Bacterial Origin

LuxR-type proteins can recognize molecules other than their cognate acyl-HSL(s). Clearly, noncognate acyl-HSLs can function as effective competitors for the correct signal, even as they serve as weak inducers. The best example of this is *V. fischeri* where octanoyl-HSL is produced by the AinS synthase (Section VIIC; and Hanzelka et. al., 1999). *V. fischeri* mutants that cannot synthesize octanoyl-HSL induce bioluminescence at lower population densities than the wild-type cells (Kuo et al., 1996). Octanoyl-HSL is thought to compete with the cognate 3-oxo-hexanoyl-HSL for its binding site on LuxR, thereby necessitating higher concentrations of the inducing pheromone to activate *lux* gene expression. It is unclear how many other systems employ additional acyl-HSLs to modulate the activity of LuxR type proteins. Many acyl-HSL synthases produce secondary acyl-HSLs, and it is plausible that these side products act to modulate the activity of the cognate LuxR-type protein (Fuqua and Eberhard, 1999; Winson et al., 1995). An intriguing extension of this general idea is that acyl-HSLs produced by different, but neighboring, bacteria can impinge upon the function of LuxR-type proteins in different types of bacteria, via cell-to-cell cross talk. In several cases cross talk among co-resident bacteria has been convincingly demonstrated (McKenney et al., 1995; Wood et al., 1997).

There are examples of non-acyl-HSL molecules interacting with LuxR proteins and affecting their function. Several bacteria have been shown to produce cyclic dipeptides (diketopiperazines) that can act as weak inducers of LuxR-type acyl-HSL receptors (Holden et al., 1999). As with noncognate acyl-HSLs, the weakly inducing cyclic dipeptides act as inhibitors of LuxR-acyl-HSL interactions. The concentration of cyclic dipeptides required for recognition by LuxR-type proteins is significantly higher than that for the cognate acyl-HSLs and a physiological role of these compounds in signaling has not been established.

A diffusible signal molecule, 2-heptyl-3-hydroxy-4-quinolone (called the *P. aeruginosa* quinolone signal, or PQS), is integrated with the Las and Rhl quorum sensing systems of *P. aeruginosa* (Pesci et al., 1999). PQS increases expression of the *rhlR* gene, and thereby indirectly affects the expression of *P. aeruginosa* quorum sensing target genes (McKnight et al., 2000). Synthesis of the PQS signal is dependent on the LasR transcription factor, although it remains uncertain whether this molecule acts in addition to or through the LasR-dependent control of *rhlR*. PQS concentrations in wild-type *P. aeruginosa* increase during late stationary phase, and it is speculated that PQS may be involved in elevating RhlR-dependent gene expression during slow growth or starvation.

B. Bacterial Proteins That Modulate Quorum Sensing

LuxR-LuxI-type regulatory proteins are sufficient to impart population-density respon-sive gene regulation in almost all quorum sensing systems that have been characterized. However, there are several examples of additional regulatory proteins that exert direct and profound influence on the quorum sensing mechanism (Fig. 4). In *A. tumefaciens* there are two separate mechanisms by which TraR-dependent transcriptional activation is modulated, both of which likely function through formation of inactive heterocomplexes. TraS is a homologue of TraR that is highly similar to the amino-terminal 181 amino acid residues of TraR but lacks the DNA binding domain (Oger et al., 1998; Zhu and Winans, 1998). TraS has a dominant negative effect on TraR, and expression of the *traS* gene is regulated by the plant tumor–released opines, thereby adjusting TraR activity in response to the specific opine composition of the environment (Fig. 4). Likewise a second inhibitory protein called TraM forms inactive heteromultimers with TraR and prevents it from activating transcription (Fig. 4; and Fuqua et al., 1995; Hwang et al., 1999; Luo et al., 2000). TraM is a small, 102 amino acid protein, and it shares no clear sequence homology with TraR. In contrast to TraS, the TraM antiactivator is required to keep the TraR protein from stimulating gene expression under noninducing conditions, and it can inhibit TraR that is in the active, acyl-HSL associated form (Luo et al., 2000; Swiderska et al., manuscript in preparation).

C. acyl-HSL Interference by Eukaryotic Signal Molecules

1. Overview

The marine red alga *Delisea pulchra* produces a family of halogenated compounds, known as furanones, that are structurally similar to acyl-HSLs. The furanones are effective antifouling agents, and they prevent the association of microorganisms as well as metazoan colonizers (de Nys et al., 1995; Reichelt and Borowitzka, 1984). At least one mechanism for furanone inhibition of microbial colonization acts through inhibition of acyl-HSL quorum sensing (Givskov et al., 1996). The furanones are thought to compete for binding sites on the LuxR-type receptor proteins. In fact, addition of a subset of the furanones at physiologically relevant concentrations to whole cell acyl-HSL binding assays with *E. coli* expressing the LuxR protein, effectively displaces the bound acyl-HSLs (Manefield et al., 1999). Although the exact mechanism of action for the furanones remains an area of intense investigation, the use of these natural compounds as well as synthetic derivatives as antifouling materials is already underway. Other metazoan organisms are likely to produce modulators of quorum sensing to control their associated bacterial populations. Recently inhibitors of quorum sensing have been identified in several varieties of pea and a number of other higher plants (Teplitski et al., 2000). Although the chemical nature of these presumptive signal molecules is not known, they are not chemically similar to known acyl-HSLs or furanones. Signal interference and possibly even signal augmentation by host organisms is an area of great interest, and there are likely many examples of this type of interkingdom communication that await discovery.

2. acyl-HSLs as immunoactive compounds

A highly specialized example of host-microbe interactions mediated directly through acyl-HSLs is the response of the mammalian immune system. Although the details remain to be elucidated, there is evidence that the 3-oxo-dodecanoyl-HSL pheromone of *P. aeruginosa* possesses immunomodulatory activity in animal infection models (Dimango et al., 1995; Telford et al., 1998). It has been suggested that the immune system interaction exhibited by 3-oxo-dodecanoyl-HSL is evidence that this signal molecule can act as a virulence factor, although it is far from clear that this response benefits the infecting bacterium. Indeed, it seems just as likely that the immune recognition of acyl-HSLs is another mechanism employed by the mammalian immune system to rid itself of bacterial

pathogens. No other acyl-HSLs have been reported to directly affect mammalian immune responses.

IX. SUMMARY

Much progress has been made toward an understanding of the molecular details of acyl-HSL quorum sensing in the last 15 years. However, a number of questions still remain to be answered. Many of these questions may be answered through the study of quorum sensing in novel systems. Each quorum sensing system studied to date has revealed unique features tailored to its host organism. Understanding how quorum sensing is integrated with other regulatory circuits in the cell could be facilitated by genomics-based research, and this should help shed light as to how quorum sensing serves a given species in its particular ecological context.

Among the obvious questions that remain are: What is the extent of the structural variation of acyl-HSLs used as signals? Is quorum sensing shut off at any point after a quorum is reached? How does growth rate affect acyl-HSL synthesis? In mixed bacterial communities, do acyl-HSLs modulate gene expression in heterologous bacterial species? How does the quorum sensing mechanism work in nonaqueous environments, such as soil? Answers to questions such as these will continue to address our understanding of the molecular mechanisms of quorum sensing and their relationship to the ecology of the organisms that use such systems.

REFERENCES

Adar YY, Simaan M, Ulitzur S (1992): Formation of the LuxR protein in the *Vibrio fischeri lux* system is controlled by HtpR through the GroESL proteins. J Bacteriol 174:7138–7143.

Adar YY, Ulitzur S (1993): GroESL proteins facilitate binding of externally added inducer by LuxR protein-containing *Escherichia coli* cells. J Biolumin Chemilumin 8:261–266.

Albus AM, Pesci EC, Runyen-Janecky LJ, West SEH, Iglewski BH (1997): Vfr controls quorum sensing in *Pseudomonas aeruginosa*. J Bacteriol 179:3928–3935.

Anderson RM, Zimprich CA, Rust L (1999): A second operator is involved in *Pseudomonas aeruginosa* elastase (*lasB*) activation. J Bacteriol 181:6264–6270.

Andersson RA, Eriksson ARB, Heikinheimo R, Mae A, Pirhonen M, Koiv V, Hyytiainen H, Tuikkala A, Palva ET (2000): Quorum-sensing in the plant pathogen *Erwinia carotovora* subsp. *carotovora*: The role of expR$_{Ecc}$. Mol Plant Micro Interact 13:384–393.

Baikalov I, Schroder I, Kaczor-Grzekowiak M, Grzekowiak K, Gunsalus RP, Dickerson RE (1996): Structure of the *Escherichia coli* response regulator NarL. Biochem 35:11053–11061.

Bainton NJ, Stead P, Chhabra SR, Bycroft BW, Salmond GPC, Stewart GSAB, Williams P (1992): *N*-(3-oxohexanoyl)-L-homoserine lactone regulates carbapenem antibiotic production in *Erwinia carotovora*. Biochem J 288:997–1004.

Bassler BL (1999): How bacteria talk to each other: Regulation of gene expression by quorum sensing. Curr Opin Microbiol 2:582–587.

Bassler BL, Wright M, Showalter RE, Silverman MR (1993): Intercellular signalling in *Vibrio harveyi*: Sequence and function of genes regulating expression of luminescence. Mol Microbiol 9:773–786.

Beck von Bodman S, Farrand SK (1995): Capsular polysaccharide biosynthesis and pathogenicity in *Erwinia stewartii* require induction by an *N*-acylhomoserine lactone autoinducer. J Bacteriol 177:5000–5008.

Beck von Bodman S, Majerczak DR, Coplin DL (1998): A negative regulator mediates quorum-sensing control of exopolysaccharide production in *Pantoea stewartii* subsp. *stewartii*. Proc Natl Acad Sci USA 95:7687–7692.

Chansey ST, Wood DW, Pierson LS III (1999): Two-component transcriptional regulation of *N*-acyl-homoserine lactone production in *Pseudomonas aureofaciens*. Appl Environ Microbiol 65:2585–2591.

Chhabra SRP, Stead P, Bainton NJ, Salmond GPC, Stewart GSAB, Williams P, Bycroft BW (1993): Autoregulation of carbapenem biosynthesis in *Erwinia carotovora* by analogues of *N*-(3-oxohexanoyl)-L-homoserine lactone. J Antibiot 46:441–545.

Choi SH, Greenberg EP (1991): The C-terminal region of the *Vibrio fischeri* LuxR protein contains an inducer-independent *lux* gene activating domain. Proc Natl Acad Sci USA 88:11115–11119.

Choi SH, Greenberg EP (1992): Genetic evidence for multimerization of LuxR, the transcriptional activator of *Vibrio fischeri* luminescence. Mol Mar Biol Biotech 1:408–413.

Choi SH, Greenberg EP (1992): Genetic dissection of DNA binding and luminescence gene activation by the *Vibrio fischeri* LuxR protein. J Bacteriol 174:4064–4069.

Costerton JW, Stewart PS, Greenberg EP (1999): Bacterial biofilms: A common cause of persistent infections. Science 284:1318–1322.

Cui Y, Chatterjee A, Liu Y, Dumenyo CK, Chatterjee AK (1995): Identification of a global repressor gene *rsmA*, of *Erwinia carotovora* subsp. *carotovora* that controls extracellular enzymes, *N*-(3-oxohexanoyl)-L-homoserine lactone, and pathogenicity in soft-rotting *Erwinia* spp. J Bacteriol 177:5108–5115.

Da Re S, Bertagnoli S, Fourment J, Reyrat J-M, Kahn D (1994): Intramolecular signal transduction within the FixJ transcriptional activator: In vitro evidence for the inhibitory effect of the phosphorylatable regulatory domain. Nucleic Acids Res 9:1555–1561.

D'Agnolo G, Rosenfeld IS, Vagelos PR (1973): β-Ketoacyl-acyl carrier protein synthase: Characterization of the acyl-enzyme intermediate. Biochim Biophys Acta 326:155–166.

De Kievit. T, Seed PC, Nezezon J, Passador L, Iglewski BA (1999): RsaL, a novel repressor of virulence gene expression in *Pseudomonas aeruginosa*. J Bacteriol 181:2175–2184.

de Nys R, Steinberg PD, Willemsen P, Dworjanyn SA, Gabelish CL, King RJ (1995): Broad spectrum effects of secondary metabolites from the red alga *Delisea pulchra* in antifouling assays. Biofouling 8:259–271.

Devine JH, Shadel GS, Baldwin TO (1989): Identification of the operator of the *lux* regulon from the *Vibrio fischeri* strain ATCC7744. Proc Natl Acad Sci USA 86:5688–5692.

Dimango E, Zar HJ, Bryan R, Prince A (1995): Diverse *Pseudomonas aeruginosa* gene products stimulate respiratory epithelial cells to produce interleukin-8. J Clin Invest 5:2204–2210.

Dunlap PV, Greenberg EP (1988): Analysis of the mechanism of *Vibrio fischeri* luminescence gene regulation by cyclic AMP and cyclic AMP receptor protein in *Escherichia coli*. J Bacteriol 170:4040–4046.

Dunny GM, Leonard BA (1997): Cell-cell communication in Gram-positive bacteria. Annu Rev Microbiol 51:527–564.

Eberhard A, Burlingame AL, Eberhard C, Kenyon GL, Nealson KH, Oppenheimer NJ (1981): Structural identification of autoinducer of *Photobacterium fischeri* luciferase. Biochem 20:2444–2449.

Eberhard A, Longin T, Widrig CA, Stranick SJ (1991): Synthesis of the *lux* gene autoinducer in *Vibrio fischeri* is positively autoregulated. Arch Microbiol 155:294–297.

Eberhard A, Widrig CA, McBath P, Schineller JB (1986): Analogs of the autoinducer of bioluminescence in *Vibrio fischeri*. Arch Microbiol 146:35–40.

Egland KA, Greenberg EP (1999): Quorum sensing in *Vibrio fischeri*: elements of the *luxI* promoter. Mol Microbiol 31:1197–1204.

Egland KA, Greenberg EP (2000): Conversion of the *Vibrio fischeri* transcriptional activator, LuxR, to a repressor. J Bacteriol 182:805–811.

Engebrecht J, Nealson KH, Silverman M (1983): Bacterial bioluminescence: isolation and genetic analysis of the functions from *Vibrio fischeri*. Cell 32:773–781.

Evans K, Passador L, Srikumar R, Tsang E, Nezezon J, Poole K (1998): Influence of the MexAB-OprM multidrug efflux system on quorum sensing in *Pseudomonas aeruginosa*. J Bacteriol 180:5443–5447.

Fuqua C, Burbea M, Winans SC (1995): Activity of the *Agrobacterium* Ti plasmid conjugal transfer regulator TraR is inhibited by the product of the *traM* gene. J Bacteriol 177:1367–1373.

Fuqua C, Eberhard A (1999): Signal generation in autoinduction systems: Synthesis of acylated homoserine lactones by LuxI-type proteins. In Winans SC, Dunny GM (eds): "Cell-Cell Signaling in Bacteria." Washington, DC: ASM Press, pp 211–230.

Fuqua C, Greenberg EP (1998): Self perception in bacteria: Quorum sensing with acylated homoserine lactones. Curr Opin Microbiol 1:183–189.

Fuqua C, Winans SC (1996): Localization of the OccR-activated and TraR-activated promoters that express two ABC-type permeases and the *traR* gene of the Ti plasmid pTiR10. Mol Microbiol 120:1199–1210.

Fuqua C, Winans SC (1996): Conserved *cis*-acting promoter elements are required for density-dependent transcription of *Agrobacterium tumefaciens* conjugal transfer genes. J Bacteriol 178:435–440.

Fuqua WC, Winans SC, Greenberg EP (1994): Quorum sensing in bacteria: The LuxR/LuxI family of cell density-responsive transcriptional regulators. J Bacteriol 176:269–275.

Fuqua WC, Winans SC, Greenberg EP (1996): Census and consensus in bacterial ecosystems: the LuxR-LuxI family of quorum-sensing transcriptional regulators. Annu Rev Microbiol 50:727–751.

Gambello MJ, Iglewski BH (1991): Cloning and characterization of the *Pseudomonas aeruginosa lasR* gene, a transcriptional activator of elastase expression. J Bacteriol 173:3000–3009.

Garwin JL, Klages AL, J.E. Cronan J Jr. (1980): Beta-ketoacyl-acyl carrier protein synthase II of *Escherichia coli*: Evidence for function in the thermal regulation of fatty acid synthesis. J Biol Chem 255:3263–3265.

Gilson L, Kuo A, Dunlap PV (1995): AinS and a new family of autoinducer synthesis proteins. J Bacteriol 177:6946–6951.

Givskov M, de Nys R, Manefield M, Gram L, Maximilien R, Eberl L, Molin S, Steinberg PD, Kjelleberg S (1996): Eukaryotic interference with homoserine lactone-mediated prokaryotic signalling. J Bacteriol 178:6618–6622.

Graf J, Ruby EG (1998): Host-derived amino acids support the proliferation of symbiotic bacteria. Proc Natl Acad Sci USA 95:1818–1822.

Gray KM, Passador L, Iglewski BH, Greenberg EP (1994): Interchangeability and specificity of components from the quorum-sensing regulatory systems of *Vibrio fischeri* and *Pseudomonas aeruginosa*. J Bacteriol 176:3076–3080.

Hanzelka BL, Greenberg EP (1995): Evidence that the N-terminal region of the *Vibrio fischeri* LuxR protein constitutes an autoinducer-binding domain. J Bacteriol 177:815–817.

Hanzelka BL, Greenberg EP (1996): Quorum sensing in *Vibrio fischeri*: Evidence that *S*-adenosylmethionine is the amino acid substrate for autoinducer synthesis. J Bacteriol 178:5291–5294.

Hanzelka BL, Parsek MR, Val DL, Dunlap PV, Cronan JEJ, Greenberg EP (1999): Acylhomoserine lactone synthase activity of the *Vibrio fischeri* AinS protein. J Bacteriol 181:5766–5770.

Hanzelka BL, Stevens AM, Parsek MR, Crone TJ, Greenberg EP (1997): Mutational analysis of the *Vibrio fischeri* LuxI polypeptide: Critical regions of an autoinducer synthase. J Bacteriol 179:4882–4887.

Holden MTG, Chhabra SR, de Nys R, Stead P, Bainton NJ, Hill PJ, Manfield M, Kumar N, Labatte M, England D, Rice S, Givskov M, Salmond GPC, Stewart GSAB, Bycroft BW, Kjelleberg S, Williams P (1999): Quorum-sensing crosstalk: Isolation and chemical characterization of cyclic dipeptides from *Pseudomonas aeruginosa* and other gram-negative bacteria. Mol Microbiol 33:1254–1266.

Hwang I, Smyth AJ, Luo Z-Q, Farrand SK (1999): Modulating quorum sensing by antiactivation: TraM interacts with TraR to inhibit activation of Ti plasmid conjugal transfer genes. Mol Microbiol 34:282–294.

James S, Nilsson P, James G, Kjelleberg S, Fagerstrom T (2000): Luminescence control in the marine bacterium *Vibrio fischeri*: An analysis of the dynamics of *lux*regulation. J Mol Biol 296:1127–1137.

Jiang Y, Camara M, Chhabra SR, Hardie KR, Bycroft BW, Lazdunski A, Salmond GPC, Stewart GSAB, Williams P (1998): In vitro biosynthesis of the *Pseudomonas aeruginosa* quorum-sensing signal molecule, *N*-butanoyl-L-homoserine lactone. Mol MIcrobiol 28:193–204.

Kahn D, Ditta G (1991): Modular structure of FixJ: Homology of the transcriptional activation domain with the −35 binding domain of sigma factors. Mol Microbiol 5:987–997.

Kaplan HB, Greenberg EP (1985): Diffusion of autoinducer is involved in regulation of the *Vibrio fischeri* luminescence system. J Bacteriol 163:1210–1214.

Kolibachuk D, Greenberg EP (1993): The *Vibrio fischeri* luminescence gene activator LuxR is a membrane-associated protein. J Bacteriol 175:7307–7312.

Kuo A, Callaghan SM, Dunlap PV (1996): Modulation of luminescence operon expression by *N*-octanoyl-L-homoserine lactone in *ainS* mutants of *Vibrio fischeri*. J Bacteriol 178:971–976.

Latifi A, Foglino M, Tanaka K, Williams P, Lazdunski A (1996): A hierarchical quorum-sensing cascade in *Pseudomonas aeruginosa* links the transcriptional activators LasR and RhlR (VsmR) to expression of the stationary-phase sigma factor RpoS. Mol Microbiol 21:1137–1146.

Luo Z-Q, Farrand SK (1999): Signal-dependent DNA binding and functional domains of the quorum-sensing activator TraR as identified by repressor activity. Proc Natl Acad Sci USA 96:9009–9014.

Luo Z-Q, Qin Y, Farrand SK (2000): The antiactivator TraM interferes with the autoinducer-dependent binding of TraR to DNA by interacting with the C-terminal region of the quorum-sensing activator. J Biol Chem 275:7713–7722.

Magnuson K, Jackowski S, Rock CO, JE Cronan J (1993): Regulation of fatty acid biosynthesis in Escherichia coli. Microbiol Rev 57:522–542.

Manefield M, de Nys R, Kumar N, Read R, Givskov M, Steinberg P, Kjelleberg S (1999): Evidence that halogenated furanones from Delisea pulchra inhibit acylated homoserine lactone (AHL)-mediated gene expression by displacing the AHL signal from its receptor protein. Microbiol 145:283–291.

McFall-Ngai MJ, Ruby EG (1991): Symbiotic recognition and subsequent morphogenesis as early events in an animal-bacterial mutualism. Science 254:1491–1494.

McKenney D, Brown KE, Allison DG (1995): Influence of Pseudomonas aeruginosa exoproducts on virulence factor production in Burkholderia cepacia: Evidence of interspecies communication. J Bacteriol 177:6989–6992.

McKnight SL, Iglewski BH, Pesci EC (2000): The Pseudomonas quinolone signal regulates rhl quorum sensing in Pseudomonas aeruginosa. J Bacteriol 182:2702–2708.

McLean RJC, Whiteley M, Stickler DJ, Fuqua WC (1997): Evidence of autoinducer activity in naturally occurring biofilms. FEMS Microbiol Lett 154:259–263.

More MI, Finger LD, Stryker JL, Fuqua C, Eberhard A, Winans SC (1996): Enzymatic synthesis of a quorum-sensing autoinducer through use of defined substrates. Science 272:1655–1658.

Nasser W, Bouillant ML, Salmond G, Reverchon S (1998): Characterization of the Erwinia chrysanthemi expI-expR locus directing the synthesis of two N-acyl-homoserine lactone signal molecules. Mol Microbiol 29:1391–1405.

Nealson KH, Hastings JW (1979): Bacterial bioluminescence: its control and ecological significance. Microbiol Rev 43:496–518.

Nealson KH, Platt T, Hastings JW (1970): Cellular control of the synthesis and activity of the bacterial luminescent system. J Bacteriol 104:313–322.

Oger P, Kim K-S, Sackett RL, Piper KR, Farrand SK (1998): Octopine-type Ti plasmids code for a mannopine-inducible dominant-negative allele of traR, the quorum-sensing activator that regulates Ti plasmid conjugal transfer. Mol Microbiol 27:277–288.

Parsek MR, Greenberg EP (2000): Acyl-homoserine lactone quorum sensing in Gram-negative bacteria: A signaling mechanism involved in associations with

higher organisms. Proc Natl Acad Sci USA 97:8789–8793.

Parsek MR, Schaefer AL, Greenberg EP (1997): Analysis of random and site-directed mutations in rhlI, a Pseudomonas aeruginosa gene encoding an acylhomoserine lactone synthase. Mol Microbiol 26:301–310.

Parsek MR, Val DL, Hanzelka BL, Cronan JE, Jr., Greenberg EP (1999): Acyl homoserine-lactone quorum-sensing signal generation. Proc Natl Acad Sci USA 96:4360–4365.

Passador L, Cook JM, Gambello MJ, Rust L, Iglewski BH (1993): Expression of Pseudomonas aeruginosa virulence genes requires cell-to-cell communication. Science 260:1127–1130.

Passador L, Tucker KD, Guertin KR, Journet MP, Kende AS, Iglewski BH (1996): Functional analysis of the Pseudomonas aeruginosa autoinducer PAI. J Bacteriol 178:5995–6000.

Pearson JP, Pesci EC, Iglewski BH (1997): Roles of Pseudomonas aeruginosa las and rhl quorum-sensing systems in the control of elastase and rhamnolipid biosynthesis genes. J Bacteriol 179:5756–5767.

Pearson JP, Van Delden C, Iglewski BH (1999): Active efflux and diffusion are involved in transport of Pseudomonas aeruginosa cell-to-cell signals. J Bacteriol 181:1203–1210.

Pesci EC, Iglewski BH (1999): Quorum sensing in Pseudomonas aeruginosa. In Dunny GM, Winans SC (eds): "Cell-Cell Signaling in Bacteria." Washington, DC: ASM Press, pp 147–155.

Pesci EC, Milbank JBJ, Pearson JP, McKnight S, Kende AS, Greenberg EP, Iglewski BH (1999): Quinolone signaling in the cell-to-cell communication system of Pseudomonas aeruginosa. Proc Natl Acad Sci USA 96:11229–11234.

Pesci EC, Pearson JP, Seed PC, Iglewski BH (1997): Regulation of las and rhl quorum sensing in Pseudomonas aeruginosa. J Bacteriol 179:3127–3132.

Piper KR, Beck von Bodman S, Farrand SK (1993): Conjugation factor of Agrobacterium tumefaciens regulates Ti plasmid transfer by autoinduction. Nature 362:448–450.

Piper KR, Beck von Bodman S, Hwang I, Farrand SK (1999): Hierarchical gene regulatory systems arising from fortuitous gene associations: controlling quorum sensing by the opine regulon in Agrobacterium. Mol Microbiol 32:1077–1089.

Puskas A, Greenberg EP, Kaplan S, Schaefer AL (1997): A quorum-sensing system in the free-living photosynthetic bacterium Rhodobacter sphaeroides. J Bacteriol 179:7530–7537.

Reichelt JL, Borowitzka MA (1984): Antimicrobial from marine algae: results of a large scale screening programme. Hydrobiol 116/117:158–168.

Reimmann C, Beyeler M, Latifi A, Winteler H, Foglini M, Lazdunski A, Haas D (1997): The global activator GacA of Pseudomonas aeruginosa PAO positively controls the production of the autoinducer N-butyryl-homoserine lactone and the formation of the virulence factors pyocyanin, cyanide, and lipase. Mol Microbiol 24:309–319.

Rhodius VA, Busby SJW (1998): Positive activation of gene expression. Curr Opin Microbiol 1:152–159.

Ruby EG (1996): Lessons from a cooperative bacterial-animal association: the *Vibrio fischeri-Euprymna scolopes* light organ symbiosis. Annu Rev Microbiol 50:591–624.

Ruby EG, McFall-Ngai MJ (1992): A squid that glows in the night: development of an animal-bacterial mutualism. J Bacteriol 174:4865–4870.

Rust L, Pesci EC, Iglewski BH (1996): Analysis of the *Pseudomonas aeruginosa* elastase (*lasB*) regulatory region. J Bacteriol 178:1134–1140.

Schaefer AL, Hanzelka BL, Eberhard A, Greenberg EP (1996): Quorum-sensing in *Vibrio fischeri*: probing autoinducer-LuxR interactions with autoinducer analogs. J Bacteriol 178:2897–2901.

Schaefer AL, Hanzelka BL, Parsek MR, Greenberg EP (2000): Detection, purification and structural elucidation of acylhomoserine lactone inducer of *Vibrio fischeri* luminescence and other related molecules. Methods Enzymol 305:288–301.

Schaefer AL, Val DL, Hanzelka BL, Cronan JE, Jr., Greenberg EP (1996): Generation of cell-to-cell signals in quorum sensing: acyl homoserine lactone synthase activity of a purified *Vibrio fischeri* LuxI protein. Proc Natl Acad Sci USA 93:9505–9509.

Schlaman HRM, Okker RJH, Lugtenberg BJJ (1989): Subcellular localization of the *nodD* gene product in *Rhizobium leguminosarum*. J Bacteriol 174:4686–4693.

Seed PC, Passador L, Iglewski BH (1995): Activation of the *Pseudomonas aeruginosa lasI* gene by LasR and the *Pseudomonas* autoinducer PAI—An autoinduction regulatory hierarchy. J Bacteriol 177:654–659.

Shadel GS, Baldwin TO (1992): Positive autoregulation of the *Vibrio fischeri luxR* gene. J Biol Chem 267:7696–7702.

Shadel GS, Young R, Baldwin TO (1990): Use of regulated cell lysis in a lethal genetic selection in *Escherichia coli*: Identification of the autoinducer-binding region of the LuxR protein from *Vibrio fischeri* ATCC 7744. J Bacteriol 172:3980–3987.

Sitnikov D, Schineller JB, Baldwin TO (1995): Transcriptional regulation of bioluminescence genes from *Vibrio fischeri*. Mol Microbiol 17:801–812.

Slock J, Kolibachuk D, Greenberg EP (1990): Critical regions of the *Vibrio fischeri* LuxR protein defined by mutational analysis. J Bacteriol 172:3974–3979.

Stevens AM, Dolan KM, Greenberg EP (1994): Synergistic binding of the *Vibrio fischeri* LuxR transcriptional activator domain and RNA polymerase to the *lux* promoter region. Proc Natl Acad Sci USA 91:12619–12623.

Stevens AM, Greenberg EP (1997): Quorum sensing in *Vibrio fischeri*: Essential elements for activation of the luciferase genes. J Bacteriol 179:557–562.

Stevens AM, Greenberg EP (1999): Transcriptional activation by LuxR. In Winans SC, Dunny GM (eds): "Cell-Cell Signaling in Bacteria." Washington, DC: ASM Press, pp 231–242.

Stickler DJ, Morris NS, McLean RJC, Fuqua C (1998): Biofilms on indwelling urethral catheters produce quorum-sensing molecules in situ and in vitro. Appl Environ Microbiol 64:3486–3490.

Telford G, Wheeler D, Williams P, Tomkins PT, Appleby P, Sewell H, Stewart GSAB, Bycroft BW, Pritchard DI (1998): The *Pseudomonas aeruginosa* quorum-sensing signal molecule N-(3-oxododecanoyl)-L-homoserine lactone has immunomodulatory activity. Infec Immun 66:36–42.

Teplitski M, Robinson JB, Bauer WD (2000): Plants secrete substances that mimic bacterial N-acyl homoserine lactone signal activities and affect population density-dependent behaviors in associated bacteria. Mol Plant Micro Interact 13:637–648.

Welch M, Todd DE, Whitehead NA, McGowan SJ, Bycroft BW, Salmond GPC (2000): N-acyl homoserine lactone binding to the CarR receptor determines quorum-sensing specificity in *Erwinia*. EMBO J 19:631–641.

White D, Hart ME, Romeo T (1996): Phylogenetic distribution of the global regulatory gene *csrA* among eubacteria. Gene 182:221–223.

Whiteley M, Lee KM, Greenberg EP (1999): Identification of genes controlled by quorum sensing in *Pseudomonas aeruginosa*. Proc Natl Acad Sci USA 96:13904–13909.

Whiteley M, Parsek MR, Greenberg EP (2000): Regulation of quorum sensing by RpoS in *Pseudomonas aeruginosa*. J Bacteriol 182:4356–4360.

Winson MK, Camara M, Latifi A, Foglino M, Chhabra SR, Daykin M, Bally M, Chapon V, Salmond GPC, Bycroft BW, Lazdunski A, Stewart GSAB, Williams P (1995): Multiple N-acyl-L-homoserine lactone signal molecules regulate production of virulence determinants and secondary metabolites in *Pseudomonas aeruginosa*. Proc Natl Acad Sci USA 92:9427–9431.

Wood DW, Gong F, Daykin MM, Williams P, Pierson III LS (1997): N-acyl-homoserine lactone-mediated regulation of phenazine gene expression by *Pseudomonas aureofaciens* 30–84 in the wheat rhizosphere. J Bacteriol 179:7663–7670.

You Z, Fukushima J, Ishiwata T, Chang B, Kurata M, Kawamoto S, Williams P, Okuda K (1996): Purification and characterization of LasR as a DNA-binding protein. FEMS Microbiol Lett 142:301–307.

Zhang L, Murphy PJ, Kerr A, Tate ME (1993): *Agrobacterium* conjugation and gene regulation by N-acyl-L-homoserine lactones. Nature (London) 362:446–448.

Zhu J, Beaber JW, More MI, Fuqua C, Eberhard A, Winans SC (1998): Analogs of the autoinducer 3-ox-ooctanoyl-homoserine lactone strongly inhibit activity of the TraR protein of *Agrobacterium tumefaciens.* J Bacteriol 180:5398–5405.

Zhu J, Winans SC (1998): Activity of the quorum-sensing regulator TraR of *Agrobacterium tumefaciens* is inhibited by a truncated, dominant defective TraR-like protein. Mol Microbiol 27:289–297.

Zhu J, Winans SC (2001): The quorum-sensing transcriptional regulator TraR requires its cognate signaling ligand for protein folding, protease resistence, and dimerization. Proc Natl Acad Sci USA 98:1507–1512.

Zhu J, Winans SC (1999): Autoinducer binding by the quorum-sensing regulator TraR increases affinity for target promoters *in vitro* and decreases TraR turnover rates in whole cells. Proc Natl Acad Sci USA 96:4832–4837.

Section 3: GENETIC EXCHANGE

17

Bacterial Transposons—An Increasingly Diverse Group of Elements

GABRIELLE WHITTLE AND ABIGAIL A. SALYERS

Department of Microbiology, University of Illinois, Urbana, Illinois 61801

I. WHAT IS A TRANSPOSON?

Transposons have been defined as genetic elements that can move from one position in a DNA segment to another by means of excision and integration reactions that are independent of homologous recombination. The first transposons to be characterized appeared to be very simple elements. Subsequently, however, more complex transposable elements have been discovered, and it has become clear that transposons are unexpectedly diverse in their structures and properties. There now are at least four modes of transposition known: conservative transposition, replicative transposition, excisive transposition, and retrotransposition. These modes of transposition are illustrated in Figure 1.

Transposons differ in site selectivity. Some, like Tn5 and Mu, integrate into many sites. Others integrate into a limited number of sites, and still others integrate site-specifically. Transposons differ greatly in size. The first transposons to be described were relatively small, less than 15 kb, and only moved from site to site within the same cell. Recently much larger-transposable elements have been discovered, elements capable of transferring themselves by conjugation to another cell (conjugative and

Modern Microbial Genetics, Second Edition. Edited by Uldis N. Streips and Ronald E. Yasbin. ISBN 0–471–38665–0
Copyright © 2002 Wiley-Liss, Inc.

I. Conservative and replicative transposition

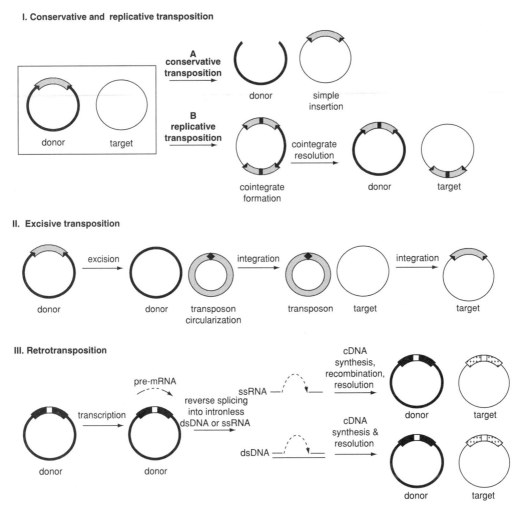

Fig. I. Four modes of transposition. Different products are generated dependent upon the transposition mechanism. In **conservative transposition** (cut-and-paste or nonreplicative transposition) (**IA**), the transposon (indicated by a gray box) excises from the donor replicon and inserts into the target replicon, generating a target containing the transposon and a gapped donor replicon. The gapped replicon can be repaired. In **replicative transposition** (**IB**), the transposon does not excise from the donor replicon; instead, the donor and target replicons form a cointegrate in a process that generates a second copy of the transposon. The cointegrate is later resolved via homologous recombination between the two copies of the transposon or by an element-encoded site-specific recombinase system (resolvase) that recognizes the *res* sites (black boxes). This yields donor and target replicons both containing a copy of the transposon. In **excisive transposition** (**II**), an integrase, together with other proteins and host factors, catalyzes the excision of the transposon from the donor replicon. This process generates a covalently closed circular transposition intermediate and reseals the donor site. The circular intermediate then integrates into the target replicon. In **retrotransposition** (**III**), an RNA copy an *orf* containing the group II intron (retrotransposon, white box) is transcribed. The intron part of the mRNA is subsequently reverse spliced into an intronless target molecule, which may be either a dsDNA or ssRNA molecule. The target molecule is an intronless *orf* that can be different from the *orf* that originally contained the intron. A cDNA copy of the inserted intron RNA molecule is then synthesized, and then the intron RNA itself is replaced by cDNA.

mobilizable transposons). The largest transposon identified to date is a conjugative transposon from *Mesorhizobium* (formerly *Rhizobium*) *loti*, which is approximately 500 kb in length (Sullivan and Ronson, 1998). The extraordinary ability of bacteria to acquire and move such large pieces of DNA no doubt has had a significant impact on their evolution.

Transposons vary greatly in function. Some are DNA segments whose sole function is to move from one DNA segment to another. Others carry genes that are useful to their bacterial host. Early on, it became clear that some lysogenic phages had transposon-like properties. Transposons called integrons create operons, and may have played a role in the evolution of operons in bacteria.

Transposons exhibit diversity in the genes they carry. The simplest transposing elements, the insertion sequences (IS elements), carry only the gene(s) involved in the transposition process itself. The larger transposons can carry a variety of genes that are not directly involved in transposition, including antibiotic resistance genes, catabolic genes, vitamin synthesis genes, nitrogen fixation genes, and heavy metal resistance genes.

Finally, transposons are widely dispersed in nature, having been found in nearly all organisms. Their wide distribution is evidence of their significant contribution to the evolution of microbial genomes and genetic diversity, since they provide a means by which new genes may be acquired, and they can promote DNA rearrangements (deletions, inversions, and replicon fusions) that are central to evolutionary processes.

The growing size and diversity of the transposon family has made it necessary to expand the old classification system in which transposable elements were divided into three classes based on transposon structure and sequence homology (Fig. 2). Class I composite transposons included all those transposons that contain IS elements. Class II (noncomposite) transposons were distinguished from Class I transposons by not containing IS elements, and by being closely related to Tn*3*. Class III transposons are the transposing bacteriophages. Today we know that there are mobile genetic elements that do not fit into these categories (Fig. 2), including retrotransposons, conjugative transposons, mobilizable transposons, and integrons. Perhaps the most useful currently used classification system is the one based on transposition mechanism (Fig. 1), but the question of how best to classify transposons remains to be answered.

For those who find this growing diversity daunting, there is some reassuring news; transposonlike elements of *Saccharomyces cerevisciae* (e.g., Ty1, Ty3) have many features in common with bacterial transposons such as Tn*5* and Tn*10*. The P elements of the fruit fly *Drosophila melanogaster* also have many characteristics in common with the bacterial transposons (Craig, 1997; Hallet and Sherratt, 1997; Polard and Chandler, 1995a). The retrotransposons are thought to be evolutionary progenitors of the introns so commonly found in the genes of eukaryotes. So there may be limits to the diversity of these elements after all.

This chapter will focus on the bacterial transposons. A number of excellent reviews have been published in recent years on specific features of bacterial transposons, such as transposition mechanism or site-specificity (Berg and Howe, 1989; Craig, 1996, 1997; Mahillon and Chandler, 1998; Merlin et al., 2000; Mizuuchi, 1992b; Plasterk, 1993; Polard and Chandler, 1995a; Saedler and Grierl, 1995). This chapter will cover, in a more general way, the traits of the various types of bacterial transposons. It will also touch upon the ecology and diversity of members of the transposon family.

II. THE PARADIGM TRANSPOSONS

Intensive work on the transposition mechanisms of a small number of transposons has provided a detailed picture now available of the various transposition mechanisms. It has also revealed some features that are common

A. Insertion sequences

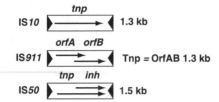

B. Class I composite transposons

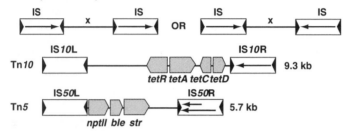

C. Non-composite transposons

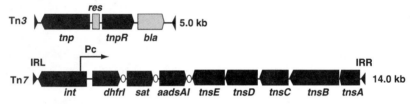

D. Class III - transposing bacteriophage

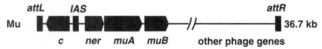

E. Integrons

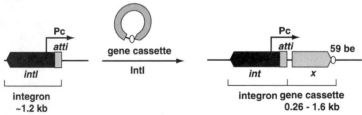

Fig. 2. The structural and functional organization of different types of bacterial transposable elements. Insertion sequences (**A**) are the simplest type of transposable element. They encode only genes required for transposition. This includes a transposase (*tnp*) which may be the product of one (e.g., IS*10*) or two genes (IS*911*). Horizontal arrows indicate the extent and the direction of transcription of the transposase genes. In the case of IS*50* a smaller transcript, *inh*, is also present and inhibits expression of the *tnp* gene. Inverted repeats that flank the ends of the IS elements are indicated by filled triangles. Class I composite transposons (**B**) contain two IS elements, oriented in the same or opposite directions, that flank a central DNA segment that contains genes whose products play no role in transposition (X). In Tn*10* and Tn*5* only one of these IS elements is active. The horizontal arrow in the IS(R) elements of Tn*10* and Tn*5* indicates an active copy of the transposase gene. The other IS(L) element has a virtually identical DNA sequence to the IS(R), but mutations in the *tnp* gene have inactivated it. Like many other composite transposons Tn*10* and Tn*5* encode antibiotic resistance genes. These include resistances to tetracycline (*tet*), kanamycin (*nptII*), bleomycin (*ble*), and streptomycin (*str*). Noncomposite transposons (**C**) are not flanked by IS elements. Tn*3* is an example of a

F. Retrotransposons (Group II introns) of bacteria

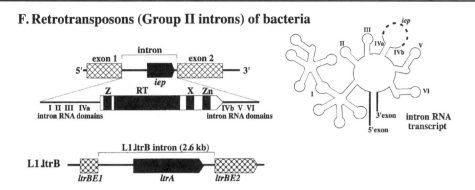

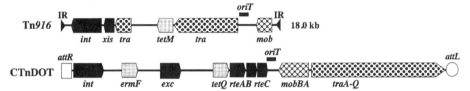

G. Conjugative transposons

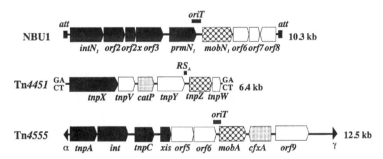

H. Mobilizable transposons

class II transposon. Class II transposons like Tn3 typically contain a transposase gene (*tnp*), a *res* site that is involved in resolution of the cointegrate transposition intermediate (Fig. 1), and a site-specific recombinase gene (also called resolvase; *tnpR*). Class II transposons may also contain antibiotic resistance genes. Tn3 encodes an ampicillin resistance gene (*bla*). Tn7, is also a noncomposite transposon, however it is very different from Tn3 in both structure and sequence, and so is not a Class II transposon. Tn7 is more complex, containing five genes that have a role in transposition (*tnsA-E*) and an integron (see category E). The integron is linked to three antibiotic resistance gene cassettes, encoding trimethoprim (*dhfrI*), streptothricin (*sat*), and streptomycin/spectinomycin (*aadsAI*) resistance, which play no role in transposition. Inverted repeats at each end of Tn7 are indicated by filled triangles and are labeled IRL and IRR. Class III transposons (**D**) are comprised of transposing bacteriophages, of which Mu is the best-studied example. The sites involved in integration of the phage are located at the ends of Mu, and are designated *attL* and *attR* respectively. The transposase is encoded by two genes *muA* and *muB*. The *c* and *ner* genes encode negative regulators. IAS is a sequence that enhances Mu transposition and is the site to which the Mu repressor encoded by *c* binds. Integrons (**E**) are portions of transposons that generate operons by integrating circular gene cassettes. An integron consists of an *intI* gene, which encodes a site-specific integrase, and a promoter (Pc), which directs transcription of genes integrated into the *attI* site. The incoming gene cassette is a DNA circle that contains an open reading frame (gray arrow) but no promoter of its own. It also contains a sequence (called the 59 bp element, or 59 be) that interacts with the *attI* site to integrated the circular DNA molecule in a direction-specific manner. The 59 bp element is indicated by a small open oval. An example of an integron that has acquired three gene cassettes is seen in the case of Tn7 in panel C. Retrotransposons (group II introns) (**F**) have an RNA intermediate. The intron is located in an exon. The *iep* (intron-encoded protein) gene product to most transposons. In general, the first step in transposition involves transposase-mediated recognition of and cleavage at the ends of the transposon. This step is followed by capture of the target DNA molecule and strand transfer reactions which begin to insert the transposon ends into the target DNA molecule. In the final step the strand transfer complex is resolved yielding the final products of transposition. In addition some transposons undergo replication during transposition.

Surprisingly, despite considerable structural and sequence differences between transposable elements, the chemical processes involved in transposition of class I, II, and III transposons are fundamentally similar. Elements for which the mechanism of transposition has been investigated in detail include bacteriophage Tn5, Tn10, Mu, IS10, Tn7, IS911, and the retrotransposons (Craig, 1995; Mahillon and Chandler, 1998; Mizuuchi, 1992a; Mizuuchi, 1992b). In all these cases, transposition begins when the transposase makes a cut that exposes the 3' ends of the transposon with the aid of a nucleophile, which is usually water (Fig. 3). In some cases, a reaction that frees the 5' ends of the transposon also occurs. This cleavage step generates a 3' hydroxyl group which acts as a nucleophile, attacking a 5' phosphate group intramolecularly to separate the transposon ends from the donor replicon. Alternatively, the 3' hydroxyl may initiate an intermolecular attack on the target DNA molecule (strand transfer). It is the number of strands cleaved, and the order in which the cleavages and strand transfers occur, that determines the reaction products formed during and at the end of the transposition process (Fig. 1 and 3).

---

is a multifunctional protein that directs transposition of the intron. The structure of *iep* gene is shown in more detail below the top diagram. The Iep proteins contain domains encoding reverse transcriptase (RT) and maturase activities (X), a zinc finger mediated DNA-binding region (Zn), and another region of unknown function (Z). The secondary structure of the mRNA that is transcribed from the intron DNA is shown at the right of the panel. A detailed map of the Ll.ltrB intron is shown at the bottom of this panel. The intron interrupts the *ltrBE* gene and contains *ltrA*, the *iep* of this intron. Conjugative transposons (**G**) have a circular transposition intermediate and are capable of transferring themselves by conjugation. The structures of the two best-studied conjugative transposons, Tn916 and CTnDOT, are shown. The *int* gene product (Int) mediates integration. Together with the proteins encoded by *xis* (Tn916) or *exc* (CTnDOT), the Int protein catalyzes excision and circularization. The *mob* genes (hatched arrows) encode proteins that nick at the transfer origin (*oriT*) to initiate conjugal transfer and the *tra* genes (spotted arrows, here shown as a region rather than as individual genes) encode proteins of the mating apparatus. Often conjugative transposons carry antibiotic resistance genes (gray) encoding tetracycline (*tet*) or erythromycin (*erm*) resistance. The *rteABC* genes of CTnDOT are regulatory genes that control excision and transfer of the element. The fill in the various boxes indicates a functional type of gene, not sequence similarity between Tn916 and CTnDOT. Not all genes on these conjugative transposons are shown. Mobilizable transposons (**H**), like conjugative transposons, have a circular intermediate and can be transferred by conjugation, but they cannot transfer themselves. The *tra* genes of a conjugative transposon must be provided *in trans*. The integrase of NBU1 is encoded by *intN1*. The *orf2, orf2x, orf3, prmN1* and *mobN1* genes are required, together with *intN1* for excision and circularization of NBU1. The *mobN1* gene product also nicks at the *oriT* to initiate transfer of the circular form. The transposase genes of Tn4451 and Tn4555 are designated *tnp*. How the proteins encoded by these genes mediate transposition is described in the text. The mobilization genes of Tn4451 and Tn4555 are designated *tnpA* and *mobA*, respectively. As in panel G, the fill in the boxes designates a functional type of gene, not sequence similarities between the mobilizable transposons shown here. Mobilizable transposons can carry antibiotic resistance genes (e.g., *catP, cfxA*). Genes with no known function are shown as white arrows. Although the diagrams are not to scale for the respective elements, their relative sizes have been indicated. In the case of phage Mu (**D**) and the conjugative transposons Tn916 and CTnDOT (**G**), not all genes are shown.

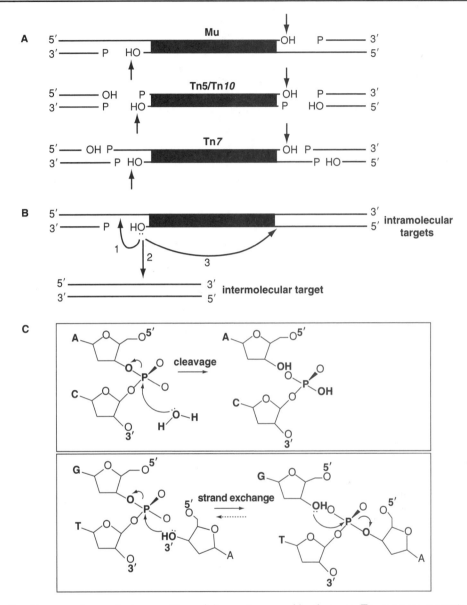

Fig. 3. Common themes in transposition of diverse transposable elements. Transposases recognize and cleave the ends of the transposon, generating an active 3'-OH group (**A**). Cleavage can occur only at the 3' end, at the 3' end and subsequently at the 5' end, or cuts may occur simultaneously at both the 3' and the 5' ends of the transposon. Examples shown include bacteriophage Mu (class III), Tn5/Tn10 (class I), and Tn7 (noncomposite transposon). Following cleavage of the transposon ends, (**B**) these activated 3'-OH groups initiate nucleophilic attack either intramolecularly on the opposite strand, which will generate the formation of hairpins (**1**); intermolecularly on a target molecule which would generate a cointegrate or deletions (**2**); or intramolecularly on the other end of the transposon which generates the formation of a figure 8 that may later be resolved to form circles (**3**). The chemistry of transposition is shown (**C**). The top panel shows the nucleophilic attack of water on the phosphate backbone of the transposon DNA, where A represents the terminal base of the transposon and C represents a base in the donor molecule phosphate backbone ("cleavage"). Nucleophilic attack on the phosphate backbone (bottom panel) generates a 3'-OH group on the transposon end, which is thereby activated to attack the phosphate backbone of the target DNA (bases G and T) that joins the 3' end of the transposon to the 5' end of the target DNA molecule ("strand transfer").

A. Tn5 and Tn10

Tn5 and Tn10 (Fig. 2) are Class I composite transposons that transpose by a conservative cut and paste mechanism (Berg, 1989; Kleckner, 1989; Kleckner et al., 1996b). The IS elements that flank Class I composite transposons carry a gene encoding a transposase protein that catalyzes the transposition reaction. The DNA located between the IS elements usually has no role in transposition. This "nonessential" DNA, however, can be important for determining the distribution and maintenance of the transposon in a bacterial population. A transposon that carries an antibiotic resistance gene may, for example, provide a new function that is useful to its bacterial host, or it may insert into a mobile element, such as a plasmid, which can then pass this trait to another host organism. If the transposon can transpose and be maintained in the new host, it has expanded its host range. Carrying selectable markers, as long as they are expressed and active in the new host, promotes the maintenance of the transposon in each new host it reaches.

Since IS elements are themselves transposable elements, the two IS elements of a composite transposon could move independently of the transposon. In the case of Tn5 and Tn10, this possibility has been eliminated because only one of the IS elements encodes an active transposase. The other IS element has a mutation in its transposase gene that renders it inactive. Thus Tn5 and Tn10 are both programmed to move as a unit. This characteristic appears not to be shared by most compound transposons, which have two active IS elements.

The transposases of Tn5 and Tn10 mediate a hairpin mechanism of strand cleavage and integration (Davies et al., 2000). This mechanism is illustrated in Figure 4. The transposition process starts with binding of transposase to each end of the transposon. The transposase makes double stranded cuts at each end of the transposon, exposing a 3'-hydroxyl group, which is able to initiate a nucleophilic attack on the 5' end of the other strand of the transposon DNA, creating a hairpin structure. The excised transposon is held in the form of a DNA circle by intramolecular interactions between the transposase molecules bound to the transposon ends, but the ends of the transposon DNA are not covalently joined to each other. During integration into a new DNA segment, a nucleophilic attack on the hairpin by water produces the exposed 3'-hydroxyl groups of the transposon so that the transposon can insert into its target site. Insertion of Tn5 or Tn10 into a new site creates a short directly repeated sequence of 9 bp at both ends of the inserted transposon. These duplications occur because the transposase makes two staggered single-stranded breaks in the target DNA molecule that are 9 bp apart, one in each strand. These staggered single-stranded breaks generate two overhanging DNA fragments, to which the transposon ends are subsequently ligated. These overhanging DNA fragments are filled in by the DNA repair machinery, hence generating target site duplications (Fig. 5).

The insertion of many other transposons into a site also generates short directly repeated sequences. Since the target site for a transposon may vary, the sequence of these repeats for a given transposon will be different for different sites. However, the size of the duplication is generally the same for a given element. The size of the duplication ranges from 2 to 14 bp depending on the transposon (Mahillon and Chandler, 1998).

The Tn5 transposase has been co-crystallized with the transposon ends, enabling scientists to check in detail whether the association between the transposase and the cleaved donor molecule would allow the mechanism deduced from genetic and biochemical experiments to occur as shown in Figure 4 (Davies et al., 2000). The crystal structure revealed that the two transposase molecules interact with each other to hold the cleaved ends of the transposon DNA close together. Each transposase protein contacts a 17 bp sequence at each of the

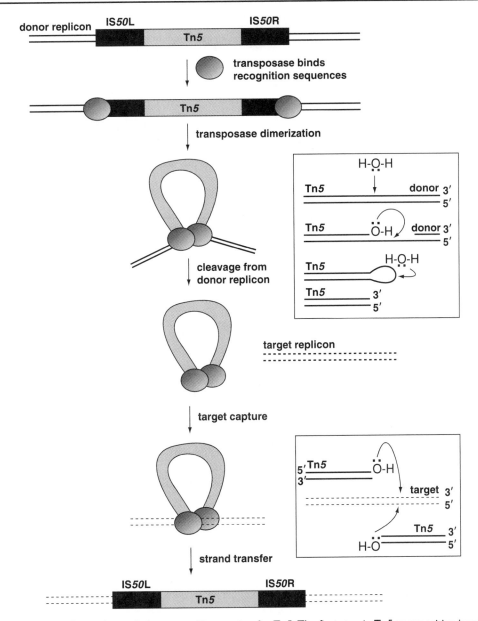

Fig. 4. Hairpin form of strand cleavage and integration for Tn5. The first step in Tn5 transposition involves the binding of the transposase (sphere) to the 19 bp recognition sequences at the left and right ends of Tn5, which is followed by dimerization of the transposase to form a synaptic complex, in which strand cleavage occurs. The first step in strand cleavage begins with the nucleophilic attack of one DNA strand by a water molecule; this results in the hydrolysis of one DNA strand, exposing a 3′ OH group. The exposed 3′ OH group then nucleophilicly attacks the opposite DNA strand, resulting in the formation of a hairpin structure. Further addition of water results in hydrolysis of the hairpin structure generating blunt transposon ends. Subsequently the synaptic complex binds the target DNA molecule, and the free 3′-OH groups at the ends of the transposon nucleophilicly attack the target DNA, resulting in strand transfer. This diagram was adapted from Davies et al., 2000.

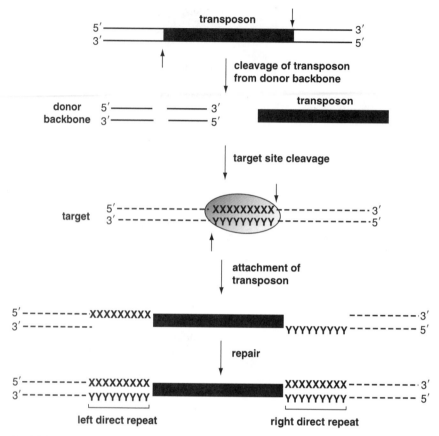

Fig. 5. Model for generation of directly repeated target sequences upon integration of a transposon. The transposase cleaves the transposon from the donor DNA molecule (solid lines) and makes two staggered single-stranded breaks in the target DNA (dashed lines) molecule, one in each strand. The transposon is then attached to the X and Y overhangs generated by the transposase. This attachment leaves two gaps opposite the target sequences (X) and (Y), which are filled in, and the nicks are sealed. Hence this process generates a directly repeated sequence that flanks the left and right ends of the inserted transposon.

transposons ends, a sequence that had been identified in genetic studies as the transposase binding site. Analysis of the crystal structure further revealed that there is one Mn^{2+} molecule in the active site of each transposase. This metal ion is thought to participate in the activation of the 3′ hydroxyl at the transposon end so that it can make the nucleophilic attack that results in the formation the hairpin structure.

The type of transposition shown in Figure 4 leaves a gap in the donor molecule. For this reason it has been called *cut-and-paste* transposition. The gap can be processed in at least three ways (Craig, 1996). First, the gap may

result in loss of the donor replicon. More commonly the donor DNA molecule may be repaired by a transposition-independent process. Repair can occur *via* a homologous recombination (gene conversion) mechanism, where a sister chromosome present in the same cell is utilized as a template for repair of the gap (Bender et al., 1991; Hagemann and Craig, 1993; Merlin et al., 2000). In this case the result may be a DNA segment that still contains a copy of the transposon, so both the target and donor DNA segments contain copies of the transposon. Alternatively, repair of the donor molecule may simply rejoin the broken regions that flanked the transposon. This requires the bacterial repair enzymes (SOS system) as well as homologous recombination, and may create a deletion in the regenerated donor site.

This later feature of conservative transposition was very helpful to scientists in the early days of bacterial genetics because it gave them a way to generate deletions in a specific region of the chromosome (Kleckner et al., 1979; Picksley et al., 1984; Raleigh and Kleckner, 1984). Also the fact that Tn5 and Tn10 integrate almost randomly into DNA has made them useful for creating mutants in which open reading frames are disrupted. Transposon mutagenesis has been the backbone of genetic analysis in a number of bacterial systems (Dyson, 1999).

The utility of Tn5 mutagenesis has now been expanded beyond those bacteria in which the Tn5 transposase gene is expressed. One approach is to introduce the transposon randomly into isolated DNA segments from the organism of interest, using an in vitro reaction. The mutagenized DNA is then introduced into the target cell by electroporation, and the mutagenized DNA segments insert into the genome by homologous recombination. A second approach is to electroporate a mixture of the transposase protein and copies of transposon DNA (Transposomes™) directly in the cell type of interest (Goryshin et al., 2000; Goryshin and Reznikoff, 1998). An advantage of this second approach is that the form of the transposon used does not carry the transposase gene and can thus not continue to transpose.

The fact that Tn5 and Tn10 integrate almost randomly has implications for the bacterium that harbors them. Obviously acquisition of antibiotic resistance genes found on such transposons can confer a selective advantage under some conditions. Also transposons can activate genes that were not previously expressed if the transposon integrates in the promoter region of the gene. Gene activation can be caused by promoters at the end of the transposon or IS element that direct transcription outward

from the ends of the element. Alternatively, insertion of an IS element may lead to the formation of new promoters upon insertion if the end of the element contains part of a promoter sequence, for example a −35 sequence (Galas and Chandler, 1989). There are numerous examples of antibiotic resistance genes that have been activated in these ways by transposons, and the phenomenon occurs with other genes as well (Charlier et al., 1982; Ciampi et al., 1982; Dalrymple, 1987; Leelaporn et al., 1994; Prentki et al., 1986; Reimmann et al., 1989; Ton-Hoang et al., 1997). Finally, there have been cases in which the insertion of a transposon activated gene expression by altering the local topology around the site of insertion (Reynolds et al., 1981, 1986; Schnetz and Rak, 1992). A detailed explanation of how these topological changes result in activation of transcription is not yet available.

The potential benefits of acquiring an active transposon can come at a cost to the bacterial host, because transposons like Tn5 and Tn10 can interrupt open reading frames that are important for bacterial survival (Berg, 1989; Reznikoff, 1993). Bacterial survival is important not only to the bacterium but also to the transposon. Perhaps this is the reason why there are some limitations on the ability of transposons to cause irreversible damage to a bacterium. For example, Tn10 can be excised precisely, in a way that regenerates the donor molecule without DNA repair (Kleckner, 1989). This type of excision is mediated by host enzymes, not the Tn10 transposase, and it occurs via the direct repeats adjacent to the ends of the transposon. The process is independent of homologous recombination, however, and may occur due to slip-strand replication in which the transposon forms a hairpin structure that allows DNA polymerase to skip from one end of the transposon to the other, effectively deleting the transposon from the site. In a variant on this type of excision, excision occurs due to direct repeat sequences within the transposon, leaving behind a 50 bp segment of transposon DNA. Although such an excision event leaves a piece of the transposon in the site, it eliminates the polar effect that an intact copy of the transposon may exert on downstream genes in the same operon.

Another way in which bacteria are protected from detrimental transposition events is that transposons like Tn5 and Tn10 do not integrate nearly as readily into DNA that is being transcribed, as into untranscribed DNA (Casadesus and Roth, 1989; Craig, 1997; Kleckner, 1989). Thus actively transcribed genes, which are presumably important for bacterial survival, are somewhat protected. This inhibition may have something to do with the altered structure of DNA in a region where RNA polymerase is separating strands of DNA and moving along the template strand. This altered structure may interfere with the assembly of the transpososome complex, and thus precludes transposition.

Still another property of transposons that may prevent adverse integration events is a phenomenon called target immunity. When a transposon like Tn10 or Tn5 is integrated into a site, secondary integration events by another copy of the same transposon in the same or nearby sites are inhibited (Kleckner et al., 1996a; Reznikoff, 1993). Protection of the region can extend for considerable distances. Other transposons such as Tn3, Mu, and Tn7 also display this property (Arciszewska et al., 1989; Darzins et al., 1985; Lee et al., 1983). Thus, if a transposon has integrated into a site, presumably in a way that did not have an adverse effect on the bacterial cell, the DNA surrounding the insertion is protected from other transposition events. The mechanism of target inhibition is poorly understood.

B. Tn3 And Tn3-Related Transposons

Tn3 and Tn3-like transposons comprise a distinct family of transposable elements, designated class II transposons. Transposons belonging to this class have been found in a diverse range of Gram-negative and Gram-positive bacteria and have been studied ex-

tensively (Abraham and Rood, 1987; Grindley, 1983; Murphy, 1989; Olson and Chung, 1988; Sherratt, 1989). Tn3 is the best characterized of this family of transposons. It is a 5.0 kb transposon that is flanked by 38 bp terminal repeats and carries an ampicillin resistance gene, bla (Fig. 2C). Although Tn3 inserts almost randomly, it shows a preference for AT-rich regions. Moreover it inserts preferentially into plasmids and inserts only rarely into chromosomal target sites (Sherratt, 1989). The reason for this preference is still unclear.

Transposition of the class II Tn3-like transposons is mediated by a replicative mechanism, in which two different recombination proteins, a transposase (Tnp) and a recombinase (also called resolvase, TnpR) participate. In the first step, the transposase generates a cointegrate transposition intermediate that contains both the donor and target replicon and two copies of the transposon (Fig. 1). In the second step, this cointegrate is resolved by the transposon-encoded resolvase (TnpR), which catalyzes a site-specific recombination reaction between the res sites present on the duplicate copies of the transposon present in the cointegrate. The enzyme name, resolvase, reflects the fact that the enzyme resolves the donor and target replicons into separate entities.

In Tn3 the res site is located between two divergently transcribed open reading frames, the transposase gene (tnp) and the resolvase gene (tnpR), but not all transposons in this Tn3 family are organized like Tn3. In a subgroup of the Tn3-family, designated the Tn501 group, the tnpR and tnp genes are expressed as an operon, and the res site is located upstream of the tnpR gene. There are yet other class II transposons that are not organized like either Tn3 or the Tn501 subgroup. For example, on Tn4651, two genes encode the resolvase activity, one on the left side of the res site and the other on the right side. The resolvase genes and the res site are separated from the transposase by a 40 kb region, which encodes proteins involved in toluene catabolism (Tsuda et al., 1989).

Some noncomposite transposons have been identified that do not belong to the class II group of transposons but preferentially insert into (70%), or adjacent to (30%), the *res* sites of Tn3-family transposons. Hence they have been dubbed "*res*-site hunters." These elements also insert preferentially into a site on plasmid RP1, which binds the resolvase-like protein encoded by the *parA* gene of RP1 (Kholodii et al., 1993). The *parA* gene product is involved in partitioning copies of the plasmid after plasmid replication. Results from a recent study suggest that for this Tn5053/Tn402 transposon family, most but not all *res*-containing plasmids may serve as target sites, as long as a related resolvase is functioning *in cis* or *in trans* (Minakhina et al., 1999; Radstrom et al., 1994; Shapiro and Sporn, 1977). The *res* site-hunting elements appear to have evolved a clever mechanism to ensure their dispersal in the microbial population, by targeting the large family of class II transposons, that are present in diverse bacterial species.

C. Bacteriophage Mu

Although Mu is a bacteriophage, it has become in effect an "honorary transposon" (class III) because of its many transposonlike features. To integrate into the bacterial chromosome during lysogeny, Mu uses a nonreplicative transposition mechanism. However, during lytic growth Mu uses a replicative transposition mechanism (Harshey, 1984; Lavoie and Chaconas, 1996; Mizuuchi, 1992b). Mu inserts into target DNA in a nearly random manner, like many transposons, and so is an excellent mutator, hence its name (Pato, 1989). The Mu genome is 36.7 kb in size but only a small part of the Mu genome is required for transposition functions (Fig. 2D). The ends of the phage (*attL* and *attR*), the transposase (MuA), and a transposase-activating protein (MuB) are the only sequences and proteins that are essential for excision. A sequence that is located in the middle of Mu, the enhancer or gyrase-binding site, is not essential but

stimulates transposition significantly. The rest of the Mu genome contains genes involved in lytic growth and packaging of the genome into the phage head.

A smaller form of Mu, called mini-Mu, has been used widely for generalized mutagenesis, just as transposons have (Haapa et al., 1999). Phage Mu was the first element for which an in vitro transposition system was developed, a major breakthrough in the study of transposition reactions (Lavoie and Chaconas, 1996; Mizuuchi, 1992b). Thus research on Mu has had benefits outside the immediate area of transposition mechanism.

The mechanism of replicative transposition as carried out by phage Mu has been extensively studied (Levchenko et al., 1997; Yamauchi and Baker, 1998). A simplified model for how Mu excision works is provided in Figure 6. Mu is not excised from its integration site as are Tn5 and Tn10. Rather, the ends of the integrated form (*attL* and *attR*) are first brought close together by a complex that includes MuA, IHF, and HU. The enhancer sequence of Mu binds to this complex in the early stage of the transposition process and may help the initial complex form. The enhancer sequence is not directly involved in later steps. The MuA complex makes strand cleavages in one strand of the DNA at each end, exposing a 3′ OH group. Mu B binds to the target DNA, in a reaction that consumes ATP. The MuB-target DNA complex then interacts with the MuA complex in such a way as to allow each of the two exposed 3′ OH groups to attack a strand of the target DNA. The result is a structure in which Mu ends hold the donor molecule and target molecule (which are both part of the same replicon, the chromosome) together in a figure eight configuration. The host chaperone ClpX interacts with the MuA-MuB-DNA complex to destabilize the complex enough so that DNA replication proteins can assemble on the complex and duplicate each strand of Mu. This replication step resolves the figure eight structure, and the

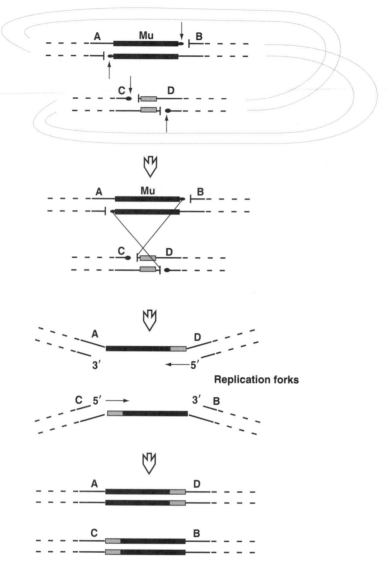

Fig. 6. Overview of Mu replicative transposition. In the first step of transposition, the MuA-host factor complex (not shown for simplicity) binds the two ends of Mu and makes two single-stranded cuts that create a free 3'-OH (small dark oval) at each end of the element. MuB binds the target site (gray rectangles), and similar staggered cuts are made in the target site. The 3'-OH ends of Mu are ligated with the free 5' ends of the nicked target site. The remaining gaps stimulate replication of the two strands of Mu DNA, resolving the joined donor and target regions. The final result is that a copy of Mu remains in the donor site and a new copy is introduced into the target site.

DNA is released from the complex. The end result is that a second copy of Mu is located elsewhere in the chromosome. Replicative transposition by other transposons, such as Tn*3,* occurs by a similar mechanism.

D. Tn7—A Site-Specific Transposon

Tn*7* utilizes a conservative transposition mechanism (Bainton et al., 1991), but in contrast to Tn*5* and Tn*10,* Tn*7* has an unusually complex system for catalyzing transposition. Transposition requires five proteins, encoded by *tnsA, tnsB, tnsC, tnsD,* and *tnsE* (Fig. 2). One combination of these proteins (TnsABCD) mediates the high-frequency insertion of Tn*7* into a single site in the *E. coli* chromosome, designated *att*Tn*7*, which is located immediately downstream of the *glmUS* operon. Another combination of proteins (TnsABCE) catalyzes the low-frequency transposition of Tn*7* into many different target sites in the *E. coli* chromosome (Craig, 1991, 1996; DeBoy and Craig, 2000; Lu and Craig, 2000).

TnsA and TnsB together form a heteromeric transposase, each protein with a distinct function. TnsB binds specifically to DNA sequences at the ends of Tn*7* and carries out cleavage and joining reactions at the 3′ end of Tn*7*. TnsA is required for cleavage of the 5′ ends of Tn*7*. Although TnsA has not been shown to bind DNA directly, it has been proposed that its ability to cleave DNA may be attributable to its interactions with the TnsB protein, since TnsA and TnsB together are required for formation of the active transposase complex (Lu and Craig, 2000). TnsA participates with TnsB to carry out the strand cleavage and ligation reactions that allow the element to transpose (May and Craig, 1996). The crystal structure of TnsA has been solved, and the structure most closely resembles that of a type II restriction enzyme (Hickman et al., 2000). This fits with the notion that TnsA participates in the catalytic reaction rather than playing some noncatalytic role in the TnsAB complex that forms at the ends of Tn*7*. It does not fit in the sense that restriction enzymes bind to specific sequences, as TnsA appears not to do. TnsB is a member of the superfamily of retroviral integrases.

Normally, during Tn*7* transposition, double-stranded breaks are made at both of the ends of the transposon and the newly liberated ends are integrated into a new site (Fig. 7). That is, transposition is an intermolecular reaction. However, in vitro, TnsA and TnsB alone are able mediate an intramolecular reaction in which a double-stranded break is made at only one end of the transposon, and the liberated end interacts with the other end of the transposon to form a lariat-shaped structure (Biery et al., 2000). This reaction requires Mn^{2+} but not ATP (Fig. 7). To obtain a true intermolecular transposition event, TnsC has to be added to the reaction. The TnsABC complex cuts at both ends of the transposon, as occurs in vivo. This reaction requires ATP and Mg^{2+}. TnsC is an ATP-binding protein that steers the complex in the direction of intermolecular transposition (Fig. 7).

Tn*7* can integrate either site-specifically or in many different sites. Which direction it takes depends on TnsE and TnsD. Site-specific integration of Tn*7* is catalyzed by the core enzyme complex, TnsABC, plus TnsD. The TnsABCD complex mediates integration of Tn*7* in a single site, *att*Tn*7*. The ability to integrate site-specifically ensures that Tn*7* will not interrupt an essential gene, killing its host. If TnsE rather than TnsD dominates the reaction, however, integration becomes more random. This type of integration may be important in hosts that do not have an *att*Tn*7* site. The integration events directed by TnsABC + TnsE are not completely random. TnsE-dominated integration events occur preferentially either near termination (*ter*) sites on the bacterial chromosome or in an actively conjugating plasmid. The presence of an actively transferring plasmid stimulates Tn*7* transposition so that over 90% of the integration events occur in the plasmid. Nonconjugal plasmids do not serve as a target, so there is something about the conjugation process that stimulates Tn*7* transposition (Wolkow et al., 1996).

Recently Peters and Craig (2000) proposed an explanation for this type of specificity. They showed that Tn*7* integrated preferentially next to single-stranded breaks in DNA. Such breaks occur commonly during termination of chromosome replication and during conjugal transfer of DNA. In the case of conjugal transfer, a single-stranded copy of the plasmid is transferred and is then copied in the recipient. During

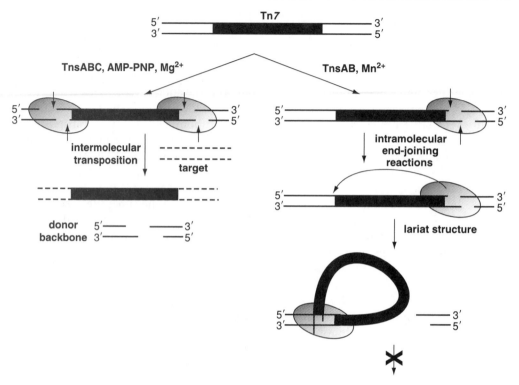

Fig. 7. The in vitro transposition pathways of Tn7. The left panel shows the steps involved in the transposition of Tn7 (black rectangle) in the presence of TnsABC, nonhydrolyzable ATP derivative AMP-PNP, and Mg^{2+}. Double-stranded breaks are made at both ends of Tn7 and is followed by the intermolecular transposition of Tn7 into the target (dashed line) replicon (Bainton et al., 1993; Gary et al., 1996). Arrows indicate points of transposase mediated cleavage. The right panel shows the products of Tn7 transposition in the presence of TnsAB and Mn^{2+}, in which the donor substrate undergoes double-stranded cleavage predominantly at one end of the transposon. Subsequently the cleaved end undergoes and intramolecular end-joining reaction with the opposite end of the transposon, which is stabilized by the transposase nucleoprotein complex (gray sphere), producing the second double-stranded break. Intermolecular transposition does not proceed beyond this point in the absence of TnsC (Biery et al., 2000).

synthesis of the second strand, which turns out to be the lagging strand, transient single-stranded gaps occur due to removal of the RNA primers.

The involvement of host factors in the integration and excision of bacteriophages has been well documented. Histonelike proteins such as integration host factor (IHF), HU and Fis play key roles. Similarly host factors play an important role in Mu transposition. Much less is known about host factors involved in transposition of other transposable elements. Some of the factors thought to play a role in transposition include the aforementioned histonelike proteins, DnaA, protein

chaperone/proteases (ClpX, ClpP, ClpA), SOS control proteins (LexA, RecAB), Dam DNA methylase, and DNA gyrase (Mahillon and Chandler, 1998). Two host factors involved in Tn7 site-specific transposition have been identified, and their identity is a surprise. Instead of being DNA-bending proteins like IHF and HU, which aid in the formation of the phage lambda recombinase complex and the Mu transpososome complex, the proteins that stimulated Tn7 transposition proved to be acyl carrier protein (ACP) and ribosomal protein L29 (Sharpe and Craig, 1998). These proteins aid the TnsABC + TnsD transposition reaction specifically. Their role is unclear. Although they are not essential, they do stimulate site-specific Tn7 transposition both in vitro and in vivo.

E. IS911—Another Maverick

IS911 is an insertion sequence than can form covalently closed circles when it excises (Polard and Chandler, 1995b; Polard et al., 1992). Similar circular transposition intermediates have been observed for other members of the IS3 family (Lewis and Grindley, 1997; Sekine et al., 1994). IS911 circle formation has been demonstrated both in vivo and in an in vitro system (Ton-Hoang et al., 1997, 1998). A model for the excision of IS911 is shown in Figure 8. During circle formation in vivo, the replicon from which the IS excises is sometimes recovered but it seems likely that resealing of the replicon during the excision reaction is not the normal course of events.

The excised circular form can integrate efficiently in vitro. Curiously integration was most efficient if the ends of IS911 were separated by 3 bp in the circular form. These 3 bps are present in the excised circles seen in the excision reaction. This is reminiscent of the coupling sequences of conjugative transposons (see later section), which consist of 4 to 6 bp of DNA adjacent to the true ends of the element and appear between the ends of the element in the excised circular form. In addition to circles, linear forms of IS911 have also been seen in vivo (Ton-Hoang et al., 1999). This raises the question of whether the true transposition intermediate is a circle or a linear DNA molecule. The transposase of IS911 can cleave the circles to form linear segments. Evidence from in vitro studies tends to support formation of the circles, then linearization of them rather than the direct transfer of the linear form by a cut-and-paste mechanism. In the *in vitro* system, however, integration by the circular form was more efficient than integration of the linear form. So, although much progress has been made toward understanding the steps in transposition of this interesting IS element, questions still remain about the identity of the true transposition intermediate.

Another unusual feature of IS911 is its transposase, which like some other members

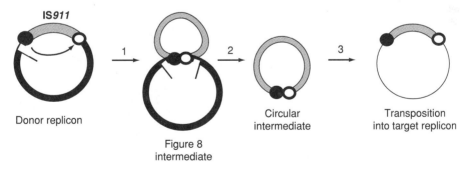

Figure 8
intermediate

Fig. 8. Schematic model for the excision and transposition of IS911. Transposon DNA circles (gray) are generated by a two-step intramolecular transposition process. In the first step (**1**), one strand end of the transposon (open circles) undergoes a single-stranded cleavage event and the exposed end transfers to the opposite end of the transposon, resulting in circularization of one strand and the formation of a figure 8 transposition intermediate. The freed 5′ end is also shown. The mechanism by which the second strand circularization (**2**) occurs has not yet been elucidated but is thought to occur either by replication (which would generate a copy of the donor molecule) or by second-strand cleavage. The formation of the circular transposition intermediate results in high-level expression of transposition proteins that cleave a single strand at both ends of IS911, and transposition into a target replicon (**3**) (Polard and Chandler, 1995b; Turlan et al., 2000).

of the IS*3* family (Chandler and Fayet, 1993) is encoded by two slightly overlapping *orfs* (*orfA* and *orfB*). They are translated into a single protein (OrfAB) by a programmed −1 translational frameshift at an A6G heptanucleotide sequence in the 3′ end of *orfA* (Fig. 2). In addition to the OrfAB fusion protein, OrfA is also made. OrfA contains a helix-turn-helix motif that may have a role in the recognition and binding of the ends of the transposon. OrfB contains a DD(35)E motif that has been determined to be the catalytic center of many transposase proteins (Andrake and Skalka, 1996; Grindley and Leschziner, 1995; Mahillon and Chandler, 1998; Polard and Chandler, 1995a). When OrfA is present along with OrfAB in vivo, the number of circle forms is reduced (Polard et al., 1992), but the OrfAB-dependent integration process was shown to be greatly enhanced by the concomitant production of OrfA (Ton-Hoang et al., 1998). Thus OrfA may act as a switch that influences OrfAB to move in the direction of insertion rather than excision.

III. INTEGRONS

Integrons are genetic elements that have an unusual property: they recruit and assemble gene cassettes into an operon structure, using a site-specific recombination mechanism (Collis et al., 1993; Collis and Hall, 1992a, 1992b). The structure of an integron and the mechanism of gene cluster assembly are illustrated in Figure 2. An integron consists of an integrase gene (*intI*) and an attachment site (*attI*) into which circular gene cassettes are inserted. Gene cassettes consist of a single promoterless open reading frame, which is covalently attached to a recombination site called the 59 bp element ("59 be" or *attC*). The 59 be sequences range in size from 57 to 141 bp. They also vary in sequence, even within a single bacterial strain, but are defined by having two core site sequences (GTTRRRY), that are central to the recombination process. The integron-encoded integrase recognizes the 59 be, and mediates site-specific recombination between the 59 be in a circularized gene cassette and the *attI*

(Hall et al., 1991; Martinez and de le Cruz, 1990; Stokes et al., 1997). The integron also provides a promoter that drives expression of the adjacent gene cassette cluster (Fig. 2).

Although integrons are unable to catalyze self-transposition, they are often associated with other mobile elements such as IS elements and transposons. Integrons have also been found in conjugative plasmids, which are able to serve as vectors for the mobilization and transfer of integrons and their associated gene cassettes (Rowe-Magnus et al., 2001).

Eight different classes of integrons have been identified, based on the sequences of their respective integrase genes (Arakawa et al., 1995; Clark et al., 2000; Mazel et al., 1998; Nield et al., 2001; Recchia and Hall, 1995; Sundstrom and Skold,). Classes 1, 2 and 3 are all associated with multidrug resistance cassettes, while classes 4 through 8 have not yet been linked to antibiotic resistance cassettes.

The first-described integrons were relatively small, containing less than five antibiotic resistance genes in an operon array (Hall and Collis, 1995; Recchia and Hall, 1995). More recently much larger integrons called superintegrons have been identified in *Vibrio* spp. (Clark et al., 2000, 1997; Mazel et al., 1998; Rowe-Magnus et al., 2001), and it has also become clear that genes other than antibiotic resistance genes, can be found in integrons. The class 4 superintegron found in *Vibrio cholerae* is 126 kb in size and contains 179 potential genes. Many of these genes have no known function, but some encode virulence factors. These superintegrons differ from the smaller integrons in a number of ways. First, the 59 be equivalents of the superintegrons (called *Vibrio cholerae* repeats— VCRs) have sequences that are very similar. Second, a number of the gene cassettes in the superintegrons contain their own promoters, and some of the *orfs* on the cassettes are not transcribed in the same direction (Rowe-Magnus et al., 2001).

The discovery of such large integrons has led scientists to speculate that such elements may have played an important role in the evolution of bacterial genomes. So far all genes carried by integrons that have an identifiable function have been genes needed for adaptation (e.g., antibiotic resistance genes) rather than essential genes. This observation further supports the notion that integrons have evolved to make bacteria better able to adapt to new conditions. That is, the bacterial chromosome has a conserved core of essential genes but acquires and rearranges other genes that increase the range of conditions under which the bacterium can grow.

The origin of gene cassettes remains a mystery. One recently advanced hypothesis is that a reverse transcriptase makes a DNA copy of an mRNA molecule (Recchia and Hall, 1995). Then a 59 be is added to one end of the DNA copy of the open reading frame, and the molecule circularizes to make the gene cassette. The notion that mRNA might be the template arises from the observation that the open reading frames in gene cassettes often have no promoters and are thus like copies of an mRNA transcript of a gene. Many bacteria have reverse transcriptase enzymes, which could make DNA copies (cDNA) of mRNA molecules. This model does not explain where the 59 be sequence comes from or how it is attached to the DNA copy of the mRNA made by reverse transcriptase.

IV. RETROTRANSPOSONS (GROUP II INTRONS)

Retrotransposons have been found widely in eukaryotic cells, from eukaryotic microbes to humans (Michel et al., 1989). At first, it appeared that bacteria lacked such elements. But some enterprising scientists reasoned that since genes in mitochondria and chloroplasts had introns, and since these organelles were probably once bacteria, it seemed likely that bacteria might have them too (Ferat and Michel, 1993). To find the putative introns, they constructed PCR primers based on known intron sequences. They first looked for PCR products in *Azotobacter vinlandii*

and in the cyanobacterium *Calothrix*, since these bacteria were closely related to mitochondria and chloroplasts, respectively. They were successful in amplifying DNA that had significant sequence similarity to group II introns of eukaryotes.

Group II introns encode a catalytic RNA molecule, which is the intron mRNA, and a multifunctional protein (*iep*, intron-encoded protein). Together with the intron RNA itself, the intron-encoded protein promotes splicing of both DNA and RNA molecules. Transcripts of group II introns are characterized by having an RNA structure with six stem-loop domains (Fig. 9), which are required for catalytic activity (Martinez-Abarca and Toro, 2000; Michel et al., 1989). The intron RNA copy inserts site-specifically in the same location in a given allele (*homing*) via a single-stranded RNA intermediate, and they are also able to transpose, at low frequency, to other novel sites that have homology to the homing site. It seems that these retrotransposons have developed a clever strategy to avoid any deleterious effects on their host that may be caused by their insertion, since if they insert into an essential gene they can be post-transcriptionally removed from the mRNA by RNA splicing.

The best-characterized bacterial group II intron, and the only one shown to be fully functional in vivo, is the retrotransposon Ll.LtrB from *Lactococcus lactis*. This retrotransposon was first identified within a gene, *ltrB*, which encodes a relaxase protein. The *ltrB* gene is located on the conjugal plasmid pRS01 and is essential for initiating plasmid transfer (Fig. 2). Conjugal transfer of pRS01 requires splicing of the intron called L1.LtrB from the exon, *ltrB* (Mills et al., 1996). L1.LtrB encodes a multifunctional intron-encoded protein called LtrA, which has several activites. These activities include reverse transcriptase activity (for cDNA synthesis), maturase activity (to stabilize catalytic RNA structure and promote splicing), and a DNA endonuclease activity (for site-specific cleavage of dsDNA recipient alleles) (Matsuura

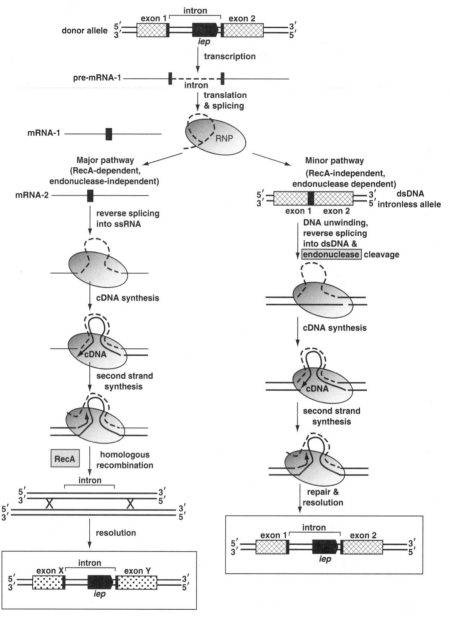

Fig. 9. Proposed pathways for transposition of bacterial retrotransposon L1.ltrB. Retrotransposition is initiated by a ribonucleoprotein complex (RNP) (gray background), which is comprised of the excised intron RNA (dashed line) and the product of intron-encoded protein gene (*iep*). The *iep* gene product has three activities: reverse transcriptase, maturase, and endonuclease. In both the major and minor retrotransposition pathways, the pre–mRNA-1 message is transcribed, spliced, and *iep* translated. Then in the major pathway (left, RecA dependent–endonuclease independent), the intron RNA is reverse spliced into the target site (black box) in a ssRNA molecule designated mRNA-2 (maturase activity), of which a cDNA copy is generated (RT activity). A second strand of cDNA is synthesized, displacing the RNA. This step is followed by RecA-dependent recombination of the integrated intron into a dsDNA target molecule. In the minor pathway (right, RecA independent–endonuclease dependent) the intron RNA molecule is reverse spliced into one strand of dsDNA target site, an endonuclease-dependent step. A cDNA copy of the inserted RNA is made in the antisense strand and then in the sense strand.

et al., 1997). LtrA also contains a another conserved domain whose function is unknown (Fig. 2).

Two pathways have been proposed for the movement of bacterial retrotransposons (Fig. 9; Cousineau et al., 2000). In the major pathway the single-stranded intron RNA molecule invades an mRNA molecule. This requires only the reverse transcriptase and maturase activities of LtrA. The reverse transcriptase activity allows the enzyme to make a double-stranded copy of the DNA segment that contains the integrated RNA intermediate. Since only a linear segment of double-stranded DNA results, homologous recombination is necessary to integrate the segment carrying the integrated retrotransposon into the genome. The need for homologous recombination might seem to violate the definition of a transposon, but technically this is not the case because the initial integration of the retrotransposon itself does not involve RecA. Only the capture and fixing of the already integrated DNA copy of the retrotransposon requires RecA.

The minor pathway, also shown in Figure 9, has no requirement for RecA, and requires all three LtrA activities. In this pathway the intron inserts into one strand of a copy of an exon in dsDNA. This pathway begins with nicks being made in both strands of the DNA target due to the endonuclease activity of the LtrA protein. The single-stranded RNA copy of the retrotransposon integrates into one of the DNA strands at the nick. The other nick serves as a starting point for a DNA copy of the integrated RNA molecule. This copy is created by the reverse transcriptase activity of LtrA. Synthesis of the single-stranded DNA copy displaces the DNA strand that was formerly in place, ensuring that new DNA molecule will contain a copy of the retrotransposon. DNA replication to form the other strand starts from a nick at the end of the integrated RNA molecule on the other strand, displacing the RNA segment and resulting in a DNA segment that contains an integrated DNA form of the retrotransposon (Cousineau et al., 2000).

In addition to the genera already mentioned, retrotransposons have now been found in diverse bacterial species including *Escherichia, Clostridium, Pseudomonas, Sinorhizobium, Bacillus Mycobacterium, Streptomyces, Yersinia, Shigella, Serratia*, and *Sphingomonas* (Martinez-Abarca and Toro, 2000; Zimmerly et al., 2001). This, together with the observation that most of these introns are associated with mobile genetic elements, suggests that retrotransposons may even be more widespread in bacteria than previously suspected. Finding retrotransposons in bacteria has provided evidence to support the hypothesis that these are very ancient elements. The first introns found in prokaryotes were found integrated into rRNA molecules. More recently, however, retrotransposons have been found in other genes as well (Cousineau et al., 1998, 2000; Martinez-Abarca and Toro, 2000; Zimmerly et al., 2001). In the future, analysis of bacterial and archaeal genome sequences will likely reveal more such elements. The function of these retrotransposons in bacteria is still a mystery. Like introns in the nuclear mRNA of eukaryotes, group II introns splice via a lariat intermediate. This has led to the suggestion that retrotransposons may be the progenitors of the nuclear spliceosomal introns of eukaryotes.

V. CONJUGATIVE AND MOBILIZABLE TRANSPOSONS

A. Conjugative Transposons

Conjugative transposons (CTns) are integrated DNA elements that can excise themselves from the chromosome (or plasmid) into which they are integrated and form a circular transfer intermediate (Salyers et al., 1995a,b; Scott, 1992). A single-stranded copy of this circular intermediate is then transferred by conjugation to a recipient cell, where it becomes a double-stranded circle and integrates into the recipient's genome. CTns, like nonconjugative transposons, are a very diverse group of elements. First found in the Gram-positive cocci, they have now been found in the Gram-negative *Bacteroides-Prevotella* phylogenetic group and in the proteobacteria. So far, in the proteobacteria, integrated transmissible elements have been found in *Vibrio cholerae, Salmonella seftenberg, Pseudomonas putida* F1, and *Mesorhizobium loti* (Table 1) (Hochhut et al., 1997; Ravatn et al., 1998; Sullivan and Ronson, 1998; Waldor et al., 1996). The fact that the existence of conjugative transposons has been slow to emerge from studies of the proteobacteria is not surprising, given that for a long time no one was looking for them. Conjugative transposons are usually found by accident because there is no simple procedure similar to a plasmid preparation procedure for identifying them. Moreover they have proved to be so diverse at the DNA sequence level that no DNA signatures have been identified. The diversity in sequence and characteristics has generated a sometimes confusing diversity in nomenclature. Some conjugative transposons, like Tn*916*, have been given a transposon designation. Others have been designated by a CTn label (e.g., CTnERL from *Bacteroides*). The conjugative transposon from *Vibrio cholerae* has been given its own special name, Constin.

Conjugative transposons range greatly in size, from 18 kb (Tn*916*, *Enterococcus faecalis*), to approximately 500 kb (*Mesorhizobium loti*), but most seem to be in the 50 to 80 kb size range (Table 1). The fact that they are generally larger than class I and II transposons is no surprise, since the conjugative transposons carry genes that encode the mating apparatus as well as genes required for excision and integration. The question of how small a conjugative transposon can be is complicated by the fact that there are small (5–12 kb) elements that are mobilized *in trans* by conjugative transposons but are not self-transmissible. These have been called mobilizable transposons (MTns). Thus a very small element can appear to be "self-transmissible," although it is actually being mobilized by a conjugative transposon element in the same strain. A criterion that has

been used in the past to determine whether a transferred element is self-transmissible is that once transferred, it can retransfer from the recipient cell to another cell. This is not a reliable criterion, at least in the case of *Bacteroides* conjugative transposons, because over 80% of recent *Bacteroides* isolates carry a conjugative transposon capable of mobilizing these smaller elements (Shoemaker et al., 2001). Also a conjugative transposon can cotransfer with the element it mobilizes.

Despite their diversity the conjugative transposons do have some features in common. First, they all have a circular transfer intermediate. These circular intermediates have been demonstrated by PCR amplification of the joined ends of the circular form, and by Southern blot analysis using the joined-end sequences as probes (Cheng et al., 2001). In the case of Tn*916*, the circular form has been isolated from donor cells and used to transform recipients (Scott, 1992; Scott and Churchward, 1995; Scott et al., 1988). When this was done, the circular form integrated efficiently in the recipient. Thus, there seems to be little question that the circular intermediate is the true transposition intermediate. However, as mentioned in an earlier section, some members of the IS*3* family of insertion sequences, such as IS*911*, do have circular transposition intermediates, although the circle is not transferred to another cell by conjugation. Thus having a circular intermediate does not necessarily make an element a conjugative transposon. A second common feature, which distinguishes conjugative transposons from other transposons that practice conservative transposition, is that when they excise, they reseal the donor molecule (Fig. 1). So far there have been no examples of conjugative transposons that transpose by a replicative mechanism.

It makes sense that there should be a circular intermediate because the next step after excision is transfer of a single-stranded copy of the conjugative transposon to the recipient. Since the transfer origin at which strand transfer begins (*oriT)* is in the middle of the

TABLE 1. Conjugative Transposons and Mobilizable Transpo Sons from Gram-Positive and Gram-Negative Bacteria

Type	Size (kb)	First Identified in:	Marker Genes	Similarities to Other CTns and MTns	References
Tn916	18.3	*Enterococcus faecalis*	*tetM*	Tn1549	Franke and Clewell, 1981
Tn918	16.0	*Enterococcus faecalis*	*tetM*	Tn916	Clewell et al., 1985
Tn919	15.4	*Streptococcus sanguis*	*tetM*	Tn916	Fitzgerald and Clewell, 1985
Tn925	16.0	*Enterococcus faecalis*	*tetM*	Tn916	Guffanti et al., 1991; Stratz et al., 1990; Torreset al.,1991
Tn1545	25.3	*Streptococcus pneumoniae*	*tetM* *aphA-3* *ermAM*	Tn916	Courvalin and Carlier, 1987; Le Bouguenec et al., 1990; Poyart-Salmeron et al., 1989
Tn1549	34.0	*Enterococcus faecalis* 268–10	*vanB*	Tn916	Garnier et al., 2000
Tn3701	60.0	*Streptococcus pyogenes*	*tetM* *ermAM*	Tn3703/Tn3951	Le Bouguenec et al., 1988; Le Bouguenec et al., 1990
Tn3702	18.0	*Enterococcus faecalis*	*tetM*	Tn916	Horaud et al., 1990
Tn3703	19.7	*Streptococcus pyogenes*	*tetM* *ermAM*	Tn916	Le Bouguenec et al., 1990
Tn3704	20.3	*Streptococcus anginosus*	*tetM* *ermAM*	Tn916	Clermont and Horaud, 1994
Tn3872	nd	*Abiotrophia defectiva* (formerly *Streptococcus defectivus*)	*ermB*	Tn916	Poyart et al., 2000
Tn3951 (*cat-erm-tet*)	67.0	*Streptococcus agalactiae* B109	*tetM* Cm^r *ermAM*	Tn3701	Inamine and Burdett, 1985
Tn4371	55.0	*Ralstonia eutropha* (formerly *Alcaligenes eutrophus*)	Bph+	none	Merlin et al., 1999
Tn5031	70.0	*Enterococcus faecium*	*tetM*	Tn916	Fletcher et al., 1989; Horn et al., 1991

TABLE 1. (*Continued*)

Type	Size (kb)	First Identified in:	Marker Genes	Similarities to Other CTns and MTns	References
Tn5032	nd	*Enterococcus faecium*	*tetM*	Tn916	Fletcher et al., 1989
Tn5033	nd	*Enterococcus faecium*	*tetM*	Tn916	Fletcher et al., 1989
Tn5251	18.0	*Streptococcus pneumoniae*	*tetM*	Tn916, Tn5253.	Ayoubi et al., 1991; Vijayakumar et al., 1986
Tn5252	47.5	*Streptococcus pneumoniae*	Cm^r	Tn916, Tn3951; Tn5253	Vijayakumar and Ayalew, 1993
Tn5253 Ω (*cat-tet*) element	65.5	*Streptococcus pneumoniae*	*tetM* Cm^r	Tn916. A Tn5251::Tn5252 composite.	Ayoubi et al., 1991
Tn5256	70.0	*Lactococcus lactis* NIZO R5	Nip$^+$, Suc$^+$	nd	Rauch and De Vos, 1992
Tn5276	70.0	*Lactococcus lactis* NIZO R5	Nip$^+$, Suc$^+$	nd	Rauch and De Vos, 1992
Tn5278	70.0	*Lactococcus lactis*	Nip$^+$, Suc$^+$	nd	Rauch et al., 1994
Tn5301	70.0	*Lactococcus lactis* NCFB894	Nip$^+$, Suc$^+$	nd	Rauch et al., 1994
Tn5301	68.0	*Lactococcus lactis*	Nip$^+$, Suc$^+$	nd	Horn et al., 1991
Tn5306	60.0	*Lactococcus lactis*	Nip$^+$, Suc$^+$, CEO$^+$	nd	Thompson et al., 1991
Tn5307	+/−70.0	*Lactococcus lactis*	Nip$^+$, Suc$^+$	nd	Broadbent et al., 1995; Thompson et al., 1991
Tn5381	18.0	*Enterococcus faecalis*	*tetM*	Tn916	Rice et al., 1992
Tn5382	27.0	*Enterococcus faecium*	*vanB*	Tn916, Tn1549.	Carias et al., 1998
Tn5383	18.0	*Enterococcus faecalis* DS16	*tetM*$^{-r}$	Tn916	Rice et al., 1992
Tn5385	65.0	*Enterococcus faecalis* strains CH19 and CH116 (clinical isolates)	*ermAM* *aac6'-aph2''* *merRAB* *aadE* *tetM* *bla*	Tn916, Tn5384, Tn4001, Tn552.	Rice and Carias, 1998
Tn5397	20.0	*Clostridium difficile*	*tetM*	Tn916 (partial), Tn4451a, Tn4453a	Mullany et al., 1996; Wang et al., 2000a

TABLE 1. (*Continued*)

Type	Size (kb)	First Identified in:	Marker Genes	Similarities to Other CTns and MTns	References
Tn*5398*	nd	*Clostridium difficile*	*ermB*	nd	Mullany et al., 1995
CTn*scr94*	100.0	*Salmonella senftenberg* 5494-57	Suc^+	nd.	Hochhut et al., 1997
clc element	105.0	*Pseudomonas putida* F1	clc^+	nd	Ravatn et al., 1998
unnamed	500.0	*Mesorhizobium loti* (formerly *Rhizobium loti*)	biotin thiamin nitrogen fixation nodulation	nd	Sullivan and Ronson, 1998
ICE*St1*	35.5	*Streptococcus thermophilus* CNRZ368	nd	Tn*916*	Burrus et al., 2000
XBU4422	60	*Bacteroides uniformis* 1001	none	CTnERL	Bedzyk et al., 1992
CTnERL	52	*Bacteroides fragilis* ERL	*tetQ*	CTnDOT	Valentine et al., 1988
CTnDOT	65	*Bacteroides thetaiotaomicron* DOT	*tetQ* *ermF* *tetX* *aadS*	CTnERL	Valentine et al., 1988
CTn*7853*	70	*Bacteroides thetaiotaomicron* 7853	*tetQ* *ermG*	none	Nikolich et al., 1994
CTn*12256* (also Tn*5030*)	150.0	*Bacteroides fragilis* 12256 (also *Bacteroides fragilis* V503)	*tetQ* *ermF* *tetX* *aadS*	composite of an unknown CTn and CTnDOT	Halula and Macrina, 1990; Valentine et al., 1998
SXT element	62.5	*Vibrio cholerae* 0139	nd -Sul^r nd -Tp^r nd -Cm^r nd -Str^r	IncJ conjugal plasmid R371	Hochhut et al., 2000; Hochhut and Waldor, 1999; Waldor et al., 1996

TABLE 1. (*Continued*)

Type	Size (kb)	First Identified in:	Marker Genes	Similarities to Other CTns and MTns	References
Tn*B123*	40.0–60.0	*Butyrivibrio fibrisolvens* 2221	nd -Tc^r	Tn*916* (weak)	Scott et al., 1997
NBU1[a]	10.3	*Bacteroides uniformis*	none		Shoemaker et al., 1993
NBU2[a]	11.1	*Bacteroides fragilis* ERL	*mefE* *linA*		Shoemaker and Salyers, 1988; Wang et al., 2000b
NBU3[a]	10.0	*Bacteroides fragilis* 12256	unknown		Bedzyk et al., 1992; Shoemaker and Salyers, 1988
Tn*4555*[a]	12.5	*Bacteroides vulgatus* CLA341	*cfxA*		Smith and Parker, 1993; Smith and Parker, 1996; Tribble et al., 1999a; Tribble et al., 1999b
Tn*4399*[a]	9.6	*Bacteroides fragilis* TM4.2321	none		Hecht and Malamy, 1989; Murphy and Malamy, 1993
Tn*5520*[a]	4.7	*Bacteroides fragilis* LV23	none		Vedantam et al., 1999
Tn*4451*[a]	6.4	*Clostridium difficile*	*catP*	Tn*4453a* and Tn*4453b*	Bannam et al., 1995; Crellin and Rood, 1997; Crellin and Rood, 1998; Wang et al., 2000a
Tn*4453a/b*[a]		*Clostridium perfringens*	*catD*	Tn*4451*	Lyras et al., 1998

Note: Resistance genes are specified where known and are as follows: cefoxitin (*cfxA*), chloramphenicol (*catD, catP*), erythromycin (*mefE*), erythromycin and MLS$_B$- group antibiotics (*ermAM, ermB, ermF, ermG*), kanamycin (*aphA-3*), gentamycin (*aac6′-aph2″*), lincomycin (*linA*), mercuric chloride (*merRAB*), penicillins (*bla*), streptomycin (*aadS, aadE*), tetracycline/minocycline (*tetM, tetQ, tetX*), and vancomycin (*vanB*). Where the source of antibiotic resistance is unknown, the resistance phenotype is indicated as follows: chloramphenicol (Cm^r), streptomycin (Str^r), sulfomethoxazole (Sul^r), tetracycline (Tc^r) and trimethoprim (Tp^r). Other phenotypes associated with these elements are as follows: Bph+: ability to catabolize biphenyl; *clc*+: ability to degrade chlorocatechol; Nip+: nisin production; Suc+: sucrose utilization; CEO+: N5-(carboxyethyl)ornithine synthetase activity.
[a] Distinguishes nonconjugative but mobilizable transposons.

element, not at the ends, it is necessary to have a circular intermediate if the entire conjugative transposon is to be transferred. The fact that conjugative transposons must first be nicked at the ends (for excision), then later nicked at an internal *oriT* (conjugal transfer) raises the question of how these different nicking reactions are coordinated. So far, there have been no reports of Hfr-type transfer of part of a conjugative transposon, a process that would be possible if nicking at the *oriT* occurred independently of nicking at the ends of the element. Tn*916* solves the coordination problem in a simple but elegant way. There is a promoter in one end of Tn*916* that is only placed upstream of the transfer genes when the excised element assumes its circular form. Thus the transfer genes are expressed only when excision is completed (Celli and Trieu-Cuot, 1998). This appears not to be the case for the *Bacteroides* conjugative transposons such as CTnDOT. For one thing, the transfer region is too far removed from either end (Bonheyo et al., 2001; Li et al., 1995a). For another, the regulation of excision and transfer functions is controlled by regulatory proteins (RteA, RteB, and RteC), which are encoded near the middle of the element (Stevens et al., 1993).

Finally, a feature that all conjugative transposons share is that they integrate into the recipient's genome after they transfer. Such integration events can be detected by pulsed field electrophoresis (Bedzyk et al., 1992). Most of the conjugative transposons are large enough to give a detectable band shift on a pulsed field gel of recipient DNA. The integrases of the conjugative transposons, which catalyze integration, have so far proved to be members of the phage lambda integrase family. They share little amino acid sequence similarity with lambda integrase throughout much of the protein, but they have the conserved residues that are located in the phage integrase catalytic site. The conjugative transposon integrases are just as distantly related to each other, sharing little more than residues in the presumed catalytic region.

A model for the excision and integration of Tn*916* is shown in Figure 10 (Bringel et al., 1992; Caparon and Scott, 1989; Marra and Scott, 1999; Rudy and Scott, 1994). Single-stranded staggered cuts are made 4 to 6 bp from the ends of the element. These short segments of chromosomal DNA are called coupling sequences. When the DNA segments are ligated to form the circular intermediate, a small region of non-base-paired sequence is created. Presumably this heterology is resolved in favor of one coupling sequence or the other, but the mechanism of resolution is not known. Transfer of a single-stranded copy of the conjugative transposon would also accomplish this. What is not clear is whether resolution actually occurs in the donor prior transfer. If the mechanism shown in Figure 10 is correct, the relationship of the conjugative transposon integrases to phage lambda is hard to understand, given that their mechanism of excision and integration appears to be so different from that of phage lambda (see Hendrix, this volume).

Conjugative transposons differ in their site-specificity. Tn*916* integrates almost randomly in most hosts, although it prefers AT-rich regions (Ike et al., 1992; Scott et al., 1994; Trieu-Cuot et al., 1993). The *Bacteroides* element, CTnDOT, integrates site-selectively into about seven sites (Bedzyk et al., 1992). Integration into these sites is orientation specific. There is a 10 bp sequence located adjacent to the integration site that is also found adjacent to one end of CTnDOT. This sequence may explain the site-selectivity and orientation-specificity of CTnDOT integration (Cheng et al., 2000).

An important breakthrough has been made in the case of Tn*916*, in the form of the first in vitro system for a conjugative transposon. Churchward and colleagues (Jia and Churchward, 1999) have shown that integrase protein encoded by Tn*916* binds to the ends of the element. Finally, they have used DNA footprinting to show exactly where the integrase binds, and they have shown that the protein encoded by a

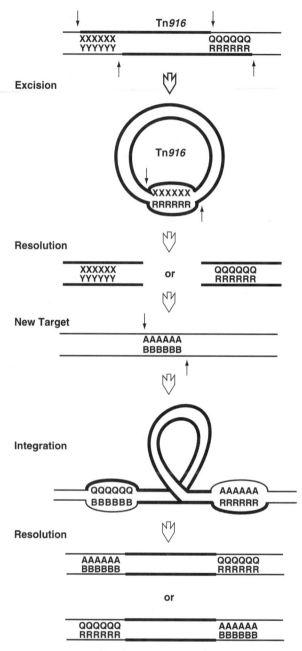

Fig. 10. Model for excision of conjugative transposon Tn916 (Scott and Churchward, 1995). Staggered cuts are made into chromosomal DNA 4 bp to 6 bp from the true end of the element. The chromosomal sequences are called coupling sequences. The coupling sequences do not base-pair with each other and hence have been indicated by X(Y)s or Q(R)s. Thus when a conjugative transposon excises from the donor backbone, a 6 bp region of heterology is formed. The heterology may be resolved before the circle integrates either by repair enzymes or during conjugation (Manganelli et al., 1997; Rudy and Scott, 1994). Integration is similar to excision in the sense that small regions of heterology are formed adjacent to the ends of the integrating element.

gene called *xis*, which is necessary for excision of Tn*916*, also binds near the ends of Tn*916* (Jia and Churchward, 1999; Rudy et al., 1997a).

B. Mobilizable Transposons

Some transposonlike mobile elements, which are much smaller than the conjugative transposons, are also transferred by conjugation. These elements do not carry the genes required to form the mating apparatus through which they are transferred, but they instead rely on transfer proteins supplied by other self-transmissible elements such as conjugative transposons. Because of this they have been called mobilizable transposons (MTns). So far conjugative transposons have been found to be the mobilizing element that supplies the transfer proteins, but in theory plasmids could also provide transfer functions (Salyers et al., 1995a, 1999). In fact IncP plasmids such as RK2 can mobilize a plasmid carrying the mobilization region of such *Bacteroides* mobilizable transposons as NBU1 (Li et al., 1995b). Although this is an artificial situation, in which a portion of a mobilizable transposon was cloned into a plasmid, it shows that plasmid mobilization of the circular forms of a mobilizable transposon is possible.

The conjugative transposons that transfer mobilizable transposons also carry genes that trigger the mobilizable transposon to excise and form the circular intermediate. In the case of the CTnDOT type of conjugative transposon, the genes necessary for triggering excision, encode regulatory proteins (RteA, RteB) which presumably interact with sequences or proteins provided by the mobilizable transposon to initiate its excision. As already mentioned, RteA and RteB also control the expression of genes that encode essential conjugative transposon functions, such as excision and conjugal transfer.

The mobilizable transposons studied to date all contain a transfer origin (*oriT*) and one or two genes that encode mobilization proteins, proteins that form the relaxosome that creates a single-stranded nick at the *oriT*

to initiate the conjugal transfer process (Fig. 2). In theory, since most mobilization proteins are capable of acting *in trans*, there could be mobilizable transposons that carry only an *oriT* and derive mobilization proteins as well as the mating apparatus proteins from another source. So far no such element has been discovered. Mobilizable transposons have been found in *Bacteroides* species and some Gram-positive bacteria (Table 1), but they may well have a wider distribution. Most of the mobilizable transposons have been given Tn designation but other names, such as NBUs, have also been used.

Mobilizable transposons exhibit considerable diversity in their excision and integration mechanisms, as will be illustrated by the examples given in this section, but they do have some common properties in addition to their dependence on the transfer machinery provided by the conjugative transposon. Like the conjugative transposons, they have a circular transfer intermediate, and their integrase proteins are members of the phage lambda integrase family. Another shared feature is that they carry one or two mobilization genes that encode the protein(s) that make a single-stranded nick at the internal *oriT* that initiates the transfer of a copy of the circular intermediate.

NBUs are 10 to 12 kb mobilizable transposons found in *Bacteroides* species. The name NBU stands for "nonreplicating Bacteroides unit," but it is now clear that they are mobilizable transposons (Li et al., 1993, 1995b; Shoemaker et al., 2000; Wang et al., 2000b). NBU-related elements are very widespread in *Bacteroides* species. In a recent survey of *Bacteroides* strains representing over 10 *Bacteroides* species, DNA that cross-hybridized with the NBUs was found in over half of the strains tested (Wang et al., 2000b). The best-studied NBUs are NBU1 and NBU2. Both of these have a single integrase gene and integrate site-specifically. Integration is the one NBU activity that is independent of the conjugative transposons. NBU1 integrates into the 3′ end of a leucine *tRNA* gene. NBU2 has two primary sites, but both of these are identical and are in serine *tRNA* genes, which are located at different positions in the *Bacteroides* chromosome. NBUs have a 14 bp *att* site that is created when the NBU ends are joined to form the circular transfer intermediate (Shoemaker et al., 1993 & 1996). This *att* site is identical to a 14 bp chromosomal target site and the site is duplicated when the NBU integrates. The mechanism of integration and the preference for the 3′ ends of tRNA genes are very similar to properties of the lambdoid phages. Moreover the integrases of NBU1 and NBU2 have amino acid sequence signatures found in members of the phage lambda integrase family. The integrases of NBU1 and NBU2 are only distantly related to the phage integrases (Shoemaker et al., 1996; Wang et al., 2000b).

The excision machinery of NBU1 is complex, involving the integrase plus at least four other proteins (Fig. 2). The *oriT* is part of the region required for excision. Excision is controlled by two genes carried on the CTnDOT family of conjugative transposons, *rteA* and *rteB*. The proteins encoded by these genes are the sensor and response regulator, respectively, of a two component regulatory system, which also controls excision and transfer of the conjugative transposon. The NBU1-encoded mobilization protein, MobN1, nicks the circular form of the NBU at its *oriT* to initiate the conjugal transfer process (Shoemaker et al., 2000).

Given that the NBUs are so dependent on the functions of the conjugative transposons for excision and transfer, it is surprising that there seems to be virtually no significant amino acid sequence similarity between the excision proteins of the two types of elements. Although the integrases of the NBUs, like those of the conjugative transposons, are members of the lambda integrase superfamily, they are only distantly related to each other. This relationship between integrases raises an additional question. Why do two members of the lambda integrase family

carry out such apparently different reactions? The NBUs integrate in a lambda-like way, site-specifically in the ends of *tRNA* genes, whereas the conjugative transposons are much less site-specific, have coupling sequences that do not create a target site duplication and do not integrate in *tRNA* genes.

Tn*4555* is another *Bacteroides* mobilizable transposon, which is 12.1 kb in size and carries a cefoxitin resistance gene (*cfxA;* Fig. 2). In the presence of a conjugative transposon, Tn*4555* excises from the *Bacteroides* chromosome to form a circular transposition intermediate whose ends are joined by 6 bp coupling sequences. That is, its excision and integration mechanism appears to be more similar to that of the conjugative transposons than that of the NBUs. As with the NBUs, integration of Tn*4555* in the recipient chromosome is a conjugative transposon-independent process (Smith and Parker, 1993). Tn*4555* appears to utilize two pathways for integration into the *Bacteroides* genome. One is site-specific integration into either of two tandem 207 bp direct repeats which comprise the primary target sites (PT-1 and PT-2). A second pathway mediates random target selection. Although there is only limited similarity between Tn*4555* sequences and the primary target sites, 80% of the time Tn*4555* integrates into the primary target site, whilst only 20% of the time does Tn*4555* integrate into other sites (Tribble et al., 1997).

The site-specific integration of Tn*4555* is catalyzed by a lambda-like integrase. However a second protein called TnpA is also involved in the integration process and is necessary for site-specificity. Tn*4555* is the first example of a lambda-like integrase that requires a targeting protein for integration. The only other transposon known to encode such a targeting protein is Tn*7* (TnsD). Another Tn*4555* protein called TnpC appears to play a role in the efficiency of Tn*4555* integration, but it is not clear how TnpC does this (Tribble et al., 1999b). Like the NBUs, Tn*4555* encodes a single mobilization protein (MobA), which shares amino acid sequence homology with the MobN1 and MobN2 proteins of the NBUs. Tn*4555* is also unusual because, based on amino acid sequence homology, it appears to be the only *Bacteroides* element to encode a lambda Xis-like protein, which has homology to several excise proteins from other genetic elements (Tribble et al., 1999a). Whether this Xis-like protein turns out to be the only protein besides the integrase to be required for excision remains to be determined.

Tn*4399* is a 9.6 kb mobilizable transposon from *B. fragilis* that is mobilizable by IncP plasmids from *E. coli* to either *Bacteroides* or *E. coli* recipients (Murphy and Malamy, 1993). The ends of Tn*4399* contain 13 bp inverted repeats. Tn*4399* inserts fairly randomly in *Bacteroides* but has a preference for AT-rich sequences. Insertion of Tn*4399* results, in most cases, in a 3 bp duplication of the target site, with an additional 5 bp at the right-hand end (Hecht and Malamy, 1989). Thus Tn*4399* combines features of both nonconjugative transposons (target site duplication) and conjugative transposons (coupling sequences). The mechanism of Tn*4399* transposition has not yet been determined. Another unique feature of Tn*4399* is that in contrast to other *Bacteroides* mobilizable transposons, Tn*4399* has two genes that are necessary for mobilization, *mocA* and *mocB*, rather than one.

Bacteroides mobilizable transposon, Tn*5520* is the smallest of the mobilizable transposons identified to date, being only 4.7 kb in size and containing only two *orfs*. One encodes an integrase protein (*bipH*), and the other encodes a mobilization protein (*bmpH*), which has amino acid sequence similarity to the mobilization proteins from the NBUs. Integration of Tn*5520* is not site-specific. Instead, Tn*5520*, like Tn*916*, integrates preferentially into AT-rich regions in an orientation-specific way (Vedantam et al., 1999). The termini of Tn*5520* contain a 22 bp imperfect inverted repeat. Transposition into a target site does not result in target site duplication.

Tn*4451* was the first mobilizable transposon to be found in a Gram-positive bacterium (*Clostridium difficile*). It is located on a tetracycline resistance plasmid pIP401 (Abraham and Rood, 1987). Since then, two other almost identical mobilizable transposons, Tn*4453a* and Tn*4453b*, have been identified in *Clostridium* species. Both are structurally and functionally related to Tn*4451* (Lyras et al., 1998). All three of these mobilizable transposons carry a chloramphenicol resistance gene, and are able to excise precisely from multicopy plasmids. They are mobilizable both in *C. perfringens* and *E. coli*. Tn*4451* is 6.4 kb mobilizable transposon that contains six genes. One gene encodes a site-specific recombinase (*tpnX*), which contains motifs from both the Tn*3* resolvase family and the lambda integrase family of recombinases. The other genes include a single mobilization gene (*tnpZ*), a chloramphenicol resistance determinant (*catP*) and three genes of unknown function (*tnpV*, *tnpY*, and *tnpW*).

Tn*4451* integrates into target sites that contain a GA dinucleotide sequence. The Tn*4451* transposase introduces a 2 bp staggered cut at the GA dinucleotides sequence (Crellin and Rood, 1997), which, after insertion and repair, results in duplication of this GA nucleotide sequence at both ends of the inserted element. Upon excision, Tn*4451* excises precisely from its integration site, leaving a GA dinucleotide sequence at the site of excision. The other GA nucleotide separates the right and left ends of the circular transposition intermediate. Tn*4451*, like the *Bacteroides* mobilizable transposons, forms a circular transfer intermediate. In contrast to the *Bacteroides* mobilizable transposons, excision of Tn*4451* appears to occur independently of the conjugal element that is required to facilitate its mobilization (Lyras and Rood, 2000).

The frequency of excision appears to depend upon the sequence of the integration site. This is may be due to this fact that the −35 region of the *tnp* promoter is provided by chromosomal or plasmid sequences flanking the integrated transposon. A strong promoter is formed when the ends of the excised element join to form the circular intermediate, and high levels of the TnpX are produced. It has been proposed that since Tn4451 forms a nonreplicating circular intermediate, transposage expression needs to be high if Tn4551 is to be maintained in the bacterial population (Lyras and Rood, 2000). Lethality due to excessive expression of transposases has been reported (Perkins-Balding et al., 1999; Ton-Hoang et al., 1998), but since high levels of TnpX are produced only transiently, this may reduce deleterious effects on the host cell. There are other examples of mobile elements in which promoters are generated via formation of a circular transposition intermediate, including Tn916 (Celli and Trieu-Cuot, 1998) and IS911 (Ton-Hoang et al., 1998). Other examples of promoters produced by formation of joined transposon ends include those for IS492 (Perkins-Balding et al., 1999), IS21 (Reimmann et al., 1989), IS30 (Dalrymple, 1987), and IS2 (Szeverenyi I, 1996). However, in these cases the stronger promoter is created by the formation of tandem dimers during transposition, and not a circular intermediate.

VI. HOW FAR CAN THE DEFINITION OF TRANSPOSON BE STRETCHED?

A. T-DNA

The T-DNA segment of the *Agrobacterium tumifaciens* Ti plasmids is usually not classified as a transposon, but T-DNA fits within the definition of a transposable element: it is capable of excising from one DNA segment and integrating into another DNA segment in a homology-independent manner (Bevan and Chilton, 1982). In this case excision occurs in a bacterial donor, and integration occurs in a plant cell recipient. Considered as a transposon, T-DNA has a couple of unusual features. First, its "transposase" genes are not interior to the excised element. Rather, these genes are located on another part of the Ti plasmid that carries the

T-DNA segment. Transposition of T-DNA could in fact be considered to be a form of conjugation. The ends of the T-DNA segment, where the nicks occur that release a single-stranded copy of the T-DNA segment, have significant similarity to *oriT* regions of conjugal plasmids (Christie, 1997; Waters and Guiney, 1993; Zhu et al., 2000). Thus T-DNA excision might be viewed as a form of mobilization, in which, in contrast to plasmid transfer, there are two rather than one *oriT* regions (see Ream, this volume).

After a single-stranded nick is made at each end of the T-DNA the linear single-stranded DNA segment is removed from the plasmid and is replaced by DNA replication. The single-stranded DNA transfer intermediate binds single-stranded DNA-binding proteins, which presumably protect it from degradation, and the DNA-protein complex is transferred by conjugation into plant cells. In the plant cell the T-DNA integrates randomly by a process that must be catalyzed by host factors, since the T-DNA does not encode any known proteins with integrase function (Christie, 1997; Waters and Guiney, 1993; Zhu et al., 2000). This latter feature might also seem to exclude T-DNA from the family of what is labeled "transposon," but this family is becoming so inclusive that a spirit of "anything goes" seems more appropriate than excluding elements.

B. Pathogenicity and Other Genomic Islands

As long as a move is being made in the direction of inclusiveness, we might as well claim pathogenicity and genomic islands as possible transposons. Pathogenicity islands (PAIs) are defined as discrete segments of DNA, which are present in the genomes of pathogenic strains of bacteria but are absent from the genomes of nonpathogenic strains of the same or related species (Hacker and Kaper, 2000; Kaper and Hacker, 1999). PAIs typically range in size from 10 kb to more than 200 kb, although some of less than 10 kb have been identified in *Salmonella* spp. (SPI-5), *Bacteroides* spp. (BfPAI), and *Listeria* (LIPI-I, LIPI-2). These smaller segments have been designated islets rather than islands. Most PAIs are located on the bacterial chromosome, although they can sometimes be part of a bacterial plasmid, a bacteriophage or a conjugative transposon. PAIs have been comprehensively reviewed recently (Hacker et al., 1997; Hacker and Kaper, 2000; Kaper and Hacker, 1999).

PAIs are distinguished from the surrounding "core" genomic sequences by several distinctive features. Since PAIs tend to be horizontally transferred elements their %G + C content and codon useage pattern often differ from those of the rest of the host genome. Also most of the pathogenicity islands are flanked by direct repeats, and they are often associated with tRNA genes, or other genes encoding small RNA molecules. Such genes are known to be target sites for the integration of other foreign DNA elements such as phages (Cheetham and Katz, 1995). PAIs also encode genes that are commonly found on mobile genetic elements, genes required for integration and transposition. Perhaps the single most distinguishing feature of a PAI is that they encode at least one virulence factor (Hacker and Kaper, 2000).

PAIs are usually assumed to be integrating mobile elements that are spread by horizontal transfer. However, there are only two cases in which transfer of these islands from one cell to another has been demonstrated. The *Vibrio cholerae* PAI (VPI) has been shown to be transmissible from one strain of *V. cholerae* to another by phage transduction (Karaolis et al., 1999). Similarly the *Staphylococcus aureus* PAI (SaPIs) is part of a defective bacteriophage that is excised, circularized, and packaged by helper phages φ13 and φ80α (Lindsay et al., 1998). Other mechanisms of PAI transfer are likely to be discovered in the future. Some may prove to be conjugative transposons. There is also one example of a PAI that moves intracellularly, like a transposon, into any of three different *asn* tRNA genes. This PAI is found in *Yersi-*

nia pseudotuberculosis (Buchrieser et al., 1998).

With the sequencing of more microbial genomes, other horizontally acquired islands, termed *genomic islands* are being identified. These DNA segments harbor genes involved in catabolism, antibiotic resistance, secretion, and fitness in general. PAIs probably constitute a subcategory of the transposon-like elements designated genomic islands (Hacker and Kaper, 2000; Kaper and Hacker, 1999).

VII. FUTURE DIRECTIONS

The elegant studies that have allowed scientists to probe the details of the transposition process—at the atomic level in the case of the paradigm transposons—will certainly continue to yield new insights into DNA rearrangements and bacterial adaptation. One hopes to see similar advances with conjugative and mobilizable transposons, which have not yet been analyzed in such depth. The existence of an in vitro system for Tn*916* (Rudy et al., 1997b) is certainly an encouraging harbinger of advances to come. It will be interesting to see whether transposable elements such as Tn*916* and the NBUs, which now look so different mechanistically from each other, are really as different as they appear to be. An understanding of the basic unities that tie these elements together may well be revealed by future studies.

An emerging area of interest is the role of host factors in transposition. Host factors like IHF and HU have long been known to play a key role in the integration and excision of bacteriophages, and are now starting to appear as players in the transposition process. The finding that ribosomal protein L29 stimulates the binding of Tn7 protein TnsD in vitro and stimulates transposition in vivo (Sharpe and Craig, 1998) expands still further the range of possible interactions between the bacterial host and the transposons it harbors. There have been many anecdotal accounts of a connection between growth phase and transposition rates, which may be linked to the production of host factors in a growth phase-dependent way (Sharpe and Craig, 1998; Talaat and Trucksis, 2000; Weinreich and Reznikoff, 1992; Ziebuhr et al., 1999). L29 may be the beginning of an answer to the question of whether growth phase affects transposition rates, and if so, how this effect is mediated. It seems reasonable that a bacterium experiencing a bout of starvation might want to throw the genetic dice more often, and increasing transposition rates is certainly one way to do that.

Target site immunity, the protection of a region around a transposon from secondary integration events, is another understudied area of transposon biology that may yield unexpected insights in the future. This trait is widespread among known transposable elements, a fact that suggests the phenomenon may be more significant than we now realize. Also the scattered reports of a connection between conjugation and the activity of transposable elements are intriguing. Many areas of research from genome sequencing to plasmid and phage transfer of DNA are pointing to the importance of horizontal gene transfer in the evolution of prokaryotes. If so, there may be coordination between the various types of elements that contribute to DNA rearrangements and DNA transfer. But the implications could spread even further. Analysis of the human genome sequence has revealed a surprising number of integrated retrovirus-like elements interspersed throughout the genome. Insights into the role of transposable elements in bacterial evolution could well provide the key to the lock that reveals the answer to what role these integrated elements have played in mammalian evolution.

REFERENCES

Abraham LJ, Rood JI (1987): Identification of Tn4451 and Tn4452, chloramphenicol resistance transposons from Clostridium perfringens. J Bacteriol 169.

Andrake MD, Skalka AM (1996): Retroviral integrase, putting the pieces together. J Biol Chem 271:19633–19636.

Arakawa Y, Murakami M, Suzuki K, Ito H, Wacharotayankun R, Ohsuka S, Kato N, Ohta M (1995): A novel integron-like element carrying the metallo-beta-lactamase gene blaIMP. Antimicro Agents Chemother 39:1612–1615.

Arciszewska LK, Drake D, Craig NL (1989): Transposon Tn7 cis-acting sequences in transposition and transposition immunity. J Mol Biol 207:35–52.

Ayoubi P, Kilic AO, Vijayakumar MN (1991): Tn5253, the pneumococcal omega (cat tet) BM6001 element, is a composite structure of two conjugative transposons, Tn5251 and Tn5252. J Bacteriol 173:1617–1622.

Bainton R, Gamas P, Craig NL (1991): Tn7 tranposition in vitro proceeds through an excised transposon intermediate generated by staggered breaks in DNA. Cell 65:805–816.

Bainton RJ, Kubo KM, Feng JN, Craig NL (1993): Tn7 transposition: Target DNA recognition is mediated by multiple Tn7-encoded proteins in a purified in vitro system. Cell 72:931–943.

Bannam TL, Crellin PK, Rood JI (1995): Molecular genetics of the chloramphenicol-resistance transposon Tn4451 from Clostridium perfringens: The TnpX site-specific recombinase excises a circular transposon molecule. Mol Microbiol 16:535–551.

Bedzyk LA, Shoemaker NB, Young KE, Salyers AA (1992): Insertion and excision of Bacteroides conjugative chromosomal elements. J Bacteriol 174:166–172.

Bender J, Kuo J, Kleckner N (1991): Genetic evidence against the intramolecular rejoining of the donor DNA molecule following IS10 transposition. Genetics 128:687–694.

Berg DE (1989): Transposon Tn5. In Berg DE, Howe MM (eds): "Mobile DNA." Washington, DC: ASM Press, pp 185–210.

Berg DE, Howe MM (eds) (1989): "Mobile DNA." Washington, DC: ASM Press.

Bevan MW, Chilton MD (1982): T-DNA of the Agrobacterium Ti and Ri plasmids. Annu Rev Genet 16:357–384.

Biery MC, Lopata M, Craig NL (2000): A minimal system for Tn7 transposition: The transposon-encoded proteins TnsA and TnsB can execute DNA breakage and joining reactions that generate circularized Tn7 species. J Mol Biol 17:25–37.

Bonheyo G, Graham D, Shoemaker NB, Salyers AA (2001): Transfer region of a Bacteroides conjugative transposon, CTnDOT. Plasmid 45:41–51.

Bringel F, Van Alstine GL, Scott JR (1992): Conjugative transposition of Tn916: The transposon int gene is required only in the donor. J Bacteriol 174:4036–4041.

Broadbent JR, Sandine WE, Kondo JK (1995): Characteristics of Tn5307 exchange and intergeneric transfer of genes associated with nisin production. Appl Microbiol Biotech 44:139–146.

Buchrieser C, Brosch R, Bach S, Guiyoule A, Carniel E (1998): The high-pathogenicity island of Yersinia *pseudotuberculosis* can be inserted into any of the three chromosomal asn tRNA genes. Mol Microbiol 30:965–978.

Burrus V, Roussel Y, Decaris B, Guedon G (2000): Characterization of a novel integrative element, ICESt1, in the lactic acid bacterium *Streptococcus thermophilus*. Appl Environ Microbiol 66:1749–1753.

Caparon MG, Scott JR (1989): Excision and insertion of the conjugative transposon Tn916 involves a novel recombination mechanism. Cell 59:1027–1034.

Carias LL, Rudin SD, Donskey CJ, Rice LB (1998): Genetic linkage and cotransfer of a novel, *vanB*-containing transposon (Tn5382) and a low-affinity penicillin-binding protein gene in a clinical vancomycin-resistant *Enterococcus faecium* isolate. J Bacteriol 180:4426–4434.

Casadesus J, Roth JR (1989): Transcriptional occlusion of transposon targets. Mol Gen Genet 216:204–209.

Celli J, Trieu-Cuot P (1998): Circularization of Tn916 is required for expression of the transposon-encoded transfer functions: characterization of long tetracycline-inducible transcripts reading through the attachment site. Mol Microbiol 28:103–117.

Chandler M, Fayet O (1993): Translational frameshifting in the control of transposition in bacteria. Mol Microbiol 7:497–503.

Charlier D, Piette J, Glansdorff N (1982): IS3 can function as a mobile promoter in *E. coli*. Nucleic Acids Res 10:5935–5948.

Cheetham BF, Katz ME (1995): A role for bacteriophages in the evolution and transfer of bacterial virulence determinants. Mol Microbiol 18:201–208.

Cheng Q, Paszkiet BJ, Shoemaker NB, Gardner JF, Salyers AA (2000): Integration and excision of a *Bacteroides* conjugative transposon, CTnDOT. J Bacteriol 182:4035–4043.

Cheng Q, Sutanto Y, Shoemaker NB, Gardner JF, Salyers AAS (2001): Identification of genes required for excision of CTnDOT, a *Bacteroides* conjugative transposon. Mol Microbiol 41:625–632.

Christie P (1997): *Agrobacterium tumefaciens* T-complex transport apparatus: A paradigm for a new family of multifunctional transporters in eubacteria. J Bacteriol 179:3085–3094.

Ciampi MS, Schimd MB, Roth JR (1982): Transposon Tn10 provides a promoter for transcription of adjacent sequences. Proc Nat Acad Sci USA 79:5016–5020.

Clark CA, Purins L, Kaewrakon P, Focareta T, Manning PA (2000): The *Vibrio cholerae* O1 chromosomal integron. Microbiol 146:2605–2612.

Clark CA, Purins L, Kaewrakon P, Manning PA (1997): VCR repetitive sequence elements in the *Vibrio cholerae* chromosome constitute a mega-integron. Mol Microbiol 26:1137–1143.

Clermont D, Horaud T (1994): Genetic and molecular studies of a composite chromosomal element (Tn3705) containing a Tn916-modified structure (Tn3704) in *Streptococcus anginosus* F22. Plasmid 31:40–48.

Clewell DB, An FY, White BA, Gawron-Burke C (1985): *Streptococcus faecalis* sex pheromone (cAM373) also produced by *Staphylococcus aureus* and identification of a conjugative transposon (Tn918). J Bacteriol 162:1212–1220.

Collis CM, Grammaticopoulos G, Briton J, Stokes HW, Hall RM (1993): Site-specific insertion of gene cassettes into integrons. Mol Microbiol 9:41–52.

Collis CM, Hall RM (1992a): Gene cassettes from the insert region of integrons are excised as covalently closed circles. Mol Microbiol 6:2875–2885.

Collis CM, Hall RM (1992b): Site-specific deletion and rearrangement of integron insert genes catalyzed by the integron DNA integrase. J Bacteriol 174:1574–1585.

Courvalin P, Carlier C (1987): Tn1545: A conjugative shuttle transposon. Mol Gen Genet 206:259–264.

Cousineau B, Lawrence S, Smith D, Belfort M (2000): Retrotransposition of a bacterial group II intron. Nature 404:1018–1021.

Cousineau B, Smith D, Lawrence-Cavanagh S, Mueller JE, Yang J, Mills D, Manias D, Dunny G, Lambowitz AM, Belfort M (1998): Retrohoming of a bacterial group II intron: Mobility via complete reverse splicing, independent of homologous DNA recombination. Cell 94:451–462.

Craig NL (1991): Tn7: A target site-specific transposon. Mol Microbiol 5:2569–2573.

Craig NL (1995): Unity in transposition reactions. Science 270:253–254.

Craig NL (1996): Transposition. In Neidhardt FC (ed): "*Escherichia coli* and *Salmonella*: Cellular and Molecular Biology", Vol 1. Washington, DC: ASM Press, pp 2339–2562.

Craig NL (1997): Target site selection in transposition. Annu Rev Biochem 66:437–474.

Crellin PK, Rood JI (1997): The resolvase/invertase domain of the site-specific recombinase TnpX is functional and recognizes a target sequence that resembles the junction of the circular form of the *Clostridium perfringens* transposon Tn4451. J Bacteriol 179:5148–5156.

Crellin PK, Rood JI (1998): Tn4451 from *Clostridium perfringens* is a mobilizable transposon that encodes the functional Mob protein, TnpZ. Mol Microbiol 27:631–642.

Dalrymple B (1987): Novel rearrangements of IS30 carrying plasmids leading to the reactivation of gene expression. Mol Gen Genet 207:413–420.

Darzins A, Simons RA, Kleckner N (1985): Bacteriophage Mu sites required for transposition immunity. Proc Natl Acad Sci USA 85:6826.

Davies DR, Goryshin IY, Reznikoff WS, Rayment I (2000): Three-dimensional structure of the Tn5 synaptic complex transposition intermediate. Science 289:77–85.

DeBoy RT, Craig NL (2000): Target site selection by Tn*7*: *att*Tn*7* transcription and target activity. J Bacteriol 182:3310–3313.

Dyson P (1999): Isolation and development of transposons. In Smith MCM, Sockett RE (eds): "Methods in Microbiology", Vol 29. Sydney: Academic Press, pp 133–167.

Ferat JL, Michel F (1993): Group II self-splicing introns in bacteria. Nature 364:358–361.

Fitzgerald GF, Clewell DB (1985): A conjugative transposon (Tn*919*) in *Streptococcus sanguis*. Infect Immun 47:415–420.

Fletcher HM, Marri L, Daneo-Moore L (1989): Transposon-*916*-like elements in clinical isolates of *Enterococcus faecium*. J Gen Microbiol 135:3067–3077.

Franke AE, Clewell DB (1981): Evidence for a chromosome-borne resistance transposon (Tn*916*) in *Streptococcus faecalis* that is capable of conjugal transfer in the absence of a conjugative plasmid. J Bacteriol 145:494–502.

Galas DJ, Chandler M (1989): Bacterial insertion sequences. In Berg DE, Howe MM (eds): "Mobile DNA." Washington, DC: ASM Press, pp 109–162.

Garnier F, Taourit S, Glaser P, Courvalin P, Galimand M (2000): Characterization of transposon Tn*1549*, conferring VanB-type resistance in *Enterococcus* spp. Microbiol 146:1481–1489.

Gary PA, Biery MC, Bainton RJ, Craig NL (1996): Multiple DNA processing reactions underlie Tn*7* transposition. J Mol Biol 257:301–316.

Goryshin IY, Jendrisak J, Hoffman LM, Meis R, Reznikoff WS (2000): Insertional transposon mutagenesis by electroporation of released Tn*5* transposition complexes. Nat Biotechnol 18:97–100.

Goryshin IY, Reznikoff WS (1998): Tn*5* in vitro transposition. J Biol Chem 273:7367–7374.

Grindley ND, Leschziner AE (1995): DNA transposition: From a black box to a colour monitor. Cell 83:1063–1066.

Grindley NDF (1983): Transposition of Tn*3* and related transposons. Cell 32:3–5.

Guffanti AA, Quirk PG, Krulwich TA (1991): Transfer of Tn*925* and plasmids between *Bacillus subtilis* and alkaliphilic *Bacillus firmus* OF4 during Tn*925*-mediated conjugation. J Bacteriol 173:1686–1689.

Haapa S, Taira S, Heikkinen E, Savilahti H (1999): An efficient and accurate integration of mini-Mu transposons in vitro: A general methodology for functional genetic analysis and molecular biology applications. Nucleic Acids Res 27:2777–2784.

Hacker J, Blum-Oehler G, Muhldorfer I, Tschape H (1997): Pathogenicity islands of virulent bacteria: Structure, function and impact on microbial evolution. Mol Microbiol 23:1089–1097.

Hacker J, Kaper JB (2000): Pathogenicity islands and the evolution of microbes. Annu Rev Microbiol 54:641–679.

Hagemann AT, Craig NL (1993): Tn*7* transposition creates a hotspot for homologous recombination at the transposon donor site. Genetics 133:9–16.

Hall RM, Brookes DE, Stokes HW (1991): Site-specific insertion of genes into integrons: role of the 59-base element and determination of the recombination cross-over point. Mol Microbiol 5:1941–1959.

Hall RM, Collis CM (1995): Mobile gene cassettes and integrons: Capture and spread of genes by site-specific recombination. Mol Microbiol 15:593–600.

Hallet B, Sherratt DJ (1997): Transposition and site-specific recombination: Adapting DNA cut-and-paste mechanisms to a variety of genetic rearrangements. FEMS Microbiol Rev 21:157–178.

Halula M, Macrina FL (1990): Tn*5030*: A conjugative transposon conferring clindamycin resistance in *Bacteroides* species. Rev Infect Dis 12:S235–242.

Harshey RM (1984): Transposition without duplication of infecting bacteriophage Mu DNA. Nature 311:580–581.

Hecht DW, Malamy MH (1989): Tn*4399*, a conjugal mobilizing transposon of *Bacteroides fragilis*. J Bacteriol 171:3603–3608.

Hickman AB, Li Y, Mathew SV, May EW, Craig NL, Dyda F (2000): Unexpected structural diversity in DNA recombination: the restriction endonuclease connection. Mol Cell 5:1025–1034.

Hochhut B, Jahreis K, Lengeler JW, Schmid K (1997): CTnscr94, a conjugative transposon found in enterobacteria. J Bacteriol 179:2097–2102.

Hochhut B, Marrero J, Waldor MK (2000): Mobilization of plasmids and chromosomal DNA mediated by the SXT element, a constin found in *Vibrio cholerae* O139. J Bacteriol 182:2043–2047.

Hochhut B, Waldor MK (1999): Site-specific integration of the conjugal *Vibrio cholerae* SXT element into *prfC*. Mol Microbiol 32:99–110.

Horaud T, Delbos F, de Cespedes G (1990): Tn*3702*, a conjugative transposon in *Enterococcus faecalis*. FEMS Microbiol Lett 60:189–194.

Horn N, Swindell S, Dodd H, Gasson M (1991): Nisin biosynthesis genes are encoded by a novel conjugative transposon. Mol Gen Genet 228:129–135.

Ike Y, Flannagan SE, Clewell DB (1992): Hyperhemolytic phenomena associated with insertions of Tn*916* into the hemolysin determinant of *Enterococcus faecalis* plasmid pAD1. J Bacteriol 174:1801–1809.

Inamine JM, Burdett V (1985): Structural organization of a 67-kilobase streptococcal conjugative element mediating multiple antibiotic resistance. J Bacteriol 161:620–626.

Jia Y, Churchward G (1999): Interactions of the integrase protein of the conjugative transposon Tn*916* with its specific DNA binding sites. J Bacteriol 181:6114–6123.

Kaper JB, Hacker J (eds) (1999): "Pathogenicity Islands and Other Mobile Virulence Elements." Washington, DC: ASM Press.

Karaolis DK, Somara S, Maneval DR, Johnson JA, Kaper JB (1999): A bacteriophage encoding a pathogenicity island, a type-IV pilus and a phage receptor in cholera bacteria. Nature 399:375–379.

Kholodii GYA, Yurieva OV, Lomovskaya OL, ZhM G, Mindlin SZ, Nikiforov VG (1993): Tn5053, a mercury resistance transposon with integron's ends. J Mol Biol 230:1103–1107.

Kleckner N (1989): Transposon Tn10. In Berg DE, Howe MM (eds): "Mobile DNA". Washington, DC: ASM Press, pp 227–268.

Kleckner N, Chalmers RM, Kwon D, Sakai J, Bolland S (1996a): Tn10 and IS10 transposition and chromosome rearrangements: mechanism and regulation in vivo and in vitro. Curr Top Microbiol Immunol 204:49–82.

Kleckner N, Chalmers RM, Kwon D, Sakai J, Bolland S (1996b): Tn10 and IS10 transposition and chromosome rearrangements: Mechanisms and regulation in vivo and in vitro. In Saedler H, Gierl A (eds): "Transposable Elements". Heidelberg: Springer, pp 49–82.

Kleckner N, Reichardt K, Botstein D (1979): Inversions and deletions of the Salmonella chromosome generated by the translocatable tetracycline-resistant element Tn10. J Mol Biol 127:89–115.

Lavoie BD, Chaconas G (1996): Transposition of phage Mu DNA. Curr Top Microbiol Immunol 204:83–99.

Le Bouguenec C, de Cespedes G, Horaud T (1988): Molecular analysis of a composite chromosomal conjugative element (Tn3701) of Streptococcus pyogenes. J Bacteriol 170:3930–3936.

Le Bouguenec C, de Cespedes G, Horaud T (1990): Presence of chromosomal elements resembling the composite structure Tn3701 in streptococci. J Bacteriol 172:727–734.

Lee C-H, Bhagwat A, Heffron F (1983): Identification of a transposon Tn3 sequence required for transposition immunity. Proc Natl Acad Sci USA 80:6765–6769.

Leelaporn A, Firth N, Byrne ME, Roper E, Skurray RA (1994): Possible role of insertion sequence IS257 in dissemination and expression of high- and low-level trimethoprim resistance in staphylococci. Antimicrob Agents Chemother 38:2238–2244.

Levchenko I, Yamauchi M, Baker TA (1997): ClpX and MuB interact with overlapping regions of Mu transposase: implications for control of the transposition pathway. Genes Dev 11:1561–1572.

Lewis LA, Grindley ND (1997): Two abundant intramolecular transposition products, resulting from reactions initiated at a single end, suggest that IS2 transposes by an unconventional pathway. Mol Microbiol 25:517–529.

Li LY, Shoemaker NB, Salyers AA (1993): Characterization of the mobilization region of a Bacteroides insertion element (NBU1) that is excised and transferred by Bacteroides conjugative transposons. J Bacteriol 175:6588–6598.

Li LY, Shoemaker NB, Salyers AA (1995a): Location and characteristics of the transfer region of a Bacteroides conjugative transposon and regulation of transfer genes. J Bacteriol 177:4992–4999.

Li LY, Shoemaker NB, Wang GR, Cole SP, Hashimoto MK, Wang J, Salyers AA (1995b): The mobilization regions of two integrated Bacteroides elements, NBU1 and NBU2, have only a single mobilization protein and may be on a cassette. J Bacteriol 177:3940–3945.

Lindsay JA, Ruzin A, Ross HF, Kurepina N, Novick RP (1998): The gene for toxic shock toxin is carried by a family of mobile pathogenicity islands in Staphylococcus aureus. Mol Microbiol 29:527–543.

Lu F, Craig NL (2000): Isolation and characterization of Tn7 transposase gain-of-function mutants: A model for transposase activation. EMBO J 19:3446–3457.

Lyras D, Rood JI (2000): Transposition of Tn4451 and Tn4453 involves a circular intermediate that forms a promoter for the large resolvase, TnpX. Molec Microbiol 38:588–601.

Lyras D, Storie C, Huggins AS, Crellin PK, Bannam TL, Rood JI (1998): Chloramphenicol resistance in Clostridium difficile is encoded on Tn4453 transposons that are closely related to Tn4451 from Clostridium perfringens. Antimicrob Agents Chemother 42:1563–1567.

Mahillon J, Chandler M (1998): Insertion sequences. Microbiol Mol Biol Rev 62:725–774.

Manganelli R, Ricci S, Pozzi G (1997): The joint of Tn916 circular intermediates is a homoduplex in Enterococcus faecalis. Plasmid 38:71–78.

Marra D, Scott JR (1999): Regulation of excision of the conjugative transposon Tn916. Mol Microbiol 31:609–621.

Martinez E, de le Cruz F (1990): Genetic elements involved in Tn21 site-specific integration, a novel mechanism for the dissemination of antibiotic resistance genes. EMBO J 9:1275–1281.

Martinez-Abarca F, Toro N (2000): Group II introns in the bacterial world. Mol Microbiol 38:917–926.

Matsuura M, Saldanha R, Wank H, Yang J, Mohr G, Cavanagh S, Dunny GM, Belfort M, Lambowitz AM (1997): A bacterial group II intron encoding reverse transcriptase, maturase, and DNA endonuclease activities: biochemical demonstration of maturase activity and insertion of new genetic information within the intron. Genes Dev 11:2910–2924.

May EW, Craig NL (1996): Switching from cut-and-paste to replicative Tn7 transposition. Science 272:401–404.

Mazel D, Dychinco B, Webb VA, Davies J (1998): A distinctive class of integron in the *Vibrio cholerae* genome. Science 280:605–608.

Merlin C, Mahillon J, Nesvera J, Toussaint A (2000): Gene recruiters and transporters: The modular structure of bacterial mobile elements. In Thomas CM (ed): "The Horizontal Gene Pool." Amsterdam: Harwood Academic pp 363–409.

Merlin C, Springael D, Toussaint A (1999): Tn*4371*: A modular structure encoding a phage-like integrase, a *Pseudomonas*-like catabolic pathway, and RP4/Ti-like transfer functions. Plasmid 41:40–54.

Michel F, Umesono K, Ozeki H (1989): Comparative and functional anatomy of group II catalytic introns—A review. Gene 82:5–30.

Mills DA, McKay LL, Dunny GM (1996): Splicing of a group II intron involved in the conjugative transfer of pRS01 in lactococci. J Bacteriol 178:3531–3538.

Minakhina S, Kholodii G, Mindlin S, Yurieva O, Nikiforov V (1999): Tn*5053* family transposons are res site hunters sensing plasmidal *res* sites occupied by cognate resolvases. Mol Microbiol 33:1059–1068.

Mizuuchi K (1992a): Polynucleotidyl transfer reactions in tranpositional DNA recombination. J Biol Chem 267:21273–21276.

Mizuuchi K (1992b): Transpositional recombination: Mechanistic insights from studies of Mu and other elements. Annu Rev Biochem 61:1011–1051.

Mullany P, Pallen M, Wilks M, Stephen JR, Tabaqchali S (1996): A group II intron in a conjugative transposon from the Gram-positive bacterium, *Clostridium difficile*. Gene 174:145–150.

Mullany P, Wilks M, Tabaqchali S (1995): Transfer of macrolide-lincosamide-streptogramin B (MLS) resistance in *Clostridium difficile* is linked to a gene homologous with toxin A and is mediated by a conjugative transposon, Tn*5398*. J Antimicrob Chemother 35:305–315.

Murphy CG, Malamy MH (1993): Characterization of a "mobilization cassette" in transposon Tn*4399* from *Bacteroides fragilis*. J Bacteriol 175:5814–5823.

Murphy E (1989): Transposable elements in Gram-positive bacteria. In Bergs DE, Howe MM (eds): "Mobile DNA". Washington, DC: ASM Press, pp 269–288.

Nield BS, Holmes AJ, Gillings MR, Recchia GD, Mabbutt BC, Nevalainen KM, Stokes HW (2001): Recovery of new integron classes from environmental DNA. FEMS Microbiol Lett 195:59–65.

Nikolich MP, Shoemaker NB, Wang GR, Salyers AA (1994): Characterization of a new type of *Bacteroides* conjugative transposon, Tc^r Em^r 7853. J Bacteriol 176:6606–6612.

Olson ER, Chung ST (1988): Transposon Tn*4556* of *Streptomyces fradiae*: sequence of the ends and target sites. J Bacteriol 170:1955–1957.

Pato ML (1989): Bacteriophage Mu. In Berg DE, Howe MM (eds): "Mobile DNA." Washington, DC: ASM Press, pp 23–52.

Pato ML, Banerjee M (2000): Genetic analysis of the strong gyrase site (SGS) of bacteriophage Mu: localization of determinants required for promoting Mu replication. Mol Microbiol 37:800–810.

Perkins-Balding D, Duval-Valentin G, Glasgow AC (1999): Excision of IS*492* requires flanking target sequences and results in circle formation in *Pseudoalteromonas atlantica*. J Bacteriol 181:4937–4948.

Peters JE, Craig NL (2000): Tn*7* transposes proximal to DNA double-strand breaks and into regions where chromosomal DNA replication terminates. Mol Cell 6:573–582.

Picksley SM, Attfield PV, Llyod RG (1984): Repair of double strand breaks in *Escherichia coli* K12 requires a functional RecN product. Mol Gen Genet 195:267–274.

Plasterk RHA (1993): Molecular mechanisms of transposition and its control. Cell 74:781–786.

Polard P, Chandler M (1995a): Bacterial transposases and retroviral integrases. Mol Microbiol 15:13–23.

Polard P, Chandler M (1995b): An in vivo transposase-catalyzed single-stranded DNA circularization reaction. Genes Dev 9:2846–2858.

Polard P, Prere MF, Fayet O, Chandler M (1992): Transposase-induced excision and circularization of the bacterial insertion sequence IS*911*. EMBO J 11:5079–5090.

Poyart C, Quesne G, Acar P, Berche P, Trieu-Cuot P (2000): Characterization of the Tn*916*-like transposon Tn*3872* in a strain of *Abiotrophia defectiva* (*Streptococcus defectivus*) causing sequential episodes of endocarditis in a child. Antimicrob Agents Chemother 44:790–793.

Poyart-Salmeron C, Trieu-Cuot P, Carlier C, Courvalin P (1989): Molecular characterization of two proteins involved in the excision of the conjugative transposon Tn*1545*: homologies with other site-specific recombinases. EMBO J 8:2425–2433.

Prentki P, Teter B, Chandler M, Galas DJ (1986): Functional promoters created by the insertion of transposable element IS*1*. J Mol Biol 191:383–393.

Radstrom P, Skold O, Swedberg G, Flensburg J, Roy PH, Sundstrom L (1994): Transposon Tn*5090* of plasmid R751, which carries an integron, is related to Tn7, Mu, and the retroelements. J Bacteriol 176:3257–3268.

Raleigh EA, Kleckner N (1984): Multiple IS*10* rearrangements in *E. coli*. J Mol Biol 173:437–461.

Rauch PJ, Beerthuyzen MM, de Vos WM (1994): Distribution and evolution of nisin-sucrose elements in *Lactococcus lactis*. Appl Environ Microbiol 60:1798–1804.

Rauch PJ, De Vos WM (1992): Characterization of the novel nisin-sucrose conjugative transposon Tn*5276* and its insertion in *Lactococcus lactis*. J Bacteriol 174:1280–1287.

Ravatn R, Studer S, Zehnder AJB, Roelof van der Meer J (1998): Int-B13, an unusual site-specific recombinase of the bacteriophage P4 integrase family, is responsible for chromosomal insertion of the 105-kilobase *clc* element of *Pseudomonas* sp. strain B13. J Bacteriol 180:5505–5514.

Recchia GD, Hall RM (1995): Gene cassettes: A new class of mobile element. Microbiol 141:3015–3027.

Reimmann C, Moore R, Little S, Savioz A, Willetts NS, Haas D (1989): Genetic structure, function and regulation of the transposable element IS*21*. Mol Gen Genet 215:416–424.

Reynolds AE, Felton J, Wright A (1981): Insertion of DNA activates the cryptic *bgl* operon in *E. coli* K12. Nature 293:625–629.

Reynolds AE, Madahadevan S, LeGrice SF, Wright A (1986): Enhancement of bacterial gene expression by insertion elements or by mutation in a CAP-cAMP binding site. J Mol Biol 191:85–95.

Reznikoff WS (1993): The Tn*5* transposon. Annu Rev Microbiol 47:945–963.

Rice LB, Carias LL (1998): Transfer of Tn*5385*, a composite, multiresistance chromosomal element from *Enterococcus faecalis*. J Bacteriol 180:714–721.

Rice LB, Marshall SH, Carias LL (1992): Tn*5381*, a conjugative transposon identifiable as a circular form in *Enterococcus faecalis*. J Bacteriol 174:7308–7315.

Rowe-Magnus DA, Guerout A-M, Ploncard P, Dychinco B, Davies J, Mazel D (2001): The evolutionary history of chromosomal super-integrons provides an ancestry for multiresistant integrons. PNAS 98:652–657.

Rudy C, Taylor KL, Hinerfeld D, Scott JR, Churchward G (1997a): Excision of a conjugative transposon in vitro by the Int and Xis proteins of Tn*916*. Nucleic Acids Res 25:4061–4066.

Rudy CK, Scott JR (1994): Length of the coupling sequence of Tn*916*. J Bacteriol 176:3386–3388.

Rudy CK, Scott JR, Churchward G (1997b): DNA binding by the Xis protein of the conjugative transposon Tn*916*. J Bacteriol 179:2567–2572.

Saedler H, Grierl A (1995): Transposable elements. Curr Top Microbiol Immunol 204:27–48.

Salyers AA, Shoemaker NB, Bonheyo G, Frias J (1999): Conjugative transposons: Transmissible resistance islands. In Kaper JB, Hacker J (eds): "Pathogenicity Islands and Other Mobile Virulence Elements." Washington, DC: ASM Press, pp 331–346.

Salyers AA, Shoemaker NB, Li LY (1995a): In the driver's seat: The *Bacteroides* conjugative transposons and the elements they mobilize. J Bacteriol 177:5727–5731.

Salyers AA, Shoemaker NB, Stevens AM, Li LY (1995b): Conjugative transposons: An unusual and diverse set of integrated gene transfer elements. Microbiol Rev 59:579–590.

Schnetz K, Rak B (1992): IS*5*: A mobile enhancer of transcription in *Escherichia coli*. Proc Natl Acad Sci USA 89:1244–1248.

Scott JR (1992): Sex and the single circle: Conjugative transposition. J Bacteriol 174:6005–6010.

Scott JR, Bringel F, Marra D, Van Alstine G, Rudy CK (1994): Conjugative transposition of Tn*916*: Preferred targets and evidence for conjugative transfer of a single strand and for a double-stranded circular intermediate. Mol Microbiol 11:1099–1108.

Scott JR, Churchward GG (1995): Conjugative transposition. Annu Rev Microbiol 49:367–397.

Scott JR, Kirchman PA, Caparon MG (1988): An intermediate in transposition of the conjugative transposon Tn*916*. Proc Natl Acad Sci USA 85:4809–4813.

Scott K, Barbosa T, Forbes K, Flint H (1997): High-frequency transfer of a naturally occurring chromosomal tetracycline resistance element in the ruminal anaerobe *Butyrivibrio fibrisolvens*. Appl Environ Microbiol 63:3405–3411.

Sekine Y, Eisaki N, Ohtsubo E (1994): Translational control in production of transposase and in transposition of insertion sequence IS*3*. J Mol Biol 235:1406–1420.

Shapiro JA, Sporn P (1977): Tn*402*: A new transposable element determining trimethoprim resistance that inserts in bacteriophage lambda. J Bacteriol 129:1632–1635.

Sharpe PL, Craig NL (1998): Host proteins can stimulate Tn*7* transposition: A novel role for the ribosomal protein L29 and the acyl carrier protein. EMBO J 17:5822–5831.

Sherratt D (1989): Tn*3* and related transposable elements. In Berg DE, Howe MM (eds): "Mobile DNA." Washington, DC: ASM Press, pp 163–184.

Shoemaker NB, Salyers AA (1988): Tetracycline-dependent appearance of plasmidlike forms in *Bacteroides uniformis* 0061 mediated by conjugal *Bacteroides* tetracycline resistance elements. J Bacteriol 170:1651–1657.

Shoemaker NB, Wang GR, Salyers AA (1996): The *Bacteroides* mobilizable insertion element, NBU1, integrates into the 3′ end of a Leu-tRNA gene and has an integrase that is a member of the lambda integrase family. J Bacteriol 178:3594–3600.

Shoemaker NB, Vlamakis H, Hayes K, Salyers AA (2001): Evidence for extensive resistance gene transfer among *Bacteroides* spp. and among *Bacteroides* and other genera in the human colon. Appl Environ Microbiol 67:561–568.

Shoemaker NB, Wang GR, Salyers AA (2000): Multiple gene products and sequences required for excision of the mobilizable integrated *Bacteroides* element NBU1. J Bacteriol 182:928–936.

Shoemaker NB, Wang GR, Stevens AM, Salyers AA (1993): Excision, transfer, and integration of NBU1, a mobilizable site-selective insertion element. J Bacteriol 175:6578–6587.

Smith CJ, Parker AC (1993): Identification of a circular intermediate in the transfer and transposition of Tn*4555*, a mobilizable transposon from *Bacteroides* spp. J Bacteriol 175:2682–2691.

Smith CJ, Parker AC (1996): A gene product related to TraI is required for the mobilization of *Bacteroides* mobilizable transposons and plasmids. Mol Microbiol 20:741–750.

Stevens AM, Shoemaker NB, Li LY, Salyers AA (1993): Tetracycline regulation of genes on *Bacteroides* conjugative transposons. J Bacteriol 175:6134–6141.

Stokes HW, O'Gorman DB, Recchia GD, Parsekhian M, Hall RM (1997): Structure and function of 59-base element recombination sites associated with mobile gene cassettes. Mol Microbiol 26:731–745.

Stratz M, Gottschalk G, Durre P (1990): Transfer and expression of the tetracycline resistance transposon Tn*925* in *Acetobacterium woodii*. FEMS Microbiol Lett 56:171–176.

Sullivan JT, Ronson CW (1998): Evolution of rhizobia by acquisition of a 500-kb symbiosis island that integrates into a phe-tRNA gene. Proc Natl Acad Sci USA 95:5145–5149.

Sundstrom L, Skold O (1990): The *dhfrI* trimethoprim resistance gene of Tn7 can be found at specific sites in other genetic surroundings. Antimicrob Agents Chemother 34:642–650.

Szeverenyi I Bodoky T, Olasz F (1996): Isolation, characterization and transposition of an (IS2)2 intermediate. Mol Gen Genet 251:281–289.

Talaat AM, Trucksis M (2000): Transformation and transposition of the genome of *Mycobacterium marinum*. Am J Vet Res 61:125–128.

Thompson J, Nguyen NY, Sackett DL, Donkersloot JA (1991): Transposon-encoded sucrose metabolism in *Lactococcus lactis*: Purification of sucrose-6-phosphate hydrolase and genetic linkage to $\underline{N}$5-(L-1-carboxyethyl)-L-ornithine synthase in strain K1. J Biol Chem 266:14573–14579.

Ton-Hoang B, Betermier M, Polard P, Chandler M (1997): Assembly of a strong promoter following IS*911* circularization and the role of circles in transposition. EMBO J 16:3357–3371.

Ton-Hoang B, Polard P, Chandler M (1998): Efficient transposition of IS*911* circles in vitro. EMBO J 17:1169–1181.

Ton-Hoang B, Polard P, Haren L, Turlan C, Chandler M (1999): IS*911* transposon circles give rise to linear forms that can undergo integration in vitro. Mol Microbiol 32:617–627.

Torres OR, Korman RZ, Zahler SA, Dunny GM (1991): The conjugative transposon Tn*925*: Enhancement of conjugal transfer by tetracycline in *Enterococcus faecalis* and mobilization of chromosomal genes in *Bacillus subtilis* and *E. faecalis*. Mol Gen Genet 225:395–400.

Tribble G, Parker A, Smith C (1997): The *Bacteroides* mobilizable transposon Tn*4555* integrates by a site-specific recombination mechanism similar to that of the Gram-positive bacterial element Tn*916*. J Bacteriol 179:2731–2739.

Tribble GD, Parker AC, Smith CJ (1999a): Genetic structure and transcriptional analysis of a mobilizable, antibiotic resistance transposon from *Bacteroides*. Plasmid 42:1–12.

Tribble GD, Parker AC, Smith CJ (1999b): Transposition genes of the *Bacteroides* mobilizable transposon Tn*4555*: Role of a novel targeting gene. Mol Microbiol 34:385–394.

Trieu-Cuot P, Derlot E, Courvalin P (1993): Enhanced conjugative transfer of plasmid DNA from *Escherichia coli* to *Staphylococcus aureus* and *Listeria monocytogenes*. FEMS Microbiol Lett 109:19–23.

Tsuda M, Minegishi K, Iino T (1989): Toluene transposons Tn*4651* and Tn*4653* are class II transposons. J Bacteriol 171:1386–1393.

Turlan C, Ton-Hoang B, Chandler M (2000): The role of tandem IS dimers in IS911 transposition. Mol Microbiol 35:1312–1325.

Valentine PJ, Shoemaker NB, Salyers AA (1988): Mobilization of *Bacteroides* plasmids by *Bacteroides* conjugal elements. J Bacteriol 170:1319–1324.

Vedantam G, Novicki TJ, Hecht DW (1999): *Bacteroides fragilis* transfer factor Tn*5520*: The smallest bacterial mobilizable transposon containing single integrase and mobilization genes that function in *Escherichia coli*. J Bacteriol 181:2564–2571.

Vijayakumar MN, Ayalew S (1993): Nucleotide sequence analysis of the termini and chromosomal locus involved in site-specific integration of the streptococcal conjugative transposon Tn*5252*. J Bacteriol 175:2713–2719.

Vijayakumar MN, Priebe SD, Pozzi G, Hageman JM, Guild WR (1986): Cloning and physical characterization of chromosomal conjugative elements in streptococci. J Bacteriol 166:972–977.

Waldor MK, Tschape H, Mekalanos JJ (1996): A new type of conjugative transposon encodes resistance to sulfamethoxazole, trimethoprim, and streptomycin in *Vibrio cholerae* O139. J Bacteriol 178:4157–4165.

Wang H, Roberts AP, Lyras D, Rood JI, Wilks M, Mullany P (2000a): Characterization of the ends and target sites of the novel conjugative transposon Tn*5397* from *Clostridium difficile*: Excision and circularization is mediated by the large resolvase, TndX. J Bacteriol 182:3775–3783.

Wang J, Shoemaker NB, Wang GR, Salyers AA (2000b): Characterization of a *Bacteroides* mobilizable transposon, NBU2, which carries a functional lincomycin resistance gene. J Bacteriol 182:3559–3571.

Waters VL, Guiney DG (1993): Processes at the nick region link conjugation, T-DNA transfer and rolling circle replication. Mol Microbiol 9:1123–1130.

Weinreich MD, Reznikoff WS (1992): Fis plays a role in Tn5 and IS50 transposition. J Bacteriol 174:4530–4537.

Wolkow CA, DeBoy RT, Craig NL (1996): Conjugating plasmids are preferred targets for Tn7. Genes Dev 10:2145–2157.

Yamauchi M, Baker TA (1998): An ATP-ADP switch in MuB controls progression of the Mu transposition pathway. EMBO J 17:5509–5518.

Zhu J, Oger PM, Schrammeijer B, Hooykaas PJJ, Farrand SK, Winans SC (2000): The bases of crown gall tumorigenesis. J Bacteriol 182:3885–3895.

Ziebuhr W, Krimmer V, Rachid S, Lossner I, Gotz F, Hacker J (1999): A novel mechanism of phase variation of virulence in *Staphylococcus epidermidis*: Evidence for control of the polysaccharide intercellular adhesin synthesis by alternating insertion and excision of the insertion sequence element IS256. Mol Microbiol 32:345–356.

Zimmerly S, Hausner G, Wu X (2001): Phylogenetic relationships among group II intron ORFs. Nucleic Acids Res 29:1238–1250.

18

Transformation

ULDIS N. STREIPS

Department of Microbiology and Immunology, School of Medicine, University of Louisville, Louisville,
Kentucky 40292

**Modern Microbial Genetics, Second Edition. Edited by
Uldis N. Streips and Ronald E. Yasbin. ISBN 0–471–38665–0
Copyright © 2002 Wiley-Liss, Inc.**

I. INTRODUCTION

Transformation, or the uptake of exogenous deoxyribonucleic acid (DNA) by living bacteria, was first described in *Streptococcus pneumoniae* by Griffith, even though he did not identify DNA as the agent for transmitting the information (Griffith, 1928). Excitement over the novelty of the system, coupled to the caution of the scientific mind, characterized the early years for the development of the pneumococcal transformation system. This can be contrasted to the explosive pace of discovery in the last decade in describing the molecular basis for the transformation systems in the pneumococcus and other bacteria, notably *Bacillus subtilis*. The repertoire of transformable organism has increased logarithmically since the last edition of this book. Although much still needs to be understood about the molecular mechanisms of the transformation process, transformation has become the workhorse of the genetic engineering field.

II. THE SEMINAL EXPERIMENTS

Streptococcus pneumoniae is virulent in both mice and humans, and this virulence is correlated to the extent of capsule formation by this organism. Pneumococcal capsules can be typed and differentiated chemically and immunologically. Mutants, which cannot produce capsules, are nonvirulent. Such mutants are designated as "R" due their "rough" colony morphology on agar, differentiating them from the virulent "smooth," "S," colonies of the encapsulated pneumococcus.

Griffith described experiments in which living "R" mutants of capsule type II *S. pneumoniae* (IIR) or heat-killed, wild-type capsule type I (IS) cells were injected separately into mice. Neither preparation caused infection in these animals, as expected. However, when a mixture of these two different capsule type pneumococci was injected into the subcutaneous tissue of mice, a proportion of the animals died, and encapsulated pneumococci of type I could be isolated from the blood of some mice and type II from others. Since the "R" mutation in type II pneumococci is not readily revertible, it is evident that information must have passed from the killed, Type IS *S. pneumoniae* to enable the nonvirulent, type IIR cells to synthesize type I capsule and in some instances to activate the mutated type II capsule synthesis. Similar experiments could also be done using capsular type III pneumococci. Griffith stated that the capsule-forming, "transforming" principle was probably "that specific protein structure of the virulent pneumococcus which enables it to manufacture a specific soluble carbohydrate" (Griffith, 1928)

For the next 10 years, even though the transformation reaction was just as effective in the test tube as in the animal, and some chemical analyses on the transforming principle had been done, this opinion was not substantially altered. These intervening years between Griffith's original observation and the appearance of the classic manuscript from Avery's laboratory (Avery et al., 1944) have been amply reviewed by many investigators intimately involved with this research at that time (Dubos, 1976; Hotchkiss; 1982; McCarty, 1980,1985). The major problems in ultimately elucidating the DNA nature of the transforming principle were (1) the unreliability of the extraction process, (2) the inability to generate transformable cells on a consistent basis, and (3) the continuing confusion surrounding the chemical properties and activity spectrum of the polysaccharide in the transforming samples. As late as 1941 MacLeod stated that an increase in pneumococcal type III polysaccharide concentration might improve transformation activity.

However, by 1942 there was already substantial evidence that the transforming principle was DNA. This work was finally published in 1944 (Avery et al., 1944). Avery and his coworkers used *S. pneumoniae*, type IIIS DNA to transform type IIR variants to produce a type III capsule. The DNA was extracted by washing the pneumocci in saline, then lysing the cells with desoxycholate. The DNA was precipitated in alcohol, dissolved in saline, deproteinized using chloroform, and reprecipitated. An enzyme to hydrolyze type III polysaccharide was added and the DNA was precipitated and deproteinized again. From this process, 10 to 25 mg of DNA was obtained from 75 liters of culture. This active preparation would induce transformation at concentration of DNA as low a 1 ng/ml. However, not all extracted preparations were this active. It is interesting that in Avery's manuscript evidence was already presented that dog or rabbit serum containing "desoxyribonucleodepolymerase" (DNase) activity appeared to destroy this transforming capability. As the authors stated, "so far as the writers are aware, a nucleic acid of the desoxyribose type has not heretofore been recovered from pneumococci nor has specific transformation been experimentally induced *in vitro* by a chemically defined substance" (Avery et al., 1944). Starting with this finding, the story of transformation, and in fact much of molecular biology, unfolds.

In 1946 McCarty applied crude DNase to the preparations and destroyed transforming activity, as did Hotchkiss three years later with crystalline DNase I. Any role for associated protein was dispelled and the protein content in active transformation preparations was calculated to be less than 0.01% (Hotchkiss, 1982).

Subsequent to this seminal research, the description of transformation has been a dissection of the phenomenon itself, rather than the intensive debate on the chemical identification of the transforming principle. For practical purposes the transformation systems of *S. pneumoniae*, *B. subtilis*, and *Haemophilus influenzae* will be used as the main models for the following discussion and will be presented sequentially in different sections in the chapter. For more detailed information on any of these organisms, the references should be consulted.

The principal steps in the transformation reaction, leading to a transformed cell, are outlined in the diagram below and discussed in that order in the following sections:

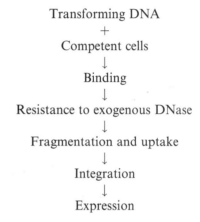

Transforming DNA
+
Competent cells
↓
Binding
↓
Resistance to exogenous DNase
↓
Fragmentation and uptake
↓
Integration
↓
Expression

III. DNA ACTIVE IN TRANSFORMATION

Native bacterial chromosomal fragments, plasmids, and bacteriophage DNA, as well as genetically constructed chimeric molecules can be used to transform bacteria. Bodmer applied a sample of transforming chromosomal DNA to neutral and alkaline sucrose gradients (Bodmer, 1966). The DNA was obtained by first lysing *B. subtilis* cells, then extracting the lysate with phenol. A representation of the results is shown in Figure 1. The obvious heterogeneity of the chromosomal fragments (ranging in molecular weight from 5.9×10^6 [9 kb] to 4.9×10^7 [70 kb], with bacteriophage T7 DNA as the size standard (40 kb), suggests that fragments of various sizes are generated during the extraction process and could participate in the transformation reaction. The average molecular weight for these fragments is 2.1×10^7 (32 kb). Moreover centrifugation of this same sample in alkaline sucrose (right panel) demonstrated that one to two

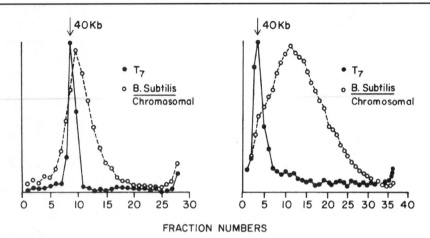

FRACTION NUMBERS

Fig. 1.   Distribution of extracted *Bacillus subtilis* and standard (bacteriophage T7) DNA on sucrose gradients (adapted from Bodmer, 1966). **Left panel:** Representation of DNA distribution in a neutral sucrose gradient. **Right panel:** The same DNA samples distributed in an alkaline sucrose gradient. A mixture of the two DNAs was layered onto 5–20% linear sucrose gradients and centrifuged for 3 hours at 38,000 rpm in a SW39 rotor. The T7 DNA (tritium-labeled) marks the 40 kb size range of the DNA. As can be seen, the extracted *B. subtilis* DNA (^{14}C-labeled) shows a heterogeneity of sizes due to fragmentation of the chromosome during extraction. The sedimentation pattern in the alkaline sucrose gradient suggested that single-strand breaks were also induced by the extraction process.

single-strand breaks were present in the extracted chromosomal DNA endogenously or due to the extraction process. Cato and Guild (1968) fractionated sheared chromosomal DNA and analyzed the resulting fragments for transforming activity. They found that transforming activity decreased with size and the smallest effective fragment was 0.5 kb in size. The transforming activity of these very small fragments was 10,000-fold lower, when compared to fragments of 32 kb. The reasons for the necessity of relatively large fragments for optimal transformation probably reside in the damage suffered by the DNA as it enters the competent cell, as well as the minimum size requirement for effective homologous pairing during recombination (see below). Conversely, bacteria can be transformed with DNA preparations from gently lysed cells or with DNA released by osmotically lysed L-forms (Horowitz et al., 1979). Such large molecular weight DNA fragments are not as efficient in transformation as fragments of 32 kb. Presumably the large molecular weight fragments interfere with the efficiency of the uptake mechanism. However, gene linkages can be established for markers far apart on the chromosome, if these large chromosomal fragments are used for transformation (Kelley, 1967).

Plasmid and bacteriophage DNAs can enter cells by transformation, but they are also subject to nuclease-mediated degradation (see below). Consequently, since these molecules cannot be expressed as fragments, it is estimated that several molecules of bacteriophage or plasmid genomes (perhaps as many as 10,000 equivalents) are necessary to cooperate and initiate the formation of one infectious center or to generate one plasmid transformant.

Transformation, in general, is dependant on DNA concentration. Exposure of bacteria to increasing concentrations of DNA produces a proportional rise in the number of transformants, until saturation is reached (Fig. 2). In most bacterial systems saturation is achieved by adding 1 µg to 10 µg of chromosomal DNA to a milliliter of competent cells. The level of uptake can also be measured by adding radioactively labeled

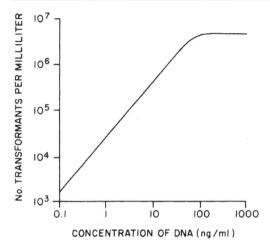

CONCENTRATION OF DNA (ng/ml)

Fig. 2. Relationship between concentration of DNA added and number of transformants recovered (Hayes, 1968). As concentration of transforming DNA is increased, the level of transformants rises until saturation of the system is achieved.

chromosomal DNA to the competent cells and quantifying the amount of DNA incorporated into the bacteria. Such measurements indicate that 100 to 200 molecules of chromosomal fragments must be added per cell to obtain a single transformant. These findings relate to the probability of the transforming gene being present in the specific fragments taken up by the competent cells, as well as to the damage to the DNA as it enters the cell and becomes incorporated. The relative concentration of DNA added becomes very critical when mapping or using congression for strain construction (see below) is attempted in transformation systems.

IV. NATURAL COMPETENCE

Competence, defined as the ability of a bacterium to take up DNA from the medium, can be naturally developed or artificially induced. We will consider these two ways for developing competence separately. Although DNA is often readily available in the environment (from lysed cells, DNA excretion, etc.), the transformation process is initiated only when competent bacteria are present.

With the exception of *Neisseria* and related bacteria, which express competence constitutively, the other commonly used, naturally competent bacteria genetically regulate (competence regulon) the development of their competence. Competence is a normal physiological development state in these bacteria. The Gram-positive bacteria usually regulate their competence by secreted external factors (competence factors or phermones), as well as the expression of specific genes and synthesis of both internal and external proteins related to the development of competence. The example we will use for a naturally competent Gram-negative organism, *Haemophilus influenzae*, regulates competence primarily internally. However, care must be taken to not relate many of the normal changes occurring in cells as they pass from the exponential growth phase toward the stationary phase of growth to development of competence. Such normal structural and regulatory alterations may have nothing to do with the ability to take up DNA. By the same token, recent research has delineated specific pathways that are initiated primarily to enable the bacterial cells to acquire exogenous DNA.

The detection of competence can be most easily achieved by the use of a DNA molecule, that requires entry into the cell but does not require subsequent integration into the resident chromosome. A nonintegrating molecule avoids problems of nonhomology or recombination deficiency in the recipient cell. Both plasmids and bacteriophage DNA are useful for this purpose. A representative assay is shown in Figure 3. Cells would be grown to various densities, and then exposed to the nonintegrating DNA. If DNA concentrations are kept constant, then the rise in the number of cells taking up bacteriophage DNA and becoming plaque-forming units (PFUs) (i.e., bacteria that will make bacteriophage from the introduced DNA and form a plaque on an indicator lawn) must indicate the increasing and decreasing ability of cells to take up DNA. As shown in the sample figure, total

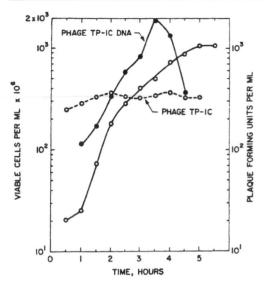

Fig. 3. Detection of optimal competence. *Bacillus stearothermophilus* cells were grown in complex media, and 2.0 μg of DNA from bacteriophage TP-1C or whole TP-1C particles (8.3×10^2 PFU) were added to the cells. The number of plaques reflected successful DNA uptake and ability to infect, respectively. Maximum DNA uptake occurred after growth for 3.5 hours, whereas phage infection was constant throughout the growth period. (Reproduced from Streips and Welker, 1969, with permission of the publisher.)

number of cells keeps increasing, but as they pass out of competence the number of PFUs falls as less and less DNA is taken up. As a control, whole bacteriophage particles will keep up a constant level of infection.

In Gram-positive cells competence has been studied most intensively in the naturally competent bacilli and streptococci.

A. *Bacillus subtilis*

Bacilli, as exemplified by *B. subtilis*, require a defined minimal medium to achieve optimal competence during late stationary phase of growth. To obtain the specific methodology for competence development, consult Anagnostopoulos and Spizizen (1961) and Bott and Wilson (1967). Anagnostopoulos and Spizizen (1961) formulated the glucose minimal salt growth medium and the two-step, shift-down regimen for the development of competence. This medium includes the divalent cations Mg^{2+} and Ca^{2+}, of which Mg^{2+} is necessary for the activation of specific nucleases involved in the uptake process. Normally 20% competence with a 1–2% transformation frequency is considered to be successful DNA introduction into *B. subtilis*. Bott and Wilson were able to achieve 50% competency and 10% transformation rate in their system (Bott and Wilson, 1967). It should be mentioned that bacilli can achieve competence in complex media, though at lower levels. Also competent cells can be purchased commercially and are highly competent. In addition Graham and Istock (1978) documented the ability of *B. subtilis* to develop competence in natural environmental conditions. Similarly marine pseudomonads (Gram-negative) become naturally competent in the environment (Stewart, 1987). Therefore competence is not only a laboratory phenomenon, which can be manipulated and optimized under specific conditions, but is also a likely natural event in the life cycle of environmental, naturally competent bacteria.

Competent *B. subtilis* cells are biosynthetically latent. Bodmer grew thymine-requiring mutants to competence and then suspended the mutants in medium lacking thymine (Bodmer, 1966). Most of the culture underwent thymineless death. However, there was no loss in the number of transformants when compared to the control, which was grown in the presence of thymine. In addition competent cells are also resistant to ampicillin and synthesize macromolecules at a fraction of the wild-type rate. Only about 2.7% of the chromosome DNA content is synthesized during the onset of competence. Most of this synthesis is accounted for by the induction of the SOB (SOS analogue) repair system in competent *B. subtilis* cells (see Yasbin, this volume). Love and coworkers demonstrated the link between competence and SOB repair (Love et al., 1985). By inserting a transposon, which was fused to a promoterless beta-galactosidase gene (Tn917-*lacZ*), into random sites in the

B. subtilis chromosome (Youngman et al., 1985), these investigators were able to show β-galactosidase induction in some mutants both during exposure to DNA damaging agents and growth to competence. Obviously in these mutants the transposon had inserted close to a promoter which was activated under both conditions—repair of DNA damage and expression of competence. Thus competence development is coordinated with DNA repair regulons (see Yasbin, this volume). The accumulation of gaps in DNA at the time of competence may induce the repair system. Also RecA protein induction during competence is greater than during DNA damage (Lovett et al., 1989).

Recently the molecular basis for the development of competence in *B. subtilis* has been at least partially elucidated. This is mostly the work of Dubnau's group and David Dubnau has written a thorough review article on DNA uptake in bacterial transformation (Dubnau, 1999). In Figure 4 is a representation of the current understanding about events at the time of competence development.

When *B. subtilis* is becoming competent, two peptides are externalized. One is the competence phermone, ComX, a modified 9 to 10 amino acid peptide (Magnuson et al., 1994), the other is the competence stimulating factor (CSF) (Solomon, et al., 1995) As shown in the figure, ComX is exported from the cell. The extracellular peptide then attaches to its own cell and others and provides a quorum sensing environment (see the Chapter by Parsek and Fuqua, this volume). It binds to ComP (a histidine kinase) which is part of a two component sensing system (see the chapter by Bayles and Fujimoto, this volume). ComP is a membrane spanning (6–8 membrane spanning domains), sensing protein that binds ComX at the N-terminus.

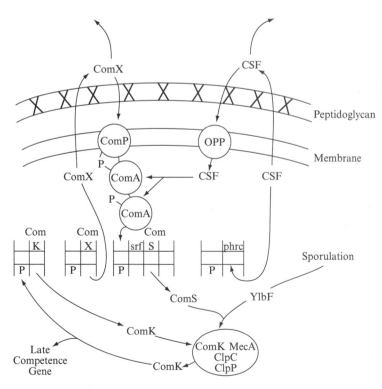

Fig. 4. Composite diagram of proteins involved in the development of competence in *Bacillus subtilis*.

Following the binding, autophosphorylated ComP donates a phosphate to ComA, the responder. The CSF is synthesized from the *phrC* gene at the same time, also exported to nearby cells as part of the quorum sensing system, then imported by the oligopeptide permease (Solomon et al., 1996). The oligopeptide permease is part of the sporulation pathway, (see the chapter by Moran this volume) and this is but one of the links competence development has to sporulation (see below). Imported CSF has the function of inhibiting dephosphorylation of ComA. Phosphorylated ComA then promotes the transcription of the *srf* operon. *Srf* consists of four genes encoding the enzymes for the lipopetide antibiotic, surfactin (Nakano, 1991). The important member of this operon is the gene *comS*, which is out of reading frame for the *srf* genes (D'Souza et al., 1993). As part of the transcription initiated by phophorylated ComA, ComS is produced. ComS has the primary function of activating ComK by releasing it from the MecA, ClpC, ClpP complex (Ogura et al., 1999; Persuh et al., 1999). During logarithmic phase growth, when *B. subtilis* is not competent, ComK is complexed with MecA, an adapter protein that targets ComK for proteolysis by the ClpC protease. It is known that ComK binds to the N-terminus of MecA, while ClpC binds to the C-terminus. The activity of MecA is modulated by ClpP in the complex. As would be expected, a *clpP*-deleted mutant increases the levels of MecA, decreases the levels of ComK, and it is interesting that this mutant also has a defect in sporulation at stage II (Msadek et al., 1998). When competence is induced and ComS is synthesized, it also binds at the N-terminus of MecA and helps to release ComK from this complex. Active ComK then is free to autoinitiate synthesis and also initiate the transcription of later competence genes (below). At this point it is interesting to note that among the genes ComK stimulates are also those for DNA recombination and repair. MecA appears to be generally present in Gram-positive cells

and may perform similar adapter functions of sequestering proteins for degradation as part of genetic regulation of several systems (Ogura et al., 1999). Tortosa and coworkers, using mini Tn*10* transposition generated a library of mutants to examine possible novel competence-regulated genes (Tortosa et al., 2000). Among these was the gene *ylbF*, which appears to increase the activity of ComS in protecting ComK in the MecA, ClpC, ClpP complex. Adding to the context that many of these genes network between regulons, *ylbF*, also appears to be necessary in spore development prior to stage II (see the chapter by Moran). Once developed, competence in bacilli is maintained for hours. Tortosa et al. (2001) have postulated that perhaps *B. subtilis* acquired this quorum sensing mechanism for competence through horizontal transfer, which would open the possibility for noncompetent bacteria to gain the ability of competence.

At this point it should be mentioned that care must be taken in evaluating mutants that are deficient in competence, or any other aspect of transformation. As explained below, there are several discrete steps related to binding, uptake, integration, and expression of donor DNA. If only the final events (i.e., successful integration and expression) are used as a criterion for noncompetence, then many mutants will be isolated that have nothing to do with competence-related events. A similar statement can be made for isolation of mutants in recombination or any other phase of transformation. In this instance careful evaluation has to be made to discern the blocked step in competence development. Competence-specific mutants are not expected to be extra sensitive to DNA-damaging agents as recombination specific mutants would be. Moreover the use of a nonintegrating DNA molecule (e.g., a plasmid or bacteriophage DNA, as described above) in the assay for competence mutants should identify cells blocked only in competence pathways and eliminate the possibility of recombination deficient mutations.

B. *Streptococcus pneumoniae*

In contrast to *B. subtilis*, *S. pneumoniae* develops competence in complex media only. This development of competence is related with the synthesis and release of a protein competence factor (Tomasz and Hotchkiss, 1964) during growth to an appropriate density of culture. Recently Havarstein and coworkers (1995) isolated this factor and determined that it is a 17 residue polypeptide. A synthetic peptide containing the same sequence could also induce competence. Once the concentration of the competence stimulating polypeptide (CSP) has reached a threshold level (usually before early stationary phase of growth), competence is induced rapidly in the whole population of pneumococci. The synthetic peptide extends the range over which competence can be obtained (Havarstein, et al., 1995). The molecular basis for this mechanism of quorum sensing to achieve competence in *S. pneumoniae* is now better understood (Lacks et al., 2000; and reviewed by Lacks, 1999) and is shown in diagrammatic form in Figure 5.

The CSP is produced from a competence operon containing *comC-E*. The gene for pre-CSP is *comC*, and the pre-CSP is a 41 amino acid peptide, which is cleaved to the 17 amino acid final molecule. In a separate operon are the genes for ComA and Com B that are the membrane-located processor and exporter of CSP into the environment, via an ABC transporter-type system. Both of these operons are transcribed constitutively by SigA, the main vegetative polymerase of *S. pneumoniae*. As CSP accumulates in the medium, it signals ComD, a membrane-located histidine protein kinase to phosphorylate ComE in a 2-component sensing system (see above for competence in bacilli and Bayles and Fujimoto, this volume). The phosphorylated ComE then autoinduces both the *comA-B* and the *comC-D-E* operons to produce more of the competence products and competence development is enhanced. It is postulated that phosphorylated ComE may also induce the production of a secondary competence-specific sigma factor, *comX* product (SigH-like), which then transcribes

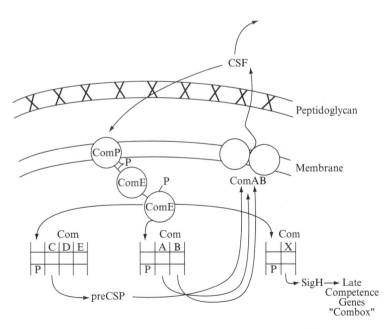

Fig. 5. Composite diagram of proteins involved in the development of competence in *Streptococcus pneumoniae*

late competence genes (Lee and Morrison, 1999) (see below).

In *S. pneumoniae* the maintenance of competence is very short lived, lasting only a few minutes. The cells are refractory to repeat competence induction for about 60 minutes (Chen and Morrison, 1987). Alloing and coworkers (1998) suggest that ComE may have a dual function-inducing competence at low doses of CSP signal, then inhibiting the process when the level of CSP becomes high, although Lee and Morrison (1999) suggest that a gene regulated by SigH may control the shutoff and postcompetence events. In the absence of the CSP the cells are noncompetent. Like bacilli, the streptococci appear to undergo a differentiation during competence development. For instance, *S. pneumoniae* cells form longer chains, growth rate is depressed, autolysins are activated and at least 16 proteins associated with competence appear (Morrison et al., 1982). The operons encoding some of the proteins share a consensus "combox" sequence in the promoter, TACGAATA, where presumably the sigH containing RNA polymerase interacts for transcription of uptake and processing systems (see below) (Pestova and Morrison, 1998). In *S. sanguis*, another competent streptococcus with a competence inducing factor, 10 proteins are transiently synthesized after the addition of the competence-inducing polypeptide (Raina and Ravin, 1980). Whatmore and coworkers (1999) have shown that CSPs are also present in *S. mitis* and *S. oralis*. Although these CSPs are quite similar to that for *S. pneumoniae*, suggesting that there may be cross-competence development and possible genetic exchange among the several streptococcal species inhabiting the respiratory tree, no "cross talk" has been observed. Indeed, the CSP receptors are highly allele specific, and even in species where there are 2 or more CSP alleles, no "cross talk" occurs (D. Morrison, personal communication). The Avery A6 strain does not respond to CSP from other strains. Also *Streptococcus pyogenes*, which is not naturally competent, carries homologues to many of the transformation-linked genes in the pneumococcus, including *comX*. So, even the nontransformable streptococci seem to have the machinery and perhaps the potential to achieve transformation.

C. General Considerations

There are also other more general physiological changes that differentiate competent Gram-positive cells from the noncompetent population (see Young and Wilson, 1972; Smith et al., 1981). In *B. subtilis* there is an increase in galactosamine content in cell wall–associated teichoic acids. In *S. pneumoniae*, choline is required in the cell wall for competence development. If ethanolamine is substituted, the cells are noncompetent. These findings probably relate to the initial attraction of the DNA by the competent cells. Autolytic enzyme content also appears to be important for both the bacilli and pneumococci for competence development. Nonautolytic mutants tend to be less competent. It is likely that the autolytic system provides easier access to the membrane for the incoming DNA. In this regard, when *B. subtilis* is grown at 45°C, a temperature that sustains good cellular growth, the secretion of autolysins and many other exported proteins is diminished. These cells grown at 45°C are also less competent.

Thus, in the Gram-positive organisms, competence is a complex correlation of physiological conditions and genetic regulatory events.

D. *Haemophilus influenzae*

Haemophilus influenzae is representative of Gram-negative bacteria that are naturally transformable. The *Neisseria* also belong to this group, and they will be mentioned at times where research has brought out some important facets of the transformation process in Gram-negatives using these organisms.

Haemophilus influenzae becomes transformable at a low level during late exponential growth phase. However, Herriott and coworkers have described a two-step technique whereby the cells become highly competent (Herrriott, et al., 1970). The cells are first grown to stationary growth phase in a chemically defined medium (MI$_c$), then shifted to a nongrowth medium (MIV). Following the downshift to nongrowth medium, competence development is rapid, high level, and independent of cell concentration. Using this method as much as 50% of the population becomes competent and 5% transformation levels can be reached (Herriott et al., 1970). In these cells competence is quite stable, as was seen with *B. subtilis*. When cells are transferred to rich growth medium, DNA synthesis resumes and competence decays. There are structural alterations associated with this development of competence. Competent *H. infuenzae* cells develop numerous membrane vesicles, also called transformasomes, that bud from the external layer of the outer membrane and contain proteins that may react specifically with conserved sequences present on *Haemophilus* DNA (Kahn, et al., 1983). Noncompetent cells have few such vesicles. Six polypeptides, ranging in size from 95 to 45 kDa, were found in the crude envelope fractions from competent *H. influenzae* cells, but not from noncompetent cells (Zoon et al., 1976).

It was also found that simply adding cAMP to cells in the early exponential growth phase results in a low level of competence, similar to that seen at late exponential growth (above). Mutations that inactivate adenylate cyclase or the cAMP binding protein result in noncompetence If cAMP is added to the adenylate cyclase mutant in MIV medium, normal competence can be restored (Gwinn et al., 1996).

A gene that appears to be one of the regulators of competence has been identified (*sxy*). Mutations in this gene result in noncompetence. Conversely, specific point mutations in this gene have been isolated that result in constitutive competence. The *sxy (tfoX)* gene is also required for the induction of later transformation genes, such as *dprABC* (Karudapuram and Barcak, 1997)

and *comF*. It is possible that this is one of the or perhaps the regulatory factor for transcription of operons that are involved in competence and transformation. It is interesting to note that a 26 bp palindromic site is found upstream of *dprA*, the *com A-F*, and the *pilA-D* genes. This could be the site where Sxy and/or another regulatory element acts. Recently it was found that *Helicobacter pylori* has a *dprA* gene and is naturally transformable (Gwinn et al., 1998). Also *Haemophilus*, *Bacillus* (the ComG proteins) and *Streptococcus* have orthologues to type IV pili components (see below in binding section). These are all necessary for competence in these strains, but they do not form pili. However, in the transformable *Neisseria*, *Pseudomonas*, *Moraxella*, and *Legionella* fully formed pili are required for transformability (Graupner et al., 2000). Nonpiliated strains are up to 3 logs lower in transformation frequency. As discussed below, these proteins, necessary for competence, function in aspects of DNA uptake.

V. BINDING OF DNA TO COMPETENT CELLS

Both Gram-positive and Gram-negative cells exhibit "loose" and "tight" DNA binding. In the Gram-positive bacteria the DNA is taken up very quickly. The negatively charged DNA is initially loosely attracted to the positively charged outside wall and associated molecules in both bacilli and pneumococci.

A. *Streptococcus pneumoniae*

In *Streptococcus pneumoniae*, much of the loosely bound DNA is converted to tightly bound DNA-cell complexes within 5 to 10 seconds at 30°C. Then there is a gradual increase in DNA binding for another 30 minutes or more. Fox and Hotchkiss (1957) analyzed this binding reaction. By treating the bacteria-DNA interaction leading to a transformant as a conventional enzyme-substrate reaction, these investigators were able to derive rate constants for the various steps leading to a transformation event.

A calculation of the ratio of these constants at high and dilute DNA concentration enabled Fox and Hotchkiss to estimate that pneumococci can adsorb approximately 30 molecules of DNA per cell. Conversely, by comparing the V_{max} values at high and low DNA concentrations in this reaction, Fox and Hotchkiss determined that there were as many as 75 binding sites per pneumococcal cell. From these experiments it is then estimated that each competent pneumococcus can bind 30 to 80 molecules of DNA. This binding is independent of molecular weight or any sequence specificity.

The bound DNA is sensitive to shear and to DNase but is resistant to gentle washing. Excess nonspecific carrier DNA will not compete for binding with the tightly bound transforming DNA. As described above, a wave of competence factor production takes place, which localizes to the membranes of both competent and noncompetent pneumococci. The association of the competence factor with the cell comes at a time when there is also demonstrated an increase in autolysis. This may allow DNA to reach binding sites more rapidly.

Lacks and his associates isolated in *S. pneumoniae* a series of mutants that lack membrane-bound major endonuclease activity (*endA* mutants) (Rosenthal and Lacks, 1980). If more than 1% of the endonuclease activity remains, transformation is normal. However, if the mutants exhibit less than 1% activity, the cells bind DNA but cannot process it further. If the bound DNA is reextracted, following exposure to alkali, it is obvious that single-stranded breaks have been introduced approximately 6 kb apart. This nicking occurs even in the presence of EDTA (ethylene diamine tetraacetic acid, a chelator of divalent cations). Lacks suggested that the incoming DNA may bind to a nicking protein, which then remains at the nick site during subsequent uptake events (Lacks and Greenberg, 1976). The mutants that have greater than 1% major endonuclease activity also bound DNA efficiently.

In the presence of EDTA the same single-stranded breaks were introduced.

However, when EDTA is not added, a second nick, caused by EndA, occurs opposite the single-stranded nicks, resulting in a double-stranded break and then entry of a single strand of the transforming DNA (see below). Thus in pneumococcus the binding reaction is a continuum of events resulting in fragmentation of the transforming DNA to double-stranded fragments, which then initiate the uptake process. The binding and subsequent entry of DNA may be facilitated by orthologues of the ComEA and ComEC bacilli proteins, which provide sites of binding and entry for the incoming DNA strand (see below).

B. Bacillus subtilis

In *Bacillus subtilis* there are about 30 to 50 sites for DNA binding during competence (Dubnau and Cirigliano, 1972), and DNA is bound rapidly. After a lag of about 1 minute the DNA becomes resistant to exogenous DNase. The DNA is also fragmented, with an average size of about 13.5 to 18 kb, and a surface-located nuclease (NucA) is postulated to interact with the bound DNA. (Provvedi, et al., 2001).

In *B. subtilis* a variety of proteins participate in binding DNA and subsequently processing it during the transformation reaction (Dubnau, 1999). The principal protein that binds the incoming DNA is ComEA. This protein may also have function in transport. It is a bitopic membrane spanning protein with the C-terminus to the outside (Inamine and Dubnau, 1995). The protein domain that binds the incoming DNA is located at the C-terminus. ComEA binds double-stranded DNA preferentially and can bind molecules as small as 22 bp (Provvedi and Dubnau, 1999). In addition a series of ComG proteins(ComGA-GG) are required for the binding reaction, though it is postulated to be more in a role of providing access by the incoming DNA to the ComEA protein. The ComG proteins are part of the class of proteins that are designated as PSTC (pilus type IV, type-2 secretion, twitching motility, competence) and have orthologues in both Gram-positive and Gram-negative cells. As such, they may help to provide a channel for the DNA through the thick peptidoglycan of bacilli (see Fig. 6). To support this idea, it is known that in membrane vesicles, where no peptidoglycan is present, ComEA is still required for DNA binding but the ComG proteins are nonessential (Provvedi and Dubnau, 1999). The expression of the *comG* genes is negatively regulated by ComZ (Ogura and Tanaka, 2000). Autolytic enzyme could facilitate this process.

C. Haemophilus influenzae

In *Haemophilus influenzae*, as well as in *Neisseria*, the binding and uptake of DNA is determined by the specificity of the DNA. Uptake sequences are present in the chromosome at a frequency of about 1 per 4000 bp. The segment that regulates this uptake is an 11 bp consensus sequence, 5'-AAGTGCG GTCA-3'. Foreign DNA, lacking the consensus sequence, is adsorbed poorly and does not compete well. The DNA is initially bound loosely to the surface of both competent and noncompetent cells and can be readily removed by high-salt concentrations or

DNase. Stuy has estimated that only 2 to 4 molecules interact with each competent cell, but the interaction is quite productive (Stuy and Stern, 1964). Deich and Smith (1980) estimate 4 to 8 uptake sites. Tight binding commences with the formation of a DNA-recipient cell surface complex that becomes resistant to DNase. Since EDTA does not affect the binding and uptake, it is likely that there is no surface nuclease that requires divalent cations, as seen in the Gram-positive systems. Binding events are difficult to measure due to the rapid uptake of bound DNA, but they are likely associated with the uptake-mediating vesicles. There is evidence that the DNA may be bound to a 29 kDa polypeptide prior to transport (see Goodgal, 1982). Also antibodies generated to competent cells will block DNA binding and uptake in competent *H. influenzae* cells, but antibodies to noncompetent cells will not (Bingham and Barnhardt, 1973). This suggests that the incoming DNA must interact with surface proteins prior to entry. Knockout of the Sxy protein results in no binding of DNA, and Sxy appears to be modulated by cAMP, reinforcing the idea that this may be a regulatory factor for transformation in this organism.

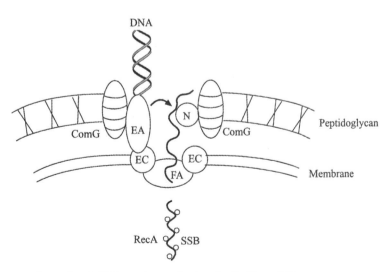

Fig. 6. Diagram of postulated DNA uptake mechanism for *Bacillus subtilis*.

D. General Considerations

DNA binding in all of the competent bacteria has an overlapping similarity, namely the presence of the PSTC proteins. As discussed above, though these proteins are found in type IV pili, they appear to serve the same function in Gram-positives, which express no pili. In fact the orthologues for the ComG proteins from *B. subtilis* include the CglA-D, CilC and CilD, D1, and D2, as well as ComYA-YD in *S. pneumoniae*, and the PilA-C, and ComE proteins in *H. influenzae* (Dubnau, 1999). Extrapolating from the involvement by these proteins in the formation of uptake channels in *B. subtilis*, it is tempting to conclude that the orthologues may function similarly in the other natural transformation systems (see Section VI). In *Neisseria gonorrheae* the ComP protein, a member of this class of proteins is required for transformation, but is not essential for pilus formation (Wolfgang, et al., 1999)

Second, in all three systems only double-stranded DNA is naturally fully active. Under standard competence conditions, binding and uptake of single-stranded DNA, RNA and RNA-DNA hybrids is very inefficient if nonexistent. In *B. subtilis* and in *S. pneumoniae* a low level of transformation can be obtained using single-stranded DNA, but only under very specific conditions (with EDTA and at low pH) when the DNA is protected from degradation by nucleases (Miao and Guild, 1970; Tevithia and Caudill, 1971).

VI. DNA UPTAKE

In the Gram-positive cells uptake is temporally concomitant with the development of resistance to DNase. This is the first point at which the transformation system no longer can be abrogated by exogenous DNase. Remember, in *S. pneumoniae*, just prior to the development of DNase resistance, double-stranded breaks have been introduced into the entering DNA by the endogenous membrane-bound endonuclease (*endA*), cleaving opposite the already present nick (see above). In the pneumococci this generates fragments of 6 to 9 kb. In *B. subtilis*, Ca^{2+} and Mg^{2+} are necessary for competence development, and Mg^{2+} activates the major endonuclease to introduce breaks and generate double stranded fragments of 13.5 to 18 kb (see above). So, at the start of uptake, the DNA is fragmented and resistant to DNase.

The elegant early experiments of Lacks and Dubnau elucidated many of the subsequent events occurring during the entry of DNA in both the pneumococcal and bacillus systems, respectively (Lacks, 1962, 1977; Lacks et al., 1967; Dubnau, 1982). More recent experiments have delineated the molecular basis for DNA uptake (Lacks et al., 2000; Dubnau, 1999). Competent cells bind large amounts of DNA when uptake is blocked with EDTA (EDTA may damage DNA receptor sites and scavenges necessary divalent cations). When EDTA is removed and divalent cations are added, uptake of DNA is initiated and can be measured. The following description of uptake is largely derived from these key experiments.

A. *Bacillus subtilis*

Since many of the proteins for DNA uptake were first described in *B. subtilis* and have orthologues in the pneumococcus system, it will be clearer if the bacillus uptake system is described first.

In *B. subtilis*, after binding to the ComEA protein, the double-stranded fragments are rapidly converted to single strands by an associated exonuclease. From the point of adding exogenous DNA to resistance to DNase and single-strand uptake is about 2 minutes in *B. subtilis*. The double-stranded fragments are digested to single-strand fragment of 7 to 15 kb. In *B. subtilis* the DNA single strand may enter either $3'-5'$ or $5'-3'$. These single strands are in the eclipse complex, where the single-stranded DNA has yet to associate with the chromosome for recombination. An equal amount of oligonucleotides can be found in the medium from the degraded strand. This suggests that the digestion is performed outside the membrane. Integrated DNA is smaller, 8.5 to 10 kb (Dubnau and Cirigliano, 1972). ComEA also plays a role in transport and may physically deliver the double-stranded DNA fragment to the nuclease. ComEC is required for transport. This is a polytopic membrane-spanning protein and may form the aqueous channel for DNA transport. ComFA binds DNA and is also required for DNA transport. It is an integral membrane protein. It has similarity to helicases that travel on DNA using energy from ATP hydrolysis. It is postulated this protein may generate the mechanism and the energy for DNA translocation. Null mutants of ComFA are decreased in transformation 1000-fold (Londono-Vallejo and Dubnau, 1993). However, since null mutations of most of the other competence-related proteins are far more restrictive to transformation, it is postulated that the cell may have an inefficient, passive way to transport the DNA in the absence of ComFA. The incoming DNA then associates with SSB and RecA for integration. The composite of this reaction is presented in Figure 6 and is based on a more complete figure seen in the review by Dubnau (1999).

B. *Streptococcus pneumoniae*

In *S. pneumoniae*, after binding, the double-stranded fragments are rapidly converted to a single strand of 7 to 8 kb that is taken up with 3′–5′ polarity, while the other strand is broken down in 5′–3′ polarity (Mejean and Claverys, 1988). Lacks showed that the sum of DNA entering and trichloroacetic acid (TCA)-soluble oligonucleotides released accounts for all the added DNA. The single strand associates with SSB proteins for stability and then is brought to complementary regions in the chromosome by RecA. Non-homologous DNA enters but is degraded over time and reutilized in endogenous DNA synthesis. The uptake and degradation of the other strand appears to be initiated by the major, membrane-bound endonuclease, EndA. The orthologues of bacillus proteins

from the ComG, ComE, and ComF operons also appear to be in the apparatus where the DNA binds and enters the cell. CelA and CelC are orthologues of ComEA and Co-mEC, while Cgl ABCD are orthologues to the ORFs 1–4 of the *comG* locus in *B. subtilis*. CflA is the orthologue to the ComFA protein of bacillus (Lacks et al., 2000; Dubnau, 1999). CelB provides the pore through the membrane for DNA uptake, while CelA participates in DNA binding. The Cgl proteins interact in the peptidoglycan and may have a similar function to that seen in *B. subtilis* (i.e., providing a channel or wall alteration to allow DNA access to membrane proteins). The CflA protein is located on the inner face of the membrane and is the putative driving force for single-strand entry. Dubnau suggests that ComF may provide the link to energy required for DNA transport. Cfl may have a similar function, though this is not yet known. All of this composite information is presented in Figure 7 (based on more detailed figure presented in Lacks et al., 2000).

C. *Haemophilus influenzae*

As described above, in *Haemophilus* and *Neisseria* only DNA carrying the specific recognition sequence is bound. In *Haemophilus* binding occurs in the tranformasome, where resistance to DNase and fragmentation occurs, and then is translocated to the cytoplasm. Uptake is rapid from the few binding sites, and it has been reported at up to 1000 nucleotides per second (Kahn and Smith, 1984). Concommitant to uptake, the DNA is converted to a single strand. The 5′ strand is completely degraded, while the 3′ strand is reduced in molecular size. Since single-stranded DNA is not detected in the cytoplasm, it is likely that the DNA is rapidly integrated. Mutants in the *rec2* gene cannot do this translocation, and the DNA remains trapped in the transformasome. Rec2 is a homologue of ComEC, and it's reasonable to assume that Rec2 forms a similar channel. Com101 is homologue to ComF, and though its function is not absolutely known, mutants in Com101 do not degrade

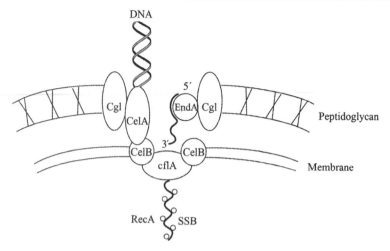

Fig. 7. Diagram of postulated DNA uptake mechanism for *Streptococcus pneumoniae*.

DNA to single strands and don't translocate. So it may serve a similar function to ComF in the Gram-positives. As mentioned above, the Sxy protein may be a regulatory element and there is some evidence that it may regulate Rec2 and Com101A expression. The 26 bp palindromic competence related regulatory region is also upstream from the *rec2* gene. Of course, the PSTC proteins are also present in both *Haemophilus* and *Neisseria*. These proteins most likely perform similar functions of channeling the DNA to appropriate receptors. A periplasmic protein, Por (protein disulfide oxidoreductase), is essential for transformation, and it may function to properly conform competence proteins for DNA uptake (Tomb, 1992). Dpr (above) is a competence protein that is common to all the systems. In *Haemophilus* it most likely helps to translocate the DNA across the inner membrane. Autolytic enzymes may aid DNA passage through the peptidoglycan, and there is a ComL orthologue in the *Haemophilus* data base that may correspond to murein hydrolase activity (Dubnau, 1999).

D. Other Considerations

I. Gram-Positives

In both *S. pneumoniae* and *B. subtilis* there is a finite time when the DNA fragment, which has entered, can be reextracted prior to inte-

gration. This suggests that incorporation into the bacterial chromosome by recombination is not concommitant to uptake. In both systems about 2 minutes or more from the time of addition of DNA are required before appearance of transformants. The *Bacillus* system appears to be slower and may again reflect the relative biosynthetic latency of the process or perhaps a mode of entry of DNA that requires additional steps from that known for the pneumococcus.

Since the appearance of a distantly linked marker (as measured by cotransformation frequency; see Appendix) commences at 2.6 minutes, 0.4 minutes *after* the initial appearance of the selected gene (see Fig. 8), it is evident that in the natural transformation systems the DNA must be taken up longitudinally, rather than en masse. Since in a similar experiment two very closely linked genes enter at the same time, from the distance between these markers, calculations have revealed that the minimum effective length of 5000 bp is necessary for successful transformation, and the uptake process proceeds at a constant rate (approximately 180 bp/s at 28°C) (Dubnau, 1999).

In *B. subtilis* there is also evidence that the proton motive force is necessary for DNA uptake (Van Nieuwenhoven et al., 1982). Van Nieuwenhoven and coworkers tested

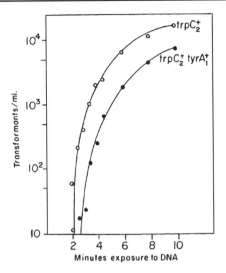

Fig. 8. Appearance of linked genes as a function of time. Competent *B. subtilis* cells (trp⁻, tyr⁻, and his⁻) were exposed to wild-type *B. subtilis* DNA. Samples were removed at various times and deoxyribonuclease was added. The mixtures then were plated on media containing histidine and tyrosine (for trp⁺ transformants) and onto histidine-supplemented plates for trp⁺, tyr⁺ transformants. (Reproduced from Strauss, 1966, with permission of the publisher.)

the effects of ionophores on both DNA binding and uptake. Although the ionophores were found to inhibit transformation, in other experiments DNA that was previously bound to the surface was not released by the ionophores. This suggests that some of the ionophores interact with DNA-binding sites on the cell. Therefore experiments were initiated to measure DNA uptake directly. When cells to which DNA had been bound in the absence of Mg^{2+} (same effect as addition of EDTA; see above) were supplemented with Mg^{2+}, DNA uptake was initiated. Addition of ionophores at this step measured uptake specifically and showed marked inhibition of the reaction, if the ionophores decreased ΔpH. Thus nigericin and monesin, which dissipate ΔpH but increase ΔY slightly, were the most effective inhibitors of DNA uptake. These results suggest that ΔpH but not $\Delta\Psi$ is involved as a driving force for DNA entry. It is quite logical that the electrochemical gradient would play a role in

transformation, since so many uptake and cell surface functions, including autolysis, are regulated by this system (Jolliffe et al., 1981).

There is evidence that proton motive force may play a role in DNA uptake in the artificial transformation system of *Escherichia coli* (see below). On the other hand, there is no evidence for involvement of this process during DNA uptake in *S. pneumoniae* (Clave and Trombe, 1987). Thus it still is not obvious what the driving force for DNA uptake may be in this organism. Lacks suggested that degradation of one strand may provide energy for the uptake of the other (Lacks, 1962), but this would not account for transformation with single-stranded DNA, which proceeds by the same uptake mechanism.

2. Gram-Negatives

As stated above, competent *H. influenzae* cells have more transformasomes than noncompetent. Mutant *com*-51 lacks competence and sheds the vesicles to the medium. These vesicles, when purified, could still bind DNA. However, in these purified vesicles protection against DNase was lost, suggesting that there must be additional processing of the incoming DNA beyond association with the transformasomes that is required for the uptake reaction. The intact vesicles extend 35 nm from the cell, are 20 nm in diameter, and "May be internalized by DNA binding", especially in *H. parainfluenzae*. The uptake reaction is so rapid in *Haemophilus* that the vesicles may deliver the incoming DNA to the proximity of the chromosome and recombination may ensue. The uptake of DNA requires active metabolism and also requires the proton motive force (Bremer et al., 1984)

As also mentioned previously, there is stringent species specificity to DNA uptake in both *H. influenzae* and *N. gonorrheae*. When *Haemophilus* DNA is cloned into an *E. coli* plasmid (pBR 322), the entire plasmid can be taken up. Normally this plasmid cannot transform *H. influenzae*. (Sisco and Smith, 1979). Danner and coworkers

(1980) calculated that there are about 600 uptake-specifiying sites on the chromosome of this organism, or roughly one site per 4000 nucleotides. Statistically this is far more than would be expected by chance. Curiously the *E. coli* chromosome should have about 2 sites but acts as if there were about 60. The ability of some nonspecific DNA to partially avoid the binding and uptake discrimination suggests that a degenerate or subset version of the sequence may also be operative, or there may be additional, yet unknown factors necessary for DNA uptake. The reasons for this specificity and the maintenance of these uptake sequences in *Haemophilus* DNA are not clear. It may be that these sequences have a secondary function, which may provide the necessary selective pressure for their maintenance. Since EDTA does not prevent DNA entry into *Haemophilus*, there are presumably no external nucleases required for the process.

VII. INTEGRATION OF TRANSFORMING DNA

It is impressive how effectively the introduced DNA fragments are integrated into the recipient chromosome. In all three of the major natural transformation systems discussed to this point, integration is accomplished via a single strand of incoming DNA.

Gurney and Fox (1968) used a heavy label (15^N)-substituted donor DNA and demonstrated that on the average, single-stranded regions of 5 to 10 kb of DNA are integrated into the chromosome of *S. pneumoniae*. In *B. subtilis* the integrated segment is on the average 8.5 kb, while in *H. influenzae* the size is 18 kb (see Dubnau, 1982). Occasionally larger than average integration may occur, and multiple single-strand fragments can also interact with the chromosome in a clustered fashion.

To study the integration process, thymine-requiring recipient cells were grown to competence in the presence of 5-BU (heavy label), then transformed with radioactive, non 5-BU containing, (light) donor DNA. After the DNA was extracted from the transformed cells and fractionated according to density, it was found that all donor activity (detected by the label) was found in the lighter fraction, suggesting that integration is a replacement reaction (Bodmer, 1966). This replacement is driven by homology of the single donor strand to the recipient chromosome and facilitated by the RecA equivalents in the recipient cells (see Levene and Huffman, this volume, and Yasbin, this volume). In bacteria and yeast, integration of exogenous DNA is driven by homology and replacement of endogenous DNA. Conversely, in most eukaryotic systems, the introduced DNA is added to the genome by nonhomologous recombination mechanisms (see Levene and Huffman, this volume).

In bacteria, limited homology decreases the probability that the DNA will be integrated. Wilson, in a series of experiments, showed that the surrounding DNA sequences of the incoming gene (the "neighborhood") are vital to integration (Wilson and Young, 1972). In Table 1 it can be seen that DNA from *B. amyloliquefasciens* poorly transformed *B. subtilis*, when two different genes were examined. Antibiotic resistance-determining genes (in this instance Rif^R) appear to be more conserved as to homology, producing higher transformation rates. This may relate to evolutionary pressure conserving the apparati for transcription and translation. Transformation for the *his* gene, on the other hand, was markedly depressed, when compared to the homologous system. However, if the DNA was reisolated from rare transformants (termed intergenotes, where even in the absence of good homology an integration event occurred) and used to transform *B. subtilis* again, the efficiency of transformation was significantly higher (almost to homologus levels). Therefore, in the second transformation, the intergenote DNA had now acquired a more compatible neighborhood through mismatch repair and recombination, which now favored successful integration. At 45°C discrimination for homology is decreased. The same results have been obtained from experiments in the natural

TABLE 1. Transformation with Heterologous DNA

Source of DNA	Rif^r Transformants per 10^8 cells	his$^+$ Transformants per 10^8 cells
Homologous	7.95×10^5	6.80×10^4
Homologous	5.50×10^5	5.28×10^4
Heterologous	8.26×10^4	2.18×10^1
Intergenote	3.66×10^5	3.96×10^4
Intergenote	7.96×10^5	8.05×10^4

Source: Data derived from Wilson and Young (1972).
Note: *Bacillus subtilis* was the transformation recipient in all experiments. Homologous DNA was from *B. subtilis*. Heterologous DNA was from *B. amyloliquifaciens* H. Intergenote DNA was obtained from *B. subtilis* strains which had been transformed with the Rif^r and his$^+$ genes from *B. amyloliquifaciens*. Transformations were for 30 minutes at 37°C and were terminated with deoxyribonuclease. DNA concentrations were adjusted to the same absorbance at 260 nm for all experiments.

transformation systems for *S. pneumoniae* and *H. influenzae*.

So transformation is a driven process that can extend into nonpaired regions as long as some homology anchors the single strand to the recipient chromosome. The single strand invades the recipient DNA duplex and, by RecA-mediated recognition, finds homology and is integrated, with the corresponding removal of the relevant recipient DNA strand. This system is identical to the Meselson-Radding D-loop recombination pathway described by Levene (this volume). This model should be applicable to all the natural transformation systems.

The integrated donor strand creates a mismatched heteroduplex, which will be resolved by DNA replication or repair. Mismatch repair is best documented in the *hex* system described in pneumococci (see Claverys and Lacks, 1986). There are low-and high-efficiency transforming markers that result from differing mismatch repair activity. The low-efficiency (LE) markers are considered to represent primarily transition mutations, while the high-efficiency markers (HE) appear to arise from transversions and deletions (Balganesh and Lacks, 1985). Consequently the *hex*-encoded mismatch repair system recognizes the mismatch for incoming LE markers and eliminates them along with bordering sequences. An asymmetry (due to the mismatch) between the

donor and recipient strands results in a specific degradation of the donor DNA by exonuclease, which specifically eliminates the donor LE markers from the transformed population. High-efficiency markers escape this mismatch correction. Isolation of *hex* mutants, in which all markers as expected would be HE, supports this scenario. These *hex* mutants, missing this part of mismatch repair, also show a high mutation rate, as is also known in other systems (notably *E. coli* and the mutator strains (Grady, 1996). There are at least two separate genes that cooperate in the recognition and elimination of mismatches in *S. pneumoniae*.

A. Integration in Gram-Positive Cells

In the naturally competent Gram-positive transformation systems integration can be viewed as a five-step process (see Smith et al., 1981; Stewart and Carlson, 1986):

1. Internalization of the DNA to the proximity of the chromosome
2. Formation of an unstable donor-recipient complex
3. Formation of stable noncovalent complex
4. Covalent complexing of donor-recipient DNA
5. Resolution of the covalently bound mismatched heteroduplex DNA

Experiments that support such a stepwise scenario are summarized here. In *B. subtilis* when the DNA is crosslinked, or if the donor DNA is exposed to intercalating agents, the donor-recipient complexes become unstable (Mooibroek and Venema, 1982). Coumermycin, a DNA gyrase inhibitor, blocked conversion from steps 2 to 3, suggesting that superhelical structures may favor recombination. Also, in *B. subtilis*, if the temperature was dropped from 37 °C to 17 °C immediately after DNA uptake, the integration process was arrested at step 2, and the transformation was still sensitive to S1 (single-stranded DNA-specific) nuclease. In this context Shoemaker and Guild (1972) used the *S. pneumoniae* transformation system to measure integration of donor genes at various temperatures. These investigators found that the rate of the reaction from step 1 to step 2 is marker dependent and has an activation energy of 50 cal/mol. This may indicate that the initial step is a localized denaturation of a small region of the recipient chromosome. On the other hand, the conversion of the unstable complex into a stable complex is largely marker independent and proceeds at a rate that suggests formation of a single covalent bond at a time. This step has an activation energy of 20 to 21 cal/mol. These are just a few of the experiments that support the stepwise donor DNA integration sequence outline above.

An interesting finding by te Riele and Venema (1984) bears on this integration sequence. These investigators found that incoming single-stranded DNA in the *B. subtilis* transformation system entered at sites in the membrane where the DNA is associated (Streips et al., 1980). Moreover they postulated that association of the incoming DNA with the recipient chromosome also occurs at that site, mediated by the RecA protein. Homologous DNA then proceeded to resolution of the complex (steps 3–5). However, heterologous DNA remained complexed to the membrane and was not stably integrated (blocked at step 2), Thus much of the continuous process leading of a stable transformant

may occur in localized regions of the cell, which may relate to the uptake scenario in Figure 6.

B. Integration in *Haemophilus influenzae*

Integration of transforming DNA into the *H. influenzae* genome also occurs via a single strand of donor DNA. As mentioned above, the incoming DNA is protected by a sequestering event, perhaps mediated by both the vesicles and associated proteins. As one strand is integrated, the other is degraded. Since the appearance of transformants is very rapid, it is reasonable to postulate that integration in this organism may occur at the site and together with uptake. Competent cells of *Haemophilus* also have single-stranded gaps in DNA, and this may facilitate the formation of donor-recipient complexes. In this organism single-stranded transforming DNA cannot be recovered from the cytoplasm, again suggesting that integration is quite rapid.

C. Recombination-Deficient Mutants

In all three natural transformation systems mutants (Rec⁻) that fail to integrate transforming DNA have been isolated (see Smith et al., 1981; Dubnau, 1982; Goodgal, 1982; Rhee and Morrison, 1988; Setlow, 1988). These mutants are sensitive to DNA-damaging agents (e.g., methyl methanesulfonate) and are found to be altered in some aspect of the recombination pathway (see Levene and Huffman, this volume). Commonly these mutants are unable to transform when chromosomal DNA fragments are added. Plasmids and bacteriophage DNA can transform such mutants, unless recombination is needed for establishment or propagation of these molecules.

VIII. ARTIFICIAL COMPETENCE

Competence can also be developed artificially. The two major artificial competence regimens are (1) the divalent cation method for transforming *Escherichia coli* and related Gram-negative cells, and (2) the protoplast transformation techniques used with a variety of bacteria, including many that cannot develop natural competence.

A. *Escherichia coli* Transformation

Escherichia coli does not develop natural competence. However, since this organism has been the cornerstone for many cloning experiments, it was necessary to develop an artificial transformation system (Mandel and Higa, 1970; Cohen et al., 1972). In initial experiments it became obvious that incoming DNA faces tremendous hazards due to powerful nucleases in the organism (Cosloy and Oishi, 1973; Clark, 1973). Ultimately it was determined that inactivation of the *recBC* genotype or the *recD* allele and activation of the minor recombination pathways, *sbcBC*, were necessary to protect linear DNA and increase transformation efficiency (Russell et al., 1989).

Plasmid molecules survive transformation better. Transformation with plasmids is a vital part of many cloning regimens in *E. coli*. A one-step method for preparing competent *E. coli* cells has been described by Chung and coworkers (1989). Hanahan found that different strains of *E. coli* respond differently to the various competence regimens (Hanahan, 1985). Consequently he developed a methodology that applies to most commonly used *E. coli* strains and also has listed the transformation protocols for many strains (Hanahan 1985). Many companies offer guaranteed competent *E. coli* cultures, obviating the need to do it yourself.

The incoming plasmids in artificial competence are thought to associate with specific channels on the outer surface of *E. coli* cells. It is estimated that each cell has at least 10 and as a many as 200 such channels used for uptake of nutrients and other materials the cell requires for growth and propagation. Competition experiments using the plasmids pBR322 and its derivative pXf1 showed that a 1000 : 1 ration of pXf1 to pBR322 (total plasmid to cell ratio of 10 : 1) did not impede transformation of pBR322 at all. Increasing the pXf1 to pBR322 ratio to 10,000 : 1 (total

plasmid to cell ratio 100 : 1) or 100,000 : 1 (total plasmid to cell ratio 1000 : 1) decreased the total pBR322 transformation to only 40% and 20% of the maximal, noncompetition rate, respectively (Hanahan, 1983). So there must not only be many independent channels for uptake but the *E. coli* cells must compete for the uptake of DNA.

It is postulated that the presence of the divalent cations and dimethyl sulfoxide (DMSO) in many of the regimens affects the integrity and organization of the lipopolysaccharide layer in *E. coli* at low temperatures. The alteration of this layer may result in unmasking or providing better access to the uptake channels. The channels could be associated with fusion regions between the outer and inner membranes (Bayer's junctions) (Bayer, 1979). Hexamine cobalt (III) chloride ($HACoCl_3$) apparently facilitates the interaction between DNA and the cell, specifically the uptake channels. A heat shock step is necessary in most protocols and increases the uptake of adsorbed DNA. These procedures yield about 4% transformation rates.

Escherichia coli is nonselective in DNA uptake. In fact most *E. coli* strains do not discriminate against foreign DNA and even use heterologous promoters to direct transcription and protein synthesis (see Hellman, this volume). This makes *E. coli* a desirable cloning system for initial constructs (see Geoghegan, this volume). The competent *E. coli* cells, like many of the naturally competent cultures, can be frozen and used at later times.

B. Polyethylene Glycol-Mediated Transformation

Gram-positive bacteria that do not develop natural competence, as well as those that do, can be transformed using polyetheylene glycol (PEG, usually PEG6000) to drive the uptake reaction (Bibb et al., 1978; Chang and Cohen, 1979). The bacteria need to have the peptidoglycan cell wall removed and then be sustained in an osmotically stabilizing environment. The resulting protoplasts are able to take up plasmid or bacteriophage DNA with high efficiency when PEG6000 is added to the mixture. The role of the PEG in this process is not clear, although PEG is known to fuse cell membranes and effectively precipitate DNA. Following the DNA uptake event, the cell wall needs to be regenerated for the transformant to survive and to assay for DNA uptake. The protoplasting and regeneration systems differ among the many Gram-positive organisms. Therefore a careful study must be initiated to assess the proper parameters before the transformation system becomes operative with a new organism (see Crawford et al., 1987). However, in well-established systems, such as *B. subtilis*, as many as 80% of viable cells can be transformed by this technique, with a yield of 10^7 transformants per microgram of plasmid DNA. Chromosomal fragments are not readily introduced using this method, though they can competitively inhibit the uptake of both plasmid and bacteriophage DNA (White and Streips, 1986). This technique has opened the possibilities for any organism to be transformed, including cell wall-deficient forms like *Mycoplasma* and L-forms, which are normally refractory to any type of genetic exchange (White et al., 1981).

IX. DNA MOLECULES THAT REQUIRE NO INTEGRATION

Two types of DNA that require no integration can enter competent cells. As already discussed above, these are plasmid and bacteriophage DNAs. The extent of success using these types of DNA molecules depends partly on whether they enter bacteria through the natural or artificial competence regimens.

A. Natural Competence Pathway

Naturally competent cells of all three systems discussed above, *S. pneumoniae*, *B. subtilis*, and *H. influenzae* take up and express plasmid and bacteriophage DNA (transfection). Chromosomal fragments compete with these molecules for uptake, so entry must occur by very similar or identical mechanisms. However, entry of plasmid and bacteriophage DNAs by natural competence regimens yield very low transformation frequencies.

In the Gram-positive cells the plasmid and bacteriophage DNA must face the same uptake pathway of nuclease attack and conversion to single strands. These single-stranded fragments cannot replicate, nor can they integrate into the chromosome due to lack of any homology. Consequently both types of DNA require the cooperation of several copies to create an active replicating plasmid or bacteriophage DNA genome. This is why as many as 1000 to 10,000 equivalents of these genomes are necessary for successful transformation. This is also why single-plasmid monomers are inactive in transformation of naturally competent Gram-positive organisms (Canosi et al., 1978; Saunders and Guild, 1981).

In *H. influenzae* there is also a 10^{-4} reduction in transformation efficiency with plasmids. Presumably a similar mechanism operates, and through the binding and sequestration steps the DNA is also degraded sufficiently to require recombination for establishment in the cytoplasm.

An exception to the low frequency of expression in competent cells by these self-replicating molecules occurs when homology for the incoming DNA is provided (see Dubnau, 1982). For instance, *B. subtilis* transformation using monomeric plasmid DNA is efficient only if the recipient harbors a plasmid, which carries homology to the incoming DNA. Recombination (RecA mediated) occurs between the two plasmids and leads to successful transformation. Second, if the incoming monomeric plasmid has sequences homologous to the chromosome, it will survive by integrating into the chromosome via the homologous region. While this process can be eliminated by eliminating the RecA protein, this method can also be used to intentionally put genes of commercial interest into the chromosomes of relevant bacteria. Following the integration of the desired plasmid chimera, the organism would then replicate the molecule along with the chromosome in a stable configuration.

Finally, as described above, the introduction of multimeric forms of plasmids into the naturally competent bacteria will result in transformation due to recombination among redundant sequences in the multimer to generate a replicating molecule. In *S. pneumoniae* transformation monomers transform only through a "two-hit" mechanism, whereby two copies of the plasmid enter and recombine (see Fig. 9). Adding homology via a chromosomal fragment also enhances plasmid transformation efficiency in the pneumococcus, as well as in other transformable species (see Stewart and Carlson, 1986).

B. Transformation after Inducing Artificial Competence

Both plasmid and bacteriophage DNA transformation is most efficient using the artificial competence system. All of the natural competence systems we have discussed are also amenable to artificial competence induction. Curiously chromosomal fragment uptake is very inefficient in the artificial competence systems. It may be the fact that entry of the DNA is unaffected by nucleases in this process and thus preserves integrity of

plasmids and bacteriophage DNA, which aids their capacity to replicate and become established in the cell. However, the integrity of chromosomal fragments may interfere with the ability to efficiently integrate into the chromosome using the RecA system.

As already described, *E. coli* is extremely capable for plasmid transformation and transfection, as is protoplasted *B. subtilis*. *H. influenzae* can also be transformed using the divalent cation artificial competence regimen. However, even in this artificial environment there is not a significant increase in plasmid transformation efficiency.

Transformation with protoplasts is also useful for eukaryotic cells, notably yeast. Therefore the artificial competence systems have opened up possibilities for introducing self-replicating DNA molecules into a wide variety of cells. As such, this method has become a cornerstone for much of the cloning work in both Gram-positive and Gram-negative bacteria.

X. TECHNOLOGICAL TRANSFER SYSTEMS

In both bacterial and eukaryotic systems, where DNA uptake by either natural or

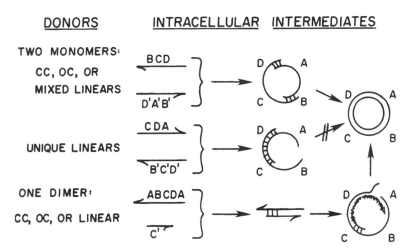

Fig. 9. Plasmid transformation of *S. pneumoniae*. Model shows that a mixture of fragments from a plasmid monomer can form an active transforming molecule; so can a dimer or multimer. However, a single monomer, or unique fragments which do not represent the entire plasmid molecule, will be inactive in transformation (see text). (Reproduced from Saunders and Guild, 1981, with permission of the publisher.)

artificial competence pathways is blocked, inefficient, or simply not expedient, novel mechanisms for introducing nucleic acid molecules into cells have been developed. The most prominent among these technological systems are electroporation and biolistic, high-velocity projectile-mediated DNA transfer.

A. Electroporation

Electroporation works for both eukaryotes and prokaryotes (see Shigekawa and Dower, 1988; Sambri and Lovett, 1990). To date both Gram-positive and Gram-negative organisms, many of which are totally refractory to DNA transformation by any technique, have been successfully transformed using electroporation.

Briefly, high-voltage electrical discharges into a suspension of cells produce localized membrane breaks or pores through which all manner of macromolecules, including DNA, can enter. In *borrelia*, PCR products and oligonucleotides have been used to transform and mutagenize the cells (Samuels et al., 1994; Samuels and Garon, 1997). In *E. coli*, the viability and efficiency of electroporation transformation are related to the electrical field strength of the pulses used (Dower et al., 1988). The cells are put on ice for 10 minutes, pelleted, suspended in cold 10–15% glycerol plus sucrose, pelleted aging, and suspended to a small volume 250 to 500 µl. DNA (1–5 µl) is added. The cells plus DNA are put in an electroporation cuvette and pulsed with 1.8 to 2.5 kV in 3 to 5 ms pulses. The resulting mixture is then incubated and plated for transformants (Samuels and Garon, 1997). In *E. coli* transformation efficiencies in the range of 10^9 to 10^{10} transformants per microgram of DNA could be obtained and electroporation routinely yielded about 10-to 20-fold more transformants than the competence regimens outlined earlier in this chapter. Larger plasmids show lower transformation frequencies. Many types of bacteria have been tried with varying levels of success (Wirth et al., 1989). Strains isolated from the environment

appear to be less susceptible than laboratory-adapted bacteria. However, from many perspectives it is worth transforming with electroporation bacteria that are refractory to other genetic exchange methodologies, even if efficiency of the process is quite low.

B. Biolistic Transformation

Biolistic transformation employs DNA precipitated onto spherical gold particles and accelerated into eukaryotic tissues with a specialized device (Sanford et al., 1993). Briefly, supercoiled DNA (about 5–25 µl of 1 mg/ml) is precipitated onto the gold spheres in the presence of divalent cations and fresh spermidine. The mix is iced and centrifuged. The pellet is suspended in ethanol and recentrifuged, then precipitated again. The target tissue is densely packed onto filters and bombarded with the gold-DNA mixture from a distance of several centimeters. If gold residue is not present on the cells, they are bombarded again. The cells need to be incubated and regenerated for selection (Fromm et al., 1990; Kemper et al., 1996).

XI. LINKAGE

Since a fairly large fragment of chromosomal DNA is integrated, there will be linkage of closely situated genes in the transformation. By definition, two genes are considered to be linked if they are carried on the same fragment of DNA. Calculations for linkage are explained in detail in the mapping appendix to this chapter. In reality, with more and more organisms having their genomes sequenced, linkage of markers appears on the physical maps and does not need to be deduced by transformation or other genetic means. On the other hand, many organisms of scientific and industrial interest are not genetically sequenced, and this exercise could be useful.

If the two genes of interest are on different DNA fragments, then the probability of them arriving simultaneously in a competent cell is a product of the independent events. Therefore **ab** competent bacteria are transformed using donor DNA which is **AB** [**ab**(bacteria) × **AB** DNA). If the **A** and **B** genes arrive on separate fragments, then **Ab** transformants appear 1/100 cells and **aB** transformants (1/100). **AB** transformants would be expected at 1/1000 cells, *if the genes arrive on separate fragments.* Any frequency of transformation to **AB** that is significantly higher than 1/1000 cells would suggest that the genes arrived on the same fragment and could be linked.

However, this assumes that the population is 100% competent. That is not true in *B. subtilis*. In this system only 20% of the population is competent. In this instance the *unlinked* frequency for **AB** transformants would be 1/400 cells rather then 1/1000 as calculated above. Consequently a transformation rate for **AB** significantly higher than 1/400 is required in *B. subtilis* transformation to infer linkage.

In *B. subtilis* calculations (as in the Appendix) can be used to determine linkage. Alternatively, dilution curves indicate linkage. If **ab** competent bacteria are exposed to **AB** donor DNA in decreasing DNA concentrations, then linked markers should decrease at the same rate as single transformant numbers since both markers should be on the same fragment repeatedly (Fig. 10). On the other hand, unlinked genes will yield curves with half the slope, because interaction of the competent cell with two separate molecules is required for **AB** transformants to appear. Such an interaction is decreasingly probable as the concentration of DNA added to cells decreases—a 10-fold reduction in DNA concentration will result in a 100-fold reduction of transformants.

Transformation is useful in establishing linkage for genetic markers in localized regions of the bacterial chromosome. If larger regions need to be mapped, generalized transduction is a more useful tool (see Weinstock, this volume).

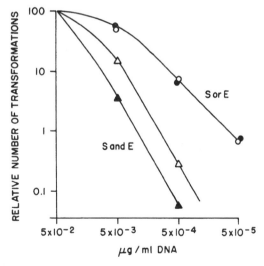

Fig. 10. Effect of DNA concentration on single and double transformations. Competent *H. influenzae* cells were exposed to DNA from an erythromycin-resistant and streptomycin-resistant strain (Goodgal, 1961). Since the resistance genes come from two DNA preparations, they cannot arrive on the same DNA fragment. Transformants for each single gene show the same increase in relative numbers as DNA concentration increases (as seen in Fig. 2). However, since two DNA molecules need to interact with each competent cell to yield a double transformant, a greater concentration is necessary to achieve equivalent transformation. The slope of the curves is indicative of nonlinkage of double transformants. If the genes were linked and could be cotransformed by the same fragment of DNA, the slope should be similar to that exhibited by the transformants for a single trait (erythromycin or streptomycin resistance). (Reproduced from Goodgal, 1961, with permission of the publisher.)

Conversely, the phenomenon of congression can be effectively used to manipulate the genetic makeup of competent bacteria. Congression is based on the assumption that given highly saturating concentrations of DNA, each competent bacterium will take up multiple fragments of DNA. In the partially competent *B. subtilis*, a bacterium that has taken up a fragment is far more likely to take up another rather than a cell that has not developed competence. This fact allows the investigator to move genes and construct strains for specific needs. For example, there is a strain of *B. subtilis* that carries the auxotrophic requirements: *purA⁻*, *leuA⁻*, *metB10⁻*. However, for research purposes you need a strain that carries the *purA* and *leuA* markers but not *metB10*. Instead, you need the strain to have the auxotrophic marker *hisA⁻*. Such needs arise when large regions of the chromosome need to be analyzed for mapping purposes or for physiological analyses of chromosome structure associations (Streips et al., 1980). The new strain could be obtained through laborious mutagenesis regimens, but instead can be constructed relatively easily by congression. Since the *hisA* and *metB10* genes are not linked by transformation, the relevant genotype cannot be expected to arrive into the recipient competent cell on the same DNA fragment. However, since saturating DNA levels are used, and each cell has many fragments available for uptake, the switch of *hisA⁻* to *metB10⁻* can be accomplished. For this purpose a donor DNA must be isolated from a strain which carries the *metB10⁺* and *hisA⁻* markers. After the transformation with this DNA, the competent portion of the population can be identified by selection for *metB10⁺* transformants. Among these transformants there should also be cells that picked up a second fragment of DNA, specifically the one that carries *hisA⁻* and incorporated it into the genome. These can be identified by transferring all the *metB10⁺* bacteria onto appropriate selective media (sterile toothpicks and eager students are wonderful for this), and

isolating any *hisA⁻* transformants. Then the new *purA⁻*, *leuA⁻*, *hisA⁻* has been isolated and can be used for experiments. The frequency of congression for any one specific exchange event will range from 1/100 to 1/1000, and how quickly you isolate the strain you need will depend on your luck in picking the right colony.

XII. CONCLUDING THOUGHTS

I trust that this discussion of DNA entry into bacteria, using transformation mechanisms, has provided the reader with up-to-date knowledge, generated some questions, and perhaps evoked some innovative ideas for future research. The research described in this chapter represents a small fraction of work published in this field. Many excellent laboratories continue research in this area, as is attested by many references within the last couple of years. Yet the discipline still has a lot of room to grow and flourish. The knowledge gained from the bacterial system has already been applied to eukaryotic cells. Of course, transformation will always be a vital step in cloning methodologies, as well as in the genetic analysis of diverse bacteria.

REFERENCES

Alloing G, Martin B, Granadel C, Claverys JP (1998): Development of competence in *Streptococcus pneumoniae*: Phermone autoinduction and control of quorum-sensing by the oligopeptide permease. Mol Microbiol 29:75–83.

Anagnostopoulos C, Spizizen J (1961): Requirements for transformation in *Bacillus subtilis*. J Bactreiol 81:741–746.

Avery OT, MacLeod CM, McCarty M (1944): Studies on the chemical nature of the substance inducing transformation of pneumococcal types: Induction of transformation by a desoxyribonucleic acid fraction isolated from pneumococcus type III. J Exp Med 79:137–158.

Balganesh TS, Lacks S (1985): Heteroduplex mismatch repair system of *Streptococcus pneumoniae*: cloning and expression of the *hexA* gene. J Bacteriol 162:979–984.

Bayer ME (1979): The fusion sites between outer membrane and cytoplasmic membrane of bacteria: Their role in membrane assembly and virus infection. In Inouye M (ed): "Bacterial Outer Membranes." New York: Wiley, pp 167–202.

Bibb MR, Ward JM, Hopwood DA (1978): Transformation of plasmid DNA into *Streptomyces* at high frequency. Nature 274:398–400.

Bingham DP, Barnhardt BJ (1973): Inhibition of transformation by antibodies against competent *Haemophilus influenzae*. J Gen Microbiol 75:249–258.

Bodmer W (1966): Integration of deoxyribonuclease treated DNA in *Bacillus subtilis* transformation. J Gen Physiol 49:233–258.

Bott KF, Wilson GA (1967): Development of competence in the *Bacillus subtilis* transformation system. J Bacteriol 94:562–570.

Bremer W, Kooistra J, Hellingwerf KJ, Konings WN (1984): Role of the electrochemical proton gradient in genetic transformation of *Haemophilus influenzae*. J Bacteriol 157:868–873.

Canosi U, Morelli G, Trautner TA (1978): The relationship between molecular structure and transformation efficiency of some *Staphylococcus aureus* plasmids isolated from *Bacillus subtilis*. Mol Gen Genet 166:259–267.

Cato A Jr, Guild WR (1968): Transformation and DNA size: I. Activity of fragments of defined size and fit to a random double cross-over model. J Mol Biol 37:157–178.

Chang S, Cohen SN (1979): High frequency transformation of *Bacillus subtilis* protoplasts with plasmid DNA. Mol Gen Genet 168:111–115.

Chen JD, Morrison DA (1987): Modulation of competence for genetic transformation in *Streptococcus pneumoniae*. J Gen Microbiol 133:1959–1967.

Chung CT, Niemela SL, Miller RH (1989): One-step preparation of competent *Escherichia coli*: Transformation and storage of bacterial cells in the same solution. Proc Natl Acad Sci USA 86:2172–2175.

Clark AJ (1973): Recombination deficient mutants of *Escherichia coli* and other bacteria. Annu Rev Genet 7:67–86.

Clave C, Trombe M-C (1987): Abstracts, Wind River Conference on Genetic Exchange, Estes Park, CO.

Claverys J-P, Lacks SA (1986): Heteroduplex deoxyribonucleic acid base mismatch repair in bacteria. Microbiol Rev 50:133–165.

Cohen SN, Chang ACY, Hsu L (1972): Nonchromosomal antibiotic resistance in bacteria:genetic transformation of *Escherichia coli* by R factor DNA Proc Natl Acad Sci USA 69:2110–2114.

Cosloy SD, Oishi M (1973): Genetic transformation in *Escherichia coli* K-12. Proc Natl Acad Sci USA 70:84–87.

Crawford IT, Greis KD, Parks L, Streips UN (1987): Facile autoplast generation and transformation in *Bacillus thuringiensis* subsp. *kurstaki*. J Bacteriol 169:5423–5428.

Danner DB, Deich RA, Sisco KL, Smith HO (1980): An eleven-base-pair-sequence determines the specificity of DNA uptake in *Haemophilus* transformation. Gene 11:311–318.

Deich RA, Smith HO (1980): Mechanism of homospecific DNA uptake in *Haemophilus influenzae* transformation. Mol Gen Genet 177:369–374.

Dower WS, Miller JF, Ragsdale CW (1988): High efficiency transformation of *Escherichia coli* by high voltage electroporation. Nucleic Acids Res 16:6127–6145.

D'Souza C, Nakano NM, Zuber P (1994) Identification of *comS*, a gene of the *srfA* operon that regulates the establishment of genetic competence in *Bacillus subtilis*. Proc Natl Acad Sci USA 91:9397–9401.

Dubnau D (1999): DNA uptake in bacteria. Ann Rev Microbiol 53:217–244.

Dubnau D, Cirigliano C (1972): Fate of transforming DNA following uptake by competent *Bacillus subtilis*: IV: The endwise attachment and Uptake of transforming DNA. J Mol Biol 64:31–46.

Dubos R (1976): "The Professor, the Institute and DNA." New York: Rockefeller University Press.

Fromm ME, Morrish F, Armstrong C, Williams R., Thomas J. Klein TM (1990): Inheritance and expression of chimeric genes in the progeny of transgenic maize plants. Bio/Technol 8:833–839.

Fox MS, Hotchkiss RD (1957); Initiation of bacterial transformation. Nature 179:1322–1324.

Goodgal SH (1961): Studies on transformation of *Haemophilus influenzae*: IV. Linked and unlinked transformation. J Gen Physiol 45:205–228.

Goodgal S (1982): DNA uptake in *Haemophilus* transformation. Annu Rev Genet 16:169–192.

Graham JB, Istock CA (1978): Genetic exchange in *Bacillus subtilis* in soil. Mol Gen Genet 166:287–290.

Grady D (1996): Quick change pathogens gain an evolutionary edge. Science 204:1081.

Graupner S, Frey V, Hashemi R, Lorenz MG, Brandes G, Wackernagel W (2000): Type IV pilus genes *pilA* and *pilC* of *Pseudomonas stutzeri* are required for natural genetic transformation, and *pilA* can be replaced by corresponding genes from nontransformable species. J Bacteriol 182:2184–2190.

Griffith F (1928): The significance of pneumococcal types. J Hyg (Lond) 27:113–159.

Gurney TJ, Fox MS (1968): Physical and genetic hybrids formed in bacterial transformation. J Mol Biol 32:83–100.

Gwinn ML, Yi, D, Smith HO, TombJ-F (1996): Role of the two-component signal transduction and the phosphoenolpyruvate:carbohydrate phosophotransferase systems in competence development of *Haemophilus influenzae* Rd. J Bacteriol 178:6366–6368.

Gwinn ML, Ramanathan R, Smith HO, Tomb, J-F (1998): A new transformation-deficient mutant of *Haemophilus influenzae* Rd with normal DNA uptake. J Bacteriol 180:746–748.

Hanahan D (1983): Studies on transformation of *Escherichia coli* with plasmids J Mol Biol 166:557–580.

Hanahan D (1985): Techniques for transformation of *E. coli*. In Glover DM (ed): "DNA Cloning, Vol 1: A Practical Approach." Oxford: IRL Press, pp 109–135.

Haverstein LS, Coomaraswamy G, Morrison DA (1995) An unmodified heptadecapeptide induces competence for genetic transformation in *Streptococcus pneumoniae*. Proc Natl Acad Sci USA 92:11140–11144.

Hayes W (1968): "The Genetics of Bacteria and Their Viruses." New York: Wiley.

Herriott RM, Meyer EM, Vogt M (1970): Refined non-growth media for stage II development of competence in *Haemophilus influenzae*. J Bacteriol 101:517–524.

Horowitz S, Doyle RJ, Young FE, Streips UN (1979): Selective association of the chromosome with the membrane in a stable L-form of *Bacillus subtilis*. J Bacteriol 138:915–922.

Hotchkiss RD (1982): The first generation of gene transfer in bacteria. In Streips UN, Goodgal SH, Guild WR, Wilson GA (eds): "Genetic Exchange: A Celebration and New Generation." New York: Dekker, pp 9–17.

Inamine GS, Dubnau D (1995): ComEA, a *Bacillus subtilis* integral membrane protein required for genetic transformation, is needed for both DNA binding and transport. J Bacteriol 177:3045–3051.

Jolliffe LK, Doyle RJ, Streips UN (1981): Energized membrane and cellular autolysis in *Bacillus subtilis*. Cell 25:753–763.

Kahn ME, Smith HO (1984): Transformation in *Haemophilus*: a problem in membrane biology. J Memb Biol 81:89–103.

Kahn ME, Barany F, Smith HO (1983): Transformasosmes: Specialized membranous structures that protect DNA during *Haemophilus* transformation. Proc Natl Acad Sci USA 80:6927–6931.

Karudapuram S, Barcak GJ (1997): The *Haemophilus influenzae dprABC* genes constitute a competence-inducible operon that requires the product of the *tfoX* (*sxy*) gene for transcriptional activation. J Bacteriol 179:4815–4820.

Kelly MS (1967): Physical and mapping properties of distant linkages between genetic markers in transformation of *Bacillus subtilis*. Mol Gen Genet 99:333–349.

Kemper EL, da Silva MJ, Arruda P (1996): Effect of microprojectile bombardment parameters and osmotic treatment on particle penetration and tissue damage in transiently transformed immature maize (*Zea Mays* L.) embryos. Plant Sci 121:85–93.

Lacks S (1962): Molecular fate of DNA in genetic transformation of pneumococcus. J Mol Biol 5:119–131.

Lacks SA (1977): Binding and entry of DNA in bacterial transformation. In Reissig JL (ed): "Microbial Interactions." London: Chapman and Hall, pp 177–232.

Lacks SA (1999): DNA uptake by transformable bacteria. In Broome-Smith JK, Baumberg S, Stirling CJ, Ward FB (eds): "Transport of Molecules across Microbial Membranes." Cambridge: Cambridge University Press, pp 138–168.

Lacks SA, Ayalew S, de la Campa AG, Greenberg B (2000) Regulation of competence for genetic transformation in *Streptococcus pneumoniae*: Expression of *dpnA*, a late competence gene encoding a DNA methyltransferase of the DpnII restriction system. Mol Microbiol 35:1089–1098.

Lacks S, Greenberg B (1976): Single-strand breakage on binding of DNA to cells in the genetic transformation of *Diplococcus pneumoniae*. J Mol Biol 101:255–275.

Lacks, S, Greenberg B, Carlson K (1967): Fate of donor DNA in pneumococcus transformation. J Mol Biol 29:327–347.

Lee MS, Morrison DA (1999). Identification of a new regulator in *Streptococcus pneumoniae* linking quorum sensing to competence for genetic transformation. J Bacteriol 181:5004–5016.

Londono-Vallejo JA, Dubnau D (1994): Membrane association and role in DNA uptake of the *Bacillus subtilis* PriA analog ComF1. Mol Microbiol 13:197–205.

Love PE, Lyle MJ, Yasbin RE (1985): DNA-damage inducible (*din*) loci are transcriptionally activated in competent *Bacillus subtilis*. Proc Natl Acad Sci USA 82:6201–6205.

Magnuson R, Solomon J, Grossman AD (1994): Biochemical and genetic characterization of a competence phermone. Cell 77:207–216.

Mandel M, Higa A (1970): Calcium-dependent bacteriophage DNA infection. J Mol Biol 53:159–162.

McCarty M (1980): The early days of transformation. Annu Rev Genet 14:1–15.

McCarty M (1985): "The Transforming Principle." New York: Norton.

Mejean V, Claverys J (1988): Polarity of DNA entry in transformation of *Streptococcus pneumoniae*. Mol Gen Genet 213:444–448.

Miao R, Guild WR (1970): Competent *Diplococcus pneumoniae* accept both single- and double-stranded deoxyribonucleic acid. J Bacteriol 101:361–364.

Mooibroek H, Venema G (1982): The effect of chlorpromazine on transformation in *Bacillus subtilis*. Mol Gen Genet 185:65–168.

Morrison DA, Mannarelli B. Vijayakumar MN (1982): Competence for transformation in *Streptococcus pneumoniae*: An inducible high capacity system for genetic exchange. In Schlessinger D (ed): "Microbiology 1982." Washington, DC: ASM Press, pp 136–138.

Msadek T, Kunst F, Rapaport G (1994): MecB of *Bacillus subtilis* is a pleiotropic regulator of the ClpC ATPase family, controlling competence gene expression and survival at high temperature. Proc Natl Acad Sci USA 91:5788–5792.

Nakano MM, Xia L, Zuber P (1991): Transcription initiation region of the *srfA* operon, which is controlled by the *comP-comA* signal transduction system in *Bacillus subtilis*. J Bacteriol 173:5487–5493.

Ogura M, Tanaka T (2000): *Bacillus subtilis comZ (yjzA)* negatively affects expression of *comG* but not *comK*. J Bacteriol 182:4992–4994.

Persuh M, Turgay K, Mandic-Mulec I, Dubnau D (1999): The N- and C-terminal domains of MecA recognize different partners in the competence molecular switch. Mol Microbiol 33:886–894.

Pestova EV, Morrison DA (1998): Isolation and characterization of three *Streptococcus pneumoniae* transformation-specific loci by use of a *lacZ* reporter insertion vector. J Bacteriol 180:2701–2710.

Provvedi R, Dubnau D (1999): ComEA is a DNA receptor for transformation of competent *Bacillus subtilis*. Mol Microbiol 31:271–280.

Provvedi R, Chen I, Dubnau D (2001) NucA is required for DNA cleavage during transformation of *Bacillus subtilis*. Mol Microbiol 40:634–644.

Raina JL, Ravin AW (1980): Switches in macromolecular synthesis during induction of competence for transformation in *Streptococcus sanguis*. Proc Natl Acad Sci USA 77:6060–6062.

Rhee D-K, Morrison DA (1988): Genetic transformation of *Streptococcus pneumoniae*: Molecular cloning and characterization of *recP*, a gene required for genetic recombination. J Bacteriol 170:630–637.

Rosenthal AL, Lacks SA (1980): Complex structure of the membrane nuclease of *Streptococcus pneumoniae* revealed by two-dimensional electrophoresis. J Mol Biol 141:133–146.

Russell CR, Thaler DS, Dahlquist FW (1989): Chromosomal transformation of *Escherichia coli recD* strains with linearized plasmids. J Bacteriol 171:2609–2613.

Sambri V, Lovett MA (1990): Survival of *Borrelia burgdorferi* in different electroporation buffers. Microbiologica 13:79–83.

Samuels DS, Garon CF (1997): Oligonucletide-mediated genetic transformation of *Borelia burgdorferi*. Microbiol 143:519–522.

Samuels DS, Marconi RT, Huang WM, Garon CF (1994): *gyrB* mutations in coumermycin A1-resistant *Borrelia burgdorferi*. J Bacteriol 176:3072–3075.

Sanford JC, Smith FD, Russell JA (1993): Optimizing the biolistic process for different biological applications. Methods Enzymol 217:483–509.

Saunders CW, Guild WR (1981): Pathway of plasmid transformation in pneumococcus: Open circular and linear molecules are active. J Bacteriol 146:517–526.

Setlow JK, Spikes, D, Griffin K (1988): Characterization of the *rec-1* gene of *Haemophilus influenzae* and behavior of the gene in *Escherichia coli*. J Bacteriol 170:3876–3881.

Shigekawa K, Dower WJ (1988): Electroporation of eukaryotes and prokaryotes: A general approach to the introduction of macromolecules into cells. Biotechniq 6:742–751.

Shoemaker NB, Guild WR (1972): Kinetics of integration of transforming DNA in pneumococcus. Proc Natl Acad Sci USA 69:3331–3335.

Sisco KL, Smith HO (1979): Sequence-specific DNA uptake in *Haemophilus* transformation. Proc Natl Acad Sci USA 76:972–976.

Smith HO, Danner DB, Deich RA (1981): Genetic transformation. Annu Rev Biochem 50:41–68.

Solomon J, Magnuson R, Srivastava A, Grossman AD (1995): Convergent sensing pathways mediate response to two extracellular competence factors in *Bacillus subtilis*. Genes Dev 11:119–128.

Stewart GJ (1987): "Abstracts." Wind River Conference on Genetic Exchange, Estes Park, CO.

Stewart GJ, Carlson CA (1986): The biology of natural transformation. Annu Rev Microbiol 40:211–235.

Strauss N (1966): Further evidence concerning the configuration of transforming deoxyribonucleic acid during entry into *Bacillus subtilis*. J Bacteriol 91:702–708.

Streips UN, Welker NE (1969): Infection of *Bacillus stearothermophilus* with bacteriophage deoxyribonucleic acid. J Bacteriol 99:344–346.

Streips UN, Horowitz S, Doyle RJ (1980): Genetic analysis of DNA-surface interactions in *Bacillus subtilis*. In Shlessinger D (ed): "Microbiology 1980." Washington, DC: ASM Press, pp 284–287.

Stuy JH, Stern D (1964): The kinetics of DNA uptake by *Haemophilus influenzae*. J Gen Microbiol 35:391–400.

te Riele HPJ, Venema G (1984): Heterospecific transformation in *Bacillus subtilis*: Protein composition of a membrane DNA complex containing a heterologous donor-recipient complex. Mol Gen Genet 197:478–485.

Tevethia MJ, Caudill CP (1971): Relationship between competence for transformation of *Bacillus subtilis* with native and single-stranded deoxyribonucleic acid. J Bacteriol 106:808–811.

Tomasz A, Hotchkiss RD (1964): regulation of the transformability of pneumococcal cultures by macromolecular cell products. Proc Natl Acad Sci USA 51:480–487.

Tomb JF (1992): A periplasmic protein disulfide oxidoreductase is required for transformation of *Haemophilus influenzae* Rd. Proc Natl Acad Sci USA 89:10252–10256.

Tortosa P, Logsdon L, Kraigher B, Itoh Y, Mandic-Mulec I, Dubnau D (2001): Specificity and genetic polymorphism of the *Bacillus* competence quorum-sensing system. J Bacteriol 183:451–460.

Van Nieuwehoven MH, Hellingwerf KJ, Venema G, Konings NN (1982): Role of proton motive force in genetic transformation of *Bacillus subtilis*. J Bacteriol 151:771–776.

Whatmore AM, Barcus VA, Dowson CG (1999): Genetic diversity of the streptococcal competence (*com*) gene locus. J Bacteriol 181:3144–3154.

White T, Doyle RJ, Streips UN (1981): Transfection of an L-form from *Bacillus subtilis* with bacteriophage deoxyribonucleic acid. J Bacteriol 145:878–883.

White T, Streips UN (1986): Methods to enhance uptake of DNA in polyethylene glycol mediated transformation. J Microbiol Methods 5:191–198.

Wilson GA, Young FE (1972): Intergenotic transformation of the *Bacillus subtilis* genospecies. J Bacteriol 111:705–716.

Wirth R, Friesinger A, Fiedler S (1989): Transformation of various species of gram-negative bacteria belonging to eleven different genera by electroporation. Mol Gen Genet 216:175–179.

Wolfgang M, van Putten JPM, Hayes SF, Koomey M (1999): The *comP* locus of *Neisseria gonorrheae* encodes a type IV prepilin that is dispensable for pilus biogenesis but essential for natural transformation. Mol Microbiol 31:1345–1357.

Young FE, Wilson GA (1972): Genetics of *Bacillus subtilis* and other Gram-positive sporulating bacilli. In Campbell LL (ed): "Spores V." Washington, DC: ASM Press, pp 77–106.

Youngman P, Perkins JB, Sandman K (1985): Use of Tn917-mediated transcriptional gene fusions to *lacZ* and *cat-86* for the identification and study of *spo* genes in *Bacillus subtilis*. In Hoch JA, Setlow P (eds): "Molecular Biology of Microbial Differentiation." Washington, DC: ASM Press, pp 47–54.

Zoon KC, Habersat M, Scocca, JJ (1976): Synthesis of envelope polypeptides by *Haemophilus influenzae* during development of competence for genetic transformation. J Bacteriol 127:545–554.

APPENDIX: GENE MAPPING BY TRANSFORMATION

Transformation has been useful for fine structure mapping in a variety of microorganisms. The following series of practical applications were developed by Dr. G.A. Wilson and illustrate mapping processes.

Cotransfer Index

Using the generic notation, donor DNA can be identified as a′, b′, c′, and so on. The genes of the recipient conversely can be designated as a°, b°, c°, and so on. There is no consideration for linkage or auxotrophy or prototrophy in these generic designations. Specific auxotrophs are listed as a⁻ and prototrophs are shown as a⁺. It is useful to recall that

fragments of DNA will carry approximately 1% of the bacterial genome.

If the following cross is made—a′b′ DNA transformed into an a°b° strain—then the following classes of transformants could be expected: a′b′, a′b°, and a°b′. The cotransfer index (CI) is defined as

$$CI = \frac{a'b'}{a'b' + a'b° + a°b'}$$
$$= \frac{a'b'}{a' + b' - a'b'}$$

The second equation is true because

$$a' = a'b° + a'b'$$
$$b' = a°b' + a'b'$$
$$a' + b' = a'b° + a°b' + 2a'b'$$

Substitute in the first equation and the second equation is derived.

So, by calculating the CI, two mapping points of information can be gained. First, as shown in Table A1, the linkage of various markers can be established. From these data it is possible to determine that *ind* (now called *trpC2*) is linked to *his2* (now designated as *hisB*) in *B. subtilis*. However, *ind* is not linked to *his1* and *his3*, which both belong to the *hisA* locus. The first and fourth experiments clearly define and confirm linkage. In the first experiment, CI = 0.51 (meaning that 51% of the *ind⁺* transformants were also transformed for the *his⁺* gene). This level of cotransfer is much higher than could be expected by random coincidence (RC). Because of only partial competence in *B. subtilis*, the calculation for RC is not entirely accurate. In the first experiment, if all of the cells were competent, then the expected frequency of unlinked double transformants by RC would be 0.5 per 10,000 cells, rather than the 48 observed. Illustrated in the fourth experiment is the control for this data. In this control, the same recipient cells were used as in the first experiment. However, the donor DNA was a mixture designed so that double transformants could only be obtained following transformation through two independent entry events. In

TABLE A1. Cotransformation of Genetic Markers

Donor DNA (ind$^+$ his$^+$)	Recipient Cells	Transformant Classes (per 10,000 Recipient Cells)			RC^a	CI^b
		ind$^+$	his$^+$	ind$^+$ his$^+$		
10 μg	ind$^-$his$_2^-$	70	72	48	0.5	0.51
10 μg	ind$^-$his$_3^-$	3.9	5.0	0.03	0.002	0.003
10 μg	ind$^-$his$_1^-$	50	100	0.8	0.5	0.005
10 μg (mixture^c)	ind$^-$his$_2^-$	46	20	0.2	0.09	0.003

Data obtained from Nester and Lederberg (1961).
^{a} Random coincidence, expressed as ind$^+$ his$^+$ transformants per 10,000 cells.
^{b} Calculated cotransfer index.
^{c} Mixture of ind$^+$ his$_2^-$ DNA and ind$^-$ his$_2^+$ DNA.

this case the cotransfer indicates that the genes are unlinked. It is important to note that in the fourth experiment the double transformant value is higher than the expected RC frequency (0.2 vs. 0.09). This is again due to the partial competence of the population and the fact that competent cells are more likely to incorporate a second molecule of DNA. The observed doubles are always greater than the predicted value. However, in the first experiment the doubles far exceed (almost 100-fold) any random transformation possibility.

The second point of information gained from calculations of CI is the calculation of relative distance between markers. As shown in Table A2 and Figure A1, determination of values for 1-CI can be used to construct a relative arrangement of genes in linear order.

Cotransfer of genetic markers can also be determined by selecting for one marker only, then testing the transformants

TABLE A2. Cotransfer Index and Mapping

Genotype of Transformants	Number of Transformants per 10,000 cells	CI^a	1-CI
A$^+$	150	—	—
B$^+$	87	—	—
C$^+$	96	—	—
D$^+$	182	—	—
E$^+$	112	—	—
A$^+$B$^+$	96	0.68	0.32
A$^+$C$^+$	87	0.54	0.46
A$^+$D$^+$	108	0.48	0.52
A$^+$E$^+$	93	0.55	0.45
B$^+$C$^+$	34	0.23	0.77
B$^+$D$^+$	37	0.16	0.84
B$^+$E$^+$	92	0.86	0.14
C$^+$D$^+$	134	0.93	0.07
C$^+$E$^+$	27	0.15	0.85
D$^+$E$^+$	14	0.05	0.95

^{a} Cotransfer index.

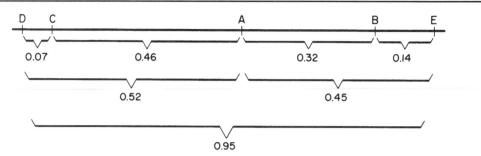

Fig. A1. Map showing the relative position (distance and order) of the loci (**A–E**) using values indicated in Table A2.

individually for the unselected marker. By testing a large number of clones, gene linkage can be determined. This method is even more reliable than the calculation of CI, though notably more laborious.

Three-Factor Crosses

Three-factor crosses are a far more accurate method to determine the relative order of three genes which have been shown to be linked. Four-or five-factor crosses can also be done. Also this principle is common to the other genetic exchange methodologies (conjugation and transduction). In generalized transduction, especially in *B. subtilis*, this technique can be used for markers separated by longer distances than in transformation.

In this example donor DNA from a'b'c' cells is designated as (III) and the recipient, a°b°c° is represented as (000). Two assumptions must be made to make this analysis valid. First, only an even number of crossovers can occur during integration events (see Levene and Huffman, this volume). Second, crossovers in any localized region will be relatively infrequent and the minimum number of crossovers should occur most of the time. So, to integrate a fragment containing three genes, two crossovers are required just to insert the single-stranded fragment or any part of it. It would then be expected that the majority of transformants will exhibit just two crossovers. Four crossovers would mean

that not only was just the fragment integrated, but that internal recombination between the introduced fragment and the chromosome also had to occur. This is shown diagramatically in Figure A2. In examples I, II, and IV, only two crossovers were required to integrate all or part of the donor markers. The first marker is always inherited, since it is selected in the transformation. In example III, however, an extra crossover pair takes place to accommodate the possibility that the outside markers are donor in genotype, while the internal marker is from the recipient.

POSSIBLE MOLECULAR BASIS

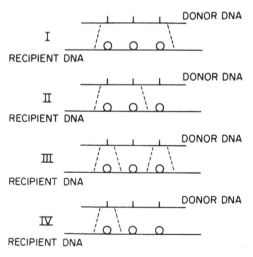

Fig. A2. Three-factor cross analysis (see text).

As an example for this model, consider the cross his^+, trp^+, tyr^- (donor) with his^-, trp^-, tyr^+ (recipient), where his$^+$ transformants are selected, then tested for *trp* and *tyr* characteristics. Four classes of transformants can be expected (Table A3, matched to examples in Fig. A2). They will all be *his*$^+$. As mentioned above and shown in Figure A2, the least frequent class will be the one requiring four crossover events. In this experiment, the least frequent transformant class is class III (his^+, trp^-, tyr^-) (IOI) and the order of the three genes must be *his trp tyr* or *tyr trp his*. If, on the other hand, the rarest class of transformants had been his$^+$, trp$^+$, tyr$^+$, then the order would have had to be *his tyr trp* or *trp tyr his*.

TABLE A3.   Three-Factor Cross Analysis

Transformation: DNA—(his$^+$, trp$^+$, tyr$^-$)
(III)
Recipient Cells—
(his$^-$, trp$^-$, tyr$^+$) (000)

Selection: his$^+$ Transformants

his$^+$ Classes Expected			Number of Transformants
his	*trp*	*tyr*	
1	1	1	103
1	1	0	180
1	0	1	12
1	0	0	240

Note: Evaluation of this analysis is described in the Appendix. The numbers correlate to expected crossover events as depicted in Figure A2.

19

Conjugation

RONALD D. PORTER

Department of Biochemistry and Molecular Biology, The Pennsylvania State University, University Park, Pennsylvania 16802

Modern Microbial Genetics, Second Edition. Edited by
Uldis N. Streips and Ronald E. Yasbin. ISBN 0–471–38665–0
Copyright © 2002 Wiley-Liss, Inc.

I. INTRODUCTION

Conjugation is the mode of gene transfer that involves the transfer of DNA between two live bacterial cells that are in direct contact. Although conjugation in nature most often simply involves the transfer of plasmid DNA from donor to recipient cell, chromosomal DNA can be transferred under certain circumstances. Much of the discussion in this chapter will focus on the F factor-mediated conjugation system of *Escherichia coli* as this system serves as a prototype for conjugation in Gram-negative bacteria. The less well-characterized Gram-positive conjugation systems will be described later in the chapter.

There are many aspects of the discovery of conjugation in *E. coli* by Lederberg and Tatum (1946) that were strongly influenced by elements of serendipity. The choice of the K-12 strain for use in the initial experiments is one of the most striking examples of such happy chance. The protocol involved mixing two isolates that each had at least two nutritional deficiencies so that cells where one marker had reverted would not be scored as recombinants. Although the K-12 strain was chosen primarily because of the availability of isolates with more than one counterselectable nutritional marker, it also happened to contain a self-transmissible plasmid. The particular self-transmissible plasmid in K-12, the F factor, was also unusual in that it both constitutively expresses conjugal transfer functions (see below) and contains several transposable elements that allow it to interact with the bacterial chromosome.

A. C. Regulation of F factor fertility

Lederberg later estimated that only about five percent of randomly selected *E. coli* isolates would have given recombinants with the selection protocol that was initially employed. It was also happy chance that the nutritional genetic markers in the isolates selected were closely clustered on the chromosome in each strain so that optimal yields of recombinants were readily obtained

without the requirement for multiple crossovers. Although the work of the Avery group with transformation in *Streptococcus pneumoniae* (Avery et al., 1944) had set the stage, there were still many scientists who were reluctant to believe that the lowly bacteria could engage in any form of sexual activity. The work done by Lederberg and others who soon followed convincingly demonstrated that bacteria were organisms where genetics could be productively practiced.

There are two distinct requirements that must be met in order for conjugation to occur. The first of these requirments is that the cells be able to engage in a specific contact cycle. The second is that some DNA in the donor cell be capable of undergoing mobilization. Plasmids that encode all of the necessary gene products to enable the potential donor cell to carry out a specific contact cycle with a suitable recipient cell are said to be conjugative. Plasmids whose DNA can be prepared for transfer to a recipient cell are called mobilizable. Both of these capabilities do not always reside on the same plasmid, however, and neither ability alone is sufficient for conjugal DNA transfer. Inability to carry out either or both of these functions classifies a plasmid as being nonconjugative and/or nonmobilizable. A plasmid may simply lack one or both of these abilities as it was originally isolated, or it may have lost one or both of these abilities through mutation. Plasmids that are mobilizable, but nonconjugative, are often efficiently transferred to recipient cells when other plasmids present in the donor cell provide the necessary cell contact functions. Plasmids that are both conjugative and mobilizable are termed self-transmissible (Clark and Warren, 1979).

It should be noted, however, that the word mobilization is also often used to describe the situation where a plasmid, generally a self-transmissible one, is able to affect the conjugational transfer of donor cell chromosomal DNA to a recipient cell. In fact the word mobilization is most often used in this sense. While plasmid mobilization refers to a plasmid's possession of the ability to transfer a copy of its own DNA to a recipient cell whenever a mating pair has been formed, chromosome mobilization in conjugation occurs as the result of some kind of physical association between the donor cell chromosome and the plasmid undergoing conjugational transfer. These plasmid/chromosome associations can be very stable in the case of Hfr strains or very transient in the case of cointegrates formed as intermediates in transposition; these examples will be discussed later in the chapter. Regardless of the degree of stability involved, the chromosome is passively carried along to the recipient cell as the result of its covalent association with the plasmid during chromosomal mobilization. The word mobilization will be used both ways in this chapter, and the student should make every effort not to interchange the two meanings of the word.

II. CONJUGATION BY THE E. COLI F FACTOR

A. Overview

E. coli cells totally lacking the presence of the F factor in any form are called F^- cells. The F factor can, however, exist in a cell in three different forms. First, cells containing an autonomously replicating F plasmid are called F^+ cells. Such cells efficiently transfer the F plasmid to a suitable recipient but rarely transfer donor cell chromosomal DNA. Second, the F factor is able to integrate into the donor cell chromosome to give rise to an Hfr (high frequency of recombination) cell that can efficiently transfer donor cell chromosomal DNA to a recipient cell by conjugation. Third, F-prime plasmids arise when the integrated F factor in an Hfr carries some chromosomal DNA with it as it is recombined out of the chromosome and returns to the autonomously replicating state. F-primes are transferred to a suitable recipient in much the same manner as a wild-type F factor. Although the establishment of the mating pair and the initiation of DNA transfer is identical in all three cases, the

ability of these three donor types to transfer chromosomal DNA to a recipient cell differs considerably. These three different types of F factor–containing cells will be discussed in more detail later in the chapter.

In the case of F factor–mediated conjugation, contact initially occurs between the tip of the donor cell's F factor–encoded sex pilus and the exterior envelope of the recipient cell. Direct contact, presumably achieved by basal disassembly of the pilus, produces an unstable mating pair. Multiple cell interactions frequently give rise to mating cell aggregates that may contain up to 20 cells (Achtman et al., 1978a). A picture and diagrammatic representation of *E. coli* mating aggregates are shown in Figure 1. Although some DNA transfer between the cells may occur at these early stages, most DNA transfer occurs between pairs of cells specifically stabilized within the mating aggregate. Cells are called exconjugants after mating pair dissociation, and recipient cells that have received DNA from donor cells are called merozygotes. These merozygotes become transconjugants after the donor DNA has become stabilized in the recipient cell. This stabilization of donor DNA can occur either by recombination with recipient DNA or, in the case of transferred plasmid DNA, by establishment of the transferred plasmid DNA as an independent replicon in the recipient cell. The various steps in this overall process will be discussed in more detail below. A number of excellent recent review articles dealing with F factor–mediated conjugation are available for additional study (Willetts and Skurray, 1980, 1987; Willetts and Wilkens, 1984; Ippen-Ihler and Minkley, 1986, Frost et al., 1994, Firth et al., 1996).

B. Structure of the F Factor

The F factor is a 100 kb plasmid that can be divided into four fairly distinct regions (see Fig. 2). The region that is labeled *inc, rep* is the portion of the F factor that is involved in the vegetative replication of the plasmid. Mini-F plasmids containing only this region can be constructed, and these

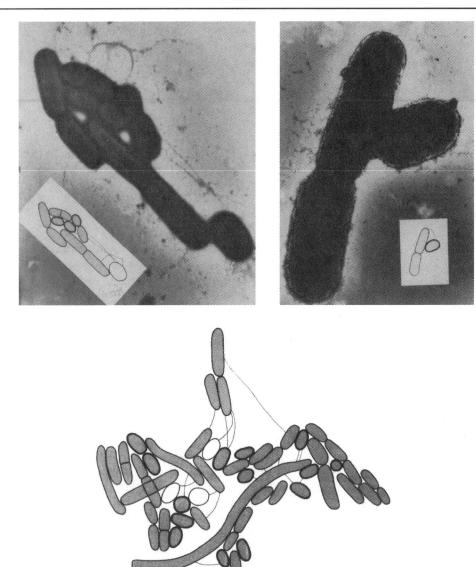

Fig. I. Hfr times F⁻ mating aggregates. An interpretative diagram is shown within each micrograph of mating *E. coli* cells. The Hfr cells in these diagrams are drawn with thin walls while the F⁻ cells are drawn with thick walls. The cells for which the Hfr versus F⁻ assignment was uncertain are shown in white, and F pili that are thought to connect cells are indicated. (Reproduced from Achtman et al., 1978a, with permission of the publisher.)

mini-F derivatives demonstrate all of the replication properties of the parent plasmid. It is this region of the plasmid that determines the F factor's incompatibility properties regarding other plasmids in the same cell (see Perlin, this volume).

The F factor also contains a region where four transposable elements are clustered. In addition to the two copies of IS3 and the copy of IS2, there is a copy of Tn*1000*, also known as γδ, whose transposition properties are very similar to those of Tn3. As

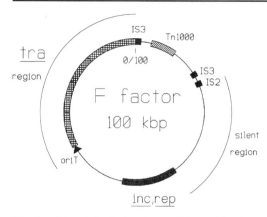

Fig. 2. Map of the *Escherichia coli* F factor. The four major regions of the F factor indicated on the figure are (1) the *inc,rep* region which determines vegetative replication and plasmid incompatibility properties, (2) the *tra* region which stretches from *oriT* to IS3 and provides conjugative and DNA mobilization functions, (3) the region containing the four transposable elements that facilitate interactions between the F factor and other DNA molecules, and (4) the "silent region" between IS2 and the *inc,rep* region.

we will see later in this chapter, it is these transposable elements that are primarily responsible for the ability of the F factor to interact with other DNA molecules, including the chromosome, in the cell. The third region of the F factor is sometimes referred to as the silent region as few distinct genetic functions have been shown to reside there.

The approximately 35 kb region of the F factor labeled *tra* is the fourth region, and it is involved in making the F factor a self-transmissible plasmid. This *tra* region is very similar in organization to the *tra* regions of many F-like R factors. It contains the *oriT* site, at which DNA transfer is initiated, and DNA sequence analysis (Frost et al., 1994) indicates the presence of 36 open reading frames with most of the likely genes designated *tra* and some *trb*. Three of the translated genes (*traM, traJ,* and *artA*) produce separate transcripts, but all of the other genes form a single operon starting with *traY*. Although this huge operon has secondary promoters, the *traY* promoter appears to be dominant under conjugative conditions.

The overall structure of the *tra* regulon is shown in Figure 3.

C. Regulation of F Factor Fertility

The main *tra* operon (starting with *traY*) is positively regulated by the product of the separately transcribed *traJ* gene (Willetts, 1977; Gaffney et al., 1983). A number of mechanisms have been proposed to explain the need for TraJ protein in the efficient expression of the *traY* promoter, but it is currently thought that TraJ protein binding works by providing sufficient superhelicity for transcription initiation (Gaudin and Silverman, 1993). In most F-like plasmids, the *traJ* gene is normally negatively regulated by the *finO* and *finP* gene products (*fin* = fertility inhibition). A virtue of the F factor in genetic studies, however, is its lack of fertility inhibition due to its lack of a functional *finO* gene. The *tra* genes and conjugal ability are therefore constitutively expressed unless *finO* is provided in-*trans* by a *fin*[+] F-like plasmid. This constitutive expression of fertility functions was an important element in the discovery of conjugation by Lederberg and Tatum (1994). The initially mysterious lack of fertility regulation in F is due to an IS3 insertion, which traditionally marks one end of the *tra* regulon, in the *finO* gene (Yoshioka et al., 1987). *finP* is transcribed from the antisense strand in the mRNA leader region of the *traJ* gene, but the *finP* transcript apparently does not code for a protein (Johnson et al., 1981). The overlap of the *finP* RNA and the leader region of the *traJ* mRNA is diagrammatically illustrated in Figure 4. The binding of these two RNA molecules places the translational start signals for *traJ* within an RNA duplex region which would presumably preclude translation initiation (Finlay et al., 1986). The presence of FinO protein has been shown to greatly extend the half-life of *finP* RNA even in the absence of *traJ* mRNA (Lee et al., 1992), and the FinO protein is capable of binding to secondary structural features of both RNA species to promote the formation of an RNA duplex (van Biesen and Frost, 1994; Ghetu et al.,

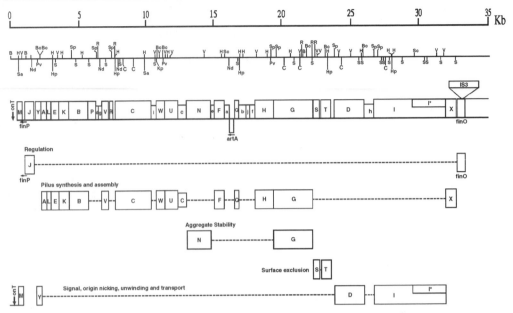

Fig. 3. Map of the F factor transfer region. The top line gives a size scale in kilobase pairs, while the second line show restriction enzyme sites that are not relevant to our discussion here. The third line represents the genes with *tra* genes shown in uppercase letter and *trb* genes shown in lowercase letters; the *oriT* site, the IS3-containing *finO* gene, and the *finP* transcript are also shown. The remaining lines show the various genes grouped by function. (Reproduced from Frost et al., 1994, with permission of the publisher.)

1999). It is only when the FinO protein stabilizes this RNA duplex that the translation of the *traJ* mRNA is precluded.

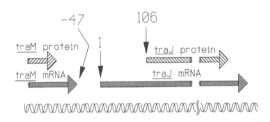

Fig. 4. Location of the F factor *finP* gene. The position of the *finP* transcript is shown relative to the *traM* and *traJ* genes. The coordinate positions indicated are relative to the start of the *traJ* mRNA. The *finP* transcript starts leftward at position 111 and extends to roughly position 34. It should be noted that the *finP* transcript is complementary to much of the leader region of the *traJ* mRNA, and overlaps slightly with the coding region for the TraJ protein.

The other separately transcribed *tra* gene, *traM*, is located very close to *oriT* and appears to be negatively autoregulated by means of binding sites for its own gene product in the dual promoter region (Penfold et al., 1996). The demonstration that *traM* is also positively regulated by TraY protein binding (Penfold et al., 1996) has indicated that the long suspected possibility of *traJ* regulation of *traM* is the result of an indirect effect.

D. Establishment of Cell Contact

A typical F factor-containing *E. coli* cell will possess one to three F-encoded (sex) pili on its surface. The F pilus is a hollow cylinder with an exterior diameter of 8 nm and an interior diameter of 2 nm (Folkhard et al., 1979). These sex pili may be up to 20 μm in length, and are often visualized microscopically by the adsorption of male-specific phages. The F sex pili consist of many molecules of a single protein, pilin, which is encoded by the *traA* gene (Frost et al., 1984). The *traQ* gene product apparently converts the initial 121 amino acid *traA* polypeptide into the functional 70 amino acid polypeptide, perhaps while acting as a chaperone during its insertion into the inner membrane (Maneewannakul et al., 1993). The N-terminal amino acid of the mature pilin is acetylated by the product of the *traX* gene, but the conjugation properties of *traX* mutants lacking this acetylation seem largely unaffected (Moore et al., 1993). Claims for the phosphorylation and glycosylation of pilin have not been substantiated. At least 13 additional genes are required for the assembly of a functional pilus, but their specific roles are not known.

The initiation of mating pair formation requires that the tip of the F pilus make contact with a specific receptor site on the surface of the recipient cell. Although the exact nature of that receptor is not known, mutations that render a cell incapable of functioning as a recipient in conjugation (Con⁻) generally map in either *ompA* or in genes involved in lipopolysaccharide (LPS) synthesis. It appears that LPS participates in the initiation of mating pair formation, and it has also been shown that Zn^{++} is required at the earliest stages of this process. The expression of the *ompA* gene in the recipient cell, on the other hand, is necessary for the stabilization of the mating pair. It remains possible, however, that the OmpA protein is also part of the receptor site, as added LPS-OmpA protein complexes block mating pair formation more effectively than LPS alone (Achtman et al., 1978b). The ability of LPS or LPS-OmpA protein complexes to prevent mating pair formation presumably results from their ability to interact with F pili and thereby prevent the pili from making contact with recipient cells.

Although DNA transfer can occur through an extended F pilus (Ou and Anderson, 1970; Harrington and Rogerson, 1990), little DNA transfer normally occurs before mating pair stabilization. The F pilus has been shown to retract upon the attachment of male-specific phages (Jacobson, 1972), and it is generally assumed that the initiation of direct envelope-to-envelope contact between donor and recipient cell involves the retraction of the pilus by disassembly at its base within the donor cell. The direct contact between cells yields an unstable mating pair where little DNA transfer occurs, and it is likely that the pilus is no longer required after unstable mating pair formation has occurred (Achtman et al., 1978a). The conversion of this unstable mating pair to a stable mating pair where DNA transfer can occur efficiently involves the participation of the *traG* and *traN* gene products, as mutations in either of these genes result in inefficient mating pair formation without reducing the extent of conjugal DNA replication (Manning et al., 1981). The exact nature of the final surface-to-surface interaction between two mating cells is not well understood, but the *traD* gene product may be involved in the formation of a pore between the two inner membranes (Panicker and Minkley, 1985).

Effective mating pair formation between two F factor-containing donor cells is prevented by a phenomenon called surface exclusion. Surface exclusion requires the *traS* and *traT* gene products which are located in the inner and outer membranes, respectively (Achtman et al., 1977, 1979; Minkley and Willetts, 1984; Cheah et al., 1986). Although pilus to envelope contacts between donor cells do occur, surface exclusion prevents mating pair stabilization between two donor cells as well as the initiation of donor conjugal DNA synthesis (see below). Mating between two donor cells can be achieved, however, by a procedure called F⁻ phenocopy mating. This involves growing the cell to be used as the recipient into stationary phase so that *tra* expression, and therefore surface exclusion, is minimized. It is also interesting to note that the *traT* gene product becomes a major component of the outer membrane and plays a role in serum resistance and in reducing the susceptibility of cells to phagocytosis (Moll et al., 1980; Aguero et al., 1984). Although serum resistance is not directly relevant to conjugation mechanism, this *tra*-dependent phenotype provides a selective advantage to F factor-containing cells in some environments.

E. DNA Mobilization and Transfer

For DNA to actually be transferred from donor to recipient cell, the plasmid DNA in the donor cell must go through a series of processing steps that we refer to as mobilization. As replacement synthesis of the transferred strand of donor cell plasmid DNA is generally concurrent with DNA transfer, the entire process (preparation for transfer, or mobilization, and transfer itself) is sometimes also referred to as donor conjugal DNA synthesis (DCDS). Four or five *tra* gene products are involved in DCDS, and several events must occur before the actual DNA synthesis and transfer begins. First, one strand of the DNA is nicked at the F factor *oriT* (origin of transfer) site. This nicking was shown by infecting cells with a bacteriophage λ derivative carrying the *oriT* site, and examining the DNA of the progeny phage (Everett and Willetts, 1980). λ *oriT* was used as a convenient means of packaging the DNA of interest as it is much easier to isolate the appropriate DNA, with and without nicks, from virions than to purify a minority species of nicked plasmid DNA from cell lysates. It was found that 5–10% of the λ *oriT* phage contained a nick within *oriT* when F*lac* was present in the infected cell; there was no nicking observed when F*lac* was not present.

These and other in vivo studies indicated that this strand-specific nicking reaction at *oriT* required the *traY* gene product plus the product of a gene called *traZ* that later turned out to be a secondary translation product of the *traI* gene (Everett and Willetts, 1980; Traxler and Minkley, 1987). Studies to determine the precise location of the nick at *oriT* and to carry out the nicking reaction in vitro (Thompson et al., 1989; Matson and Morton, 1991; Reygers et al., 1991) revealed that the complete *traI* gene product possesses both the *oriT* nicking activity and the strand separation activity known as *E. coli* DNA helicase I (Abdel-Monem et al., 1983) that is responsible for

separating the two DNA strands during transfer. The *traI* nickase-helicase apparently becomes covalently linked to the 5′ end of the nicked DNA strand during the nicking reaction and may play a role in plasmid recircularization after transfer is complete (Reygers et al, 1991; Matson et al., 1993). The *traY* gene product and the integration host factor (IHF) of *E. coli* both have binding sites in the *oriT* region, and both are needed for efficient *oriT* nicking by the TraI protein in vitro under more physiologically relevant conditions (Nelson et al., 1995). The nicking reaction at *oriT* is constitutive in that it occurs in the absence of either mating pair formation or DCDS (Everett and Willetts, 1980), so a nick at *oriT* does not automatically lead to the initiation of DCDS. It has been suggested that the binding of TraY protein and IHF at *oriT* permit the binding of TraI protein to form a complex referred to as a relaxosome (Howard et al, 1995). The presence of such relaxosomes is a common attribute of many plasmids whose DNA can be self-mobilized.

The role of the *traM* gene product has not been well defined. It is not required for piliation or nicking at *oriT*, but it is required for DNA transfer and replacement strand synthesis in the donor cell. It thus seems to trigger the start of DCDS at a nicked *oriT* site in response to a signal arising after the tip of the F pilus contacts a suitable recipient cell (Everett and Willetts, 1980). There are three TraM protein binding sites near *oriT*, but these are on the 3′ side of the nick site and therefore do not involve the leading end of the transferred strand. These TraM binding sites clearly play a role in *traM* autoregulation, and the ability of TraM protein to also bind TraD protein (Disque-Kochem and Dreiseikelmann, 1997) may serve to indicate a role for TraM protein in positioning the nicked DNA at the transfer portal of which TraD protein is thought to be a part (Panicker and Minkley, 1985).

A single strand of F factor DNA is transferred to the recipient cell starting with the 5′ end from the nicked *oriT* site (Rupp and

Ihler, 1968; Ohki and Tomizawa, 1968). The transfer of the displaced strand to the recipient cell is normally accompanied by DNA polymerase III-mediated replacement synthesis in the donor cell, but transfer can occur in the absence of this synthesis (Sarathy and Siddiqi, 1973; Kingsman and Willetts, 1978). This replacement DNA synthesis requires priming by RNA polymerase (Kingsman and Willetts, 1978), and this requirement may reflect blockage of the 3′ end by one of the proteins involved in DCDS.

A new complementary strand for the entering donor DNA is synthesized by the recipient cell's normal DNA synthesis machinery. It now appears that the necessary priming for this complementary strand synthesis is achieved by a special promoter called F*rpo* that allows host cell RNA polymerase to initiate a transcript on single-stranded DNA that can be continued by DNA polymerase III holoenzyme (Masai and Arai, 1997). This special promoter and possibly others like it also appear to allow for the rapid expression of a number of genes from the leading region of the transferred F factor DNA in the recipient cell before complementary DNA strand synthesis has been accomplished. While the role of many of these leading region genes has not been determined, this group includes *ssf*, the F factor's SSB or single-stranded DNA-binding protein gene, and the *psiB* gene whose product acts to prevent SOS induction by the entering single-stranded DNA in the recipient cell during DNA transfer (Bailone et al., 1988; Bagdasarian et al., 1992).

The stabilization of the F factor in the recipient cell is a *recA*-independent process (Clark, 1967) that typically requires the transfer of both ends of *oriT* (Everett and Willetts, 1982). Despite earlier evidence for the transfer of single-stranded concatemers of F factor DNA (Ohki and Tomizawa, 1968; Matsubara, 1968), the observed requirements for recircularization favor the transfer of unit length DNA strands (Willetts and Skurray, 1987). The TraI protein bound to the 5′ end of the transferred strand could simply re-ligate that 5′ end to the 3′ end when it arrives in a direct reversal of the original *oriT* nicking reaction if the 3′ end remains unobstructed during the course of replacement DNA synthesis in the donor cell. The fact that transfer can occur in the absence of replacement DNA synthesis in the donor cell indicates that this reaction can most likely occur. While the distinct priming event for initiation of replacement strand synthesis in the donor (Kingsman and Willetts, 1978) indicates that the 3′ end is not initially involved in rolling circle replication, that result does not rule out a later extension event involving that 3′ end. If the free 3′ end in the donor cell is not preserved throughout the transfer event, then circularization to complete transfer would presumably involve another nicking reaction at the reconstituted *oriT* site. While there is no experimental evidence that directly supports or contradicts this second possibility, the fact that some *oriT* mutations yield reduced nicking efficiency without reducing termination or circularization efficiency (Gao et al., 1994) makes it seem unlikely that this is the primary mechanism. The student is referred to Wilkens and Lanka (1993) for a more extensive discussion of this subject. Figure 5 shows a model for the transfer of F factor DNA during conjugation.

F. Separation of the Mating Pair

The destabilization of the mating pair and its separation are poorly understood. Mechanical disruption of the mating pairs leaves little apparent lasting damage (Low and Wood, 1965), and it is therefore possible that mating pair disruption is sometimes a spontaneous and random process. In Hfr by F⁻ matings, where the transfer of the *tra* regulon to the recipient occurs only after 100+ minutes of DNA transfer (see below), the mating pairs (aggregates) do not show detectable levels of separation for at least 60 minutes (Achtman et al., 1978a). When F⁺ or F-prime cells are used as donors, however, an intact *tra* regulon is quickly transferred to the recipient and mating aggregates

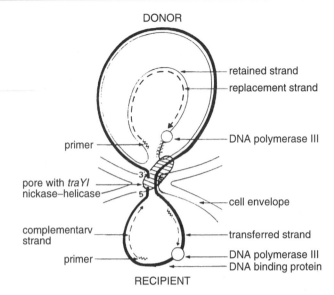

Fig. 5. Model for conjugative transfer of F. A specific strand of the plasmid (thick line) is nicked at *oriT* by the TraYI nickase-helicase and transferred in the 5′-to-3′ direction through a pore formed between juxtaposed donor and recipient cells. The plasmid strand retained in the donor is shown as a thin line. The termini of the transferred strand are attached to the cell membrane. DNA helicase I (from the *traI* gene) is bound to the membrane near the pore, and its migration along the transferred strand provides the motive force for displacing the strand into the recipient cell. New F factor DNA (broken lines) is synthesized in both donor and recipient cells by DNA polymerase III. The model assumes that a primer is required for the DNA synthesis and that single-stranded DNA-binding protein (small circles) coats the single-stranded DNA. (Reproduced from Willetts and Skurray, 1987, with permission of the publisher.)

rapidly decay (Achtman et al., 1978a). Although there is no firm evidence that initial mating pair stability is the same in these two situations, it seems reasonable to speculate that the expression of transferred *tra* genes in the recipient cell may play an active role in mating pair disaggregation.

III. HFR CONJUGATION AND CHROMOSOMAL TRANSFER

A. How Hfr Strains Arise

The integration of the F factor into the *E. coli* chromosome gives rise to Hfr strains that efficiently transfer, or mobilize, donor cell chromosomal markers to recipient cells. The first Hfr strains were isolated by Cavalli-Sforza (HfrC) and Hayes (HfrH), and many other Hfr strains have subsequently been isolated. Although Hfr's representing a minimum of 20 clearly distinct sites of F factor

integration have been described (Low, 1972), it appears that a limited number of integration sites are highly favored. There are four transposable elements on the F factor (Davidson et al., 1975)—two copies of IS*3*, one copy of IS*2*, and one copy of Tn*1000* also known as γδ—and some Hfr formation definitely involves *recA*-dependent recombination between an F factor-borne transposable element and a homologue in the cell's chromosome. It has in fact been shown that the sites of Hfr formation largely correlate with known positions of IS elements in the *E. coli* chromosome (Umeda and Ohtsubo, 1989). Hfr's arise in *recA⁻* cells at considerably lower frequencies (Deonier and Mirels, 1977; Cullum and Broda, 1979), but the mechanistic basis for this *recA*-independent Hfr formation is not known. Hfr's vary greatly in their stability; excision of the integrated F factor from some chromosomal locations is essentially never observed, but F⁺ cells arise at high frequency with some Hfr strains (Low, 1973). The most unstable Hfr's are generally those whose integration involved γδ, but the basis for the variations in stability of other Hfr's is not known. The relative position and orientation of the integrated F factor for many of the commonly used Hfr's is shown in Figure 6.

The limited ability of an autonomous (nonintegrated) F factor (F⁺) to transfer the donor chromosome cannot be explained solely by the frequency with which Hfr's arise in an F⁺ culture. A second component of F⁺-mediated chromosome transfer involves a process called conduction. Conduction is a type of passive mobilization that can involve any replicon, including the chromosome, present in an F⁺ cell. When the Tn*1000* present on the F factor initiates replicative transposition (see Whittle and Salyers, this volume) to another replicon in the donor cell, an intermediate step in the transposition process is a cointegrate structure where the two copies of Tn*1000* serve as the boundaries between the F factor and the other replicon. Normally such a cointegrate structure is rapidly resolved into separate replicons in the F⁺ cell by the Tn*1000*-encoded resolvase. The F factor is unchanged by this process, but the other replicon has a newly added copy of Tn*1000*. If, however, DNA transfer is initiated during the cointegrate or replicon fusion stage of this transposition, the replicon that is covalently linked to the F factor by Tn*1000* will also be involved in DNA transfer.

When all or part of another replicon is transferred to a conjugal recipient by such a series of events, we say that it has been "conducted" by the F factor. This process was first described when it was shown that the low frequency transfer of pBR322 from donor to recipient was always accompanied by the addition of a copy of Tn*1000* to the transferred pBR322 plasmid (Guyer, 1978).

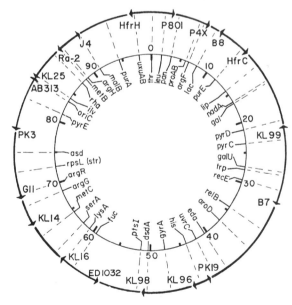

Fig. 6. Map positions of integrated sex factors for some *E. coli* Hfr strains. Each arrowhead indicates the position and orientation of integration of the sex factor on the 100 minute map of the *E. coli* chromosome. The location of some chromosomal genetic markers is also shown. The sequence of markers transferred from a given strain begins behind the arrowhead. Thus HfrC (located at about 13 minutes) transfers counter clockwise from the point of origin (*purE* then *lac* then *argF*) while HfrH (located at about 98 minutes) transfers clockwise. (Reproduced from Low, 1987, with permission of the publisher.)

The transposition of Tn*1000* from the F factor to the chromosome can similarly result in the transfer of donor cell chromosomal DNA to a recipient cell where it is available for recombination with the recipient cell chromosome. Although the rapid resolution of any such transposition intermediates in the donor cell precludes their identification as Hfr's, such replicon fusions between the F factor and the donor cell chromosome temporarily function as Hfr's before they are resolved. The Tn*1000*-based conduction of a plasmid by the F factor is shown diagrammatically in Figure 7. Other transient associations between the F factor and the chromosome may also promote chromosome transfer by F$^+$ donor cells (Goto et al., 1984).

B. Properties of Hfr's

In any typical Hfr strain, the integrated F factor resides at a particular location within the chromosome, and *oriT* is pointed in one of the two possible directions. This fact leads to one of the two most important descriptive properties of Hfr conjugation: orientation of transfer. The orientation of transfer depends on whether oriT is pointed clockwise or counterclockwise on the *E. coli* genetic map, and determines the order in which chromosomal markers will be transferred by the donor. For example, one Hfr strain

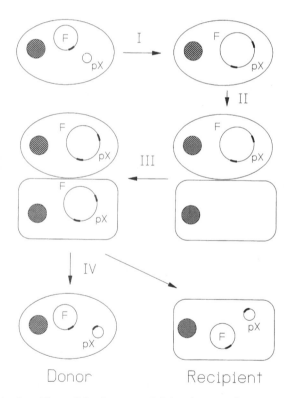

Fig. 7. Plasmid conduction. The cell in the upper left-hand corner has a copy of the F factor and a nonmobilizable plasmid called pX. The F factor copy of γ-δ is indicated by the thicker portion of the line and the cellular nucleoid is shown as a cross-hatched circle. In step I, γ-δ transposition to pX is initiated with the formation of a cointegrate. Step II shows that this donor cell has formed a mating pair with a suitable recipient cell before resolution of the cointegrate, and step III shows transfer of the plasmid cointegrate to the recipient cell. In step IV, resolution of the cointegrate occurs independently in both donor and recipient cell; pX now has a copy of γ-δ in both cases. By using a double selection scheme that allows the growth of only those recipient cells that express a pX-borne gene (typically a drug resistance determinant), a collection of γ-δ insertion mutants can be obtained for any DNA fragment carried by pX.

might transfer a set of hypothetical markers in the order A then B then C then D, while another Hfr strain would transfer D then C then B then A.

The second important descriptive property of Hfr conjugation is referred to as the gradient of transfer. The gradient of transfer was originally thought to occur because mating pairs undergo spontaneous random disruption (Jacob and Wollman, 1961), but subsequent work has indicated that the time dependence of the marker transfer gradient is not correlated with the time dependence of mating pair disaggregation (Wood, 1968; Achtman et al., 1978a). Whatever the actual mechanism, the net result is that markers transferred early are transferred at a higher frequency than markers that are transferred later. The gradient of transfer dictates that marker transfer efficiency will depend on the marker's position relative to that of the integrated F factor.

One of the initial questions that arose during the characterization of Hfr conjugation dealt with the nature of the transfer event. The initial data could be explained by assuming that a uniformly sized piece of donor DNA was always transferred and that recombination with the recipient chromosome began at the proximal end of the transferred donor DNA segment. In that situation the gradient of transfer would result from a cessation of recombination with time and length as processing continued along the fragment. It could also be explained by assuming that different sized segments of donor DNA were transferred and that recombination was limited by the size of the piece that had been transferred. The latter hypothesis was shown to be correct on the basis of experiments involving a phenomenon called zygotic induction (Wollman et al., 1956).

When an Hfr that is lysogenic for bacteriophage λ is mated with an F$^-$ recipient, λ DNA is transferred to the recipient cell without the λ cI-encoded repressor. If the recipient cell is nonlysogenic, there is no λ cI-encoded repressor present, and the entering λ DNA therefore undergoes induction to

yield a burst of phage in the recipient cell or zygote without the need for any recombination. This was initially observed in experiments where an HfrH(λ^+) strain was mated with an F$^-$ (λ^-) strain when it was found that Gal$^+$ transconjugants were essentially undetectable. This was the result of the tight linkage between the *E. coli gal* operon and the λ prophage location on the chromosome; rarely were *gal* genes transferred to the recipient without the simultaneous transfer of λ as the *gal* operon and the primary bacteriophage λ integration site are very closely linked (0.3 minutes on a 100 minute scale) on the *E. coli* chromosome. At the same time, however, recombinants involving markers closer to the HfrH point of origin, such as Thr$^+$ (17 minutes earlier) or Lac$^+$ (9 minutes earlier), were readily detectable in those crosses. It was concluded that those transconjugants resulted from the transfer of shorter pieces of donor DNA that did not involve the transfer of the λ prophage.

The nearly uniform rate of transfer of DNA from each Hfr makes conjugation a powerful tool for genetic mapping over very long distances. Transfer is initiated at the *oriT* site of the integrated F factor and proceeds in the direction dictated by the orientation of its integration. Transfer initiates rapidly (within about 3 minutes of mixing cells) and proceeds at a reasonably uniform rate (Wood, 1968) of about 45,000 base pairs per minute, making time of entry a good criterion for determining the distance of a marker from the Hfr origin (Low, 1987). Since transfer is initiated in the middle of the integrated F factor, the recipient cells remain F$^-$ (Hayes, 1953) unless the entire donor chromosome is transferred and the F factor is subsequently integrated. Complete transfer is rare, but can be detected by selecting recombinants for a marker transferred late. Chromosomal DNA transferred by an Hfr can also recombine with homologous plasmid-borne DNA in the recipient (Porter, 1982). Chromosomal mapping by conjugation will be discussed in more detail in the Appendix at the end of this chapter.

C. Recombination after Hfr Conjugation

Once the transfer of variable portions of the Hfr chromosome had been established by the zygotic induction experiments described above, it became possible to estimate the efficiency of recombination events after donor DNA transfer. Among recombinants selected for a distal marker, more than 50% inherit any given nonselected proximal marker from the donor. This serves to indicate that there is a greater than 50% probability of a donor marker being recombined into the recipient cell chromosome when the presence of a more distal donor marker serves to clearly show that the proximal marker has been transferred. Inheritance of more distal markers in these selected recombinants is less frequent, however, probably due at least in part to subsequent interruption of DNA transfer.

Very long linkage groups are typically observed by genetic criteria in Hfr conjugation. One study of marker linkage in Hfr conjugation estimated a 20% probability of a crossover per "minute" (one "minute" is 1% of the *E. coli* chromosome—about 45 kilobase pairs) of transferred DNA (Low, 1965), while another study estimated an even lower frequency of crossovers (Pittard and Walker, 1967). The net result of these long linkage groups is a low frequency of recombination between two closely linked proximal markers. As an example, you might determine the frequency of Thr$^+$, Leu$^+$, and Pro$^+$ transconjugants among those selected for Gal$^+$ from a cross between HfrH and a multiply marked F$^-$ recipient. Although Gal$^+$ transconjugants that are plus for all three of these non selected proximal markers would be common, recombinants that are plus for one or two of these markers and minus for the others would be considerably less frequent. Although you would find that more than 50% of the Gal$^+$ transconjugants were "plus" for any of the three proximal markers scored individually, such classes of recombinant would show considerable overlap for these three markers because the more closely linked markers appear to be frequently recombined into the recipient chromosome as a group.

Markers very near the Hfr origin, however, are not frequently inherited. The rare inheritance of very early markers (less than one to two minutes from the origin) from the donor was proposed to be due to a length exclusion effect whereby the earliest sequences transferred were somehow not available for recombination (Low, 1965). However, crossovers do occur frequently in this very early region (Pittard and Walker, 1967), suggesting that increased crossover frequency leads to the reduced recovery of these markers. An anti-pairing effect of the leading F factor DNA has also been suggested (Pittard and Walker, 1967). In contrast, the probability of crossover per minute of transferred DNA is somewhat less for very late markers; this effect leads to physically larger linkage groups (Verhoff and DeHaan, 1966).

The long linkage groups observed genetically are at variance with the results of physical studies of recombination following conjugation. Differentially labeled donor and recipient DNA become covalently associated, but only short pieces (mostly about 0.4 kb) of single-stranded donor DNA appeared to be integrated (Siddiqi and Fox, 1973). Incorporation of double-stranded donor DNA was not detected (Siddiqi and Fox, 1973), even though the transferred single-stranded DNA is rapidly converted to the double-stranded state in the recipient cell. The method used for detection of inserted double-stranded DNA, however, required that the light density donor DNA initially be found in association with heavy density recipient DNA so as to distinguish it from unrecombined donor DNA. That criterion would be valid if the double-stranded insertions were short relative to the broken fragments of recipient DNA after cell lysis, but it would not be valid for double-stranded insertions whose length might be comparable to or greater than the recipient DNA fragments produced by cell lysis and subsequent sample manipulations.

It is now accepted that large segments of double-stranded donor DNA are incorporated into the recipient chromosome (see below), and the single-stranded insertions seen by Siddiqi and Fox (1973) may simply represent heteroduplex regions generated by branch migration and resolution of Holliday junctions at the crossover sites.

Smith (1991) reevaluated a great deal of published linkage data in light of an improved understanding of recombination mechanism. He suggested that most recombination events in *E. coli* cells with a functional RecBCD pathway occur by means of RecBCD enzyme entry at both the leading end of the transferred DNA and the broken end generated by termination of transfer. The RecBCD enzyme then processes through the DNA in DNA helicase mode until it encounters a Chi recombinational hotspot (an asymmetric 8 base sequence that occurs about every 5 kbp in *E. coli* DNA). Nicking at those Chi sites results in the displacement of a single-stranded DNA tail by continued RecBCD enzyme helicase action, and the resulting single-stranded tail allows the binding of RecA protein for recombination initiation. This "long chunk" mechanism produces a crossover at each end of the donor DNA fragment so that essentially the entire donor sequence is incorporated into the recipient cell genome. Smith then proposes that there is also a "short chunk" mechanism that accounts for situations where most of the donor DNA sequence proximal to the selected marker is not integrated into the recipient cell chromosome. The action of RecBCD enzyme at the broken or distal end of the donor DNA fragment is envisioned to be the same in this "short chunk" case, but the second crossover event does not involve RecBCD-dependent recombination at the leading end. The speculation is that recombination within the transferred donor DNA fragment is promoted by the RecF recombination pathway by which recombination may be initiated by means of the inherent partial single-strandedness of recently transferred donor DNA. This speculation is consistent with published observations that strains with an active RecF pathway (*recBCD⁻ sbcBC⁻*) show considerably reduced linkage following Hfr conjugation as compared to *recBCD⁺* strains where the RecBCD pathway is active and thought to predominate (Mahajan and Datta, 1979; Lloyd and Thomas, 1983).

Lloyd and Buckman (1995) subsequently carried out a study of recombinants formed after Hfr conjugation that involved analyzing the effect of both distance from the origin of transfer and numerous recombination genes. Their results were consistent with a mechanism such as Smith's RecBCD-dependent "long chunk" model for most recombinants where the amount of Hfr donor DNA transferred was in the range of 500 kbp or less. While Smith's model would require termination of DNA transfer to allow entry of the RecBCD enzyme at the leading or *oriT* end, they suggest that at least some of these "long chunk" events may involve non-RecBCD-dependent initiation events utilizing transient single-strandedness near the leading end while DNA transfer is still occurring as originally proposed by Lloyd and Thomas (1984). They also observed, however, that many of the so-called short chunk recombinants where donor markers proximal to the selected marker were not incorporated arose within sectored colonies that also contained recombinants showing the much longer "long chunk" linkage groups. They propose that these sectored colonies may have arisen from secondary recombination events involving the displaced recipient DNA sequence after the recombined donor sequence has undergone one round of chromosomal replication. As nonsectored short chunk recombinants might have arisen from secondary recombination events occurring prior to recipient cell chromosome replication, they regard such secondary recombination events are being the probable source for many of the short chunk recombinants.

A somewhat different story emerged when the selected marker was such that donor DNA segments of 1000 kbp or more had to be transferred (Lloyd and Buckman,

1995). While very long linkage groups still predominated under those conditions, it appeared that fewer of the proximal or leading end crossover events involved donor DNA sequence transferred at the earliest times. There is no simple, straightforward explanation for this phenomenon, but it was suggested that recombinants involving longer transfer times may more often involve proximal end initiation at single-stranded gaps that may occur further from the proximal/leading end than RecBCD-dependent events occurring after DNA transfer has been terminated. The student is referred to the original work (Lloyd and Buckman, 1995) for a discussion of how a number of *rec* gene mutants affect linkage parameters. All of the preceding discussion assumes a need for an even number of crossover events to produce a viable recombinant, and most of that discussion has been focused on scenarios involving the minimum number of two such events. It should be noted, however, that none of the data rules out at least the occasional appearance of recombinants involving a larger, but still even, number of such crossover events.

When recombination between closely linked markers in the *lacZ* gene was measured in transconjugants, it was found that there was very little correlation between recombination frequency and the map order of the alleles as determined by deletion mapping (Norkin, 1970). This phenomenon (lack of correlation between recombination frequency and physical distance for closely linked markers) is referred to as marker effects. Such marker effects are thought to result from gene conversion events involving the nonrandom correction of nucleotide base mismatches in heteroduplex DNA produced by recombination. The dramatic marker effects in Hfr conjugation (Norkin, 1970) provide a strong argument for the generation of heteroduplex DNA during conjugational recombination, and this is strengthened by the fact that such marker effects subsequent to Hfr conjugation have been shown to be dependent on the mismatch correction genes in *E. coli* (Feinstein and Low, 1986). These heteroduplex regions contain one strand from each of two different parental DNA molecules, and would only occur where a single strand of donor DNA became integrated into the recipient DNA homoduplex (perhaps as part of a recombination initiation event) or where heteroduplex DNA had been generated by branch migration of a Holliday junction. As the equivalent of a crossover event must involve one of two closely linked markers for recombinants to be observed, it is not surprising that such marker effects are observed.

Although the recombination that occurs after Hfr conjugation is classified as homologous or general, there are other considerations that may have a bearing on the nature and distribution of recombination events that occur. *fre* (frequent recombination exchange) regions where genetic exchanges by the RecF pathway are clustered on the *E. coli* genome have also been suggested (Bressler et al., 1978, 1981). As already discussed above, the location of Chi sites may affect the distribution of genetic exchanges when the RecBCD pathway is involved in conjugational recombination. Therefore genetic exchanges between donor and recipient DNA are probably not entirely random following Hfr conjugation.

IV. F-PRIME CONJUGATION AND MERODIPLOIDS

A. The Generation of F-primes

F-prime factors (see Low, 1972; Holloway and Low, 1987 and 1996 for reviews) can arise from Hfr's by a number of different mechanisms. Those mechanisms include illegitimate recombination events (those with no known mechanistic basis), recombination between IS elements that flank the integrated F factor, recombination between an IS element within an integrated F factor and another copy of the same IS element in flanking chromosomal sequence, recombination between homologous chromosomal sequences flanking the integrated F factor, or abortive intramolecular transposition where the resolution step of Tn*1000* transposition does not occur. A schematic representation of the abortive intramolecular transposition event that gave rise to a particular F-prime called F42*lac* is shown in Figure 8. Early studies with F13 (a particular F-prime that arose in an Hfr13 strain) demonstrated that the chromosomal genes present on the F-prime were missing from the chromosome (Scaife and Pekhov, 1964). This observation indicates that F-prime formation is mechanistically equivalent to a chromosomal deletion involving the production of two smaller circles from one larger circle in a manner that is roughly analogous to prophage excision by the Campbell model (Low, 1972). Such deletion events would normally result in the loss of the smaller circle, but it is preserved as a newly generated F-prime when the deleted

sequence includes F factor replication functions. Type I F-primes incorporate host DNA from only one side of the integrated F factor and leave behind part of the F factor. Type II F-primes incorporate host DNA from both sides of the integrated F factor and retain the complete F factor (see Fig. 9).

Although the strain in which the F-prime arises, the primary F-prime strain, may initially contain both an intact and a deleted version of the chromosome, the cells remain functionally haploid and may require the F-prime for viability when only the deleted version of the chromosome is present (Scaife and Pekhov, 1964). Such primary F-prime strains do not promote efficient chromosome mobilization because the chromosome lacks the sequences needed for high-frequency homologous recombination with the F-prime

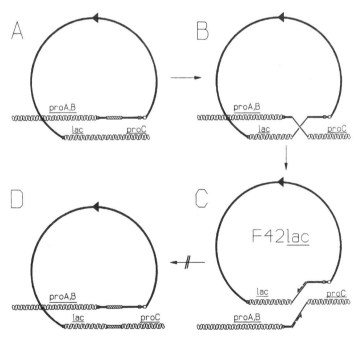

Fig. 8. Formation of F42*lac* by abortive transposition. The strain in which F42*lac* arose contained an F factor that had integrated at an IS*3* between *proA,B* and *lac* in the *E. coli* chromosome. **A:** The integrated F factor, shown as a heavy line, has looped around to bring its copy of γ-δ into close proximity with the chromosome at a point between *lac* and *proC*. **B:** γ-δ has begun transposition by breakage of the DNA at the chromosomal target site and ligation of single strands of γ-δ at each end of the target site break. **C:** A hypothetical intermediate where the molecules have been realigned and replication of γ-δ is indicated by the arrows. **D:** Completed transposition event where the replicated copies of γ-δ have undergone site-specific recombination at their internal resolution sites. F42*lac* formation, however, occurred when the final resolution step (panel C to panel D) failed to happen.

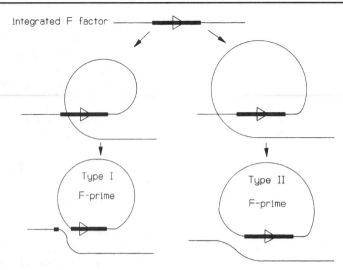

Fig. 9. Generation of type I and type II F-primes. The insertion of the F factor (heavy line with arrow to indicate *oriT*) within the chromosome (thin line) as found in an Hfr is shown at the top of the figure. On the left side of the figure, a recombination event between the flanking chromosomal DNA and a site within the integrated F factor gives rise to a type I F-prime and leaves some F factor DNA in the chromosome. On the right side of the figure, recombination between flanking chromosomal DNA on either side of the integrated F factor gives rise to a type II F-prime with the concurrent production of a chromosomal deletion.

(Pittard and Ramakrishnan, 1964; Scaife and Pekhov, 1964; Berg and Curtiss, 1967). Transfer of an F-prime to a recipient with an intact chromosome produces a secondary F-prime strain which is merodiploid, or partially diploid, for the host chromosomal DNA carried by the F-prime. Such merodiploids are frequently used for genetic complementation experiments, but *recA* mutants should be used to preclude recombination, which can lead to confusing results. Recombination between the chromosomal and F-prime copies of host DNA, and subsequent segregation (which may not be necessary if the recombination event is nonreciprocal) can convert an initial heterogenote into a homogenote. Such "homogenotization" is commonly used to move alleles between strains (Jacob and Wollman, 1961; Low, 1972). With the use of an appropriate selection or screening procedure, an F-prime can be used to pick up a mutant allele from the chromosome of one strain by homogenotization. The F-prime carrying that particular mutant allele can then be mated into another strain where it is transferred to the chromo-

some by a reversal of the homogenotization protocol. After spontaneous or acridine orange promoted curing of the F-prime (Miller, 1972), the second strain contains the desired mutant chromosomal allele from the first strain with very little perturbation of the surrounding chromosomal region. The *tra*-dependent enhanced recombination properties of F-primes (see Section IVB below) cause them to demonstrate homogenotization more frequently than most other plasmids (Yancey and Porter, 1985).

Although the isolation of primary F-prime strains is not straightforward (Berg and Curtiss, 1967), a number of methods have been described for directly obtaining secondary F-prime strains (Holloway and Low, 1996). These methods all involve Hfr times F-matings, and the first requirement is an Hfr strain where the F factor is integrated near the chromosomal segment that is to be carried by the F-prime being sought. The underlying assumption is that F-primes will have arisen at a low frequency in any given Hfr strain, and that one merely needs a means of selecting for the transfer of such spontaneously arising F-primes into a suitable recipient cell.

When the marker to be selected for incorporation into an F-prime is one that is transferred late (late-situated) by the Hfr strain being used, early interruption of a mating often yields F-primes in the selected recipients. As the interruption of the mating precludes the transfer of a complete donor chromosome, the late-situated marker can only be transferred to the recipient as part of an F-prime. A more general method involves the use of $recA^-$ recipient strains in such matings (Low, 1968). No recombination can occur between donor and recipient chromosomes in the $recA^-$ recipient unless $recA^+$ is introduced from the donor, and this is usually prevented by interruption of the mating. Therefore the transferred marker can only be selected in the recipient if it is part of a functional replicon such as an F-prime. Double male strains with two integrated copies of the F factor have also been used in F-prime isolation (Clark et al., 1969). When such a strain is mated with a suitable recipient, F-primes containing the chromosomal region between the two F factor insertion sites can be isolated at a relatively high frequency.

Although F-primes involving essentially every region of the *E. coli* chromosome have been described (Low, 1972), the process is far from random and certain F-primes repeatedly appear (Hadley and Deonier, 1979, 1980). This preferential formation of certain specific F-primes appears to be due to the presence and positioning of transposable elements (Timmons et al., 1983; Umeda and Ohtsubo, 1989), and recombination involving the multiple copies of ribosomal RNA operons (*rrn*) present in the *E. coli* chromosome have also been shown to sometimes participate in F-prime formation (Blazey and Burns, 1983). When transposable elements are involved, the necessary circularization can occur by general recombination between two copies of the transposable element or by incomplete replicative transposition events initiated by a single copy (Hadley and Deonier, 1979, 1980; see the chapter by Streeps, this volume and Figs. 8 and 9). These same types of recombination events may also have a major impact on the stability of F-primes, particularly large ones. Although F-primes containing up to 30% of the *E. coli* chromosome have been reported (Low, 1972), deletion derivatives frequently appear in such cultures because the presence of such large F-primes slows the growth rate of the host cell (Simons et al., 1980). It is probable that such deletions also frequently occur via recombination events involving transposable elements.

B. Conjugation Properties of F-primes

Due to general recombination in Rec$^+$ secondary F-prime strains, there is an equilibrium between the integrated and autonomous state of the F-prime. Because of their ability to reintegrate by homologous recombination, F-primes were initially described as F factors that could "remember" where they had been integrated (Richter, 1957). F-primes recombine both into and out of the chromosome more frequently than an unaltered F factor because the recombination events involve extensive stretches of homology; dozens to hundreds of kilobases as compared to the one to two kilobases of IS (insertion sequence) homology typically involved in F factor integration and excision. In its autonomous state the F-prime promotes conjugation and self-transfer like the F$^+$ factor. Such conjugation establishes the F-prime in the recipient cell as an independent replicon (repliconation), but the host DNA carried by the F-prime can recombine with recipient DNA even when complete repliconation does not occur (e.g., if part of the F DNA is not transferred, as frequently happens with large F-primes). The integrated form of the F-prime behaves like an Hfr, and thereby serves as the basis of F-prime mobilization of the chromosome. This is the usage of the word mobilization where a plasmid transfers part or all of the donor cell chromosome to the recipient cell by means of a physical association between plasmid and chromosome. F-primes generally mobilize the chromosome of a Rec$^+$ secondary donor cell quite effectively, and transfer efficiencies that are only down 5- to 100-fold from Hfr-mediated chromosome transfer are often achieved. F-prime mobilization of the chromosome can be used to move specific markers by conjugation when a suitable Hfr is not readily available.

The usefulness of F-primes for both chromosome mobilization and homogenotization is accentuated by their enhanced recombination properties. This phenomenon was first shown with λplac5 specialized transduction where the transducing phage recombines with F42lac (see Fig. 8) about 30 times more efficiently than with a chromosomal lac gene (Porter et al., 1978). This recombination enhancement depends on both constitutive expression of the F factor tra regulon (Porter, 1981) and the presence of a functional RecBCD enzyme (exonuclease V) in the recipient cell (Porter et al., 1978, 1982). Only the oriT site must be in-cis (on the same DNA molecule as) to the recombining lac genes (Seifert and Porter, 1984), and it appears that the action of trans-acting tra gene products (specifically traY and traI) at oriT is crucial in allowing enhanced recombination involving F42lac or any oriT-containing plasmid (Carter and Porter, 1991; Carter et al., 1992). The exact role of the RecBCD enzyme in promoting enhanced recombination is not known, but it has been shown that this same recombination enhancement phenomenon participates in recombination between F42lac and chromosomal lac genes in lac merodiploids (Yancey and Porter, 1985). This tra- and RecBCD enzyme-dependent enhancement therefore facilitates both chromosome mobilization and homogenotization by promoting increased recombinational interaction between fertile F-primes and the host cell chromosome.

V. CONJUGATION OF FERTILITY-INHIBITED F-LIKE PLASMIDS

Although the conjugation properties of the F factor are representative of those seen with a variety of self-transmissible plasmids that have been examined, it is atypical in that its fertility inhibition system is inoperative. As mentioned previously, this is because the finO gene of the F factor has been inactivated by an IS3 insertion. This was fortunate for the development of bacterial genetics, but it does not appear to be common in nature.

There are a large number of plasmids, however, whose tra regulon structure is similar to that of the F factor while their fertility inhibition systems remain functional. R factor plasmids such as R1 and R100 are among the well-characterized members of this F-like group. This situation poses two questions regarding the general subject of fertility inhibition. The first is why fertility inhibition is so commonly found in F-like plasmids. The second is how significant levels of conjugal transfer are achieved under fertility inhibition conditions.

There is no definitive answer to the first question, but there is a reasonably good speculation that can be advanced. The production of pilin and all of the other tra proteins that are synthesized when tra expression is derepressed must exert a significant energy drain on the plasmid-containing cell. Although this is of little consequence in a laboratory setting, it could represent a significant selective disadvantage to a cell that is attempting to compete in a natural setting. The conferral of serum resistance by the expression of the surface exclusion proteins could represent a significant selective advantage for tra expression in some situations, but the surface exclusion genes of the F factor are expressed from a secondary promoter even when the main tra operon promoter (for traY to traI; see fig. 3) is not functioning (Cheah et al., 1986).

The answer to the second question stems from the nature of biological regulation. Such regulation is essentially never absolute as demonstrated by the fact that a low level of lacZ expression occurs even when lacI repression is functioning. This is also the case with the fin system, and a minority of the cells in a plasmid-bearing population, are always expressing the *tra* regulon even when fertility inhibition is fully functional. This minority of conjugally proficient cells is capable of initiating conjugal plasmid transfer whenever a population of suitable recipient cells in encountered.

The functional donor cells quickly establish mating pairs and achieve plasmid transfer to a limited number of recipient cells. In those transconjugant recipient cells, an expression race between *fin* and *tra* ensues. Initially *tra* expression wins this race, and the primary recipient cells (the cells that just recieved a copy of the plasmid) readily become functional donor cells. These primary recipient cells then function as secondary donor cells. As this process continues, a mating cascade develops that results in transfer of the plasmid to a majority of the suitable recipient cells that are available. Eventually *fin* expression "catches up," and the newly expanded population of plasmid-bearing cells returns to the state where only a small minority of the cells are conjugally proficient at any given time. This scenario provides a means of increasing genetic variation in the overall population without excessive energy expenditure by the plasmid-bearing cells.

This whole process is analogous to the situation observed in the classic PAJAMA experiment (Pardee et al., 1959). In the PAJAMA experiment (the name comes from PArdee, JAcob, and Monod), there is an expression race between the *lacI* gene and the structural genes of the *lac* operon (*lacZ*, *Y*, and *A*) when a *lacI⁺ lacZ⁺* version of the *E. coli lac* region is transferred to a *lacI⁻ lacZ⁻* recipient cell by Hfr conjugation. The burst of β-galactosidase synthesis seen in that situation indicates that the *lacZ* gene was expressed more rapidly than the *lacI* gene in the recipient cell. Eventually *lacI* expression catches up, and repression of the *lac* structural genes is established.

The net result in the *fin-tra* case is a situation where a moderate reduction in mating proficiency is probably more than compensated for by the energy savings experienced by the general population. The particular subset of cells that will be expressing *tra* at any given time under fertility inhibition conditions is random, so no particular cell experiences the selective growth disadvantage conferred by *tra* expression for an extended period of time.

VI. NONCONJUGATIVE, MOBILIZABLE PLASMIDS

Many smaller plasmids possess a system that frequently provides them with the ability to achieve efficient conjugal transfer without carrying the considerable number of genes that are required for the specific cell contact cycle. Whenever these plasmids are present in the same cell as an appropriate conjugative plasmid, the ability to mobilize their DNA allows efficient DNA transfer after the formation of a stable mating pair has been achieved by the other plasmid. Note that here we are using the word mobilization in the sense of preparing plasmid DNA for transfer, not in the sense of transferring the donor cell chromosome.

While numerous nonconjugative but mobilizable plasmids are found both in nature and in the scientific literature, the colicin E1 (ColE1) plasmid (see Perlin, this volume) will serve as a good example of such a plasmid for our purposes here. This 6.8 kb plasmid achieves high copy number, and its *dnaA*-independent origin of vegetative replication has been used in the construction of many of the common *E. coli* cloning vectors. The plasmid contains a *nic* or *bom* (basis of mobility) site that functions much like the *oriT* site of the F factor. Three specific proteins were found to be associated with purified ColE1 DNA, and treatment of this DNA with ionic detergents or proteases caused a relaxation of the superhelical DNA to an open circular form (Clewell and Helinski, 1969, 1970). The nick responsible for the relaxation occurs at a unique site (Lovett et al., 1974), and one of the proteins becomes covalently associated with the 5′ end of the broken strand during relaxation (Guiney and Helinski, 1975). It was later inferred that the nick introduced during relaxation occurs at the *nic* site, as plasmid derivatives with *cis*-acting mutations incapable of relaxation are also incapable of mobilization (Warren et al., 1978).

ColE1 also contains a *mob* (mobilization) region that contains four genes designated *mbeA* through *mbeD* where "mbe" specifies mobilization for ColE1 (Boyd et al., 1989). Three of these *mbe* gene products are the proteins involved in the relaxation complex, and it is the MbeA protein that becomes covalently bound to the nicked DNA as is seen with TraI protein at the F factor *oriT* site. The role of the fourth Mbe protein is unknown. The proteins encoded by these *mob* (the generic term) genes are specifically required for mobilization, however, as their function(s) cannot be provided by the conjugative plasmid that is responsible for mating pair formation. The variation in *mob*-type gene specificities among a number of characterized mobilizable and self-transmissible plasmids is thought to be the result of differences in the *oriT*-like sites carried by these plasmids (Willetts and Wilkens, 1984).

The ability of *traI⁻* derivatives of the F factor to mobilize ColE1 (Willetts and Maule, 1979) serves to make two additional important points. The student will recall (see above) that the *traI* gene of the F factor encodes the DNA helicase involved in strand separation during the transfer of F factor DNA (Abdel-Monem et al., 1983), and *traI⁻* F factors cannot transfer their DNA despite the fact that they can produce stable mating pairs. The first point is that the conjugative plasmid does not have to be capable of transferring its own DNA in order to permit transfer of the mobilizable plasmid. The second point is that the mobilizable plasmid must use a non-F factor host cell function or provide its own function for strand separation during DNA transfer.

The complexity of the interactions between conjugative and mobilizable plasmids is illustrated by the variations in mobilization efficiency achieved in the presence of different conjugative plasmids. ColE1, for example, is efficiently mobilized in the presence of IncF, IncI, or IncP conjugative plasmids, while mobilization is inefficient in the presence of IncW plasmids (Reeves and Willetts, 1974; Warren et al., 1979). RSF1010 is another mobilizable plasmid that is efficiently mobilized by IncP conjugative plasmids while it is not mobilized at all by IncF plasmids (Willetts and Crowther, 1981). It therefore appears that interaction between *mob*-type proteins and an *oriT*-like site is a necessary, but not a sufficient, condition for achieving mobilization. A systematic study involving all combinations of six distinct conjugative plasmids and four mobilizable plasmids revealed that interactions between the relaxosome of the mobilizable plasmid, the *traD*-like portal or coupling protein(s) available, and the particular pilus type provided by the conjugative plasmid all play an important role in mobilization efficiency (Cabezon et al., 1997).

VII. BROAD HOST RANGE SELF-TRANSMISSIBLE PLASMIDS

While the F factor is capable of conjugal transfer to and maintenance in *E. coli* and its close relatives (see Section VIIA below), a variety of self-transmissible plasmids have a much broader host range for both conjugation and replication. The best-characterized plasmid of this type is known as RK2, RP1, and RP4 as the three independently isolated plasmids are thought to be essentially identical. This plasmid can move between and be stably maintained in a large number of different Gram-negative bacteria (Guiney, 1984). In addition to differences in replication competence in different organisms, however, plasmids differ in their ability to promote mating pair formation and DNA transfer across boundaries involving species, genus and kingdom. RP4 can promote DNA transfer by conjugation to a variety of different Gram-positive bacteria (Guiney et al., 1985; Trieu-Cuot et al., 1987), but selection of transconjugants in those experiments requires the use of a shuttle plasmid containing an origin or replication that is functional in the Gram-positive recipients (Trieu-Cuot et al., 1987). ColE1-derived plasmids can also be mobilized from *E. coli* to filamentous cyanobacteria of the genus *Anabaena* when RK2-promoted mating pair formation occurs (Wolk et al., 1984). Several plasmids including the F factor can promote conjugal DNA transfer from *E. coli* to the yeast *Saccharomyces cerevisiae* when the plasmid being transferred contains a yeast origin of replication (Heinemann and Sprague, 1989; Bates et al., 1998), but RP4 does this much more efficiently than the F factor (Bates et al., 1998). Thus there are also plasmid-dependent differences in mating pair formation and DNA transfer efficiency that influence such DNA transfers across classification boundaries as well as in host range for DNA replication.

The conjugation properties of these broad host range plasmids show both similarities and differences with the F factor conjugation system that are worthy of note. First, RP4 contains a *nic* or *oriT*-like site where plasmid-specific proteins act to form a relaxosome as part of the DNA mobilization process. It also carries the genes involved in producing specialized conjugative pili and promoting mating pair formation and stabilization. However, RK2 encodes a pilus that is thinner and more rigid than the F pilus (Bradley, 1980). One obvious result of this difference in pilus structure is that RK2 is only capable of mating pair formation on a solid surface while the F factor mates equally well under either solid or liquid media conditions (Bradley et al., 1980). Also, the genes of the *tra* regulon of RK2 are not contiguous as they are with the F factor in that they are divided between two regions called Tra1 and Tra2 that constitute almost half of the 60 kbp plasmid (Pansegrau et al., 1994). Not all of the genes in these two Tra regions are essential for plasmid transfer between two *E. coli* cells (those required for such transfer are called *core genes*), and it seems possible that at least some of the other genes

in these regions are involved in transfers involving more distantly related organisms.

Self-transmissible broad host range plasmids such as RK2 must overcome several barriers in order to cross genus and species boundaries. The initial barrier to be overcome is stable mating pair formation and DNA mobilization. The presence of a restriction enzyme system in the recipient cell that degrades incoming plasmid DNA potentially constitutes a second barrier to successful plasmid transfer and establishment. As already mentioned, the third possible barrier is a replication defect that would preclude the autonomous replication of the transferred plasmid in the new recipient host cell. All of these steps have been shown to play a role in determining effective plasmid host range.

One obvious way in which a surface barrier could operate would be to prevent the donor pilus tip from making suitable contact with the surface of a potential recipient cell. Purified F sex pili bind to *E. coli* cells but not to *Pseudomonas aeruginosa* cells (Helmuth and Achtman, 1978), and it has been shown that the F factor cannot promote plasmid transfer to *P. aeruginosa* (Guiney, 1982). Experiments with a well-characterized series of LPS mutants of *Salmonella minnesota* have been used to demonstrate how this surface barrier phenomenon can operate. Although the progressive reduction in the length of the LPS sugar side chain had no effect on the ability of the F factor or RK2 to transfer to *S. minnesota* recipients, the ability of R64 (an IncI plasmid) to transfer dropped off in parallel to the reduction in sugar side chain length (Guiney, 1984). It is also possible that future studies of surface phenomena will uncover situations where mating pair formation and stabilization are inhibited despite efficient contact by the tip of the pilus.

Restriction endonuclease degradation in the recipient cell can also potentially limit the host range of a self-transmissible plasmid. The resident DNA (chromosomal and plasmid) already in a restriction endonucleave-containing cell is resistant to restriction because of specific methylation by a corresponding modification enzyme, and newly replicated cellular DNA is temporarily hemimethylated (methylated on one strand —the parental one) and still protected. Although plasmid DNA entering as a single-strand would not be subject to degradation by most restriction endonucleases, the double-stranded DNA resulting from complementary strand synthesis would be unmodified and susceptible to degradation. One solution for the plasmid would be to eliminate possible restriction enzyme recognition sites. If RK2 did actually go through a process of "shedding" restriction enzyme recognition sites, the mechanism whereby it was accomplished remains a mystery. Digestion of RK2 with restriction enzymes recognizing a hexanucleotide sequence revealed that only one of 18 enzymes tested cut it at the expected frequency, while many of these enzymes cut this 60 kb plasmid only once or twice (Lanka et al., 1983). The single *Eco*RI site present in RK2 resulted in a 10-fold reduction in its ability to transfer into a strain expressing *Eco*RI, and the introduction of several additional *Eco*RI sites into the plasmid caused transfer into an *Eco*RI expressing strain to drop to nearly undetectable levels (Guiney, 1984).

An alternative mechanism for dealing with restriction activity in the recipient cell is exemplified by the *ardA* gene originally described for the ColIb-P9 plasmid and subsequently found to be present on a variety of self-transmissible plasmids from enteric bacteria (Chilley and Wilkens, 1995). When expressed in the recipient cell, the product of this gene interferes with the action of the type I restriction endonucleases commonly found in enteric bacteria and thus provides protection for the entering plasmid.

Inability of a conjugally transferred plasmid to replicate in its new host comprises the third major host range barrier. The fact, however, that self-transmissible plasmids seemingly often promote mating pair formation at low efficiency with organisms where plasmid replication is not possible, can be exploited in experiments utilizing nonreplicating "suicide" plasmids to introduce genetic information into an organism by conjugation. The introduction of a transposon by this method allows efficient transposon mutagenesis of organisms where no other method for introducing DNA is available. This method can also be used to make very specific mutants in such organisms by reintroducing cloned fragments of chromosomal DNA that have been modified in vitro. Recombination with the conjugally introduced DNA fragment can result in the replacement of the corresponding sequences originally present in the chromosome of the recipient cell.

VIII. CHROMOSOME MOBILIZATION BY NON-F PLASMIDS

Gram-negative bacteria commonly contain plasmids, many of which have conjugative or mobilizable properties, or both. Some of these plasmids possess a transfer system that is highly analogous to that of the F factor (e.g., overlapping *fin* specificity or specific *tra* gene complementation), but transfer systems quite distinct from F are also found. The interrupted *tra* regulon structure of RK2 has already been mentioned (Pansegrau et al., 1994), and the narrow host range IncP plasmid, R91, has also been shown to have two separate *tra* regions (Moore and Krishnapillai, 1982). *tra* regulons significantly smaller than the 30^+ kb regions seen with F and F-like plasmids have also been described. The entire tra regulon of R46, the IncN plasmid on which the SOS-enhancing *muc* genes were originally found (see DNA repair chapter by Yasbin, this volume), has been cloned on a 22.5 kb fragment (Brown and Willetts, 1981), and a 19.4 kb fragment from the closely related pCU1 plasmid was sufficient to confer self-transmissibility on the cloning vector (Thatte and Iyer, 1983). Tn5 mutagenesis studies with pCU1 have suggested that the *tra* regulon may not even occupy the entire 19.4 kb fragment (Thatte and Iyer, 1983).

Although most such plasmids rarely affect the transfer of donor chromosomal DNA to the recipient, there are three possible mechanisms by which chromosomal DNA mobilization or transfer can theoretically be achieved. Here we are using the word mobilization to refer to plasmid-mediated transfer of donor cell chromosomal DNA; we are not using it to describe the preparation of plasmid DNA for transfer. First, donor DNA may be transferred to a recipient by an Hfr-like mechanism after a self-transmissible plasmid has stably integrated into the donor cell chromosome. Second, donor DNA can be transferred by an F-prime-like mechanism. If the self-transmissible plasmid carries substantial amounts of DNA sequence homologous to the donor cell chromosome (as exemplified by *E. coli* F-primes), the plasmid can recombine in and out of the donor cell chromosome. A *recA*-dependent recombinational equilibrium is established between the autonomous and integrated states of the plasmid, and transfer of donor chromosomal DNA is achieved when conjugation initiates while the plasmid is in the integrated state. Third, the chromosome can be "conducted" (Clark and Warren, 1979) to the recipient cell when the plasmid has become involved in the formation of a transposition cointegrate with the donor cell chromosome. This third situation generally represents a transient intermediate in replicative transposition to or from the plasmid, and is therefore typically a low-frequency event.

Although low-level chromosome mobilization often occurs without firm evidence for plasmid integration, bonafide Hfr formation has only rarely been documented (Holloway, 1979). Hfr-like strains can be derived with a number of different plasmids, however, by a phenomenon called *integrative suppression*. When strains containing a $dnaA_{ts}$ mutation are grown at high temperature, various plasmids integrate at low frequency into the chromosome at apparently random locations to form Hfr-like strains in which the integrated plasmid provides the basis for chromosomal replication. The DNA replication defect of the $dnaA_{ts}$ mutation is suppressed, hence the term integrative suppression. The F factor, a number of F-like plasmids that normally do not integrate into the chromosome, and some I-like plasmids (Datta and Barth, 1976) have all been shown to be capable of forming Hfr-like strains by integrative suppression.

Chromosome mobilization by non-F plasmids often appears more akin to mobilization by F-primes (a recombinational equilibrium between the autonomous and integrated states of a homology bearing plasmid) than stable Hfr-like mobilization. The same plasmid often gives considerably different results in different bacterial species. This may often be due to differing degrees and amounts of homology; for example, several derivatives of plasmid RK2 carrying differing amounts of sequence homologous to the *E. coli* chromosome were tested for their ability to mobilize the donor cell chromosome. The amount of homology between the plasmid and the chromosome was directly correlated with the ability of the plasmid to mobilize the chromosome by a $recA^+$-dependent mechanism (Grinter, 1981).

The third mechanism for chromosome mobilization is conduction by the formation of transposition cointegrates between a self-transmissible plasmid and the donor cell chromosome. The actual mechanistic basis of conduction has been most clearly demonstrated for the conduction of pBR322 (a mob^-, and therefore nonmobilizable, derivative of ColE1) to recipient cells by the F factor. In this situation, Tn*1000* (γδ) forms cointegrates between the F factor and pBR322, and pBR322 transferred by this mechanism contains a copy of the transposable element left by resolution of the cointegrate in the recipient (Guyer, 1978; see Fig. 7). Mutations in genes on nonmobilizable plasmids can thus be readily obtained (Sancar et al., 1981). It is widely assumed that the donor cell chromosome can be conducted by the same mechanism.

This mechanism has been shown to be implicated in the mobilization (actually conduction) of the chromosome of a variety of gram-negative bacteria by a plasmid called R68.45 (Willetts et al., 1981). The parent R68 plasmid (another name for RK2) mobilizes the chromosome very inefficiently, but the R68.45 derivative was isolated on the basis of efficient mobilization of the *Pseudomonas aeruginosa* chromosome. R68 contains a single copy of IS*21* that transposes very infrequently, but the two tandem copies of IS*21* in R68.45 allow very efficient transposition to occur. As there was no detectable sequence homology between R68.45 and the *P. aeruginosa* chromosome, it was concluded that efficient chromosome mobilization was the result of high-frequency cointegrate formation between the plasmid and the chromosome. Other self-transmissible plasmids containing a suitable transposable element may be able to act similarly. The student is referred to a review article by Holloway (1979) for a detailed discussion of chromosome mobilization by plasmids.

IX. PLASMID-BASED CONJUGATION IN OTHER BACTERIA

While much of the effort expended on our understanding of conjugation systems has involved work with *E. coli*, conjugational phenomena have been described in a wide variety of other organisms. Some of these additional systems involve conjugative and/ or mobilizable plasmids, but conjugative transposons have emerged in recent years as an increasingly important aspect of conjugation. This section will deal with plasmid-based conjugation in other bacteria while the conjugative transposons will be treated separately in the next section.

A. Salmonella

Although a variety of natural isolates of *Salmonella* sp. contain self-transmissible plasmids, these plasmids have not been extensively used in the development of *Salmon-*ella genetics. Instead, the F factor from *E. coli* K-12 has been imported and utilized to develop Hfr strains in *S. typhimurium* and *S. abony*. Many of the initial F-containing strains of *Salmonella* were largely infertile because the strains also contained Fin$^+$ plasmids, but the curing of those plasmids or the use of F factor fertility mutants resulted in F-containing *Salmonella* strains that were fully as fertile as their *E. coli* counterparts (Sanderson et al., 1983).

In some strains of *S. typhimurium* (now more correctly referred to as *S. enterica* serovar Typhimurium), the F factor integrated at only one position on the chromosome, and it was postulated that this was due to the presence of a sex factor affinity site where the F factor could readily recombine at that location in the chromosome (Sanderson et al., 1972). Although the basis for this sex affinity site in some strains of *S. typhimurium* is not known, a sex affinity site in the *E. coli* chromosome was shown to be due to the presence of γδ (Guyer et al., 1980). The Hfr collection for *S. typhimurium* and *S. abony* is not as extensive as that for *E. coli*, but it has been used extensively in chromosomal mapping in these organisms (Sanderson, 1996).

The power of chromosomal mapping by conjugation in *Salmonella* has been greatly extended by the use of a technique employing Tn*10* insertions positioned around the chromosome (Chumley et al., 1979). An F*lac* with a temperature-sensitive replication defect and containing a copy of Tn*10* is introduced into a strain carrying Tn*10* at the desired location on the *Salmonella* chromosome. When the temperature is then raised, Lac$^+$ survivors will result from the integration of the F*lac* by recombination between the two copies of Tn*10*. Although these pseudo-Hfr's are fairly unstable, a fresh isolate functions quite effectively in mobilizing donor chromosomal DNA to a recipient cell. Variations on this technique have also been used in *Pseudomonas* (see below), and may well be developed in a number of additional organisms.

B. *Pseudomonas*

The pseudomonads are a diverse group of microorganisms, and that diversity is reflected in the variety of conjugation systems that have been developed for use with various members of this group. A large number of self-transmissible and mobilizable *Pseudomonas* plasmids have been described, and some of these have been extensively studied because they carry groups of genes involved in the metabolism of complex organic compounds that bacteria normally cannot metabolize. Although conjugation has been reported in a number of *Pseudomonas* species, the discussion here will be restricted to chromosome mobilization in *P. aeruginosa* and *P. putida* where conjugation has been utilized most extensively for genetic mapping. Even with these two species, extensive variations in conjugative behavior have been described, and plasmids that may mobilize the chromosome of one strain are often unable to function effectively in other strains of the same species. The student is referred to a review article (Holloway, 1986) for a more detailed discussion of conjugation in pseudomonads. The student should note that the following discussion of chromosome mobilization in *Pseudomonas* deals with plasmid-mediated transfer of the donor cell chromosome rather than the preparation of plasmid DNA for transfer.

I. Chromosome mobilization in *P. aeruginosa*

Although a wide variety of chromosome mobilizing plasmids have been found in various pseudomonads, the beginnings of a defined conjugation system in a pseudomonad occurred when it was found that a plasmid from *P. aeruginosa* strain PAT could mobilize the chromosome of *P. aeruginosa* strain PAO (Holloway and Jennings, 1958). The plasmid, called FP2, can mobilize the host cell chromosome unidirectionally from a single chromosomal site. Subsequent usage of this plasmid has allowed time-of-entry style conjugational mapping (see Appendix) to be done quite effectively for about 25% of the *P. aeruginosa* chromosome. Other plasmids of the FP series with different origin sites for chromosome mobilization have been described, but their use in mapping studies has not been extensive. The mechanism by which the FP plasmids mobilize the *P. aeruginosa* chromosome remains unclear.

Transposon insertions in the chromosome have also been used to provide the basis for chromosome mobilization in *P. aeruginosa*. This is mechanistically similar to the system described for *Salmonella* above, and involves a self-transmissible plasmid containing a transposon recombining with another copy of the same transposon in the donor cell chromosome. R18 and R91–5 are self-transmissible plasmids that carry Tn*1* in opposite orientations relative to the plasmid's origin of transfer, and it has been shown that these plasmids can mobilize the *P. aeruginosa* chromosome (each of the plasmids in the opposite direction) starting from a variety of chromosomal Tn1 insertions (Krishnapillai et al., 1981). This methodology contributed to the development of a genetic map for *P. aeruginosa*.

Genetic mapping in *P. aeruginosa* has also been greatly facilitated by the use of a group of ECM (enhanced chromosome-mobilizing) plasmids that is typified by an IncP1 plasmid called R68.45 which contains a tandem duplication of an insertion sequence called IS*21*. The transposition of IS*21* is greatly increased when in this tandem duplication arrangement, and this allows for reasonably frequent conduction of the donor cell chromosome when DNA transfer initiates while a transposition cointegrate exists. R68.45 can mobilize all regions of the *P. aeruginosa* strain PAO chromosome, and time-of-entry experiments have been done for the entire chromosome by this methodology (Haas and Holloway, 1978; Royle et al., 1981).

2. Chromosome mobilization in *P. putida*

Although plasmids analogous to the FP plasmids of *P. aeruginosa* have not been found in *P. putida*, a number of genetic mapping systems based on chromosomal mobilization have been described. The earliest system involved chromosome mobilization by a hybrid of the *Pseudomonas* OCT and XYL plasmids, and the combining of data from this system with transduction data allowed the generation of a primitive chromosomal map for *P. putida* PpG1 (Mylroie et al., 1977). Other conjugational mapping work has involved the use of ECM plasmids such as R68.45 with *P. putida* strain PPN, but a plasmid called pMO22 has been used to develop a much more powerful mapping system.

R91–5 is a Fin⁻ derivative of R91 that has been used to mobilize the chromosome of Tn*1*containing strains of *P. aeruginosa* (see above). This plasmid cannot replicate in *P. putida*, and transconjugants with this organism cannot be achieved in matings with *P. aeruginosa* PAO donors. It was found, however, that plasmid markers could be recovered at a low frequency in *P. putida* recipients when pMO22, an R91–5 derivative containing Tn*501*, was mated from a *P. aeruginosa* PAO donor. Such transconjugants turn out to have the entire pMO22 plasmid integrated into the *P. putida* chromosome, and subsequently function effectively as Hfr-like donors in matings with other *P. putida* strains (Dean and Morgan, 1983). A number of distinct chromosomal insertion sites for pMO22 were found, and the resulting collection of Hfr-like strains has been instrumental in the development of a genetic map for *P. putida* strain PPN. Although the insertion of either Tn*501* or Tn*7* into R91–5 allows such Hfr-like strains to be derived, it appears that the Tn*1* already present in R91–5 provides the means for the actual chromosomal insertion event. The basis for the coordinate involvement of Tn*501* or Tn*7* with Tn*1* in the integration process remains a mystery.

3. R-primes in pseudomonads

R-primes are derivatives of R factor plasmids that have acquired additional DNA from the chromosome of a host bacterium.

These R-primes are largely analogous to the F-primes of *E. coli* discussed above, and the extended host range capabilities of many R factors has allowed R-primes to be generated and utilized in a variety of bacterial genera (Holloway and Low, 1987, 1996). They have been particularly useful in pseudomonads where the extensive use of self-transmissible R factors has provided a framework for the isolation of a variety of R-primes.

R68.45 and other similar ECM plasmids have been extensively used for generating R-primes containing *P. aeruginosa* chromosomal DNA. The resulting R-primes contain the chromosomal DNA between the two copies of IS*21* present in the original R68.45 plasmid. These can be selected for by genetic complementation after transfer into suitably marked Rec⁻ strains of *P. aeruginosa*, or by transfer into *P. putida*. Genetic complementation still occurs in *P. putida*, but a lack of sufficient nucleotide base homology prevents recombination between the R-prime-borne *P. aeruginosa* DNA and the *P. putida* chromosome. This is analogous to the situation that exists when *E. coli*/*S. typhimurium* merodiploids are made. R-primes can also be obtained from strains of *P. putida* containing integrated derivatives of R91–5 by mechanisms that appear to be highly analogous to F-prime production in Hfr strains of *E. coli*.

C. *Streptomyces*

The importance of members of the genus *Streptomyces* in commercial antibiotic production has led to extensive efforts to develop genetic systems for these organisms. As a result the first conjugative plasmid from a Gram-positive organism was found in *S. coelicolor* A3(2) (Vivian, 1971). The plasmid, SCP1, was characterized on the basis of its ability to influence the exchange of chromosomal genes and to produce the antibiotic methylenomycin. Although it does not appear to carry much in the way of important genetic information for the cell, it is unusual in that it is a linear double-stranded DNA plasmid of approximately

350 kbp (Kinashi and Shimaji-Murayama, 1991).

The various conjugative behavior patterns of the SCP1 plasmid of *S. coelicolor* are somewhat analogous to those of the F factor of *E. coli*. Strains lacking SCP1 (SCP1⁻; originally called UF for ultrafertility) correspond to F⁻ strains *E. coli*. Attempted matings between two such SCP1⁻ strains that also lack all other known fertility plasmids yield verifiable recombinants for chromosomal markers at extremely low levels (Bibb and Hopwood, 1981). Such recombinants were not detectable in attempted crosses between two strains of *S. lividans* that were totally lacking in fertility plasmids (Hopwood et al., 1983). SCP1⁺ strains, also called IF for initial fertility, appear to correspond to F⁺ strains of *E. coli* in that transfer of the presumably autonomous plasmid is several orders of magnitude (roughly 10^3 to 10^5 times) more frequent than the detectable transfer of chromosomal markers. The fact that recombinants from SCP1⁺ times SCP1⁻ crosses tend to preferentially inherit chromosomal markers from the donor strain has led to the conclusion that most of the transfer occurs as the result of Hfr-like cells, such as NF (see below), present in the SCP1⁺ culture (Hopwood et al., 1973).

A variety of Hfr-like derivatives of SCP1⁺ strains of *S. coelicolor* have been described. The original such strain was called NF, for normal fertility, and it has been determined that the SCP1 plasmid integrates by means of recombination between a copy of an insertion sequence (now called IS*446*) on the plasmid and one of two closely positioned copies of IS*446* on the *S. coelicolor* chromosome (Kendall and Cullum, 1986). The integration mechanism also involves the deletion of both chromosomal and plasmid sequence, but exactly how these deletions arise during integration remains unclear. Such strains can effectively transfer chromosomal genes to a suitable recipient, but the donor properties of these strains differ from those of *E. coli* Hfr's in two principal ways (Hopwood et al., 1969). First, the transfer appears to be bidir-

ectional in that markers on either side of the integrated SCP1 plasmid are transferred with equal efficiency. Second, all of the recombinants from such NF times UF crosses become NF, whereas the vast majority of recombinants from an Hfr times F⁻ cross in *E. coli* remain F⁻. Although unidirectional donors that appear to involve the integration of SCP1-prime plasmids (see below) at alternate chromosomal sites have been described, these also involve transfer of the donor character to the recipient in all cases (Vivian and Hopwood, 1973).

It has been proposed that these two variations from Hfr-style chromosomal mobilization could be explained by either of two models (Hopwood et al., 1985). The first would have a unidirectional transfer of DNA, perhaps single-stranded as in *E. coli*, emanating from the integrated SCP1 with the proviso that transfer normally involves a DNA length greater than that of the chromosome. With this model the unidirectional transfer observed in some situations might be explained on the basis of a recombination hotspot of the SCP1 integration site in that particular donor strain. The second model is that passive conduction to the recipient of the entire chromosome occurs as an extension of the fact that the SCP1 plasmid itself is transferred to the recipient as an intact double-stranded DNA molecule. Although experimental evidence to discriminate between these two models is not available, the second model is made attractive by the types of marker transfer gradients that are observed in NF times UF crosses.

Various SCP1-prime plasmids have also been described for *S. coelicolor* (Hopwood and Wright, 1976). Although the mechanistic details of their formation have not been determined, their conjugative properties generally correspond to those of F-primes in *E. coli* and R-primes in a variety of Gram-negative bacteria. These SCP1-prime plasmids can provide for the mobilization of donor cell chromosomal markers that are not actually present on the SCP1-prime, and both stable and unstable Hfr-like strains appear to be formed by the integration of these plasmids into the chromosome (Vivian and Hopwood, 1973).

In addition to SCP1, a variety of other fertility plasmids from *S. coelicolor* have been characterized. The first such plasmid described was a 31 kb self-transmissible plasmid called SCP2 (Bibb et al., 1977). This plasmid promotes little, if any, transfer of chromosomal markers, but variants called SCP2* spontaneously arise that are capable of low frequency chromosomal transfer. Unlike the case with SCP1, neither Hfr-like nor F-prime-like donors involving SCP2 have been found in vivo. The smaller size of SCP2, however, has allowed the in vitro construction of SCP2-primes that can mediate the transfer of the host cell chromosome (Hopwood et al., 1985).

One of the more interesting phenotypes associated with *Streptomyces* conjugation is called pocking. These matings typically involve the mixing of two strains, often starting with spores, on an agar surface, followed by mycelial growth, mating, and the subsequent transfer of resulting spores to selective media to obtain recombinants. The regions of the mating plate where plasmid transfer has occurred are visible as regions of reduced growth or pocks—it may be useful to think of pocks on a *Streptomyces* mating plate as being similar in appearance to turbid bacteriophage plaques. Plasmid mutants that give rise to small pocks also show less extensive plasmid transfer, and it has been hypothesized that such small pocks arise when the plasmid is incapable of intramycelial transfer in the recipient strain, referred to as *spread*, while retaining the ability to achieve intermycelial transfer between the donor and recipient strains. Although a definitive relationship between between plasmid transfer and pocking has not been demonstrated, the weight of evidence suggests that the two phenomena are intimately associated (Hopwood et al., 1985).

A clue as to the basis of pocking appeared when it was found that insertions into a certain nonreplicative region of pIJ101, a small self-transmissible plasmid also used as a *Streptomyces* cloning vector, gave derivatives that were incapable of transforming *S. lividans* (Kieser et al., 1982). It was originally thought that pIJ101 must contain one or more *kil* genes whose expression results in cell death unless they are repressed by plasmid-borne *kor*, or kill override, genes (Kendall and Cohen, 1987). It was thought that the original insertion mutant of pIJ101 had inactivated a *kor* gene, and it seemed reasonable to conclude that this had made it impossible for *S. lividans* cells containing that particular plasmid derivative to survive. One might then explain pocking as cell growth inhibition by a temporary expression of *kil* genes in recipient cells before establishment of *kor* repression allows cell growth to resume. It has not been possible to document the presence of bonafide *kil* genes, however, and it is now thought that the *kor* effect stems from its role in transcriptional regulation of plasmid genes involved in transfer (Stein, et al., 1989). The burst of plasmid gene expression when this *kor* repression is temporarily alleviated immediately following transfer might contribute to both increased conjugal proficiency and reduced growth rate on the part of new recipient cells.

Space considerations do not allow a discussion of the numerous small self-transmissible *Streptomyces* plasmids that have been characterized, but the pIJ101 plasmid serves to illustrate a couple of additional important features of conjugation in *Streptomyces*. First, only a single *tra* gene found on pIJ101 is required for conjugal function on the part of the plasmid (Kendall and Cohen, 1987), so it seems likely that *trans*-acting host cell functions must be involved in the conjugal transfer of at least some of the smaller *Streptomyces* plasmids. Second, a mutational analysis of the pIJ101 *tra* gene has shown that plasmid transfer and chromosome mobilizing ability are differentially dependent on the gene product as some mutations affect chromosome mobilizing ability to a much greater extent than plasmid transfer (Pettis and Cohen, 2000). In summary, a great deal remains to be learned about conjugation mechanism in these interesting organisms. For further details, the student is referred to Hopwood and Kieser (1993).

D. Gram-Positive Cocci (*Streptococcus*, etc.)

Although the streptococci as a group are more noted for the transformation systems that have been developed in *S. pneumoniae* and *S. sanguis*, three distinctive types of conjugation system have also been described for this group of organisms. Two of these systems appear to involve plasmid conjugation and associated phenomena while the third involves a unique type of conjugal transfer of chromosomal elements called conjugative transposons (Section X). One type of plasmid conjugation system normally utilizes sex pheromones and operates at high efficiency, but the other type of plasmid conjugation system operates at low efficiency phenomena and does not involve the participation of sex pheromones. The pheromone-enhanced conjugation systems seem largely limited to the enterococci (see Perlin, this volume).

Sex pili have not been observed in conjugative streptococci, and conjugal transfer frequently depends on the co-precipitation of donor and recipient cells onto a solid surface. High-efficiency mating in liquid culture can be obtained, however, with a few plasmid-containing *Enteroccoccus faecalis* strains, whose mating is controlled by small peptide sex pheromones (see Clewell, 1993 and 1999 for overviews). In this type of plasmid conjugation system, the recipient cells produce a pheromone that elicits a mating response from donor cells. The induction of the donor mating response by pheromones increases the level of plasmid transfer by about 10,000-fold as compared to the level observed in the uninduced state (Dunny et al., 1979; Ike and Clewell, 1984). There are a number of these pheromones, and each appears to be keyed to a specific plasmid (Dunny et al., 1979). Genomic analysis has recently revealed that a number of the common pheromones are processed from the signal sequence segments of surface lipoproteins (Clewell et al., 2000), and a chromosomal gene of *Enterococcus faecalis* called *eep* (for enhanced expression of pheromone) has been identified that may be involved in the specific processing of these signal sequence peptides (An et al., 1999).

After receiving the plasmid, the recipient cell ceases production of the particular pheromone associated with that plasmid, but continues to produce other pheromones (Dunny et al., 1979; Ike et al., 1983). The transconjugant, as well as the initial donor cell, also produces a new plasmid-encoded peptide that functions as a competitive inhibitor of the corresponding pheromone. It has been speculated that the inhibitor prevents the plasmid-containing cells from responding to a concentration of pheromone that is too low to result in effective mating aggregate formation. The nomenclature used for the components of these systems is standardized. As an example, cells containing plasmid pAD1 respond to a pheromone called cAD1 and produce an inhibitor that is called iAD1.

A pheromone-induced donor cell produces a surface adhesion protein, referred to as the aggregation substance (Ehrenfeld et al., 1986). This donor aggregation substance binds to a surface receptor called the enterococcal binding substance to promote clumping, and the ability of low concentrations of lipoteichoic acid to prevent this clumping suggests that it may be a component of the binding substance (Ehrenfeld et al., 1986). This enterococcal binding substance is present on potential donors and potential recipients alike, and pure cultures of donor cells exhibit clumping when treated with exogenous pheromone.

Although successful pheromone-mediated plasmid transfer has thus far only been observed between *E. faecalis* cells, one of the *E. faecalis* sex pheromones, called cAM373, is also produced by a variety of *Staphylococcus aureus* isolates and by a few isolates of other species of streptococci (Clewell et al., 1985). These cAM373-producing isolates can induce clumping with *E. faecalis* strains bearing the corresponding plasmid, pAM373, but potential recipients containing a functionally replicating plasmid were not obtained. DNA transfer apparently occurred, however, as conjugative transposons (see Section X) carried by pAM373 were found to have successfully integrated into the *Staphylococcus aureus* chromosome after such matings. Thus the sex pheromone-mediated clumping phenomenon initially described for *E. faecalis* may have a much broader role in genetic exchange between Gram-positive bacteria.

Initial genetic analysis with one of these *E. faecalis* plasmids, pAD1, involved the use of Tn*917* insertions to delineate a 31.2 kb region of the plasmid that is necessary for high-efficiency plasmid conjugation involving pheromones (Ehrenfeld and Clewell, 1987). Insertions in part of this region reduced or eliminated plasmid transfer without affecting the clumping response, while insertions in other regions prevented cell aggregation (as well as plasmid transfer) in liquid cultures. While a number of regulatory genes and the gene producing the aggregation substance have been studied in some detail for several of these plasmids (see Clewell, 1999), very little is still known about the genes and gene products that are directly involved in plasmid transfer. Interestingly, the *oriT* analogue for pAD1 is located within a group of genes thought to function in the vegetative replication of the plasmid (An and Clewell, 1997).

Considerably less is known about the plasmid conjugation system that does not involve sex pheromones (see Macrina and Archer, 1993, for a review). The lack of inducible aggregation results in much lower plasmid transfer efficiencies, but many of these plasmids demonstrate a much broader host range. One example is the *E. faecalis* plasmid pAMβ1 which transfers to *Lactobacillus casei*, *Staphylococcus aureus*, and *Bacillus subtilis*, as well as to a variety of other species of streptococci. The size of pAMβ1, approximately 26 kb, is much less than that of the conjugative plasmids that are phero-

mome responsive, but the nature of the genes whose products mediate the actual transfer event has not been determined. Some of these conjugative streptococcal plasmids of both types (pheromone induced and non-pheromone induced) can mobilize non-conjugative plasmids (Dunny and Clewell, 1975; Smith et al., 1980), and others chromosomal DNA (Franke et al., 1978). The actual basis of mobilization is not yet characterized, however, for any of these plasmids.

E. Other Plasmid-Based Conjugation Systems

It is not our purpose here to catalog all bacterial conjugation systems, and many that have been described in less well-known organisms have not been characterized in any detail. The intriguing Ti plasmid conjugation system from *Agrobacterium* is covered separately by Ream (this volume), and a discussion of plasmid conjugation systems in anaerobes, particularly the genera *Bacteroides* and *Clostridium*, may be found in the article by Macrina (1993) as well as by Whittle and Salyers, this volume. We do want to note, however, the existence of conjugation in thermophiles and hence in the *Archaea* as well. Since the first report of a conjugative plasmid in an archaeobacterium of the genus *Sulfolobus* (Schleper, et al., 1995), it has been shown that conjugative plasmids are quite common in that organism (Prangishvili et al, 1998; Stedman et al. 2000) and that transfer of chromosomal markers can be promoted (Schmidt, et al., 1999). Plasmid conjugation has also been demonstrated in a eubacterial thermophile of the genus *Thermus* with the additional possibility of Hfr-like transfer suggested (Ramirez-Arcos, et al., 1998).

X. CONJUGATIVE TRANSPOSONS

These interesting elements can both transpose to new locations in the cell in which they reside and insert into a replicon in another cell after mediating their own transfer by conjugation without any stabilized existence as a plasmid. They will not be dealt with extensively here because they are covered in considerable detail by Whittle and Salyers, this volume. The student is also referred to review articles by Salyers et al. (1995), Scott and Churchward (1995), and Rice (1998).

The first two of these elements described were Tn916 from *Enterococcus faecalis* (Franke and Clewell, 1981) and Tn5253, originally known as the Ω(*cat-tet*) element, from *Streptococcus pneumoniae* (Shoemaker et al., 1980). The much larger Tn5253 element turns out to contain a copy of the 18 kbp Tn916 element, and such composite structures seem to be relatively common among conjugative transposons. The intensively studied Tn916 element seems to be the most common such element among the Gram-positive cocci, and many of the other elements that have been described are related to it. Many of these conjugative transposons carry multiple drug-resistance determinants, but tetracyline resistance is almost always observed to be present in such elements.

Intercellular transfers by these elements only occur at low frequency during filter (solid surface) matings in the laboratory, but interspecies transfer is quite common (Courvalin, 1994). Tn916 or Tn916-like elements presumably of Gram-positive origin have been found in a number of Gram-negative species, but distinctively different conjugative transposons are quite commonly found in the Gram-negative genus *Bacteriodes*.

A model for the excision, conjugation, and insertion of Tn916 is shown in Figure 10 (from Scott and Churchward, 1995). An element-encoded integrase protein is needed for the excision and integration steps while the element-encoded excision protein only plays a role in the excision step. These proteins show some interesting functional similarities to the well-known bacteriophage λ intregase and excision proteins. The fact that some strains of *Lactococcus lactis* can serve as a recipient but not a donor for the conjugative transposition of Tn916 has led to the speculation that an IHF-like host factor is also required for excision (Bringel et al., 1991). The six base pair staggered cuts

made to initiate excision are called coupling sequences, and the excised element is circularized even though base-pairing within these coupling sequences is unlikely. The integration reaction does not duplicate the target site as generally seen for more conventional transposons, but it does generate flanking mismatched heteroduplexes that can be eliminated either by mismatch correction or by a round of recipient replicon replication. A role for the double-stranded circular intermediate in the integration reaction is supported by the observation that such circular intermediates purified from *E. coli* strains can integrate subsequent to protoplast transformation in *Bacillus subtilis* (Scott et al., 1988). Tn916 does not seem to be very site specific with regard to the integration reaction, but this is not the case with all conjugative transposons.

The conjugation part of the process is much less well understood than the excision and integration reactions. One study exploring concomitant transfer of chromosomal markers suggested the occurrence of something approximating a complete cell fusion event (Torres et al., 1991), but a direct correlation of chromosomal marker transfer with conjugative transposition was not established. At the same time the indirect mobilization of donor cell plasmids by conjugative transposons in several systems (see Salyers et al., 1995) suggests that at least pore formation between donor and recipient cell is mediated by conjugative transposons. There is no evidence for either the involvement of sex pili as seen with plasmid conjugation in the Gram-negative bacteria or the involvement of adhesion proteins such as are seen with pheromone-mediated plasmid conjugation in *E. faecalis*. It has been suggested that a single-strand of the excised double-stranded circular element is transferred and then duplicated in the recipient cell before integration (Scott et al., 1994), and *oriT*-like function has been demonstrated for a 466 bp segment of Tn916 (Jaworski and Clewell, 1995). Tn5 insertion analysis indicated that about 11 kbp located largely towards one

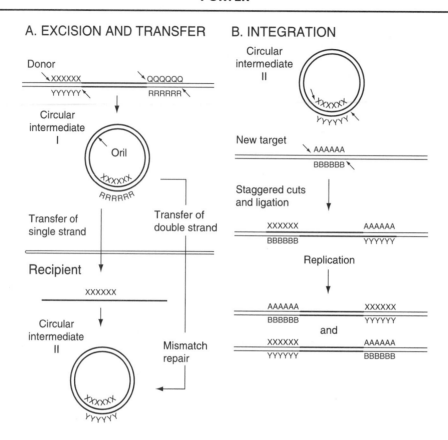

Fig. 10. Proposed model for Tn*916* transposition and conjugation. **a:** Staggered nicks in the sequences flanking the initial position of the element result in the production of nonreplicating circular intermediate I with six potentially mismatched bases. While this double-stranded circular form may be transferred directly to the recipient and then possibly undergo correction of the mismatched bases, it is perhaps more likely that nicking at the presumptive *oriT*-like site within the element allows transfer of a single strand to the recipient cell where complementary strand synthesis then occurs; circular intermediate II results in either case. **b:** Cleavage of circular intermediate II and the target is followed by ligation to produce an integrated element that is flanked by mismatched coupling sequences until one round of replication has occurred. (Reproduced from Scott and Churchward, 1995 with permission from the "Annual Review of Microbiology" Volume 49 © 1995 by Annual Reviews.)

end of Tn*916* consists of genes required for intercellular transfer but not intracellular transfer of the element (Senghas et al., 1988), and there are 11 putative open reading frames in that region. One of those open reading frames possesses significant similarity to the *mbeA* gene of ColE1 whose product acts at the ColE1 *nic* site (Flannagan et al., 1994), but nothing else is known about these putative conjugation genes. The *traA* gene of Tn*916* is located at the other end of the element with the genes for the integase

and excisase, and *traA* seems to be a positive regulator of other genes involved in the conjugative process (Jaworski et al., 1996). It has been suggested that concerted regulation of *traA* and the *int-Tn* and *xis-Tn* genes may serve to coordinate the transposition and conjugation functions of the element (Jaworski et al., 1996).

XI. CONCLUSION

Bacterial conjugation systems are both diverse and complex. In the Gram-negative bacteria, conjugation is generally a plasmid phenomenon that only occasionally results in the transfer of some portion of the donor cell chromosome. F factor-mediated transfer of the chromosome can be systematically achieved with *E. coli* and some of its close relatives, but it is achieved much less frequently in a variety of Gram-negative bacteria when plasmids other than the F factor are utilized. Although conjugation is also known to often involve plasmids in Gram-positive organisms, the conjugative transposons found in both Gram-positive and Gram-negative bacteria provide what seems to be an exception to the requirement for plasmid involvement in all conjugation systems. Conjugative donor cell chromosome transfer is certainly achieved in the streptomycetes, but there are certainly mechanistic differences as compared to the *E. coli* F factor system. Nevertheless, this most ascetically appealing of the bacterial gene transfer processes undoubtedly plays a very important role in the production and maintenance of genetic variability in bacteria.

XII. APPENDIX: CONJUGATIONAL MAPPING

A variety of techniques have been developed for genetic mapping in bacteria. Both transformation and generalized transduction are very useful for fine-structure mapping in organisms where these gene transfer modes operate at reasonably high efficiency. However, conjugation using Hfr-type donors remains the single most powerful technique for mapping large segments of the bacterial chromosome. Despite the power of this method, conjugational mapping has only been done in a handful of organisms where such Hfr-type donors have been available. The prototype system for such chromosomal mapping involves the use of Hfr's based on the *E. coli* F factor. This mapping system has been used in some *Salmonella* species as well as in *E. coli* itself. The principles are the same for the Hfr-like systems that have been derived for *Pseudomonas aeruginosa* and *P. putida*. The conjugation system in *Streptomyces coelico-*

lor has also been used for mapping purposes, but the mapping strategy in that organism differs considerably from that employed in the Gram-negative organisms. The discussion here will be limited to the principles used in the *E. coli*-like systems.

The mapping strategy often used in Hfr-style conjugation is based on a principle called *time of entry*. The rate at which DNA is transferred from donor to recipient cell is fairly constant for any given *E. coli* Hfr strain, even though some variation between Hfr's can be observed. The time after the initiation of mating pair formation at which any given chromosomal marker enters the recipient depends on its position relative to the integrated F factor in terms of both distance and direction. The *E. coli* chromosomal map is calibrated in minutes to reflect this time-of-entry phenomenon. The map was originally 90 minutes in length, but it was re-calibrated to 100 minutes in the mid-1970s in order to more accurately reflect numerous mapping data on various parts of the chromosome. Map positions given in the older literature have to be evaluated in light of this re-calibration.

The actual time-of-entry data obtained reflect both the gradient of transfer and orientation of transfer that were discussed in the chapter. An actual experiment involves the following steps. First, the Hfr and F⁻ strain are grown to a specified cell density (generally $1–2 \times 10^8$ cells per ml) at 37°C. Temperature has a strong effect on fertility (F factor *tra* expression is minimal at 30°C and below) and transfer velocity, so it is important to ensure that both strains are kept at 37°C both prior to and during mating. This means that you must adjust relative cell density at an early stage of growth so that both strains reach the desired level at the same time. Second, the two strains are mixed, generally at a ratio of one Hfr cell per 10 F⁻ cells, and incubation is continued at 37°C. Third, samples of the mating culture are taken periodically for mating interruption. This is called the *blendor* technique, but a real blendor is seldom used. Although vigorous vortex mixing is sometimes used to disrupt mating pairs, the method of choice involves a special tube-holding attachment for a sabre saw. The underlying principle is that only so much donor DNA transfer can occur in any such defined time period. Plating agar is frequently present during the interruption step, and further initiation of mating is prevented by both dilution and cell immobilization in the agar after plating.

The interrupted samples of mating culture are plated on the appropriate media for scoring the desired recombinants. This involves both a selection for the recombinant F⁻ cells (e.g., omitting arginine from minimal media to select Arg⁺ recombinants) and a counterselection for the Hfr donor. Most matings done for mapping purposes involve Str^s Hfr strains (sensitive to the antibiotic streptomycin) and Str^r recipients so that streptomycin can be used as the counterselective agent. The data obtained are then plotted as recombinants per Hfr or recombinants per ml versus minutes of mating before blending. Some hypothetical time-of-entry data are shown in Figure 11.

For each genetic marker scored, the time-of-entry curve can be broken down into three phases as shown for marker A in Figure 11. The initial phase is the baseline portion of the curve where the scored marker has not yet reached any recipients and no recombinants can scored. The second phase involves the part of the curve where the recombinants are beginning to appear and the slope of the curve is rapidly changing. The third phase of the curve occurs when the slope has stabilized and the number of scorable recombinants is rapidly rising. This is the part of the curve that is used to determine an actual time of entry for a given marker. Eventually such curves would show a shoulder and plateau as the recombinant level maximizes, but this portion of the curve is not terribly informative and is generally ignored.

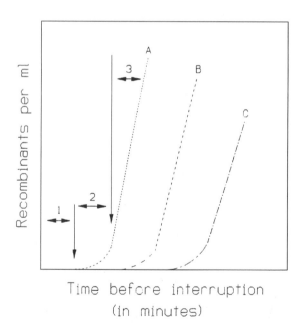

Fig. 11. Time-of-entry curve for a hypothetical mating interruption experiment. An Hfr times F⁻ mating is initiated, and samples of the mating culture are periodically interrupted and plated on plates that select for F⁻ recipient cells that have received genetic marker A, B, or C from an Hfr donor cell. The time of entry for any given marker correlates to its chromosomal location, and the distance of any selected marker from the Hfr origin of transfer (*oriT*) can be determined as described in the text. The three phases of the time-of-entry curve for marker A, as described in the text, are shown by numbers.

The time of entry for a given marker is derived by an extrapolation to the abscissa from the earliest part of phase three of the curve where recombinants are just beginning to appear in significant numbers. The value obtained, in minutes of mating before interruption, provides a good initial indication of the distance between the marker and the site of F factor integration. This value is often compared to the time of entry achieved in similar matings for known markers in that region of the chromosome. The use of more than one Hfr is frequently used to firm up the approximate position before using P1 transduction to map the marker with precision.

Although the initiation of DNA transfer is not actually synchronous, the rapid rise in recombinants at the time-of-entry point often closely approximates a step function for genetic markers that are transferred early by the Hfr. The degree of apparent synchrony with which transfer occurs is lessened with markers further from the F factor site of integration. This is reflected in an increase in the phase two region of the curve and a reduction in the slope observed in phase three of the time-of-entry curve. These effects can be seen for markers B and C in Figure 11. Although useful data can be obtained for distances up to and beyond 50 minutes, the use of an Hfr that transfers the marker of interest prior to 30 minutes is preferable for mapping purposes.

All of the discussion to this point has assumed that a direct selection is available for the marker being transferred to the F^- recipient. When such a direct selection is not available, a select-then-screen strategy must be employed. Recombinants for an available selectable marker are obtained at various time points, and these are tested for the nonselectable marker by replica plating or some other suitable screening procedure. If the marker of interest is between the site of F factor integration and the selected marker, transfer of the marker of interest will be observed in roughly half of the recombinants. High levels of co-transfer ($> 75\%$) indicate tight linkage between the selected marker and the screened marker. When the nonselected marker of interest is transferred after the selected marker, co-transfer of the two markers will seldom be observed. This type of procedure allows the nonselectable marker to be assigned to the particular region of the chromosome between two such selectable markers. The power and ease of drug-resistance selection has made use of Hfr strains with mapped transposon insertions a very popular method of carrying out this type of select-then-screen protocol.

When the marker of interest is a conditional lethal (e.g., auxotrophy or temperature sensitivity), a print mating procedure can be employed to make the initial assignment of marker position to a defined region of the chromosome. A plate is prepared with about 20 Hfr strains patched on the plate at positions corresponding to the appropriate F factor integration site on the chromosome. The clockwise-transferring Hfr's and counterclockwise-transferring Hfr's are generally arranged in two concentric circles to simplify interpretation of the actual print mating plate. A Str^r F^- strain carrying the conditional lethal marker to be mapped is grown up and spread on a streptomycin-containing plate. This plate must either select against the conditional lethal marker or be incubated under conditions, such as high temperature, that provide the necessary selection. Then a fresh replica of the Hfr plate is printed on the lawn of F^- cells by replica plating. After incubation under suitable conditions, this print mating plate is read. For each ring of Hfr's, the gradient of transfer should be apparent. There will be little or no growth at positions where the corresponding Hfr is transferring the marker late while strong patches will result from Hfr's that transfer the marker early. For each set of Hfr's (clockwise and counterclockwise) a sharp breakpoint between early and late transfer will be evident. This allows the marker to be assigned to the region of the chromosome between the F factor integration site for the earliest clockwise-transferring and the earliest counterclock-wise-transferring Hfr. Data of this type greatly facilitates the choice of Hfr's for subsequent interrupted mating experiments.

When possible, the use of a print mating and a subsequent interrupted mating experiment provides the most effective means of localizing a marker on the *E. coli* chromosome. The availability of numerous Hfr strains containing well-mapped transposon insertions makes it possible to map essentially any genetic marker on the basis of time of entry. The student is referred to an article by Low (1991) for a more extensive discussion of conjugational mapping methods.

LITERATURE CITED

Abdel-Monem M, Taucher-Scholz G, Klinkert M-Q (1983): Identification of *Escherichia coli* DNA helicase I as the *traI* gene product of the F sex factor. Proc Natl Acad Sci USA 80:4659–4663.

Achtman M, Kennedy N, Skurray R (1977): Cell-cell interactions in conjugating *Escherichia coli*: Role of *traT* protein in surface exclusion. Proc Natl Acad Sci USA 74:5104–5108.

Achtman M, Morelli G, Schwuchow S (1978a): Cell-cell interactions in conjugating *Escherichia coli*: Role of F pili and fate of mating aggregates. J Bacteriol 135:1053–1061.

Achtman M, Schwuchow S, Helmuth R, Morelli R, Manning G (1978b): Cell-cell interactions in conjugating *Escherichia coli*: Con⁻ mutants and stabilization of mating aggregates. Mol Gen Genet 164:171–183.

Achtman M, Manning PA, Edelbluth C, Herrlich P (1979): Export without proteolytic processing of inner and outer membrane proteins encoded by F sex factor *tra* cistrons in *Escherichia coli* minicells. Proc Natl Acad Sci USA 76:4837–4841.

Aguero CW, Aron L, DeLuca AG, Timmis KN, Cabello FC (1984): A plasmid-encoded outer membrane protein, TraT, enhances resistance of *Escherichia coli* to phagocytosis. Infect Immun 46:740–746.

An FY, Clewell DB (1997): The origin of transfer (*oriT*) of the enterococcal, pheromone-responding, cytolysin plasmid pAD1 is located within the *repA* determinant. Plasmid 37:87–94.

An FY, Sulavik MC, Clewell DB (1999): Identification and characterization of a determinant (*eep*) on the *Enterococcus faecalis* chromosome that is involved in the production of the peptide sex pheromone cAD1. J Bacteriol 181:5915–5921.

Avery OT, MacLeod CM, McCarty M (1944): Studies on the chemical nature of the substance inducing transformation of pneumococcal types: Induction of transformation by a desoxyribonucleic acid fraction isolated from pneumococcus type III. J Experimental Med 79:137–158.

Bagdasarian M, Bailone A, Angulo JF, Scholz P, Bagdasarian MM, Devoret R (1992): PsiB, and anti-SOS protein, is transiently expressed by the F sex factor during its transmission to an *Escherichia coli* K-12 recipient. Mol Microbiol 6:885–893.

Bailone A, Backman A, Sommer S, Celerier J, Bagdasarian MM, Bagdasarian M, Devoret R (1988): PsiB polypeptide prevents activation of RecA protein in *Escherichia coli*. Mol Gen Genet 214:389–395.

Bates S, Cashmore AM, Wilkins BM (1998): IncP plasmids are unusually effective in mediating conjugation of *Escherichia coli* and *Saccharomyces cerevisiae*: Involvement of the Tra2 mating system. J Bacteriol 180:6538–6543.

Berg CM, Curtiss III R (1967): Transposition derivatives of an Hfr strain of *Escherichia coli* K-12. Genetics 56:503–525.

Bibb MI, Freeman RF, Hopwood DA (1977): Physical and genetical characterization of a second sex factor, SCP2, for *Streptomyces coelicolor* A3(2). Mol Gen Genet 154:155–166.

Bibb MI, Hopwood DA (1981): Genetic studies of the fertility plasmid SCP2 and its SCP2* variants in *Streptomyces coelicolor* A3(2). J Gen Microbiol 126:427–442.

Blazey DL, Burns RO (1983): *recA*-dependent recombination between rRNA operons generates type II F′ plasmids. J Bacteriol 156:1344–1348.

Boyd AC, Archer JAK, Sherratt DJ (1989): Characterization of the ColE1 mobilization region and its protein products. Mol Gen Genet 217:488–498.

Bradley DE (1980): Morphological and serological relationships of conjugative pili. Plasmid 4:155–169.

Bradley DE, Taylor DE, Cohen DR (1980): Specification of surface mating systems among conjugative drug resistance plasmids in *Escherichia coli* K-12. J Bacteriol 143:1466–1470.

Bresler SE, Krivonogov SV, Lanzov VA (1978): Scale of the genetic map and genetic control of recombination after conjugation in *Escherichia coli* K-12. Mol Gen Genet 166:337–346.

Bresler SE, Goryshin I-Yu, Lanzov VA (1981): The process of general recombination in *Escherichia coli* K-12: structure of intermediate products. Mol Gen Genet 183:139–143.

Bringel F, Van Alstine GL, Scott JR (1991): A host factor absent from *Lactococcus lactis* subspecies *lactis* MG1363 is required for conjugative transposition. Mol Microbiol 5:2893–2993.

Brown AMC, Willetts NS (1981): A physical and genetic map of the IncN plasmid R46. Plasmid 5:188–201.

Cabezon E, Sastre JI, de la Cruz F (1997): Genetic evidence of a coupling role for the TraG protein family in bacterial conjugation. Mol Gen Genet 254:400–406.

Carter JR, Porter RD (1991): *traY* and *traI* are required for *oriT*-dependent enhanced recombination between *lac*-containing plasmids and λp*lac5*. J Bacteriol 173:1027–1034.

Carter JR, Patel DR, Porter RD (1992): The role of *oriT* in *tra*-dependent enhanced recombination between mini-F-*lac-oriT* and λp*lac5*. Genet Res Camb 59:157–165.

Cheah K-C, Ray A, Skurray R (1986): Expression of F plasmid *traT*: Independence of *traY* → *Z* promoter and *traJ* control. Plasmid 16:101–107.

Chilley PM, Wilkens BM (1995): Distribution of the *ardA* family of antirestriction genes on conjugative plasmids. Microbiol Reading 141:2157–2164

Chumley FG, Menzel R, Roth JR (1979): Hfr formation directed by Tn*10*. Genetics 91:639–655.

Clark AJ (1967): The beginning of a genetic analysis of recombination-proficiency. J Cell Physiol 70 (Suppl 1):165–180.

Clark AJ, Maas WK, Low B (1969): Production of a merodiploid strain from a double male strain of *E. coli* K12. Mol Gen Genet 105:1–15.

Clark AJ, Warren GJ (1979): Conjugal transmission of plasmids. Ann Rev Genet 13:99–125.

Clewell DB (1993): Sex pheromones and the plasmid-encoded mating response in *Enterococcus faecalis*. In Clewell DB (ed): "Bacterial Conjugation" New York: Plenum Press, pp 349–367.

Clewell DB (1999): Sex pheromone systems in enterococci. In Dunny GM, Winans SC (eds): "Cell-Cell Signaling in Bacteria." Washington, DC: ASM Press, pp 42–65.

Clewell DB, Helinski DR (1969): Supercoiled circular DNA-protein complex in *Escherichia coli*: Purification and induced conversion to an open circular DNA form. Proc Natl Acad Sci USA 62:1159–1166.

Clewell DB, Helinski DR (1970): Properties of a supercoiled deoxyribonucleic acid-protein relaxation complex and strand specificity of the relaxation event. Biochem 9:4428–4440.

Clewell DB, An FY, White BA, Gawron-Burke C (1985): *Streptococcus faecalis* sex pheromome (cAM373) also produced by *Staphylococcus aureus* and identification of a conjugative transposon (Tn*918*). J Bacteriol 162:1212–1220.

Clewell DB, An FY, Flannagan SE, Antiporta M, Dunny GM (2000): Enterococcal sex pheromone precursors are part of signal sequences for surface lipoproteins. Mol Microbiol 35:246–247.

Courvalin P (1994): Transfer of antibiotic resistance genes between Gram-positive and Gram-negative bacteria. Antimicrob Agents Chemother 38:1447–1451.

Cullum J, Broda P (1979): Chromosome transfer and Hfr formation by F in *rec*[+] and *recA* strains of *Escherichia coli* K-12. Plasmid 2:358–365.

Datta N, Barth PT (1976): Hfr formation by I pilus-determining plasmids in *Escherichia coli* K-12. J Bacteriol 125:811–817.

Davidson N, Deonier RC, Hu S, Ohtsubo E (1975): Electron microscope heteroduplex studies of sequence relations among plasmids of *Escherichia coli*: X. Deoxyribonucleic acid sequence organization of F and F-primes, and the sequences involved in Hfr formation. In Schlessinger D (ed): "Microbiology 1974". Washington, DC: ASM Press, pp 56–65.

Dean HF, Morgan AF (1983): Integration of R91-5 :: Tn*501* into the *Pseudomonas putida* PPN chromosome and genetic circularity of the chromosomal map. J Bacteriol 153:485–497.

Deonier RC, Mirels L (1977): Excision of F plasmid sequences by recombination at directly repeated insertion sequence 2 elements: involvement of *recA*. Proc Natl Acad Sci USA 74:3965–3969.

Disque-Kochem C, Dreiseikelmann B (1997): The cytoplasmic DNA-binding protein TraM binds to the inner membrane protein TraD in vitro. J Bacteriol 179:6133–6137.

Dunny GM, Clewell DB (1975): Transmissible toxin (hemolysin) plasmid in *Streptococcus faecalis* and its mobilization of a noninfectious drug resistance plasmid. J Bacteriol 124:784–790.

Dunny GM, Craig RA, Carron RL, Clewell DB (1979): Plasmid transfer in *Streptococcus faecalis*. Production of multiple sex pheromones by recipients. Plasmid 2:454–465.

Ehrenfeld EE, Kessler RE, Clewell DB (1986): Identification of pheromone-induced surface proteins in *Streptococcus faecalis* and evidence for a role for lipoteichoic acid in formation of mating aggregates. J Bacteriol 168:6–12.

Ehrenfeld EE, Clewell DB (1987): Transfer functions of the *Streptococcus faecalis* plasmid pAD1: Organization of plasmid DNA encoding response to sex pheromone. J Bacteriol 169:3473–3481.

Everett R, Willetts N (1980): Characterization of an *in vivo* system for nicking at the origin of conjugal DNA transfer of the sex factor F. J Mol Biol 136:129–150.

Everett R, Willetts N (1982): Cloning, mutation, and location of the F origin of conjugal transfer. EMBO J 1:747–753.

Feinstein SI, Low KB (1986): Hyper-recombining recipient strains in bacterial conjugation. Genetics 113:13–33.

Finlay BB, Frost LS, Paranchych W, Willetts NS (1986): Nucleotide sequences of five IncF plasmid *finP* alleles. J Bacteriol 167:754–757.

Firth N, Ippen-Ihler K, Skurray RA (1996): Structure and function of the F factor and mechanism of conjugation, p.2377–2401. In Neidhardt FC, Curtiss III R, Ingraham JL, Lin ECC, Low KB, Magasanik B, Reznikoff WS, Riley M, Schaechter M, Umbargar HE (eds): "*Escherichia coli* and *Salmonella*: Cellular and Molecular Biology". Washington, DC: ASM Press.

Flannagan SE, Zitzow LA, Su YA, Clewell DB (1994): Nucleotide sequence of the 18 kp conjugative transposon Tn*916* from *Enterococcus faecalis*. Plasmid 32:350–354.

Folkhard W, Leonard KR, Malsey S, Marvin DA, Dubochet J, Engel A, Achtman M, Helmuth R (1979): X-ray diffraction and electron microscope studies of the structure of bacterial F pili. J Mol Biol 130:145–160.

Franke AE, Dunny GM, Brown BL, An F, Oliver DR, Damle SP, Clewell DB (1978): Gene transfer in *Streptococcus faecalis*: Evidence for the mobilization of chromosomal determinants by transmissible plasmids, p. 45–47. In Schlessinger D (ed): "Microbiology 1978". Washington, DC: ASM Press, pp 45–47.

Franke AE, Clewell DB (1981): Evidence for a chromosome-borne resistance transposon (Tn*916*) in *Streptococcus faecalis* that is capable of "conjugal" transfer in the absence of a conjugative plasmid. J Bacteriol 145:494–502.

Frost LS, Paranchych W, Willetts NS (1984): DNA sequence of the F *traALE* region that includes the gene for F pilin. J Bacteriol 160:395–401.

Frost LS, Ippen-Ihler K, Skurray RA (1994): Analysis of the sequence and gene products or the transfer region of the F sex factor. Microbiol Rev 58:162–210.

Gaffney D, Skurray R, Willetts N (1983): Regulation of the F conjugation genes studied by hybridization and *tra-lacZ* fusion. J Mol Biol 168:103–122.

Gao Q, Luo Y, Deonier RC (1994): Initiation and termination of DNA transfer at F plasmid *oriT*. Mol Microbiol 11:449–458.

Gaudin HM, Silverman PM (1993): Contribution of promoter context and structure to regulated expression of the F plasmid *traY* promoter in *Escherichia coli* K-12. Mol Microbiol 8:335–342.

Ghetu AF, Gubbins MJ, Oikkawa K, Kay CM, Frost LS, Glover JNM (1999): The FinO repressor of bacterial conjugation contains two RNA binding regions. Biochem 38:14036–14044.

Goto N, Shoji A, Horiuchi S, Nakaya R (1984): Conduction of nonconjugative plasmids by F′ *lac* is not necessarily associated with transposition of the γ-δ sequence. J Bacteriol 159:590–596.

Grinter NJ (1981): Analysis of chromosome mobilization using hybrids between plasmid RP4 and a fragment of bacteriophage λ carrying IS*1*. Plasmid 5:267–276.

Guiney DG (1982): Host range of conjugation and replication functions of the *Escherichia coli* sex plasmid F*lac*: Comparison with the broad host range plasmid RK2. J Mol Biol 162:699–703.

Guiney DG (1984): Promiscuous transfer of drug resistance in Gram-negative bacteria. J Infect Dis 149:320–329.

Guiney DG, Helinski DR (1975): Association of protein with the 5′ terminus of the broken DNA strand in the relaxed complex of plasmid ColE1. J Biol Chem 250:8796–8803.

Guiney DG, Chikami G, Deiss C, Yakobson E (1985): The origin of plasmid DNA transfer during bacterial conjugation. In Helinski DR, Cohen SN, Clewell DB, Jackson DA, Hollaender A (eds): "Plasmids in Bacteria". New York: Plenum Press, pp 521–534.

Guyer MS (1978): The γ-δ sequence of F is an insertion sequence. J Mol Biol 125:233–247.

Guyer MS, Reed RR, Steitz JA, Low KB (1980): Identification of a sex-factor-affinity site in *E. coli* as γ-δ. Cold Spring Harbor Symp Quant Biol 45:135–140.

Haas D, Holloway BW (1978): Chromosome mobilization by the R plasmid R68.45: A tool in *Pseudomonas* genetics. Mol Gen Genet 158:229–237.

Hadley RG, Deonier RC (1979): Specificity in formation of type II F′ plasmids. J Bacteriol 139:961–976.

Hadley RG, Deonier RC (1980): Specificity in the formation of Δ*tra* F-prime plasmids. J Bacteriol 143:680–692.

Harrington LC, Rogerson AC (1990): The F pilus of *Escherichia coli* appears to support stable DNA transfer in the absence of wall-to-wall contact between cells. J Bacteriol 172:7263–7264.

Hayes W (1953): The mechanism of genetic recombination in *Escherichia coli*. Cold Spring Harbor Symp Quant Biol 18:75–93.

Heinemann JA, Sprague Jr GF (1989): Bacterial conjugative plasmids mobilize DNA transfer between bacteria and yeast. Nature 340:205–209.

Helmuth R, Achtman M (1978): Cell-cell interactions in conjugating *Escherichia coli*: Purification of F pili with biological activity. Proc Natl Acad Sci USA 75:1237–1241.

Holloway BW (1979): Plasmids that mobilize the bacterial chromosome. Plasmid 2:1–19.

Holloway BW (1986): Chromosome mobilization and genomic organization in *Pseudomonas*. In Gunsalus IC, Sokatch JR, Ornston LN (eds): "The Bacteria: A Treatise on Structure and Function", Vol 10. "The Biology of Pseudomonas". Orlando, FL: Academic Press, pp 217–249.

Holloway BW, Jennings PA (1958): An infectious fertility factor for *Pseudomonas aeruginosa*. Nature 181:855–856.

Holloway B, Low KB (1987): F-prime and R-prime factors. In Neidhardt FC, Ingraham JL, Low KB, Magasanik B, Schaechter M, Umbargar HE (eds): "*Escherichia coli* and *Salmonella typhimurium*: Cellular and Molecular Biology". Washington, DC: ASM Press, pp 1145–1153.

Holloway B, Low KB (1996): F-prime and R-prime factors. In Neidhardt FC, Curtiss III R, Ingraham JL, Lin ECC, Low KB, Magasanik B, Reznikoff WS, Riley M, Schaechter M, Umbargar HE (eds): "*Escherichia coli* and *Salmonella*: Cellular and Molecular Biology". Washington, DC: ASM Press, pp 2413–2420.

Hopwood DA, Harold RJ, Vivian A, Ferguson HM (1969): A new kind of fertility variant in *Streptomyces coelicolor*. Genetics 62:461–477.

Hopwood DA, Chater KF, Dowding JE, Vivian A (1973): Advances in *Streptomyces coelicolor* genetics. Bacteriol Rev 37:371–405.

Hopwood DA, Wright HM (1976): Genetic studies on SCP1-prime strains of *Streptomyces coelicolor* A3(2). J Gen Microbiol 95:107–120.

Hopwood DA, Kieser T, Wright HM, Bibb MJ (1983): Plasmids, recombination and chromosome mapping in *Streptomyces lividans* 66. J Gen Microbiol 129:2257–2269.

Hopwood DA, Lydiate DJ, Malpartida F, Wright HM (1985): Conjugative sex plasmids of *Streptomyces*. In Helinski DR, Cohen SN, Clewell DB, Jackson DA, Hollaender A (eds): "Plasmids in Bacteria". New York: Plenum Press, pp 615–634.

Hopwood DA, Kieser T (1993): Conjugative plasmids of *Streptomyces*. In Clewell DB (eds): "Bacterial Conjugation". New York: Plenum Press, pp 293–311.

Howard MT, Nelson WC, Matson SW (1995): Stepwise assembly of a relaxosome at the F plasmid origin of transfer. J Biol Chem 270:28381–28386.

Ike Y, Craig RC, White BA, Yagi Y, Clewell DB (1983): Modification of *Streptococcus faecalis* sex pheromones after acquisition of plasmid DNA. Proc Natl Acad Sci USA 80:5369–5373.

Ike Y, Clewell DB (1984): Genetic analysis of the pAD1 pheromone response in *Streptococcus faecalis*, using transposon Tn917 as an insertional mutagen. J Bacteriol 158:777–783.

Ippen-Ihler K, Minkley Jr EG (1986): The conjugation system of F, the fertility factor of *Escherichia coli*. Ann Rev Genet 20:593–624.

Jacob F, Wollman EL (1961): Sexuality and the genetics of bacteria. New York: Academic Press.

Jacobson A (1972): Role of F pili in the penetration of bacteriophage f1. J Virol 10:835–843.

Jaworski DD, Clewell DB (1995): A functional origin of transfer (*oriT*) on the conjugative transposon Tn916. J Bacteriol 177:6644–6651.

Jaworski DD, Flannagan SE, Clewell DB (1996): Analysis of *traA*, *int-Tn*, and *xis-Tn* mutations in the conjugative transposon Tn916. Plasmid 36:201–208

Johnson D, Everett R, Willetts N (1981): Cloning of F DNA fragments carrying the origin of transfer *oriT* and the fertility inhibition gene *finP*. J Mol Biol 153:187–202.

Kendall K, Cullum J (1986): Identification of a DNA sequence associated with plasmid integration in *Streptomyces coelicolor* A3(2). Mol Gen Genet 202:240–245.

Kendall KJ, Cohen SN (1987): Plasmid transfer in *Streptomyces lividans*: Identification of a *kil-kor* system associated with the transfer region of pIJ101. J Bacteriol 169:4177–4183.

Kieser T, Hopwood DA, Wright HM, Thompson CJ (1982): pIJ101, a multi-copy broad host-range *Streptomyces* plasmid: Functional analysis and development of DNA cloning vectors. Mol Gen Genet 185:223–238.

Kinashi H, Shimaji-Murayama M (1991): Physical characteristics of SCP1, a giant linear plasmid from *Streptomyces coelicolor*. J Bacteriol 173:5123–5129.

Kingsman A, Willetts N (1978): The requirements for conjugal DNA synthesis in the donor strain during F*lac* transfer. J Mol Biol 122:287–300.

Krishnapillai VN, Royle P, Lehrer J (1981): Insertions of the transposon Tn1 into the *Pseudomonas aeruginosa* chromosome. Genetics 97:495–511.

Lanka E, Lurz R, Furste JP (1983): Molecular cloning and mapping of *Sph*I restriction fragments of plasmid RP4. Plasmid 10:303–307.

Lederberg J, Tatum EL (1946): Gene recombination in *Escherichia coli*. Nature 158:558.

Lee SH, Frost LS, Paranchych W (1992): FinOP repression of the plasmid involves extension of the half-life of FinP antisense RNA by FinO. Mol Gen Genet 235:131–139.

Lloyd RG, Thomas A (1983): On the nature of the RecBC and RecF pathways of conjugal recombination in *Escherichia coli*. Mol Gen Genet 190:156–161.

Lloyd RG, Thomas A (1984): A molecular model for conjugational recombination in *Escherichia coli* K12. Mol Gen Genet 197:328–336.

Lloyd RG, Buckman C (1995): Conjugational recombination in *Escherichia coli*: Genetic analysis of recombinant formation in Hfr x F-crosses. Genetics 139:1123–1148.

Lovett MA, Guiney DG, Helinski DR (1974): Relaxation complexes of plasmids ColE1 and ColE2: Unique site of the nick in the open circular DNA of the relaxed complexes. Proc Natl Acad Sci USA 71:3854–3857.

Low B (1965): Low recombination frequency for markers very near the origin in conjugation in *E. coli*. Genet Res 6:469–473.

Low B (1968): Formation of merodiploids in matings with a class of Rec⁻ recipient strains of *Escherichia coli* K12. Proc Natl Acad Sci USA 60:160–167.

Low KB (1972): *Escherichia coli* K-12 F-prime factors, old and new. Bacteriol Rev 36:587–607.

Low B (1973): Rapid mapping of conditional and auxotrophic mutations in *Escherichia coli* K-12. J Bacteriol 113:798–812.

Low KB (1987): Mapping techniques and determination of chromosome size. In Neidhardt FC, Ingraham JL, Low KB, Magasanik B, Schaechter M, Umbargar HE (eds): "*Escherichia coli* and *Salmonella typhimurium*: Cellular and Molecular Biology". Washington, DC: ASM Press, pp 1184–1189.

Low KB (1991): Conjugational methods for mapping with Hfr and F-prime strains. Methods Enzymol 204:43–62.

Low B, Wood TH (1965): A quick and efficient method for interruption of bacterial conjugation. Genet Res 6:300–303.

Macrina FL (1993): Conjugal transfer in anaerobic bacteria. In Clewell DB (ed): "Bacterial Conjugation". New York: Plenum Press, pp 331–348.

Macrina FL, Archer GL (1993): Conjugation and broad host range plasmids in streptococci and staphylococci. In Clewell DB (ed): "Bacterial Conjugation". New York: Plenum Press, pp 313–329.

Mahajan SK, Datta AR (1979): Mechanisms of recombination by the RecBC and the RecF pathways following conjugation in *Escherichia coli*. Mol Gen Genet 169:67–78.

Maneewannakul K, Maneewannakul S, Ippen-Ihler K (1993): Synthesis of F pilin. J Bacteriol 175:1384–1391.

Manning PA, Morelli G, Achtman M (1981): *traG* protein of the F sex factor of *Escherichia coli* and its role in conjugation. Proc Natl Acad Sci USA 78:7487–7491.

Masai H, Arai K (1997): F*rpo*: A novel single-stranded DNA promoter for transcription and for primer RNA synthesis of DNA replication. Cell 89:897–907.

Matson SW, Morton BS (1991): *Escherichia coli* DNA helicase I catalyzes a site-and strand-specific nicking reaction at the F plasmid *oriT*. J Biol Chem 266:16232–16237.

Matson SW, Nelson WC, Morton BS (1993): Characterization of the reaction product of the oriT nicking reaction catalyzed by *Escherichia coli* DNA helicase I. J Bacteriol 175:2599–2606.

Matsubara K (1968): Properties of sex factor and related episomes isolated from purified *Escherichia coli* zygote cells. J Mol Biol 38:89–108.

Miller JH (1972): "Experiments in Molecular Genetics." Cold Spring Harbor, NY: Cold Spring Harbor Laboratory Press.

Minkley EG Jr, Willetts NS (1984): Overproduction, purification, and characterization of the F *traT* protein. Mol Gen Genet 196:225–235.

Moll A, Manning PA, Timmis K (1980): Plasmid-determined resistance to serum bactericidal activity: A major outer membrane protein, the *traT* gene product is responsible for plasmid-specified serum resistance in *Escherichia coli*. Infect Immun 28:359–367.

Moore D, Hamilton CM, Maneewannakul K, Mintz Y, Frost LS, Ippen-Ihler K (1993): The *Escherichia coli* K-12 F plasmid gene *traX* is required for acetylation of F pilin. J Bacteriol 175:1375–1383.

Moore RJ, Krishnapillai V (1982): Tn*7* and Tn*501* insertions into *Pseudomonas aeruginosa* plasmid R91-5: mapping of two transfer regions. J Bacteriol 149:276–283.

Mylroie JR, Friello DA, Siemens TV, Chakrabarty AM (1977): Mapping of *Pseudomonas putida* chromosomal genes with a recombinant sex factor plasmid. Mol Gen Genet 157:231–237.

Nelson WC, Howard MT, Sherman JA, Matson SW (1995): The *traY* gene product and Integration Host Factor stimulate *Escherichia coli* DNA helicase I-catalyzed nicking at the F plasmid *oriT*. J Biol Chem 270:28374–28380.

Norkin LC (1970): Marker-specific effects in genetic recombination. J Mol Biol 51:633–655.

Ohki M, Tomizawa J (1968): Asymetric transfer of DNA strands in bacterial conjugation. Cold Spring Harbor Symp Quant Biol 33:651–657.

Ou JT, Anderson TF (1970): Role of pili in bacterial conjugation. J Bacteriol 102:648–654.

Panicker MM, Minkley EG Jr (1985): DNA transfer occurs during a cell surface contact stage of F sex factor-mediated bacterial conjugation. J Bacteriol 162:584–590.

Pansegrau W, Lanka E, Barth PT, Figurski DH, Guiney DG, Haas D, Helinski DR, Schwab H, Stanisich VA, Thomas CM (1994): Complete nucleotide sequence of Birmingham IncPα plasmids: Compilation and comparative analysis. J Mol Biol 239:623–663.

Pardee AB, Jacob F, Monod J (1959): The genetic control and cytoplasmic expresssion of inducibility in the synthesis of β-galactosidase by *E. coli*. J Mol Biol 1:165–178.

Penfold SS, Simon J, Frost LS (1996): Regulation of the expression of the *traM* gene of the F sex factor of *Escherichia coli*. Mol Microbiol 20:549–558.

Pettis GS, Cohen SN (2000): Mutational analysis of the *tra* locus of the broad-host-range *Streptomyces* plasmid pIJ101. J Bacteriol 182:4500–4504.

Pittard J, Ramakrishnan T (1964): Gene transfer by F′ strains of *Escherichia coli*: IV. Effect of a chromosomal deletion on chromosome transfer. J Bacteriol 88:367–373.

Pittard J, Walker EM (1967): Conjugation in *Escherichia coli*: recombination events in terminal regions of transferred deoxyribonucleic acid. J Bacteriol 94:1656–1663.

Porter RD (1981): Enhanced recombination between F42*lac* and λ*plac*5: Dependence on F42*lac* fertility functions. Mol Gen Genet 184:355–358.

Porter RD (1982): Recombination properties of P1d*lac*. J Bacteriol 152:345–350.

Porter RD, McLaughlin T, Low KB (1978): Transduction versus "conjuduction": Evidence for multiple roles for exonuclease V in genetic recombination in *Escherichia coli*. Cold Spring Harbor Symp Quant Biol 43:1043–1047.

Porter RD, Welliver RA, Witkowski TA (1982): Specialized transduction with λ*plac*5: Dependence on *recB*. J Bacteriol 150:1485–1488.

Prangishvili D, Albers S-V, Holz I, Arnold HP, Stedman K, Klein T, Singh H, Hiort J, Schweier A, Kristjansson JK, Zillig W (1998): Conjugation in *Archaea*: Frequent occurrence of conjugative plasmids in *Sulfolobus*. Plasmid 40:190–202.

Ramirez-Arcos S, Fernandez-Herrero LA, Marin I, Berenguer J (1998): Anaerobic growth, a property horizontally transferred by an Hfr-like mechanism among extreme thermophiles. J Bacteriol 180:3137–3143.

Reeves P, Willetts N (1974): Plasmid specificity of the origin of transfer of sex factor F. J Bacteriol 120:125–130.

Reygers U, Wessel R, Muller H, Hoffman-Berling H (1991): Endonuclease activity of *Escherichia coli* DNA helicase I directed against the transfer origin of the F factor. EMBO J 10:2689–2694.

Rice LB (1998): Tn*916* family conjugative transposons and dissemination of antimicrobial resistance determinants. Antimicrob Agents Chemother 42:1871–1877.

Richter A (1957): Complementary determinants on an Hfr phenotype in *E. coli* K12. Genetics 42:391.

Royle PL, Matsumoto H, Holloway BW (1981): Genetic circularity of the *Pseudomonas aeruginosa* PAO chromosome. J Bacteriol 145:145–155.

Rupp WD, Ihler G (1968): Strand selection during bacterial mating. Cold Spring Harbor Symp Quant Biol 33:647–650.

Salyers AA, Shoemaker NB, Stevens AM, Li L-Y (1995): Conjugative transposons: An unusual and diverse set of integrated gene transfer elements. Microbiol Rev 59:579–590.

Sancar A, Wharton RP, Seltzer S, Kacinski BM, Clarke ND, Rupp WD (1981): Identification of the *uvrA* gene product. J Mol Biol 148:45–62.

Sanderson KE, Ross H, Zeigler L, Makela PH (1972): F$^+$, Hfr, and F′ strains of *Salmonella typhimurium* and *Salmonella abony*. Bacteriol Rev 36:608–637.

Sanderson KE, Kadam SK, MacLachlan PR (1983): Derepression of F factor function in *Salmonella typhimurium*. Can J Microbiol 29:1205–1212.

Sanderson KE (1996): F-mediated conjugation, F$^+$ strains, and Hfr strains of *Salmonella typhimurium* and *Salmonella abony*. In Neidhardt FC, Curtiss III R, Ingraham JL, Lin ECC, Low KB, Magasanik B, Reznikoff WS, Riley M, Schaechter M, Umbarger HE (eds): "*Escherichia coli* and *Salmonella*: Cellular and Molecular Biology". Washington, DC: ASM Press, pp 2406–2412.

Sarathy PV, Siddiqi O (1973): DNA synthesis during bacterial conjugation: II. Is DNA replication in the Hfr obligatory for chromosome transfer? J Mol Biol 78:443–451.

Scaife J, Pekhov AP (1964): Deletion of chromosomal markers in association with F-prime factor formation in *Escherichia coli* K12. Genet Res 5:495–498.

Schleper C, Holz I, Janekovic D, Murphy J, Zillig W (1995): A multicopy plasmid of the extremely thermophilic archaeon *Sulfolobus* effects its transfer to recipients by mating. J Bacteriol 177:4417–4426.

Schmidt KJ, Beck KE, Grogan DW (1999): UV stimulation of chromosomal marker exchange in *Sulfolobus acidocaldarius*: Implications for DNA repair, conjugation and homologous recombination at extremely high temperatures. Genetics 152:1407–1415.

Scott JR, Kirchman PA, Caparon MG (1988): An intermediate in transposition of the conjugative transposon Tn*916*. Proc Natl Acad Sci USA 85:4809–4813.

Scott JR, Bringel F, Marra D, Van Alstine G, Rudy CK (1994): Conjugative transposition of Tn*916*: Preferred targets and evidence for conjugative transfer of a single strand and for a double-stranded circular intermediate. Mol Microbiol 11:1099–1108.

Scott JR, Churchward GG (1995): Conjugative transposition. Annu Rev Microbiol 49:367–397.

Shoemaker NB, Smith MD, Guild WR (1980): DNA se-resistant transfer of chromosomal *cat* and *tet* insertions by filter mating in pneumococcus. Plasmid 3:80–87.

Seifert HS, Porter RD (1984): Enhanced recombination between λp*lac*5 and F42*lac*: Identification of *cis*- and *trans*-acting factors. Proc Natl Acad Sci USA 81:7500–7504.

Senghas E, Jones JM, Yamamoto M, Gawron-Burke C, Clewell DB (1988): Genetic organization of the bacterial conjugative transposon Tn*916*. J Bacteriol 170:245–249.

Siddiqi O, Fox MS (1973): Integration of donor DNA in bacterial conjugation. J Mol Biol 77:101–123.

Simons RW, Hughes KT, Nunn WD (1980): Regulation of fatty acid degradation in *Escherichia coli*: Dominance studies with strains merodiploid in gene *fadR*. J Bacteriol 143:726–730.

Smith MD, Shoemaker NB, Burdett V, Guild WR (1980): Transfer of plasmids by conjugation in *Streptococcus pneumoniae*. Plasmid 3:70–79.

Smith GR (1991): Conjugational recombination in *E. coli*: Myths and mechanisms. Cell 64:19–27.

Stedman KM, She Q, Phan H, Holz I, Singh H, Prangishvili D, Garrett R, Zillig W (2000): pING family of conjugative plasmids from the extremely thermophilic archaeon *Sulfolobus islandicus*: Insights in recombination and conjugation in *Crenarchaeaota*. J Bacteriol 182:7014–7020.

Stein DS, Kendall KJ, Cohen SN (1989): Identification and analysis of transcriptional regulatory signals for the *kil* and *kor* loci of *Streptomyces* plasmid pIJ101. J Bacteriol 171:5768–5775.

Thatte V, Iyer VN (1983): Cloning of a plasmid region specifying the N transfer system of conjugation in *E. coli*. Gene 21:227–236.

Thompson TL, Centola MD, Deonier RC (1989): Location of the nick at *oriT* of the F plasmid. J Mol Biol 207:505–512.

Timmons MS, Bogardus AM, Deonier RC (1983): Mapping of chromosomal IS*5* elements that mediate type II F-prime plasmid excision in *Escherichia coli* K-12. J Bacteriol 153:395–407.

Torres OR, Korman RZ, Zahler SA, Dunny GM (1991): The conjugative transposon Tn*925*: Enhancement of conjugal transfer by tetracycline in *Enterococcus faecalis* and mobilization of chromosomal genes in *Bacillus subtilis* and *E. faecalis*. Mol Gen Genet 225:395–400.

Traxler BA, Minkley EG Jr (1987): Revised genetic map of the distal end of the F transfer operon: Implications for DNA helicase I, nicking at *oriT*, and conjugal DNA transport. J Bacteriol 169:3251–3259.

Trieu-Cuot P, Carlier C, Martin P, Courvalin P (1987): Plasmid transfer by conjugation from *Escherichia coli* to gram-positive bacteria. FEMS Microbiol Lett 48:289–294.

Umeda M, Ohtsubo E (1989): Mapping of insertion elements IS*1*, IS*2* and IS*3* on the *Escherichia coli* K-12 chromosomes: Role of the insertion elements in formation of Hfrs and F' factors and in rearrangement of bacterial chromosomes. J Mol Biol 208:601–614.

van Biesen T, Frost LS (1994): The FinO protein of IncF plasmids binds FinP antisense RNA and its target, *traJ* mRNA, and promotes duplex formation. Mol Microbiol 14:427–436.

Verhoff C, DeHaan PG (1966): Genetic recombination in *Escherichia coli*: I. Relation between linkage of unselected markers and map distance. Mutation Res 3:101–110.

Vivian A (1971): Genetic control of fertility in *Streptomyces coelicolor* A3(2): Plasmid involvement in the interconversion of UF and IF strains. J Gen Microbiol 69:353–364.

Vivian A, Hopwood DA (1973): Genetic control of fertility in *Streptomyces coelicolor* A3(2): New kinds of donor strains. J Gen Microbiol 76:147–162.

Warren GJ, Twigg AJ, Sherratt DJ (1978): ColE1 plasmid mobility and relaxation complex. Nature 274:259–261.

Warren GJ, Saul MW, Sherratt DJ (1979): ColE1 plasmid mobility: Essential and conditional functions. Mol Gen Genet 170:103–107.

Wilkens B, Lanka E (1993): DNA processing and replication during plasmid transfer between Gram-negative bacteria. In Clewell DB (ed), "Bacterial Conjugation". New York: Plenum Press, pp 105–136.

Willetts N (1977): The transcriptional control of fertility in F-like plasmids. J Mol Biol 112:141–148.

Willetts N, Maule J (1979): Investigations of the F conjugation gene *traI*: *traI* mutants and λ*traI* transducing phages. Mol Gen Genet 169:325–336.

Willetts NS, Skurray R (1980): The conjugative system of F-like plasmids. Ann Rev Genet 14:41–76.

Willetts N, Crowther C (1981): Mobilization of the nonconjugative IncQ plasmid RSF1010. Genet Res 37:311–316.

Willetts NS, Crowther C, Holloway BW (1981): The insertion sequence IS*21* of R68.45 and the molecular basis for mobilization of the bacterial chromosome. Plasmid 6:30–52.

Willetts N, Skurray R (1987): Structure and function of the F factor and mechanism of conjugation. In Neidhardt FC, Ingraham JL, Low KB, Magasanik B, Schaechter M, Umbarger HE (eds), "*Escherichia coli* and *Salmonella typhimurium*: Cellular and Molecular Biology". Washington, DC: ASM press, pp 1110–1133.

Willetts NS, Wilkins B (1984): Processing of plasmid DNA during bacterial conjugation. Microbiol Rev 48:24–41.

Wolk CP, Vonshak A, Kehoe P, Elhai J (1984): Construction of shuttle vectors capable of conjugative transfer from *Escherichia coli* to nitrogen fixing filamentous cyanobacteria. Proc Natl Acad Sci USA 81:1561–1565.

Wollman E-L, Jacob F, Hayes W (1956): Conjugation and genetic recombination in *Escherichia coli* K-12. Cold Spring Harbor Symp Quant Biol 21:141–162.

Wood TH (1968): Effects of temperature, agitation, and donor strain on chromosome transfer in *Escherichia coli* K-12. J Bacteriol 96:2077–2084.

Yancey SD, Porter RD (1985): General recombination in *Escherichia coli* K-12: In vivo role of RecBC enzyme. J Bacteriol 162:29–34.

Yoshioka Y, Ohtsubo H, Ohtsubo E (1987): Repressor gene *finO* in plasmids R100 and F: Constitutive transfer of plasmid F is caused by insertion of IS*3* into F *finO*. J Bacteriol 169:619–623.

20

The Subcellular Entities a.k.a. Plasmids

MICHAEL H. PERLIN

Department of Biology, University of Louisville, Louisville, Kentucky 40292

I. INTRODUCTION

A striking discovery from the genetic examination of bacteria was that these microscopic organisms, originally ignored by geneticists, can actually exchange genetic information by cell-to-cell contact. Yes, bacteria have sex! This ability was discovered to depend on a relatively small "extra" chromosome, called F (for "fertility factor") in *Escherichia coli* (Jacob and Wollman, 1956). This sex factor primarily encodes the apparatus for the sexual process and the ability to transfer a copy of itself to cells that lack the factor. On the face of it, such a piece of DNA would seem little more than a promiscuous entity, providing no apparent benefit to its bacterial hosts (F^+ cells) save the uncertain pleasure of close contact with other bacterial cells (F^- cells). However, it was learned that recombination between the chromosome and F could allow F^+ cells to transfer other potentially useful genetic information to their bacterial cohorts. Furthermore Jacob and Wollman (1958) found that F can exist in alternative states in the cell—as an autonomously replicating molecule and as inserted into the bacterial chromosome. In this latter state it is possible to transfer large portions of the host chromosome to F^- bacteria (see Porter, this volume).

We will soon see that there are many varieties of DNAs that resemble the *E. coli* sex factor. The *E. coli* F factor can be transferred to and propagated in *Shigella* species. However, in this organism it does not integrate into the chromosome. In addition there are a number of autonomously replicating DNA molecules that are unable to promote sexual transfer. Despite such differences all these molecules belong to the same general class, called *plasmids* by Lederberg (1952).

Plasmids, that is to say, are extrachromosomal self-replicative heredity units. When existing autonomously in the cytoplasm, most plasmids take the form of double-stranded, closed circular DNAs. Plasmids, such as the F factor (discussed extensively by Porter, this volume), which can also integrate into the bacterial chromosome, are called *episomes*. Multiple copies of plasmids may be present within a single cell. A particular plasmid may contain as few as one gene or as many as several hundred. The bacterial chromosome is roughly 5×10^6 bp; plasmids range in size from as small as 1×10^3 bp to as large as one-third the size of the *E. coli* chromosome.

Interest in plasmids has increased rapidly since the early 1970s with the discovery of transposable elements and restriction endonucleases (see Gingeras, this volume). The ability to manipulate these relatively small self-replicative DNA molecules is the cornerstone for the widely growing recombinant DNA technology, used both for cloning and genetic modification of bacterial strains (see Geoghegan and Haller and DiChristina, this volume).

But are these previous descriptions of plasmids adequate? Is a plasmid just a piece of nucleic acid? Some plasmid features support their characterization as subcellular organisms. A plasmid may be considered as a replicon, meaning an autonomously replicating unit of nucleic acid; moreover the plasmid is a replicon that is independent of the cellular chromosome. Plasmids possess a number of characteristics that suggest a more autonomous nature (Novick, 1980). First of all, they control their own number of copies per host cell. Next, plasmids can in many cases provide for their own transfer between cells via conjugation; this form of transfer may even allow their exchange between different genera where chromosomal exchange is never observed (Roberts and Kenny, 1987). Finally, plasmids may transfer from one species to a competing species. This may provide a benefit to the competing species at the expense of the donor. Thus, if a plasmid were just a part of a cell, such exchanges would seem to run counter to the notion of evolution via natural selection. In fact, if all plasmids did were to provide for their own propagation, one might think of them as selfish, promiscuous DNA. But plasmids do provide useful genetic information for their host bacteria—genes for resistance to antibacterial agents such as drugs and heavy metals genes, for production of toxins to make the bearer more competitive, genes for attachment to intestinal mucosa, and genes for novel degradative pathways. Under appropriate conditions almost any chromosomal gene from the host may recombine onto a plasmid and thus be transferred. Plasmids can then be thought of as endosymbionts; they exist stably within bacteria and other microbes, and they often provide useful functions. They are dependent on the host cell for all life-support systems, except those functions related to plasmid autonomy.

A. Requirements for Autonomy: Replication

A first requirement for all plasmids is their ability to *replicate*. Replication itself is a process with certain general requirements for all DNA molecules. Usually this involves beginning at a specific point on the DNA, the origin, and proceeding in a linear fashion to another point, the terminus. An initiator substance is often found to act at the origin to trigger replication. Replication of all plasmids to date has been found to be semiconservative; that is each strand serves as a template for the synthesis of complementary strand. In addition three alternative inter-

mediates have been observed by electron microscopy (EM):

1. *The Cairns intermediate-circular form with two Y forks.* At least one fork is a replication site; the other is either a second replication fork (if replication is bidirectional) or the origin. This is the intermediate observed for P1.

2. *Cairns intermediate with supercoils.* It is the same as above, but the unreplicated origin remains supercoiled (e.g., ColE1, SV40, and mitochondrial DNA); nicking and closing occur as replication proceeds.

3. *Rolling circle* (Gilbert and Dressler, 1968). One strand is attached at the origin and is elongated using the other strand as a template; the circle may be supercoiled, and the daughter monomers are covalently linked in tandem. This type of replication is seen with γ phage lytic infection and with replication of F for conjugation.

Different plasmids have different requirements for host enzymes to aid in replication. For instance, chain growth of the F plasmid requires *E. coli* DNA polymerase III, whereas ColE1 uses DNA polymerase I for the initial stages of replication and polymerase III only for later stages. Some plasmids can use host products exclusively; ColE1 can be replicated by cell-free extracts from *E. coli* strains that were plasmid-free. Plasmids like F, which can be transferred from one cell to another via conjugation, have special requirements that are lacking in plasmids, which are never transferred via conjugation (for a review, see Willetts and Wilkins, 1984; Ingraham et al., 1987). Some of those requirements are directly related to transfer and are discussed for the F plasmid elsewhere (see Porter, this volume). DNA synthesis for conjugation involves transfer of a single strand. This poses certain unique tasks for replication. The most prominent of these are (1) synthesis of a replacement strand in the donor cell, (2) complementary strand synthesis in the recipient, and (3) circularization in the recipient.

Much of the information on these processes has come from F plasmid. This plasmid promotes efficient conjugation in liquid culture. DNA synthesis for conjugation requires the formation of a 3'-OH primer end and an elongation reaction involving a DNA polymerase. There are also participating enzymes in *E. coli*. These include the products of the *dnaB* (a prepriming protein), *dnaE* (encoding the catalytically active subunit of DNA polymerase III), and *dnaG* (primase) genes. Strains with temperature-sensitive mutations in these genes cease vegetative synthesis at the restrictive temperature (Willetts and Wilkins, 1984).

Some plasmids, although they are nonconjugative, are mobilized by certain conjugative plasmids. The mobilized plasmids are small (<10 kb), naturally occurring, and include such plasmids as ColE1, CloDF13, and the broad host range IncQ plasmid RSF1010. A site for recognition by a protein factor n' (involved in the *E. coli* prepriming complex) has been detected on both strands of ColE1 (Nomura et al., 1982) and of pBR322 (Marians et al., 1982). Short DNA segments associated with initiation of primosome-dependent DNA synthesis have been characterized for a number of these plasmids. DNA sequence comparisons of these plasmids revealed that inverted repeat structures may be part of the process of recognition by the n' protein and there may be a consensus sequence of 5'-AAGCGG-3' for these sites (Van der Ende et al., 1983).

Although linear plasmids have been described for fungi (Blackburn, 1985; Tudzynski and Esser, 1985), prokaryotes were thought to contain only circular plasmids. Linear DNA molecules pose a unique problem for replication, since free ends are highly recombinogenic (Blackburn, 1985) and such molecules may thereby be lost due to recombination with other cellular replicons. This would make plasmids highly unstable due to a high probability for recombination with the chromosome. Later we discuss linear plasmids and some aspects of how they are maintained despite these apparent problems.

B. Further Requirements for Autonomy

1. Plasmid replication control

Plasmids, in order to ensure their independence from the host chromosome, must encode the means necessary to control their copy number within the host cell. This control might take several forms. All of the plasmids that have been examined to date use *negative control* either directly or indirectly to regulate their replication. The mechanisms for such control can be divided into two major categories. The first control mechanism is called the *inhibitor-target* mechanism (Novick, 1987). For this mechanism a small countertranscript RNA molecule is used as the main inhibitor to control initiation of replication. The targets for these countertranscript RNAs are RNAs transcribed from the opposite strand which either (1) code for a Rep protein needed to initiate replication or (2) serve as a primer or primer precursor to initiate replication. The RNAs base pair due to the complementarity of the unpaired loops formed by secondary structure folding in both RNAs; this RNA-RNA duplex formation is what inhibits replication. This type of control is found in small plasmids such as ColE1, staphylococcal plasmids like pT181, and large F-like conjugative plasmids.

The second category of control is the *iteron-binding* mechanism. In this case a series of direct DNA repeats, each about 20 bp long, serve as the main replication inhibitor (and incompatibility determinant, see below). Plasmids under this type of control have these repeats (iterons) near their origins of replication and near a *rep* gene (encoding a positive factor, Rep, needed for initiation of plasmid replication). The iterons are supposed to control replication by competing for binding of the *trans*-acting Rep protein. This type of control is found in such replicons as F, P1, R6K, and pSC101.

The outcome of control of the rate of initiation of replication is that each plasmid has its own characteristic number of copies per host cell. Some plasmids may be present in one to two copies per host chromosome, while others may be found in 10 to 100 copies per chromosome. Replication is allowed to initiate when the negative control element is diluted out by cell growth in preparation for cell division.

2. Conservation of space

Space plays several roles for the plasmid molecule. For one, the plasmid must balance the size it requires to encode all its functions for maintenance within the host cell with the burden that plasmid size places on the host. Second, interconnected control regions on the plasmid must be kept close together in order to ensure that they are coinherited.

3. Incompatibility

Different closely related plasmids usually are not able to be stably maintained in the progeny of a single cell. This is what we call *incompatibility*. Two plasmids are said to be in different *incompatibility groups* if they can stably reside within the same cell. A thorough review of the theoretical aspects of plasmid incompatibility and its contents has been provided by Novick (1987). If two plasmids are so similar at an essential stage of maintenance that they cannot be distinguished from one another, then one or the other will be lost when the cell divides. For example, if two plasmids have identical or nearly identical negative control elements (repressors), then the repressor of one could regulate expression of the other, and vice versa. The choice of which plasmid will be allowed to replicate is random, so that as the DNA in the cell doubles in preparation for cell division, just prior to division the cell *could* have a ratio of 3 : 1 for one of the plasmids to the other. After cell division, one of the plasmids would then be lost to one of the daughter cells. In the case of multicopy plasmids a similar phenomenon occurs, but in this case the *distribution* of plasmid copies into daughter cells is also random. A comparison of plasmid distribution for plasmids showing segrega-

tional incompatibility as opposed to that for two compatible plasmids is shown in Figure 1.

C. Requirements for Autonomy: Partitioning

Another requirement for plasmid autonomy is *partitioning*—the assurance that each daughter cell gets a copy of the plasmid. In theory, each autonomous replicon could be attached to the cell membrane for distribution near the place where the cell will divide (Jacob et al., 1963). Growth of a septum along the equatorial plane would allow for proper distribution. This is in fact the case for a number of replicons including the *E. coli* chromosome (see Firshein, this volume). Alternatively, plasmids may be distributed to the daughter cells by random diffusion, as appears to be the case for the high copy number plasmids, such as pBR322.

II. EXAMINATION OF SPECIFIC PROKARYOTIC PLASMIDS

Generally, plasmids are classified according to their incompatibility group or according to the genetic information specified by their DNA. The F factor, as described by Porter (this volume), is a fertility factor and contains genes necessary for sex pilus formation, transfer, and replication. Other plasmids contain genes conferring resistance to antibiotics (*R plasmids*), heavy metals, and ultraviolet light, and leading to the production of toxins (*Col plasmids*).

In the following sections, the mechanisms of autonomy are described for a variety of non-F plasmids. These sections are divided according to (1) modes of replication, copy number control, incompatibility, (2) modes of partition and (3) sex-pheromone plasmids of Gram-positive bacteria. The discussion will point out peculiarities of these plasmids, including whether they are limited to Gram-negative or Gram-positive bacteria and aspects of their characteristic numbers per cell that might influence their strategies for survival.

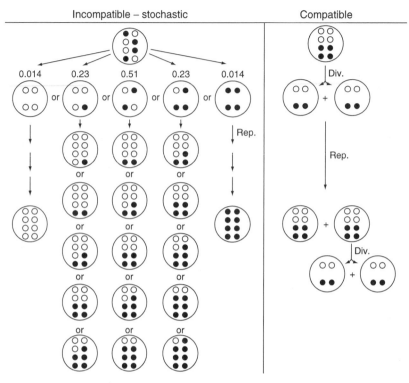

Fig. I. Segregational incompatibility. Black-and-white symbols represent plasmids with different selectable genetic traits that replicate in a 4–8–4 cycle. In the left half, the two plasmids have similar control mechanisms and show segregational incompatibility. At the top, the various possible combinations that can arise by random assortment with equipartitioning are modeled; the numbers represent the binomial probabilities for each combination. Beneath the partitioning diagram are illustrated the possible combinations that can result from random selection for replication. In the right half, the two plasmids have different control mechanisms and are compatible. These models apply to incompatibility for multicopy plasmids. (Reproduced from Novick, 1987, with permission of the publisher and the author.)

A. Replication Control

1. Control that employs a countertranscript

IncFII Plasmids. The IncFII plasmids are all self-transmissible (i.e., conjugative) plasmids. These plasmids are found in the Enterobacteriaceae only. All that have been studied so far maintain a low copy number (ca. 2) relative to the chromosome, and they are all nearly 100 kb. These plasmids require certain *E. coli* (i.e., not plasmid encoded) products for replication. Among these are the products of the *dnaB, C, E, F*, and *G* genes (Rownd, 1978); they do not require the product of the *polA* gene (DNA polymerase I) (Kolleck et al., 1978). Originally these

plasmids were thought not to require the product of the *dnaA* gene (Goebel, 1974; Frey et al., 1979), since they replicate in DnaA mutants; however, since DnaA temperature-sensitive mutants are "leaky," it was thought that except for the ColE1 plasmids, all other *E. coli* plasmids, including the IncFII plasmids, require *dnaA* product (T. Atlung, personal communication). Nevertheless, with the use of *dnaA* null mutants that contain no detectable DnaA protein, Tang et al. (1989) have shown that DnaA protein is not essential for replication of R100. From EM analyses of heteroduplexes between them, it was determined that three members of this group share a great deal of homology: R-1, R6–5, and R100. The replication of these three has been studied in detail, so they will serve as the basis for the discussion of replication which follows. These plasmids are too large to study normally so that in vitro techniques have been employed to reduce their sizes. The *origins* of the miniplasmids (Syneki et al., 1979) have been determined and observed by EM and replication was found to be unidirectional from a unique origin. For R100 it was determined that the miniplasmid origin is the one used in vivo; this was determined by a variety of methods, including autoradiography and EM (Silver et al., 1977). Also, when R100 is transferred to *Proteus mirabilis* there are other origins which are activated.

Replication control: What is the evidence from which an understanding of IncFll replication can be derived? Replication of IncFll plasmids is under negative control. What types of mutants might we expect if replication were under negative control? We might expect *conditional* (e.g., temperature-sensitive) or *trans-recessive* mutations characteristic of an altered regulatory molecule; also we should see *cis-dominant* mutations associated with an altered target for regulatory action. Actually these are the types of increased copy number mutants which have been isolated (Nordström et al., 1972; Gustafsson and Nordström, 1978; Molin et al., 1981). Two loci were mapped for these mutations: *copA* and *copB*. The promoters in these regions were characterized by RNA protection and binding experiments to restriction fragments from R6–5 and R100 (Easton et al., 1981; Lurz et al., 1981).

Those mutations that mapped to *copB* had no effect on incompatibility. Rather, *copB* mutants show a greater than 10-fold increase in mini R1 copy number. These mutants can be complemented in *trans* by introducing a wild-type R1 plasmid into the same cell with a mini R1 copy number mutant. The result is a return to normal (low) copy number. The *copB* mutants still express R1 incompatibility, so whatever the *copB* product is, it is not essential for incompatibility (Molin et al.,

1981). R1 *copB* mutants are not complemented by wild-type R100, but there is an analogous gene in R100 that *is* complemented in *trans* by wild-type R100. The product of this gene, RepA2, would be basic (as predicted from the DNA sequence) and as such would be expected to bind to DNA. This protein would have a difference in sequence from that of an analogous CopB protein from R1 (Rosen et al., 1980); since these two plasmids are mutually incompatible, this means that the CopB protein (or RepA2) is *not* involved in incompatibility.

What are the relative roles of *copB* compared with *copA*? Addition of extra *copB* product has no observable effect on wild-type plasmid. In fact R1 copy number is not affected when we place *copB* into the same cell on a different plasmid. Also, if we use wild-type R1 to integratively suppress a cell (see Appendix B, Integrative Suppression), replication is not shut down when we introduce the plasmid bearing *copB*. No effect on incompatibility would be expected if the *copB* target is already saturated in the wild-type state. This appears to be the case. The expression of *copB* (as measured by assaying for β-galactosidase activity from *copB-lacZ* fusions) is not regulated by other plasmid-encoded factors functioning in *trans*. The expression of *copB* is growth-rate and gene-dosage dependent. That would seem to suggest that *copB* plays no significant role in FII plasmid replication under normal conditions for this group of plasmids (Light and Molin, 1982).

What about the role of *copA*? The *copA* product controls incompatibility of the IncFII plasmids. It does so by specific interaction with the *copB* transcript. The product of *copA* is a countertranscript for a portion of *copB*. The function of *copA* has been analyzed using *repA1-lacZ* (=*copB-lacZ* for R1) fusions (and assaying for β-galactosidase activity). Such studies have shown that *repA1* expression is inhibited by *copA* product (Light and Molin, 1981). However, this is *not* transcriptional inhibition; why? In one *copA* mutant of R100, *repA1* transcription is the same as in wild-type; but there is an increase in *repA*1 gene expression from this mutant plasmid and an increase in copy number (Miki et al., 1980). Therefore the *copA* product regulates *repA*1 expression after transcription has already taken place.

How could the *copA* product affect *repA*1 repression posttranscriptionally? What is the target? The target for the *copA* product (RNA I; see Fig. 2) is in a countertranscript (i.e., the *copB* transcript) made from the same region of DNA. Mutations in IncFII plasmids that reduce incompatibility with wild-type plasmids are located in the *copA* gene and affect interaction of RNA I with target (CopT RNA) (Miki et al., 1980). But two different plasmids with the same target-type mutation are mutually incompatible; that is to say, they have an alteration of both the *copA* product *and* its target! Thus the target for the *copA* product must be within the coding region for *copA* (Miki et al., 1980). A point mutation can affect both the *copA* RNA and its target at the same time: using translational RepA1–LacZ fusions, one can assay for β-galactosidase activity from wild-type and mutant *copA* genes in the presence of extra copies of the respective *copA* gene in *trans* (Light and Molin, 1982). When the *copT* DNA is present in *trans* on a high copy number plasmid, the copy increases. Why? Because of the increase in

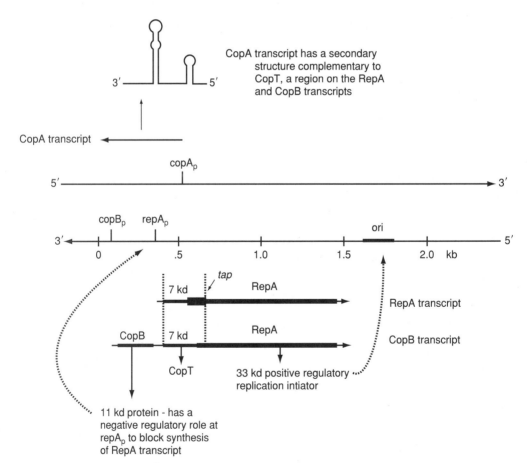

Fig. 2. Plasmid R1 control circuits for replication. Diagram shows transcription, translation, and regulatory loops.

target sites that compete with the test plasmid's RNA I binding site (Danbara et al., 1980). It turns out that the *copT* region is located *within* the *copA* gene, 1.6 kb upstream from *ori*. Is the RNA I target then its own DNA template? No. The extent to which RNA I interacts depends on the amount of transcription through the *copT* DNA proceeding toward the *repA*1 gene (Light and Molin, 1983). Thus it would appear that RNA I binds directly to the *repA*1 mRNA rather than to the DNA itself. The *copA* product does *not* block transcription from the *repA* gene; *copA* product provided in *trans* has very little effect on *lac* expression from transcriptional fusions of *lacZ* to the *repA*1 promoter (*repA_p*) (Light and Molin, 1983). But *copA* strongly inhibits production of RepA-LacZ fusion products (Light and Molin, 1982). Thus *copA* product interferes with translation of the *repA*1 mRNA.

How could such an interference come about? Well, there is currently no evidence that the interaction between CopA and CopT introduces a change in secondary structure that hides the ribosome binding site for *repA*. Unlike the case with other plasmids (see pT181-like plasmids, below), no compelling data have been presented that would suggest that the interaction causes premature termination of transcription of *repA*, à la attenuation in the *trp* operon. There is evidence that the CopA/CopT interaction could help target the *repA* transcript for decay. Complete duplex formation between these two RNAs is a substrate for digestion by the cellular RNase III; however, cleavage by this enzyme is not an absolute requirement for inhibition of *repA* expression (Wagner et al., 1992). The region that spans the CopT target encodes a 7 kDa protein that is expressed at low levels. Translation of this protein was originally proposed as necessary for expression of *repA* by some sort of translational coupling mechanism, or translation of the message for the 7 kDa protein might control the interaction between *copA* and *copT* RNAs (Wagner and Nordström, 1986). Yet, a nonsense mutation

in the 7 kDa reading frame did not prevent expression of RepA. Thus this ORF is not likely involved in translational coupling.

Instead, a second ORF, located between the *copA* and *repA* reading frames, was found that coded for a 24 amino acid peptide, Tap (translational activator peptide; Blomberg et al., 1992). The Shine-Delgarno (SD, or ribosome binding) region for *tap* was only two nucleotides downstream of the region where CopA and CopT interact, and its stop codon was two nucleotides after the GUG start codon for the *repA* gene. Synthesis of Tap was inhibited by CopA. Moreover translation of the *tap* gene was required for expression of *repA* (Blomberg et al., 1994). Thus the antisense RNA, CopA, controlled expression of *repA* indirectly, by regulating translation of *tap*. The question was, Why can't *repA* be translated directly?

Interestingly, if one compares IncFII plasmids, one finds that they each carry a highly conserved inverted repeat that, in the transcribed RNA, should fold to form a stem-loop structure that hides both the GUG start codon and the Shine-Delgarno region from the ribosomes. The predicted lollipop structures are almost identical between the different plasmids, even though there are a large number of differences in sequence between them. It is expected that translation of *tap* is required to make available the nearby ribosome binding site for *repA*. Blomberg et al. (1994) have provided compelling evidence that supports this model. First, they demonstrated that when *tap* cannot be translated (due to an introduced premature stop codon), disruption of the proposed stem-loop by a second mutation allows translation of *repA*. However, in this instance expression was approximately 20-fold lower than wild-type, and these authors attributed the difference to greater efficiency in utilizing the SD sequence when translation of *tap* and *repA* are coupled. In other instances where translational coupling has been seen, ribosomes terminating one reading frame apparently increase initiation at the next SD region because of a ribosome loading effect (de Smit and van Duin, 1990). A similar mutation leading to independent *repA1* expression was seen for plasmid NR1 (Wu et al., 1993), where the region corresponding to *copA* and to the leader peptide gene was identical to that found in plasmid R1. Blomberg et al. (1994) also found that for R1, mutations that more optimized the spacing between the SD region and the start codon for *repA* resulted in additional increased expression of that gene.

Summary of IncFII replication control and incompatibility. A diagram of the locus involved in plasmid R1 replication and copy number control is shown in Figure 2. Table 1 serves as a dictionary to translate the R1 designations into the comparable genes, gene products, and sites for R100. In general, the R1 terminology will be used unless facts were specifically obtained with R100. The *repA* gene encodes a positive regulatory protein whose function is rate-limiting for R1 replication and that works only in *cis*. This gene is transcribed from two promoters, $repA_p$ and $copB_p$. But the CopB protein (from the $copBp$ transcript) negatively regu-

lates $RepA_p$ transcription. Therefore, under normal circumstances in the cell, RepA is made almost exclusively from the CopB transcript. So, you say, why have transcription from $repA_p$ at all? Possibly it is so that when this plasmid is transferred to another cell during conjugation, there can be an overshoot in replication to ensure that the plasmid is maintained at the proper level in the new cell (see below) (Gerdes et al., 1986b).

There is a second mechanism that is normally more important for copy number control. The *copA* gene encodes a 90 nucleotide RNA that blocks translation of the RepA mRNA (or CopB mRNA at the *repA* gene). How? The CopA RNA has a high degree of secondary structure that contains two stems and loops. These interact with a region of the CopB or RepA RNAs that is referred to as CopT. This CopT region has secondary structure such that base pairing can occur between its loops and those of the CopA RNA. In addition a gene for a short protein, Tap, exists just prior to the *repA* gene. Translation of this part of the RepA or

TABLE 1.   Translation of Designations Used in R1 and R100 Replication Control Systems

R1	R100	Function
RepA	RepA1	Positive regulatory protein for initiation of replication
CopB	RepA2	Negative regulatory protein that blocks RepA transcription
$repA_p$	P_A	Promoter of *repA*
$copB_p$	P_C	Promoter of *copB*
$copA_p$	P_E	Promoter of *copA*
RepA RNA	RNA-A	RNA that encodes positive regulator for initiation
CopB RNA	RNA-CX	RNA that encodes positive regulator and its negative regulator
RNA for only *copB*	RNA-C	(A countertranscript) 90 nucleotide RNA
gene CopA RNA	RNA-E	that blocks translation of the *repA* gene by base-pairing with its RNA[a]
tap		

[a] Translation of this region of RNA changes RepA mRNA to make available an otherwise "hidden" ribosome binding site for the *repA* gene. This also improves the efficiency of *repA* expression via translational coupling.

CopB transcripts helps *repA* expression by translational coupling. CopA/CopT interaction interferes with this positive control.

Why have two types of control for replication? The function of *copB* is to make the switch between low and high copy number replication; *copA* is to maintain a constant copy number in either mode. These differences are analogous to the transcription of the cI gene of λ initially from P_{re} and subsequently from P_{rm}, once lysogeny has been established (see Hendrix, this volume).

Col Plasmids. The ColE1 plasmid produces colicin E1 (see Appendix A). This plasmid and related plasmids are small and found in many copies per chromosome. They replicate in the absence of de novo protein synthesis (Staudenbauer, 1978). Chloramphenicol (CM) or other protein synthesis inhibitors cause chromosomal replication to stop, although ColE1 replication continues. This fact allows the amplification of the ColE1 plasmid up to 50-fold relative to the chromosome (Donoghue and Sharp, 1978). Why is this possible? Because ColE1 does not require plasmid-encoded proteins for replication in vivo, the necessary host proteins are stable and available long after host chromosomal replication has stopped. This feature of the ColE1-derived plasmids makes them useful for purposes of cloning.

Using a derivative of ColE1 (mini ColE1) as a template. DNA replication was shown to start at any of three consecutive bases which we term the origin (*ori)* (Itoh and Tomizawa, 1980). In order for *initiation* of ColE1 replication to occur, transcription is required. Transcription of an RNA preprimer begins about 500 bp upstream of the replication *ori*. Initiation requires that the RNA preprimer hybridize to the DNA template in the origin region (Itoh and Tomizawa, 1980). The DNA-RNA hybrid is recognized by RNase H, which cleaves the preprimer to yield the primer (to which the dNTPs will be added).

This system is under negative control, as follows: RNA I is an antisense RNA or *countertranscript* (ca. 100 nucleotides long)

(Oka et al., 1979) that is transcribed in the opposite direction from the RNA primer. This RNA has been shown genetically to be involved in copy number control (Itoh and Tomizawa, 1980). When the RNA I hybridizes with the preprimer, the preprimer cannot be used to form the primer since this interaction blocks RNase H (see Figure 3). In addition to this mechanism of control, the *rop* gene (repressor of primer (Twigg and Sherratt, 1980); also know as *rom*, RNA on modulator) encodes a 63 amino acid polypeptide that inhibits ColE1 replication. It is an RNA binding protein which exists as a dimer in solution and can accelerate the association of RNA I with certain types of primer precursors (Polisky, 1988). It accelerates RNA I interaction with primers shorter than 85 nucleotides and longer than 135 nucleotides; it can inhibit interactions with intermediate-size primers. These characteristics of the Rop suggest its sensitivity to RNA conformation. In vitro, Rop stabilizes interactions between the preprimer and RNA I (Li et al., 1997). Although the three-dimensional structure of the Rop protein has been solved by X-ray crystallography, it is not yet clear how its structure contributes to its highly specific biochemical properties.

What is the evidence which supports this picture of ColE1 replication? ColE1 replication requires the following host functions: DNA polymerase I (Kingsbury and Helinski, 1973), RNA polymerase (Staudenbauer, 1975), DNA polymerase III (Staudenbauer, 1976), the products of the *dnaB, C, G,* and *Z* genes. These requirements are shared by other ColE1-like plasmids (e.g., RSF 1030, pMB1, CloDF13, which encodes a cloacin, an antibacterial toxin; see Appendix A). We know that *regulation occurs at initiation* since increased copy mutants of CloDF13 do not replicate any faster than wild-type under the same conditions (Veltkamp and Nijkamp, 1976). The involvement of RNA I in negative control is implied from the fact that the sequences of RNA I and the preprimer can each form three stem-and-loops (Lacatena and Cesareni, 1983), and there is potential

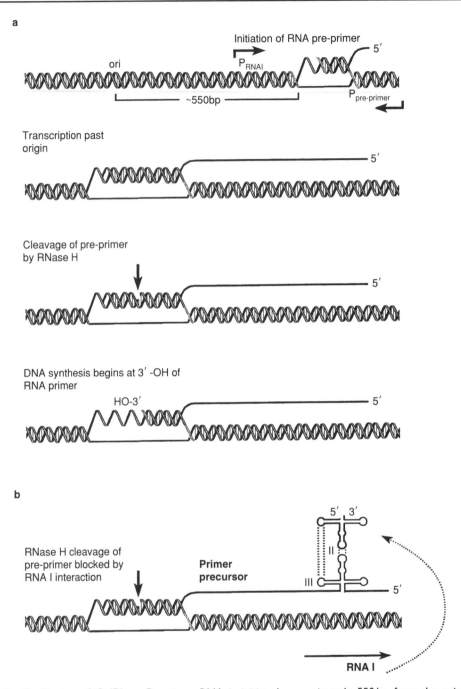

Fig. 3. Replication of ColEI. **a:** Preprimer RNA is initiated approximately 550 bp from the origin of replication (*ori*). Transcription proceeds past *ori*, where the preprimer is cleaved by RNase H into the primer used to initiate DNA replication. Replication begins at the 3'-OH end of the primer. **b:** RNAI is synthesized from the opposite DNA strand as the preprimer. The RNAI secondary structure allows it to base pair with the preprimer, which interferes with RNase H cleavage and thereby prevents initiation of replication.

for base pairing between them. High copy number mutants have demonstrated that these structures are important for processing the primer and for RNA I-preprimer interaction. The RNA I-preprimer complex is sensitive (since it is double stranded) to RNase III (Tomizawa, 1984).

This RNA I-preprimer interaction also seems to be involved in incompatibility. If one blocks interaction of the preprimer with the DNA in vitro, primer formation can be prevented, probably by preventing cleavage of the preprimer by RNase H. RSF1030 is compatible with ColE1 and has a different sequence in the primer-RNA I region; however, there is analogous folding of the RNAs, indicating a similar form of regulation.

Lacatena and Cesareni (1983) have found that the actual sequence in loop II (see Fig. 3) is not important, but the interaction in this area determines that incompatibility for the plasmid. In analyzing a large number of mutants they examined the RNA I-preprimer interaction. They found that point mutations in the target (preprimer) produce inhibitors that are unable to interact with the wild-type target (Lacatana and Cesareni, 1983). Two types of target mutants were found. One type had a mutant inhibitor which can only interact with the mutant target; this type of mutant predominated. These target mutations defined new incompatibility groups, and they were therefore RNA I mutations. These mutations helped show that changes in loops I and II of RNA I alter the incompatibility; however, these changes do not prevent RNA I from functioning as an inhibitor. Isolated mutations of the second type were rarer. In these mutants the RNA I did not interact with either the wild-type or its own mutant target. These were mutations in the stability of the secondary structure and destroyed loops I and II. For these mutants the copy number increased at least 2.5-fold. Temperature-sensitive copy number mutants of ColE1 were examined by fusing the promoters of the mutant RNA I or preprimer genes to the galactokinase gene; such muta-

tions had no effect on galactokinase production, implying that the effect seen was due not to transcriptional changes but to a change in the effect of RNA I on preprimer secondary structure. Further evidence for inhibition by RNA I is that if one provides additional copies of RNA I genes the copy number of the plasmid decreases proportionately (Wong et al., 1982).

The other mode of negative control which has been characterized (Twigg and Sherratt, 1980) is the *rop* locus, which is distant from *ori*. Mutants that map there have increased copy number. The mutants are recessive and can be complemented by a compatible plasmid ColK. Therefore the *rop* mutation does not affect incompatibility. As mentioned earlier the Rop protein appears to be a repressor of the primer (Twigg and Sherratt, 1980; Polisky, 1988), possibly by preventing formation of the necessary preprimer precursor. In transcriptional fusions of RNA $I_p lacZ$, β-galactosidase is not affected by ColE1 or pMB1 (ColE1-like). On the other hand, in fusions of preprimer$_p$-*lacZ*, β-galactosidase *is* repressed by either plasmid. Therefore Rop is the same for both plasmids. There are other results which suggest that Rop may play a role that affects the normal termination of transcription in *E. coli*. In *E. coli*, normal termination involves a protein called Rho that binds to nascent RNAs and interacts with paused RNA polymerase to stop transcription. There are *rho* mutants in which plasmids like ColE1 and its derivatives are unable to reside. If one tries to transform *rho* mutant cells with these plasmids, no transformants appear. Moreover residence of such plasmids inhibits the growth of such mutant cells. In one particular mutant, *rho-15*, this has been shown specifically to be due to expression of Rop. This appears to be due to some, as yet unexplained, effect that Rop has on normal transcription termination, and this effect is exacerbated in *rho-15* mutants. In fact, if one deletes the *rop* gene, ColE1-derivatives are again able to transform the *E. coli rho-15* mutants (Li et al., 1997).

Conjugative transfer of Col plasmids. There are models to explain DNA synthesis for conjugation-mediated transfer of ColE1 and other nonconjugative plasmids. These methods are described in detail by Porter (this volume).

Synthesis of the complementary strand in the recipient. Two events are necessary in the recipient after transfer of the single strand from the donor: synthesis of a complementary strand and circularization. For the nonconjugative plasmids priming is most likely to use the priming machinery of the recipient cell since the ColE1 H strand, which is the strand transferred during conjugation (Warren et al., 1978), contains the *rrlb* locus. This locus promotes rifampin-resistant initiation of DNA synthesis by the primosome (Nomura et al., 1982), is recognized by the n' protein (for primosome assembly), and contains *nic*. This locus is nonessential for vegetative DNA replication (i.e., normal replication of the plasmid in its host cell). The same seems to be true for pBR322 (Zipursky and Marians, 1981). The mechanism of circularization is, at present, unknown. One possibility is that circularization of the donated linear strands follows a mechanism similar to that for the single-stranded phage, φX174 (Eisenberg et al., 1976; also LeClerc, this volume).

pT181-like Plasmids. pT181 is part of a large group of plasmids originally discovered in *Staphylococcus aureus* for which rolling circle replication is used (for a review of rolling circle replication of plasmids, see Khan, 1997). The mechanism employs a Rep initiator protein that creates a site-specific nick on the leading strand of the plasmid (double-strand) origin. This nick is extended by DNA polymerase from the free 3'-OH. The lagging strand is synthesized using host enzymes and factors and begins at a separate origin only "visible" once exposed in its single-strand form. Since their original discovery in *S. aureus*, these types of plasmids have also been found in many other Gram-positive bacteria (e.g., *Bacillus, Clostridium, Streptomyces*), and more recently in

Gram-negatives, other prokaryotes such as *Treponema denticola* (spirochete), *Mycoplasma mycoides*, and even in the hyperthermophilic bacterium, *Thermatoga* (Yu and Noll, 1997), and the archaeon, *Pyrococcus abyssi* (Erauso et al., 1996). With the large amount of data available now about them, the rolling circle plasmids can be grouped based on homologies in leading strand origins and Rep proteins, biochemical properties of such proteins, functional domains of the double-strand origin, the mechanism of lagging-strand synthesis, and the ways in which replication and copy number are controlled.

Replication control for pT181-like plasmids. Work with a family of *Staphylococcus aureus* plasmids has shown that there may be more involved with plasmid stability than just replication control and partition (Novick et al., 1986). There may be other plasmid loci involved with host-plasmid interaction that also play a role in stability. The prototypes from which this knowledge was derived form a family of small plasmids that occur in Gram-positives, especially in *S. aurlus aureus* and *Bacillus subtilis*. Many of these have been characterized and several have been sequenced in their entirety (Novick et al., 1986). In Table 2 various plasmids in the overall family are broken down into three groups, those derived from or similar to pT181, those from pUB110, and those from pSN2. There is much that has been conserved in this plasmid family, and the organization of pT181 may thus serve as a model for the family. The plasmid contains a tetracycline resistance gene. In addition this 4.4 kb plasmid has its *ori* region located between nucleotides 43 and 151 on the pT181 map; this is also where the amino portion of the RepC protein is encoded. The *repC* gene encodes a *trans*-acting positive regulatory protein (RepC) necessary to initiate replication. It appears that bending of the DNA is involved in initiation, in order to make recognition of the binding site in *ori* easier for the RepC protein (Koepsel and Khan, 1986). This binding site contains a

TABLE 2. Representative Small Plasmids of Gram-Positive Bacteria

Family[a]	Plasmid	Resistance Marker[b]	Size
pT181	pT181	TE	4.4 kb
	pC221	CM	4.6 kb
	pS194	SM	4.3 kb
	pC223	CM	4.7 kb
	pUB112	CM	4.4 kb
	pCW7	CM	4.2 kb
	pHD2	—	2.1 kb
	pLUG10	CD	3.1 kb
	pOg32	—	2.5 kb
	pT127	TE	4.4 kb
	pTZ12	CM	2.5 kb
	pE194	EM	3.7 kb
pUB110/pC194	pUB110	KM	4.5 kb
	pC194	CM	2.9 kb
	pAMα1	TE	9.6 kb
	p353–2	—	2.4 kb
	pBC16	TE	4.5 kb
pSN2	pSN2	—	1.3 kb
	pE12	EM	2.5 kb
	pIM13	EM	2.5 kb

[a] Complete family members listed for pT181; representative members for the other families; information from Khan (1997).
[b] TE, tetracycline; CD, cadmium; CM, chloramphenicol; SM, streptomycin; EM, erythromycin; KM, kanamycin.

bend which is enhanced by RepC binding (Koepsel and Khan, 1986). The RepC protein binds to a 32 bp sequence in *ori* and makes a nick in one strand of the DNA within it in order to initiate rolling circle replication. The *ori* contains a stretch of 21 nucleotides with the sequence 5' (Pu-Py)$_5$ PU(Pu-Py)$_5$ which is equivalent to a dodecamer with alternating purines and pyrimidines found to be involved in initiation of replication in various replicons (Calladine, 1982).

For the plasmids in the pT181 group, each Rep protein molecule is inactivated after it has initiated one round of replication. The protein functions as a RepC/RepC dimer (Jin et al., 1997), and subsequently one of the pair is modified by the addition of a short oligonucleotide to Tyr-191. This modification produces the RepC/RepC* heterodimer (Jin et al., 1997). The heterodimer can still bind to DNA, but unlike its predecessor, it has very little nicking activity and cannot denature the DNA in the region of the origin. Moreover a regulatory role for the heterodimer has been suggested by its in vitro inhibition of RepC-supported replication (Jin et al., 1997).

Within the *pre* coding sequence is a *cis*-specific region of DNA called *cmp*. If this region is rearranged to produce a Cmp$^-$ phenotype, a plasmid is found to have 1/20 its normal copy number when coresident in a cell with a wild-type plasmid; at the same time, the wildtype plasmid has a corresponding 20-fold increase in copy number. This implies that there is a competition for a rate-limiting RepC protein and that *cmp* enhances the efficiency of RepC utilization. The *cmp* element thus acts as an enhancer or stimulator of replication, acting up to 1 kb away from the double-strand origin. *cmp* is approximately 100 bp and is involved in

DNA bending and enhancing the interaction between RepC and the origin. A host factor, CBF1, binds to *cmp*, and its role in this process is currently being investigated (Khan, 1997).

Copy number control for pT181-like plasmids. As with the IncFII and ColE1 plasmids, countertranscripts play a role in regulating copy number for pT181 and other plasmids within the overall group that replicates by rolling circle. Unlike what we have seen before, however, the antisense RNAs appear to act by causing premature termination of the Rep mRNA. One or more countertranscripts are produced by such plasmids (85 nt and 150 nt, for pT181; 75 nt and 250 nt, for p353–2 [in the pC194/pUB110 family]; Pouwels et al., 1994) and these have been found to interact with the leader portions of their respective Rep mRNAs, as the latter are being synthesized. This interaction eventually results in the refolding of the Rep mRNA to create a stem-and-loop structure recognized as a signal for termination of transcription. The counterstranscripts are constitutively expressed, limiting full-length synthesis of the Rep mRNA to only about 5% of all such transcripts initiated. Thus Rep protein levels are kept low.

A streptococcal plasmid which replicates via rolling circle, pLS1, has been shown to have two mechanisms for controlling replication. It too has a countertranscript, RNA II, that binds to the Rep mRNA, and this blocks translation of the mRNA, likely by blocking the ribosome binding site (del Solar et al., 1997). In addition there is a transcriptional repressor, a small 5.1 kD CopG protein. This binds to the promoter region for the *rep* gene and blocks transcription (del Solar et al., 1997).

Self-correction of plasmid copy number. Using temperature-sensitive replication mutants (Tsr) or high-frequency transduction (Hft) systems, Novick et al. (1986) examined re-population kinetics for pT181. What they found was that initially replication begins quite rapidly, causing a two- to threefold overshoot compared to the steady-state plas-

mid copy number. Then replication is stopped to allow the copy number to reach steady state. These authors concluded that during the approach to steady state there is overshoot because initiator and inhibitor are not made at the same rates. In cells where the interaction between initiator and inhibitor does not occur, autoregulation by the initiator in high concentrations appears to influence the new steady state. The maximum rate the plasmid could replicate would be its doubling time during exponential cell growth. In the absence of effective inhibitor functions, this represents the plasmid's maximum self-correction rate.

Another region affecting copy number. Another important region on the pT181 map is the *palA* region. This region is suspected by Novick et al. (1986) to be the initiation site for lagging strand synthesis during rolling circle replication. The region is required for normal replication and maintenance. Plasmids that have PalA⁻ mutations are very unstable and have a 10-fold decrease in copy number. One can exchange *palA* segments between different plasmids, but they must be in their normal orientation with respect to the origin in order to be effective (Gruss et al., 1986). Further the orientation of *palA* relative to the *ori* is important, suggesting asymmetrical initiation of replication. A similar locus, originally termed *par*, was identified in the *B. subtilis* plasmid, pLS11 (Chang et al., 1987). This locus appeared to govern faithful partition of the plasmid to daughter cells, but its function, like that of *palA* for pT181, was found to be dependent on the orientation of the *par* region with respect to the replication origin on the same plasmid. It appears that the *par* site acts only as the lagging strand conversion signal. If so, this could explain the dependence of *par* function on orientation relative to the origin.

2. Control that relies on iteron binding

P1. A comprehensive review of P1 has been prepared by Yarmolinsky and Sternberg (1988). P1 is a temperate phage that infects

E. coli and whose prophage is a 100 kb plasmid (instead of being integrated). It can replicate as a low copy number plasmid prophage or exclusively as a lytic phage. As a plasmid, P1 is a member of incompatibility group Y (Hedges et al., 1975). Also included in this group are plP231, which encodes H_2S production and tetracycline resistance (Briaux et al., 1979), and plasmid prophage P7. P7 shows 90% homology to P1 (Yun and Vapnek, 1977), but both behave as if they have different immunities. This is because the two phages require the expression of two repressors to confer immunity. Their c1 immunity repressors are identical in specificity and almost identical in DNA sequence; the other repressor is of different specificity, and this is what makes their immunities appear to differ.

Since P1 exists as both a lytic phage as well as a lysogenic prophage, one can ask if there are two replicons for P1, both a lytic form and a lysogenic form? As a prophage, the requirements for maintaining itself at one copy per chromosome are the same as for other plasmids with a fine-tuned mechanism for regulation and partition. As a lytic phage, a separate origin is required for initiation of rolling circle replication so that the concatamers may be packaged. P1 can thus be considered as an infectious plasmid (as the virion).

Replication and partition are two independent processes for the P1 plasmid (prophage). How do we know this? If the *par* region is deleted from mini P1, the plasmid still replicates, but it is not properly partitioned. Also, if a region is to function as an analogue of a centromere (*CEN*), then one would expect stabilization by *parS* of a plasmid in *cis* (i.e., when *parS* is on that plasmid); on the other hand, in *trans* one would expect destabilization (incompatibility) of a plasmid (i.e., another plasmid) by *parS*. This explains why *parS* is also referred to as *incB* (Friedman et al., 1986).

P1 replication control. There is also a portion of the origin (*oriR*) that exhibits incompatibility. The effect of this region is only seen when *oriR* is cloned on a multicopy plasmid and is placed in the same cell with P1. The replicon required for replication of P1 as a plasmid prophage requires only about 1.5 kb of DNA (see Fig. 4). The required DNA consists of (1) the approximately 245 bp *oriR* (Chattoraj et al., 1985); (2) the *repA* gene, which encodes an essential 286 amino acid replication initiator protein; and (3) *incA*, a 285 bp replication controlling sequence (Abeles et al., 1984). It turns out that the *incA* region is actually dispensable for replication (Pal and Chattoraj, 1986).

Comparison of the *oriR* replicon with analogous regions from other low copy number plasmids shows sequence homology which seems to be a consensus sequence for *E. coli* DnaA protein binding (Hansen and Yarmolinsky, 1986). DnaA protein can in fact be assumed to bind to this conserved sequence in P1, since replication of P1 depends on *dnaA* function. In the origin the DnaA-binding site is next to an A-T-rich region that includes a repeat of five GATC sequences. The origin is completed with sets of repeated DNA sequences (referred to as *iterons*). There are five 19 bp iterons in *oriR*. These have the same orientation and are spaced close together so that similar portions are two turns of the double helix apart; this section of iterons is *incC*. In addition there are nine iterons downstream of *repA*; this is *incA*.

Replication of the P1 plasmid in vitro is unidirectional away from the *repA* gene and takes place in the Cairn's mode (θ); this was determined with a mini-plasmid containing *oriR*. Evidence suggests that the RepA protein required for initiation at *oriR* represses its own transcription from its promoter: We can construct a fusion between the *repA* gene and the *lacZ* gene (encoding β-galactosidase) and thereby generate a protein fusion (see Appendix B). Let's say we make such a fusion present in one copy in the cell (e.g., by using λ as a vehicle for integration of the fusion into the chromosome). In such a case *lacZ* expression is reduced more than 99% by

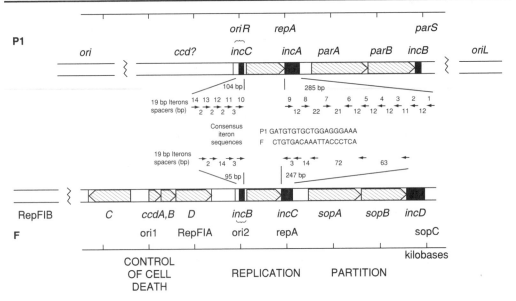

Fig. 4. Map of the region of P1 required for plasmid replication compared with that for F plasmid. The shaded arrows represent open reading frames of the indicated structural genes. Black boxes represent segments that can exert incompatibility. Small arrows indicate the organization of the 19 bp iterons. *Ori* and *oriL* are two other replication origins used by P1. (Reproduced from Yarmolinsky and Sternberg, 1988, with permission of the publisher and author.)

the RepA protein encoded on a P1 plasmid resident in the cell.

The specific binding of RepA protein to the iterons in *oriR* and *incA* is suggested by the strategic placement of iterons on either side of *repA*, so the RepA protein can be involved in replication *and* in replication control. How is control exerted in *incA*? The binding of the essential RepA protein to *incA* means that RepA in unavailable for initiation of DNA synthesis at the origin. Using immunoblotting techniques (Western blots), Swack et al. (1987) determined that in vivo there are 20 RepA dimers per P1 plasmid per cell. Since there are 14 binding sites for the RepA protein per P1 replicon, the concentration of repA would thus appear sufficiently low to be rate limiting. The copy number of the plasmid is inversely proportional to the number of iterons in the same cell (Chattoraj et al., 1985). There is competition for RepA protein between *oriR* and *incA*. A similar competition by iterons as targets for the Rep protein is found in plas-

mid pSC101. However, in this case a second region, normally associated with plasmid partitioning, *par*, may also have an effect on plasmid copy number (Ohkubo and Yamaguchi, 1995).

The RepA protein actually has two roles. It serves as a positive regulator to initiate replication; it also serves in a negative role, to repress expression from the *repA* gene. The concentration of RepA protein is controlled by binding to *incA* and by repression of the *repA* gene by RepA.

But doesn't this pose a conflict? Why doesn't the autoregulatory circuit counteract the effect of *incA* (i.e., by increasing RepA production)? One possibility is that the RepA dimer is asymmetric (perhaps as a result of DNA binding) so that one part has an affinity for any iteron; the other part of the dimer has greater affinity for the repA promoter. Any RepA dimer bound at *incA* is unavailable for initiation. However, by looping of the DNA, the repressor portion of the dimer could bind to the RepA operator site. This model seems to be supported by EM visualization (D.K. Chattoraj, R.J. Mason, and S.H. Wickner, cited in Yarmolinsky and Sternbert, 1988).

Similar control mechanisms are also found in F, pSC101, R6K, Rts1, and RK2. Only in F, Rts1, and P1 is there a second region containing the repeats that bind the Rep protein nonproductively so that Rep is unavailable to bind to the origin (Scott, 1984).

R6K. An IncX plasmid, R6K (38 kb), is a self-transmissible *E. coli* plasmid that encodes bacterial resistance to streptomycin and ampicillin (Datta and Kontomichalou, 1965). It is found in roughly 15 to 20 copies per cell (Kontomichalou et al., 1970), and three distinct origins are used for replication (Crosa, 1980); α, β, and γ. The γ origin is located between α and β, 2 kb from each. Replication of R6K appears to be dependent on several *E. coli* gene products. In vitro replication is sensitive to novobiocin, suggesting a requirement for DNA gyrase. Also replication of R6K is sensitive to streptolydigin and to rifampin, implying an involvement of RNA polymerase. R6K replication takes place in *polB⁻* mutants, so DNA polymerase II is *not* required. However, DNA polymerase III appears to be required, since R6K replication is sensitive to arabinosyl-CTP (which inhibits both DNA polymerases III and II).

Although 90% of replication for R6K is initiated at the α and β origins, this initiation is dependent on a region within the γ origin which is recognized by a positive initiator protein, π. The π protein is a 35 kDa product of the *pir* gene, located between γ and β (Stalker et al., 1982). This protein binds to seven iterons or direct repeats (22 bp each) located within a 277 bp segment of the γ origin (Stalker et al., 1979) and to an eighth; π also binds to a pair of inverted repeats that overlap the promoter for *pir*. Thus π binding may be involved in both γ replication and autoregulation from *pir* (Kelley and Bastia, 1985, cited in McEachern et al., 1985). There is also evidence that π is later deployed to the α and β origins by a looping mechanism

after first binding to γ (Miron et al., 1994). The iterons normally found in the γ origin can also act to inhibit R6K replication if provided in *trans*. The model explaining this phenomenon, the "handcuffing" model (McEachern et al., 1989), proposes that the π protein can cause different sets of iterons to cluster at different plasmid origins of replication and that such intermolecularly coupled plasmids are consequently prevented from initiating replication. This was supported by EM analyses that demonstrated that π protein can associate two DNAs that bear the γ origin sequences (McEachern et al., 1989). Further experiments showed that mutants of the initiator protein, π113 and π108, both caused increases in plasmid copy number of a replicon with a γ origin (Miron et al., 1994). Both mutants apparently retained their abilities to bind the iterons in the γ origin. However, one of these, π113, was impaired in its ability to loop out DNA and cause pairing of γ replicons, whereas the other was unaffected in these functions. These results suggested that handcuffing was, indeed, one mechanism of copy number control for R6K but that additional mechanisms must exist, such as interaction of π with host proteins.

From either the α or the β origins, replication proceeds to a terminus, *ter,* which is not located directly opposite the origin. This phenomenon is somewhat unique for plasmids. The *ter* region is 216 bp and appears similar to the terminus for the *E. coli* chromosome (Louarn et al., 1977) in that it delays, but does not stop, the replication forks. When recombinant plasmids are constructed bearing both the ColE1 and R6K origins, the *ter* region from R6K delays replication originating from either origin (Kolter and Helinski, 1982). The *ter* region contains no dyad symmetry and no extensive open reading frame (Bastia et al., 1981). Apparently, no additional protein is required for termination, since the effect of the *ter* region is chloramphenicol resistant. However, the *ter* region does not appear to be essential for replication, since it may be deleted and R6K replication will still occur.

3. Transcriptional controls: IncP and IncQ Plasmids

We have described some aspects of transcriptional controls above, in the cases of CopB protein action as a negative regulatory protein for the initiation of RepA transcription (IncFII plasmids) and in the production of transcription termination signals via interaction with countertranscripts (pT181). Below we describe an example that relies primarily on negative control of transcriptional initiation.

Plasmids that are able to replicate in a variety of hosts might be expected to have complicated modes of replication and replication control. Examples of such plasmids are the IncP and IncQ plasmids. These are able to replicate in several different Gram-negative hosts. Plasmid RSF1010 is a wide host range IncQ plasmid, about 8.7 kb, normally found in 10 to 12 copies per cell in *E. coli* (Bagdasarian et al., 1986). There are at least three essential proteins involved in replication of this plasmid (Scherzinger et al., 1984) RepC is a positive regulatory protein that is responsible for initiating replication by binding to *oriV* (the origin). RepA is a DNA-dependent ATPase that may be similar in function to DnaB protein of *E. coli.* RepB is a primase that is found in two forms: a 38 kDa RepB, that is essential for replication, and a large, 70 kDa protein, RepB*, which is translated from the same reading frame as RepB and which cross-reacts with anti-RepB antibody. The function of this larger form is unknown. The genes for these essential proteins are arranged in somewhat of an operon. These genes are transcribed in the order *repB* repB repA repC*, from promoter p1. In addition another promoter, p2, may be used to transcribe a RepA-RepC transcript independently. For each of these promoters there is a negative regulatory protein (10 kDa for p1; 7 kDa for p2) that is encoded just downstream of the respective promoter; each has been shown to bind to its respective promoter in vitro (Scholtz et al., 1985) and thus to control the copy number by regulating the amount of the positive factor, RepC. Mutations that alter control at p1 affect plasmid stability in *Pseudomonas.* This suggests that proper regulation of copy number is important in this plasmid's ability to switch species.

4. Linear plasmids

In eukaryotes, remember, linear plasmids had been identified. However, linear plasmids have to deal with the problem that their ends are highly recombinogenic and thus that plasmids might continually be lost due to recombination with the chromosomes. This problem is solved for stable linear DNA (e.g., chromosomes) by repetitive ends called telomeres. This role is accomplished by virtue of rendering the ends generally inert in interactions with other chromosome ends. In contrast to unprotected ends or broken chromosomes whose ends are "sticky" and readily rejoin to other such ends, telomeric regions do not fuse with broken ends or with one another. Despite these expectations Barbour and Garon (1987) reported the discovery of a double-stranded linear plasmid in *Borrelia burgdorferi*, a spirochete that causes Lyme disease. This 49 kb plasmid encoded *ospA* and *ospB* genes for major outer membrane proteins of the strain and the plasmid's ends were protected from recombination by being covalently closed.

Since their original discovery, other linear plasmids have been identified and characterized in both *Borrelia* species and in the filamentous soil bacteria *Streptomyces* species (Wu and Roy 1993; Redenbach et al., 1998). For *Borrelia burgdorferi* Sensu Lato isolates, which include five species of the genus *Borrelia*, the chromosome is linear and individual isolates have variable numbers and sizes of circular and linear plasmids. Marconi et al. (1996) have identified from these strains unusually large linear plasmids, ranging from 92 to 105 kb. Most of these were dimers of the 50 kb plasmid pAB50, which bears the *ospA* and *ospB* genes; however, one isolate was unique and showed no homology to pAB50 or to total DNAs from any other isolates tested. The other large plasmids apparently were dimers of the smaller, more common plasmids. These authors have proposed that dimer formation may have occurred during replication which employs a monomeric circular intermediate. Alternatively, it has been proposed (Hinnebusch et al., 1990) that such multimers may have arisen by recombination, in particular, after horizontal exchange between *Borrelia* and African swine fever virus, both of which are found in the same arthropod vector.

Streptomyces species frequently contain linear plasmids (Redenbach et al., 1998), ranging in size from 12 kb to as large as 1 Mb! These plasmids contain terminal inverted repeats 44 to 95 kb in length, and all the plasmids so far studied are blocked at the 5′ ends by a protein. Like *Borrelia* species described above, several *Streptomyces* species have also been shown to have linear chromosomes (Redenbach et al., 1996).

B. Plasmid Segregation

I. Random diffusion and/or maintenance of monomers

What about mechanisms for maintaining stability of ColE1? Plasmids ColE1, CloDF13, pMB1, and ColK have the ability to ensure that they remain as monomers; this ability is important for their stability (Makkaart et al., 1984; Summers and Sherratt, 1984, 1985). This is because multicopy plasmids are normally distributed at random when the cell divides. The system that controls their copy number per cell keeps track of the number of origins instead of actually counting the number of plasmid copies (Summers and Sherratt, 1984). If a plasmid occurs as multimers, it will be found at a lower copy number than the monomers; such multimeric plasmids will be improperly segregated into the daughter cells. Many ColE1 derivatives form multimers in *recA*$^{+}$ *E. coli* and as a result they are lost. ColE1, on the other hand, maintains itself as monomers

and is not lost. How? ColE1 has a site for recombination, *cer*, that is dominant and acts in *cis* as a site for an *E. coli*-encoded recombinase, Xer. Multimers of ColE1 have multiple direct repeats of *cer*, and these are converted to monomers by Xer-mediated recombination in order to maximize the number of copies of ColE1 available for segregation. Deletion of *cer* results in unstable inheritance. ColE1 lacking *cer* form multimers via the *recA* recombination system; these do not form in RecA⁻ strains. The *cer* locus is located in a 280 bp region (Chan et al., 1985); recombination takes place within a 35 bp stretch at the end of this region. For the ColE1-related plasmids pMB1, ColK, and CloDF13, if these regions are deleted, the plasmids become unstable inherited. Several plasmids unrelated to ColE1 also employ a similar mechanism. For example, the partition regions for the broad–host range plasmids RK2 (*parCBA*) and RP4 (*par*) each encode "resolvases" that participate in resolving plasmid multimers back into monomers (Easter et al., 1997; Gerlitz et al., 1990). There are specific sites for recombination in pT181-like plasmids. Two sites on pT181 are involved in site-specific recombinations, RS$_A$ and RS$_B$. The RS$_A$ site is found immediately after t$_1$, while RS$_B$ is much closer to *ori*. The RS$_A$ site contains the *pre* promoter and is used as a site for recombination between different plasmids and within cointegrates. This site is shared by pT181, pE194, and the pSN2 family.

2. *Par* (*CEN*-like) regions

Many low-copy plasmids have a partition locus, *par*, that is required for segregation of the plasmid to daughter cells. These loci usually encode two proteins, A and B, and a *cis*-acting site on which these proteins act. The B protein has, in most cases, been shown to bind with the site and is involved in attaching the original and replicated copies of the plasmid to each other. The A protein has recently been found to be associated with the bacterial membrane, at least in the case of the F episome and plasmid QpH1 of *Coxiella burnetti* (Lin and Mallavia, 1998). The A proteins, then, appear to tether the plasmid copies to a position on the cellular membrane prior to division of the cell. In this sense, the *par* systems would seem to be analogous to centromeres in eukaryotes.

IncFII. *Partition of plasmid R1*. As evidenced by the preceding sections, much is known about the mechanisms by which plasmid copy number and incompatibility are controlled. On the other hand, what about partition of the plasmid? How is a low copy plasmid, like R1, stably segregated to the daughter cells. Such stable partitioning is of particular importance for these plasmids. One might expect that a mechanism exists analogous to the *CEN* regions (centromeres) for eukaryotic chromosomes. In fact such loci have been detected on several plasmids. For F, one region has been identified that functions in this capacity (Lane et al., 1987). Regions homologous to such *par* loci have been identified on bacterial chromosomes. Some of these are known to participate in the segregation of the respective chromosomes (e.g., *soj* and *spoOJ* in *B. subtilis*, Ireton et al., 1994; *parA* and *parB* in *Caulobacter crescentus*, Mohl and Gobers, 1997). As we will see for R1, though, several alternatives may be employed simultaneously.

For R1, two loci have been identified as being involved in partition: the *parA* and *parB* loci. The *parA* locus (Gerdes and Molin, 1986) is a 1500 bp region whose nucleotide sequence contains two open reading frames capable of encoding both a 320 residue (36 kDa) protein, ParM, and a 60 residue (7 kDa) protein, ParR. The *parA$_p$-lacZ* fusions demonstrate that *parA* encodes a negative regulatory factor for the 36 kDa protein, which is possibly the 36 kDa protein itself (Gerdes and Molin, 1986). There is also ParC, a 160 bp region covering the *parA$_p$*. In this region *parA$_p$* is flanked by two sets of iterons which are bound by ParM. All 10 of these repeated sequences are required for partitioning and for regulation of the promoter by ParR. The ATPase activity of ParM is required for this protein to mediate efficient pairing of plasmids for partitioning. The ParC site thus acts like a centromere (*CEN*) in eukaryotes, and there could be competition in *trans* for a limited number of *CEN* sites. If one clones the *parA* locus to an unstable mini F plasmid devoid of its own partition genes, there is a 100-fold increase in stability for the plasmid. There is a sequence downstream that is rich in dyad symmetry and could be the site of adherence to a cellular structure involved in partition. Although there are no sequence similarities at the amino acid or nucleotide levels between *parA* and the corresponding systems in P1 and F, there are similarities between these *par* loci, in that both of these possess a 35–45 kDa protein that is autoregulated (Ogura and Hiraga, 1983). But there is also an important difference—the incompatibility determinants for these other plasmids are located downstream of their respective protein-coding regions. In studies to examine how the plasmids are localized during partitioning, the F plasmid and SopB protein were co-localized to the mid-cell or $\frac{1}{4}$ and $\frac{3}{4}$ positions (Kim and Wang, 1998). Jensen and Gerdes (1999) have shown that ParM co-localizes with R1 during the cell cycle. They used fusions with the green fluorescence protein (GFP) and immunofluorescence microscopy to show that ParM and plasmids containing ParC are found together at specific sites near the cell poles or mid-cell. There is pairing of newly replicated plasmids at mid-cell, and then they separate and move to the cell poles in preparation for cell division.

P1. *P1 partition*. P1 is a low copy number plasmid. As such, it requires an efficient system to ensure that each daughter cell receives a copy of the plasmid upon cell division. This probably requires cell membrane involvement and a specific binding site on P1. The sequences required for proper plasmid partition of P1 are found within a 3.0 kb region next to the *rep* gene (Friedman et al., 1986). This 3.0 kb *par* region does not appear to be directly involved with replication; *par* deletions allow replication but make the plasmid unstable—they are lost at a high frequency because they are partitioned randomly into the daughter cells. P1 *par* can stabilize several other types of replicons, including F and pBR322, as well as P1 (Austin et al., 1986). P1 *par* seems analogous to a *CEN* region for eukaryotic chromosomes. Abeles et al. (1985) have determined the complete sequence of the *par* region. Two genes, *parA* and *parB*, are predicted from the existence of two open reading frames. The products of both of these genes are essential. The *parB* protein is hypothesized to bind to *parS* (a 49 bp region in which there is a 20 bp AT-rich inverted repeat). This is thought to be the plasmid partition site. The binding of the *parA* and *parB* products to *parS* serves as a unit for interaction with the host components involved in the segregation process. The *par* region might also be involved in incompatibility; if the sites recognized by the plasmid partitioning mechanism are the same on different plasmids, competition for the cellular segregation components should prevent plasmids from segregating independently.

The ParB product (38 kDa) binds to ParS and, along with the ParA product (44 kDa), and is essential for partitioning. There is normally a higher concentration of ParB than ParA in the cell due to a differential translation efficiency for the transcripts of the two genes. Expression of the *parA* and *parB* genes is autoregulated by the ParA product alone or by ParA and ParB products acting together. Why is regulation of these genes necessary? Excess ParB product causes a decrease in partitioning efficiency, possibly by obscuring parS; in addition, ParB destabilizes even high copy number stable plasmids bearing *parS*, presumably by causing them to aggregate (M. Yarmolinsky, personal communication). The effect of excess *parB* is even more severe than if there is no *parS* at all—segregation is less stable than that observed in random distribution to daughter cells (Funnell, 1988). It appears that excess *parB* sequesters *parS*-containing plasmids so that they are prevented from

properly segregating, even by random diffusion.

There is incompatibility associated with *parS*. For this reason *parS* is also referred to as *incB*. A 175 bp restriction fragment causes incompatibility relative to mini P1 plasmids. This fragment is 63% A+T and has a 13 bp perfect inverted repeat; the inverted repeat is followed by a stretch of C and T nucleotides and the sequence AAAA AAATAAAAAAA. It was determined that *par* was required in *cis* for partitioning by cloning fragments into pBR322. *Par* activity was assayed by a heterodimer pickup assay developed by Friedman et al. (1986) (see Appendix B). *parS* regions of different sizes (produced by deletions) affect plasmid incompatibility but may still allow *par* function. This can be explained if we assume that heterologous pairs (i.e., between a deleted and a wild-type *par*) will not serve as a substrate for partitioning, and only homologous pairs will.

How does the *par* apparatus function in the host cell? There must be an interaction between *par* and some cellular components. The likely model is as follows. The exact role of ParA protein is unknown, but there are many similarities between it and the corresponding A proteins from the *par* loci of other low copy number plasmids, such as F and QpH1 (Lin and Mallavia, 1998). First, the active DNA binding-form of ParA is a homodimer, and second, the protein has ATPase activity, which is enhanced by interaction with ParB. The ParB protein and host cell integration host factor (IHF) bind the *parS* to cause a specific formation of plasmid pairs, each of which attaches to one of several possible membrane sites; when the sites split, they segregate to separate cells and the plasmids detach from each other. *Par* specific pairing of the daughter plasmids precedes membrane attachment. Also *incB* could be involved, leading to competition for *par* action—heterologous pair formation would cause the loss of one of the daughter plasmids. This process is distinct from the mechanism used to partition the *E. coli* chromosome. This is known from experiments that examine partitioning of P1 in mutants defective for normal chromosome partitioning. MukB is a protein required for normal chromosomal positioning prior to cell division; if this protein is defective, at each division cells are produced that lack chromosomes. Funnell and Gagnier (1995) found that in *mukB* mutants P1 partitions effectively into both the chromosome-containing and chromosome-free cells, indicating that P1 utilizes a separate membrane attachment mechanism than that used by the bacterial chromosome. Also, in contrast to the *E. coli* chromosome, both P1 and F were found to localize to the center of the cell, which is the site of septum formation (Gordon et al., 1997). After replication, the copies move quickly to the quarter poles, the sites where the next septa will form.

In addition to the *par* mechanism described above, it appears that a recombination system exists for P1 (the *cre-lox* system), which serves the same function for P1 as the *cer* system discussed for ColE1 (see above). This recombination system is used to resolve multimers of P1 produced by host *rec* systems (Austin et al., 1981, cited in Yarmolinsky and Sternberg, 1988).

pSC101. Plasmid pSC101 is a plasmid that encodes bacterial drug resistance. It was isolated originally from *Salmonella* and is also functional in *E. coli*. The pSC101 plasmid contains a *par* locus necessary for partition. This is a *cis*-acting locus that does not encode a protein (Biek and Cohen, 1986). This region consists of three PR segments. These three sequence elements are related in such a way that (considering each strand of DNA separately) the central sequence could form a hairpin by base-pairing with either of the flanking sequences on that strand. Deletion of any one of these sequences decreases stability of the plasmid (probably by impeding base-pairing with the others). Deletion of one PR leads to a decrease in competitiveness compared to wild-type pSC101; if both plasmids are in the same cell, the mutant is lost more fre-

quently. There are other mutations that affect stability. The superPar⁻ phenotype occurs when all three PRs are deleted; in this case the plasmid shows even less stability then that seen by random distribution. There are also *E. coli* chromosomal mutations that affect stability of pSC101: (1) mutations in the *himA* and *himD* loci responsible for integration host factor (IHF; Kikuchi et al., 1985)—the mechanism of this effect is unknown; (2) the *recD* locus (Biek and Cohen, 1986) between *recB* and *argA*—mutations in this locus lead to increased multimerization. (This may be contrasted with the need of Xer function to reduce multimers of ColE1.)

Conley and Cohen (1995) have also identified secondary sites that, when mutated, may be used by the plasmid for partition in the absence of the *par* locus. There are seven sites in the 5' half of the *repA* gene (which encodes the protein necessary to help initiate replication). Combinations of some of these mutations allowed replication of pSC101 in the absence of IHF. It was suggested by this work that mutations in these sites enable the formation of RepA-DnaA-DNA complexes that play a role in plasmid stability separate from their normal role in replication of the plasmid.

3. Postsegregational killing

IncFII. The *parB* locus for R1 is a 600 bp region that greatly enhances the stability of plasmids into which it has been inserted (Gerdes et al., 1986a). Several methods were employed to characterize the way in which this locus functions; these include *lacZ* protein fusions and DNA sequencing (see Fig. 5), and mRNA secondary structural analyses (Fig. 6). A model has been developed by Gerdes and coworkers (1986b) for the effect of *parB* on partition. A strategy for a population of cells to maintain a plasmid might occur without the development of a complicated partition apparatus. What if, instead of ensuring proper segregation of plasmid copies into daughter cells, the *par* locus ensured that only plasmid-bearing cells

```
  1   AACAAACTCC GGGAGGCAGC GTGATGCGGC AACAATCACA CGGATTTCCC GTGAACGGTC
      TTGTTTGAGG CCCTCCGTCG CACTACGCCG TTGTTAGTGT GCCTAAAGGG CACTTGCCAG

                                        -35            phok
 61   TGAATGAGCG GATTATTTTC AGGGAAAGTG AGTGTGGTCA GCGTGCAGGT ATATGGGCTA
                          ^CTTTCAC TCACACCAGT CGCACGTCCA TATACCCGAT
```

```
                         TCTGCC TCATGACGTG AAGGTGGTTT GTTGCCGTGT
                         AGACGG AGTACTGCAC TTCCACCAAA CAACGGCACA
```

```
                         CGTAGT AAGTTAATTT TCATTAACCA CCACGAGGCA
                         GCATCA TTCAATTAAA AGTAATTGGT GGTGCTCCGT
```

```
                         ATAGCC TCTTACCGCG CTTTGCGCAA GGAGAAGAAG
                         TATCGG AGAATGGCGC GAAACGCGTT CCTCTTCTTC
                         10         psok         -35
```

```
                         :CTTGTC TGGTGTGTGT TGATCGTGTG TCTCACACTG
                         :GAACAG ACCACACACA ACTAGCACAC AGAGTGTGAC
```

```
                         :AAATCG CTGTGCGAGA TTCGTTACAG AGACGGACAC
                         :TTTAGC GACACGCTCT AAGCAATGTC TCTGCCTGTG
```

```
                                                       ───────────→ IV
                         :TACGAA TCCGGTAAGT AGCAACCTGG AGGCGGGCGC
                         :ATGCTT AGGCCATTCA TCGTTGGACC TCCGCCCGCG
                                      mok hok
                                    stop stop
```

```
                         GCTGGTC TGACTACTGA AGCGCCTTTA TAAAGGGGCT
                         CGACCAG ACTGATGACT TCGCGGAAAT ATTTCCCCGA
```

```
                    CTCCTTG CTGATGTTGT
                    GAGGAAC GACTACAACA
                    *       *
                    ion of full
                    ok transcripts
```

us from plasmid R1. Shown are the locations of the *mok*, *hok*, ing gene, toxin and antisense RNAs, respectively. −35 and −10 icated, and the beginning of each transcript is indicated by an and stop codons for the *mok* and *hok* genes are underlined. and V, which are likely rho-independent termination signals for pectively. (Reproduced from Gerdes et al., 1990, with permis- ...thor.)

survived? Thus the *parB* locus would not influence the rate of formation of plasmid-free cells; rather, the locus would ensure the postsegregational killing of plasmid-free cells (see Fig. 7A).

How does this occur for *parB*? This locus encodes two products. The first is encoded by a gene termed *hok*, for "host killing." The *hok* gene encodes a 52 amino acid protein that causes rapid cell killing; the viability drops to a 30 second half-life. The proton gradient across the membrane collapses, and oxygen uptake ceases (Gerdes et al., 1986a).

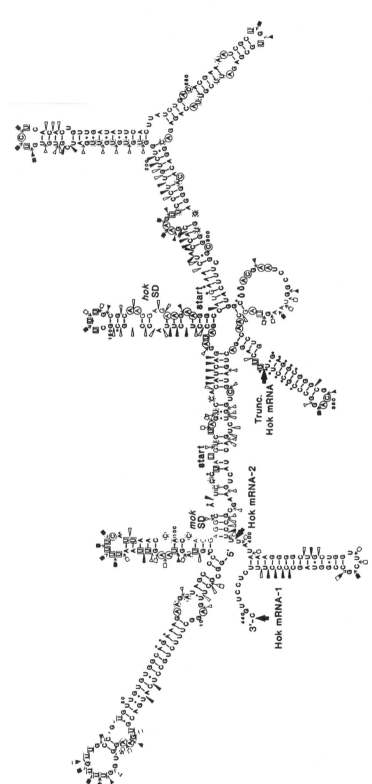

Fig. 6. Secondary structural analysis of the Hok mRNAs. The numbers indicate the nucleotide positions relative to the 5′ end. The *mok* and *sok* Shine-Delgarno regions and stop codons are indicated. The portions of the 3′ ends of the full length Hok mRNA-1 (at +441), full-length mRNA-2 (at +398) and truncated Hok mRNA-3 (at +361) are indicated with arrows. (Reproduced from Thisted et al., 1995, with permission of the publisher.)

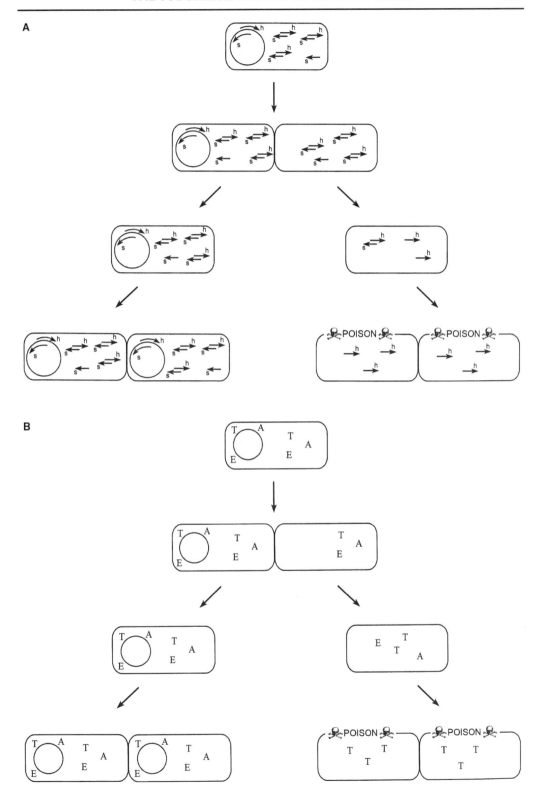

C

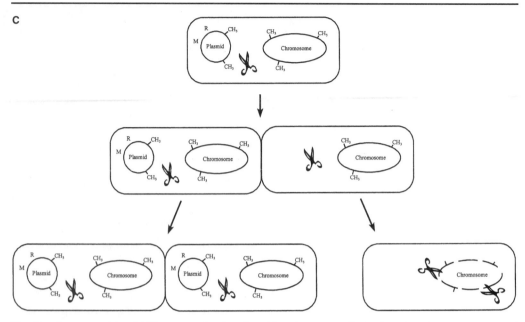

Fig. 7. Postsegregational killing strategies for plasmid maintenance. **A:** *parB*-mediated postsegregational killing of plasmid-free cells. The complex of *hok* mRNA–*sok* RNA prevents expression of the *hok* gene product. When the *sok* RNA is degraded in the plasmid-free cell, the *hok* mRNA can be translated. This produces the *hok* protein, which results in cell death. **B:** Toxin-antidote system modeled after the *pas* system from pTF-FC2. In plasmid-bearing cells two-to-three proteins are produced, a long-lived toxin, T, an antidote to the toxin, A, and, in some cases, a protein that enhances the neutralizing effects of the antidote, E. The antidote and enhancer are short-lived and need to be constantly resynthesized. In cells that do not receive a copy of the plasmid, they are degraded, leaving the toxin available to kill the plasmid-free segregant. **C:** Plamids bearing restriction-modification systems. The plasmid bears a gene, M, encoding a methyltransferase, and a gene, R, encoding a restriction endonuclease (indicated by the scissors). The methyltransferase adds a methyl group to specific bases present in the cell's DNA (plasmid and chromosome). Modified DNA cannot be cut by the restriction enzyme, but when the plasmid is lost, there ceases to be a source of methylase. The resulting unmodified DNA is then cleaved by the remaining endonuclease. This causes double-strand breaks in the chromosome and cell death.

The second gene is the *sok* ("suppression of killing") gene; it encodes an untranslated antisense RNA (at least 90 nucleotides of which are required for function), which reduces or eliminates *hok*-mediated killing if *sok* is highly expressed. The *sok* promoter is much stronger than that for *hok*. This was shown by the fact that *sok* RNA greatly reduces the production of *hok-lacZ* transcriptional fusions.

According to the model, in a cell carrying a *parB*⁺ plasmid, *hok* mRNA is made and so is *sok* RNA. When the cell divides, any cells that do not receive a copy of the plasmid will cease to produce new *hok* or *sok* RNAs. Nevertheless, *hok-sok* RNA hybrids remain in the plasmid-free daughter cells. If *sok* RNA is less stable than *hok* mRNA, the *hok* mRNAs could be translated in these plasmid-free cells once the *sok* RNAs have been degraded. Thus only daughter cells that do *not* receive a copy of the plasmid will be killed. In fact "ghost" cells characteristic of *hok* protein killing are not viable and contain no plasmid; all normal viable cells contain plasmid. There is a locus homologous to *parB* on F and two on the *E. coli* chromosome (Gerdes et al., 1986b). This model predicts, and the data have shown (Gerdes et al., 1986a), that *parB* can stabilize other plasmids. Moreover evidence that the Hok and Sok RNAs actually interact in vivo comes from experiments in which Sok RNA is provided to cells in *trans* from another plasmid and expression of *sok* is placed under control of a temperature-sensitive repressor protein (Gerdes et al., 1992). These data, added to evidence that Sok RNA binding to the Hok RNA did not overlap the translation initiation site for the *hok* gene (Gerdes et al., 1988), suggested that Sok RNA might block translation of Hok by a more indirect mechanism. The interaction between these RNAs is unlike several we have already observed (e.g., CopA/CopT in R1; RNA I/Preprimer RNA in ColE1); in this case the interaction between Sok RNA and Hok RNA is initiated not between loops on the respective RNAs but instead by pairing of the 5′ tail of the Sok RNA.

In addition to *hok* and *sok*, a third gene was discovered to also reside in the *parB* locus. This gene, *mok* ("mediation of killing"), is transcribed onto the same mRNA as *hok*. Gerdes et al. (1992) first identified this gene on R1, after a similar gene had been discovered for the F episome (Loh et al., 1988). What does *mok* have to do, if anything, with the mechanism outlined above for *hok/sok* mediated postsegregational killing? Well, it turns out that the secondary structure of the Mok/Hok mRNA is such that the translation initiation region for Hok is normally unavailable to ribosomes. It becomes available only after the upstream *mok* gene is first translated. Normally, however, even this is not possible because the 3′ end of the Mok/Hok mRNA contains a structure which acts as an anti–Shine-Delgarno sequence. This structure folds back and blocks the *mok* initiation region, preventing ribosome binding and translation (Thisted et al., 1995). Of course, even if this region were exposed, at least one round of Mok translation is required to cause a change in the secondary structure of the RNA that would allow translation of Hok.

The first blockage to *hok* expression mentioned above is removed by slow processing of the 3′ end of the Mok/Hok mRNA, relieving the inhibition of Mok translation by

eliminating the fold-back inhibitory loop. This begins the process whereby Hok eventually could be translated. However, Sok RNA plays several roles in preventing this from happening. It binds to all three forms of the Mok/Hok RNA that have been identified: full-length mRNA, RNA truncated at its 3′ end and ready for Mok translation, and mRNA whose secondary structure has already been altered by translation of Mok. It is this last form for which Sok RNA has the greatest affinity for binding. This is important since the duplex is a target for disposal by RNase III ribonuclease digestion. It makes sense that Sok would have the highest affinity for the most potentially lethal Hok mRNA form.

So long as a constant supply of Sok RNA is provided (i.e., in cells which maintain a copy of the plasmid), equilibrium can be maintained with regard to potentially lethal Mok/Hok mRNAs, while at the same time the most dangerous forms are rapidly degraded. On the other hand, when a cell fails to get a copy of the plasmid, there is no longer a source of Sok, and as it degrades and is no longer replenished, the sequence of events leading to Hok expression and cell death rapidly follows.

Molin et al. (1987) have characterized expression of the *hok* protein while under control of the *lac* or *trp* promoters. They have shown that such controlled expression of *hok* can be used to selectively eliminate bacterial cells from a population. This locus may thus be useful as a stabilizing cassette or as a tool in constructing a plasmid where biological containment is important. Several groups have developed such "conditional-lethal" cassettes in which the lethal *hok* gene, a related gene, or a series of lethal genes has been fused to a promoter from which expression can be carefully controlled. Cells bearing such constructs commit "suicide" by expressing the toxic gene under appropriate conditions (Bej et al., 1989, 1992; Knudsen and Karlström, 1991; Knudsen et al., 1995). For example, the *hok* gene could be placed under control of the promoter for a xenobiotic degradation pathway and placed into a aromatic compound-degrading microbe to be released into the environment. Then expression of the gene would begin when the level of pollutant was reduced below a critically low concentration in a contaminated sample (Contreras et al., 1991).

Poison/Antidote. Systems analogous to the one described above have been discovered that still employ a toxin or poison, but control of its lethal activity is moved to the level of the toxin protein itself. In these systems, a toxin with a long half-life is expressed at low levels, while another protein, the antidote, is expressed at high levels. However, the latter protein is unstable, thus requiring its constant production. When a plasmid-bearing cell fails to yield a daughter containing a copy of the plasmid, the toxin is able to have its way, since no further synthesis of the unstable antidote will occur in this now plasmid-free cell. Thus, once again, failure to segregate a copy of the plasmid results in rapid cell death for the hapless offender bacterial progeny. This type of partition system has now been identified in a variety of plasmids, including F (*ccd*; Jaffe et al., 1985), P1 (the *phd*/*doc* system; Lehnherr et al., 1993), plasmid RK2 (*parDE*; Roberts and Helinski, 1992; Easter et al., 1997), and the IncFII plasmids (identical *parD*/*pem* and *kis*/*kid* systems; Ruiz-Echevarria et al., 1991; Tsuchimoto et al, 1988).

Smith and Rawlings (1998a, b) have recently described work with a 12.2 kb mobilizable plasmid originally identified in *Thiobacillus ferrooxidans*. This broad–host range plasmid, pTF-FC2, appears to be a natural hybrid between IncQ plasmids (e.g., RSF1010) and IncP plasmids (e.g., RK2). This plasmid has a replicon with similarity to the former and a mobilization region with low but clear relatedness to the latter. The plasmid has a poison/ antidote maintenance system, called *pas*, for "plasmid addiction system" (see Fig. 7B). However, this system has an interesting twist: it contains three genes, *pasA*, encoding the antidote, *pasB*, encoding a bactericidal toxin, and *pasC*,

which encodes a protein that enhances the neutralizing effects of the antidote. In examining the ability of this locus to stabilize heterologous plasmids in *E. coli*, these authors found that both PasA and PasB were involved in autorepression of the operon and that this regulation was necessary for the stabilization of such heterologous plasmids (Smith and Rawlings, 1998b). Moreover the efficiency of the *pas* system was dependent on which *E. coli* host was examined, with host proteases such as Lon playing a role in antidote degradation and thus in the plasmid maintenance function (Smith and Rawlings, 1998a).

As with the *hok* system described above, such toxin-antitoxin systems have been employed as containment systems, this time in yeasts (Kristoffersen et al., 2000).

Restriction-Modification. The recognition that bacterial restriction-modification systems bear many of the same characteristics as the postsegregational killing systems just outlined above has led some authors to explore their possible roles in plasmid maintenance (Pomiankowski, 1999; Kulakauskas et al., 1995). These systems, thought to primarily play a role in bacterial protection from foreign DNAs, including phage and plasmids, contain, in the case of the best-studied group, the type II systems, two-protein functions. The first is a methyltransferase, which modifies specific nucleotides on resident DNAs in the cell; the second component of the system is a restriction endonuclease, which is capable of making blunt or staggered double-strand cuts in DNAs that have not been modified by the methyltransferase. Over 200 type II enzyme systems are now known, and about half their genes have been cloned and sequenced. Much is also known about the structure of the proteins themselves and their interactions with their respective DNA substrate targets. What was particularly intriguing was the discovery that several of these systems reside on naturally occurring plasmids. Some are encoded by small plasmids (*Eco*RI and *Pvu*II), while others have been found on large plasmids (*Pae*RI and

*Eco*RII) (Kulakauskas et al., 1995). Other types of restriction-modification systems have also been found on plasmids (*Sty* R124I (Type I), *Eco* P15 (Type III), and *Eco* P1 (Type III); Kulakauskas et al., 1995). In fact *Eco* P1 is encoded by the P1 prophage, and thus it may provide an additional role in stability of this plasmid, as we will see below.

How might such a system be used to promote plasmid maintenance? Well, once again, we have a system where a potentially lethal product is produced by the plasmid. In this case it is an enzyme that causes double-strand breaks in DNA. Such breaks in the chromosome are extremely difficult to repair and severely diminish cell survival (Handa et al., 2000). When *Eco*RI was used in a conditional lethal containment system, even low-level expression of the enzyme was sufficient to reduce cell viability (Molin et al., 1993). In the normal setting, the methyltransferase protects cells, with modification of only one strand of a target providing sufficient protection from the restriction endonuclease. But here again, the methyltransferase is significantly less stable than the endonuclease and must continually be synthesized in order to afford protection. So, if a plasmid bears such a system, should a cell fail to acquire a copy of the plasmid, such daughter cells would inherit endonuclease in the absence of a renewable source of protection for their DNA, and they would subsequently be killed (see Fig. 7C). Kulakauskas et al. (1995) have shown that this type of system can be used experimentally to enhance the maintenance of unstable plasmids. The fact that restriction-modification systems are not normally as important in providing bacterial protection against phage infection as supposed originally suggests that their residence on plasmids may provide a mutualistic benefit. When present on the chromosome alone, it is possible that their expression was not sufficiently high to overcome the invasion by phage DNAs. However, on plasmids, their increased expression might protect the cell more effectively, while at the same time pro-viding the benefit of a plasmid maintenance function for the plasmids.

C. Sex Pheromone Plasmids—A Peculiarity of Gram-Positive Conjugation

Plasmids are about as common in strepto-cocci as they are in the Gram-negative en-teric bacteria (Kempler and McKay, 1979). They are easier to detect in some species than in others. For example, in *Enterococcus fae-calis* and *Streptococcus lactis* it is common to find five to six plasmids per strain; it is rare to find strains with no plasmids (Kempler and McKay, 1979). In *Streptococcus mutans* and *Streptococcus pneumoniae* plasmids are found less frequently (Dunny et al., 1978; Buu-hoi and Horodniceanu, 1980).

In *E. faecalis* there exist conjugative plas-mids that bear fertility factors (Clewell, 1981, 1985) and novel transposons that can func-tion in other Gram-positives. In addition these plasmids confer bacterial resistance on those bacteria that contain them. Some of these plasmids transfer in broth and confer a mating response to peptide sex phero-mones excreted by the recipients (Dunny et al., 1978, 1995). Others only transfer on solid surfaces (i.e., filter matings) and have a broad host range among Gram-positives. For instance, one novel transposon, Tn916 (encoding resistance to tetracycline), is trans-ferred on a plasmid between *E. faecalis* and *Mycoplasma hominis* at approximately 10^{-6} and 10^{-7} per recipient (Roberts and Kenny, 1987). So far *Enterococcus faecalis* is the only species in which plasmids that transfer in response to pheromones have been detected. Although most *Staphylococcus aureus* strains excrete compounds similar to one *E. faecalis* pheromone, cAM373, these compounds do *not*, in most cases, appear to function as pheromones (Clewell et al., 1985; but see below). Those strains of *E. faecalis* that serve as recipients can excrete five or more differ-ent pheromones; each of these is specific for a donor strain that contains a related plas-mid. A sharp drop in production and excre-tion of the particular pheromone occurs as soon as the recipient bacterium receives a copy of the plasmid; the cells then excrete a peptide that competitively blocks the action of the pheromone (Ike et al., 1983). Other unrelated pheromones are still made. Upon exposure to pheromone, there is a 10,000-fold increase in transfer frequency of the plasmid from donors cells (Ike and Clewell, 1984) (see also Porter, this volume).

Aggregation substance (a protein-like material) is produced by donor cells induced with a pheromone; this substance coats the donor cell surface and helps donors and recipients remain together for conjugation once they have come into contact (Yagi et al., 1983). There appears to be a direct interaction between the aggregation substance and a *binding substance* on the recipient cell surface. There is some evidence that donor cells contain binding substance on their cell surfaces as well, since pheromone-induced donor cells will also aggregate with each other (Yagi et al., 1983). These substances also appear to be involved with binding of the bacteria to host tissues (Muscholl-Silberhorn, 1998).

One *E. faecalis* conjugative plasmid is pAD1, which encodes a hemolysin-bacteriocin and UV resistance. This 57 kb plasmid also encodes a mating response to recipient-produced sex pheromone cAD1. Clewell et al. (1986) detected four surface proteins produced in strains bearing pAD1 after induction by pheromone cAD1. At least one of these proteins was thought to be the aggregation substance. The binding substance might be lipoteichoic acid (LTA), since low concentrations of LTA from *E. faecalis* were found to inhibit aggregation.

Another *E. faecalis* plasmid, pAM373, is unusual in that it encodes an adhesin that mediates clumping. This adhesin, Asa373, is unrelated to the other aggregation substances, which are highly conserved at the amino acid level (Muscholl-Silberhorn, 1999). Controlled expression of the gene for this protein resulted in constitutive clumping of cells, while expression in cells defective in production of binding substance was just as effective. This indicated that this adhesin has a separate mechanism for cell-cell interaction than that normally used by other aggregation substances with their targets. Some amino acid regions in Asa373 resembled motifs in adhesins from oral streptococci. Moreover the leader sequence contained a conserved domain that might be involved in regulation of Asa373 expression.

E. faecalis pheromones have been purified and sequenced (see Table 3; Clewell et al., 1986; Nakayama et al., 1995). They have been found to be hydrophobic octa-or heptapeptides. Also cells harboring sex pheromone plasmids excrete a competitive inhibitor of the respective pheromone produced by plasmid-free cells. For example, cells bearing plasmid pAD1 produce an inhibitor of cAD1, called iAD1, that is also a hydrophobic octapeptide, in which four of the amino acids are the same as in cAD1. This inhibitor is probably plasmid encoded (Clewell et al., 1986).

Although the *S. aureus* "pheromone" staph-cAM373 does not appear to have an effect on other *E. faecalis* plasmid-bearing strains, for those containing pAM373, there is an effect. Muscholl-Silberhorn et al. (1997) found that 85/100 coagulase[+] *S. aureus* strains produced this compound, which induced pAM373 to produce adhesin and subsequent clumping of *E. faecalis* cells. This effect was not observed for other sex pheromone plasmids, pAD1 and pPD1; moreover not a single coagulase[−] staphylococcal strain was observed to produce this effect. These authors postulated that the function of staph-cAm373 was to stimulate transfer of pAM373 from *E. faecalis* to *S. aureus* strains.

Several of the pheromone-responding plasmids described above can mobilize bacterial drug resistance markers unlinked to a particular plasmid. The mating response and subsequent aggregation in broth has resulted in the transfer of the nonconjugative R plasmids pAMβ1 (Dunny and Clewell, 1975) and pAD2 (Tomich et al., 1979). Also plasmids that transfer poorly in broth (pAMβ1) (LeBlanc et al., 1978) have improved efficiencies of transfer (as high as 10^{-4} per donor) if they

TABLE 3.   Sex Pheromones and Related Inhibitors

Pheromone/Inhibitor	Peptide Sequence[a]	Plasmids Affected
cAD1	H-Leu-Phe-Ser-Leu-Val-Leu-Ala-Gly-OH	pAD1, pAMα1, pJH2, pBEM10
cAM373	H-Ala-Ile-Phe-Ile-Leu-Ala-Ser-OH	pAM373
cCF10	H-Leu-Val-Thr-Leu-Val-Phe-Val-OH	pCF10
cOB1	H-Val-Ala-Val-Leu-Val-Leu-Gly-Ala-OH	pOB1, pYI1
cPD1	H-Phe-Leu-Val-Met-Phe-Leu-Ser-Gly-OH	pPD1
iAD1	H-Leu-Phe-Val-Val-Thr-Leu-Val-Gly-OH	
iPD1	H-Ala-Leu-Ile-Leu-Thr-Leu-Val-Ser-OH	

[a] Data presented in Clewell, 1993, and Nakayama et al., 1995.

are in donors along with plasmids involved in the pheromone response (Clewell et al., 1982). Finally plasmids involved in the mating response can cause transfer of chromosomal markers and can greatly increase the transfer frequency of conjugative transposons, even between species (Franke and Clewell, 1981; Roberts and Kenny, 1987). This may be true even if the transferred plasmid itself is not able to replicate; in such cases, upon entry into the new cell, the transposon is induced to transpose and inserts into the chromosome. The spread of antibiotic resistance may thus be greatly facilitated by sex pheromone–mediated conjugation in this species (Clewell et al., 1986).

III. EUKARYOTIC PLASMIDS

Naturally occurring plasmids in eukaryotes have been most frequently observed and studied in fungi. Some have also been reported in other microbes. These autonomously replicating nucleic acids can be best divided into those found in organelles, those found in the cytoplasm, and those residing and maintained in the nucleus. Those plasmids found in fungal mitochondria have been thoroughly reviewed (see Griffiths, 1995); in addition circular dsRNAs have been found in *Ustilago maydis* that appear to be viral sequences associated with a "killer" phenotype in this fungus (Koltin and Day, 1976). These two types of molecules will not be discussed in this chapter. Instead, we will focus on plasmids that reside

in the nucleus. These include the 2μm and 2μm-like plasmids of yeasts, the *Dictyostelium* plasmids, and plasmids found in the parasitic protozoans. Although these latter plasmids are widespread in protozoans (Gardner et al., 1991; Hightower et al., 1988), only the *Saccharomyces cerevisiae* 2μm and *Dictyostelium* plasmids are well studied in terms of their replication.

A. Replication Control

1. 2μm plasmids in *S. cerevisiae*

In most strains of *S. cerevisiae* there exists a double-stranded circular plasmid DNA found in the nucleus at 30 to 100 copies per cell. This DNA may constitute up to 5% of the total nuclear DNA (Broach and Volkert 1991). Cells that contain these plasmids are referred to as cir[+], whereas those that are plasmid free are cir°. The only observable phenotypic difference between isogenic cir[+] and cir° strains is that the cir[+] strains have a slightly reduced growth rate (Mead et al., 1986). Thus the 2μm plasmids would appear to be relatively benign subcellular parasites (although at artificially high copy numbers, unusual cell phenotypes do occur). This contrasts with other fungal plasmids which appear to encode molecules necessary for the organism. For example, a circular extrachromosomal DNA has been observed in the mucoraceous fungus *Absidia glauca* (Hänfler et al., 1992). In this case the plasmid encodes a surface protein present only in the mycelia of mating-type (+) strains.

As with the other plasmids we have already discussed, persistence of 2μm plasmids in populations of yeast requires appropriate assurance of replication, and control of copy number and plasmid partitioning to daughter cells. A conventional ARS (autonomously replicating sequence) is used for initiation of replication by host proteins. Each ARS contains an 11 bp core called an ACS (ARS consensus sequence: 5′-A/TTTTA(A)TA/GTTTA/T-3′; Blackburn, 1985) and an AT-rich stretch of nucleotides to the 3′ of the T-rich strand of the ACS. Just as host chromosome replication follows strict cell cycle limitations, 2μm plasmid replication is limited to the S phase of cell growth; other requirements of chromosomal DNA replication also are followed (Zakian et al., 1979). This means that the only four open reading frames (ORFs) present on the plasmid must determine both the copy number control and plasmid maintenance functions. One of these ORFs, FLP, encodes a recombinase that causes copy number increase under certain circumstances by allowing production of multiple copies, even though there is only one replication initiation per plasmid in any one S phase. Two other genes, REP1 and REP2, encode proteins that repress FLP, while RAF encodes an antirepressor whose expression is, in turn, negatively regulated by the Rep1p/Rep2p complex. Murray et al. (1987) have presented a model in which increased FLP expression counteracts decreases in plasmid copy numbers—as the expression of REP1 and/or REP2 falls, there is a direct effect of reducing the repression on FLP; there is also the indirect effect of increased RAF expression.

2. Linkage of transcription and replication?

There is a developing body of information that suggests that replication in eukaryotes may be linked to transcription. This is not unreasonable. Remember that we have already seen such linkage for prokaryotic plasmids, such as ColE1, where transcription through the origin is required to produce a primer for replication. Regulating whether the resulting RNA pre-primer is cleaved within the origin is a way to control copy number for such plasmids (see Section IIA.1b above). In S. cerevisiae, studies of ARS1 have shown that this replication origin consists of multiple binding elements, and one of these is a binding site for ABF1, an activator of transcription. The ARS1 origin is bound in vitro and in vivo by a protein complex, the ORC (origin recognition complex); one member of this complex is linked to transcriptional repression (Patnaik et al., 1994). The ORC also binds each of the four silencers at the silent mating-type loci HML and HMR (Patnaik et al., 1994) and silencing in S. cerevisiae is dependent on DNA replication. Together these facts suggest that origins and their associated proteins may play a role in transcriptional domains on eukaryotic chromosomes and possibly on plasmids.

Transcription likewise plays an important role in the replication of plasmids in Trypanosoma brucei, one of the organisms that causes African trypanosomiasis. This organism is among the most ancient of the eukaryotes (Patnaik et al., 1994). Transcription is polycistronic, which is unusual for eukaryotes, and there is a trans-splicing of a 39 nt spliced leader onto each mature mRNA. The PARP (procyclic acidic repetitive protein) promoter is used during the stage in which this protozoan develops within the insect midgut. This promoter has been shown to play a role in plasmid DNA replication, while other T. brucei promoters do not support stable extrachromosomal maintenance of DNAs (Patnaik et al., 1994). In contrast to what we saw for ColE1 plasmids, however, transcription into or through the origin may not be the critical factor (Patnaik et al., 1994). So how does it play a role? Well, proteins that bind to transcriptional regulatory elements in the DNA could have an indirect effect—by preventing binding of core histones, thus keeping the DNA open for assembly of the replication initiation complex; alternatively, by binding with or stabilizing the initiation complex at the origin, their effect could be indirect.

3. *Entamoeba histolytica* rDNA plasmid

Entamoeba histolytica is a protozoan which possesses a 24.5 kb nuclear plasmid, EhR1, encoding the rRNA genes. This plasmid is found in multiple copies and its amount in the cell comprises about 10% of the genome. Within the plasmid there are two rRNA-encoding genes found as inverted repeats, separated by upstream and downstream spacers. In microbial DNAs, and in microbial plasmids in particular, initiation has been found to occur at a specific site or sites. In the *D. discoideum* plasmid 2, for example, initiation occurs at three copies of a 49 bp element that is an essential part of the origin (Dhar et al., 1996). Strikingly, the EhR1 rDNA plasmid of *E. histolytica* was found instead to initiate at multiple dispersed sites, much like what is seen in metazoan chromosomes. Replication of the plasmid was examined (Dhar et al., 1996) by neutral/neutral 2D gel electrophoresis (Brewer and Fangman, 1987) and by electron microscopy. Initiation was found to occur more frequently from within the transcription units than from the intergenic spacers. There was also a direct observation of multiple bubbles within the same molecule.

Such a mechanism of initiation is in contrast with what has been observed for other rDNA plasmids and for other rDNAs. For example, for the yeast rDNA cluster initiation occurs in the intergenic spacer at intervals of 5 repeat units. The rDNA plasmids of other unicellular eukaryotes provide further contrasts: *Naegleria gruberi* (Clark and Cross, 1987) and *Euglena gracilis* (Ravel-Chapuis et al., 1985), where rolling circle intermediates have been observed. For the EhR1 rDNA plasmid of *E. histolytica*, the two transcription units were inverted. If the replication forks were moving in the direction opposite to that of transcription and they were stalled, replication of the intergenic region could be blocked. But this was not the case. So either replication of these plasmids was not affected by rDNA transcription or the plasmids used for rRNA synthesis were not the templates used for replication.

B. Partitioning

1. 2μm plasmids in *S. cerevisiae*

As mentioned above, stable extrachromosomal maintenance of DNAs requires both a mechanism for replication and a mechanism of partitioning. Epstein Barr virus, that maintains itself in infected cells extrachromosomally, requires two regions, one that serves as an origin and one that causes the viral DNA to be maintained in the nucleus (Patnaik et al., 1994). In yeast, if a plasmid contains only the replicative functions of an ARS, then such a plasmid is highly unstable. In fact, when investigators attempt to use such plasmids to transform yeast, they must maintain a constant selection for traits encoded by the plasmid, or the plasmid will be lost. This will be discussed in greater detail in Appendix A, under Plasmids used for Cloning and Expression in Yeast.

Why are such plasmids unstable in yeast? This is due to a phenomenon known as MIB (maternal inheritance bias)—such plasmids replicate efficiently but are lost in 10–50% of cells per cell division, and they collect at very high levels in what are predominantly mother cells. This occurs despite a relatively equal division of the nuclear volume during mitosis. Although the basis for MIB is not yet understood, it does play a role in the "aging" of yeast cells. This occurs when plasmids bearing an ARS but lacking partitioning functions are used in transformation or naturally, when excised circular molecules bearing the rDNA region(s) and an ARS accumulate (Sinclair and Guerente 1997).

So how do plasmids like the 2μm circle overcome this problem? The *STB* locus is required in *cis* and thus would seem to provide a *CEN*-like function we observed for the *parA* locus of the IncFII plasmids, as well as others discussed previously (in Section IIB.2). This locus contains five copies of imperfect tandem repeats of a 62 bp sequence and three if the four proteins encoded by the plasmid bind to these repeats in a sequence-specific manner (Bogdonova et al., 1998). Moreover two of these proteins, Rep1p and

Rep2p, are required for plasmid maintenance and correct partitioning. As just mentioned, both proteins bind to *STB*, and there
is a wealth of data that suggests that both
proteins are localized to discrete nuclear
sites. First of all, Rep1p co-purifies with the
nuclear matrix-pore complex-lamina fraction and solubilization of both Rep1p and
Rep2p requires urea (Hadfield et al., 1995).
Immunofluorescence experiments show that
these nuclear proteins are dispersed throughout the nucleus, although Rep2p appears in
the nucleus only when co-expressed with
Rep1p. Both proteins can bind independently to *STB*, but they apparently require
additional yeast-encoded components for efficient binding and function (Hadfield et al.,
1995). Moreover fusion proteins between the
green fluorescence protein (GFP) and the
Rep1p and Rep2p proteins provided further
evidence about binding and localization of
the proteins (Ahn et al., 1997). The fluorescent fusion proteins were localized best in
cir$^+$ strains, whereas in cir$^\circ$ strains, they
were not well localized to the nucleus. Together with *REP1*, the GFP-Rep2p promoted stability of 2μm-derived plasmids;
similarly, when present with a functional
REP2 gene, the GFP-Rep1p promoted maintenance of such plasmids as well. In both
cases the proteins localized to the nucleus
(Ahn et al., 1997). It was found that localization did not require FLP or RAF1. Immu-
noprecipitation and protein-binding assays
(similar to the yeast two-hybrid assays)
showed the following interactions: Rep1p-
Rep1p, Rep2p-Rep2p, and Rep1p-Rep2p.

Scott-Drew and Murray (1998) also provided direct evidence for Rep1p/Rep2p interactions. They confirmed that these proteins
form homocomplexes and heterocomplexes
in vitro. They also used anti-Rep1p and
anti-Rep2p antisera to demonstrate the localization of these proteins to specific nuclear
sites. Further the sites were found to duplicate during the cell cycle and segregate so that
both the mother and daughter cells possessed
them. The experiments also showed that
Rep2p was required for proper localization

of Rep1p to the nucleus. Ultimately, these
findings support a model for 2μm partitioning in which plasmid molecules associate with
specific segregated sites in the nucleus (Reynolds et al., 1987; Scott-Drew and Murray
1998). The amount and position of the nuclear sites for the Rep proteins is similar to
that observed for Sir2-4p (the silencing proteins, Palladino et al., 1993; see below) and
for Rap1p, the telomere-associated protein
(Gotta et al., 1996). Several groups are actively investigating whether the 2μm DNAs
are in fact interacting with such regions.

2. Transcription plays a role in partitioning

A process known as *transcriptional repression*
or *silencing* affects genes placed near the silent
mating-type loci in *S. cerevisiae* or adjacent to
telomeres (telomere position effect). Similar
cis and *trans* factors act at both locations, so
the mechanisms for silencing at both locations must be related. The *HML* and *HMR*
loci each contain two silencers, E and I. These
silencers have binding sites for three proteins
or complexes: Rap1p, Abf1p, and ORC. Telomeres contain many Rap1p binding sites
within a 300 to 600 bp $(TG_{1-3})_n$ /$(C_{1-3}A)_n$
repeated sequence. Binding sites for Abf1
and ORC (which, remember, binds to
ARSs) are located within groups of larger
subtelomeric repeats. These repeats are not
necessary for either chromosome maintenance or telomeric silencing.

Interestingly both the *HMR* E silencer and
telomeric sequence tracts (TG_{1-3} arrays)
also have been shown to function in plasmid
partitioning (Kimmerly and Rine, 1987;
Longtine et al., 1992). Why should this be
true for telomeres? From cytological evidence it has been found that telomeres play
an important role in the transport of
chromosomes to or keeping chromosomes
at nuclear positions. During meiotic pro-
phase in *S. pombe* telomeres lead movement
of the chromosomes; the telomeres then cluster near the spindle pole body. In contrast,
during mitotic growth, centromeres are used
for chromosome movement and telomeres
group at the outskirts of the nucleus. So

perhaps the telomeric sequences are guiding the plasmids to specific nuclear locations for proper segregation.

And what about the components of silencing? Three *SIR* gene products, Sir2p, Sir3p, and Sir4p are needed to form a complex with a repressive effect on chromatin that blocks access to DNA. Sir3p and Sir4p interact with peptides in histone tails and with DNA-bound Rap1p. In this way the silencing complex is brought to silencers and to telomeres. Sir4p, together with Orc1p and Sir1p, help collect other components at silencers. But, if Sir3p and Sir4p are already bound to the DNA directly, say via fusion of another protein's DNA-binding domain to Sir4p, there is no longer a requirement for Rap1p (Lustig et al., 1996).

There are similar requirements for partitioning of plasmids bearing the *HMR* E silencer or telomeric sequence tracts, i.e., Rap1p and the Sir protein partitioning is blocked if a telomere-based plasmid is present in a *rap1* mutant or if the Rap1p binding site is removed from an *HMR* E plasmid. There are two models to explain this partitioning effect. In the first model, Sir proteins dislodge plasmids from the mother cell to allow exchange between the mother and bud nuclei. An alternative model is that partition occurs by attachment of chromatin-bound Sir proteins to a nuclear target that is divided evenly between the cells.

Evidence supports the second model, some evidence stronger than others. The Sir4p contains a motif homologous to regions in lamins A and C in the human nuclear envelope, and this might link telomeres to the lamin-like shell at the boundary of the nucleus. The importance of this evidence is diminished by the fact that no true lamins have yet been found in yeast. Moreover the coiled-coil motif of Sir4p is not sufficient for plasmid partitioning or DNA anchoring; instead, the PAD4 domain (partitioning and anchoring domain, amino acids 950–1262) is sufficient for both these activities (Ansari and Gartenburg, 1997). Stronger evidence for the model is as follows. Overexpression

of a C-terminal Sir4p fragment releases Sir3p and full-length Sir4p from nuclei. Also Sir4p co-localizes with Sir3p and telomeric DNA in a pattern of distinct dots in the peripheral shell nuclear volume.

Ansari and Gartenburg (1997) examined the role of Sir4p in plasmid partitioning. Sir4p was attached to the DNA via a heterologous DNA-binding domain. Mitotic segregation of a plasmid bearing LexA-binding sites was examined when Sir4p was fused to the LexA protein. Such constructs were able to partition normally unstable plasmids. Targeting of Sir4p to DNA stabilized plasmids and bypassed the need for silencers or telomeric sequences in partitioning. How does this occur? Via its association with an immobile nuclear component, the Sir4p divides between the progeny at mitosis. However, despite its importance and obvious involvement, the Sir4p is not likely the sole component directing telomeric DNA movements, and telomeric plasmids are still able to partition in *sir* mutants.

3. 2μm plasmid-like plasmids in other fungi

What has been seen for 2μm plasmids in *S. cerevisiae* is also true for several related yeasts, including *Zygosaccharomyces* and *Kluyveromyces*. As seen below, there may also be some components of this system that allow for plasmids to be maintained in yeasts for which naturally occurring plasmids have not heretofore been observed.

2μm-like plasmids in *Zygosaccharomyces rouxii*. *Z. rouxii* is an osmotolerant yeast often found associated with food spoilage. Strains have been shown to possess the 2μm-like plasmids, pSR1 and pSB3 (Toh-e et al., 1982; Jearnpipatkul et al., 1987; Toh-e and Utatsu 1985). These plasmids contain ORFs whose presence are necessary for a *trans*-acting stability function. There are also *cis*-acting components that include both direct and inverted repeats; these elements are apparently sufficiently conserved to allow the stability of pSR1-like plasmids in *S. cerevisiae*, where they presumably interact with *S. cerevisiae* components.

Action of 2μm Components in *Hansenula polymorpha*. Recall that in the *S. cerevisiae* 2μm plasmid there are five copies of imperfect tandem repeats of a 62 bp sequence; three out of four proteins encoded by this plasmid bind to the *STB* repeats in a sequence-specific manner: Rep1, Rep2, RAF. Moreover the yeast chromosomes encode specific STB-binding components. *Hansenula polymorpha* is a methylotrophic yeast for which endogenous plasmids have not been observed. Nevertheless, plasmids have been constructed for use in this organism by cloning endogenous *H. polymorpha* ARS sequences into bacterial plasmids. These plasmids are not stably inherited in transformants. However, insertion of the 2μm *STB* locus and some surrounding DNA into *H. polymorpha* replicative plasmids leads to decreased copy number and an increase in both meiotic and mitotic stability (Bogdanova et al., 1998). This increase appeared to be associated with several copies of 24–28 bp imperfect repeats. Deletion of these repeats lead to a decrease in plasmid stability and indicated that downstream sequences may affect chromatin structure of the plasmid region(s) involved in stability.

4. *CEN* plasmids

Naturally occurring plasmids bearing *CEN* sequences have not been observed. Nevertheless, when such sequences have found their way onto plasmids in the laboratory, they provide an increased stability for the plasmids. (A discussion of their usage in gene cloning and expression experiments in yeast in provided in Appendix A.)

IV. COMPARISONS AND CONCLUSIONS

A. Plasmid Replication Control in Prokaryotes

Plasmids, if they are to be considered as endosymbionts, must depend to a certain degree on host factors encoded by the host chromosome. But the replication maintenance and copy number control functions must be encoded by the plasmid. There are two types of control mechanisms:

1. *Positive control,* where the addition of factors is needed for initiation. Sometimes these factors are *proteins,* such as the RepA protein of R1 (Danbara et al., 1980) or of P1 (Chattoraj et al., 1985), the RepC protein of pT181 (Novick et al., 1986), and (protein of R6K (Filutowicz et al., 1986). Sometimes the controlling factor is an *RNA molecule,* such as the preprimer RNA II for ColE1.

2. *Negative control,* where factors directly or indirectly inhibit replication. There are *proteins,* such as the CopB protein of R1 (Wagner and Nordström, 1986), *RNAs,* such as the *cop* RNA molecules of ColE1 and the IncFII plasmids (Wagner and Nordström, 1986), and *direct repeats* of DNA in the origins of several plasmids (Stalker et al., 1982; Chattoraj et al., 1985). Negative control appears to be the most important form of control, since there are no plasmids known that lack this type of control (Nordström, 1985).

Folding of RNA is known to be important for a number of control systems. We can compare this type of control for ColE1, the IncFII plasmids, and pT181. In all three systems a positive regulatory protein is required for initiation. For all three, refolding of a large regulatory RNA with a positive regulatory effect (direct for ColE1, indirect for pT181) is induced by a small negative regulatory RNA. Further comparisons are given in Table 4.

The use of a small *trans*-acting RNA for negative regulation is *not* found in large low copy plasmids such as F and P1, which use a positive regulatory protein that autoregulates its expression.

Regarding conservation of space, there is an overlap in the sequences encoding two replication functions—the countertranscripts and their target RNAs for ColE1 and the IncFII plasmids (as well as the *hok* and *sok* functions of R1, see below). Why should there be this conservation of space? The fact that two

TABLE 4. Comparisons among ColE1, IncFII Plasmids, and pT181

ColE1	IncFII	pT181
The large regulatory RNA is a preprimer that hybridizes to the template DNA and is cleaved to the primer by RNase H.	Large RNA is a message for the RepA1 positive regulatory protein for initiation.	Large RNA is the transcript for the RepC positive regulatory protein
A cloverleaf forms that exposes key single-stranded regions.	Folding of the large RNA exposes a Shine-Delgarno sequence for the *repA*1 transcript.	Large RNA folding exposes the *repC* transcript's Shine-Delgarno sequence.
The role of the small RNA is to base-pair with the preprimer to prevent RNase H from forming the primer.	Small RNA role: refold the large RNA so that the Shine-Delgarno sequence is no longer in the single-stranded region.	The small RNA binds with large RNA to cause refolding: this produces signal to terminate transcription of the large RNA.
The negative regulatory RNA is a countertranscript for the large RNA.	Negative regulatory RNA is a countertranscript for the large RNA.	Negative regulatory RNA is countertranscript for the large RNA.
The effect of successful regulation is that no primer is formed. No initiation comes from the individual plasmid.	RepA protein (which only acts in *cis*) is not formed; replication cannot be initiated from the individual plasmid.	RepC protein (which can function in *trans*) is not made, since its transcript is not completed.

functions are tightly linked means that they are likely to be inherited together. Also there should be a decrease in genetic drift (accumulation of random mutations), since the regulatory sequence is also part of the origin or initiator sequence. In such a circumstance one mutation could be lethal. This arrangement decreases the likelihood of mutation in the region of the regulatory gene.

Are there copy number limitations besides the normal negative regulatory circuit? Copy number mutations can only be isolated if they are not lethal. This means there must be some other type of control at work. Copy number might be limited by some new factor that then becomes limiting in the copy number mutant. For example, the pVH51 mini ColE1 plasmid has about the same amount of plasmid DNA per cell as ColE1 (300 megadaltons in exponentially growing

cells; Hershfield et al., 1976). This is so even though pVH51 has a fivefold increase in copy number relative to ColE1. Availability of some positive regulatory protein may also play a role. For P1 and other low copy plasmids, RepA is rate-limiting for initiation of replication. It appears that these proteins may have two active domains, one for replication initiation, and the other for autoregulation. The autoregulatory domain would bind to the operator to block transcription. However, since there are several such sites on the plasmid strategically placed for cooperativity, there is a competition between autoregulation and initiation of replication. Thus the amount bound at any time is critical to the rate of replication. In R6K the π protein is a positive regulatory protein that is required for replication but is *not* rate limiting. This system does use repeated DNA sequences for its positive effect.

B. Partition in Prokaryotes

There are two basic strategies for partitioning of the plasmid copies to the daughter cells. The high copy number plasmids exploit the fact that there is a high probability of copies being transmitted to daughter cells by random distribution. Controls for high copy number plasmids ensure a rapid recovery in cells receiving few copies. There are also special mechanisms that ensure that there will be a high number of free copies of the plasmid. For example, we have seen this for the *cer* site in ColE1, and possibly this true for the *par* site in pSC101 (Biek and Cohen, 1986). It takes a great deal of energy to make so many copies, so high copy number plasmids tend to be small in size. This means that their maintenance tends to be less of a drain on the host; they must therefore depend on more host functions for their survival.

The low copy plasmids, such as P1 and F, cannot be properly distributed by random diffusion. They require an *active-partition system*. For P1, ParA, and ParB are two plasmid-encoded proteins that work at a *cis*-acting partition site, ParS, in a process that seems analogous to mitosis in eukaryotes. This allows partition in the absence of extensive host burden in the process and goes along a general design for the large low copy plasmids. Because they are large, they tend to encode a considerable amount of information for their own maintenance (e.g., P1 lytic functions and F conjugation). The large size of these plasmids allows acquisition of new functions that enhance their ability to fine-tune partition (e.g., the ability to resolve dimers) (Austin et al., 1986). Another partition mechanism is found for R1. Here we have a *parA* locus that is similar to the partition system for P1 (without incompatibility). In addition R1 (and possibly the other IncFII plasmids) has a *parB* locus. This locus ensures that a plasmid is maintained in a population of cells, not by ensuring proper segregation, but by *killing* all those cells that do not receive a copy of the plasmid! This mechanism of partition may be quite common among low copy number plasmids. Similar mechanisms also occur in F (*ccdA* and *ccdB*). Alternative postsegregational killing mechanisms involving poison-antidote or restriction-modification add to the arsenal of strategies available to plasmids to ensure proper maintenance in cells. Such mechanisms lend credence to the view of plasmids as subcellular organisms. It is likely that by ensuring proper segregation of plasmids, the *par* loci are responsible for maintaining a pool of antibiotic-resistant bacteria even in the absence of selection.

C. What about Eukaryotic Plasmids?

Plasmids found in eukaryotic organisms face many of the same problems and requirements as those we have just discussed for prokaryotes. For example, in both prokaryotes and eukaryotes a place to initiate replication is required for the plasmid—*ori* for prokaryotes, ARS for eukaryotes.

Control of replication is just as important for eukaryotic plasmids as for prokaryotic plasmids. In eukaryotes, as in prokaryotes, recombination is used as one tool in regulating plasmid copy number. In eukaryotes, however, this is employed as a mechanism to *increase* the copy number when normal constraints of the cell cycle serve as an impediment to repeated initiations of replication. So we have the Flp recombinase which provides multiple copies of the 2μm plasmids, whereas recombinase systems like *cer*/Xer (ColE1) and *cre*/*lox* (P1) are used to resolve multimers of their respective plasmids so that the copy number control system is not confused into allowing a drop in plasmid copy number below what would be acceptable for proper maintenance.

And how about plasmid maintenance? Remember, we just said above that the prokaryotic high copy number plasmids exploit the fact that there is a high probability of copies being transmitted to daughter cells by random distribution. However, in yeast this strategy does not appear to be sufficient for stable segregation of ARS plasmids lacking a *CEN* region (Murray and Szostak, 1983).

Partition strategies for eukaryotic and prokaryotic plasmids do share some common features, but there are more similarities between maintenance mechanisms of eukaryotic plasmids and those of the lower copy plasmids of prokaryotes. The most notable similarity is the requirement of a specific site on the plasmid (*par* loci in prokaryotes, *STB* on 2μm plasmids, *CEN* on *CEN*-based plasmids) that is recognized by plasmid-encoded proteins and/or cellular components. In prokaryotes, the proteins are the plasmid-encoded (Par) A and B proteins that bind the *par* site on the plasmid and serve as liaison between these and the corresponding cellular sites on the membrane of dividing bacteria. The analogous proteins in eukaryotes are the Rep1p and Rep2p for the 2μm plasmid. In both cases the mode of segregation is independent from that of the resident cellular chromosome(s). In contrast, for the *CEN*-based plasmids, the proteins that mediate segregation are the very spindle fibers used by the cell to normally distribute the replicated chromosomes into the daughter cells.

In conclusion, we have seen a myriad of strategies that have evolved to provide plasmids with the ability to propagate themselves and to be maintained within populations of bacteria or other microbes. Although these strategies may differ in specifics, a number of similarities are found. If one steps back to view the bigger picture, plasmid survival is tied to and dependent on the cellular machinery of the host and the subsequent survival of hosts bearing the plasmid. Therefore plasmid survival depends on adapting to the host's way of doing things and providing a useful, selectable advantage to the host.

ACKNOWLEDGMENTS

The author is indebted to Diane Ashby for typing the first daft of this chapter and to David Smith for help in production of several of the figures. I would also like to thank J. Seth Strattan for helpful discussions regarding recent work on the mechanism of *mok/hok/sok* interactions.

REFERENCES

Abeles AL, Friedman SA, Austin SJ (1985): Partition of unit-copy miniplasmids to daughter cells III. The DNA sequence and functional organization of the P1 partition region. J Mol Biol 185:261–272.

Abeles AL, Snyder KM, Chattoraj DK (1984): P1 plasmid replication: Replicon structure. J Mol Biol 173:307–324.

Ahn YT, Wu XL, Biswal S, Velmurugan S, Volkert FC, Jayaram M (1997): The 2 μm-plasmid-encoded Rep1 and Rep2 proteins interact with each other and colocalize to the *Saccharomyces cerevisiae* nucleus. J Bacteriol 179:7497–7506.

Alexeyev MF (1999): The pKNOCK series of broad-host-range mobilizable suicide vectors for gene knockout and targeted DNA insertion into the chromosome of Gram-negative bacteria. BioTechniques 26:825–828.

Ansari A, Gartenberg MR (1997): The yeast silent information regulator Sir4p anchors and partitions plasmids. Mol Cell Biol 17:7061–7068.

Austin S, Friedman S, Ludtke D (1986): Partition functions of unit-copy plasmids can stabilize the maintenance of plasmid pBR322 at low copy number. J Bacteriol 168:1010–1013.

Bagdasarian MM, Scholz P, Frey J, Bagdasarian M (1986): Regulation of the *rep* operon expression in the broad host range plasmid RSF1010. In Levy SB, Novick R (eds): "Antibiotic Resistance Genes: Ecology, Transfer, and Expression," Banbury Report 24. Cold Spring Harbor, NY: Cold Spring Harbor Laboratory.

Barbour AG, Garon CF (1987): Linear plasmids of the bacterium *Borrelia burgdorferi* have covalently closed ends. Science 237:409–411.

Bastia D, Germino J, Crosa JN, Ram J (1981): The nucleotide sequence surrounding the replication terminus of R6K. Proc Natl Acad Sci USA 78:2095–2099.

Biek DP, Cohen SN (1986): Identification and characterization of *recD*: A gene affecting plasmid maintenance and recombination in *E. coli*. J Bacteriol 167:594–603.

Bittner M, Vapnek D (1981): Versatile cloning vectors derived from the runaway-replication plasmid pKN402. Gene 15:319–329.

Blackburn EH (1985): Artificial chromosomes in yeast. Trends Genet (January): 8–12.

Blomberg P, Nordström,K, Wagner EGH (1992): Replication control of plasmid R1: RepA synthesis is regulated by CopA RNA through inhibition of leader peptide translation. EMBO J 11:2675–2683.

Blomberg P, Engdahl HM, Malmgren C, Romby P, Wagner EGH (1994): Replication control of plasmid R1: Disruption of an inhibitory RNA structure that sequesters the *repA* ribosome-binding site permits *tap*-

independent RepA synthesis. Mol Microbiol 12:49–60.

Bogdanova AI, Kustikova OS, Agaphonov MO, Ter-Avenesyan MD (1998): Sequences of the *Saccharomyces cerevisiae* 2 μm DNA improving plasmid partitioning in *Hansenula polymorpha*. Yeast 14:1–9.

Brantl S, Wagner GHW (1997): Dual function of the *copR* gene product of plasmid pIP501. J Bacteriol 179:7016–7024.

Brewer BJ, Fangman WL (1987): The localization of replication origins on ARS plasmids in *S. cerevisiae*. Cell 51:463–471.

Briaux S, Gerband G, Jaffe-Brachet A (1979): Studies of a plasmid coding for tetracycline resistance and hydrogen sulfide production incompatible with the prophage P1. Mol Gen Genet 170:319–325.

Broach JR, Volkert FC (1991): Circular DNA plasmids of yeasts: genome dynamics, protein synthesis and energetics. In Broach JR, Pringle JR, Jone EW (eds) "The Molecular and Cellular Biology of the Yeast *Saccharomyces*". Cold Spring Harbor, NY: Cold Spring Harbor Press, pp 297–331.

Buckle TA, Krüger DH (1993): Biology of DNA restriction. Microbiol Rev 57:434–450.

Buu-hoi A, Horodniceanu T (1980): Conjugative transfer of multiple antibiotic resistance markers in *Streptococcus pneumoniae*. J Bacteriol 143:313–320.

Calladine C (1982): Mechanics of sequence-dependent stacking of bases in B-DNA. J Mol Biol 161:343–352.

Chakrabarty AM (1972): Genetic basis of the biodegradation of salicylate in *Pseudomonas*. J Bacteriol 112:815–823.

Chan PT, Ohmori H, Tomizawa J, Leibowitz J (1985): Nucleotide sequence and gene organization of ColE1 DNA. J Biol Chem 260:8925–8935.

Chang ACM, Slade MB, Williams KL (1990): Identification of the origin of replication of the eukaryote *Dictyostelium discoideum* nuclear plasmid Ddp2. Plasmid 24:208–217.

Chang S, Chang S-Y, Gray D (1987): Structural and genetic analyses of a *par* locus that regulates plasmid partition in *Bacillus subtilis*. J Bacteriol 169:3952–3962.

Chatterjee DK, Chakrabarty AM (1983): Genetic homology between independently isolated chlorobenzoate-degradative plasmids. J Bacteriol 153:532–534.

Clark CG, Cross GAM (1987): rRNA genes of *Naegleria gruberi* are carried exclusively on a 14-kilobase-pair plasmid. Mol Cell Biol 7:3027–3031.

Clewell DB (1981): Plasmids, drug resistance and gene transfer in the genus *Streptococcus*. Microbiol Rev 45:409–436.

Clewell DB (1985): Sex pheromones, plasmids, and conjugation in *Enterococcus faecalis*. In Halvorson HO, Monroy A (eds): "The Origin and Evolution of Sex." New York: Alan R. Liss, pp 13–28.

Clewell DB (1993): Sex pheromones and the plasmid-encoded mating response in *Enterococcus faecalis*. In Clewell DB (ed): "Bacterial Conjugation" New York: Plenum Press, pp 349–367.

Clewell DB, An FY, White BA, Gawron-Burke C (1985): *Enterococcus faecalis* sex pheromone (cAM373) also produced by *Staphylococcus aureus* and identification of a conjugative transposon (Tri916). J Bacteriol 162:1212–1220.

Clewell DB, Ehrenfeld EE, Kessler RE, Ike Y, Franke AE, Madion M, Shaw JH, Wirth R, An F, Mori M, Kitada C, Fujino M, Suzuki A (1986): Sex pheromone systems in *Enterococcus faecalis*. In Levy SB, Novick RP (eds): "Antibiotic Resistance Genes: Ecology, Transfer, and Expression." Banbury Report 24. Cold Spring Harbor, NY: Cold Spring Harbor Laboratory.

Clewell DB, Yagi Y, Ike Y, Carig RA, Brown BL, An F (1982): Sex pheromones in *Enterococcus faecalis*: Multiple pheromone systems in strain D55, similarities of pAD1 and pAMγ1, and mutants of pAD1 altered in conjugative properties. In Schlessinger D (ed): "Microbiology 1982." Washington, DC: ASM Press.

Conley D, Cohen SN (1995): Isolation and characterization of plasmid mutations that enable partitioning of pSC101 replicons lacking the partition (*par*) locus. J Bacteriol 177:1086–1089.

Contreras A, Molin S, Ramos, JL (1991): Conditional-suicide containment system for bacteria which mineralize aromatics. Appl Environ Microbiol 57:1504–1508.

Couturier M, Bex F, Bergquist PL, Mass WK (1988): Identification and classification of bacterial plasmids. Microbiol Rev 52:375–395.

Crosa JH (1980): Three origins of replication are active in vivo in R-plasmid RSF1040. J Biol Chem 255:11075–11077.

Danbara H, Timmis JK, Lurz R, Timmis KN (1980): Plasmid replication functions: Two distinct segments of plasmid R1, RepA and RepD, express incompatibility and are capable of autonomous replication. J Bacteriol 144:1126–1138.

Danilevskaya ON, Gragerov AI (1980): Curing of *Escherichia coli* K12 plasmids by coumermycin. Mol Gen Genet 178:233–235.

Datta N, Kontomichalou P (1965): Penicillinase synthesis controlled by infectious R factors in Enterobacteriaceae. Nature 208:239–241.

Davies JK, Reeves P (1975): Genetics of resistance to colicin in *Escherichia coli*: Cross-resistance among colicins of group B. J Bacteriol 123:96–101.

del Solar D, Acebo P, Espinosa M (1997): Replication control of plasmid pLS1: The antisense RNA II and the compact *rnaII* region are involved in translational regulation of the initiator RepB synthesis. Mol Microbiol 23:95–108.

Dhar SK, Choudhury NR, Mittal V, Bhattacharya A, Bhattacharya S (1996): Replication initiates at multiple dispersed sites in the ribosomal DNA plasmid of the protozoan parasite *Entamoeba histolytica*. Mol Cell Biol 16:2314–2324.

Donoghue DJ, Sharp PA (1978): Replication of ColE1 plasmid DNA in vivo requires no plasmid-encoded proteins. J Bacteriol 133:1287–1294.

Dunny GM, Brown BL, Clewell DB (1978): Induced cell aggregation and mating in *Enterococcus faecalis*. Evidence for a bacterial sex pheromone. Proc Natl Acad USA 75:3470–3483.

Dunny GM, Clewell DB (1975): Transmissible toxin (hemolysin) plasmid in *Enterococcus faecalis* and its mobilization of a non-infectious drug resistance plasmid. J Bacteriol 124:784–790.

Dunny GM, Leonard BAB, Hedberg PJ (1995): Pheromone-inducible conjugation in *Enterococcus faecalis*: Interbacterial and host-parasite chemical communication. J Bacteriol 177:871–876.

Easter CL, Sobecky PA, Helinski DR (1997): Contribution of different segments of the *par* region to stable maintenance of the broad-host-range plasmid RK2. J Bacteriol 179:6472–6479.

Easton AM, Sampathkuman P, Rownd RH (1981): Incompatibility of IncFll R plasmid NR1. In Ray DS (ed): "The Initiation of DNA Replication." New York: Academic Press, pp 125–141.

Eisenberg S, Scott JF, Kornberg A (1976): Enzymatic replication of viral and complementary strands of duplex DNA of phage ϕX174 proceeds by separate mechanisms. Proc Natl Acad Sci USA 73:3151–3155.

Erauso G, Marsin S, Benbouzid-Rollet N, Baucher M-F, Barbeyron T, Zivanovic Y, Prieur D, Forterre P (1996): Sequence of the plasmid pGT5 from the Archaeon *Pyrococcus abyssi*: Evidence for rolling-circle replication in a hyperthermophile. J Bacteriol 178:3232–3237.

Filutowicz M, Uhlenhopp E, Helinski DR (1986): Binding of purified wild-type and mutant (initiation proteins to a replication origin region of plasmid R6K. J Mol Biol 187:225–239.

Franke A, Clewell DB (1981): Evidence for a chromosome-borne resistance transposon in *Enterococcus faecalis* capable of "conjugal" transfer in the absence of a conjugative plasmid. J Bacteriol 145:494–502.

Frey J, Chandler M, Caro L (1979): The effects of an *Escherichia coli dnaAts* mutation on the replication of the plasmids ColE1, pSC101, R100.1 and RTF-TC. Mol Gen Genet 174:117–126.

Friedman S, Martin K, Austin S (1986): The partition system of the P1 plasmid. In Levy SB, Novick RP (eds): "Antibiotic Resistance Genes: Ecology, Transfer, and Expression." Banbury Report 24. Cold Spring Harbor, NY: Cold Spring Harbor Laboratory.

Funnell BE. (1988): Mini-P1 plasmid partitioning: Excess *parB* protein destabilized plasmids containing the centromere *parS*. J Bacteriol 170:954–960.

Funnell BE, Gagnier L (1995): Partition of P1 plasmids in *Escherichia coli mukB* chromosomal partition mutants. J Bacteriol 177:2381–2386.

Gardner NJ, Williamson DH, Wilson RJ (1991): A circular DNA in malaria parasite encodes an RNA polymerase like that of prokaryotes and chloroplasts. Mol Biochem Parasitol 44:115–123.

Gennaro ML, Kornblum J, Novick RP (1987): A site-specific recombination function in *Staphylococcus aureus* plasmids. J Bacteriol 169:2601–2610.

Gerdes K, Bech FW, Jorgensen ST, Lobner-Olesen A, Rasmussen PB, Atlung T, Boe L, Karlström O, Molin S, von Meyerburg K (1986a): Mechanisms of post-segregational killing by the *hok* gene product of the *parB* system of plasmid R1 and its homology with the *relF* gene product of the *E. coli relB* operon. EMBO J 5:2023–2039.

Gerdes K, Boe L, Anderson P, Molin S (1986b): Plasmid stabilization in populations of bacterial cells. In Levy SB, Novick RP (eds): "Antibiotic Resistance Genes: Ecology, Transfer and Expression." Banbury Report 24. Cold Spring Harbor, NY: Cold Spring Harbor Laboratory.

Gerdes K, Molin S (1986c): Partitioning of plasmid R1. Structural and functional analysis of the *parA* locus. J Mol Biol 190:269–279.

Gerdes K, Thisted T, Martinussen J (1990): Mechanism of post-segregational killing by the *hok /sok* system of plasmid R1: *sok* antisense RNA regulates formation of a *hok* mRNA species correlated with killing of plasmid-free cells. Mol Microbiol 4:1807–1818.

Gerlitz M, Hrabak O, Schwab H (1990): Partitioning of broad-host-range plasmid RP4 is a complex system involving site-specific recombination. J Bacteriol 172:6194–6203.

Gilbert W, Dressler D (1968): DNA replication: The rolling circle model. Cold Spring Harbor Symp Quant Biol 33:473–484.

Goebel W (1974): Integrative suppression of temperature-sensitive mutants with a lesion in the initiation of DNA replication. Replication of autonomous plasmids in the suppressed state. Eur J Biochem 43:120–125.

Gordon GS, Sitnikov D, Webb CD, Teleman A, Straight A, Losick R, Murray AW, Wright A (1997): Chromosome and low copy plasmid segregation in *E. coli*: Visual evidence for distinct mechanisms. Cell 90:1113–1121.

Gotta M, Laroche T, Formenton A, Maillet L, Scherthan H, Gasser SM (1996): The clustering of telomeres and colocalization with Rap1, Sir3, and Sir4 proteins in wildtype *Saccharomyces cerevisiae*. J Cell Biol 134:1349–1363.

Griffiths AJF (1995): Natural plasmids of filamentous fungi. Microbiol Rev 59:673–685.

Gruss AD, Ross H, Novick RP (1987): Functional analysis of a palindromic sequence required for normal replication of several staphylococcal plasmids. Proc Natl Acad Sci USA 84:2165–2169.

Gustafsson P, Nordström K (1978): Temperature-dependent and amber copy mutants of plasmid R1drd-a9 in *Escherichia coli*. Plasmid 1:134–144.

Hadfield C, Mount RC, Cashmore AM (1995): Protein binding interactions at the STB locus of the yeast 2μm plasmid. Nucleic Acids Res 23:995–1002.

Hakkaart MJJ, van den Elzen PJM, Veltkamp E, Nijkamp HJJ (1984): Maintenance of multicopy plasmid CloDF13 in *E. coli* cells: Evidence for site-specific recombination at *parB*. Cell 36:203–209.

Handa N, Ichige A, Kusano K, Kobayashi I (2000): Cellular responses to postsegregational killing by restriction-modification genes. J Bacteriol 182:2218–2229.

Hänfler J, Teepe H, Weigl C, Kruft V, Lurz R, Wöstemeyer J (1992): Circular extrachromosomal DNA codes for a surface protein in the positive mating type of the zygomycete *Absidia glauca*. Curr Genet 22:319–325.

Hansen EB, Yarmolinsky MB (1986): Host participation in plasmid maintenance: Dependence upon *dnaA* of replicons derived from P1 and F. Proc Natl Acad Sci USA 83:4423–4427.

Hedges RW, Jacob AE, Barth PT, Ginter NJ (1975): Compatibility properties of P1 and φAMP prophages. Mol Gen Genet 141:263–267.

Heffron H (1983): Tn3 and its relatives. In Shapiro JA (ed): "Mobile Genetic Elements." New York: Academic Press, pp 223–260.

Hershfield U, Boyer HW, Chow L, Helinski DR (1976): Characterization of a mini-ColE1 plasmid. J Bacteriol 126:447–453.

Hightower RC, Ruiz-Perez LM, Wong ML, Santi DV (1988): Extrachromosomal elements in the lower eukaryote *Leishmania*. J Biol Chem 263:16970–16976.

Holmes W, Platt T, Rosenberg M (1983): Termination of transcription in *E. coli*. Cell 32:1029–1032.

Ike Y, Clewell DB (1984): Genetic analysis of the pAD1 pheromone response in *Enterococcus faecalis*, using Tn*917* as an insertional mutagen. J Bacteriol 158:777–783.

Ike Y, Craig RC, White BD, Yagi Y, Clewell DB (1983): Modification of *Enterococcus faecalis* sex pheromones after acquisition of plasmid DNA. Proc Natl Acad Sci USA 80:5369–5373.

Ingraham JL, Low KB, Magasanik B, Schaechter M, Umbarger HE (eds)(1987): "*Escherichia coli* and *Salmonella typhimurium:* Cellular and Molecular Biology," Vol 2. Washington, DC: ASM Press.

Ireton K, Gunther NW IV, Grossman AD (1994): spoOJ is required for normal chromosome segregation as well as initiation of sporulation in *Bacillus subtilis*. J Bacteriol 176:5320–5329.

Itoh T, Tomizawa J (1980): Formation of an RNA primer for initiation of replication of ColE1 DNA by ribonuclease H. Proc Natl Acad Sci USA 77:2450–2454.

Jacob F, Brenner S, Cuzin F (1963): On the regulation of DNA replication in bacteria. Cold Spring Harbor Symp Quant Biol 28:329–348.

Jacob F, Wollman EL (1956): Sur le processus de conjugaison et de recombinaison génétique sur *E. coli* l. L'induction par conjugaison ou induction zygotique. Ann Inst Pasteur 91:486.

Jacob F., Wollman EL (1958): Genetic and physical determinations of chromosomal segments in *E. coli*. Symp Soc Exp Biol 12:75.

Jaffé A, Ogura T, Hiraga S (1985): Effects of the *ccd* function of the F plasmid on bacterial growth. J Bacteriol 163:841–849.

Jearnpipatkul A, Araki H, Oshima Y (1987): Factors encoded by and affecting the holding stability of yeast plasmid pSR1. Mol Gen Genet 206:88–94.

Jensen RB, Gerdes K (1999): Mechanism of DNA segregation in prokaryotes: ParM partitioning protein of plasmid R1 co-localizes with its replicon during the cell cycle. EMBO J 18:4076–84.

Jin R, Rasooly A, Novick RP (1997): In vitro activity of RepC/RepC*, the inactivated form of the plasmid pT181 initiation protein, RepC. J Bacteriol 179:141–147.

Kelley W., Bastia D (1985): Replication initiator protein of plasmid R6K autoregulates its own synthesis at the transcription step. Proc Natl Acad Sci USA 82:2574–2578.

Kellogg ST, Chatterjee DK, Chakrabarty AM (1981): Plasmid assisted molecular breeding: New technique for enhanced biodegradation of persistent toxic chemicals. Science 214:1133–1135.

Kempler GM, McKay LL (1979): Characterization of plasmid deoxyribonucleic acid in *Streptococcus lactis* subsp. *diacetylactis:* Evidence for plasmid-linked citrate utilization. Appl Environ Microbiol 37:316–323.

Khan SA (1997): Rolling-circle replication of bacterial plasmids. Microbiol Mol Biol Rev 61:442–455.

Kikuchi A, Flamm E, Weisberg RA (1985): An *Escherichia coli* mutant unable to support site-specific recombination for bacteriophage λ. J Mol Biol 183:129–140.

Kimmerly WJ, Rine J (1987): Replication and segregation of plasmids containing *cis*-acting regulatory sites of silent mating-type genes in *Saccharomyces cerevisiae* are controlled by the *SIR* genes. Mol Cell Biol 7:4225–4237.

Kingsbury DT, Melinski DR (1973): Temperature-sensitive mutants for the replication of plasmids in *Escherichia coli:* Requirement for deoxyribonucleic acid polymerase I in the replication of plasmid ColE1. J Bacteriol 114:1116–1124.

Knudsen SM, Karlström OH (1991): Development of efficient suicide mechanisms for biological containment of bacteria. Appl Environ Microbiol 57:85–92.

Knudsen S, Saadbye P, Hansen LH, Collier A, Jacobsen BL, Schlundt J, Karlström OH (1995): Development and testing of improved suicide functions for biological containment of bacteria. Appl Environ Microbiol 61:985–991.

Koepsel RR, Khan SA (1986): Static and initiator protein-enhanced bending of DNA at a replication origin. Science 233:1316–1318.

Kollek R, Oertel M, Goebel W (1978): Isolation and characterization of the minimal fragment required for autonomous replication of a copy mutant (pKN102) of the antibiotic resistance factor R1. Mol Gen Genet 162:51–57.

Kolter R, Melinski DR (1982): Plasmid R6K DNA replication. II. Direct nucleotide sequence repeats are required for an active origin. J Mol Biol 161:45–56.

Koltin Y, Day PR (1976): Inheritance of killer phenotypes and double-stranded RNA of the killer phenotype in Ustilago maydis. Proc Nat Acad Sci USA 594–598.

Kontomichalou R, Mitani M, Clowes RC (1970): Circular R-factor molecules controlling penicillinase synthesis, replicating in Escherichia coli under either relaxed or stringent control. J Bacteriol 104:34–44.

Kristoffersen P, Jensen GB, Gerdes K, Piskur J (2000): Bacterial toxin-antitoxin gene system as containment control in yeast cells. Appl Environ Microbiol 66:5524–26.

Kulakauskas S, Lubys A, Ehrlich, SD (1995): DNA restriction-modification systems mediate plasmid maintenance. J Bacteriol 177:3451–3454.

Lacatena RM, Cesareni G (1983): The interaction between RNA1 and the primer precursor in the regulation of ColE1 replication. J Mol Biol 170:635–650.

Lane D, Rothenbuehler R, Merrillat AM, Aiken C (1987): Analysis of the F centromere. Mol Gen Genet 207:406–412.

LeBlanc D, Hawley R, Lee L, St. Martin E (1978): "Conjugal" transfer of plasmid DNA among oral streptococci. Proc Natl Acad Sci USA 75:3484–3487.

Lederberg J (1952): Cell genetics and hereditary symbiosis. Physiol Rev 32:403.

Lehnherr H, Maguin E, Jafri S, Yarmolinsky MB (1993): Plasmid addiction genes of plasmid P1: doc, which causes cell death on curing of the prophage and phd, which prevents host death when the prophage is retained. J Mol Biol 233:414–428.

Levy SB (1985): Ecology of plasmids and unique DNA sequences. In Halverson HO, Pramer D, Rogul M (eds): "Engineered Organisms in the Environment: Scientific Issues." Washington, DC: ASM Press, p 180–190.

Li T-K, Panchenko YA, Drolet M, Liu LF (1997): Incompatibility of Escherichia coli rho mutants with plasmids is mediated by plasmid-specific transcription. J Bacteriol 179:5789–5794.

Light J, Molin S (1981): Replication control functions of plasmid R1 act as inhibitors of expression of a gene required for replication. Mol Gen Genet 184:56–61.

Light J, Molin S (1982): The sites of action of the two copy number control functions of plasmid RL. Mol Gen Genet 187:486–493.

Light J, Molin S (1983): Post-transcriptional control of expression of the repA gene of plasmid R1 mediated by a small RNA molecule. EMBO J 2:93–98.

Lin Z, Mallavia LP (1998): Membrane association of active plasmid partitioning protein A in Escherichia coli. J Biol Chem 18:11302–11312.

Longtine MS, Enomoto S, Finstad SL, Berman J (1993): Telomere-mediated plasmid segregation in S. cerevisiae involves gene products required for transcriptional repression at silencers and telomeres. Genetics 133:171–182.

Louarn J, Patte J, Louarn JM (1977): Evidence for a fixed termination site of chromosome replication in Escherichia coli K12. J Mol Biol 115:195–214.

Lurz R, Danbara H, Rucker B, Timmis KN (1981): Plasmid replication functions. VII. Electron microscopic localization of RNA polymerase binding sites in the replication control region of plasmid R6-5. Mol Gen Genet 183:490–496.

Lustig AJ, Liu C, Zhang C, Hanish JP (1996): Tethered Sir3p nucleates silencing at telomeres and internal loci Saccharomyces cerevisiae. Mol Cell Biol 16:2483–2495.

Maas R, Wang C, Maas WK (1997): Interaction of the RepA1 protein with its replicon targets: two opposing roles in control of replication. J Bacteriol 79:3823–3827.

Marconi RT, Casjens S, Munderloh UG, Samuels DS (1996): Analysis of linear plasmid dimers in Borrelia burgdorferi sensu lato isolates: implications concerning the potential mechanism of linear plasmid replication. J Bacteriol 178:3357–3361.

Marians KJ, Soeller W, Zipursky SL (1982): Maximal limits of the Escherichia coli replication factor Y effector site sequences in pBR322 DNA. J Biol Chem 257:5656–5662.

McEachern MJ, Filutowicz M, Helinski DR (1985): Mutations in direct repeat sequences and in a conserved sequence adjacent to the repeats result in a defective replication origin in plasmid R6K. Proc Natl Acad Sci USA 82:1480.

McEachern MJ, Filutowicz M, Yang S, Greener A. Mukhopadhyay P, Helinski DR (1986): Elements involved in the copy number regulation of the antibiotic resistance plasmid R6K. In Levy SB, Novick RP (eds): "Antibiotic Resistance Genes: Ecology, Transfer, and Expression." Banbury Report 24. Cold Spring Harbor, NY: Cold Spring Harbor Laboratory.

McEachern MJ, Bott MA, Tooker PA, Helinski DR (1989): Negative control of plasmid R6K replication: Possible role of intermolecular coupling of replication origins. Proc Natl Acad Sci USA 86:7942–7946.

McHugh GL, Swartz MN (1977): Elimination of plasmids from several bacterial species by novobiocin. Antimicrob Agents Chemother 12:423–426.

Mead DJ, Gardner DCJ, Oliver SG (1986): The yeast 2μm plasmid: strategies for the survival of a selfish DNA. Mol Gen Genet 205:417–421.

Miki T, Easton AM, Rownd RH (1980): Cloning of replication, incompatibility, and stability functions of R plasmid NR1. J Bacteriol 141:87–99.

Miron A, Patel I, Bastia D (1994): Multiple pathways of copy control of γ replicon of R6K: mechanisms both dependent on and independent of cooperativity of interaction of π protein with DNA affect the copy number. Proc Natl Acad Sci USA 91:6438–6442.

Mohl DA, Gober JW (1997): Cell cycle-dependent polar localization of chromosome partitioning proteins in *Caulobacter crescentus*. Cell 88:675–684.

Molin S, Klemm P, Poulsen LK, Biehl H, Gerdes K, Andersson P (1987): Conditional suicide system for containment of bacteria and plasmids. Biotechnol 5:1315–1318.

Molin S, Staugaard P, Light J, Nordström M, Nordström K (1981): Isolation and characterization of new copy mutants of plasmid R1, and identification of a polypeptide involved in copy number control. Mol Gen Genet 181:123–130.

Molin S, Boe L, Jensen LB, Kristensen CS, Givskov M, Ramos JL, Bej AK (1993): Suicidal genetic elements and their use in biological containment of bacteria. Ann Rev Microbiol 47:139–166.

Murray JAH, Scarpa M, Rossi N, Cesareni E (1987): Antagonistic controls regulate copy number of the yeast 2μm plasmid. EMBO J 6:4205–4212.

Murray AW, Szostak JW (1983): Pedigree analysis of plasmid segregation in yeast. Cell 34:961–970.

Muscholl-Silberhorn A (1998): Analysis of the clumping-mediating domain(s) of sex pheromone plasmid pAD1–encoded aggregation substance. Eur J Biochem 258:515–520.

Muscholl-Silberhorn A (1999): Cloning and functional analysis of Asa373, a novel adhesin unrelated to other sex pheromone plasmid-encoded aggregation substances of *Enterococcus faecalis*. Mol Microbiol 34:620–630.

Muscholl-Silberhorn A, Samberger E, Wirth R (1997): Why does *Staphylococcus aureus* secrete an *Enterococcus faecalis* pheromone? FEMS Microbiol Lett 157:261–266.

Nakayama J, Abe Y, Ono Y, Isogai A, Suzuki A (1995): Isolation and structure of the *Enterococcus faecalis* sex pheromone, cOB1, that induces conjugal transfer of the hemolysin/bacteriocin plasmids, pOB1 and pYI1. Biosci Biotechnol Biochem 59:703–705.

Nishimura Y, Caro L, Berg CM, Hirota Y (1971): Chromosome replication in *E. coli*: IV. Control of chromosome replication and cell division by an integrated episome. J Mol Biol 55:441–456.

Nomura N, Low RL, Ray DS (1982): Identification of ColE1 DNA sequences that direct single strand-to-double strand conversion by a φX174 type mechanism. Proc Natl Acad Sci USA 79:3153–3157.

Nordström K (1985): Control of plasmid replication: Theoretical consideration and practical solutions. In Helinski DR (ed): "Plasmids in Bacteria." New York: Plenum Press.

Nordström K, Ingram LC, Lundback A (1972): Mutations in R factors of *Escherichia coli* causing an increased number of R-factor copies per chromosome. J Bacteriol 110:562–569.

Novick R (1980): Plasmids. Sci AM 243:102–122.

Novick RP (1987): Plasmid incompatibility. Microbiol Rev 51:381–395.

Novick RP, Gruss AD, Highlander SK, Gennaro ML, Projan SJ, Ross HF (1986): Host-plasmid interactions affecting plasmid replication and maintenance in Gram-positive bacteria. In Levy SB, Novick RP (eds): "Antibiotic Resistance Genes: Ecology, Transfer, and Expression," Banbury Report 24. Cold Spring Harbor, NY: Cold Spring Harbor Laboratory.

Novick RP, Projan SJ, Kumar C, Carleton S, Gruss A, Highlander SK, Kornblum J (1985): Replication control for pT181, an indirectly regulated plasmid. In Helinski DR (ed): "Plasmids in Bacteria." New York: Plenum Press.

Novick RP, Projan SJ, Rosenblum W, Edelman I (1984): Staphylococcal plasmid cointegrates are formed by host-and phage-mediated *rec* systems that act on short regions of homology. Mol Gen Genet 195:374–377.

Ogura T, Hiraga S (1983): Partition mechanism of F plasmid: Two plasmid gene-coded products and a *cis*-acting region are involved in partition. Cell 32:351–360.

Ohkubo S, Yamaguchi K (1995): Two enhancer elements for DNA replication of pSC101, *par* and a palindromic sequence of the Rep protein. J Bacteriol 177:558–565.

Ohtsubo E, Rosenbloom M, Schrempf H, Goebel W, Rosen J (1978): Site specific recombination involved in the generation of small plasmids. Mol Gen Genet 159:131–141.

Oka A, Nomura N, Morita M, Sugisaki II, Sugimato K, Takanani M (1979): Nucleotide sequence of small ColE1 derivatives: Structure of the regions essential for autonomous replication and colicin G1 immunity. Mol Gen Genet 172:151–159.

Pal SK, Chattoraj DK (1986): RepA is rate-limiting for P1 plasmid replication. In Kelly T, McMacken R (eds): "Mechanisms of DNA Replication and Recombination." New York: Alan R. Liss, pp 441–450.

Palladino F. Laroche T, Gilson E, Axelrod A, Pillus L, Gasser SM (1993): SIR3 and SIR4 proteins are required for the positioning and integrity of yeast telomeres. Cell 75:543–555.

Patnaik PK, Fang X, Cross GAM (1994): The region encompassing the procyclic acidic repetitive protein (PARP) gene promoter plays a role in plasmid DNA replication in *Trypanosoma brucei*. Nucleic Acids Res 22:4111–4118.

Polisky B (1988): ColE1 replication control circuitry: Sense from antisense. Cell 55:929–932.

Pomiankowski A (1999): Intragenomic conflict. In L. Keller (ed): "Levels of Selection in Evolution." Princeton: Princeton University Press, pp 121–152.

Pouwels PH, Van Luijk N, Leer RJ, Posno M (1994): Control of replication of the *Lactobacillus pentosus* plasmid p353-2: Evidence for a mechanism involving transcriptional attenuation of the gene coding for replication protein. Mol Gen Genet 242:614–622.

Ravel-Chapuis P, Nicolas P, Nigon V, Neyret O, Freyssinet G (1985): Extrachromosomal circular nuclear rDNA in *Euglena gracilis*. Nucleic Acids Res 13:7529–7537.

Redenbach M, Ikeda K, Yamasaki M, Kinashi H (1998): Cloning and physical mapping of the giant linear plasmid SCP1. J Bacteriol 180:2796–2799.

Reynolds AE, Murray AW, Szostak JW (1987): Roles of the 2 μm gene products in stable maintenance of the 2 μm plasmid in *Saccharomyces cerevisiae*. Mol Cell Biol 7:3566–3573.

Roberts RC, Helinski DR (1992): Definition of a minimal plasmid stabilization system from the broad-host-range plasmid RK2. J Bacteriol 174:8119–8132.

Roberts MC, Kenny GE (1987): Conjugal transfer of transposon Tn916 from *Streptococcus faecalis* to *Mycoplasma hominis*. J Bacteriol 169:3836–3839.

Rosen J, Ryder T, Inokuchi H, Ohtrubo H, Ohtsubo E (1980): Genes and sites involved in replication and incompatibility of an R100 plasmid derivative based on nucleotide sequence analysis. Mol Gen Genet 179:527–537.

Rownd RH (1978): Plasmid replication. In Molineux O, Kohoyima M (eds): "DNA Synthesis: Present and Future." New York: Plenum Press, pp 751–772.

Rownd RH, Womble DD, Dong X (1986): IncFll plasmid replication control and stable maintenance. In Levy SB, Novick RP (eds): "Antibiotic Resistance Genes: Ecology, Transfer, and Expression," Banbury Report 24. Cold Spring Harbor, NY: Cold Spring Harbor Laboratory.

Ruiz-Echevarria MJ, Berzal-Herranz A, Gerdes K, Díaz-Rejas R (1991): The *kis* and *kid* genes of the parD maintenance system of plasmid R1 form an operon that is regulated at the level of transcription by the co-ordinated action of the Kis and Kid proteins. Mol Microbiol 5:2685–2693.

Scherzinger E, Bagdasarian MM, Scholz P, Lurz R, Ruckert B, Bagdasarian M (1984): Replication of the broad host range plasmid RSF1010: Requirement for three plasmid-encoded proteins. Proc Natl Acad Sci USA 81:654–658.

Scholtz P, Haring E, Scherzinger R, Lurz R, Bagdasarian MM, Schuster H, Bagdasarian M (1985): Replication determinants of the broad host range plasmid RSF1010. In Helinski DR (ed): "Plasmids in Bacteria." New York: Plenum Press.

Scott JR (1984): Regulation of plasmid replication. Microbiol Rev 48:1–23.

Scott-Drew S, Murray JA (1998): Localisation and interaction of the protein components of the yeast 2 μm circle plasmid partitioning system suggest a mechanism for plasmid inheritance. J Cell Sci 111:1779–1789.

Silhavy TJ, Beckwith JR (1985): Uses of *lac* fusions for the study of biological problems. Microbiol Rev 49:398–418.

Silver L, Chandler M, Boy de la Tour E, Caro L (1977): Origin and direction of the drug resistance plasmid R100.1 and of a resistance transfer factor derivative in synchronized cultures. J Bacteriol 131:929–942.

Sinclair DA, Guarente L (1997): Extrachromosomal rDNA circles-a cause of aging in yeast. Cell 91:1033–1042.

Smith ASG, Rawlings DE (1998a): Efficiency of the pTF-FC2 *pas* poison-antidote stability system in *Escherichia coli* is affected by host strain, and antidote degradation requires the Lon protease. J Bacteriol 180:5458–5462.

Smith ASG, Rawlings DE (1998b): Autoregulation of the pTF-FC2 proteic poison-antidote plasmid addiction system (*pas*) is essential for plasmid stabilization. J Bacteriol 180:5463–5465.

So M, Boyer HW, Betlach M, Falkow S (1976): Molecular cloning of an *Escherichia coli* plasmid that encodes for the production of heat-stable enterotoxin. J Bacteriol 128:463–472.

Stalker DM, Kolter R, Helinski DR (1979): Nucleotide sequence of the region of an origin of replication of the antibiotic resistance plasmid R6K. Proc Natl Acad Sci USA 76:1150–1154.

Stalker DM, Kolter R, Helinski DR (1982): Plasmid R6K DNA replication: 1. Complete nucleotide sequence of an autonomously replicating segment. J Mol Biol 161:33–43.

Staudenbauer WL (1975): Novobiocin—A specific inhibitor of semi-conservative DNA replication in permeabilized *Escherichia coli* cells. J Mol Biol 96:201–205.

Staudenbauer WL (1976): Replication of small plasmids in extracts of *Escherichia coli*. Mol Gen Genet 145:273–280.

Staudenbauer WL (1978): Structure and replication of the colicin E1 plasmid. Curr Top Microbiol Immunol 83:93–156.

Summers DK, Sherratt DJ (1984): Multimerisation of high copy number plasmids causes instability: ColE1 encodes a determinant essential for plasmid monomerisation and stability. Cell 36:1097–1103.

Summers DK, Sherratt DJ (1985): Bacterial plasmid stability. Bioassays 2:209.

Swack JA, Pal SK, Mason RJ, Abeles AL, Chattoraj DK (1987): P1 plasmid replication: Measurement of initiator protein concentration in vivo. J Bacteriol 169:3737–3742.

Syneki RM, Nordheim A, Timmis KN (1979): Plasmid replication functions: III. Origin and direction of replication of a mini-plasmid derived from R6–5. Mol Gen Genet 168:27–36.

Tang XB, Womble DD, Rownd RH (1989): DnaA protein is not essential for replication of IncFll plasmid NR1. J Bacteriol 171:5290–5295.

Thisted T, Sorensen NS, Gerdes K (1995): Mechanism of post-segregational killing: Secondary structure analysis of the entire Hok mRNA from plasmid R1 suggests a fold-back mechanism that prevents translation and antisense RNA binding. J Mol Biol 247:859–873.

Timmis K, Cabello F, Cohen SN (1975): Cloning, isolation, and characterization of replication origins of complex plasmid genomes. Proc Natl Acad Sci USA 72:2242–2246.

Timmis K, Cabello F, Cohen SN (1978): Cloning and characterization of EcoRI and HindII restriction endonuclease generated fragments of antibiotic resistance plasmids R6-5 and R6. Mol Gen Genet 162:121–137.

Toh-e A, Tada S, Oshima Y (1982): 2 μm-like plasmids of the osmophilic haploid yeast *Saccharomyces rouxii*. J Bacteriol 151:1380–1390.

Toh-e A, Utatsu I (1985): Physical and functional structure of a yeast plasmid, pSB3, isolated from *Zygosaccharomyces bisporus*. Nucleic Acids Res 13:4267–4283.

Tomich PK, An FY, Damle SP, Clewell DB (1979): Plasmid-related transmissibility and multiple drug resistance in *Streptococcus faecalis* subsp. *zymogenes* strain DS16. Antimicrob Agents Chemother 15:828–830.

Tomizawa J (1984): Control of ColE1 replication: The process of binding of RNAI to the primer transcript. Cell 38:861–870.

Tsuchimoto S, Ohtsubo H, Ohtsubo E (1988): Two genes, *pemK* and *pemI*, responsible for stable maintenance of resistance plasmid R100. J Bacteriol 170:1461–1466.

Tudzynski P, Esser K (1985): Mitochondrial DNA for gene cloning in eukaryotes. In Bennett JW, Lasure LL (eds): "Gene Manipulation in Fungi." New York: Academic Press, pp 403–416.

Twigg A, Sherratt D (1980): Trans-complementable copy number mutants of plasmid ColE1. Nature 283:216–218.

Uhlin BE, Molin S, Gustafsson P, Nordström K (1979): Plasmids with temperature-dependent copy number for amplification of cloned genes and their products. Gene 6:91–106.

Van der Ende A, Teertstra R, van der Avoort HGAM, Weisbeek PJ (1983): Initiation signals for complementary strand DNA synthesis on single-stranded plasmid DNA. Nucleic Acids Res 11:4957–4975.

Vapnek D, Lipman MB, Rupp WD (1971): Physical properties and mechanism of transfer of R factors in *Escherichia coli*. J Bacteriol 108:508–514.

Veltkamp E, Nijkamp HJJ (1976): Characterization of a replication mutant of the bacteriocinogenic plasmid CloDF13. Biochim Biophys Acta 425:356–367.

Wagner EGH, Blomberg P, Nordström K (1992): Replication control of plasmid R1: duplex formation between the antisense RNA, CopA, and its target, CopT, is not required for inhibition of RepA synthesis. EMBO J 11:1195–1203.

Wagner EGH, Nordström K (1986): Structural analysis of a molecule involved in replication control of plasmid R1. Nucleic Acids Res 14:2523–2538.

Warren GJ, Twigg AJ, Sherratt DJ (1978): ColE1 plasmid DNA during bacterial conjugation. Nature 274:259–261.

Willetts N, Wilkins B (1984): Processing of plasmid DNA during bacterial conjugation. Microbiol Rev 48:24–41.

Womble DD, Dong X, Luckow VA, Wu RP, Rownd RH (1985): Analysis of the individual regulatory components of the IncFll plasmid replication control system. J Bacteriol 161:534–543.

Wong EM, Muesing MA, Polisky B (1982): Temperature-sensitive copy number mutants of ColE1 are located in an untranslated region of the plasmid genome. Proc Natl Acad Sci USA 79:3570–3574.

Yagi Y, Kessler RE, Shaw JH, Lopatin DE, An FY, Clewell DB (1983): Plasmid content of *Streptococcus faecalis* strain 39.5 and identification of a pheromone (cPD1)-induced surface antigen. J Gen Microbiol 129:1207–1215.

Yarmolinsky M, Sternberg N (1988): Bacteriophage P1. In Calendar R (ed): "The Bacteriophages," Vol 1. New York: Plenum Press.

Yu J-S, Noll KM (1997): Plasmid pRQ7 from the hyperthermophilic bacterium *Thermatoga* species strain TQ7 replicates by the rolling-circle mechanism. J Bacteriol 179:7161–7164.

Yun T, Vapnek D (1977): Electron microscopic analysis of bacteriophage P1, phage P1Cm and phage P7. Determination of genome sizes, sequence homology and location of antibiotic resistance determinants. Virol 77:376–385.

Zakian VA, Brewer BJ, Fangman BL (1979): Replication of each copy of the yeast 2 micron DNA plasmid occurs during the S phase. Cell 17:923–934.

Zipursky SL, Marians KJ (1981): *Escherichia coli* factor Y sites of plasmid pBR322 can function as origins of DNA replication. Proc Natl Acad Sci USA 78:6111–6115.

SELECTED BIBLIOGRAPHY

Ahn YT, Wu XL, Biswal S, Velmurugan S, Volkert FC, Jayaram M (1997): The 2 μm-plasmid-encoded Rep1 and Rep2 proteins interact with each other and colocalize to the *Saccharomyces cerevisiae* nucleus. J Bacteriol 179:7497–7506.

Barbour AG, Garon CF (1987): Linear plasmids of the bacterium *Borrelia burgdorferi* have covalently closed ends. Science 237:409–411.

Blackburn EH (1985): Artificial chromosomes in yeast. Trends Genet, January, pp 8–12.

Dunny GM, Leonard BAB, Hedberg PJ (1995): Pheromone-inducible conjugation in *Enterococcus faecalis*: interbacterial and host-parasite chemical communication. J Bacteriol 177:871–876.

Khan SA (1997): Rolling-circle replication of bacterial plasmids. Microbiol Mol Biol Rev 61:442–455.

Meynell E, Datta N (1967): Mutant drug resistant factors of high transmissibility. Nature 214:885–887.

Novick RP (1987): Plasmid incompatibility. Microbiol Rev 51:381–395.

Silhavy TJ, Beckwith JR (1985): Uses of *lac* fusions for the study of biological problems. Microbiol Rev 49:398–418.

Smith ASG, Rawlings DE (1998a): Efficiency of the pTF-FC2 *pas* poison-antidote stability system in *Escherichia coli* is affected by host strain, and antidote degradation requires the Lon protease. J Bacteriol 180:5458–5462.

Thisted T, Sorensen NS, Gerdes K (1995): Mechanism of post-segregational killing: secondary structure analysis of the entire Hok mRNA from plasmid R1 suggests a fold-back mechanism that prevents translation and antisense RNA binding. J Mol Biol 247:859–873.

Tomizawa J (1984): Control of ColE1 replication: The process of binding of RNAI to the primer transcript. Cell 38:861–870.

Uhlin BE, Molin S, Gustafsson P, Nordström K (1979): Plasmids with temperature-dependent copy number for amplification of cloned genes and their products. Gene 6:91–106.

Yasuda S, Takagi T (1983): Overproduction of *Escherichia coli* replication proteins by the use of runaway-replication plasmids. J Bacteriol 154:1153–1161.

APPENDIX A: SPECIAL-USE PLASMIDS

R Plasmids

R plasmids are naturally occurring plasmids which encode genes for bacterial resistance. Such genes may confer resistance to antibiotics, heavy metals, UV light, and may specify restriction-modification systems. Among the best-studied R plasmids are R100 (Womble et al., 1985), R1 (Nordström et al., 1972), pSC101 (Biek and Cohen, 1986), and the pT181 family (Novick, et al., 1986). (For a review, see Levy, 1985).

Pathogenicity

In *E. coli* there is an enterotoxin-producing plasmid (which causes cholera-like symptoms) and a plasmid that encodes the pili K88 antigen. The K88 antigen is a fimbrial protein that allows adhesion to host mucosa. *Both* this plasmid and an enterotoxin-producing plasmid are required for pathogenicity. Two types of enterotoxin are produced in *E. coli:* (1) LT, a heat-labile toxin, is immunologically related to cholera toxin. It functions by activating the adenylate cyclase enzyme of the gut epithelial host cells to increase greatly the levels of cAMP, which in turn causes secretion of fluid and electrolytes into the gut lumen. (2) ST, heat-stable toxin, stimulates guanylate cyclase in host cells. It is found on a transposon (So et al, 1976). Individual cells may have either or both of these toxin-bearing plasmids. They can be conjugative.

Bacteriocins

Some plasmids encode production of bacteriocins. These are compounds that are produced by bacteria to kill other bacteria. Only a small percentage of the cell population produces the bacteriocin. Each type of bacteriocin is named according to the bacteria of origin. For example, there are megacins (from *B. megaterium*) and colicins (from *E. coli*).

The colicins are the best-studied bacteriocins (Davies and Reeves, 1975). They are specified by plasmids. They are encoded on conjugative plasmids or on small multicopy nonconjugative plasmids (e.g., ColE1 and ColI). The cellular colicin receptors are outer membrane proteins. Different colicins have different mechanisms of action. Colicins E1 and K uncouple energy-dependent processes in the cytoplasmic membrane, colicin E2 causes inhibition of cell division and DNA degradation, and colicin E3 prevents protein synthesis by cleavage of the 16S rRNA. This cleavage can be inhibited in vitro by a ColE3–associated immunity protein specified by the plasmid.

The *cloacins* are derived from *Enterobacter cloacae*. These proteins bind selectively to receptor sites on sensitive cells. Their action is similar to that of ColE3. In addition cloacin from CloDF13 binds tightly to a unique site on this plasmid and makes two staggered single-strand nicks in the DNA. This activity is inhibited by the immunity protein. Does this play a role in replication?

Degradative Plasmids in Pseudomonads

A number of plasmids have been characterized (Chakrabarty, 1972), or engineered (Kellogg et al., 1981), that are found in *Pseudomonas* spp. and encode unusual degradative pathways. The first reported was found in *P. putida* and encoded a function called SAL (salicylate utilization) (Chakrabarty, 1972). The criterion used in determining if this function was plasmid encoded was that the SAL$^+$ phenotype was transmissible at 10^{-5} per donor, compared with 10^{-9} for chromosomal markers. Other pathways that have since been reported include CAM (camphor utilization), OCT (octane utilization), NAH (naphthalene utilization), and TOL (benzoate and toluene utilization). In addition *P. cepacia* strains have been engineered that are capable of degrading pesticides such as 2, 4, 5-T (2, 4, 5-trichlorophenoxyacetic acid) and chlorobenzoate (Chatterjee and Chakrabarty, 1983).

Tumorogenic Plasmids in *Agrobacterium tumefaciens*

Plasmid-encoded abilities to cause tumorogenesis in plants are discussed by Ream (this volume).

Cloning Vectors

The use of plasmids for cloning strategies is described by Geoghegan et al. (this volume). Below are listed some special cloning vectors not discussed in that chapter.

Runaway replication plasmids

Cloning vectors have been designed from mutant miniplasmid derivatives of plasmid R1 (Uhlin et al., 1979; Bittner and Vapnek, 1981). These vectors are temperature-sensitive replicons, whose copy number at 30°C is 20 to 50 copies per *E. coli* cell; above 35°C the copy number of these plasmids is uncontrolled, such that after 2 to 3 hours of additional growth the plasmid DNA represents up to 75% of the total cellular DNA. Various versions of these vectors have been designed to contain genes conferring resistance to ampicillin, chloramphenicol, tetracycline, and aminoglycosides, while some derivatives also bear the *lacZ* gene. In the last case, conveniently placed restriction sites allow detection of inserts via growth of cells on X-gal plates. These vectors can be used for production of very high levels of recombinant protein.

pKNOCK

It is important to be able to determine the function of genes, particularly in light of the genome sequencing projects that have been completed and those currently underway. One way to evaluate the importance and/or function of a gene is to disrupt it or make a gene "knock-out." To do this in the chromosome of Gram-negative bacteria, there are two methods commonly employed: (1) allelic exchange of wild-type for mutant via a plasmid-borne mutant copy, and (2) insertions into the gene-of-interest via a "suicide" plasmid that is unable to replicate. The former approach is the more approved method, but it is time-consuming and can only be used when there is effective counterselection against the donor plasmid. The latter approach has been efficiently improved upon by the construction of the pKNOCK series of vectors (Alexeyev, 1999; see Fig. 8). These are small plasmids that contain a multiple cloning site and that can be introduced into recipients via transformation, electroporation, or conjugation. The ability to transfer via conjugation is particularly attractive, since nearly 100% efficiency can be achieved and since the form of the DNA that enters the recipient is single stranded. This is an advantage due to the increased rate with which such DNA can recombine with the target in the chromosome. The pKNOCK series contains both the RP4 *ori*T (for transfer) and R6K γ-ori (see Section IIA.2b above). The R6K γ-ori makes these vectors suicidal, since it can be used only in those *E. coli* strains that provide the R6K π protein in *trans*. Thus they are propagated in *E. coli* and are then used when transferred into a wide range of Gram-negative bacteria via conjugation or by electroporation, usually the most efficient form of transformation.

Yeast cloning vectors

Cloning vectors for *S. cerevisiae* have been designed with one or more of the components discussed discussed in Section III. These plasmids can be divided according to the relative level of stability for the vector in the transformed cells.

YIP. Yeast integrative plasmid (YIP) vectors are intended for the integration of a cloned

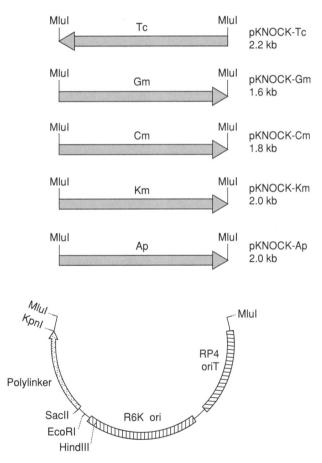

Fig. 8.   The pKNOCK series of cloning vectors. Vector sizes are below vector designations. The following refer to resistance conferred by the particular gene: Tc, tetracycline; Gm, gentamicin; Cm, chloramphenicol; Km, kanamycin; Ap, ampicillin. Polylinker: *Eco*RI-*Sac*II-*Bst*XI-*Not*I-*Eag*I-*Xba*I-*Spe*I-*Bam*HI-*Sma*I-*Pst*I-*Eco*RI-*Eco*RV-*Hind*III-*Cla*I-*Sal*I-*Hinc*II-*Acc*I-*Xho*I-*Apa*I-*Dra*II-*Kpn*I. The *Eag*I site is not unique in either pKNOCK-Km or pKNOCK-Tc; the *Eco*RV site is not unique in pKNOCK-Gm and pKNOCK-Tc; the *Dra*II site is not unique in pKNOCK-Tc and pKNOCK-Ap. (Reproduced from Alexeyev, 1999, with permission of the publisher.)

gene into the *S. cerevisiae* genome. Thus the vector itself does not replicate as an independent entity. Rather, it contains an origin for replication in *E. coli* and a selectable marker for expression in *S. cerevisiae*.

The following vectors are *E. coli* -*S. cerevisiae* shuttle vectors. This means that they can be propagated in both organisms. As such they each contain an origin of replication for both organisms and selectable markers (usually ampicillin resistance for bacteria) for both organisms.

YRP. Yeast replicating plasmids (YRP) contain an ARS as their origin of replication in yeast. These have high copy numbers but are extremely unstable in the absence of selection for the yeast marker; they are typically lost at $> 1/10$ per cell per generation.

YEP. Yeast episomal plasmids (YEP) are vectors based on the yeast $2\,\mu m$ plasmid. They have very high copy number and are much more stably inherited than the YRP plasmids.

YCP. Yeast centromeric plasmids (YCP) vectors contain both ARS and *CEN* sequences. They are approximately 10 to 100 times more stable without selection than isogenic plasmids lacking the *CEN* sequences.

YAC. Yeast artificial chromosomes (YAC) are linear DNAs containing ARS, *CEN*, and TEL (telomeric) sequences. For YACs $> 50\,kb$, the average copy number is 1 per cell, and they are only lost at 10^{-3} per cell per generation without selection. As the overall length of such vectors increases to at least $140\,kb$, loss of such plasmids occurs at approximately that of natural yeast chromosomes.

APPENDIX B: COMMONLY EMPLOYED METHODS IN CHARACTERIZING PLASMID FUNCTIONS

Complementation Analysis

The structure and function of various plasmid-encoded genes may be studied by constructions generated either in vitro or in vivo in order to test for complementation.

Cloned fragments

Fragments may be cloned from plasmids under study using recombinant DNA techniques. Such cloned fragments may be used in complementation analysis with mutant plasmids in order to determine if the mutations may be corrected in *trans* by the cloned fragment when co-resident in the cell with the mutant plasmid. Alternatively, mutations in plasmid genes may be mapped relative to one another by measuring the frequency of recombinants between two plasmids bearing temperature-sensitive lethal mutations.

Integrative suppression

Integrative suppression takes advantage of the fact that an *E. coli* strain with a temperature-sensitive defect in initiation of DNA replication (DnaA mutant) may be rescued at the higher temperature by the formation of a cointegrate between the host chromosome and the plasmid. In this case the plasmid provides the *ori* and the cointegrates require plasmid functions at the higher temperature. Using this technique, deletions in R100 making the plasmid unable to cause integrative suppression were mapped to two locations (the Rep functions). One of these was found to be absolutely required for replication (Nishimura et al., 1971).

Fusions (protein and operon)

Two types of in vivo–generated fusions are commonly employed in order to study genes at the level of transcription, translation, structure, and function. The *lacZ* gene (which encodes β-galactosidase) is most commonly employed for these fusions due to the versatility of this gene and its product (for a review, see Silhavy and Beckwith, 1985). *Operon fusions* yield constructs in which the *lacZ* gene (devoid of its promoter) is fused to a gene of interest. The result is that *lacZ* expression comes under control of the new gene's promoter. The amount of β-galactosidase activity under varying conditions may then be used as a measure of transcription from the promoter in question. *Protein fusions* result from translational gene fusions in which the

N-terminal portion of the *lacZ* gene has been replaced by all or part of the gene of interest. Both types of fusions were used to analyze the effect of the *copB* product on its target, the *repAl* gene (required for IncFll replication) (Light and Molin, 1982).

Heterodimer pickup assay for P1 *par* function in pBR322

It is difficult to assay for *par* function in a unit copy plasmid like P1. Characterization would be easier if the *par* locus were cloned into pBR322. But pBR322 has no need of this *par* function due to the plasmid's high copy number. To overcome this problem, Friedman et al. (1986) developed the following assay. Cells are cotransformed with pBR322 containing the *par* locus and pBR322 containing the P1 or F *rep* region. The drug resistance markers for both recombinant plasmids are selected for during the transformation. A small proportion of the transformants will contain heterodimers, in which the two plasmids have recombined with each other. The plasmid DNA is then isolated from the transformants and used to transform a PolA⁻ strain. Plasmid pBR322 cannot replicate in a PolA⁻ strain, so replication must come from the P1 origin. The degree of such heterodimers is a measure of whether or not *par* was cloned.

Miniplasmid Derivatives

To reduce the size of a plasmid for better study, miniplasmid derivatives of large plasmids may be selected by in vitro or in vivo methods. Miniplasmids have been used to examine *origins of replication* and *plasmid-encoded functions*. For IncFll plasmids such as R100, EM analysis, denaturation, mapping, and autoradiography were used to show that a miniplasmid origin was the same origin used in vivo for R100 (Silver et al., 1977)

Miniplasmids are also used in order to measure the minimum required for plasmid replication. In plasmid p1258 of *Staphylococcus aureus* up to two-thirds of the plasmid can be deleted without affecting autonomy,

but there is a limit. A miniplasmid can be constructed by ligating a *bla* gene (AP resistance) to an *Eco*RI digest of an FII plasmid (Timmis et al., 1975) to yield a 2.6 kb minireplicon. This constructed minireplicon has the normal incompatibility and approximately normal copy number, and can replicate in PolA⁻ hosts (Timmis et al., 1978). There is a slight instability for this minireplicon due to a loss of the *par* region. These findings agree with heteroduplex analyses of R1 and R100 which show a common 2.5 kb essential region (Ohtsubo et al., 1978).

Incompatibility Assays

One normally tests for plasmid incompatibility by introducing a plasmid into a strain that already contains another plasmid. These plasmids must contain different genetic markers so as to follow their segregation. Usually selection is for the entering plasmid, and then daughter cells are examined for presence of the original plasmid. If the original plasmid is eliminated, then the two plasmids are considered *incompatible*.

There are some pitfalls to such assays for incompatibility. There may be problems with a particular replicon being tested. For instance, it may have no selectable marker, or there may be some surface barrier to entry of the donor plasmid. In order to alleviate some of these problems, a series of marker miniplasmids with known incompatibility groups have been designed to use as the resident plasmids in crosses. These plasmids contain a Gal⁺ marker. Incompatibility of a donor plasmid is then measured as the appearance of Gal⁻ segregants on indicator media (Couturier et al., 1988).

There are other problems associated with grouping plasmids based on incompatibility. The two primary ones are (1) the presence of several replicons in a plasmid and (2) mutational alterations that affect both a replication inhibitor and its target. In the former case determination of incompatibility can be confused, since one would expect that a plasmid with two different replicons would simply rely on the second control mechanism if a plasmid entered the cell that was incompatible with the first control mechanism. The latter situation involving mutational alterations would lead to altered incompatibility determinants even among closely related replicons. To alleviate these problems, Couturier et al. (1988) developed a plasmid classification system that relies on specific DNA probes and DNA-DNA hybridization to test for the presence of basic replicons. According to these authors, this method is more direct than incompatibility grouping, and it is easier to do.

Curing of Plasmids

Plasmids may be eliminated selectively from cells by treatment of the cells with agents that inhibit DNA gyrase, since DNA supercoiling is required for initiation of replication of some plasmids and elongation of others. *Novobiocin* has been used to treat cells to inhibit replication of low copy number plasmids (McHugh and Swartz, 1977). Concentrations of novobiocin (50–175 µg/ml) sufficient to inhibit cell multiplication or produce minimal killing bring about 15–37% elimination of such plasmids. Elimination of multicopy plasmids such as pBR322, pMB9, and other ColE1-derived plasmids is more difficult and less success is found using novobiocin. *Coumermycin* is more effective in eliminating such plasmids and seems to result in plasmid degradation in addition to inhibition of replication (Danilevskaya and Gragerov, 1980). Such treatment can lead up to 72% loss of pMB9.

21

Transduction in Gram-Negative Bacteria

GEORGE M. WEINSTOCK

Department of Biochemistry and Molecular Biology, University of Texas Medical School,
Houston, Texas 77225

I. INTRODUCTION

Transduction is the mode of genetic exchange between cells that is mediated by viruses. The term transduction refers to the transfer of cellular DNA or RNA from one cell to another by infection with a viral vector. Virus particles that carry and introduce cellular nucleic acids on infection are called transducing particles. Bacteriophages are transducing viruses if they carry parts of bacterial or plasmid chromosomes from one bacterial host to another. For some phages this DNA is joined to the viral chromosome in the particle (as in specialized transduction by phage λ), while for other transducing phages only DNA of host origin is present in the particle (as in generalized

**Modern Microbial Genetics, Second Edition. Edited by
Uldis N. Streips and Ronald E. Yasbin. ISBN 0–471–38665–0**

transduction by phages P1 or P22). The formation of transducing particles is a part of the normal DNA metabolism of viral infection. In addition, when foreign DNA is introduced into a virus chromosome by recombinant DNA techniques, the resulting phages that carry the cloned foreign DNA are transducing particles. Finally, RNA viruses, such as retroviruses, that carry host RNA and introduce it into a recipient cell are transducing viruses.

The process of transduction is divided into two types. In generalized transduction, the virus can introduce any region from the donor chromosome into a recipient. In specialized transduction, the virus always carries the same segment into the recipient. A phage that randomly packages host DNA into particles is a generalized transducing phage. A phage that has had a particular gene stably introduced into its chromosome is a specialized transducing phage. For detailed reviews of these processes, see Margolin (1987) and Weisberg (1987); for a summary of the early literature, see Low and Porter (1978) (also see Hendrix, this volume).

II. GENERALIZED TRANSDUCTION

In 1952 Zinder and Lederberg first reported generalized transduction in *Salmonella tymphimurium*. They were able to show that genetic exchange occurred in *Salmonella* by mixing different auxotrophic mutants together and isolating prototrophic recombinants. To differentiate this mechanism of genetic exchange from conjugation, they showed that transfer of the genetic material did not require physical contact between the donor and recipient bacteria. The agent responsible for transfer was able to pass through the pores of a filter, pores that were too small to allow bacteria through. Moreover the genetic exchange was not due to transformation (see Streips, this volume), since the process was resistant to DNase treatment. It was eventually shown that the agent that carried the genes from the donor to the recipient bacterium was the temperate bacteriophage P22. This dis-

covery was made possible because fortuitously one of the *Salmonella* strains that Zinder and Lederberg had used in the mixing experiment was a lysogen for P22. During their experiments the P22 prophage had spontaneously induced, forming particles that infected the second *Salmonella* strain. The infection of the second strain then produced particles that carried genes of the second host. When these particles infected the original auxotrophic bacterium, they introduced the wild-type gene, which could recombine and replace the defective segment of the chromosome, forming a prototrophic transductant bacterium.

Subsequently this process of bacteriophage-mediated genetic transduction was shown to occur in many other Gram-negative bacteria, including *Escherichia coli, Myxobacteria, Rhizobium, Caulobacter*, and *Pseudomonas*. In *E. coli*, for example, transduction was shown to occur with bacteriophage P1. In addition it was possible to transfer any host gene from one bacterium to another with these phages, and hence this process came to be known as generalized transduction. That this was due to physical transfer of DNA from the donor bacterium to the recipient was demonstrated for both P22 (Ebel-Tsipis et al., 1972) and P1 (Ikeda and Tomizawa, 1965a) transduction. To show this (Fig. 1), the donor bacteria were grown in a medium containing heavy isotopes of nitrogen and hydrogen (^{15}N and ^{2}H) to make the donor DNA more dense. Then the bacteria were shifted to a normal medium (^{14}N and ^{1}H) and infected with the phage. During growth, the newly synthesized phage DNA was light, while the preexisting bacterial DNA was heavy. It was then shown that the phage prepared in this way contained heavy DNA; that is, they had packaged the donor bacterial DNA into particles, and when they were used to infect a recipient bacterium that had light DNA, some of the heavy DNA became incorporated into the recipient chromosome.

These experiments illustrate the overall process of generalized transduction. They show that there are two parts to the mechanism: the packaging of donor DNA into a phage

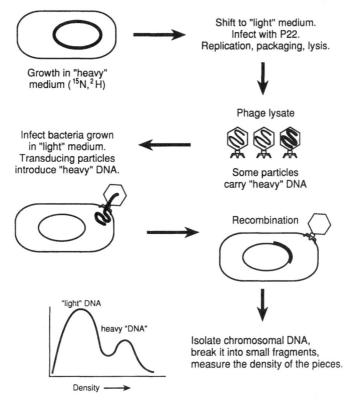

Fig. 1. Physical transfer of DNA in generalized transduction. An example of generalized transduction by phage P22 is shown. The donor bacteria are grown in a medium containing heavy isotopes of nitrogen and hydrogen so that their DNA has a heavier density than normal (the heavy DNA is represented by the thick circular line in the cell at the upper left). Following P22 infection of this cell, some heavy chromosomal DNA is packaged into phage particles. These can then be used to infect a cell grown in medium with light isotopes of nitrogen and hydrogen, and thus containing DNA with a lighter density than the donor's. When heavy donor DNA is introduced, it can recombine and replace some light DNA in the recipient's chromosome. When the chromosome is broken into small pieces, the heavy DNA can be detected by equilibrium gradient centrifugation because it bands at a different density than the bulk of the light DNA.

particle and the stable introduction of this packaged DNA into the recipient cell, usually through genetic recombination with the recipient chromosome (Fig. 2). The ability of a phage to perform generalized transduction thus depends on the mechanism of packaging DNA into phage particles. If this mechanism allows host as well as phage DNA to be encapsidated, generalized transduction invariably occurs. As we will see, there are numerous types of DNA metabolism that lead to generalized transduction. In each case the capability for generalized transduction is a result of the mode of packaging phage DNA.

A. Events in the Donor: Bacteriophage P22

A well-studied example of generalized transduction is that of the *Salmonella* phage P22. This is a useful paradigm for DNA metabolism leading to generalized transduction, since it is not unique and is representative of a number of other phages (see Guttman and Kutter, this volume, for phage T4). The DNA metabolism of phage P22 has been reviewed by Susskind and Botstein (1978).

The phage DNA in a P22 particle is terminally redundant and circularly permuted (Fig. 3). Each phage chromosome has the

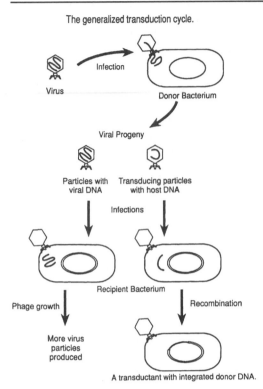

The generalized transduction cycle.

Fig. 2. Overview of generalized transduction. The scheme shown applies to phages like P22 and P1. During normal productive infection of bacteria two kinds of particles are produced: those with a viral chromosome and those containing a fragment of the host chromosome. When the phage with a viral chromosome infects a cell, another round of productive infection occurs. However, when the particle with the host DNA injects this DNA into a cell, the fragment can recombine with the recipient's chromosome, resulting in a transductant.

leftmost sequence of its chromosome repeated at the right end (terminally redundant). However, this repetitious segment (about 2% of the chromosome) is not unique and differs for each chromosome (circularly permuted). This chromosome structure is a result of the mechanism of packaging DNA employed by P22, the headful packaging mechanism (Fig. 3).

1. P22 DNA metabolism

When P22 infects a cell, its DNA must circularize for the infection to be productive and the chromosome to replicate. This obligatory

circularization occurs by homologous recombination between the repetitious ends of the chromosome. This is the significance of the terminal redundancy of the chromosome—it provides the substrate for the circularization reaction. The circular molecule replicates and ultimately produces a concatemer, a long DNA molecule containing multiple copies of the genome. This probably arises from rolling circle replication, an efficient process where a single replication fork continually circles a circular chromosome synthesizing a long concatemeric tail (see Fig. 3). Packaging of this DNA initiates at

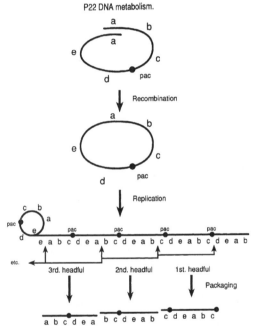

P22 DNA metabolism.

Fig. 3. Scheme of phage P22 DNA metabolism. At the top is shown the linear DNA molecule carried by the phage particle. This chromosome is terminally redundant, containing a copy of the "a" region of the genome at each end. The two copies of the "a" region can recombine to circularize the chromosome. Then a concatemer is formed by rolling circle replication. This concatemeric DNA is packaged into phage particles by sequential cuts starting at a *pac* site. Each packaging event measures out a "headful" of DNA, which is slightly longer than the size of the genome. As a result a series of chromosomes is produced. Each chromosome is terminally redundant, but the redundant sequence is different for each one. Thus the chromosomes are said to be circularly permuted.

a specific region in the genome (*pac*) and proceeds processively along the concatemer from this site. The first packaging cut, by a phage-encoded enzyme, is site specific, at *pac*, and a defined length of DNA is encapsidated. When this length (a "head-ful") is packaged, the second cut is made and a phage head containing DNA is formed. This second cut also defines the start of packaging for the next phage head to be formed. The same headful length of DNA is encapsulated starting at the second cut, and when this length is reached a third cut is made. The third cut plays the same role as the second. It terminates packaging of the second headful and starts the packaging into the next head. This process continues. Thus, although DNA packaging starts at a defined sequence, the subsequent cuts are not sequence specific. Rather, it is the same length of DNA that is packaged each time. For phage P22, the size of the genome is slightly less than the length of the headful. Thus, each time a head is formed, the encapsidated DNA contains a short terminal repeat. Since each packaging event starts at a different sequence, the repeats are different, producing permuted chromosomes. Headful packaging thus provides the mechanism for generating the essential terminal repeats.

2. Formation of P22 transducing particles

During P22 infection the host DNA is not degraded (in contrast to infections with virulent phages like T4; see Guttman and Kutter, this volume), and thus is potentially also a substrate for DNA packaging. As just described, the characteristics of DNA packaging are that a concatemeric molecule is encapsidated in processive headfuls starting from a unique site. Exactly the same process operating on the host DNA generates transducing particles (Fig. 4). When a sequence similar to the *pac* sequence occurs in the host chromosome, the phage packaging apparatus uses this analogous *pac* site and the host chromosome can be packaged. When a *pac* site occurs in the host DNA, packaging

A. Wild type chromosome

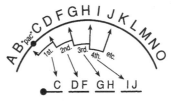

B. Chromosome deleted for E

Fig. 4. Packaging of host DNA by the phage P22 headful packaging mechanism. **A:** A wild-type bacterial chromosome contains a site, *pac*, that is recognized by the phage DNA packaging apparatus. The bacterial chromosome is packaged into phage particles starting at *pac* and proceeding sequentially in one direction. Each packaging event cuts a headful of DNA and each of the resulting transducing fragments carries a characteristic set of genes. Genes on the same fragment are cotransducible because they can be introduced into a cell together. **B:** The packaging of a bacterial chromosome that is deleted for gene E is shown. The *pac* site is still recognized and the overall process of forming transducing particles is the same as in A. However, because of the deletion, the cotransduction of genes is altered. This is because each packaging event measures the same amount of DNA and the deletion has changed the distance of distal genes from the *pac* site.

initiates and proceeds processively, generating a series of phage particles that carry different parts of the host chromosome. These transducing particles carry DNA fragments that are all the same size but obviously do not have the terminal repetition that the phage chromosomes have, since the genome size of the host is much greater than the headful size that is packaged. For P22, although there are roughly equal amounts of phage and host DNA in the cell during infection, only a few percent of the host chromosome is packaged and only a few percent of the particles are transducing particles. What limits formation of transducing particles is the number of *pac* sites in the host DNA, as well as the fact that these chromosomal sites are probably not identical in sequence to the phage *pac* site and are thus inefficiently used. It has been estimated (Chelala and Margolin, 1974) that there are 10 to 15 of these *pac*-like sites in the *Salmonella* chromosome. Other studies (reviewed in Margolin, 1987) indicate that at least 5–10% of the bacterial chromosome is packaged in the sequential headfuls from each *pac* site. However, the efficiency of packaging declines the more distant a gene is from a *pac* site, and this contributes to the wide range in transduction frequencies seen for different genes transduced with P22. This variation in frequency can be two or three orders of magnitude between different genes.

One experimental approach (Chelala and Margolin, 1974) that provided support for this view of transducing particle formation comes from the study of the effects of deletions in the bacterial chromosome on transduction (Fig. 4). One effect of deletions is on the cotransduction of genes, the frequency at which two nearby genes are transduced together on the same fragment. Some deletions were found to alter this frequency, presumably by altering the register of subsequent packaging events. Effects on transduction have also been observed for insertions; these are due to an analogous mechanism. The studies imply that host DNA packaging initiates from a limited number of sites and occurs processively with a characteristic register for each region.

Another important approach to understanding this mechanism of generalized transduction was the isolation and analysis of HT mutants of P22: mutants that perform generalized transduction at a higher frequency than a wild-type phage (Schmieger, 1972). HT mutants, which can increase the frequency of transduction of some genes by as much as 10,000-fold, have an alteration in the phage protein involved in recognizing the *pac* site (Raj et al., 1974). As a consequence of this mutation, the HT mutants show an altered specificity in making the first cut:

packaging initiates at a different, more common site (Casjens et al., 1987). With respect to packaging of the phage chromosome, it is observed that in HT mutants there is a different distribution of cut sites. With respect to generalized transduction, in the HT mutants other, more numerous sites in the bacterial chromosome can be used to initiate packaging and hence the sites are no longer as limiting a factor in the initiation of packaging. This means that the host DNA competes more effectively with phage DNA for the packaging apparatus, and there is a higher proportion (up to 50%) of phage particles that carry host DNA, resulting in a higher frequency of transduction. In addition there is less variation in the transduction frequency of different genes. Because of these properties the HT mutants are extremely useful tools when transduction is needed for genetic manipulations.

B. Formation of Generalized Transducing Particles by Other Phages

Phage P22 is a useful model, both to illustrate a common mode of DNA metabolism as well as its consequences for transduction. However, among the phages that perform generalized transduction, there are both minor variations on this theme as well as completely different modes of transduction. Some examples are given below.

1. Phage P1

The *E. coli* phage P1 is in many ways as well studied as P22 (see Sternberg and Hoess, 1983, for a review). Phage P1 is noteworthy because it is the major generalized transducing phage used in *E. coli* and can infect a broad range of hosts, making it an important tool for genetic manipulation. In general, P1 is similar to P22 in using a processive headful packaging process initiating at a specific site to produce terminally redundant, circularly permuted phage DNA. The P1 *pac* region has been analyzed and is complex, comprising a number of short sequence elements in a 161 bp region (Sternberg and Coulby, 1987). The size of the P1 headful is about twice that of P22, thus larger segments of the bacterial chromosome can be transduced. There is less variation in transduction frequency between different genes with P1 than seen with P22. This could result from the larger headful as well as there being more packaging sites in the bacterial chromosome. These results have also been interpreted to mean that initiation of packaging of the bacterial chromosome by phage P1 does not use *pac* sites like those found in the phage chromosome (Sternberg, 1986).

2. Phage T4

As detailed by Guttman and Kutter (this volume), phage T4 is not usually thought of as a generalized transducing phage. T4 is a virulent phage that degrades the host DNA during infection and normally packages only phage DNA by a headful packaging mechanism. This degradation is a result of another feature of T4 DNA metabolism. T4 DNA is modified and contains glycosylated hydroxymethylcytosine, a modification that results from phage-encoded enzymes. The degradation is due to a phage-encoded nuclease that cuts DNA at unmodified cytosines, generating substrates for other DNases in the cell. A multiple mutant phage was constructed (Wilson et al., 1979) with mutations to prevent degradation of host DNA. This phage can perform generalized transduction, presumably packaging host DNA as well as phage DNA because the mechanism to distinguish these two has been removed.

3. Phage λ

Although λ is not usually thought of as a generalized transducing phage, it can package host DNA (Sternberg, 1986). This phage does not use a headful packaging mechanism. During infection, replication of λ DNA generates concatemers from a rolling circle, like the phages described above. However the packaging mechanism is site specific. A particular sequence, *cos*, is recognized and the DNA is cut (discussed below; Fig. 9). Two *cos* sites must be present in the chromo-

some for packaging to occur. Cutting at *cos* produces a double-stranded break that is not blunt ended; a 12 nucleotide single-strand end is produced that is necessary for the chromosome to circularize in the next infection. This DNA processing system, although suited for processing of the phage's DNA, is apparently only loosely related to the mechanism of formation of generalized transducing particles. A study of the formation of transducing particles suggests that λ packages host DNA without recognition of specific *cos* sites (Sternberg, 1986). Moreover formation of transducing particles is not seen until 60 to 90 minutes after infection and is inhibited by the phage exonuclease synthesized during infection. Because of these properties, λ transducing particles are difficult to detect in a wild-type infection, where cell lysis occurs after about 60 minutes. Lysis defective mutants (also mutant for the exonuclease) must be used for maximal transducing particle formation.

4. Phage Mu

Mu has quite a different DNA metabolism from the phages discussed so far (see Whittle and Salyers, this volume). Mu is a transposable element that replicates by copying its genome and integrating this copy at random sites in bacterial DNA (replicative transposition; Fig. 5). During infection the cell thus accumulates a number of different insertions of the Mu chromosome. Packaging of these Mu chromosomes is by a headful mechanism. In this case the *pac* site is at the *c* end and one cut is made in bacterial DNA about 100 nucleotides outside of this end while the other cut is a fixed length in the direction of the *S* end, from 500 to 3200 nucleotides outside of this end for the wild-type phage. Hence the cuts are made in the bacterial DNA that flanks the integrated Mu genome. This genome structure is important for the initial transposition event in subsequent infections. Genetic studies (Howe, 1973) showed that Mu is capable of generalized transduction. Presumably *pac* sites in the

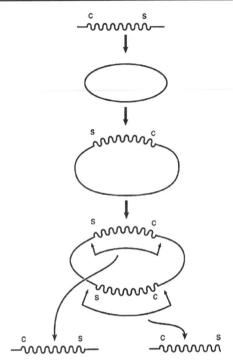

Fig. 5. Phage Mu DNA metabolism. Shown at the top is the linear phage Mu chromosome as it is carried in the phage particle. The central (wavy) portion represents phage genes while the terminal (straight) portions flanking the phage c and s regions are bacterial DNA segments. Upon infection, the Mu segment of the chromosome transposes into the circular bacterial chromosome. This integrated genome then replicates and transposes to additional sites in the chromosome. Finally, the integrated genomes are packaged by a process that involves cutting in the bacterial DNA that flanks the integrated Mu genomes. This produces chromosomes that contain different regions of the bacterial genome.

host chromosome are also recognized, leading to the formation of transducing particles.

C. Events in the Recipient

As stated earlier, generalized transduction can be thought of as a two-stage process. The varied ways in which the first stage, packaging of host DNA into phage particles, can occur has been described. In this section the second stage of transduction is described—the fate of the host DNA after it has been introduced into a recipient cell. When the packaged host DNA is injected into a recipient cell, it can become stably integrated into the recipient cell's chromosome by homologous recombination, or it can remain unintegrated in the cytoplasm. In this latter unintegrated case the DNA will eventually be lost if it cannot replicate in the cytoplasm (abortive transduction), whereas it will be maintained as an extrachromosomal element if it can replicate as a plasmid.

1. Homologous recombination with the recipient's chromosome

Most of what is known about integration of transducing fragments by recombination comes from studies of phages P22 and P1. In the case of P1 for example, linear doublestranded DNA molecules are injected into the recipient where they may assume a circular conformation due to the presence of a bound protein. Studies with isotopically labeled DNA showed that there is a physical replacement of the recipient's DNA by the transduced fragment, as described earlier. In the case of transduction by both P1 and P22, it was determined that this is a double-strand replacement. That is, a double-stranded segment of the recipient's chromosome is replaced by the homologous double-stranded segment from the transducing fragment. Such an event requires two crossovers, one at each end of the replaced segment (Fig. 6). It has been estimated that only about 5% of the DNA introduced as transducing fragments ends up being incorporated into the bacterial chromosome. This is in part because some fragments do not recombine at all and remain in the cytoplasm. In addition, for those fragments that do recombine, the entire transducing fragment that is introduced into the recipient need not integrate into the recipient chromosome. The portion that is integrated is determined by where the two crossovers occur.

D. Measuring Transduction

In discussing the various fates of the DNA in the recipient, it is also useful to consider the practical problems in assaying for

Recombination in transduction

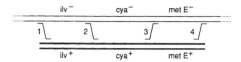

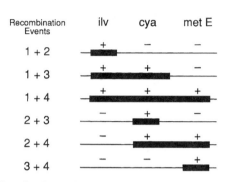

Fig. 6. Recombination in the recipient: possible outcomes of a three-factor transductional cross. Shown at the top is a recipient chromosome (thin lines) carrying mutations in the *ilv, cya,* and *metE* genes and a donor transducing DNA fragment (thick lines) that is wild-type at these three loci. In order to integrate donor DNA into the recipient chromosome, two crossovers must occur, and these can be in any of the four regions defined by the three genes. At the bottom is shown the genotypes resulting from crossovers in different regions. The donor DNA (thick line) substituted for that of the recipient (thin line) is shown for each case.

transduction. Imagine an experiment to measure the transduction of a specific gene, *metE*, that encodes the enzyme homocysteine methylase, required for methionine biosynthesis. The donor is *metE*⁺ and the recipient is *metE*⁻. After growth of a phage like P1 or P22 on the donor strain, the resulting phage lysate contains two kinds of particles (Fig. 2). The majority of particles carry phage DNA and will cause phage production and cell death upon subsequent infection. The remaining particles carry host DNA from all regions of the bacterial chromosome, so only a small fraction of these carry the *metE*⁺ gene. For P22, only a few percent of the particles carry host DNA, and of these, at most only a few percent carry *metE*. In the transduction experiment, the phage lysate is

mixed with the recipient cells and, after the phages have adsorbed to the bacteria, the mixture is plated on a medium that selects for *metE*⁺ cells. Most of the cells fail to grow on this medium because they remain *metE*⁻, but a few cells can grow and these form colonies. These are the cells that have received a transducing fragment carrying the *metE*⁺ gene and this wild-type gene has become integrated into the recipient chromosome by recombination, replacing the *metE*⁻ gene and making the cell *metE*⁺ (Fig. 6). The problem is to quantitate the frequency of transducing the *metE*⁺ gene.

One factor that complicates this quantitation is the multiplicity of infection (MOI), the ratio of phage particles to bacterial cells (Fig. 7). At low multiplicities (MOI ≪ 1) most bacteria receive no phage particles and those that are infected receive only a single particle. Very few cells receive more than one particle and thus the number of *metE*⁺ transductants observed is simply related to the number of particles carrying *metE*⁺ DNA. The number of transductants will be less than the number of particles carrying *metE*⁺ gene because, as mentioned above, not all of

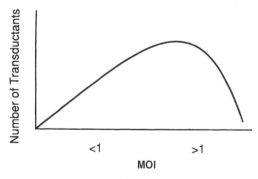

Fig. 7. Effect of multiplicity of infection (MOI) on transduction frequency. This curve illustrates the fact that at high MOI, where all cells are infected by a particle containing phage DNA, there will be killing of cells receiving transducing DNA, leading to a decrease in frequency. Note that this effect will not be as great if the phage can form lysogens. In that case there will be some lysogen formation at high MOI and correspondingly less killing. However, mutations are often introduced into transducing phages to prevent lysogenization.

these fragments recombine with the bacterial chromosome and those that do will sometimes not integrate the *metE* region of the fragment. At high multiplicities (MOI $\gg$ 1) all cells become infected and most cells receive more than one particle. In this case, when a cell is infected by a particle that carries a *metE*$^+$ gene, it is simultaneously infected by other particles, and these will carry phage DNA since this is the major class of particles present. The particles introducing phage DNA will usually cause the cell to produce more phage and lyse, thus making it impossible for the cell to become a transductant. At high MOI the only cells that can become transductants are those that receive a transducing fragment and either do not become infected by particles carrying phage DNA, or if they do, the phage infection leads to lysogeny rather than lytic growth. As a result of this, fewer transductants may be observed when large amounts of lysate are used for transduction (high MOI) than when lower amounts of lysate are used (low MOI). The optimum amount of lysate is somewhere near an MOI of 1, where the largest amount of lysate (and transducing particles) has been added that does not cause significant loss of transductants by killing due to multiple infection with particles containing phage DNA.

In measuring transduction frequencies, two approaches can be used. In one method, after the growth of the phage on the donor strain, the resulting lysate is titered for particles containing phage DNA by measuring the number of plaque-forming units (PFUs) on a suitable indicator bacterium. Next the phage lysate is titered for transducing particles by measuring the number of transductants in the volume of lysate added to recipient cells. The ratio of transductants to PFUs is then calculated, using the titer of transductants determined in the low MOI portion of the titration. This method allows the frequency of transduction of two different markers to be compared, by normalizing them both to the total number of phage DNA-containing particles. The alternative method is to measure the frequency of transduction of several different markers by always using the same volume of lysate (i.e., the same MOI) in the transductions and ignoring the titer of phage DNA-containing particles. In this case the ratio of transductants for each marker can be normalized to that for a common marker, allowing them to be compared.

I. Cotransduction of markers

One of the most important uses for generalized transduction is in measuring how far apart markers are or in determining the order of three or more markers in the chromosome. An important concept for this analysis is that of contransduction, the ability of two markers to be simultaneously integrated into the recipient's chromosome on the same transducing fragment (Figs. 4, 6). Since phages like P1 and P22 package a discrete length of DNA in their particles, for two markers to be contransduced at all requires that they not be farther apart in the bacterial chromosome than the headful length of DNA that the transducing phage will package. Furthermore the closer together two markers are located, the higher is the probability that they can be integrated into the recipient chromosome together. This probability is manifested operationally as the contransduction frequency. In the example of Figure 6, the donor strain is *ilv*$^+$ *cya*$^+$ *metE*$^+$, while the recipient is *ilv*$^-$ *cya*$^-$ *metE*$^-$. When *metE*$^+$ transductants are selected, more of these will also be *cya*$^+$ (cotransduction of *cya* and *metE*) than will be *ilv*$^+$ (cotransduction of *ilv* and *metE*) because *cya* is closer to *metE* than *ilv* is. Moreover the *metE*$^+$ transductants that are also *ilv*$^+$ will almost always be *cya*$^+$ because this marker is in the middle (only rare double recombinants will be *ilv*$^+$ *cya*$^-$ *metE*$^+$). In contrast, of the *metE*$^+$ transductants that are also *cya*$^+$, some will be *ilv*$^+$ and some will be *ilv*$^-$. These results would suggest the gene order *ilv-cya-metE*. This mapping technique is very similar to the three-factor cross procedure using bacterial transformation (see Streips, this volume).

There are actually two factors that contribute to the frequency of cotransduction of two markers. One is the probability that they will reside on the same transducting fragment formed in the donor. The closer two markers are to each other, the greater the chance that they will be packaged together on a single fragment. In addition, since the two crossovers resulting in integration in the recipient can occur anywhere along the transducing fragment, the closer together the markers are, the greater the chance that they will be integrated together. A formula relating distance to cotransduction frequency has been derived, and the relation is nonlinear, reflecting these complexities. A theoretical treatment of this mapping is presented in Mandecki et al. (1986).

2. Marker effects

Although generalized transduction is "general" because any gene in the donor chromosome can be introduced into the recipient chromosome, there are quantitative differences in the frequency with which different genes can be transduced. This variation is as much as 1000-fold for P22 transduction and 10-fold for P1. Based on the previous discussion of transduction, there are two principal sources for variation. The first is variation due to differences in the packaging efficiency of different regions of the chromosome. This results from the requirement for specific *pac* sequences, which would not be expected to be placed periodically around the chromosome. This appears to be the major source of variation in P22 transduction. It is estimated that there are only 10 to 15 *pac* sites in the *Salmonella* chromosome, and thus the transduction frequency of a gene is strongly dependent on how close it is to a *pac* site. In P22 HT mutants, which can package from other sites, the variation in transduction frequencies is only about 10-fold, and markers that are poorly transduced by wild-type P22 often show the greatest increases with HT. The second

source of variability is due to differences in the efficiency of recombination of transducing fragments from different regions of the chromosome. This appears to be the primary cause of variation in P1 transduction. Treatments that stimulate recombination, such as UV irradiation (see chapters by Yasbin and Levene and Huffman, this volume), reduce the variability of P1 transduction frequencies.

As a result of these factors, quantitating transduction frequencies can be complex. An example of this is the asymmetry of cotransduction frequencies. In an experiment where an ilv^+ $metE^+$ donor and an ilv^- $metE^-$ recipient are used, when ilv^+ is selected, about 30% of these transductants are also found to be $metE^+$. In contrast, when $metE^+$ is selected, about 70% of these are found to be ilv^+. Thus, although the distance is the same between these markers in the two crosses, different cotransduction frequencies are obtained depending on the selection. This could reflect either asymmetries in packaging or recombination. In the case of a packaging asymmetry (Fig. 8), it is possible that there are two types of transducing fragments carrying the ilv^+ gene: a major

Classes of transducing fragments

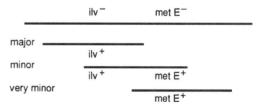

Fig. 8. Hypothetical classes of transducing fragments in the *ilv-metE* region and their relative abundance. When an $ilv^-/metE^-$ recipient is transduced to ilv^+, about 30% of these transductants are also $metE^+$. In contrast, when $metE^+$ is selected, about 70% of the transductants are ilv^+. To account for the asymmetry by differential packaging in the donor, it is proposed that there are three types of transducing fragments packaged from this region: a major one containing only *ilv*, a minor one carrying both *ilv* and *metE*, and a very minor one with only *metE*.

one that does not contain the *metE* gene and a minor one that carries *metE*. Hence the majority of the recombinants formed when *ilv*⁺ is selected will retain the recipient's *metE*⁻ marker. On the other hand, there may be more *metE*⁺ transducing fragments that carry the *ilv*⁺ marker than those that do not carry *ilv*, resulting in the higher frequency of *ilv*⁺ *metE*⁺ transductants. It is also possible to account for this difference in cotransduction based on hot spots in recombination. From these considerations it is clear that care must be taken in interpreting data based solely on transduction frequencies. At the minimum, reciprocal crosses must be performed to determine if there are marker effects. Ideally multiple-factor crosses should be used (like the *ilv cya metE* example above) to determine gene order.

3. Abortive transduction

Not all of the DNA fragments introduced into a cell by a generalized transducing phage recombine and integrate into the bacterial chromosome; the fraction that do recombine are in fact the minority. The majority of transducing fragments remain in the cytoplasm of the recipient cell, although the efficiency of recombination depends on the region (Schmieger, 1982). These fragments do not replicate, since they do not contain an origin of replication and thus are inherited by only one cell at each division. Eventually these fragments are lost from the culture because very few cells carry an unintegrated, nonreplicating fragment and because they can be degraded. This phenomenon of creating unstable transductants with unintegrated fragments is called abortive transduction.

The genes in an unintegrated fragment can be expressed and thus the cell is functionally diploid for this region of its chromosome. If the cell is mutant for a gene in the diploid region, while the fragment is wild-type, complementation will be observed. At each cell division the functional product generated in the diploid can be distributed into the two

daughter cells, and thus the daughter cell that does not receive the fragment may remain phenotypically wild-type until the functional product is degraded or diluted through subsequent cell divisions. Typically, in a transduction of an auxotrophic recipient to prototrophy, a large number of microcolonies will be observed on minimal medium. These are abortive transductants that grow by virtue of the small amount of nutrient made by the diploid. In contrast, the stable transductants (where the fragment has recombined into the bacterial chromosome) form normal-sized colonies. Another classic and beautiful example of this phenomenon is the behavor of nonmotile mutants (Lederberg, 1956; Stocker, 1956). In this case the diploid formed by abortive transduction is motile. On a plate with a low agar concentration the diploid can move, but at each cell division a nonmotile daughter is produced that will grow into a colony. Eventually a trail of these nonmotile daughters is made, called a flare, which marks the path of the original abortive transductant.

The unintegrated DNA introduced by transduction is quite stable in the cell. In the case of P1 transduction, when this DNA is isolated from a recipient it is found to be circular, even though the DNA was linear in the phage particle (Sandri and Berger, 1980). The circles can be disrupted by treatments with detergent or protease, implying that the circle is held together by a protein linker. Evidence for protein attached to the DNA in P1 transducing particles has in fact been reported (Ikeda and Tomizawa, 1965b). Normally, a linear double-stranded DNA molecule is rapidly degraded in vivo by the action of the RecBCD nuclease. Thus it appears that the protein at the DNA end may be a special mechanism to protect transducing DNA from this degradation.

4. Transduction of plasmids

Just as the bacterial chromosome is a substrate for packaging by generalized transducing phages, so is any other DNA that is inside an infected cell, including plasmid chromosomes. When the plasmid is larger than a phage headful, only a part of the plasmid can be packaged. This is the case with R factors, which are much larger than a headful of a phage like P22. When this occurs, the fragment of the plasmid that is introduced into the recipient cannot circularize efficiently because there is no homology between the fragment ends or with cellular sequences to allow recombination. Thus the majority of transduced plasmid fragments will be lost from the recipient. Nevertheless, at a low frequency it is possible to recover deleted versions of the plasmid following transduction. These represent fragments that retain the plasmid origin of replication. It is possible that these are plasmids that acquired a deletion in the donor that made them small enough to fit completely into a phage head. Alternatively, these may be fragments that were introduced into the recipient where they circularized by some inefficient "illegitimate" recombination event. This process is called transductional shortening (Shipley and Olsen, 1975; also see the review by Low and Porter, 1978).

A different situation is found when the plasmid is much smaller than the phage headful. In this case, the plasmid is packaged inefficiently because the chromosome is too small to form a stable head. However, multimeric forms of the plasmid can be packaged efficiently, as shown for transduction of plasmids by P22HT mutants (for a review see Margolin, 1987) and T4 transducing phages (Takahashi and Saito, 1982). The multimeric plasmid chromosomes may be preexisting in the donor or may be caused by the phage infection, either by phage-promoted recombination between plasmids or rolling circle replication of the plasmid. In these cases the multimers that are introduced into the recipient circularize by recombination between the repeated plasmid units.

A final case of plasmid transduction involves plasmids that are smaller than a phage headful but contain sequences that are homologous to the bacterial chromosome. Examples of this are small plasmid vectors carrying cloned genes. These can re-combine with the donor's chromosome and become integrated. This integrated plasmid may then be packaged and transduced by the usual mechanism (Trun and Silhavy, 1987). Once in the recipient, the plasmid can recombine out of the chromosome by the reverse of the integration process, leading to a transductant containing the free plasmid. This is a useful mechanism for selecting for recombination between a plasmid and the host chromosome.

E. Uses for Generalized Transduction

In the preceding sections several uses for generalized transduction have been described. These include genetic mapping, complementation analysis, transduction of plasmids, selection for deletions in plasmids, and selection for recombination between cloned genes and the bacterial chromosome. Other common uses include strain constructions, delivery of transposons, and isolation of chromosome rearrangements such as duplications (see Margolin, 1987, for a review).

III. SPECIALIZED TRANSDUCTION

Specialized transduction is the second major class of virus-mediated genetic exchange. It differs from generalized transduction in two respects. First, in this mode of transduction the transduced genes are covalently joined to the viral chromosome, allowing them to be replicated, packaged, and introduced into a recipient with the rest of the viral chromosome. Second, a specialized transducing particle carries a specific chromosome segment, and consistently only introduces this set of genes into the recipient. Hence the name, specialized transduction (see Hendrix, this volume, for a more complete discussion).

A. Bacteriophage λ DNA Metabolism

The temperate phage λ (Fig. 9 and Appendix) is the classic example of a specialized transducing phage. The λ phage particle contains a linear double-stranded DNA molecule with complementary 12 nucleotide single-stranded ends (see Furth and

λ DNA Metabolism

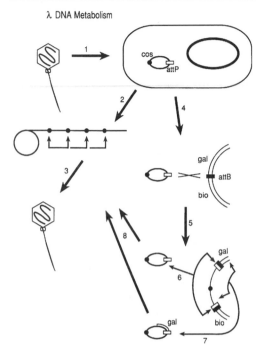

Fig. 9. Phage λ DNA metabolism. When λ infects a cell (1) the phage DNA circularizes by virtue of the single-stranded cohesive ends (cos) and the cell can have two fates. If the phage grows lytically (2), the chromosome replicates to form concatemers, which are then packaged to form new phage (3). Alternatively (4), the phage chromosome can integrate into the host chromosome during the lysogenic response. This is accomplished by site-specific recombination between the *attP* site of the phage chromosome and the *attB* site of the host (5). When a lysogen induces, the phage genome excises by a reverse of the integration process (6), leading to a wild-type chromosome that can replicate to form concatemers (8) and normal phage. However, on rare occasions the excision process is aberrant (7) and the recombination does not occur between the *att* sites. As a result a chromosome is formed that has lost some phage material and acquired some host sequences. As long as none of the missing phage genes were essential, this hybrid chromosome is capable of replicating normally (8) and producing new phage. These phages are all specialized transducing phages since they carry host genes in their chromosomes.

Wickner, 1983, for a review of λ DNA metabolism). When λ infects a cell, its chromosome circularizes by hybridization of these single-strand extensions (*cos* sites), and the phage then chooses between two alternative

life cycles. In the lytic response, the chromosome replicates to form concatemers that are packaged into particles by cutting at the *cos* sites, reforming the single-stranded ends. This packaging differs from the headful mechanism in that the size of the DNA that is encapsidated is determined by the distance between the *cos* sites and can vary considerably. This is the basis for cosmid cloning vectors (see Geoghegan, this volume). However, there are limits on the size of the packaged chromosome. DNA molecules that are larger than about 50,000 bp, or smaller than about 35,000 bp, produce unstable phage particles. The alternative life cycle is the lysogenic response. In this case the phage chromosome integrates into the bacteria chromosome and a repressor is produced that prevents expression of the lytic gene products. The phage chromosome is replicated and passively carried by the bacterium as a latent prophage. The integration reaction is performed by a phage-encoded site-specific recombination system that breaks the phage chromosome at a specific sequence, *attP*, and joins it to the host at another specific sequence, *attB* (see Weisberg and Landy, 1983, for a review). When the prophage becomes induced, the phage repressor is inactivated and the reverse reaction occurs. The site-specific recombination system acts at the *attP/attB* joints to reform a circular chromosome that undergoes the lytic cycle.

Specialized transducing particles are formed in vivo when an integrated λ prophage excises from the bacterial chromosome imprecisely (Fig. 9). In this case the breaking and joining events do not occur at the *attP/attB* joints, but are the result of "illegitimate" recombination events occurring at other sites in the vicinity of the prophage. This process occurs at low frequency and has little sequence specificity. Markers either to the left or right of the prophage can be transduced. The result is a circular chromosome that has lost some of the phage DNA sequences from one end of the prophage and in their place carries bacterial DNA that was adjacent to the other prophage end. This hybrid chromosome can replicate and be packaged, provided that the required lytic functions and *cos* site are present, and results in the formation of transducing particles that carry and can introduce the incorporated bacterial genes. Some specialized transducing phages carry the complete array of genes and sites necessary for lytic growth. These are "plaque-forming" phages, and this property is usually designated by including a "p" in their name, such as λ p*gal*, a plaque-forming *gal* transducing phage. There can also be defective transducing phages. These are phages that have lost essential genes from their chromosome in the process of excision. The minimum requirement for a phage to be packaged and maintained is that it contain the *cos* site and the origin of replication. All other functions can be provided by helper phages. To designate that a transducing phage is defective and requires a helper for growth, a "d" is included in its name, such as λp*gal*, a defective *gal* transducing phage.

From this mechanism it is clear that only a restricted set of bacterial genes can be transduced. This is a consequence of the fact that only genes near the *attB* site can be incorporated, as well as the limits on the size of the chromosome that can give a stable phage particle. However, a number of methods have been used to allow a broader range of genes to be transduced by λ. Rearrangements of the bacterial chromosome, such as deletions, can bring different genes close enough to the *attB* site to be transduced. A second method relies on the observation that when the *attB* site itself is deleted, λ now integrates at a large number of secondary attachment sites. These sites are not used as efficiently as *attB*, presumably because their sequences deviate from *attB*. However, lysogens with prophages located at these secondary sites can be used to isolate specialized transducing phages carrying the neighboring genes. In a third method, the site of integration of the phage chromosome does not depend on the site-specific recombination system at all, but rather utilizes transposons (see Whittle and Salyers, this volume). If a transposon insertion can be isolated in the bacterial chromosome near a gene of interest, then a phage chromosome that also carries an insertion of the transposon can integrate at this site by homologous recombination. This is similar to generating F plasmid integration sites using transposon Tn*10* insertions (see Porter, this volume). Finally, derivatives of λ have been developed that contain the transposition system from phage Mu (see Whittle and Salyers, this volume) instead of the λ site-specific recombination system (Bremer et al., 1984). These phages behave like λ except that integration during lysogenization uses the Mu system and there is no sequence specificity. As a result prophages can be isolated anywhere around the bacterial chromosome. This is the most general in vivo method, and it is as general as the in vitro method for forming specialized transducing phages: molecular cloning using recombinant DNA methodology.

B. Events in the Recipient

When a specialized transducing phage like λ stably transduces a recipient cell, it forms a lysogen that is, it integrates into the chromosome and expression of lytic functions is repressed. There are several different ways this can occur. If the phage has an intact *attP* site and *int* gene (the gene needed for site-specific integration at *attB*), and the repression system of the phage is intact, lysogeny can follow the same steps as for a wild-type λ phage. Often, however, the transducing phage does not have *attP* or *int*, these having been lost during formation of the transducing phage genome. In this case integration can occur through homologous recombination between the bacterial chromosomal sequences carried by the phage and its counterpart in the host chromosome. In some cases, if the recipient is lysogenized by another wild-type helper λ phage, this prophage can also provide a homologous region for recombination with the transducing phage. Finally, under some conditions, the transducing phage can be maintained as an un-integrated plasmid in the cell (see Weisberg, 1987).

C. Uses for Specialized Transduction

Specialized transducing phages provide one of the most convenient ways of manipulating genes. They represent the original method for gene cloning, using in vivo genetic techniques. Because these phages carry a small region of the chromosome, genes carried on phages can be mapped with great precision, as can mutations within these genes. Moreover mutagenesis can be targeted to specific genes. These phages also provide a convenient means for constructing strains that are diploid for a very limited region of the chromosome, useful for complementation analysis.

REFERENCES

Bremer E, Silhavy TJ, Weisemann JM, Weinstock GM (1984): λ placMu: A transposable derivative of bacteriophage lambda for creating lacZ protein fusions in a single step. J Bacteriol 158:1084–1093.

Casjens S, Huang WM, Hayden M, Parr R (1987): Initiation of bacteriophage P22 packaging series. Analysis of a mutant that alters the DNA target specificity of the packaging apparatus. J Mol Biol 194:411–422.

Chelala CA, Margolin P (1974): Effects of deletions on cotransduction linkage in Salmonella typhimurium: Evidence that bacterial chromosome deletions affect the formation of transducing DNA fragments. Mol Gen Genet 131:97–112.

Ebel-Tsipis J, Botstein D, Fox MS (1972): Generalized transduction by phage P22 in Salmonella typhimurium. I. Molecular origin of transducing DNA. J Mol Biol 71:433–448.

Furth M, Wickner S (1983): Lambda DNA replication. In Hendrix R, Roberts J, Stahl F, Weisberg R (eds): "Lambda II." Cold Spring Harbor, NY: Cold Spring Harbor Laboratory, pp 145–174.

Hendrix R, Roberts J, Stahl F, Weisberg R (eds) (1983): "Lambda II." Cold Spring Harbor, NY: Cold Spring Harbor Laboratory.

Howe M (1973): Transduction by bacteriophage Mu-1. Virology 55:103–117.

Ikeda H, Tomizawa J (1965a): Transducing fragments in generalized transduction by phage P1. I. Molecular origin of the fragments. J Mol Biol 14:85–109.

Ikeda H, Tomizawa J (1965b): Transducing fragments in generalized transduction by phage P1. II. Association of protein and DNA in the fragments. J Mol Biol 14:110–119.

Lederberg J (1956): Linear inheritance in transductional clones. Genetics 41:845–871.

Low KB, Porter DD (1978):Modes of gene transfer and recombination in bacteria. Annu Rev Genet 12:249–287.

Mandecki W, Krajewska-Grynkiewicz K, Klopotowski T (1986): A quantitative model for nonrandom generalized transduction, applied to the phage P22-Salmonella typhimurium system. Genetics 114:633–657.

Margolin P (1987): Generalized transduction. In Ingraham JL, Low KB, Magasanik B, Neidhardt FC, Schaechter M, Umbarger HE (eds): "Escherichia coli and Salmonella typhimurlum: Cellular and Molecular Biology." Washington, DC: American Society for Microbiology, pp 1154–1168.

Ptashne M (1985): "A Genetic Switch." Boston: Blackwell Scientific Publications.

Raj AS, Raj AY, Schmieger H (1974): Phage genes involved in the formation of generalized transducing particles in Salmonella-phage P22. Mol Gen Genet 135:175–184.

Sandri RM, Berger H (1980): Bacteriophage P1-mediated generalized transduction in Escherichia coli: Structure of abortively transduced DNA. Virology 106:30–40.

Schmieger H (1972): Phage P22 mutants with increased or decreased transduction abilities. Mol Gen Genet 119:75–88.

Schmieger H (1982): Packaging signals for phage P22 on the chromosome of Salmonella typhimurium. Mol Gen Genet 187:516–518.

Shipley PL, Olsen RH (1975): Isolation of a nontransmissible antibiotic resistance plasmid by transductional shortening of R factor RP1. J Bacteriol 123:20–27.

Sternberg N (1986): The production of generalized transducing phage by bacteriophage lambda. Gene 50:69–85.

Sternberg N, Coulby J (1987): Recognition and cleavage of the bacteriophage P1 packaging site (pac). II. Functional limits of pac and location of pac cleavage termini. J Mol Biol 194:469–479.

Sternberg N, Hoess R (1983): The molecular genetics of bacteriophage P1. Annu Rev Genet 17:123–154.

Stocker BAD (1956): Abortive transduction of motility in Salmonella: A nonreplicated gene transmitted through many generations to a single descendant. J Gen Microbiol 15:575–598.

Susskind MM, Botstein D (1978): Molecular genetics of bacteriophage P22. Microbiol Rev 42:385–413.

Takahashi H, Saito H (1982): High-frequency transduction of pBR322 by cytosine-substituted T4 bacteriophage: Evidence for encapsulation and transfer of head-to-tail plasmid concatemers. Plasmid 8:29–35.

Trun NJ, Silhavy TJ (1987): Characterization and *in vivo* cloning of *prlC*, a suppressor of signal sequence mutations in *Escherichia coli* K12. Genetics 116:513–521.

Weisberg RA (1987): Specialized transduction. In Ingraham JL, Low KB, Magasanik B, Neidhardt FC, Schaechter M, Umbarger HE (eds): "*Escherichia coli* and *Salmonella tymphimurium*: Cellular and Molecular Biology." Washington, DC: American Society for Microbiology, pp 1169–1176.

Weisberg R, Landy A (1983): Site-specific recombination. In Hendrix R, Roberts J, Stahl F, Weisberg R (eds): "Lambda II." Cold Spring Harbor, NY: Cold Spring Harbor Laboratory, pp 211–250.

Wilson GG, Young KKY, Edlin GJ, Konigsberg W (1979): High frequency generalized transduction by bacteriophage T4. Nature 280:80–82.

Zinder ND, Lederberg J (1952): Genetic exchange in *Salmonella*. J Bacteriol 64:679–699.

APPENDIX: BACTERIOPHAGE λ

The bacteriophage λ is one of the most thoroughly studied systems in biology. The DNA sequence of the entire phage λ chromosome has been determined (over 48,500 bp), and this organism has served as a paradigm for many phenomena in addition to genetic transduction. Much of the information about phage λ is summarized in the monograph *Lambda II* (Hendrix et al., 1983). Another good reference is the book *A Genetic Switch* (Ptashne, 1985). The metabolism of λ DNA during infection has been described above. Here we present an overview of the genetic organization of the phage and some of the important aspects of the control of phage λ gene expression to complement the chapter by Hendrix, this volume.

Genetic Organization

The λ chromosome encodes about 50 genes that are organized in functionally related clusters. At the left end of the chromosome (Fig. A1) is a block of 10 genes that is required for production of the phage head followed by 11 genes needed to make the phage tail. The central region of the chromosome contains 19 nonessential genes that are also functionally clustered. Included in this region are the genetic elements of the site-specific (*int, xis,* and the *att* site) and homologous (*exo* and *bet*) recombination systems of phage λ as well as a number of genes whose products are involved in interactions with the host. Next is a short stretch containing the important control genes *N, cl, cro,* and *cll,* followed by the DNA replication region (genes *O* and *P* and the origin of replication), the controller of late gene expression, *Q,* and the genes for cell lysis, *S, R,* and *Rz.* This overall functional arrangement along the chromosome is followed by a number of other phages of *E. coli* (e.g., phages 434 and 21) as well as phage P22 of *Salmonella.* Because of this relatedness, these phages are collectively referred to as the lamboid phages, and it has been possible to create hybrids in which functions in one lambdoid phage are replaced with the analogous genes from another phage in this race.

Transcriptional Units

There are four major transcriptional units in the phage (Fig. A1). The promoter P_L is used for the early leftward operon and the promoter P_R is used for the early rightward operon. The genes in these two operons are made early in infection. The $P_{R'}$ promoter (sometimes called the late promoter) drives a large operon consisting of genes that are expressed late in infection, such as those for head and tail production and cell lysis. Last is the transcription unit that contains the *cl* gene, encoding the λ repressor. There are two promoters for this purpose: P_{RE}, which is used early in infection, and P_{RM}, which is used in a repressed lysogenic cell. In addition to these is the P_1 promoter, which produces a small RNA to augment the expression of the *int* gene, and the P_o promoter, which makes another small RNA of unknown function.

Lytic Growth

In the lytic response, λ replicates its DNA, packages the new chromosomes into phage heads, adds phage tails to these, and finally lyses the cell. The genes for head and tail production and cell lysis are contained in the later operon, expressed from the $P_{R'}$ promoter. However, essential control genes and

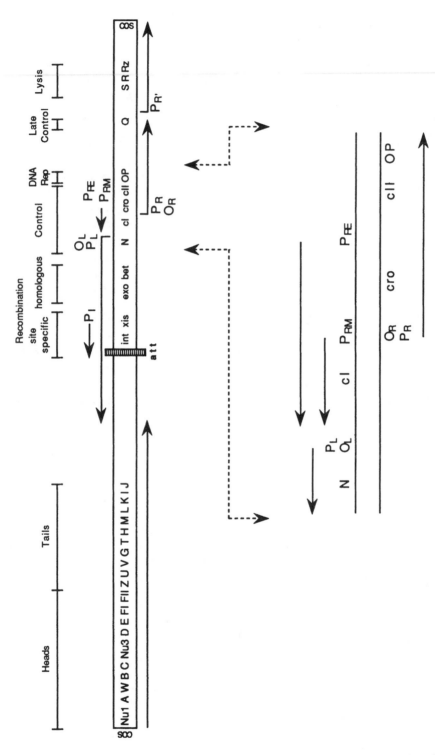

Fig. A1. Phage λ genetic map. The details of the map are described in the appendix. Arrows indicate direction of transcription of the different operons.

the DNA replication functions, located in the early leftward and rightward operons, must be expressed first.

The leftward operon contains the essential N gene, which is an antiterminator of transcription. Transcription termination signals are located throughout the leftward and rightward operons. There is a terminator immediately after the N gene as well as at other downstream sites in this operon. In the rightward operon there are terminators after the *cro* and P genes. The N gene product prevents transcription from terminating at these sites and thus allows expression of downstream genes. This provides a timing mechanism. Genes preceding the terminators (N and *cro*) are early genes, while the others are sometimes called delayed-early genes, since their expression is delayed until N protein is available.

In the rightward operon there are several essential functions for lytic development. First are the DNA replication genes O and P and the origin of replication. These are located after the first transcription termination signal in this operon. In addition is the Q gene, located after the second terminator, which thus delays its expression further. The Q gene product is required for the expression of all late proteins (i.e., production of heads, tails, and cell lysis). This is because there is a transcription terminator just after the late promoter P'_R and immediately before all of the late genes. The Q protein is a transcription antiterminator that is specific for this site and thus controls late gene expression. The delay in expression of Q contributes to the delay in expression of the late genes and ensures that plenty of DNA has accumulated before the packaging and lysis processes begin.

The Lysogenic Response

In the lysogenic response λ turns off the lytic genes by the action of a single repressor and integrates its chromosome into the bacterial chromosome. It is passively carried in this latent state as a prophage until it becomes induced. The repressor that turns off all of the lytic genes is the product of the *cl* gene. It binds to the phage chromosome at two regions: O_L and O_R (Fig. A1). Each of these is a cluster of three operator sites and binding of the λ repressor causes repression of transcription from the P_L and P_R promoters. This in turn prevents all lytic gene expression by preventing expression of the control genes N and Q.

Expression of the *cl* gene involves two promoters. The P_{RE} promoter (for the establishment of repression) is used early in infection when the decision between the lytic and lysogenic choices has not been resolved. This is a strong promoter that is stimulated by the product of the *cll* gene, located in the early rightward operon. This promoter allows a lot of repressor to be made quickly. Once a lysogen has been established, P_{RE} is not used and expression switches to the P_{RM} promoter (for maintenance of repression). This is a weaker promoter that produces enough repressor to maintain the repressed state. It can be either stimulated or inhibited by λ repressor bound at O_R (depending on the repressor concentration) and thus allows for buffering of the amount of repressor produced in the lysogen.

The other important control protein in the λ system is the *cro* gene product, which serves to counteract the effect of *cl*. The *cro* protein is also a repressor and it also binds at the sites at O_R and O_L. However, its binding characteristics are different from *cl* and as a result it can antagonize *cl* binding and lead to a lytic response. The decision between the lytic and lysogenic response involves the competition for control of the O_R and O_L sites by *cl* and *cro*. This competition is ultimately decided by the physiology of the host, which influences expression of these genes in ways that are not at present clear.

22

Genetic Approaches in Bacteria with No Natural Genetic Systems

CAROLYN A. HALLER AND THOMAS J. DICHRISTINA

School of Biology, Georgia Institute of Technology, Atlanta, Georgia 30332

I. INTRODUCTION

Recent advances in molecular genetic techniques have provided powerful tools for elucidating the molecular basis of uncharacterized bacterial systems. Previous chapters in this text have examined genetic approaches in highly characterized bacteria such as *Escherichia coli* and *Baccilus subtilis*. While many molecular genetic protocols have been designed for use in *E. coli* and related organisms, most do not operate outside non-enteric bacteria. An inability to examine a wide range of bacterial processes at the genetic level has severely hampered fundamental and applied research efforts. The objective of this chapter is to present a variety of state-of-the-art molecular genetic approaches in bacteria with no natural genetic systems.

Modern Microbial Genetics, Second Edition. Edited by Uldis N. Streips and Ronald E. Yasbin. ISBN 0–471–38665–0
Copyright © 2002 Wiley-Liss, Inc.

II. GENERAL STRATEGY

Design of a new genetic system is generally initiated with a mutant (genetic) complementation strategy. In the first step, mutants deficient in a specified activity are isolated. Mutagenesis may take place spontaneously, or it may be induced by chemical agents (e.g., ethyl methane sulfonate, hydroxylamine, 5-bromouracil, 2-aminopurine), UV radiation, or transposon insertions. Theoretical details of these mechanisms have been covered in previous chapters. The second step, and perhaps the most difficult, is to devise a selection technique that readily identifies a mutant phenotype. An effective selection technique clearly differentiates wild-type and mutant phenotypes, and thus facilitates screening tens of thousands of potential mutants quickly and accurately. The third step is to transform each putative mutant strain with a wild-type gene clone bank. This clone bank is typically mobilized from an *E. coli* host. Wild-type DNA fragments containing the genes of interest are identified by their ability to restore wild-type activity to transformants. The smallest complementing DNA fragment is isolated by subcloning and individual genes are then identified via nucleotide sequence analysis. An efficient gene transfer system is therefore critical to isolating, identifying, and manipulating genes from uncharacterized bacterial strains. The beginning of this chapter will focus on developing mutagenesis, mutant screening, and gene transfer techniques for uncharacterized bacterial strains. The end of the chapter will describe alternative ways to study bacterial systems that have proved intractable to some of the more common cloning technologies.

III. BROAD HOST RANGE GENE CLONING SYSTEMS

Broad host range cloning systems are usually based on transformation and conjugation mechanisms. Transformation is often the most attractive first option in designing a new gene transfer system. In Gram-negative bacteria, DNA uptake is artificially induced by electroporation or by treatment with a combination of divalent cations and heat shock. Gram-positive strains may display natural competence for DNA uptake, or transformation is artificially induced either by electroporation or by treatment of protoplasts with polyethylene glycol. Conjugal gene transfer may be required for bacterial strains that are not successfully transduced or transformed by standard protocols. Promiscuous plasmids that replicate in a broad range of hosts have provided the structural backbone for cloning vectors in conjugation. The transfer, replication, and maintenance functions of a variety of broad host range cloning vectors are described in the next section.

A. Broad Host Range Plasmid Transfer and Replication/Partitioning Functions

Broad host range plasmids are classified according to their transfer and replication/partitioning functions. In Gram-negative bacteria, the sex pilus (or conjugation bridge) of broad host range plasmids is encoded by the transfer (*tra*) genes. Tra functions determine the range of organisms into which the plasmid may be conjugally transferred. Broad host range Tra functions are often able to transfer DNA across species and, in some cases, domain lines (Knauf and Nester, 1982). Broad host range plasmids may be classified as either self-transmissible (Tra$^+$, Mob$^+$) or mobilizable but not self-transmissible (Tra$^-$, Mob$^+$). All functions required for conjugation are encoded on self-transmissible plasmids, while mobilizable plasmids lack *tra* genes that must be supplied *in trans* for mobilization. Tra functions are also able to recognize and activate the origin of transfer (*ori*T) contained on a mobilizable plasmid. For containment purposes in the laboratory, it is desirable to work with cloning vectors that are mobilizable but not self-transmissible. Cloning vectors are typically mobilized by Tra functions housed either on the chromosome of a mobilizing strain or on a separate plasmid in a helper strain (see below for discussion of bi- and tri-parental matings). Mobilizable plasmids lacking *tra* genes are generally smaller and thereby confer an advantage in transformation experiments where small plasmid size increases transformation frequency (see Perlin, this volume).

The replication and partitioning functions of plasmids are key to developing a cloning system for uncharacterized strains. Plasmids are divided into incompatibility (Inc) groups based on replication and partitioning functions. Two plasmids able to coexist in the same cell are assigned to different Inc groups. In other words, two plasmids are termed incompatible if they share the same replication and/or partitioning functions. Vegetative replication functions encoded by *ori*V are also involved in defining plasmid host range. A host range includes all species of bacteria in which the plasmid is replicated and stably maintained. Narrow host range plasmids, such as those derived from the ColE1 *ori*V, may only replicate in *E. coli* and related strains such as *Salmonella* and *Klebsiella*. Narrow host range plasmids are found in all Inc groups and include the well-known families of cloning vectors pBR322 and pUC (Sambrook et al., 1989). Broad host range *ori*V may permit replication in unrelated bacterial species. Broad host range plasmids are divided into the Inc C, N, P, Q, and W groups and may be used in constructing new broad host range cloning vectors (see below). In Gram-negative strains, vectors derived from RK2 or RP4 (Inc P) and RSF1010 (Inc Q) plasmids have well-characterized broad host range capabilities (Blatny et al., 1997). In Gram-positive strains, pUB110, pC194, and pIJ101 are common broad host range vectors. Several of the more common broad host range cloning vectors are listed in Table 1.

A critical step in developing a gene transfer protocol for an uncharacterized strain is to optimize conditions for transformation or conjugation. Chemical treatments that induce transformation competence include treatment with divalent cations and heat shock.

Electroporation is optimized by varying cell density, growth phase, plasmid DNA concentration, and electric field strength. Conjugation is optimized by varying mating time, mating temperature, and the ratio of donor, recipient, and helper cells. Matings are carried out directly on a solid growth medium, or cells are first captured on a filter that is then placed on a solid growth medium. Bi-parental (donor and recipient strains) and tri-parental (donor, recipient and helper strains) matings are possible options (Fig. 1). In bi-parental matings, mobilizable cloning vectors contained in an *E. coli* donor strain are mobilized via Tra functions housed either on the *E. coli* donor chromosome (Simon et al., 1983) or on a second plasmid maintained in the donor strain. In tri-parental matings, Tra functions are housed on a helper plasmid in a (helper) strain and the mobilizable cloning vector is housed in a separate donor strain.

Transconjugants are identified by plating the mating mix on solid growth medium amended with an antibiotic (e.g., tetracycline), the resistance for which is encoded on the transferred vector. An antibiotic resistance marker encoded on the recipient chromosome (e.g., rifampicin) is used to select against the *E. coli* donor and helper cells. Antibiotic effectiveness may vary from species to species and therefore antibiotic concentrations should be optimized before selection. Rifampicin-resistant (spontaneous) mutants are isolated by plating dense liquid cultures (e.g., 10^{11} cells/ml) on rifampicin-containing solid growth medium. After mating, growth of the *E. coli* donor and helper strains is inhibited by rifampicin, while growth of the untransformed (potential recipient) strains is inhibited by tetracycline (Fig. 1). Host strains housing a broad host range cloning vector (or gene library) have also been engineered to contain chromosomally encoded broad host range *tra* genes and are denoted broad host range mobilizing strains. Steps for constructing broad host range mobilizing strain *E. coli* S17-1 are shown in Figure 2.

TABLE 1. Broad Host Range Vectors

Vector	Size (Kb)	Selection	Remarks (Reference)
Inc P1			
Cloning vectors			
pRK310	20.4	Tc	Mob$^+$, Tra$^-$ contains *lacZ'* (Ditta et al., 1985)
pRK415	10.5	Tc	Mob$^+$, Tra$^-$ (Keen et al., 1988)
pFF1	5.9	Ap, Cm	Mob$^+$, Tra$^-$ (Durland et al., 1990)
pJB3Cm6	6.2	Ap, Cm	Mob$^+$, Tra$^-$ (Blatny et al., 1997)
Cosmid vector			
pVK100	23.0	Tc, Km	Mob$^+$, Tra$^-$, *cos* (Knauf and Nester, 1982)
Inc Q			
Cloning vectors			
pKT210	11.8	Cm, Sm	Mob$^+$, Tra$^-$ (Bagdasarian et al., 1979)
pDSK509	9.1	Km	Mob$^+$, Tra$^-$ (Keen et al., 1988)
pGSS15	11.3	Ap, Tc	Mob$^+$, Tra$^-$ (Barth et al., 1981)
pMMB7	8.9	Sm, Su	Mob$^+$, Tra$^-$ (Frey et al., 1983)
Cosmid vector			
pMMB31	14.75	Km, Sm	Mob$^+$, Tra$^-$ (Frey et al., 1983)
Inc W			
Cloning vectors			
pGV1106	8.7	Km, Sm	Mob$^+$, Tra$^-$ (Leemans et al., 1982)
pSa151	13.3	Km, Sm	Mob$^+$, Tra$^-$ (Tait et al., 1983)
pSa152	15.0	Cm, Km, Sm	Mob$^+$, Tra$^-$ (Tait et al., 1983)
pUCD2	13.0	Ap, Km, Sm, Tc	Mob$^+$, Tra$^-$ joint replicon with pBR322 (Close et al., 1984)
Cosmid vector			
pSa747	15.0	Km, Sm	Mob$^+$, Tra$^-$, *cos* (Tait et al., 1983)
Unclassified			
pBBR1MCS	4.7	Cm	Mob$^+$, Tra$^-$ (Kovach et al., 1994)
pME260	6.3	Ap, Km	Mob$^+$, Tra$^-$ (Itoh and Haas, 1985)
Special Vectors			
Promoter cloning			
pBBR1-GFP	6.6	Ap	Mob$^+$, Tra$^-$, GFP or CAT as reporter (Ouahrani-Bettache et al., 1999)
pAYC37	9.4	Ap	Mob$^+$, Tra$^-$, Streptomycin resistance as reporter (Tsygankov and Chistoserdov, 1985)
Expression vectors			
pJN105	6.0	Gm	Mob$^+$, Tra$^-$ araC-P-BAD controlled expression (Newman and Fuqua, 1999)
pML122	13.5	Gm	Mob$^+$, Tra$^-$ pNm controlled expression (Labes et al., 1990)

Table 1. (*Continued*)

Vector	Size (Kb)	Selection	Remarks (Reference)
Suicide vectors			
pSUP1011		Cm, Km	pACYC184 derivative carrying RP4 oriT (Mob$^+$) delivers Tn5 (Simon et al., 1983)
pSUP5011		Ap, Cm	pBR325 derivative carrying RP4 oriT (Mob$^+$)
Helper plasmids			delivers Tn5 (Simon, 1984)
pRK2013	48.0	Km	ColE1 replicon with RK2 Tra (Figurski and Helinski, 1979)
pRK2073	61.0	Sm, Tp	ColE1 replicon with RK2 Tra (Leong et al., 1982)

Abbreviations: Ap, ampicillin; Cm, chloramphenicol; Gm, gentamycin; Km, kanamycin; Sm, streptomycin; Su, sulfonamide; Tc, tetracycline; Tp, trimethoprim

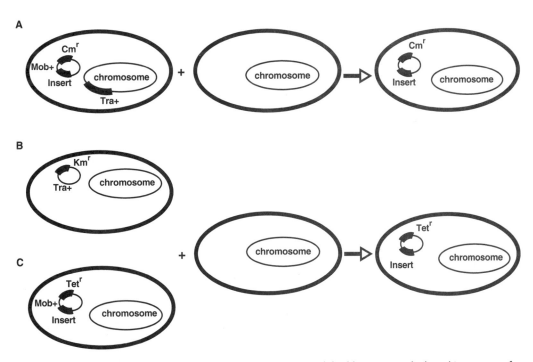

Fig. I **A:** Bi-parental mating. (**1**) *E. coli* donor strain housing mobilizable vector with cloned insert, transfer functions integrated into the chromosome; (**2**) Rifamycin resistant recipient strain; (**3**) Rifamycin/chloramphenicol resistant transconjugant. **B:** Tri-parental mating. (**1**) *E. coli* helper strain with transfer functions on helper plasmid; (**2**) *E. coli* donor strain housing mobilizable vector with cloned insert; (**3**) Rifamycin resistant recipient strain; (**4**) Rifamycin/tetracycline resistant kanamycin sensitive transconjugant.

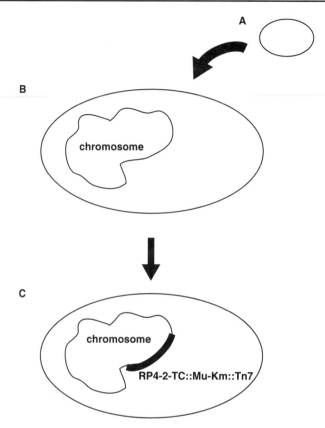

Fig. 2. Construction of mobilizing *E. coli* strain S17-1. (For experimental details, see Simon et al., 1983.) **A:** Broad host range plasmid RP4-2-TC::Mu-Km::Tn7; **B:** *E. coli* S49-20 (*recA*-); **C:** integration of RP4-2-TC::Mu-Km::Tn7 into chromosome of S49-20 provides transfer functions generating *rec* A-mobilizing strain *E. coli* S17-1.

Genetic engineering techniques have been used to construct broad host range cloning vectors from naturally occurring plasmids. For example, broad host range cloning vector pBBR1MCS (Kovach et al., 1994) has been engineered to encode chloramphenicol resistance as well as a multiple cloning site, while pSUP1011 (Simon et al., 1983) has been engineered to contain a broad host range *mob* site, a narrow host range (ColE1) replicon and a Tn5 insertion (encoding resistance to kanamycin). pSUP1011 is mobilizable to a broad range of hosts, yet it is unable to replicate outside of *E. coli* and acts as a suicide vector delivering Tn5 to recipient strains. Tn5 recipients have acquired resistance to kanamycin,

yet remain sensitive to antibiotics, the resistance for which is encoded by functions on the pSUP1011 backbone. Steps in construction of pBBR1MCS and pSUP1011 are described in Figures 3 and 4, respectively.

B. Construction of New Cloning Vectors

If the current suite of broad host range plasmids are not transmissible to or do not replicate in a bacterial strain, new cloning vectors may be constructed using endogenous plasmids. A variety of naturally occurring plasmids have received attention as potential cloning vectors, including those isolated from the extremophiles *Sulfolobus, Pyrococcus,* and *Thermotoga* (Noll and Vargas, 1997).

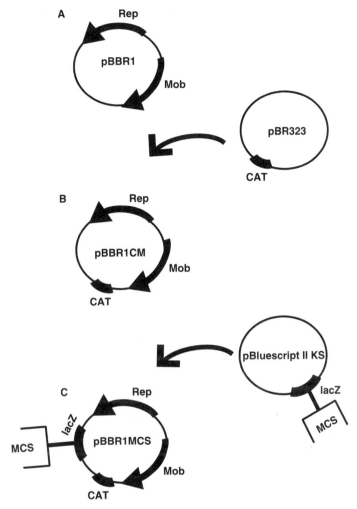

Fig. 3. Construction of broad host range cloning vector pBBR1MCS. (Adapted from Kovach et al., 1994 with permission of the publisher). **A:** 2.6 kb mobilizable, nonconjugative cryptic plasmid with broad host range replication functions isolated from *Bordetella bronchiseptica*. **B:** Chloramphenicol antibiotic resistance cassette (CAT) from pBR323 is cloned into PBBR1 to generate pBBR1CM. **C:** Multiple cloning site (MCS) containing additional unique restriction sites within *lacZ* gene from pBluescript II KS (Stratagene, La Jolla, CA, USA) cloned into pBBR1CM to generate the broad host range cloning vector pBBR1MCS.

Plasmid pNOB8 from *Sulfolobus* is especially attractive, since it is conjugally transferred between members of the genus *Sulfolobus* (Schleper et al., 1995). A cryptic mini-plasmid isolated from a *Thermotoga* species has also shown promise as a potential cloning vector. Shuttle vectors have been constructed by ligating an origin of replication recognized by a surrogate host to an endogenous plasmid.

For example, cloning vector pHS15 has been constructed by ligating a ColE1 *oriV*, a RK2 *oriT* and a streptomycin resistance marker to a cryptic plasmid isolated from *Halomonas elongata* (Vargas et al., 1997). pHS15 is mobilizable and may replicate in a broad range of hosts including *E. coli, Deleya halophila, Volcaniella eurihalina*, and various species of the genera *Halomonas*.

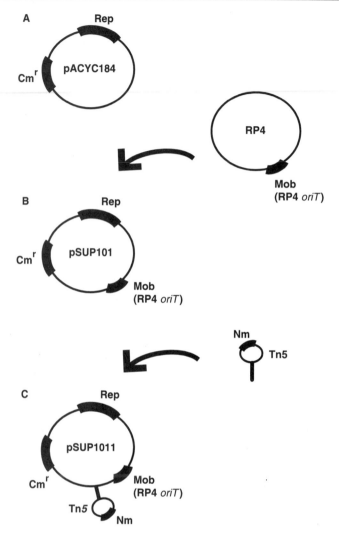

Fig. 4. Construction of Tn5 suicide delivery vector pSUP1011. (Adapted from Simon et al., 1983, with permission of the publisher.) **A:** *E. coli* vector pACYC184, nonmobilizable with chloramphenicol resistance (Cm^r) and narrow host range replication functions (Rep). **B:** Mob functions from broad host range plasmid RP4 are cloned into pACYC184 to generate pSUP101. **C:** Tn5 containing Neomycin resistance cassette cloned into pSUP101 to generate suicide vector pSUP1011. If RP4 transfer functions are provided in *trans*, pSUP1011 is mobilizable into broad range of gram-negative bacteria but is not able to replicate outside of *E. coli*. Tn5 is delivered to the recipient strain chromosome by a transposition event after mobilization.

IV. APPLICATIONS

The construction of broad host range gene transfer systems has provided a wealth of opportunities for elucidating the molecular basis of previously uncharacterized bacterial systems. Genes and predicted gene products involved in a wide range of biochemical and physiological pathways have been identified. Promoter elements have been characterized and gene expression studies have been carried out. With novel nucleotide sequence data in hand, gene replacement, site-directed mutagenesis, and nucleotide sequence analyses have been applied to study previously uncharacterized systems. The following section illustrates several of these applications, including a case study of genetic approaches used to elucidate the molecular details of metal reduction by *Shewanella putrefaciens*, a previously uncharacterized bacterial process.

A. Development of a Gene Cloning Strategy

A gene clone bank is a collection of cloned DNA fragments in which every gene from a single genome is represented at least once. To construct a gene clone bank, wild-type DNA fragments are ligated to plasmid cloning vectors and transferred to a host strain. Several methods are available to identify individual clones contained within a clone bank (Sambrook et al., 1989; Ausubel et al., 1997). If the target gene sequence is known, oligonucleotide probes can be synthesized from conserved regions. The clone carrying the desired DNA fragment is then identified via DNA/DNA (colony) hybridization analysis. Genes may also be identified by testing individual clone bank members for functional activity of the cloned gene product or by immunological assays with an antibody targeted against the gene product in other bacteria. A genetic complementation strategy may also be used to identify specific genes. In this method the gene clone bank is transferred to a mutant strain deficient in a specified activity. The complementing clone is identified by its ability to restore wild-type activity to the mutant strain. The complementing DNA fragment is subcloned to identify the smallest complementing fragment. In the absence of restriction sites, PCR is required to subclone individual open reading frames (ORFs). PCR depends on nucleotide sequence information to design forward and reverse PCR primers flanking the individual ORFs (see also Geoghegan, this volume).

B. Case Study: Anaerobic Metal Reduction by *Shewanella putrefaciens*

S. putrefaciens is extremely versatile in terms of its respiratory capability and is able to respire on oxygen, nitrate, nitrite, trimethy-lamine-*N*-oxide, sulfite, thiosulfate, elemental sulfur, fumarate, Mn(IV), Se(IV), U(VI), and Fe(III) as sole terminal electron acceptor (DiChristina and DeLong 1994). Bacterial Fe(III) reduction is a dominant terminal electron accepting process in anaerobic natural waters and sediments and can serve as a major pathway for degrading natural and hazardous organic carbon compounds. Bacterial Fe(III) reduction may also have been one of the first respiratory processes to have evolved on early earth. Ironically the molecular mechanism of bacterial Fe(III) reduction is poorly understood. Soluble terminal electron acceptors such as oxygen, nitrate, sulfate, and carbon dioxide are able to diffuse into cells and contact inner membrane-associated respiratory chains. Fe(III), on the other hand, is sparingly soluble at neutral pH, and is presumably not able to diffuse into the cell and contact the respiratory chain. To identify the genes and gene products required for Fe(III) reduction by *S. putrefaciens*, a mutant complementation strategy was developed (DiChristina and DeLong, 1994).

Attempts at directly transforming *S. putrefaciens* with broad host range plasmids have proved unsuccessful. The broad host range cosmid vector pVK100 (Knauf and Nester, 1982), however, is conjugally transferred to and stably maintained in *S. putrefaciens* at a low copy number. A *S. putrefaciens* gene clone bank was constructed by ligating partially digested *S. putrefaciens* genomic DNA into pVK100, packaging the recombinant cosmid in lambda bacteriophage particles, and transfecting *E. coli* mobilizing strain S17-1 (DiChristina and DeLong, 1994). Ethyl methane sulfonate (EMS) was used as a chemical mutagen to generate an array of *S. putrefaciens* (Fer) mutants deficient in Fe(III) reduction activity. Liquid cultures were exposed to EMS for a predetermined time period to achieve 90% kill, and surviving cells were screened for Fer mutant phenotypes via application of three Fe(III) reduction-dependent plate assays. The plate assays have circumvented problems associated with the inability of *S. putrefaciens* to form anaerobic colonies on Fe(III)- supplemented solid growth media, and they are based on the finding that the interior of aerobic colonies is essentially anaerobic. The first plate assay is based on the observation that wild-type colonies produce Fe(II) during aerobic growth on Fe(III)-supplemented solid medium, while Fer mutants do not. Fe(II) is detected by spraying colonies with a fine mist of ferrozine, a chelating agent that turns magenta-colored after complexing Fe(II) (DiChristina et al., 1988). The second plate assay is based on the observation that wild-type *S. putrefaciens* colonies produce a black FeS precipitate after aerobic growth on Fe(III)- and $S_2O_3^{2-}$-containing triple sugar iron agar, while Fer mutants are colorless (DiChristina and DeLong, 1994). The third plate assay is based on the observation that wild-type *S. putrefaciens* colonies liquefy agar during aerobic growth on Fe(III)-supplemented solid growth media, while Fer mutants do not (DiChristina and DeLong, 1994; McKinzi and DiChristina, 1999). Similar mutant screening techniques have recently been designed to identify *S. putrefaciens* respiratory mutants that are unable to respire anaerobically on Mn(IV) (Burnes et al., 1998), Se(IV) (Taratus et al., 2000), or U(VI) (Wade and DiChristina, 2000) as sole terminal electron acceptor.

Approximately 20,000 colonies (see below for statistical data) arising from EMS-treated cells were screened for Fer mutant phenotypes and 72 putative Fer mutants were identified. Each putative Fer mutant was tested for anaerobic growth in liquid culture on Fe(III) and the 10 alternate compounds as sole terminal electron acceptor. A subset of 10 Fer mutants was identified by the ability to respire on all compounds except Fe(III). Genetic analysis of the Fer mutants was carried out via conjugal mating experiments with the previously constructed *S. putrefaciens* wild-type gene clone bank. The three screening techniques were again employed to identify recipient (Fer$^+$) transconjugates that regained wild-type Fe(III) reduction activity. All Fer$^+$ trans-

conjugates were found to contain a common 23.3 kb fragment (DiChristina and DeLong, 1994). pBBR1MCS (see Fig. 3; Kovach et al., 1994) was used as a cloning vector to subclone the 23.3 kb fragment, and a 4.2 kb fragment was identified as the smallest complementing subclone (DiChristina et al., 2001). Nucleotide sequence analysis revealed that the 4.2 kb fragment contained three ORFs. *orf*1 (*ferE*) was PCR-amplified and identified as the gene required for wild-type Fe(III) reduction activity. Sequence analysis revealed that *ferE* displayed high (87%) sequence homology to the family of *pulE* homologs of type II protein secretion. The link between type II protein secretion and Fe(III) reduction is unknown, but it may involve targeting of respiration-linked Fe(III) terminal reductases to the outside face of the *S. putrefaciens* outer membrane where they are able to contact insoluble Fe(III) and complete the anaerobic electron transfer pathway (DiChristina et al., 2001). Studies on anaerobic metal reduction by *S. putrefaciens* have provided a working model of a genetic strategy that may be used to identify genes and gene products involved in previously uncharacterized bacterial processes.

Several important statistical considerations were taken into account during *S. putrefaciens* gene clone bank construction and mutagenesis experiments. The Clarke-Carbon equation (Clarke and Carbon, 1976), $P = 1 - (1 - f)^N$, where P is a desired probability value, f represents the fraction of events and N defines the number of iterations, is used to determine the number (N) of clones that are required to ensure that all genes represented in a genomic library at a desired probability value (P) with a known insert and genome size. For example, given that the cosmid vector pVK100 can house a DNA insert of 27 kb and that the *S. putrefaciens* genome size is 4000 kb, the number (N) of clones required for a 99% probability of having all sequences represented in the genomic library is calculated by $N = \ln(1 - P)/\ln(1 - f)$, where $P = 0.99$, $f = 27/4000$, and $N = 680$. As the insert size decreases, the number of required clones required increases. For example, assuming again a 4000 kb genome size, P set to 99%, and a 4 kb DNA insert, the number of clones required increases to 4603. Although the number of clone bank members that must be screened has increased nearly seven fold, the smaller insert size facilitates subsequent subcloning experiments. The benefits of a smaller insert size should be weighed against the number of clones that must be screened prior to selecting a broad host range cloning vector. The Clarke-Carbon equation may also be used to determine the number (N) of mutants (assuming a single 1 kb gene is mutated in a genome of 4000 genes) that must be screened to have a 99% probability of identifying one mutant genotype in a genome of 4000 genes: $N = \ln(1 - P)/\ln(1 - f)$, where $P = 99\%$, $f = 1/4000$, and $N = 18, 421$. In addition the number (N) of transconjugates that must be screened to have a 99% probability of complementing a strain with a 27 kb insert (assuming one mutation per gene in a 4000 kb genome; $f = 27/4000$) is 680 (see also Geoghegan, this volume).

C. Promoter Identification and Characterization

Broad host range vectors containing promoter-less reporter genes provide an attractive tool for identifying and characterizing promoters. Expression of promoterless reporter genes is detected if a DNA fragment containing an active promoter is inserted in the correct orientation upstream of the reporter. Structural reporter genes commonly used in gene fusions include β-galactosidase (*lacZ*), alkaline phosphatase (*phoA*), luciferase (*lux*), and green fluorescent protein (*gfp*) from the jellyfish *Aequoria victoria*. Promoter regions are identified by cloning random DNA fragments into a broad host range vector that contains the promoterless reporter gene. The recombinant vector is then transferred back into the host strain where reporter gene expression is measured. pBBR1-GFP is an example of a promoter-less reporter vector (Ouahrani-Bettache et al., 1999) containing two reporter genes (GFP and chloramphenicol

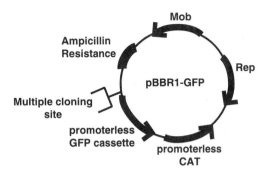

Fig. 5. Broad host range promoter-selection vector pBBR1-GFP. (Adapted from Ouahrani-Bettache et al., 1999, with permission of the publisher.)

acetyl transferase) downstream of a multiple cloning site (Fig. 5). Transposons carried on broad host range plasmids have also been utilized to insert promoterless reporter genes into bacterial chromosomes (Matthysse et al., 1996). This strategy requires delivery of the transposon on a suicide vector that is unable to replicate in the recipient strain (see above). Transposons carrying the reporter gene are transposed from the suicide vector to random locations of the chromosome. Insertion sites are identified with molecular techniques such as restriction site PCR (rsPCR). Primers targeted for internal regions of the transposon are paired with those targeted for random (upstream or downstream) restriction sites on the chromosome to amplify chromosomal DNA flanking the transposon insertion.

A Mu*d-lac* reporter gene fusion has been used to identify FNR (fumarate-nitrate regulator)-dependent gene expression in *Salmonella typhimurium* (Strauch et al., 1985). FNR regulatory elements are able to sense oxygen and either activate or deactivate gene transcription. The consensus FNR-binding site is generally located at position −41 relative to the transcription start site. Mu*d-lac* insertions have been used to generate random chromosomal insertions with *lacZ* as a reporter gene (the Mu*d* transposon is derived from the broad host range phage Mu, which will transpose randomly on the *S. typhimurium* chromosome). Transposon recipients are screened for β-galactosidase activity on MacConkey-lactose agar. Wild-type *S. typhimurium* is not able to metabolize lactose. If the *lacZ* fusion is expressed, however, lactose is broken down by β-galactosidase and a red colony is produced. FNR-dependent gene expression is identified in the Mu*d-lac* recipients by noting a fisheye colony phenotype (dark red colony center and a white periphery). Presumably cells in the colony interior (red) have become anaerobic while nascent cells in the colony periphery (white) have remained aerobic. Random chemical mutagenesis is used in a second screen to confirm the oxygen sensitivity of the fused wild-type promoter. Loss of the fisheye phenotype (entire colony turns red) in colonies arising from mutagenized cells has indicated that the original fusion is under control of an oxygen-sensitive regulator. Further mutant characterization and linkage analysis has identified a set of anaerobically regulated, FNR-dependent *lac* fusions. The anaerobically regulated genes have been identified by standard cloning and PCR techniques (Strauch et al., 1985; Wei and Miller, 1999). These studies have provided an example of a genetic strategy to identify promoter regions with randomly inserted promoter-less reporter genes.

Reporter genes have also been used to quantify promoter strength. The strength of the *S. putrefaciens* fumarate reductase promoter was measured using a *lacZ* fusion (Gordon et al., 1998). To determine conditions for optimal fumarate reductase gene expression, *lacZ* was fused immediately downstream of the fumarate reductase promoter and the construct placed in the *S. putrefaciens* chromosome via gene replacement strategy. Gene expression studies with the chromosomally integrated fumarate reductase promoter-*lacZ* fusion revealed that the fumarate reductase gene was transcribed maximally during anaerobic growth on fumarate and was regulated hierarchically in response to the redox potential of the terminal electron acceptor used for growth. Promoter regions required for high-level transcription

of the fumarate reductase gene were identified by fusing a nested set of promoter deletions to the *lacZ* reporter. These constructs have been integrated into the *S. putrefaciens* chromosome and β-galactosidase activity has been measured to identify promoter regions required for high-level transcription.

D. Regulated Gene Expression

In the fumarate reductase promoter study described above, a reporter gene on a broad host range vector is placed under control of an unregulated promoter. Broad host range vectors have also been used to clone genes downstream of tightly regulated promoters. Tightly regulated gene expression is especially important in industrial applications where cloned gene products are often toxic to host cells or gene expression is desired only at a specified cell density. The broad host range vector pMMB24 has been constructed to contain cloning sites downstream of a tightly regulated hybrid regulatory system consisting of a tryptophan biosynthesis gene promoter under the control of *lac* regulatory elements (Fig. 6). Although the *lac* promoter is not efficiently recognized in hosts other than *E. coli*, the *lac* regulatory region can be utilized to place other promoters under control of an inducer such as isopropylthiogalactoside (IPTG) (Bagda sarian et al., 1983). In another example the broad host range expression plasmid pML122 has been constructed to increase production of ethylene, a compound produced by plant pathogens such as

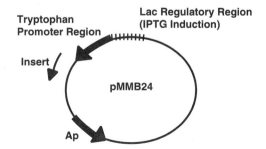

Fig. 6. Broad host range expression vector pMMB24 (tightly regulated). (Adapted from Bagda-sarian et al., 1983, with permission of the publisher).

Pseudomonas syringae. PML122 has been genetically engineered to contain broad host range (RSF1010-derived) *mob* and *rep* regions and a multiple cloning site downstream of the constitutively expressed neomycin phosphotransferase promoter (Labes et al., 1990). The *P. syringae* gene encoding the ethylene-forming enzyme has been inserted at the multiple cloning site, and *P. syringae* recipient strains harboring the resulting recombinant plasmid pMEFE1 are able to produce ethylene (under control of the neomycin phosphotransferase promoter) at levels up to 40 times higher than the original *P. syringae* strain (Ishihara et al., 1996).

E. Site-Specific Mutagenesis and Gene Replacement

Site-specific mutagenesis has been combined with gene replacement strategies to replace wild-type genes with mutated genes that have been either insertionally inactivated or altered at specific nucleotide positions. Such strategies are particularly useful for identifying enzyme active sites or altering enzyme activity. The altered gene is generally cloned on a broad host range suicide vector in *E. coli* and is mobilized to the original host where the wild-type chromosomal gene is replaced with the altered gene via homologous recombination. Gene replacement strategies have generally used antibiotic resistance cassettes to disrupt wild-type genes as well as to serve as a marker for successful recombination. For example, *S. putrefaciens* is known to contain two proteins expressing fumarate reductase activity (Myers and Myers, 1992), and a gene replacement strategy has been used to identify the respiration-linked enzyme (Gordon et al., 1998). The gene encoding *S. putrefaciens* flavocytochrome c$_3$ (*fccA*) has been insertionally inactivated with an antibiotic resistance cassette, cloned in a suicide vector and mobilized to the wild-type strain. The insertionally inactivated *fccA* gene has been homologously recombined into the *S. putrefaciens* chromosome, and the resulting *fccA* mutant is unable to grow anaerobically on fumarate as sole terminal electron acceptor.

F. Potential Problems

In many broad host range cloning systems the gene clone bank is housed in a foreign host. Cloned genomic DNA can be modified by foreign host-encoded restriction, methylation, or recombination. Such modifications have been prevented by using modification-deficient *E. coli* strains. A variety of modification- deficient host strains are available to house cloning vectors safely (Sambrook et al., 1989; Ausubel et al., 1997). *E. coli* restriction systems are encoded by the *hsdRMS*, *mcrA*, *mcrB*, and *mrr* genes. Unmethylated DNA is recognized by the *Eco*K restriction system encoded by *hsdRMS*, while unmethylated DNA is recognized by the *E.coli* restriction system encoded by *mcrAB* and *mrr*. *E. coli* methylases are encoded by *Eco*K, *dam* and *dcm*, while homologous recombination in *E. coli* is encoded by *recA*. To overcome problems associated with these modification systems, a variety of *E. coli* mutant strains have been constructed to serve as host strains (Table 2). After passage through an alternate host, cloned DNA fragments may be analyzed by restriction digests or DNA/DNA hybridization with genomic DNA to ensure that modifications have not taken place.

Application of *E. coli*-based plasmid isolation protocols to non-enteric bacteria often results in low plasmid yield. Problems with low plasmid yield have been exacerbated by low copy number cloning vectors. Several strategies have been successful in resolving these problems. Recombinant cloning vectors that have been mobilized from an *E. coli* host to a recipient (transconjugate) strain may be remobilized from the transconjugate to a recipient (plasmid-less) *E. coli* strain where standard *E. coli*-based plasmid isolation protocols can be used (DiChristina and DeLong, 1994). Recombinant plasmids may also be transferred from recipient (transconjugate) strains by electroporating (plasmid-less) *E. coli* strains with transconjugate mini-preps. In many cases the recombinant plasmid yield from the transconjugate mini-prep cannot be visualized via standard

TABLE 2. *E. coli* Strains Used in Cloning

Strain	Genotype	Remarks	Reference
DH5α	supE44ΔlacU169 ($_\phi$80 lacZ ΔM15) hsd R17 recA1 end A1 gyrA96 thi-1 relA1	Recombination-deficient, suppressing strain, for plasmid maintenance	Hanahan, 1983; Bethesda Research Laboratories, 1986
JM109	recA1 supE44 end A1 hsdR17 gyrA96 relA1 thi Δ(lac-proAB) F'[traD36proAB$^+$lacI^q lacZΔM15]	Recombination-deficient, suppressing strain	Sambrook et al., 1989
S17-1	thi pro hsdR hdsM$^+$ recA, chromosomal insertion of RP4-2 (Tc::Mu Km::Tn7)	Recombination-deficient, for plasmid maintenance, mobilizing strain	Simon et al., 1983
SM10	thi-1 thr leu tonA lacY supE recA::RP4 – 2 – Tc:: Mu Km^R	Recombination-deficient, for plasmid maintenance, mobilizing strain	Simon et al., 1983
HB 101	supE44 hsdS20($r_B^- m_B^-$) recA13 ara-14 proA2 lacY1 galK2 rpsL20 xyl-5 mtl-1	Suppressing strain, for plasmid maintenance	Sambrook et al., 1989

agarose gel electrophoresis techniques, yet is of sufficient quantity for electroporation (DiChristina et al., 2001).

V. SURROGATE GENETIC APPROACHES

The design of a broad host range gene cloning system is generally the preferred initial approach in developing a genetic system for uncharacterized bacterial strains. Broad host range gene cloning systems are not always feasible, however, and alternative genetic approaches are required. Surrogate genetic approaches have received renewed interest, largely due to the advent of bacterial artificial chromosome-based technologies. Surrogate genetic approaches have played an important role in industrial and pharmaceutical applications where increased production of commercially relevant gene products is desired.

A. Genomic Mapping

Genomic mapping in *E. coli* has traditionally involved isolation of a large number of auxotrophic mutants and transfer of chromosomal DNA on F' plasmids. F' plasmids are self-transmissible plasmids capable of integrating into the *E. coli* chromosome (see Porter, this volume). Imprecise F' excision events can result in recombinant plasmids carrying flanking chromosomal DNA. Chromosomal DNA transferred into a surrogate host harboring well-characterized mutations is scored for restoration of wild-type function and a partial genetic map is constructed. If a gene transfer system is not available, a clone bank may be used in genomic mapping. In this case clone bank fragments are used to complement mutations in a surrogate host, and the relative position of markers is determined by Southern blot analysis of overlapping clones. Alternatively, the relative position of complementing clones is determined by Southern blot analysis of large chromosomal fragments that are separated with specialized electrophoresis techniques such as field inversion gel electrophoresis (FIGE) or contour-clamped homogeneous-field gel electrophoresis (CHEF) (Dale, 1998). Unlike conventional agarose gel electrophoresis, these specialized techniques separate large DNA fragments by periodic changes in electrical field direction.

Fragments are separated by their relative ability to reorient with the electric field which is a function of fragment length.

B. Identification of Similar Genes in Other Bacteria

Genetic (mutant) complementation analysis may be used with a surrogate host to identify similar genes in a specified strain. This strategy entails mobilizing a genomic clone bank of the specified strain to a surrogate host with a known mutation and screening for restoration of (at least partial) activity to the mutant strain. Although such approaches have employed surrogate hosts that are phylogenetically related, distantly related surrogate hosts have also been used. For example, type II protein secretion in closely related Gram-negative bacteria has been examined by surrogate genetic strategies. Protein secretion-deficient mutants of a variety of *Erwinia* species (*E. carotovora* subsp. *carotovora, E. carotovora* subsp. *atroseptica*, and *E. chrysanthemi*) have been isolated. An *E. carotovora* subsp. *carotovora* gene clone bank has been used to restore activity and identify analogous genes in the suite of mutants from all *Erwinia* species (Murata et al., 1990). If a specified gene is well conserved across genus and domain lines, it is possible to restore function to a mutant strain of a phylogenetically unrelated microorganism. For example, the gene encoding *Saccharomyces cerevisiae* biotin- apoprotein ligase has been identified via genetic complementation analysis of an *E. coli* biotin-apoprotein ligase mutant (Cronan and Wallace, 1995).

C. Expression in Alternate Hosts

Surrogate genetic approaches have overcome problems associated with low-level gene expression in wild-type hosts and difficulties in cultivating environmental isolates under extreme conditions. For example, genes encoding thermally stable enzymes of hyperthermophiles and secondary metabolites of soil microorganisms have been expressed in alternate hosts. Members of the genus *Thermus* isolated from thermal springs grow optimally at 70 to 80 °C and express proteins that are both heat-stable and resistant to chemical denaturants (Alldread et al., 1992). The heat stability of *T. aquaticus* polymerase is particularly useful in PCR applications. The heat stability of *Thermus* proteins has enormous economic potential, and alternate strategies for expressing *Thermus* genes in surrogate hosts have been developed. The *T. aquaticus* gene encoding malate dehydrogenase (*mdh*) has been expressed in *E. coli* by first overcoming problems associated with codon usage and poor recognition of transcription and translation signals. High-expression clones for *T. aquaticus mdh* have been constructed by changing the start codon from GTG to ATG and shortening the distance between the promoter and the translation start codon (resulting in a 30-fold increase in MDH yield per cell). *T. aquaticus mdh* expression is increased an additional 50% by replacing the *T. aquaticus mdh* native promoters with the *E. coli trp* or *mdh* promoters (Alldread et al., 1992). Similar techniques have been applied to over express recombinant *T. aquaticus* DNA polymerase in *E. coli*. The gene encoding the *T. aquaticus* DNA polymerase has been cloned in *E. coli* with an in-frame ATG start codon and placed under control of the phage lambda P-R promoter and the regulatory elements of the *E. coli* tryptophan operon (Patrushev et al., 1993).

To avoid problems associated with expression in distantly related microorganisms, closely related strains may sometimes be used as surrogate hosts. For example, members of the genus *Thermus* have traditionally been classified as strict aerobes; *T. thermophilus* HB8, however, has recently been found to respire anaerobically on nitrate as sole terminal electron acceptor. Several cryptic plasmids have been found in *T. thermophilus* HB8, while none have been observed in any strictly aerobic *T. thermophilus* strains. Southern hybridization analysis has shown that strain HB8 contains an *E. coli*–like *nar* operon (encoding nitrate reductase), while the strict aerobic strains do not. It was subsequently hypothesized that the *nar* genes may be plasmid encoded. To test this hypothesis, the cryptic plasmids were conjugally transferred from *T. thermophilus* HB8 to the closely related (strictly aerobic) strain *T. thermophilus* HB27 and transconjugants were identified by the aquired ability to grow anaerobically on nitrate. Genes encoding other anaerobic respiratory functions may also be identified via analysis of plasmids transferred horizontally among members of the genus *Thermus* (Ramirez-Arcos et al., 1998).

Cloned gene products are often difficult to purify in surrogate gene expression systems. Foreign proteins often form polar inclusion bodies that limit active product yield. To overcome such problems, *E. coli* surrogate host strains have been engineered to express protein secretion systems that secrete cloned gene products to the cell exterior rather than allowing them to accumulate in the cytoplasm or periplasmic space. For example, *E. chrysanthemi* Type II protein secretion (*out*) genes have been engineered in *E. coli* surrogate hosts for bioconversion of lignocellulose to ethanol. Optimization of ethanol production has required extracellular secretion of lignocellulose-hydrolyzing enzymes, a function not carried out by wild-type *E. coli*. Recombinant *E. coli* expressing *E. chrysanthemi out* genes are able to secrete the tandomly cloned *E. chrysanthemi celZ* gene product that aids in cellulose solubilization, a key step in the production of ethanol (Zhou et al., 1999). Furthermore random fragments of *Zymomonas mobilis* DNA have been screened, and a surrogate promoter has been identified, for increasing expression of the *E. chrysanthemi celZ* gene product sixfold in *E. coli* (Zhou et al., 1999).

D. Bacterial Artificial Chromosome-Based Systems

An exciting advance in surrogate genetics has involved cloning and expressing extremely large DNA fragments in bacterial artificial chromosomes (BAC). F factor–based BAC vectors (Fig. 7) can carry DNA inserts as large as 300 kb and are stably maintained in *E. coli* (Shizuya et al., 1992).

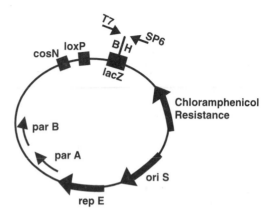

Fig. 7. Bacterial artificial cloning (BAC) vector pBeloBAC 11. Replication of the BAC vector is controlled via *oriS* and *repE* genes; low copy number is maintained by *parA* and *parB*. *Hind*III and *Bam*HI cloning sites are noted as H and B, respectively. The bacteriophage lambda *cosN* and P1 *loxP* sites provide sites for specific cleavage. The T7 and SP6 promoters are utilized for chromosome walking, DNA sequencing, and generation of RNA probes. (Adapted from Shizuya et al., 1992, and Kim et al., 1996.)

In theory, gene clusters encoding complete biochemical pathways may be cloned on BAC vectors and expressed in surrogate hosts. In addition the collective genomes of all microorganisms in a natural sample (metagenome; Handelsman et al., 1998) may be cloned on BAC vectors and transferred to *E. coli* surrogate hosts. The recombinant *E. coli* are subsequently screened for expression of genes encoding medically or industrially important gene products. In this manner BAC-based cloning systems have provided a means for isolating environmental DNA fragments without cultivation.

In a recent study a gene library from *Bacillus cereus* (a Gram-positive microorganism for which no natural genetic tools are available) was constructed in a BAC vector housed in an *E. coli* host, and hemolysin- and esculin-hydrolyzing (recombinant) clones were identified (Rondon et al., 1999). In a second study a metagenomic library from soil samples was constructed in a BAC vector housed in a surrogate *E. coli* host and clones expressing DNase, antibacterial, lipase, and amylase activity were identified. The genes encoding enzymes with DNase and antibacterial activity do not display nucleotide sequence homology to any known genes in the current sequence databases. Analysis of the 16S rRNA genes harbored on the BAC metagenomic soil library has indicated that the majority of sequences are not represented in the Ribosomal Database (Rondon et al., 2000). These findings have demonstrated the potential for BAC-based systems to uncover novel biochemical activities in bacteria with no natural genetic systems or in bacteria that resist cultivation under laboratory conditions.

VI. REVERSE GENETIC APPROACHES AND PROTEOMICS

A reverse genetics approach is based on amino acid sequence information, which guides isolation of specific genes. This approach contrasts genetic complementation strategies in which knock-out mutants are generated and wild-type genes are identified via restoration of function by a mobilized wild-type clone bank. In reverse genetics, a protein is isolated and its amino acid sequence determined via N-terminal sequencing or mass spectrometry. The amino acid sequence is translated to its corresponding nucleotide sequence, and degenerative nucleotide primers are designed to PCR-amplify the original gene from wild-type genomic DNA (Fig. 8).

If protein activity is detectable, it may be possible to design traditional protein purification or gel-based protocols to identify and isolate specific proteins in crude cell extracts. For example, a simple colorimetric assay was used to identify Fe(III) reduction activity in crude cell extracts of Fe(III)-reducing *S. putrefaciens*. The subcellular location of Fe(III)-reducing proteins was identified by incubating subcellular fractions with an electron donor, Fe(III) and ferrozine, which forms a deep magenta colored tris-complex with Fe(II). An Fe(III)-reducing *c*-type cytochrome in the outer membrane fraction was isolated by standard protein purification techniques and N-terminal sequencing provided amino acid sequence data. A degenerative primer set was designed from the N-terminal sequence, and the entire gene was amplified by PCR. Subsequent sequence

Amino Acid Sequence:

Met Gly Val Trp Phe Asp Cys Trp Gly Try Glu Pro

Codon Sequence:

AUG	GGU	GUU	UGG	UUU	GAU	UGU	UGG	GGU	UAU	GAA	CCU
	GGU	GUC		UUC	GAC	UGC		GGC	UAC	GAG	CCC
	GGA	GUA						GGA			CCG
	GGG	GUG						GGG			CCG

Degenerative Primer:

$$\text{UGG} \quad \text{UUU}^{U}_{C} \text{ GA}^{U}_{C} \text{ UG}^{U}_{C} \text{ UGG} \quad \text{UU}$$

Fig. 8. Design of a degenerative primer from amino acid sequence. Primers are designed from the region with the least amount of third-position wobble. For the amino acid sequence given above, 8 different primers are synthesized (each 17 nucleotides long) accounting for all possible sequence combinations. Codon usage bias should be considered to narrow possibilities. For example, *Thermus thermophilus* is 68.23% GC and codon possibilities are as follows: 1st letter GC 69.39%, 2nd letter GC 42.89%, 3rd letter GC 92.4%. (Adapted from Codon Usage Database, *www.kazusa.or.jp/codon/*; (Nakamura et al., 2000 with permission.)

analysis of the PCR product confirmed the presence of heme c binding sites, yet revealed no significant homology to previously characterized proteins (Myers and Myers, 1998).

Two-dimensional gel electrophoresis is used in proteome analysis to identify the suite of proteins expressed in cells under specific growth conditions. Maximum polypeptide separation is attained with two independent separation parameters, isoelectric focusing and SDS-PAGE. The phosphate starvation-inducible proteins of *B. subtilis* were recently identified with a proteomics approach (Antelmann et al., 2000). Phosphate starvation in *B. subtilis* induces expression of genes under control of the PhoR and σ^B regulons. Two-dimensional polypeptide profiles were generated from wild-type *B. subtilis* and compared to profiles generated from *sigB* and *phoR* mutant strains. Subtractive analysis software was used to identify polypeptides expressed in each strain under conditions of phosphate starvation. Proteins excised from 2-D gels were identified by mass spectrometry. Several previously identified phosphate starvation-inducible proteins were identified as well as two new proteins, glycerophosphoryl diester phosphodiesterase GlpQ and lipoprotein YdhF.

VII. RECENT ADVANCES: GENOMICS

The microbial genome sequencing initiative has revolutionized techniques for studying the molecular basis of uncharacterized bacterial systems. Currently the complete genome sequence of 28 microorganisms have been published and the sequencing of several others is in progress (Table 3). Published DNA sequence information is accessible via GenBank, the National Institute of Health genetic database (*www.ncbi.nlm.nih.gov*). Links to a wide range of other sequence databases is found at Amos's WWW links page (*www.expasy.ch/alinks.html*). These databases facilitate studies on comparative sequence homologies. For example, a Gapped-BLAST sequence similarity search (Altschul et al., 1997) enables comparison of specified nucleotide or protein sequences with all sequences contained in the database. If homologous genes are identified, the uncharacterized gene is assigned a putative function. Additional programs are available to perform other functions, such as identifying open reading frames (GeneMark; Borodovsky and McIninch, 1993), performing multiple sequence alignments (CLUSTAL W; Thompson et al., 1994), designing PCR primers

TABLE 3.   Completely Sequenced Microbial Genomes

Aeropyrum pernix	*Helicobacter pylori* strain J99
Aquifex aeolicus	*Methanobacterium thermoautotrophicum*
Archaeoglobus fulgidus	*Methanococcus jannaschii*
Bacillus subtilis	*Mycobacterium tuberculosis*
Borrelia burgdorferi	*Mycoplasma genitalium*
Campylobacter jejuni	*Mycoplasma pneumoniae*
Chlamydia pneumoniae strain CWL029	*Neisseria meningitidis* strain MC58
Chlamydia pneumoniae strain AR39	*Neisseria meningitidis* strain Z2491
Chlamydia trachomatis serovar D	*Pyrococcus horikoshii*
Chlamydia trachomatis strain MoPn	*Rickettsia prowazekii*
Deinococcus radiodurans	*Saccharomyces cerevisiae*
Escherichia coli	*Synechocystis* sp.
Haemophilus influenzae Rd	*Thermotoga maritima*
Helicobacter pylori strain 26695	*Treponema pallidum*

Note: For more information and a list of partially sequenced microbial genomes, see The Institute for Genomic Research (TIGR) Web site at *http://www.tigr.org*.

(Primer3; Rozen and Skaletsky, 1996), and constructing phylogenetic trees (PHYLIP; Felsenstein, 1993).

In a pioneering study that illustrates the power of genomics (Stein et al., 1996), environmental DNA fragments were isolated from marine picoplankton samples and archaeal clones were amplified by PCR with archaeal-biased small subunit rRNA- targeted primers. Subclone fragments generated from the resulting archaeal clones were sequenced, and the flanking genes were identified. Two open reading frames were identified, with each encoding a putative protein not previously found in archaea (RNA helicase and glutamate semialdehyde aminotransferase). A similar genomics strategy was recently used to demonstrate that archaeal-like rhodopsins are broadly distributed among different taxons, including uncultivated members of the domain bacteria (Beja et al., 2000). As sequence databases expand, genomics will become an increasingly more attractive means for identifying the physiological potential and genetic composition of bacteria in a wide variety of environments.

Nucleotide and amino acid sequence databases also provide the foundation for evolutionary studies of metabolic traits. For example, the evolutionary history of dissimilatory sulfate reduction was recently examined via comparative sequence analysis. Sulfate-reducing bacteria (SRB) are able to catalyze the six-electron reduction of sulfite to sulfide via the dissimilatory sulfite reductase (DSR) enzyme. Comparative sequence analysis of DSR genes isolated from archaeal (*Archaeoglobus fulgidus*) and bacterial (*Desulfovibrio vulgaris*) SRB revealed a high degree of DSR sequence similarity across domain lines. A DSR-biased PCR primer set was designed to amplify DSR genes from individual strains of SRB as well as bulk environmental DNA and RNA, and phylogenetic trees were constructed from nucleotide sequence analysis of the resulting DSR-biased PCR products. A comparison of DSR- and 16S rRNA-based phylogenies

has led to the hypothesis that the evolutionary history of sulfate reduction is rooted in a single ancestral progenitor (Wagner et al., 1998). In future studies, DSR-targeted DNA and RNA probes may be used as genetic tools for detecting the presence and activity of SRB in situ without biases associated with traditional cultivation techniques (Minz et al., 1999).

High-throughput analysis of DNA and RNA (cDNA) is now possible with DNA microchip array technology. DNA microchip arrays consist of a series of single-stranded oligonucleotide probes immobilized on a solid support (microchip). Fluorescently labeled DNA or cDNA is passed over the chip, and hybridization to target oligonucleotides is monitored via fluorescent signal detection. DNA microchip array technology is designed to operate on the same principles as traditional Southern or Northern hybridization techniques. Instead of targeting single genes, however, microchip arrays target entire genomes, and thus genotyping or genome-wide expression studies are possible. DNA microchip array technology was recently used to identify *E. coli* genes differentially expressed in rich and minimal media. Growth in minimal medium resulted in elevated expression of genes required for biosynthesis of amino acids, vitamins, cofactors, and prosthetic groups, a result consistent with the required de novo synthesis of macromolecular building blocks. Growth in rich medium resulted in elevated expression of genes involved in macromolecule biosynthesis and translation apparatus, a result consistent with the higher growth rates observed in rich media (Tao et al., 1999).

VIII. SUMMARY

This chapter has outlined a variety of approaches that may be followed in developing a genetic system for uncharacterized bacterial strains. Previously constructed broad host range cloning vectors have provided opportunities for genetic complementation strategies, while endogenous plasmids isolated from previously uncharacterized bacteria have allowed development of a new generation of cloning vectors. While genetic complementation systems are often attractive, newer proteomic- and genomic-based technologies have succeeded in dissecting the more intractable bacterial systems when traditional gene cloning methodologies have failed. As nucleotide and amino acid sequence databases continue to expand, comparative sequence analysis will provide unprecedented opportunities for studying novel processes in bacteria for which no natural genetic system is available.

ACKNOWLEDGMENTS

Financial support for this study was provided by the National Science Foundation Environmental Geochemistry and Biogeochemistry Program (Grant No. EAR-97085350) and the Nelson and Bennie Abell Professorship in Biology at Georgia Tech.

REFERENCES

Alldread RM, Nicholls DJ, Sundaram TK, Scawen MD, Atkinson T (1992): Overexpression of the *Thermus aquaticus* B malate dehydrogenase encoding gene in *Escherichia coli*. Gene 114:139–143.

Altschul SF, Madden TL, Schaffer AA, Zhang J, Zhang Z, Miller W, Lipman DJ (1997): Gapped BLAST and PSI-BLAST: A new generation of protein database search programs. Nucleic Acids Res 25:3389–3402.

Antelmann H, Scharf C, Hecker M (2000): Phosphate starvation- inducible proteins of *Bacillus subtilis*: Proteomics and transcriptional analysis. J Bacteriol 182:4478–4490.

Ausubel RM, Brent R, Kingston RE, Moore DD, Seidman JG, Smith JA, Struhl K (eds) (1997): "Current Protocols in Molecular Biology." New York: Wiley.

Bagdasarian M, Bagdasarian MM, Coleman S, Timmis KN (1979): In: Timmis KN, Puhler A (eds): "Plasmids of Medical, Environmental, and Commercial Importance." Amsterdam: Elsevier/North-Holland Biomedical Press, pp 411–422.

Bagdasarian MM, Amann E, Lurz R, Ruckert B, Bagdasarian M (1983): Activity of the hybrid trp-lac (tac) promoter of *Escherichia coli* in *Pseudomans putida*. Construction of broad-host- range, controlled-expression vectors. Gene 26:273–282.

Barth PT, Tobin L, Sharpe GS (1981): In: Levy SB, Clowes RC, Koenig EL (eds): "Molecular Biology, Pathogenicity, and Ecology of Bacterial Plasmids." New York: Plenum Press pp 439–448.

Beja O, Aravind L, Koonin EV, Suzuki MT, Hadd A, Nguyen LP, Jovanovich SB, Gates CM, Feldman RA,

Spudich JL, Spudich EN, DeLong EF (2000): Bacterial Rhodopsin: Evidence for a new type of phototrophy in the sea. Science 289:1902–1906.

Bethesda Research Laboratories. (1986): BRL pUC host: *E. coli* DH5α™ competent cells. Bethesda Res. Lab. Focus 8:9.

Blatny JM, Brautaset MT, Winther-Larsen HC, Haugan K, Valla S (1997): Construction and use of a versatile set of broad-host- range cloning and expression vectors based on the RK2 replicon. AEM 63:370–379.

Borodovsky M, McIninch J (1993): GeneMark: Parallel gene recognition for both DNA strands. Comput Chem 17:123–133.

Burnes BS, Mulberry MJ, DiChristina TJ (1998): Design and application of two rapid screening techniques for the isolation of Mn(IV) reduction-deficient mutants of *Shewanella putrefaciens*. AEM 64:2716–2720.

Clarke L, Carbon J (1976): A colony bank containing synthetic col E1 hybrid plasmids representative of the entire *E. coli* genome. Cell 9:91–99.

Close T, Zaitlin JD, Kado CI (1984): Design and development of amplifiable broad-host-range cloning vectors: Analysis of the vir region of *Agrobacterium tumefaciens* plasmid pTiC58. Plasmid 12:111–118.

Cronan JE, Wallace JC (1995): The gene encoding the biotin- apoprotein ligase of *Saccharomyces cerevisiae*. Fems Microbiol Lett 130:221–229.

Dale JW, (ed) (1998): "Molecular Genetics of Bacteria." New York: J Wiley.

DiChristina TJ, Arnold RG, Lidstrom ME, Hoffmann MR (1988): Dissimilative iron reduction by the marine eubacterium *Alteromonas putrefaciens* strain 200. Water Sci Tech 20:69–79.

DiChristina TJ, DeLong EF (1994): Isolation of anaerobic respiratory mutants of *Shewanella putrefaciens* and genetic analysis of mutants deficient in anaerobic growth on Fe^{3+}. J Bacteriol 176:1468–1474.

DiChristina TJ, Moore CM, Haller CA (2001): Dissimilatory Fe(III) and Mu (IV) reduction by *Shewanella putrefaciens* Requires *ferE*, a *pulE* homolog of type II Protein Secretion. J Bacteriol, submitted for review.

Ditta GT, Schmidhauser T, Yakobson E, Lu P, Liang X-W, Finlay DR, Guiney D, Helinski DR (1985): Plasmids related to the broad host range vector pRK290 useful for gene cloning and for monitoring gene expression. Plasmid 13:149–153.

Durland RH, Toukdarian A, Fang F, Helinski DR (1990): Mutantions in the *trfA* replication gene of the broad-host-range plasmid RK2 result in elevated plasmid copy numbers. J Bacteriol 172:3859–3867.

Felsenstein J (1993): PHYLIP (Phlyogeny Inference Package). Department of Genetics, University of Washington, Seattle.

Figurski DH, Helinski DR (1979): Replication of an origin- containing derivative of plasmid RK2 dependent on a plasmid function provided *in trans*. PNAS USA 76:1648–1652.

Frey J, Bagdasarian M, Feiss D, Franklin FCH, Deshusses J (1983): Stable cosmid vectors that enable the introduction of cloned fragments in a wide range of gram-negative bacteria. Gene 24:299–308.

Gordon EJJ, Pealing SL, Chapman SK, Ward FB, Reid GA (1998): Physiological function and regulation of flavocytochrome c_3, the soluble fumarate reductase from *Shewanella putrefaciens* NCIMB 400. Microbiol 144:937–945.

Hanahan D (1983): Studies on transformation of *Escherichia coli* with plasmids. J Mol Biol 166:557.

Handelsman J, Rondon MR, Brady SF, Clardy J, Goodman RM (1998): Molecular biological access to the chemistry of unknown soil microbes: A new frontier for natural products. Chem Biol 5:R245–R249.

Ishihara K, Matsuoka M, Ogawa T, Fukuda H (1996): Ethylene production using a broad-host-range plasmed in *Pseudomonas syringae* and *Pseudomonas putida*. J Ferment Bioeng 82:509–511.

Itoh Y, Haas D (1985): Cloning vectors derived from the *Pseudomonas* plasmid pVS1. Gene 36:27–36.

Keen NT, Tamaki S, Kobayashi D, Trollinger D (1988): Improved broad-host-range plasmids for DNA cloning in Gram-negative bacteria. Gene 70:191–197.

Kim U, Birren BW, Slepak T, Mancino V, Boysen C, Kang H, Simon MI, Shizuya H (1996): Construction and characterization of a human bacterial artificial chromosome library. Genomics 34:213–218.

Knauf VC, Nester EW (1982): Wide host range cloning vectors: cosmid clone bank of *Agrobacterium* Ti plasmids. Plasmid 8:45–54.

Kovach ME, Phillips RW, Elzer PH, Roop RM, Peterson KM (1994): pBBR1MCS: A broad-host-range cloning vector. BioTechniq 16:800–802.

Labes R, Puhler A, Simon R (1990): A new family of RSF1010–derived expression and *lac*-fusion broad-host- range vectors for Gram negative bacteria. Gene 89:37–46.

Leemans J, Langenakens J, DeGreve H, Deblaere R, Van Montagu M, Schell J (1982): Broad-host-range cloning vectors derived from the W-plasmid Sa. Gene 19:361–364.

Leong SA, Ditta GS, Helinski DR (1982): Heme biosynthesis in rhizobium: identification of a cloned gene coding for X-aminolevulinic acid synthetase from *Rhizobium meliloti*. J Biol Chem 257:8724–8730.

Matthysse AG, Stretton S, Dandie C, Mcclure NC, Goodman AE (1996): Construction of GFP vectors for use in Gram-negative bacteria other than *Escherichia coli*. FEMS Microbiol Lett 145:87–94.

McKinzi AM, DiChristina TJ (1999): Microbially driven fenton reaction for transformation of pentachlorophenol. Environ. Sci. Technol 33:1886–1891.

Minz D, Flax JL, Green SJ, Muyzer G, Cohen Y, Wagner M, Rittmann BE, Stahl DA (1999): Diversity of sulfate-reducing bacteria in oxic and anoxic regions of a microbial mat characterized by comparative analysis of dissimilatory sulfite reductase genes. AEM 65:4666–4671.

Murata H, Fons M, Chatterjee A, Collmer A, Chatterjee AK (1990): Characterization of transposon insertion out-mutants of *Erwinia carotovora* subsp. *carotovora* defective in enzyme export and of a DNA segment that complements *out* mutations in *E. carotovora* subsp. *carotovora*, *e. carotovora* subsp. *atroseptica* and *Erwinia chrysanthemi*. J Bacteriol 172:2970–2978.

Myers JM, Myers CR (1992): Fumarate reductase is a soluble enzyme in anaerobically grown *Shewanella putrefaciens* MR-1. FEMS Microbiol Lett 98:13–20.

Myers JM, Myers CR (1998): Isolation and sequence of *omcA*, a gene encoding a decaheme outer membrane cytochrome *c* of *Shewanella putrefaciens* MR-1, and detection of omcA homologs in other strains of *S. putrefaciens*. Biochimi Biophys Acta 1373:237–251.

Nakamura Y, Gojobori T, Ikemura T (2000): Codon usage tabulated from the international DNA sequence databases: Status for the year 2000. Nucleic Acids Res 28:292.

Newman JR, Fuqua C (1999): Broad-host-range expression vectors that carry the L-arabinose-inducible *Escherichia coli ara BAD* promoter and the *ara C* regulator. Gene 227: 197–203.

Noll KM, Vargas M (1997): Recent advances in genetic analyses of hyperthermophilic archaea and bacteria. Archives Microbiol 168:73–80.

Ouahrani-Bettache S, Porte F, Teyssier J, Liautard JP, Kohler S (1999): pBBR1-GFP: A broad-host-range vector for prokaryotic promoter studies. BioTechniq 26:620–622.

Patrushev LI, Valyaev AG, Golovchenko PA, Vinogradov SV, Chikindas ML, Kiselev VI (1993): Cloning and expression in *Escherichia coli* of the gene of the thermostable DNA polymerase of *Thermus aquaticus* YT1. Mol Biol 27:681–688.

Ramirez-Arcos S, Fernandez-Herrero LA, Marin I, Berenguer J (1998): Anaerobic growth, a property horizontally transferred by and Hfr-like mechanism among extreme thermophilies. J Bacteriol 180:3137–3143.

Rondon MR, August PR, Bettermann AD, Brady SF, Grossman TH, Liles MR, Loiacono KA, Lynch BA, Macneil IA, Minor C, Tiong CL, Gilman M, Osburne MS, Clardy J, Handelsman J, Goodman RM (2000): Cloning the soil metagenome: A strategy for accessing the genetic and functional diversity of uncultured microorganisms. AEM 66:2541–2547.

Rondon MR, Raffel SJ, Goodman RM, Handelsman J (1999): Toward functional genomics in bacteria: Analysis of gene expression in *Escherichia coli* from a bacterial artificial chromosome library of *Bacillus cereus*. PNAS USA 96:6451–6455.

Rozen S, Skaletsky HJ (1996,1997): Primer3.Code available at *http://www-genome.wi.mit.edu/genome_software/other/primer3.html*.

Sambrook J, Fritsch EF, Maniatis T (1989): Molecular Cloning: A Laboratory Manual Cold Spring Hator, NY: Cold Spring Harbor Laboratory Press.

Shizuya H, Birren B, Kim U, Mancino V, Slepak T, Tachiiri Y, Simon M (1992): Cloning and stable maintenance of 300-kilobase-pair fragments of human DNA in *Escherichia coli* using an F-factor-based vector. PNAS USA 89:8794–8797.

Simon R (1984): High frequency mobillization of Gram-negative bacterial replicons by the in vitro constructed Tn5-Mob transposon. Gen Genet 196:413–420.

Simon R, Priefer U, Puhler A (1983): A broad host range mobilization system for in vivo genetic engineering: Transposon mutagenesis in Gram negative bacteria. Biotechno 1:784–791.

Stein JL, Marsh TL, Wu KY, Shizuya H, DeLong E (1996): Characterization of uncultivated prokaryotes: isolation and analysis of a 40-kilobase-pair genome fragment from a planktonic marine archaeon. J Bacteriol 178:591–599.

Strauch KL, Lenk JB, Gamble BL, Miller CG (1985): Oxygen regulation in *Salmonella typhimurium*. J Bacteriol 161:673–680.

Tait RC, Close TJ, Lundquist RC, Hagiya M, Rodriguez RL, Kado CI (1983): Construction and characterization of a versatile broad range DNA cloning system for Gram-negative bacteria. Biotechnol 1:269–275.

Tao H, Bausch C, Richmond C, Blattner FR, Conway T (1999): Functional genomics: Expression analysis of *Escherichia coli* growing on minimal and rich media. J Bacteriol 181:6425–6440.

Taratus EM, Eubanks SG, DiChristina TJ (2000): Design and application of a rapid screening technique for isolation of selenite reduction-deficient mutants of *Shewanella putrefaciens*. Microbiol Res 155:79–85.

Thompson JD, Higgins DG, Gibson TJ (1994): CLUSTAL W: Improving the sensitivity of progressive multiple sequence alignment through sequence weighting, position specific gap penalties and weight matrix choice. Nucleic Acids Res 22:4673–4680.

Tsygankov YD, Chistoserdov AY (1985): Specific-purpose broad-host-range vectors. Plasmid 14:118–125.

Vargas C, Coronado MJ, Ventosa A, Nieto JJ (1997): Host-range, stability and compatibility of broad-host-range plasmids and a shuttle vector in moderately halophilic bacteria—Evidence of intrageneric and intergeneric conjugation in moderate halophiles. Systematic and App Microbiol 20:173–181.

Wade R, DiChristina TJ (2000): Isolation of U(VI) reduction-deficient mutants of *Shewanella putrefaciens*. FEMS Microbiology lett 184:143–148.

Wagner M, Roger AJ, Flax JL, Brusseau GA, Stahl DA (1998): Phylogeny of dissimilatory sulfite reductases supports an early origin of sulfate respiration. J Bacteriol 180:2975–2982.

Wei Y, Miller CG (1999): Characterization of a group of anaerobically induced *fnr*-dependent genes of *Salmonella typhimurium*. J Bacteriol 181:6092–6097.

Zhou S, Yomano LP, Saleh AZ, Davis FC, Aldrich HC, Ingram LO (1999): Enhancement of expression and apparent secretion of *Erwinia chrysanthemi* endoglucanase (encoded by *celZ*) in *Escherichia coli B*. AEM 65:2439–2445.

SELECTED BIBLIOGRAPHY

The Chipping Forecast (1999): Nature Genet (Microarray Suppl) 21:1–560.

Dale JW (1998): "Molecular Genetics of Bacteria." New York: Wiley.

del Solar G, Alonso JC, Espinosa M, Diaz-Orejas R (1996): Broad-host-range plasmid replication: An open question. Mol Microbiol (MicroRe) 21:661–666.

Greene JJ, Rao VB (eds) (1998): "Recombinant DNA Principles and Methodologies". New York: Dekker.

Guiney DG (1993): Broad host range conjugative and mobilizable plasmids in Gram-negative bacteria. In Clewell D (ed): "Bacterial Conjugation." New York: Plenum Press, 75–102.

Holloway BW (1993): Genetics for all bacteria. Annu Rev Microbiol 47:659–684.

Miller JH (1992): "A Short Course in Bacterial Genetics". Cold Spring Harbor, NY: Cold Spring Harbor Laboratory Press.

Rodriguez RL, Denhardt DT, (eds) (1988): Vectors: A survey of molecular cloning vectors and their uses. Boston: Butterworths.

Rousset M, Casalot L, RappGiles BJ, Dermoun Z, dePhilip JP, Wall JD (1998): New shuttle vectors for the introduction of cloned DNA in *Desulfovibrio*. Plasmid 39(2):114–122.

Salyers AA, Shoemaker NB (1994): Broad host range gene transfer: Plasmids and conjugative transposons. FEMS Microbiol Ecol 15:15–22.

Index

*Eco*R II system, transcription regulation in, 212
*Eco*R V endonuclease
  catalysis of, 203
  in stimulating DNA recombination, 188
  structure of, 202
*Eco*R124 I systems, specificity subunits in, 207–208
Edgar, R. S., 97
EDTA (ethylene diamine tetraacetic acid)
  DNA uptake and, 442, 446
  Haemophilus influenzae binding and, 441
  Streptococcus pneumoniae binding and, 440
eep (enhanced expression of pheromone) gene, in plasmid-based conjugation, 493
8-oxoG lesions, BER repair of, 35
Electroporation
  of *Myxococcus xanthus*, 299–301
  in plant genetic engineering, 337
  of *Stigmatella aurantiaca*, 299–300
  transformation via, 452
  transposons and, 397
Electrostatic forces, in DNA replication, 4
ELISA display, 170, 254
Ellis, E. L., bacteriophage studies by, 87–88
Elongation
  bacteriophage T4 and, 110
  in replicon model, 5, 7–12
  in RNA synthesis, 48, 50, 51
  in RNA translation, 55–56, 72–74
  in transcription regulation, 66–67
Elongation complex, in RNA synthesis, 49–50
Elongation factor G (EF-G), 54, 55–56
Elongation factor Tu (EF-Tu), 54, 55–56
  in suicide systems, 185–186
endA endonuclease, in transformation, 442
endA mutants, *Streptococcus pneumoniae* binding and, 440
Endogenous replication, DNA-membrane interaction and, 21
Endonuclease activity, L1.LtrB retrotransposon and, 405–407
Endonuclease II, in bacteriophage T4 infection, 108–109
Endonuclease IV, in bacteriophage T4 infection, 109
Endonucleases
  in addiction modules, 186–187
  in cloning, 244–246
  divalent cations and, 203
  homing, 205–206
  methylation-dependent, 203–205

processivity and, 211
protease control of, 211–212
restriction-modification systems and, 183, 185, 191, 192, 193–194, 201–206
in stimulating DNA recombination, 188
in transcription regulation, 212–213
in translation, 212
Endospores, 273–274
  anti-sigma factor and, 277–278
  in *Bacillus*, 273
  in *Bacillus subtilis*, 273–280
  cell developmental fates and, 277–278
  in *Clostridium*, 273
  gene expression in formation of, 274–275
  morphological development of, 273–274
  RNA polymerase and, 275
  σ factors and, 275, 278–280
  sporulation initiation for, 276–277
  structure of, 280
  unanswered questions concerning, 280
Enol form, of DNA bases, 29
Entamoeba histolytica, plasmids in, 541
Enterobacter cloacae, bacteriocins from, 556
Enterobacteriaceae, type I restriction-modification systems in, 192–193
Enterococcus, site-specific recombination system in, 233
Enterococcus faecalis
  conjugative transposons in, 408, 409, 410, 495, 496
  plasmid-based conjugation in, 493–494
  sex-pheromone plasmids in, 537–538
Enterococcus faecium, conjugative transposons in, 409–410
Enterotoxins, in *Escherichia coli*, 555
Environment
  bacterial adaptability to, 47
  mutations and, 27–28, 30–31
  in quorum sensing, 363–364
Environmental cleanup, molecular cloning in, 244
Environmental signals, for *Myxococcus xanthus* motility, 309–310
EnvZ protein, in osmolarity regulation, 353–354
Enzyme substrates, restriction-modification systems and, 196–197
Enzymes. *See also* Restriction enzymes
  of bacteriophage T4, 106–107
  in BER systems, 34–35
  in competence, 438
  in DNA elongation, 7–8, 11–12
  DNA precursors and, 17–18